THE NEW

GENERAL AND MINING

TELEGRAPH CODE

BY

C. ALGERNON MOREING, M.Inst. C.E.,

MINING ENGINEER,

AND

THOMAS NEAL,

SECRETARY OF THE MONTANA COMPANY, LIMITED, AND THE
MINES COMPANY, LIMITED.

ALPHABETICALLY ARRANGED.

FOR THE USE OF

*MINING COMPANIES, MINING ENGINEERS,
STOCKBROKERS, FINANCIAL AGENTS, AND TRUST
AND FINANCE COMPANIES.*

EIGHTH EDITION.

1901.

LONDON: WILLIAM CLOWES & SONS, LIMITED.

NEW YORK: D. VAN NOSTRAND CO., 23, MURRAY STREET.
WESTERN AUSTRALIA: PERTH, FREEMANTLE, &c.: E. S. WIGG & SON.
BROKEN HILL (N.S.W.): E S. WIGG & SON.
MELBOURNE, SYDNEY, AND BRISBANE: G. ROBERTSON & CO.
ADELAIDE: E. S. WIGG & SON AND G. ROBERTSON & CO.
CAPE TOWN: J. C. JUTA & CO.
DURBAN AND MARITZBURG: J. C. JUTA & CO. AND P. DAVIS & SON.
JOHANNESBURG: P. DAVIS & SON.

Price One Guinea.

LONDON:

PRINTED BY WILLIAM CLOWES AND SONS, LIMITED,
STAMFORD STREET AND CHARING CROSS.

PREFACE.

In the preface to the Telegraphic Mining Code, published in 1888, the author stated that, although he had done his best to make it as complete as possible, the subject dealt with was too vast for him to think that it would be regarded as perfect, but he ventured to hope that it would form the foundation of a thoroughly complete and practical Code. He invited the co-operation of Mining Engineers and others interested in collecting data, with a view of compiling a more perfect work. This invitation has been widely responded to, and the work of arranging the data so collected has been performed by Mr. Thomas Neal, the Secretary of the Montana Company, Limited, and of the Mines Company, Limited, who has had large experience in cabling.

The Telegraphic Mining Code contained 18,318 words; the New General and Mining Code contains 36,898, or more than double.

The number of sentences has been increased from 15,110 to 29,632, and the additions have been selected from such as have occurred, and been found useful, in general practice.

The tables for Money have been largely extended—

for instance, English Money has been increased from 600 to 1650 words. Numerous new Tables have been added, which will be found extremely useful, such as those for Weights, Assays, Percentages, Dates, and Points of the Compass.

A most important feature is the complete revision of the Code Words to satisfy the rules and requirements of the Telegraph Convention and the various Cable Companies. For this purpose the right has been acquired, at considerable expense, from the Proprietors of the "Telegraphic Convention Code" to use the words in the English section of that work.

The Code Words are alphabetical from beginning to end, thus avoiding the waste of time and the possibility of mistakes sometimes arising from not knowing at once where to look for a word.

The Author believes that this will now be found the most complete, the most useful, and the simplest Mining Code in existence, and that it will save much trouble and expense to Mining Companies and others adopting it.

The Author is greatly indebted to his partner, Mr. T. Burrell Bewick, for the great care he has taken in reading and revising the proofs; and he believes that the work will be found unusually free from errors.

C. ALGERNON MOREING.

SUFFOLK HOUSE,
  LAURENCE POUNTNEY HILL,
    LONDON, E.C.,
      May, 1891.

CONTENTS.

MORSE ALPHABET.

A	— —
B	— — — —
C	— — — — —
D	— — —
E	—
F	— — — —
G	— — —
H	— — — —
I	— —
J	— — — —
K	— — —
L	— — — —
M	— — —
N	— —
O	— — —
P	— — — —
Q	— — — —
R	— — —
S	— — —
T	—
U	— — —
V	— — — —
W	— — —
X	— — — —
Y	— — — —
Z	— — — —

TELEGRAPHIC CODE.

ABA

No.	Code Word.	
1	Aaronic . . .	**Abandon**
2	Aaronsrod . .	Do not abandon
3	Abaciscus . .	Likely to abandon
4	Abacist . . .	Not likely to abandon
5	Abaft . . .	Can I abandon work **on**
6	Abaiser . . .	If —— abandon(s)
7	Abalienate .	Will abandon
8	Abandoned .	Abandon work for the present on
9	Abandoning .	Will not abandon
10	Abandum . .	You must abandon
11	Abanga . . .	Abandon to insurance
12	Abaptiston .	You must abandon the idea
13	Abashed . .	Had better abandon the idea
14	Abashment .	Before we abandon
15	Abasing . . .	Shall I (we) abandon
16	Abatable . .	Have decided to abandon
17	Abatements .	You must not abandon
18	Abatial . . .	Must abandon the works unless
19	Abattised . .	If you resolve to abandon
20	Abattoir . . .	Will not be prudent to abandon
21	Abatude . .	If we are forced to abandon
22	Abbacinate .	**Abandoned**
23	Abbacy . . .	Has been abandoned
24	Abbajeer . .	The mine was abandoned
25	Abbatical . .	Was abandoned on account **of**
26	Abbess . . .	The mine would be considered **abandoned**
27	Abbeyland . .	Has (have) abandoned
28	Abbotship . .	Why was it abandoned
29	Abbreviate . .	Consequently was (were) **abandoned**
30	Abbroch . .	Has (have) —— abandoned
31	Abbwool : .	Had better be abandoned
32	Abdevenham .	The lower workings were abandoned
33	Abdicant . .	The works were abandoned (in consequence of)
34	Abdicating . .	Has (have) —— been abandoned
35	Abdicative . .	The works were abandoned in consequence of an
36	Abdicator . .	Tried and abandoned long ago [influx of water
37	Abdomen . .	Was abandoned
38	Abdominal . .	Will be abandoned
39	Abdominous .	After the mine was abandoned
40	Abduced . .	The action has been abandoned
41	Abduction . .	**Abandonment**
42	Abeam . . .	After abandonment (by)

B

ABA

No.	Code Word.	
43	Abecedary . .	**Abate**
44	Abeigh . . .	Will abate
45	Abelmosk . .	Will not abate
46	Abeltree . .	Will he (they) abate
47	Aberdevine .	Will you abate
48	Aberrancy . .	Cannot abate
49	Aberrant . .	You must not abate
50	Aberrating . .	You must not abate your exertions
51	Aberration .	**Abated**
52	Aberuncate .	Has (have) abated
53	Abettors . .	The sickness has abated
54	Abeyance .	**Abatement**
55	Abgregate . .	Agree(s) to an abatement of
56	Abhorrency .	No abatement in
57	Abhorreth . .	Will —— agree to an abatement
58	Abhorrible . .	If —— do (does) not agree to an abatement
59	Abhorring . .	If —— agree(s) to an abatement
60	Abideth . . .	An abatement of
61	Abiding . . .	In abatement of
62	Abidingly . .	In abatement of his claim
63	Abietic . . .	In abatement of our claim
64	Abigail . . .	In abatement of the damage
65	Abiliate . . .	A material abatement
66	Abiliments . .	What abatement will he make
67	Abilities . .	**Abeyance**
68	Ability . . .	The matter is in abeyance
69	Abiogeny . .	Has been in abeyance until
70	Abjectly . . .	Why is the matter in abeyance
71	Abjugating . .	(To) hold in abeyance
72	Abjunctive . .	In abeyance until we hear from (you)
73	Abjurement .	**Abide**
74	Abjuring . .	Cannot abide by
75	Ablaqueate . .	Will abide by
76	Ablaze . . .	Will not abide by
77	Ablebodied . .	Will —— abide by
78	Ablegate . .	Will you abide by
79	Ableness . .	Must abide by
80	Ablepharus . .	If —— will not abide by
81	Ablepsy . . .	Abide by our decision
82	Ableseaman .	Abide by our offer
83	Ablocated . .	Abide at all risks by
84	Abnegation .	Must we abide by
85	Abnegative . .	Refuse to abide by
86	Abnormal . .	If he (they) refuse to abide by
87	Abnormity .	**Able**
88	Abnormous .	Able to
89	Aboard . . .	Not able to
90	Abococked . .	Am (are) not able to
91	Abodance . .	Am (are) able to
92	Abolish . . .	Are you able to

No.	Code Word.	**Able** (*continued*)
93	Abolisheth .	Have you been able **to**
94	Abolishing .	Have been able to
95	Abolition . .	Have not been able to
96	Abominable .	If —— is (are) able to
97	Abominate .	If —— is (are) not able **to**
98	Aboriginal .	If I am able (to)
99	Aborigines .	If I am not able (to)
100	Aborticide .	Shall be able to
101	Abortient . .	Shall not be able to
102	Abortively .	Will —— be able to
103	Aboveboard .	When will —— be able to
104	Abovedeck .	Telegraph if you are able to
105	Abradant . .	Telegraph when you are able **to**
106	Abrahamic .	As soon as you are able to
107	Abrasions. .	As soon as we are able **to**
108	Abraxas . .	Unless we are able to
109	Abrazitic . .	Unless you are able to
110	Abreast .	Unless —— is able to
111	Abrenounce .	Must be able to
112	Abridged . .	May perhaps be able **to**
113	Abridging .	. **Abolish**
114	Abridgment .	It is proposed to abolish
115	Abrogable .	. **Abolished**
116	Abrogating .	Has been abolished
117	Abrooding .	Has not been abolished
118	Abrook . .	Will be abolished
119	Abrotanoid .	Will not be abolished
120	Abrupt. . .	Must be abolished
121	Abrupteth .	Is going to be abolished
122	Abruption .	Has been abolished for some time
123	Abruptly .	. **Abound(ing)**
124	Abruptness .	To abound in
125	Abscess . .	The district abounds **in**
126	Abscession .	Abounding in
127	Abscond . .	. **About**
128	Absconding .	About the same
129	Absence . .	Price will be about
130	Absenters. .	Profits will be about
131	Absently . .	Expenses will be about
132	Absentment .	About how much
133	Abseybook .	Write at once about
134	Absinthic . .	Please see at once about
135	Absolute . .	See —— about
136	Absolutely .	Have seen —— about
137	Absolution .	Have not been able to **see about**
138	Absolvable .	Will probably be about
139	Absolving. .	Telegraph at once about
140	Absolvitor. .	Should like to know about
141	Absonant . .	Know(s) all about
142	Absonous . .	Know(s) nothing about

No.	Code Word.	**About** (*continued*)
143	Absorbent	. What do you think will be about (the)
144	Absorption	. Can hear nothing about
145	Absorptive	. **Above**
146	Abstain	. Do not go above
147	Abstaining	. Not above
148	Abstemious	. Above suspicion
149	Abstention	. Above the present workings
150	Absterged	. Above the old workings
151	Absterging	. Above the datum level
152	Abstersive	. Above the water level
153	Abstinency	. Will be above
154	Abstinent	. Will it be above
155	Abstorted	. Expenses will be above
156	Abstract	. Cost will be above
157	Abstractly	. **Absconded**
158	Abstruse	. Has absconded (taking)
159	Abstrusely	. Has absconded, leaving heavy default
160	Abstrusion	. **Absence**
161	Absume	. During your absence
162	Absurdity	. During my (our) absence
163	Absurdness	. During the absence of
164	Abthane	. On leave of absence
165	Abundant	. Wish(es) for leave of absence till
166	Abundantly	. Can I have leave of absence till
167	Abusage	. Grant leave of absence to
168	Abuseful	. Do not grant leave of absence to
169	Abusive	. Owing to the absence of
170	Abusively	. In the absence of instructions from (you)
171	Abutilon	. Leave of absence until [right
172	Abuttal	. Can have leave of absence, if matters can go all
173	Abutting	. Cannot grant leave of absence at present
174	Abuzz	. **Absent**
175	Abysm	. I shall be absent from —— for the next —— days
176	Abysmal	. Is (are) absent
177	Abysses	. If —— is absent apply to
178	Acacia	. Will be absent
179	Acaciatree	. Is absent through sickness
180	Academical	. Expect not to be absent more than
181	Academism	. **Absolute**
182	Academy	. Has (have) absolute power to
183	Acalephan	. Granting absolute power to
184	Acalephoid	. To gain absolute control of
185	Acalycine	. Must have absolute control
186	Acantha	. Has (have) absolute control of
187	Acanthice	. **Absolutely**
188	Acanthodes	. Absolutely untrue
189	Acanthoid	. The statement is absolutely untrue
190	Acanthurus	. The report is absolutely untrue, and you can at
191	Acanthylis	. Absolutely without foundation [once contradict it
192	Acanticone	. Absolutely worthless

No.	Code Word.	Absolutely (*continued*)
193	Acaricide .	Absolutely powerless (to act)
194	Acarpous .	. **Abstract**
195	Acatelepsy .	Send me (us) an abstract of
196	Acatharsy . .	Will send an abstract by first mail
197	Acathistus .	Have sent you abstract of
198	Acatry . . .	Cannot send you abstract before
199	Acaulose . .	Send abstract of ore extracted
200	Accable . .	Send abstract of profits (for)
201	Accapitum .	Abstract of title
202	Accelerate .	Abstract of accounts
203	Accent. . .	Abstract of returns
204	Accentors . .	Abstract of monthly yield
205	Accentual . .	Abstract of expenses
206	Acceptably .	Abstract of report
207	Acceptance .	Prepare abstract (of)
208	Accepteth . .	Have prepared abstract (of)
209	Accepting .	. **Abstain**
210	Acception . .	Abstain from
211	Acceptress .	You must abstain from
212	Accessible .	Will not abstain from
213	Accessory . .	Abstain from trespassing
214	Accidence .	. **Absurd**
215	Accidental .	It would be absurd
216	Accipiter . .	The story is absurd
217	Accismus . .	——'s demands are quite absurd
218	Acclaim . .	The amount asked is absurd
219	Acclaiming .	**Abundance**
220	Acclamate .	There is an abundance of
221	Acclinal . .	Are found in abundance
222	Acclivity . .	Fuel and timber in abundance
223	Acclivous . .	There is an abundance of water
224	Accloy . . .	Fuel, timber, and water in abundance [cheap
225	Accoast . .	Fuel, timber, and water in abundance, and labour
226	Accoasting .	Fuel and timber in abundance, water scarce
227	Accol . . .	Water in abundance, timber scarce
228	Accolade . .	Is there an abundance of
229	Accolent .	. **Abundant**
230	Accompany .	Are fuel, timber, and water abundant
231	Accomplish .	Not very abundant
232	Accompt .	. **Abut**
233	Accordable .	Abuts upon
234	Accordancy .	Abutting on
235	Accorded .	. **Accede**
236	According .	Will the vendor accede to these terms
237	Accordions .	Will the company accede to these terms
238	Accountday .	Will he (they) accede to
239	Accourage .	Will you accede to
240	Accoupled .	Accede to the proposition
241	Accoupling .	Will accede to
242	Accredit . .	Will not accede to

No.	Code Word.	Accede (*continued*)
243	Accrescent .	If —— accede(s) to these terms
244	Accretions .	Should —— not accede to these terms
245	Accretive . .	I hope to induce the vendors to accede to
246	Accrue. .	. **Accept(s)**
247	Accruing . .	I (we) accept
248	Accrument .	I (we) cannot accept
249	Accumb . .	Am convinced vendors would accept
250	Accumbency .	Accept under protest
251	Accumbing .	Will you accept
252	Accumulate .	Will accept
253	Accuracy . .	Will not accept
254	Accurse . .	Will not accept less than
255	Accursing . .	Shall I (we) accept
256	Accusable .	Upon what terms will you accept
257	Accusant . .	Shall I (we) accept his (their) offer
258	Accusation .	You had better accept his (their) offer
259	Accusative .	Accept ——'s offer, if nothing better can be done
260	Accusatory .	You had better not accept
261	Accusement .	Accept subject to
262	Accustom . .	Accept subject to approval of condition(s)
263	Aceldama . .	Accept subject to conditions named
264	Acephala . .	Decline to accept without
265	Acephalist .	Shall I (we) accept for your account
266	Acephalous .	Please accept on my (our) account
267	Acepoint . .	Do not accept
268	Acerbate . .	Do not accept bill for
269	Acerbating .	Please accept bill for
270	Acerbitude .	I (we) cannot accept your bill(s)
271	Acerbity . .	Accept in full settlement
272	Acervose . .	Accept, but for cash only
273	Acescency .	Telegraph if you accept
274	Acescent . .	Telegraph if you do not accept
275	Acetabulum .	Telegraph if I (we) may accept
276	Acetamide .	How much will he (they) accept
277	Acetarious .	How much will you accept
278	Acetary . .	Provided he (they) accept(s)
279	Acetified . .	Provided we accept
280	Acetifying .	Will accept, provided
281	Acetimetry .	In case you accept
282	Acetopathy .	In case he accepts (they accept)
283	Acetosity . .	Accept on terms named
284	Acetum . .	Decline to accept on terms named
285	Acetyle . .	Will not accept on any terms
286	Achatina . .	You should not accept (his) (their) statements with-
287	Ached . . .	**Acceptance** [out confirmation
288	Acheilary . .	Acceptance is conditional (upon)
289	Acherontia .	Unconditional acceptance
290	Acherset . .	Acceptance has been returned
291	Achetidae . .	Upon acceptance of
292	Acheweed . .	Draft for acceptance by

No.	Code Word.	
293	Achievable .	**Accepted**
294	Achievance .	The offer is accepted
295	Achieved . .	The offer has not yet been accepted
296	Achieving .	The offer is not accepted
297	Achilleid . .	Has (have) —— accepted
298	Achimenes .	Has (have) not accepted owing to
299	Achirite . .	Has (have) accepted all conditions imposed
300	Achlya . . .	Will be accepted
301	Achorion . .	Has (have) accepted draft
302	Acicularly . .	Is accepted (by)
303	Acidifier . .	Not yet accepted
304	Acidify . .	Not accepted in consequence of
305	Acidimeter .	**Access**
306	Acidness . .	(To) gain access to
307	Acidulate . .	Cannot gain access to
308	Aciform . .	Can you gain access to
309	Acinous .	**Accident**
310	Acipenser . .	A bad accident has happened to
311	Aciurgy . .	Has met with an accident
312	Acknow . .	Owing to an accident to
313	Aclinic . . .	Cannot ascertain cause of accident
314	Acnestis . .	Telegraph cause of accident
315	Acolyte . .	The accident was caused by
316	Acolythist . .	Works stopped, owing to accident to
317	Acombered .	Mill shut down, owing to accident to
318	Acondylous .	Accident easily repaired
319	Aconitum . .	Accident will take some time to repair
320	Acontiadae .	Owing to serious railway accident
321	Acontias . .	Was sent by accident (to)
322	Acoraceae .	Have just heard by accident that
323	Acorn . . .	Delayed by accident
324	Acorncup . .	How long will it take to repair accident (to)
325	Acornoil . .	A fatal accident has occurred
326	Acornshell .	Killed by accident
327	Acosmistic .	In case of accident
328	Acosmium .	**Accommodate**
329	Acotyledon .	Will —— accommodate
330	Acouchy . .	Will accommodate
331	Acousmatic .	Will not accommodate
332	Acoustic . .	Done to accommodate
333	Acoustical .	Can accommodate
334	Acquiesce .	Can you accommodate
335	Acquirable .	Cannot accommodate
336	Acquired . .	Will not accommodate more than
337	Acquiring .	**Accommodated**
338	Acquisitor . .	Have accommodated
339	Acquit . . .	**Accommodation**
340	Acquitment .	Will be a great accommodation
341	Acquittal . .	Any accommodation
342	Acrasies . .	Will give all accommodation in his (their) power

8 ACC

No.	Code Word.	
		Accommodation (*continued*)
343	Acredale	Accommodation for ——
344	Acrefight	Without any accommodation (for)
345	Acreshot	Has every accommodation (for)
346	Acridian	The accommodation has been very opportune
347	Acrisy	The accommodation will be very useful
348	Acrobat	**Accompany**
349	Acrobatic	Think —— is the best man to accompany
350	Acrocarpi	Will probably accompany you
351	Acrocinus	Will accompany me (us)
352	Acrocomia	Cannot accompany me (us)
353	Acrodont	Must accompany me (us) [do the work alone
354	Acrodus	Must have some one to accompany me as I cannot
355	Acrogen	Do you require any one to accompany you
356	Acrogenous	If —— cannot accompany me (you)
357	Acrography	**Accomplish**
358	Acroleine	Can you accomplish
359	Acrolithan	Hope to accomplish
360	Acromion	Cannot accomplish
361	Acronycal	Must accomplish
362	Acropetal	Should you not be able to accomplish
363	Acropodium	We hope you can accomplish this
364	Acropolis	If you can accomplish these results
365	Acrosaurus	Hardly possible to accomplish
366	Acrospire	Can easily accomplish what you require
367	Across	To accomplish these results
368	Acrostics	To accomplish the desired results we shall have to adopt measures which we do not think judicious
369	Acroterium	Can you accomplish this without seriously over-
370	Acrotism	Cannot accomplish this without [taxing
371	Acrotomous	**Accomplished**
372	Acrylic	Has been accomplished
373	Acting	Has not been accomplished
374	Actiniadae	Can be accomplished (if)
375	Actinism	Cannot be accomplished (unless)
376	Actinoid	**Accord**
377	Actinolite	Act in accord with
378	Actinology	Will accord with
379	Actinosoma	In accord with
380	Actinozoa	**Accordance**
381	Action	In accordance with
382	Actionable	Not in accordance with
383	Actionist	In accordance with your instructions
384	Actionless	In accordance with his (their) instructions
385	Activate	In accordance with my (our) views already ex-
386	Actively	**According** [plained
387	Activement	According to ——
388	Activeness	According to the agreement
389	Activities	According to the reports
390	Actresses	According to instructions received
391	Actual	According to the mining laws here

No.	Code Word.	According (*continued*)
392	Actualist	According to private advices
393	Actuality	According to plans and specification
394	Actualness	According to circumstances
395	Actuarial	According to our view of the matter
396	Actuated	According to your view of the matter
397	Actuating	According to instructions already given
398	Actuation	According to instructions to be given
399	Actuose	You must act according to your instructions
400	Actuosity	Act according to your own discretion
401	Acturience	Acted according to instructions
402	Aculeated	**Account(s)**
403	Aculeiform	On your account
404	Aculeolate	On my (our) account
405	Aculeous	On no account
406	Acumen	On account of
407	Acuminated	No account of
408	Acuminous	Forward the accounts as quickly as you can
409	Acutely	The accounts are unsatisfactory
410	Adage	The most reliable accounts to be had
411	Adagial	Forwarding by to-day's post the accounts
412	Adamant	Send account of bullion shipped
413	Adamantean	Account is overdrawn
414	Adamantoid	Account is overdrawn, remit us by telegraph at once
415	Adamitic	Get all accounts you can of the mine
416	Adansonia	The accounts are very favourable
417	Adaptable	The accounts are very unfavourable
418	Adaptation	Have opened an account with
419	Adapting	Open an account with
420	Adaptness	It must be taken into account
421	Adaptorial	To account for
422	Adarcon	Our account at bank is very low
423	Adaunted	Accounts will be sent as soon as possible
424	Addable	No accounts have been kept
425	Addecimate	Account(s) is (are) to be depended upon
426	Addeemed	Account(s) is (are) not to be depended on
427	Addeeming	Accounts are contradictory
428	Addendum	Have you had any account of
429	Adderbolt	Have you had any account with
430	Adderflies	I (we) have not any account of
431	Adderfly	I (we) have no account with
432	Addergrass	I (we) have an account of
433	Adderpike	I (we) have an account with
434	Adderstone	On whose (what) account
435	Adderswort	Will send accounts by next mail
436	Addeth	If account(s) is (are) favourable
437	Addibility	If account(s) is (are) unfavourable
438	Addict	Can you account for
439	Addition	Cannot account for
440	Additional	That accounts for
441	Addlepated	Not on account of

No.	Code Word.	Account(s) (*continued*)
442	Addleplot . .	Entirely on account of
443	Addoom . .	Account of all expenditure
444	Addressing .	Account of all expenses incurred
445	Adducing . .	Account of all expenses on Revenue Account
446	Adelaster . .	Account of all expenses on Capital Account
447	Adelite . .	Account of ore extracted
448	Adelopod . .	Account of all ore milled
449	Adelphia . .	Account of all shipments of ore
450	Adelphous .	Account of all stores consumed
451	Adenalgy . .	Credit account (with)
452	Adeniform .	Debit account (with)
453	Adenitis . .	Account shows a balance to credit (of)
454	Adenoncus .	Account shows a debit balance of
455	Adenophyma .	Accounts for last month
456	Adenose . .	Accounts for this month
457	Adenotomy .	Accounts for current half year
458	Adephagia .	Accounts for last half year
459	Adeptist . .	Accounts for half year show profit of
460	Adequacy . .	Accounts for half year show loss of
461	Adequate . .	Sent accounts by last mail
462	Adequately .	Sent accounts on
463	Adequation .	Send accounts by next mail
464	Adhere . .	Accounts made up to
465	Adherents .	Up to what date are accounts made up
466	Adhesions .	Must have accounts made up to
467	Adhesive . .	Must have accounts not later than
468	Adhesively .	The account has been forwarded
469	Adhibit . .	We do not understand the account
470	Adhibiting .	How do you account for
471	Adhibition .	What is the state of your Bank Account
472	Adiabatic . .	Accounts are unreliable
473	Adiantites .	Accounts do not tally with
474	Adiantum . .	According to accounts
475	Adiaphory .	The accounts we have received
476	Adipate . .	The accounts do not show
477	Adipocere .	Accounts for freight
478	Adipsous . .	The accounts include
479	Adjacent . .	The accounts do not include
480	Adjacently .	Included in the accounts
481	Adjectival .	Include in the accounts
482	Adjoin . .	Why did you not include in the accounts
483	Adjoinant . .	Send us approximate accounts
484	Adjournal . .	Have sent approximate accounts
485	Adjourning .	To be charged to Revenue Account
486	Adjudge . .	To be charged to Capital Account
487	Adjudging .	Has been charged to Revenue Account
488	Adjudgment .	Has been charged to Capital Account
489	Adjudicate .	Profit and Loss Account
490	Adjunct . .	Profit as shown by the accounts
491	Adjunction .	Revenue and Expenditure Account

No.	Code Word.	Account(s) (*continued*)
492	Adjunctly . .	Stores Account
493	Adjure . .	**Accountant**
494	Adjustable .	An experienced accountant
495	Adjustage . .	**Accumulate**
496	Adjusting . .	Has been allowed to accumulate
497	Adjustive . .	Accumulate a sufficient stock of
498	Adjustment .	Which will accumulate
499	Adjutancy .	**Accumulated**
500	Adjutator .	**Accumulating**
501	Adjutrix . .	We are now accumulating
502	Adjuvant .	**Accumulation**
503	Admeasure .	A large accumulation of
504	Adminicle .	We have now a large accumulation of
505	Administer .	**Accurate**
506	Admirably .	Is (are) report(s) accurate
507	Admiral . .	Is (are) quite accurate
508	Admirance .	Is (are) not at all accurate ,
509	Admirative .	Must be accurate
510	Admirers . .	Must be more accurate
511	Admiringly .	**Accused**
512	Admissible .	Has (have) been accused of
513	Admissions .	Has (have) been unjustly accused of
514	Admissive .	Has (have) —— ever been accused of
515	Admissory .	**Acknowledge**
516	Admittable .	Acknowledge receipt of this
517	Admittance .	Will not acknowledge
518	Admitting . .	Acknowledge(s) to have
519	Admixture .	Acknowledge(s) to have received
520	Admonish .	**Acquainted**
521	Admonition .	Are you acquainted with
522	Admonitive .	Acquainted with
523	Admonitory .	Not acquainted with
524	Admoved .	**Acquire**
525	Admoving .	Acquire all the information you can about
526	Adnascent .	By this we acquire
527	Adnate . .	We acquire a most valuable property
528	Adnoun . .	Will acquire a most valuable mine
529	Adolescent .	Acquire entire control of
530	Adolode . .	Must acquire entire control
531	Adonean . .	Can you acquire
532	Adonic . .	Can acquire
533	Adopted . .	Cannot acquire
534	Adoptedly .	Should you not be able to acquire
535	Adoptious .	**Acquitted**
536	Adoptive . .	Has (have) been acquitted
537	Adorable . .	Cannot be acquitted of all blame
538	Adornated .	Has (have) —— been acquitted
539	Adornments .	Acquitted of all blame
540	Adossed . .	**Acre(s)**
541	Adpressed .	The property comprises —— acres

No.	Code Word.	Acre(s) (*continued*)
542	Adread . .	Equal to —— per acre
543	Adrift . . .	Extending over —— acres
544	Adroit . . .	How many acres are there in the property
545	Adroitly .	. Across
546	Adroitness .	Across the
547	Adscript . .	Running across the
548	Adsignify . .	Half across the
549	Adularia . .	Have come across
550	Adulating .	. Act
551	Adulation . .	Do not act except on written instructions
552	Adulatory. .	Act(s) in the most honourable way
553	Adulterate .	Act(s) in the most underhand way
554	Adulterous .	Act upon advice
555	Adultery . .	Do not act without consulting
556	Adumbrant .	Do not act till you hear from
557	Aduncity . .	Do not act
558	Aduncous. .	Do not act on my (our) letter of
559	Adusk . . .	Act promptly
560	Adustion . .	Act cautiously
561	Advance . .	Act accordingly
562	Advancing .	Act according to agreement
563	Advantage .	Act according to instructions
564	Advenient .	Act as you think best
565	Adventive. .	How shall I act
566	Adventual .	Will act on behalf of
567	Adverb . .	Who will act for us when you leave
568	Adverbial . .	Act(s) on behalf of
569	Adversable .	Will not act for
570	Adversity . .	Will act
571	Advertence .	Shall I (we) act
572	Advertised .	Cannot act
573	Adviceboat .	Cannot act until we hear from
574	Advigilate. .	For whom do(es) —— act
575	Advisably. .	Act on our behalf
576	Advise. . .	Before you act
577	Advisement .	Act under legal advice
578	Advisings . .	Continue to act
579	Advisory . .	Continue to act as you have been doing
580	Advocacy. .	Act on the defensive
581	Advocate . .	Act as proposed
582	Advocatess .	Will he (they) act
583	Advocation .	. Acting
584	Advoutress .	Who is acting for
585	Advoutry . .	Acting on behalf of
586	Advower . .	Is (are) not acting for
587	Advowson .	There is no one acting for
588	Adynamon .	For whom are you acting
589	Aerating . .	Am acting for
590	Aerations . .	You are acting right
591	Aerial . . .	You are acting wrong

No.	Code Word.	**Acting** (*continued*)
592	Aerially . .	Be cautious in acting with
593	Aerides . .	Is (are) not acting with sufficient caution
594	Aerified . .	Acting a double part
595	Aerifying . .	Acting straightforwardly
596	Aerocyst . .	Not acting in good faith
597	Aerognosy .	Acting contrary to instructions
598	Aerologist .	Acting according to instructions
599	Aeromancy .	Acting under orders from
600	Aerometers .	Before acting, take legal opinion
601	Aerometric .	Before acting, consult
602	Aeronaut . .	Before acting
603	Aeronautic .	**Action**
604	Aerophane .	Action for damages
605	Aerophobia .	Owing to the action of
606	Aerophytes .	So far they have taken no action in the matter
607	Aeroscopy .	Has (have) commenced an action against
608	Aerostatic . .	Commence an action against
609	Aerylight . .	Avoid an action if possible
610	Aesthetic . .	Before taking any action in the matter
611	Affability . .	Threaten him (them) with an action
612	Affably . .	Shall I (we) begin an action (against)
613	Affabrous . .	If necessary take action against
614	Affair . . .	Stop further action
615	Affamish . .	Abandon the action
616	Affatuate . .	Defend the action
617	Affect . . .	Push forward the action [present
618	Affectate . .	Shall I (we) suspend all further action for the
619	Affectedly . .	Has (have) abandoned the action
620	Affectible . .	Get the action postponed
621	Affecting . .	Action has been deferred
622	Affections . .	Approve(s) of your action
623	Affeerment .	Do(es) not approve of your action
624	Affianced . .	Compromise the action if possible
625	Affiancing .	Withdraw the action
626	Affidation . .	Action withdrawn
627	Affidavit . .	If action has been commenced
628	Affiliated . .	If action has not been commenced
629	Affinage . .	Has (have) taken action
630	Affinities . .	Has (have) not taken action
631	Affinity . .	We are waiting till we see what action **is taken**
632	Affirm . . .	Take immediate action
633	Affirmably .	If action is taken
634	Affirmance .	No time must be lost in taking **action**
635	Affirming . .	**Adapted**
636	Affixed . . .	Well adapted for
637	Affixture . .	**Adaptability**
638	Afflatus . .	Proved its adaptability
639	Afflicteth . .	**Add** (to)
640	Afflicting . .	This will add greatly to
641	Afflictive . .	This will not add greatly to

No.	Code Word.	Add (*continued*)
642	Affluent . .	Will this add to
643	Affluxion . .	If we add this to
644	Afford . . .	Must add
545	Affordeth . .	Will add to our reserves
646	Affordment .	Add to reserve
647	Afforest .	. **Added** (to)
648	Affray . . .	Added to amount previously sent
649	Affrayers . .	Added to reserve
650	Affrayment .	. **Addition**
651	Affreight . .	In addition to
652	Affriended .	Is this in addition to
653	Affront . .	What addition has been made
654	Affronting .	Must make some addition to
655	Affrontive . .	To make any addition to
656	Aflame . .	. **Additional**
657	Aflaunt .	Will there be any additional expense
658	Aflighted . .	Causing an additional expense (of)
659	Aforehand .	Additional information
660	Afresh . . .	These will be additional
661	Africanism .	Is (are) there any additional
662	Africanize .	. **Address.** (See Registered.)
663	Aftcastle . .	To what address shall I forward
664	Afterage . .	The registered address is
665	Afterbody . .	Telegraph at what address I shall find
666	Aftercabin .	Cannot find —— at the address given
667	Afterclaps .	Address communications direct to
668	Aftercost . .	In future address to
669	Aftercrop . .	Telegraph your address (at)
670	Afterdamp .	Telegraph the address of
671	Afterfeed . .	Address to me, care of
672	Aftergame .	. **Adequate**
673	Aftergrass .	An adequate remuneration
674	Aftergrief . .	Not an adequate remuneration
675	Afterguard .	Is (are) not adequate to
676	Afterhelp . .	Do you consider it adequate
677	Afterhind . .	Do not consider it adequate
678	Afterholds .	Without adequate
679	Afterhope . .	Without adequate means to go on
680	Afterlife . .	Adequate milling power
681	Aftermath . .	Think it is fully adequate
682	Afternoon . .	Adequate compensation
683	Afterpains .	Adequate supply of
684	Afterpeak .	. **Adhere**
685	Aftersail . .	Must adhere to
686	Aftershaft . .	Do you adhere to
687	Afterstudy .	I (we) do not adhere to
688	Afterswarm .	Do(es) adhere to
689	Aftertaste . .	Do(es) not adhere to
690	Aftertime . .	Shall I (we) adhere to
691	Afterwards .	Do you adhere to the opinions expressed

No.	Code Word.	Adhere (*continued*)
692	Afterwise . .	I (we) adhere to the letter of
693	Against .	. **Adit**
694	Agallochum .	From the adit
695	Agalwood . .	Above the adit
696	Agapemone .	Below the adit
697	Agaphite . .	Proposed deep adit
698	Agaricus . .	The length of the adit is
699	Agathosma .	Mine worked entirely by adits
700	Agathotes . .	Samples taken from the adit
701	Agedly . . .	The only development is an adit
702	Agencies . .	Adit has cut the vein —— feet below the outcrop
703	Agentship .	Adit has cut the vein after driving
704	Ageratum . .	We are driving the adit
705	Aggrace . .	Are you driving the adit
706	Aggrandize .	Adit is being driven to cut the vein
707	Aggravable .	By means of the adit
708	Aggravated .	The adit has been driven
709	Aggregator	. **Adjacent**
710	Aggression .	The adjacent mine
711	Aggressors .	Adjacent properties
712	Aggrieve . .	Ought to buy the adjacent claim
713	Aggrieving .	Vein runs into the adjacent claim
714	Aggroup . .	Adjacent location
715	Aghast . . .	Adjacent location to the north
716	Agilely . . .	Adjacent location to the east
717	Agiotage . .	Adjacent location to the south
718	Agistor . .	Adjacent location to the west
719	Agitable . .	If the adjacent location is bought
720	Agitate . .	If the adjacent location is sold
721	Agitating . .	Adjacent property can be bought
722	Agitative . .	Adjacent property cannot be bought
723	Agitators .	. **Adjunct**
724	Agletbaby . .	A most valuable adjunct
725	Aglow . . .	An undesirable adjunct
726	Aglutition .	. **Adjust**
727	Agmatology .	Can you adjust matters amicably
728	Agminated .	Try to adjust matters amicably
729	Agnail . . .	Can adjust matters amicably
730	Agnostics . .	Cannot adjust matters amicably
731	Agonies . .	Hope to adjust matters amicably
732	Agonized .	. **Adjusted**
733	Agonizing . .	Matter has not yet been adjusted
734	Agony . . .	Matters have been satisfactorily adjusted
735	Agraff . . .	Has —— matter been adjusted yet
736	Agraphis .	. **Adjustment**
737	Agrarian . .	In course of adjustment
738	Agree . . .	Adjustment of our claim
739	Agreeable . .	Adjustment of his (their) claim
740	Agreeing .	. **Admit**
741	Agreeingly	Is (are) willing to admit

No.	Code Word.	Admit (*continued*)
742	Agreements .	Will not admit
743	Agrestial . .	Must admit
744	Agrestic . .	Must not admit
745	Agricolist . .	Admit nobody
746	Agricolous .	Admit(s) of no delay
747	Agricultor . .	Cannot admit
748	Agriology . .	Will —— admit
749	Agriopus . .	If —— will not admit
750	Agronomial .	Refuse(s) to admit any one to see the mine
751	Agronomy .	Has (have) refused to admit me (us) to see the mine
752	Agrope . .	Admit —— to see the mine
753	Agrostemma .	Admit —— to see the works
754	Aground . .	Do not admit —— to the mine
755	Agroupment .	If we admit
756	Agrypnotic .	Admit nothing
757	Aguatoad . .	Cannot admit that the fault is here
758	Aguecake .	. **Advance**
759	Aguedrop . .	Will advance the money required
760	Aguefit . .	Is (are) ready to advance the money
761	Agueproof .	Am (are) ready to advance the money
762	Aguespell . .	Can you advance what is required
763	Aguetree . ..	Will advance
764	Aguishness .	Will not advance
765	Agynous . .	Cannot advance any money until (or unless)
766	Aidmajor . .	On what terms will you advance
767	Aiguille . .	What advance will be given
768	Ailanthus . .	Advance wanted of
769	Ailments . .	What can I (we) advance
770	Ailurus . .	You must be very careful about advances
771	Aimcrier . .	Will require an advance
772	Aimless . .	Must advance more
773	Airballoon .	Cannot advance more
774	Airbath . .	Do not think it would be prudent to advance more
775	Airbed . . .	Do not make any further advance(s) to
776	Airbladder .	Decline making advance before searching inspection
777	Airbones . .	Must not advance
778	Airbrake . .	You may advance up to
779	Airbraving .	What has caused this advance in
780	Airbricks . .	Advance money for examination of property
781	Airbuilt . .	What will ——advance towards
782	Aircane . .	The advance made is
783	Aircasing . .	What advance has been made in
784	Aircells . .	Will he (you) advance
785	Airchamber .	Any advance made
786	Aircourse . .	In advance of
787	Aircushion .	In advance of calls
788	Airdrain . .	Payment to be made in advance
789	Airdrill . . .	Will not pay in advance
790	Airengine . .	Advance in price
791	Aircscape . .	Owing to the advance in price of

No.	Code Word.	**Advance** (*continued*)
792	Airfilter . .	Wants us to advance
793	Airfunnel .	. **Advanced**
794	Airfurnace .	Has (have) advanced
795	Airgrating .	Has (have) not advanced
796	Airguns . .	The money thus advanced
797	Airholder . .	The works have much advanced
798	Airhole . .	The works have not advanced much
799	Airiest . . .	Matters are no further advanced
800	Airjacket . .	If —— has (have) not advanced
801	Airlevel . .	Has (have) —— advanced
802	Airlock . .	The price asked has been advanced to
803	Airmachine .	Wages have advanced
804	Airpassage .	The price has advanced
805	Airpipe . .	Shares have advanced
806	Airplant . .	Has (have) been advanced
807	Airpoise . .	Has (have) been advanced to a point
808	Airpump . .	Has (have) materially advanced
809	Airscuttle . .	Has (have) already advanced
810	Airshaft .	. **Advancing**
811	Airslacked .	Are advancing
812	Airstove . .	Are rapidly advancing
813	Airthread . .	Prices are advancing
814	Airtight . .	Rate of wages is advancing
815	Airtrap . .	We are now advancing
816	Airtrunk . .	We are now advancing in a direction
817	Airtube . .	Are you advancing favourably
818	Airvalve .	. **Advantage**
819	Airvessel . .	Is there any advantage in
820	Airwards . .	It will be a great advantage
821	Aisled . .	It will be to your advantage
822	Aitchbone .	There is no advantage
823	Aitchpiece .	It will be to our mutual advantage
824	Aizoon . . .	What would be the advantage of
825	Akerstaff . .	Taking advantage of
826	Akimbo . .	Take advantage of
827	Alabandine .	The advantage to us is that
828	Alabaster . .	The advantage to be derived from
829	Alabastrus .	What advantage is there
830	Alackaday .	What advantage do you expect from
831	Alacrious .	. **Advantageous**
832	Alacrity . .	Most advantageous offer
833	Aladinist . .	Most advantageous for our interests
834	Alarm . . .	Most advantageous for
835	Alarmbell .	Will be in an advantageous position to
836	Alarmclock .	More advantageous
837	Alarmgauge .	Will be very advantageous
838	Alarmgun .	. **Advertise**
839	Alarming . .	May I advertise
840	Alarmingly .	You may advertise
841	Alarmists . .	Will advertise

No.	Code Word.	**Advertise** (*continued*)
842	Alarmpost .	Will you advertise
843	Alarmwatch .	Will —— advertise
844	Alasmodon	. **Advertised**
845	Alaternus . .	Advertised for
846	Albatross . .	Advertised to be sold on
847	Albeit . . .	Have you advertised for
848	Albescent . .	Has (have) advertised
849	Albicore . .	Has (have) not advertise
850	Albigenses	. **Advice(s)**
851	Albinism . .	Advices from the mine are satisfactory
852	Albuginea .	Advices from the mine are unsatisfactory
853	Albumen . .	May favourable advices be expected
854	Albumenize .	Send advices immediately
855	Albuminoid .	Waiting for advice
856	Alburnitas .	No advice to hand
857	Alcahest . .	Let me (us) have at once by telegraph any favourable advices you can possibly send
858	Alchemic . .	My (our) advice is
859	Alchemical .	Take legal advice before you do anything
860	Alchemilla .	Have you had legal advice
861	Alchemized .	By advices which we have received
862	Alchymists .	Upon receipt of advices
863	Alchymy . .	As soon as we receive advices (of)
864	Alcmanian .	Have taken legal advice
865	Alcohol . .	Are daily expecting advices
866	Alcoholate .	Follow our advice
867	Alcoholism .	By following ——'s advice
868	Alcoholize .	Recent advices inform us
869	Alcoran . .	Advice received too late to act
870	Alcoranish .	What are the latest advices from
871	Alcove . .	. **Advise**
872	Alcyonaria .	Advise you to sell out your holding
873	Alcyonite . .	What do you advise
874	Alcyonoid .	Advise you to
875	Aldebaran .	Do not advise (you to)
876	Aldehyd . .	Do you advise
877	Aldehydic .	We advise you not to
878	Alderman . .	Advise by telegraph
879	Aldermancy .	What do (does) —— advise
880	Aldermanic .	Advise early what to do
881	Aleatory . .	Advise if any change occurs
882	Aleavement .	Strongly advise(s)
883	Alebench . .	We advise you to have nothing to do with
884	Aleberry .	. **Advisable**
885	Alebrewer .	Will it be advisable (to)
886	Alecampane .	It would be advisable (to)
887	Aleconner .	It would not be advisable (to)
888	Alectoria . .	Do you consider it advisable (to)
889	Aledraper .	. **Advised**
890	Alefed . . .	We have been advised (that) (to)

No.	Code Word.	**Advised** (*continued*)
891	Alegill . . .	Keep us advised
892	Aleglass . .	Keep them advised
893	Alehouse . .	Until we are further advised
894	Aleknight . .	As at present advised
895	Alembic . .	Has (have) been badly advised
896	Alembroth .	Was (were) well advised
897	Alert . . .	Have advised
898	Alertness . .	Have you advised
899	Aleshot . .	Have not advised
900	Alesilver .	. **Affair(s)**
901	Alestake . .	In the present state of affairs
902	Alewashed .	Affairs are in a hopeless muddle
903	Alewife . .	How are affairs going on
904	Alexiteric . .	Affairs are in a bad way
905	Aleyard . .	Do you think affairs can be satisfactorily arranged
906	Alfagrass . .	Affairs are progressing satisfactorily
907	Algaroth . .	On account of political affairs
908	Algates . .	With the affair
909	Algebra . .	Affair settled satisfactorily
910	Algebraics .	Is the affair settled
911	Algebraize .	Can you settle the affair
912	Algidity . .	Can settle the affair
913	Algidness .	Cannot settle the affair
914	Algific . .	**Affect**
915	Algoid . . .	Will affect the market
916	Algorithm . .	Will not affect the market
917	Alhambraic .	Will affect the sale
918	Aliases . . .	Will not affect the sale
919	Alien . . .	Will not affect me (us) in any way
920	Alienable . .	How will this affect
921	Alienage . .	Does it affect
922	Alienating .	It does not affect
923	Alienation .	Will affect
924	Alienators .	Will it affect
925	Alienism . .	Will not affect
926	Aligerous . .	It is not likely to affect
927	Alimental . .	**Affected**
928	Alimonious .	Has affected the market
929	Alimony . .	Has affected us considerably
930	Aliquot . .	Has not affected us much
931	Alisander . .	Has it affected
932	Alisma . . .	Has been affected (by)
933	Alismaceae .	Has been seriously affected (by)
934	Alitrunk . .	Has (have) not been affected (by)
935	Alkahestic .	**Affidavit(s)**
936	Alkali . . .	Affidavit(s) will be sent
937	Alkalified . .	Shall want affidavit
938	Alkalify . .	Affidavit to the following effect
939	Alkalimide .	**Affirm**
940	Alkalinity . .	If what —— affirm(s) is true

No.	Code Word.	**Affirm** (*continued*)
941	Alkalious .	. As —— affirm(s)
942	Alkalizate.	. I cannot endorse what —— affirm(s)
943	Alkalizing	. What —— affirm(s) is true
944	Alkaloids .	. He affirms (they affirm) that
945	Alkanet .	. We affirm that
946	Alkarsine .	. **Affirmed**
947	Alkermes .	. It has been affirmed that
948	Allabreve.	. Has been affirmed
949	Allagite .	. **Affirmation**
950	Allantoic .	. We shall require affirmation
951	Allatrate .	. Must make affirmation
952	Allay . .	. **Affirmative**
953	Allaying .	. In the affirmative
954	Alldreaded	. If I (we) get an answer in the affirmative
955	Allective .	. If you get an answer in the affirmative
956	Alledge .	. **Afford**
957	Allegation	. Can you afford
958	Allegeable	. Can afford
959	Allegement	. Cannot afford
960	Allegiance	. This would afford an opportunity of
961	Allegorist .	. **Afraid**
962	Allegory .	. Am afraid that
963	Allerion .	. Am not afraid that
964	Alleviator .	. Is (are) afraid
965	Allfours .	. Is (are) not afraid
966	Allgood .	. Is (are) —— afraid
967	Allhail . .	. Afraid of
968	Allhailing .	. Afraid to
969	Allhallow .	. **After**
970	Alliaceous	. After what has taken place
971	Alliance .	. After what has been done
972	Alliciated .	. After hearing from
973	Alliciency.	. After seeing
974	Allicient .	. After the mine is unwatered
975	Alligature .	. After visiting —— shall go to
976	Allocating	. After you have finished
977	Allochrous	. After finishing at —— shall go to
978	Allodially .	. After the arrival of
979	Allodials .	. After I (we) have
980	Allodium .	. After you have
981	Allograph .	. After they have (he has)
982	Allopathy .	. After the mail
983	Allophane .	. After that date
984	Alloquy .	. Look after
985	Allot . .	. On and after
986	Allotment .	. If, after
987	Allotropic.	. If, after all
988	Allotted .	. **Afternoon**
989	Allotting .	. This afternoon
990	Alloverish	. To-morrow afternoon

No.	Code Word.	
991	Allowably	**. Again**
992	Allowances .	Again as soon as you can
993	Alloxan . .	Shall have to go again
994	Alloxanate .	Cannot go again
995	Alloxanic .	Could you go again
996	Alloyage .	Must go again
997	Allperfect .	Shall have to see —— again about
998	Allseed . .	See —— again
999	Allseeing .	Cannot see —— again
1000	Allsorts . .	Till I hear again from
1001	Allspice . .	Till —— hear(s) again from
1002	Alluringly .	Has (have) been again
1003	Allusive . .	Has (have) not been again
1004	Allusively .	Again in the market
1005	Alluvial . .	If the property is again offered
1006	Alluvious .	The mine is again
1007	Allylamine .	The water has again
1008	Allylene . .	Has (have) again
1009	Almanac . .	Has (have) not again
1010	Almanrivet .	If the matter comes up again
1011	Almighty .	Has (have) gone up again
1012	Almondcake	Has (have) fallen again
1013	Almondoil .	Has (have) not gone up again
1014	Almondtree.	When shall you again
1015	Almonry . .	(Have) again started
1016	Almsdeed .	It must not occur again
1017	Almsdrink .	It shall not occur again
1018	Almsfolk . .	If it occurs again
1019	Almsgiver	**. Against**
1020	Almsgiving .	Against my (our) wishes
1021	Almshouse .	Against the hanging wall
1022	Almsman .	Against the foot wall
1023	Almucantar.	Against the adjoining claim
1024	Alnagar . .	Against advice
1025	Aloeswood .	Has (have) gone against us
1026	Aloetic . .	Is (are) against our
1027	Aloetical .	**. Agent(s)**
1028	Aloexylon .	Cannot find your agent(s)
1029	Alogians . .	You must appoint an agent
1030	Alongshore .	Act(s) as agent(s) for
1031	Alongside .	Have appointed —— as my agent (our agent)
1032	Aloofness .	Will not act as agent(s)
1033	Alopecurus .	——'s agent is away
1034	Alopecy . .	Have left it in the hands of the agent to arrange
1035	Aloud . .	Apply to the agent(s)
1036	Aloysia . .	Is acting as our agent
1037	Alpaca . .	Wish you to act as our agent
1038	Alpenhorn .	Will you act as our agent
1039	Alpenstock .	Cannot act as your agent
1040	Alpestrine .	Are our agents

No.	Code Word.	**Agent(s)** *(continued)*
1041	Alpha . .	Care of our agents
1042	Alphabet .	Address of our agents
1943	Alphabetic .	Send through our agents
1044	Alphenic . .	Our agent at
1045	Alphonsin .	Our agent at —— will attend to
1046	Alpigene	. **Agency**
1047	Alpinery . .	Will you accept local agency of
1048	Already . .	Through his agency (their agency)
1049	Alsophila .	Through your agency
1050	Altaic . .	Through the agency of
1051	Altarbread .	Decline the agency
1052	Altarcards .	If it is by the agency of
1053	Altarcloth	. **Aggregate**
1054	Altardues .	Amounting in the aggregate to
1055	Altarhorn .	The aggregate amount
1056	Altarledge .	The aggregate tonnage
1057	Altarpiece	. **Agree**
1058	Altarrail . .	Will agree
1059	Altarside .	Will agree to
1060	Altarstole .	Agree(s) to
1061	Altartable .	Will not agree to
1062	Altarthane .	The owners agree
1063	Altartomb .	Will you agree to
1064	Altarvase .	Would you agree to these terms
1065	Altazimuth .	Cannot agree to these terms
1066	Alterably .	Agree(s) to the conditions
1067	Alterative .	Agree(s) to do all that is necessary
1068	Altercate .	If —— agree(s)
1069	Alternally .	If —— do (does) not agree
1070	Alternity .	Cannot agree
1071	Alternized .	You ought to agree
1072	Altheine . .	Agree to the proposition
1073	Although .	Agree to the terms
1074	Altify . .	Do you agree to or decline
1075	Altimetry .	Cable promptly if you agree to or decline
1076	Altincar . .	If we agree to
1077	Altiscope .	If you cannot agree (to)
1078	Altisonous .	If you can agree (to)
1079	Altivolant .	If you can agree with
1080	Altoclef . .	If he does (they do) not agree
1081	Altogether .	If you do not agree
1082	Altruistic	. **Agreeable**
1083	Aludel . .	It would be agreeable
1084	Aluminite .	Would it be agreeable
1085	Aluminium .	Most agreeable information
1086	Alumish . .	It is the most agreeable information which has [reached us
1087	Alumrock	. **Agreed**
1088	Alumroots .	Has (have) agreed to, on condition that
1089	Alumschist .	Has (have) agreed to
1090	Alumslate .	Has (have) not agreed to

No.	Code Word.	**Agreed** (*continued*)
1091	Alumstone .	Has (have) agreed to all except
1092	Alunogen .	Unless —— has (have) agreed to
1093	Alutaceous .	**Agreement** (See also Sign.)
1094	Alveated. .	The agreement is signed by
1095	Alveolary .	According to the present agreement
1096	Alveolate .	Agreement has been signed
1097	Alveolus. .	According to (his) their agreement with
1098	Always . .	Not according to the agreement
1099	Alyned . .	The original agreement
1100	Alyssum . .	On the same conditions as the original agreement
1101	Amacratic .	No agreement has yet been made
1102	Amalgam .	It would be wise to make a special agreement
1103	Amalgamize	An agreement between
1104	Amalphitan .	Willing to enter into an agreement
1105	Amanuensis.	Cancel the agreement
1106	Amaranthus.	The agreement is cancelled
1107	Amaritude .	By a special agreement
1108	Amaryllis .	According to the original agreement
1109	Amarythrin .	Have made an agreement (with)
1110	Amassment .	By this agreement
1111	Amasthenic.	By the terms of the agreement
1112	Amateur. .	Will you modify the agreement
1113	Amateurish .	Cannot modify the agreement
1114	Amative . .	Agreement must be modified
1115	Amatorial .	Agreement has been modified
1116	Amatorious .	Consent to alter the agreement
1117	Amaurotic .	Has any agreement been arrived at
1118	Amazed . .	**Ahead**
1119	Amazedness	Shall be able to go ahead
1120	Amazeful .	Cannot go ahead owing to
1121	Amazingly .	To get ahead of
1122	Amazonian .	To go ahead
1123	Ambages .	**Aim**
1124	Ambagitory .	His (their) aim has been to
1125	Ambassador	What is his (their) aim
1126	Ambered .	Our aim should be (to)
1127	Ambering .	**Air**
1128	Ambergris .	Air in the mine is bad
1129	Amberseed .	Air is too bad to permit progress in the mine
1130	Ambertree .	Air-compressor
1131	Ambidexter.	Air-pressure —— pounds per square inch
1132	Ambient. .	Air-pipes
1133	Ambigenal .	An air-shaft is necessary
1134	Ambiguity .	An accident to the air-shaft
1135	Ambiguous .	**Alien**
1136	Ambilevous.	The Alien Act
1137	Ambilogy .	**Alienate**
1138	Ambitioned.	**Alienated**
1139	Ambler . .	**All**
1140	Amblingly .	All you can

No.	Code Word.	**All** (*continued*)
1141	Amblotic .	All you have
1142	Amblygon .	All right
1143	Amblygonal.	**Allot**
1144	Amblyopsis .	Wish you to allot
1145	Amblyopy .	Will you allot
1146	Ambreada .	Cannot allot
1147	Ambrosial .	**Allotted**
1148	Ambrotype .	Allotted to you
1149	Ambsace .	Allotted to me
1150	Ambulacrum	Will be allotted
1151	Ambulances	How many shares are there already allotted
1152	Ambulated .	The company have allotted —— shares
1153	Ambulating.	To be allotted to me or my nominees
1154	Ambulative.	Allotted to you —— shares
1155	Ambulatory.	Allotted to you or your nominees
1156	Amburbial .	In whose name do you want shares allotted to you
1157	Ambury . .	**Allotment** [to be placed
1158	Ambuscade .	The allotment will take place
1159	Ambush . .	No allotment has been made
1160	Ambushing .	Must proceed to allotment
1161	Ambushment	The company has gone to allotment
1162	Amelcorn .	Will go to allotment on
1163	Amenably .	Pending the allotment
1164	Amendatory	The directors refuse to go to allotment
1165	Amendful .	The company goes to allotment on
1166	Amendments	All allotments
1167	Amends . .	Will join after allotment
1168	Amenity . .	If we do not go to allotment
1169	Amercement	If we do not go to allotment immediately, the thing
1170	Ametabola .	**Allow(ing)** [will fail
1171	Amethyst .	Do not allow
1172	Amherstia .	Cannot allow
1173	Amiability .	Will you allow
1174	Amianth . .	It is impossible to allow
1175	Amianthoid.	(To) allow easy access to
1176	Amicable .	(To) allow the men to
1177	Amidogen .	It would be wise to allow
1178	Amidships .	Will not allow
1179	Amidward .	You must not allow
1180	Ammochryse	You must allow
1181	Ammodytes.	Will allow us to
1182	Ammonalum	Will not allow us to
1183	Ammonia ..	Why did you allow
1184	Ammoniacal	Why did he allow (they allow)
1185	Ammoniuret	Unless we allow
1186	Ammophila.	Unless you allow
1187	Ammunition	If you will allow
1188	Amnestied .	If you will allow us to
1189	Amnesty. .	If you will not allow
1190	Amongst. .	If you will not allow us to

ALL

No.	Code Word.	Allow(ing) *(continued)*
1191	Amorous .	Will only allow on condition that
1192	Amorously .	By allowing
1193	Amorphism .	Allowing for
1194	Amorphous .	**Allowance**
1195	Amorphozoa	Making ample allowance for
1196	Amorphy .	Not making any allowance for
1197	Ampelis . .	What allowance has been made (for)
1198	Ampelopsis .	After making reasonable allowance
1199	Ampersand .	Some allowance will have to be made for
1200	Amphibia .	No allowance has been made for
1201	Amphibious.	Allowance for travelling expenses
1202	Amphibolic .	**Allowed**
1203	Amphibrach	Has been allowed (for)
1204	Amphicome.	Has not been allowed (for)
1205	Amphicyon .	You have not allowed (for)
1206	Amphidisc .	We have fully allowed (for)
1207	Amphigean .	What have you allowed (for)
1208	Amphilogy .	Have allowed —— (for)
1209	Amphimacer	**Alluvial**
1210	Amphioxus .	Payable alluvial
1211	Amphipoda .	Payable alluvial has been found at
1212	Amphisarca .	Alluvial extends over —— acres, and averages ——
1213	Amphiscian .	Alluvial contains —— per cubic yard [feet
1214	Amphithura.	Rich alluvial
1215	Amphitrite .	Alluvial deposit(s)
1216	Amphitype .	Alluvial of no value
1217	Amphoral .	Alluvial deposits have been found
1218	Amphoteric.	Alluvial deposits are rich in gold
1219	Amplectant.	Alluvial deposits are rich in
1220	Ampleness .	Depth of the alluvial
1221	Ampliated .	If alluvial is found
1222	Ampliating	**Almost**
1223	Ampliative .	Almost all
1224	Amplifier .	Almost all the work has been done
1225	Amplify . .	Almost all the ore
1226	Amplifying .	Almost all the rock
1227	Amplitude .	Have almost finished
1228	Ampulla . .	**Along**
1229	Amputate .	Along the course of the vein
1230	Amputation.	Along the east wall
1231	Ampyx . .	Along the west wall
1232	Amuck . .	Along the —— level
1233	Amulet . .	Along the face
1234	Amurcosity .	Driving along
1235	Amurcous .	Along the hanging wall
1236	Amusable .	Along the footwall
1237	Amusing. .	Ranging along
1238	Amyelous .	**Already**
1239	Amygdalate.	Have already indications of the vein
1240	Amygdalic .	Has (have) already begun

No.	Code Word.	**Already** (*continued*)
1241	Amygdaloid.	Already completed
1242	Amygdalus .	Already in vein matter
1243	Amylaceous.	**Alter**
1244	Amyloid . .	Have had to alter
1245	Amyraldism.	Cannot alter
1246	Amyris . .	Will you alter
1247	Anabaptist .	Will alter
1248	Anabaptize .	Will not alter
1249	Anabasis. .	This will alter
1250	Anableps .	Will —— alter
1251	Anacamptic.	If —— will not alter
1252	Anacanth .	It will be necessary to alter
1253	Anacardium	Too late to alter
1254	Anachorism .	If you can alter
1255	Anaclastic .	If you cannot alter
1256	Anaclisis. .	**Alteration**
1257	Anadromous	The necessary alteration has been made
1258	Anagallis .	No alteration can be made
1259	Anaglyph .	Has any alteration been made
1260	Anaglyptic .	Great alteration has taken place
1261	Anagogic .	Make the necessary alteration
1262	Anagraphs .	No alteration has been made
1263	Analecta. .	No alteration has taken place
1264	Analepsis .	Alteration in milling
1265	Analgesia .	Alteration in the management.
1266	Analogical .	What alteration will be necessary
1267	Analogism .	Make no alteration
1268	Analogized .	No alteration is needed
1269	Analogous .	Much alteration is needed
1270	Analysable .	**Alternative**
1271	Analysing .	The only alternative is
1272	Analysis . .	Is there no other alternative
1273	Analytical .	Is there any alternative
1274	Anamesite .	**Altitude**
1275	Anamirta .	At an altitude of
1276	Ananchytes .	Owing to the high altitude
1277	Anapest . .	**Altogether**
1278	Anaplastic .	Taken altogether
1279	Anarchical .	Taking it altogether
1280	Anarchism .	**Alum**
1281	Anarchized .	**Alumina**
1282	Anarchy . .	**Aluminium**
1283	Anarrhexis .	**Amalgam**
1284	Anasarca .	Gold amalgam
1285	Anasarcous .	Silver amalgam
1286	Anastatica .	Sodium amalgam
1287	Anastrophy .	Robbery of amalgam
1288	Anathema .	Percentage of gold in the amalgam
1289	Anatocism .	Percentage of silver in the amalgam
1290	Anatomical .	**Amalgamate** (**with**)

No.	Code Word.	**Amalgamate** (*continued*)
1291	Anatomized .	Amalgamate interests
1292	Anatomy. .	Can you amalgamate
1293	Anatreptic .	Amalgamate companies as
1294	Anatripsis .	Will not amalgamate
1295	Anatropal .	Amalgamate your claims with the adjoining company
1296	Anatropous .	**Amalgamated**
1297	Ancestor. .	Have amalgamated
1298	Ancestral .	Cannot be amalgamated
1299	Ancestress .	Can only be amalgamated in pans
1300	Ancestry. .	Cannot be amalgamated in pans.
1301	Anchilops .	**Amalgamating**
1302	Anchithere .	Amalgamating plates
1303	Anchor . .	Amalgamating pans
1304	Anchorable .	**Amalgamation**
1305	Anchorage .	Amalgamation successfully carried through
1306	Anchorball .	The amalgamation process
1307	Anchordrag .	Pan amalgamation
1308	Anchoretic .	Amalgamation on plates
1309	Anchorgate .	Discontinue pan amalgamation
1310	Anchorhold .	Resume pan amalgamation
1311	Anchorice .	Ore not suitable for pan amalgamation
1312	Anchovies .	Amalgamation in tortas on the Mexican system
1313	Anchovy. .	Barrel amalgamation
1314	Anchusa. .	Before amalgamation
1315	Anchylose .	After amalgamation
1316	Anchylotic .	Amalgamation of the companies
1317	Ancient . .	Amalgamation of claims
1318	Anciently .	**Amalgamator**
1319	Ancillary .	An experienced amalgamator
1320	Ancipital. .	Engaged as amalgamator
1321	Ancipitous .	Send thoroughly competent amalgamator
1322	Ancanoid .	**Amount(s)**
1323	Ancorist . .	To what amount
1324	Ancylotome.	Amount on hand
1325	Andabatism .	Telegraph the amount
1326	Andalusite .	Amount extracted per month is
1327	Andreolite .	No record has been kept of amount extracted
1328	Androecium	A large amount has been extracted
1329	Androgynal .	Impossible to estimate the amount
1330	Androides .	What is the right amount
1331	Andromeda .	The right amount is
1332	Androphagi .	This amount will be deposited subject to your order
1333	Andropogon	Telegraph the amount of ore in sight
1334	Androspore .	Does not amount to much
1335	Androtomy .	A large amount has been lost
1336	Anecdotage .	This amount will be deposited with
1337	Anecdote .	Estimate the amount of ore in sight at
1338	Anecdotist .	Cannot estimate the total amount
1339	Anelectric .	What is the amount of capital
1340	Anemograph	Will guarantee this amount

No.	Code Word.	**Amount** (*continued*)
1341	Anemology .	What amount is there to my (our) credit
1342	Anemometry	The amount to your credit is
1343	Anemone .	What is the amount in hand
1344	Anemoscope	Cable the amount you require
1345	Anemosis .	By increasing this amount
1346	Anentera .	By reducing this amount
1347	Anenterous .	A large amount
1348	Anesthesia .	Half the amount
1349	Aneurism .	Double the amount
1350	Aneurismal .	The gross amount
1351	Anfracture .	The net amount
1352	Angelage .	The total amount
1353	Angelbed .	Only a small amount
1354	Angelfish .	What is the amount of
1355	Angelgold .	For what amount
1356	Angelhood .	A further amount
1357	Angelical .	Amount of indebtedness
1358	Angelicize .	The amount must be reduced
1359	Angelify . .	Cannot reduce amount
1360	Angelized .	If the amount is reduced
1361	Angelology .	Unless the amount is reduced
1362	Angelshot .	Unless the amount is increased
1363	Angelwater .	Amounts to less than
1364	Angioscope .	Amounts to more than
1365	Angiosperm .	**Amounting**
1366	Anglebar. .	Amounting to
1367	Anglebrace .	Not amounting to
1368	Anglefloat .	Amounting to upwards of
1369	Angleiron .	**Ample**
1370	Anglemeter .	This will be ample for
1371	Angleplane .	Ample for all demands
1372	Anglesite .	**Analysis**
1373	Angleties .	Analysis shows —— percentage of
1374	Anglicify. .	Analysis shows
1375	Anglicism .	Analysis shows only traces of
1376	Angola . .	Analysis of copper ore
1377	Angolacat .	Analysis of lead ore
1378	Angolapea .	Analysis of iron ore
1379	Angoragoat .	Analysis of the ore
1380	Angorawool.	Result of analysis
1381	Angrily . .	Having a complete analysis made
1382	Angriness .	**And**
1383	Anguineal .	And if
1384	Anguished .	And if not
1385	Angularly .	And if so
1386	Angulosity .	And when
1387	Angulous .	**Angle(s)**
1388	Angustate .	At an angle of ——
1389	Anhelose .	At right angles
1390	Anhydride .	Vein dips at an angle of ——

No.	Code Word.	
1391	Anhydrous .	**Angry**
1392	Animadvert.	Very angry with (that)
1393	Animalcule .	**Animation**
1394	Animalisms .	Considerable animation in
1395	Animality .	There is less animation in
1396	Animalized .	The animation has subsided
1397	Animalness .	The animation in the mining market
1398	Animated .	**Announce**
1399	Animation .	Enabling you to announce
1400	Animetta .	(Which) will enable us to announce
1401	Animism. .	When can we announce a dividend
1402	Animistic .	Cannot announce a dividend
1403	Animosity .	**Announced**
1404	Animus . .	It is officially announced
1405	Anisomeric .	Not yet officially announced
1406	Anisotrope .	Not yet announced
1407	Ankerite. .	Will be announced
1408	Anklebone .	The directors have announced
1409	Ankylosed .	A dividend of —— per share has been announced
1410	Annalistic .	Will not be announced (until)
1411	Annelide. .	As previously announced
1412	Annex . .	It is announced (that)
1413	Annexary .	**Announcement**
1414	Annexation .	The announcement is incorrect
1415	Annexment .	The announcement is correct
1416	Annihilate .	The announcement is incorrect, and you can con-
1417	Annotated .	Announcement is premature [tradict it
1418	Annotating .	**Annoyed**
1419	Annotation .	The directors are annoyed
1420	Annotatory .	I am (we are) much annoyed (that)
1421	Announce .	**Annoying**
1422	Announcing.	It is very annoying
1423	Annoy . .	The delay is very annoying
1424	Annoyances.	**Annual**
1425	Annoyful .	The annual meeting is fixed for
1426	Annoying .	Will be decided at the annual meeting
1427	Annoyous .	Must stand over till the annual meeting
1428	Annuities .	The annual meeting
1429	Annuity . .	At the annual meeting
1430	Annumerate	In time for the annual meeting
1431	Annunciate .	The annual report
1432	Anodynes .	Your annual report
1433	Anointed .	With the annual report
1434	Anointment.	Annual report sent
1435	Anomalous .	Annual report will be sent
1436	Anomaly. .	Annual report cannot be ready (till)
1437	Anonyme .	Send with the annual report
1438	Anonymity .	Annual statement
1439	Anonymous.	Annual statement for registration
1440	Anophyta .	Annual returns have been

No.	Code Word.	
1441	Anoplura	. **Annually**
1442	Anorexy .	. Returned annually
1443	Anorthura	. **Answer**
1444	Anoxoluin	. Telegraph ——'s answer
1445	Anserous.	. Answer my telegram of
1446	Answer .	. In consequence of receiving no answer
1447	Answering	. Answer the questions about
1448	Answerably	. Must have a definite answer by
1449	Answerless	. Answer immediately
1450	Antacid .	. Answer this by telegraph at once
1451	Antagonism.	Will communicate as soon as answer comes
1452	Antagonize	. Delay giving an answer (till)
1453	Antagony	. Answer quickly to these questions
1454	Antalgic .	. Expect to get an answer by
1455	Antalkali.	. Answer by post
1456	Antarctic	. Cannot get an answer
1457	Antbear .	. Cannot give an answer yet
1458	Antbirds.	. Cannot answer by telegraph
1459	Antcatcher .	Cannot answer by telegraph, will write you fully
1460	Anteater.	. It does not answer
1461	Antecedent	. It answers very well
1462	Antecessor	. Does it answer
1463	Antechapel	. How does it answer
1464	Antechoir	. We await your answer
1465	Antedated	. When shall we have an answer
1466	Antefixes	. Will answer when we get
1467	Antegg .	. Answer fully and in detail
1468	Antelopes	. Will give definite answer
1469	Antelucan	. Have not answer yet from
1470	Antemosaic	. **Answered**
1471	Antemural	. Has (have) answered in the affirmative
1472	Antenumber	Has (have) answered in the negative
1473	Antepenult	. Has (have) already answered
1474	Anteponed	. Has (have) not answered any of the questions
1475	Anteponing	. **Anticipate**
1476	Anteriorly	. Anticipate(s) some difficulty
1477	Anteroom	. Anticipate(s) no difficulty
1478	Antetemple	. Do you anticipate any difficulty
1479	Anthelix.	. Anticipate rise (in)
1480	Anthem .	. Anticipate fall (in)
1481	Anthemwise.	**Anticipated**
1482	Antherdust	. Have anticipated
1483	Anthericum	. **Anticipation**
1484	Antheroid	. In anticipation of
1485	Anthill .	. In anticipation of your wishes
1486	Anthillock	. **Antimony**
1487	Anthobians	. Oxide of antimony
1488	Anthodium	. Sulphide of antimony
1489	Antholite	. Antimony mine
1490	Antholysis	Antimony ore

No.	Code Word.	**Antimony** (*continued*)
1491	Anthomyia .	Largely mixed with antimony
1492	Anthotaxis .	Antimony occurs in
1493	Anthozasia .	Owing to the amount of antimony (in the)
1494	Anthozoic .	**Anxiety**
1495	Anthracite .	There is great anxiety
1496	Anthrax . .	There is great anxiety and annoyance
1497	Anthrenus .	Is there any cause for anxiety
1498	Anthriscus .	There is no cause for anxiety
1499	Anthropic .	Absolutely no cause for anxiety or alarm
1500	Anthropoid .	There is great cause for anxiety
1501	Anthurium .	In alleviation of the anxiety
1502	Anthus .	**Anxious**
1503	Anthyllis. .	Am (are) anxious
1504	Antiaditis .	Is (are) anxious
1505	Antichlor .	Anxious to have your report quickly
1506	Antichrist .	Anxious to have the matter settled
1507	Anticipant .	Anxious to hear how things are going on
1508	Anticivism .	Anxious to hear from you
1509	Anticlimax .	Anxious to get off at once
1510	Anticlinic .	Anxious for you to go to
1511	Anticmask .	Anxious to go to
1512	Anticness .	Anxious to come home
1513	Anticourt .	Anxious for you to see
1514	Antidactyl .	Anxious to have your opinion
1515	Antidesma .	You need not feel anxious
1516	Antidorcas .	Extremely anxious
1517	Antidotal .	Shareholders are very anxious
1518	Antidoting .	Shareholders are very anxious and desponding
1519	Antiemetic .	**Any**
1520	Antiface . .	Is (are) there any
1521	Antigraphy .	There is (are) not any
1522	Antilobium .	If there is (are) any
1523	Antiloquy .	There has (have) never been any
1524	Antimason .	Not any of them
1525	Antimonial .	Not any reason
1526	Antinomist .	There are hardly any
1527	Antiochian .	If there are any other
1528	Antipapal .	Any other
1529	Antipathy .	In any event
1530	Antiphonal .	**Anything**
1531	Antiphony .	Can anything be done
1532	Antipodean .	Have you done anything
1533	Antipoison .	Have not done anything
1534	Antipsoric .	Is there anything
1535	Antiptosis .	If you can do anything
1536	Antiquated .	**Apex**
1537	Antiquely .	The apex of the lode is not in our property
1538	Antiquist .	The apex of the lode is in our property
1539	Antirenter .	The apex of the lode
1540	Antiscians .	At the apex

No.	Code Word.	**Apex** (*continued*)
1541	Antiseptic .	Apex of lode is within surface boundaries
1542	Antispasis .	Apex of lode is not within surface boundaries
1543	Antitheist .	Apex of lode will be found
1544	Antithetic .	Apex of lode found
1545	Antitrade .	Apex of lode not found
1546	Antitragus .	Has apex of lode been found
1547	Antitype. .	Apex of lode intersects end lines
1548	Antitypous .	Apex of lode intersects side lines
1549	Antlered. .	Apex of lode intersects one end and one side line
1550	Antlermoth .	Apex of lode can be traced
1551	Antlion .	. **Apart**
1552	Antorbital .	Apart from this (these)
1553	Antozone .	Apart from this (these) cause(s)
1554	Antthrush	. **Apparent**
1555	Anvil . . .	It is quite apparent
1556	Anvilled . .	It is not apparent
1557	Anxieties	. **Appeal**
1558	Anxietude .	You had better appeal
1559	Anxiety . .	Intend(s) to appeal
1560	Anxious . .	Cannot appeal
1561	Anxiously .	Notice of appeal has been given
1562	Anybody. .	Shall I (we) appeal
1563	Anyhow . .	Do not appeal
1564	Anyone . .	Will —— appeal
1565	Anyrate . .	If we lose the case we shall appeal
1566	Anytime . .	Appeal at once
1567	Anywhere .	If we appeal
1568	Anywhither .	If you appeal
1569	Aortic . .	If he appeals
1570	Aortitis .	. **Appear**
1571	Apagogical .	I (we) do not want my (our) name to appear
1572	Apagynous .	Do (does) not want his (their) name to appear
1573	Apanage. .	It must not appear
1574	Apanthropy.	It should appear
1575	Apartments .	It appears to have been
1576	Apastron. .	Does not appear to be
1577	Apathetic .	It appears to be
1578	Apathist .	. **Appearance**
1579	Apathy . .	There is no appearance of
1580	Apebearer .	Judging from the appearance of (the)
1581	Apecarrier .	There is every appearance of
1582	Apedom . .	The appearance of the mine is good
1583	Apehood. .	The appearance of the mine is discouraging
1584	Apennine .	The appearance of
1585	Apepsy . .	In appearance
1586	Aperient. .	**Appertain**
1587	Aperture. .	All that appertains to
1588	Apetalous	. **Appliance(s)**
1589	Apexes . .	Complete appliances for opening up
1590	Aphanite .	What hoisting appliances have you

No.	Code Word.	**Appliance(s)** (*continued*)
1591	Aphelion .	Have you the necessary appliances for
1592	Aphnology .	Have no appliances for
1593	Aphorisms .	Require appliances for
1594	Aphoristic .	There is every appliance for
1595	Aphorized .	What appliances are there for
1596	Aphorizing .	(To) provide all appliances for
1597	Aphrizite. .	Well equipped with every appliance
1598	Aphrodites .	We have all needful appliances
1599	Aphthong .	**Application(s)**
1600	Aphyllous .	Has (have) made application to me (us)
1601	Apiarist . .	If —— make(s) an application to you
1602	Apiary . .	Will make application to you
1603	Apicillary .	To whom is application to be made
1604	Apiculture .	Has (have) sent an application
1605	Apiece . .	Do not close the list of applications till
1606	Apishly . .	Application arrived too late
1607	Apishness .	Make immediate application to
1608	Aplacental .	Application to be made by
1609	Aplanatic .	Application cannot be entertained
1610	Aplomb . .	Application will be attended to
1611	Aplotomy .	Notice has been given that the list of applications
1612	Aplustre . .	On application to [for the —— will close on
1613	Aplysia . .	If application is made
1614	Apocalypse .	Application(s) has (have) been made
1615	Apocarpous .	If no applications are made
1616	Apocopated .	Shall we make application
1617	Apocope. .	Will you make application
1618	Apocrisary .	Application has been made on your behalf
1619	Apocrustic .	When will application be made
1620	Apocryphal .	**Applied**
1621	Apocynum .	To be applied in part payment
1622	Apodeictic .	Has (have) applied for
1623	Apodema .	Has (have) not applied for
1624	Apodosis .	**Apply(ies)**
1625	Apogean . .	Apply all your energies to
1626	Apogiatura .	Does this apply to both
1627	Apograph .	Does this apply to all
1628	Apologetic .	To whom shall I (we) apply
1629	Apologist .	Apply for funds to
1630	Apologized .	Apply for information to
1631	Apologuer .	It will be necessary to apply to
1632	Apology . .	You had better apply to
1633	Apopemptic .	Apply it exclusively to
1634	Apophasis .	It is useless to apply to
1635	Apophthegm	Will not apply to
1636	Apophyge .	Does not apply to
1637	Apoplectic .	Applies to
1638	Apoplexy .	This applies only to
1639	Aporosa . .	This applies to all
1640	Aporrhais .	This applies to both

C

No.	Code Word.	
1641	Aposepidin	. **Appoint**
1642	Apostasy	. Must appoint some one to
1643	Apostated	. Do not appoint
1644	Apostatize	. Whom shall I (we) appoint
1645	Apostaxis	. You had better appoint —— (to)
1646	Apostolate	. **Appointed**
1647	Apostolic	. Has any one been appointed
1648	Apostrophe	. Who has (have) been appointed
1649	Apotactite	. Has (have) appointed
1650	Apothecary	. No one has been appointed
1651	Apothecium	Has (have) been appointed
1652	Apotheosis	. I (we) have appointed —— as my (our) attorney
1653	Apotrepsis	. **Apprehend**
1654	Apozemical	. We apprehend
1655	Appalling	. Apprehend there will be some trouble
1656	Appalment	. Apprehend there will be some difficulty
1657	Appanagist	. Need not apprehend any trouble
1658	Apparatus	. **Apprehension**
1659	Apparel .	. No cause for apprehension
1660	Apparelled	. **Approve**
1661	Apparency	. Hope you approve of the course we have taken
1662	Apparition	. The Board quite approve the steps you have taken
1663	Appeach .	. We approve the proposal
1664	Appeaching	. If you approve
1665	Appearance	. We do not approve the proposal
1666	Appeasable	. **Approved**
1667	Appeased	. Is approved
1668	Appellancy	. Is not approved
1669	Appellant	. Your course is approved
1670	Appendages	. Your course is not approved
1671	Appendicle	. Has (have) been approved by the Board
1672	Appendix	. Has (have) not been approved by the Board
1673	Appertain	. The proposition has been approved
1674	Appetence	. If the proposition is approved
1675	Appetite .	. **Approval**
1676	Appetizing	. Subject to approval (of)
1677	Applaud .	. Does not meet with the approval of (the)
1678	Applauding	. Has this your approval
1679	Applausive	. It has my (our) approval
1680	Appledrink	. The approval of
1681	Applefaced	. Awaiting approval
1682	Applegraft	. Subject to approval within —— days
1683	Applejack	. **Approximate**
1684	Applejohn	. What is the approximate value of
1685	Applemoth	. The approximate value (of)
1686	Appleparer	. The approximate amount of
1687	Applescoop	. The approximate tonnage
1688	Appletrees	. The approximate tonnage and value
1689	Applewine	. The approximate value of ore in sight
1690	Appliable	. The approximate yield

No.	Code Word.	**Approximate** (*continued*)
1691	Applicants .	The approximate assay value
1692	Applicator .	The approximate receipts
1693	Appliedly .	The approximate expenditure
1694	Applying .	**Approximation**
1695	Appoint . .	**Arbitrate**
1696	Appointeth .	Who will arbitrate (for)
1697	Appointing .	Get a competent person to arbitrate
1698	Apporters .	Cannot arbitrate
1699	Appraisal .	Will —— arbitrate
1700	Appraising .	**Arbitration**
1701	Appreciate .	Try to get the matter settled by arbitration
1702	Apprehend .	Will you agree to arbitration
1703	Apprentice .	Agree(s) to have the matter settled by arbitration
1704	Approached .	Will not agree to have the matter settled by arbi-
1705	Approbator .	Decline arbitration [tration
1706	Apprompt .	Shall I (we) submit the matter to arbitration
1707	Approof . .	The arbitration is against
1708	Approving .	The arbitration is in our favour
1709	Appulsion .	The arbitration is in favour of
1710	Appulsive .	Agree to arbitration
1711	Apricot . .	**Are**
1712	Aprilfool . .	Are they
1713	Apron . .	Are you
1714	Aproning .	Are they not
1715	Apronman .	Are you not
1716	Apronpiece .	Are not
1717	Apropos . .	If you are
1718	Apsidal . .	If you are not
1719	Apsis . . .	They are
1720	Aptable . .	There are
1721	Apteral . .	If there are
1722	Apterous . .	You are
1723	Apteryx . .	When you are
1724	Aptness . .	When we are
1725	Apyrexy . .	When they are
1726	Aquamarine .	**Area**
1727	Aquarium .	Covering an area of
1728	Aquatic . .	Extending over the entire area
1729	Aquatical .	A large area
1730	Aquatinta .	A small area
1731	Aqueduct .	The property covers an area of —— acres
1732	Aqueous . .	The property covers an area of —— square yards
1733	Aquiferous .	The property covers an area of —— square miles
1734	Aquilegia .	What is the area of the property (or concession)
1735	Aquiline . .	A larger area than
1736	Aquosity .	A smaller area than
1737	Arabical . .	A much larger area than
1738	Arabically .	**Argentiferous**
1739	Arabists . .	Argentiferous ore
1740	Arachis . .	In argentiferous

No.	Code Word.	Argentiferous (*continued*)
1741	Arachnida .	Argentiferous galena
1742	Arachnoid .	Argentiferous copper ore
1743	Araneiform .	**Argillaceous**
1744	Araneous .	In argillaceous
1745	Arapaima .	**Arid**
1746	Arapunga .	Situated in an arid district
1747	Araucaria .	The country around is very arid
1748	Arbalister .	**Arrange**
1749	Arbitrable .	You must arrange as soon as possible **to**
1750	Arbitrage .	If you cannot arrange
1751	Arbitrary .	To enable me (us) to arrange
1752	Arbitrator .	Enabling you to arrange
1753	Arbitress. .	I (we) cannot arrange to
1754	Arblasts . .	I (we) cannot arrange with
1755	Arboreal .	If you can arrange
1756	Arboretum .	You must arrange (with)
1757	Arboriform .	Endeavour to arrange
1758	Arborists .	Arrange as best you can
1759	Arborized .	Arrange as best you can, but subject to our approval
1760	Arborous .	We hope you will be able to arrange
1761	Arborvitae .	Arrange as soon as possible
1762	Arbourvine .	**Arranged**
1763	Arbuscle. .	I have arranged to go to —— on
1764	Arbuscular .	Telegraph when you have arranged
1765	Arcade . .	Everything has been satisfactorily arranged
1766	Archaical .	Nothing has been yet arranged
1767	Archaist . .	Has (have) arranged with
1768	Archangel .	To be arranged hereafter
1769	Archbands .	Has (have) arranged to
1770	Archbishop .	Have everything properly arranged for inspection
1771	Archboard .	When you have arranged
1772	Archbrick .	Why have you not arranged
1773	Archbutler .	Everything will be arranged
1774	Archchemic.	If you have arranged
1775	Archcount .	Has (have) not yet arranged
1776	Archdeacon.	Arranged as stated in our letter of
1777	Archdruid .	Arranged as proposed
1778	Archducal .	Has anything been arranged about
1779	Archduchy .	Must be arranged before
1780	Archegony .	Must be arranged before you leave
1781	Archenemy .	Must be arranged before I leave
1782	Archeology .	Can be arranged after
1783	Archeress .	**Arrangement(s)**
1784	Archerfish .	No time to make other arrangements
1785	Archery . .	What arrangement has been made (with regard to
1786	Archetypal .	Make whatever arrangements may be necessary
1787	Archiater .	By this arrangement
1788	Architects .	An arrangement has been made
1789	Architrave .	No final arrangement has yet been made
1790	Archive . .	Will not make any arrangement

No.	Code Word.	**Arrangement(s)** (*continued*)
1791	Archivists	Make an arrangement by which
1792	Archivolt	Have made arrangements to
1793	Archlute	According to arrangement
1794	Archmock	According to arrangements made (with)
1795	Archonship	Arrangement must be modified
1796	Archontic	Make other arrangements
1797	Archpastor	Have not been able to make arrangements
1798	Archpillar	Confirm arrangements
1799	Archpoet	Arrangements approved
1800	Archpriest	If arrangements are not approved
1801	Archrebel	An arrangement has been made for you to inspect
1802	Archstone	A temporary arrangement [and report upon
1803	Archtype	Arrangement(s) only temporary
1804	Archtyrant	Cannot approve such an arrangement
1805	Archways	If the arrangement is made
1806	Archwife	When arrangement is made
1807	Arcograph	**Arrastre(s)**
1808	Arctitude	The ore has been ground in arrastres
1809	**Arctogeal**	The only appliances on the mine for treating the ore are arrastres
1810	Arctomys	It will be as well to continue using arrastres (till)
1811	Arcturus	**Arrival**
1812	Arcubalist	On arrival of
1813	Ardassine	On your arrival in (at)
1814	Ardentness	Advise —— of your arrival
1815	Arduous	Wait for arrival of my written report
1816	Arduously	Cannot wait for arrival of your written report
1817	Areasneak	Telegraph on your arrival at
1818	Arefaction	Await arrival of
1819	Arenaceous	Shall await your arrival in order to discuss with you
1820	Arendalite	**Arrive**
1821	Arenicola	Must arrive here by —— or will be too late
1822	Arenilitic	Will probably arrive about
1823	Areolar	Is (are) expected to arrive about
1824	Areolation	Is (are) not expected to arrive before
1825	Areopagist	When do you expect --—— to arrive
1826	Areopagy	**Arrived**
1827	Areostyle	Arrived here safely
1828	Argandlamp	Has (have) not yet arrived
1829	Argillitic	Arrived too late
1830	Argillous	Arrived here to-day
1831	Argosies	Arrived here yesterday
1832	Argosy	Telegraph when —— has (have) arrived
1833	Argued	**Arsenic**
1834	Arguing	Arsenical pyrites
1835	Argument	Containing arsenic
1836	Argumentum	**Articles**
1837	Arguseyed	According to the articles of association
1838	Argusshell	Require the following articles
1839	Argutation	**Asbestos**

No.	Code Word.	**Asbestos** (*continued*)
1840	Arhizous .	Asbestos mine
1841	Arianize . .	Asbestos occurs in
1842	Arianizing .	**Ascertain**
1843	Aristarch .	Ascertain who is
1844	Aristocrat .	Ascertain what is owner's price
1845	Aristology .	Can you ascertain
1846	Arithmancy .	Cannot ascertain anything about
1847	Arithmetic .	Ascertain as soon as possible and telegraph
1848	Arkshell . .	Ascertain from
1849	Arlepenny .	Ascertain what is doing (at)
1850	Armband .	Could only ascertain
1851	Armchair .	Ascertain the financial condition of
1852	Armental .	To ascertain the extent of
1853	Armgaunt .	Ascertain the value of
1854	Armholes .	Ascertain if he is reliable
1855	Armillated .	Ascertain if he is experienced and reliable
1856	Armillett .	**Ascertained**
1857	Armipotent .	Have you ascertained (that)
1858	Armisonant .	(Has) have ascertained (that)
1859	Armisonous .	**Ask(s)**
1860	Armistice .	Whom shall I (we) ask
1861	Armouries .	You must ask
1862	Armozine .	I will ask
1863	Armpit . .	Ask about
1864	Armrack . .	Ask —— what he has done
1865	Armsweep .	Ask him (them) if (or whether)
1866	Armycorps .	Ask who
1867	Armylist . .	Ask what has been done
1868	Arnica . .	**Asked**
1869	Arnicine . .	Has (have) asked about
1870	Arnoldist .	Has (have) asked all who have been there
1871	Aroma . .	Have asked him (them)
1872	Aromatic .	Have asked if
1873	Aromatical .	We have been asked (to)
1874	Aromatizer .	Have asked what has been done
1875	Arpentator .	**Asking**
1876	Arquerite .	Has (have) been asking about
1877	Arraign . .	Is (are) asking a great deal too much
1878	Arraigners .	What is (are) —— asking for
1879	Arranged .	**Aspect**
1880	Arranging .	The mine has a most favourable aspect
1881	Arraswise .	I (we) do not like the aspect of affairs
1882	Arrectary .	A most promising aspect
1883	Arrest . .	A most discouraging aspect
1884	Arresteth .	Judging from the aspect of (the)
1885	Arresting .	The mine now presents a more favourable aspect
1886	Arrestment .	What is the present aspect of the mine
1887	Arrestor . .	What is the present aspect of
1888	Arrhythmy .	Presents a more promising aspect
1889	Arrival . .	The political aspect (is)

No.	Code Word.	**Aspect** (*continued*)
1890	Arrivance .	The financial aspect (is)
1891	Arrogant .	The commercial aspect (is)
1892	Arrogantly .	**Assay(s).** (See Table at end.)
1893	Arrogation .	Please have it (them) assayed for
1894	Arrogative .	From the result of these assays
1895	Arrowgrass .	~ An assay of this sample gave
1896	Arrowheads.	Assays satisfactory
1897	Arrowing .	Assays average —— per ton
1898	Arrowlet . .	Assays disappointing
1899	Arrowroot .	Assays made of the vein matter give
1900	Arrowstone .	Assays from borings in face of
1901	Arsenals . .	Assays of ore from —— foot level show
1902	Arsenical .	Assays from diamond-drillings
1903	Arsenious .	Assays made from surface ore show
1904	Arsenuret .	A rough assay showed
1905	Arson . .	Please assay for
1906	Artemisia .	There is a good assay office at the mill
1907	Arteriac . .	Have had samples assayed with following result
1908	Artesian . .	Pulp assays at mill average
1909	Artful . .	Ore on dump assays
1910	Artfully . .	An assay of this sample gave —— in gold
1911	Artfulness .	An assay of this sample gave —— ozs. in silver
1912	Arthritis . .	An assay of this sample gave —— % in lead
1913	Arthrodia .	An assay of this sample gave —— % in copper
1914	Arthropoda .	An assay of which showed [—— ozs. in silver
1915	Artichoke .	An assay of this sample gave —— in gold and
1916	Articled . .	An assay of this sample gave —— in gold and —— ozs. in silver, and —— percentage of lead
1917	Articling . .	In consequence of there being no assay office the samples have to be sent to —— to be assayed
1918	Articulate .	Average assay of ore milled for the month is
1919	Artificial . .	Pulp assay value of ore
1920	Artilize . .	Assay value of ore (from)
1921	Artilizing .	Assay of tailings
1922	Artillery . .	Assay of concentrates
1923	Artistic . .	Assay from battery
1924	Artistical .	What is the assay value of ore you are now working
1925	Artistlike .	Assay value of ore now being milled
1926	Artlessly . .	What is the assay value of ore
1927	Artocarpad .	**Assert**
1928	Artotyrite .	Do (does) —— assert the statement to be true
1929	Artunion ,	Assert(s) the statement to be true
1930	Arundelian .	Assert(s) the statement is not true
1931	Arytenoid .	**Assessment**
1932	Asadulcis .	How much will assessment be on
1933	Asafetida .	Assessment has been levied on
1934	Asaphus . .	Assessment is soon to be levied on
1935	Asbestos . .	Assessment will be levied on
1936	Asboline . .	Unless this is done, assessment will be levied on
1937	Ascendable .	They have only done the necessary assessment work

No.	Code Word.	**Assessment** (*continued*)
1938	Ascendancy .	Will not levy any more assessment
1939	Ascending .	In doing the assessment work
1940	Ascension .	Will have to levy assessment
1941	Ascertain .	Have not done the necessary assessment work
1942	Ascessant .	Will there be an assessment on
1943	Ascetic . .	Stock has been sold for assessment
1944	Asceticism .	To save the expense of assessment work
1945	Ascham . .	Assessment paid
1946	Ascidiform .	Assessment unpaid
1947	Ascidium .	**Assets**
1948	Ascigerous .	Assets reported to be about
1949	Ascitical . .	What are the expected assets
1950	Asclepiad .	There are no assets
1951	Ascospore .	The assets will be very small
1952	Ascribable .	Large assets
1953	Ascribe . .	Assets not yet ascertained
1954	Ascribing .	Assets and liabilities
1955	Ascription .	Net assets
1956	Aseptic .	Balance of assets
1957	Ashamed .	Cash assets
1958	Ashamedly .	Assets, including stores, etc.
1959	Ashbuds . .	Realized assets
1960	Ashcolour .	Unrealized assets
1961	Ashfire . .	Assets will take some time to realize
1962	Ashflies .	**Assent**
1963	Ashfly . .	Do you assent to this
1964	Ashfurnace .	We assent to the proposal
1965	Ashleach .	**Assist**
1966	Ashlering .	Can you assist us (to)
1967	Ashore . .	Can assist us
1968	Ashypale .	Is disposed to assist us
1969	Asiaticism .	Will he assist
1970	Asinine . .	Will assist you
1971	Asiphonata .	In order to assist
1972	Askance .	**Assistance**
1973	Askew . .	Must have assistance in examining the —— mine
1974	Askingly .	Must have assistance
1975	Aslant . .	Get —— to give the necessary assistance
1976	Asleep . .	Will send assistance
1977	Aslope . .	Will give you every assistance
1978	Aspalathus .	Immediate assistance
1979	Asparagus .	Give all the assistance required
1980	Aspartic . .	What assistance do you want
1981	Aspasia . .	Unable to render assistance
1982	Aspect . .	Can you render assistance
1983	Aspectable .	Further assistance
1984	Asperate . .	Cost of assistance
1985	Asperating .	**Assistant**
1986	Aspergill .	We want an assistant in the
1987	Asperities .	Get an assistant

No.	Code Word.	**Assistant** (*continued*)
1988	Asperity . .	Get an assistant if absolutely necessary
1989	Aspermous .	Will you pay cost of assistant
1990	Aspersed .	**Associated**
1991	Aspersions .	Is (are) associated with
1992	Aspersive .	Who is (are) —— associated with
1993	Aspersory .	Is (are) not any longer associated with
1994	Asphalt . .	Who is (are) —— associated with in this matter
1995	Asphaltic .	Avoid being associated with
1996	Asphodelus .	**Association**
1997	Asphyxial .	Memorandum of Association
1998	Asphyxy . .	The Articles of Association
1999	Aspidium .	In association with
2000	Aspirants .	**Assorted**
2001	Aspiration .	Assorted samples assayed
2002	Aspiratory .	Carefully assorted
2003	Aspirement .	Assorted samples of ore from the lode
2004	Aspiring . .	Assorted samples of ore from the dump
2005	Aspiringly .	**Assortment**
2006	Asplenium .	A good assortment
2007	Assailable .	**Assure**
2008	Assailants .	I (we) can assure you that
2009	Assailed . .	Assure(s) me (us) that
2010	Assapan . .	**At**
2011	Assapanic .	Shall be at —— on
2012	Assassins .	Is at
2013	Assaulter .	Will be at —— about
2014	Assaulting .	At about
2015	Assayed . .	Or at
2016	Assaying . .	And at
2017	Assecured .	And at what
2018	Assecution .	If you are at
2019	Assemblage .	If not at
2020	Assemble .	**Atmosphere**
2021	Assembling .	The effects of the atmosphere [sphere
2022	Assentator .	In consequence of the rarified state of the atmo-
2023	Assented .	**Attached(ment)**
2024	Assentient .	Bullion is attached
2025	Assertory .	Mine is attached for
2026	Asservile . .	Has (have) levied an attachment
2027	Assessably .	To get the attachment removed
2028	Assessment .	Levy an attachment
2029	Assessors . .	Cannot levy an attachment
2030	Asseverate .	**Attempt**
2031	Assiduity .	No practical attempt has been made (to)
2032	Assiduous .	Have made an attempt (to)
2033	Assientist .	Cannot attempt (to)
2034	Assignable .	No use to attempt (to)
2035	Assignment .	Shall attempt (to)
2036	Assignors .	An attempt was made (to)
2037	Assimilate .	**Attend**

No.	Code Word.	Attend (*continued*)
2038	Assistance	Please attend to this matter
2039	Assistless .	Will attend
2040	Assize . .	Cannot attend to —— till after
2041	Assizeball .	Must attend at once to
2042	Associable .	Attend immediately to
2043	Associator .	Will attend to (it)
2044	Assoil . .	Please attend promptly to
2045	Assoilize . . **Attention**	
2046	Assortment .	Give great attention to
2047	Assuage . .	No attention has been paid to saving the tailings
2048	Assuaging .	The first thing requiring attention is
2049	Assuasive .	Pay no attention to
2050	Assuetude .	Will give every attention
2051	Assumings .	This requires your greatest attention
2052	Assumpsit .	Is (are) well worthy of your attention
2053	Assumpted . **Attorney** (See Power.)	
2054	Assumption .	Power of attorney
2055	Assumptive .	Require(s) power of attorney to be produced
2056	Assurable .	Who holds power of attorney
2057	Assurances .	Must have power of attorney for
2058	Assuredly .	Is power of attorney required
2059	Assurgency .	Do not require power of attorney
2060	Astacus . .	Send power of attorney for
2061	Asteriated .	Impossible to do anything without power of attorney
2062	Asteridian . **Auriferous**	
2063	Asterisk . .	The district is highly auriferous
2064	Asteroid . .	The ore is highly auriferous
2065	Asteroidal .	Auriferous ore
2066	Astheny . .	Auriferous sand
2067	Asthma . .	The beach is auriferous
2068	Asthmatic .	The sands are auriferous
2069	Astir . . .	Auriferous gravel
2070	Astomata .	Auriferous deposits
2071	Astonish . . **Auspices**	
2072	Astound . .	Under good auspices
2073	Astounding .	Under bad auspices
2074	Astragal . . **Authorities**	
2075	Astragalus .	Authorities require
2076	Astrakhan .	Authorities strongly opposed to
2077	Astrantia . .	Authorities favourable to
2078	Astraught .	Authorities favourable to concession for
2079	Astray . .	Authorities not very favourable to concession for
2080	Astricted . **Authority**	
2081	Astringeth .	You have full authority to
2082	Astringing .	You have no authority
2083	Astrofell . .	Must have full authority
2084	Astrogeny .	By what (whose) authority
2085	Astrognosy .	Who is a great authority here on mines
2086	Astrolabe .	Cannot give the necessary authority
2087	Astrolatry .	Must have written authority

No.	Code Word.	Authority *(continued)*
2088	Astrologer .	Telegraph authority (to)
2089	Astrometer .	What authority have you for your statements
2090	Astronomic.	Has (have) full authority to
2091	Astroscope .	Has (have) no authority to
2092	Astucious .	My authority for the report is
2093	Astucity . .	Had no authority whatever to
2094	Astute . .	Without authority
2095	Astuteness .	Cannot be done without authority (of)
2096	Astylar . .	My authority is unimpeachable
2097	Asunder . .	**Authorize**
2098	Asylum . .	The Board authorize you
2099	Asymmetral.	The Board authorize you to do as you propose
2100	Asymtotic .	The Board authorize you to act as you think expe-
2101	Asynartete .	The Board authorize you to negotiate [dient
2102	Asyndetic .	The Board authorize you to purchase
2103	Atacamite .	Will the Board authorize me to
2104	Ataraxia . .	Please authorize me (us)
2105	Atavism . .	I (we) authorize
2106	Ataxic . .	Cannot authorize
2107	Ateuchus .	Do not authorize
2108	Athalamous.	**Authorized**
2109	Athalia . .	Has (have) authorized —— to
2110	Athanasian .	Has (have) been authorized to
2111	Athanor . .	Is (are) not authorized to
2112	Atheist . .	**Autumn**
2113	Atheistic . .	During the autumn
2114	Atheized . .	Last autumn
2115	Atheneum .	Next autumn
2116	Atheology .	**Available**
2117	Atheous . .	Will not be available
2118	Athericera .	Available for the purpose
2119	Atherina . .	The amount of available
2120	Athermancy	Every available means
2121	Atherome .	Every available means must be used
2122	Athlete . .	Every available man
2123	Athletics . .	Utilize all available
2124	Athrob . .	Every ton of ore available
2125	Athwart . .	Will be available (for)
2126	Athymia . .	Is —— available (for)
2127	Atlantean .	**Average**
2128	Atlasfolio .	Above the average
2129	Atmolysis .	Below the average
2130	Atmosphere.	Certain to average
2131	Atomical .	Likely to average
2132	Atomicism .	What is the average
2133	Atomicity .	Average assay value of ore in the mine is
2134	Atomistic .	Average width of the vein is
2135	Atomizer .	The average assays for last week were
2136	Atonement .	The average assays for last month were
2137	Atrabiliar .	Average about

No.	Code Word.	Average (*continued*)
2138	Atramental .	An average sample
2139	Atrocious .	Will average
2140	Atrocities .	Will not average
2141	Atrocity . .	Average assay value of ore
2142	Atropine . .	Average assay value of tailings
2143	Attach . .	Average price paid for
2144	Attachable .	Average rate of wages
2145	Attaching .	Average cost per ton
2146	Attachment .	Average weight
2147	Attacked . .	**Avert**
2148	Attacottic .	In order to avert
2149	Attagas . .	Cannot avert
2150	Attaghan .	In order to avert the damage
2151	Attainable .	In order to avert future loss
2152	Attainder .	It is desirable to avert
2153	Attaining .	To avert a catastrophe
2154	Attainment .	**Avoid**
2155	Attainture .	Avoid unnecessary expense
2156	Attaminate .	Try to avoid
2157	Attempt . .	Avoid delay as much as possible
2158	Attempteth .	Will try to avoid
2159	Attempting .	You must avoid
2160	Attemptive .	Avoid legal proceedings
2161	Attendance .	Is (are) trying to avoid
2162	Attending .	Cannot avoid
2163	Attendress .	Avoid communicating with
2164	Attentive .	**Avoided**
2165	Attenuated .	If it can be avoided
2166	Attested . .	Could not be avoided
2167	Atticism . .	Must be avoided
2168	Atticized . .	This will be avoided in future
2169	Atticizing .	Cannot be avoided
2170	Attitude . .	Has been avoided
2171	Attollent . .	Can it be avoided
2172	Attorneyed .	**Aware**
2173	Attract . .	Are you aware that
2174	Attraction .	No one was aware
2175	Attractive .	Have (has) just become aware of the fact
2176	Attrahent .	**Away**
2177	Attributed .	Is (are) away for a long time
2178	Attrite . .	Is (are) away, but will shortly return
2179	Attuned . .	When do you expect to get away
2180	Atwixt . .	Expect to get away
2181	Aubade . .	Was (were) sent away on
2182	Auburn . .	Get away as soon as possible
2183	Auctionary .	Could not get away on
2184	Auctioneer .	Has (have) been away
2185	Audacious .	Cannot get away till
2186	Audacity .	Whilst you are away
2187	Audiences .	Whilst I am (we are) away

No.	Code Word.	**Away** (*continued*)
2188	Audient . .	Whilst —— is (are) away
2189	Audiometer .	How long will he (you) be away
2190	Audiphone .	How long will —— be away
2191	Auditday .	Will be away
2192	Audited . .	Will be away some time
2193	Audithouse .	Expect to be away
2194	Auditing . .	**Await**
2195	Auditorial .	Await arrival of
2196	Auditory . .	Await my letter of
2197	Augerbit . .	Await further telegram
2198	Augergauge .	Await instructions
2199	Augerhole .	**Back**
2200	Augershell .	When will you be back at
2201	Augitic . .	Back out (from)
2202	Augmented .	If you can go back
2203	Augmenting	Can go back
2204	Augurated .	Is (are) trying to back out
2205	Augurist . .	Try and back out
2206	Augurizing .	Try and get back
2207	Augurship .	Hold back
2208	Augury . .	Shall have to go back to
2209	Augustly . .	In order to get back
2210	Augustness .	Cannot get back
2211	Aularian . .	At the back of
2212	Aulnage . .	Can you go back
2213	Aulostoma .	Has (have) gone back to
2214	Auntsally .	Wish (wishes) to get back
2215	Aurelian . .	Back from
2216	Auricular .	Is (are) holding back
2217	Auriferous .	**Backed**
2218	Aurigation .	Has (have) backed out (of)
2219	Auriscalp .	Has (have) been backed up by
2220	Auriscope .	Has (have) backed up
2221	Aurochs . .	Who has (have) backed up
2222	Auspex . .	**Backs**
2223	Auspicated .	How many feet of backs have you above —— level
2224	Auspices .	How many feet of backs have you above cross-cut
2225	Auspicial .	We have —— feet of backs
2226	Austere . .	**Bad**
2227	Austerity .	Very bad indeed
2228	Australize .	Not so bad as we were told
2229	Authentic .	The mine is looking very bad
2230	Authorial .	Owing to bad management
2231	Authority .	Owing to the bad state of the roads
2232	Authorized .	In bad condition
2233	Authorless .	From bad to worse
2234	Authorling .	Things are in a very bad state
2235	Authorship .	Owing to bad
2236	Authotype .	Has been ruined by bad management
2237	Autochthon .	**Badly**

No.	Code Word.	**Badly** (*continued*)
2238	Autoclave	Badly managed
2239	Autocracy	The mine has been badly managed
2240	Autocrat.	Badly worked
2241	Autocratic	Has (have) been badly worked
2242	Autogeneal	Has (have) been badly treated
2243	Autographs	Turning out badly
2244	Automatism	Turned out badly
2245	Automatize . **Bags**	
2246	Automatous	Packed in bags
2247	Automolite	Send us —— bags (of)
2248	Autonomic	Bags of
2249	Autophagi . **Balance(s)**	
2250	Autophoby	There will be a considerable balance
2251	Autophon	There will be only a small balance
2252	Autoplasty	What balance will there be
2253	Autopsical	Transfer the balance to
2254	Autopsies	Balance all right
2255	Autopsy .	The balance of the money to be paid
2256	Autoptic.	Telegraph the probable balance when expenses have
2257	Autumn .	Cannot remit balance until [been paid
2258	Autumnal	This will leave sufficient balance for
2259	Autumnity	The balance is of no account
2260	Auxesis .	Will the balance be sufficient for
2261	Auxiliary	The balance of the shares
2262	Availably	There is a balance at bank (of)
2263	Availeth .	What balance have you in hand
2264	Availment	Balance in hand
2265	Avalanche	Balance in our favour
2266	Avantfosse	On balance of account
2267	Avanturine	Due as balance
2268	Avarice .	The balance due to
2269	Avaricious	It will about balance the
2270	Avaunt .	Pay the balance
2271	Avauntance.	In order to balance the account
2272	Avauntry	The balance of probabilities (is)
2273	Avellane. . **Balanced**	
2274	Avenary .	The account(s) is (are) now balanced
2275	Avenged.	If you have balanced the accounts
2276	Avengeress . **Balance sheet**	
2277	Aventayle	Send the balance sheet without delay
2278	Average .	The balance sheet will be forwarded by post on
2279	Averagely . **Ballast**	
2280	Averaging	Will be taken as ballast
2281	Avercake	For ballast
2282	Avercorn	Will send —— tons as ballast
2283	Averdupois	Is in ballast
2284	Averments	Can be sent as ballast
2285	Averpenny . **Bank(s)**	
2286	Averrhoa	This sum to be deposited in the bank (of)
2287	Averrhoist	Have deposited the scrip at the bank (of)

No.	Code Word.	Bank(s) (*continued*)
2288	Aversant . .	Have deposited —— with bank (of)
2289	Aversation .	Less bank commission
2290	Averseness .	I cannot arrange with the bank (of)
2291	Aversively .	Can you arrange with the bank (of)
2292	Averters . .	Have arranged with the bank (of)
2293	Aviaries . .	Can arrange with the bank (of) [the bullion
2294	Avicennia .	The bank will credit us at once with the value of
2295	Avicularia .	No bank will discount your paper ; you must deposit
2296	Avidiously .	Send bullion to bank (of) [—— with
2297	Avifauna . .	The security is at the bank of
2298	Avisand . .	Please pay into the bank
2299	Avizandum .	Has (have) paid into the bank of
2300	Avocatory .	Will pay —— into the bank (of)
2301	Avocet . .	Banks are calling in their money
2302	Avoidable .	Banks are calling in their money, as there is likely
2303	Avoidances .	Bank will remit [to be a panic
2304	Avoided . .	Bank will allow an overdraft of
2305	Avoiding .	Will bank allow an overdraft of
2306	Avoidless .	There has been a run on the bank of
2307	Avouchable .	Bank will not advance any more, unless overdraft
2308	Avouched .	Bank refuses (to) [now due is paid
2309	Avouching .	Banks will not advance unless secured by un-
2310	Avowals . .	Bank agrees (to) [doubted collaterals
2311	Avowedly .	Badly in want of money ; telegraph to the bank of
2312	Avowries .	Bank holds the deeds [—— immediately
2313	Avowry . .	Bank holds the documents
2314	Avulsed . .	Bank refuses to give up the documents
2315	Avuncular .	Bank has given up the deeds
2316	Await . .	Bank will hold the deeds (until)
2317	Awaiting .	Bank holds deeds in escrow
2318	Awakement .	Bank holds the bullion
2319	Awaken . .	Bank pays us
2320	Awakening .	If bank refuses to advance
2321	Awarded .	Bank advance
2322	Awaygoing .	Bank will advance
2323	Awesome .	Bank will advance only
2324	Awestruck .	Account at bank
2325	Awhile . .	Bank account overdrawn
2326	Awkward .	Overdraft at bank
2327	Awkwardly .	Surplus at bank
2328	Awlshaped .	To be sent to bank
2329	Awlwort . .	Bank refused to cash draft
2330	Awnings . .	Bank refused to discount
2331	Axaycatl . .	Bank refused to advance
2332	Axeheads .	**Bankers**
2333	Axeman . .	Bankers to the Company
2334	Axeshaped .	Bankers to the Vendor(s)
2335	Axestone .	**Banket**
2336	Axinomancy	Banket reef
2337	Axiomatic .	Banket easily worked

No.	Code Word.	**Banket** (*continued*)
2338	Axioms . .	Banket is rich in gold
2339	Axiopisty .	Banket very fully developed
2340	Axlebar . .	Banket not developed
2341	Axleboxes .	Banket does not carry much gold
2342	Axleclip . .	Banket is widely distributed and developed on the
2343	Axleguard .	From the banket [whole property (district)
2344	Axlepins .	Is the banket
2345	Axleskein .	Banket assays
2346	Axlesleeve .	**Bankrupt(cy)**
2347	Axletree . .	The owner(s) is (are) bankrupt
2348	Axolotl . .	On the verge of becoming bankrupt
2349	Axotomous .	The company is bankrupt
2350	Axunge . .	Is (are) not bankrupt as reported
2351	Ayegreen .	Is (are) reported bankrupt
2352	Ayenbite .	Is it true that —— is (are) likely to become
2353	Ayenward .	Will become bankrupt unless [bankrupt
2354	Azalea . .	Bankruptcy imminent
2355	Azimuth . .	Proceedings in bankruptcy
2356	Azimuthal .	**Bar(s)**
2357	Azobenzene.	Have struck a bar of hard ground
2358	Azobenzol .	Shall ship —— bars of bullion on
2359	Azolitmine .	Have shipped this week —— bars
2360	Azotous . .	Have shipped this month —— bars
2361	Azured . .	Estimated value of bars shipped
2362	Azurestone .	When will the next shipment of bars be made
2363	Azymite . .	The shipment of bars delayed by
2364	Azzletooth .	Since the last shipment of bars
2365	Baalism . .	There has been a regular weekly shipment of ——
2366	Babblishly .	Number of bars [bars
2367	Babiana . .	Bars of gold
2368	Babiroussa .	Value of bars
2369	Babishly . .	Bars shipped
2370	Babishness .	Weight of bar(s)
2371	Baboons . .	**Bargain(s)**
2372	Babyfarmer .	A very good bargain
2373	Babyhood .	A very poor bargain
2374	Babyhouse .	Try and make a better bargain
2375	Babyish . .	Cannot make a better bargain
2376	Babyjumper	Have struck a bargain
2377	Babylonian .	Bargain is off
2378	Babylonish .	Is bargain off
2379	Babypin . .	**Barometer**
2380	Babywalker .	Mercurial barometer
2381	Baccarat . .	Aneroid barometer
2382	Bacchanal .	**Barren**
2383	Baccharic .	Barren-looking ground
2384	Bacciform .	Barren quartz
2385	Bacheleria .	Vein is barren
2386	Bachelor . .	**Basalt**
2387	Bacillaria .	**Basaltic**

No.	Code Word.	**Basaltic** (*continued*)
2388	Backarack .	In basaltic rock
2389	Backbands .	**Base**
2390	Backbiteth .	At the base of the
2391	Backbiting .	From the base of the
2392	Backbone .	To the base of the
2393	Backcarry .	**Battery**
2394	Backcast . .	What do battery assays average
2395	Backcentre .	Battery assays average
2396	Backcomb .	Battery assays for the week average
2397	Backdoors .	Battery assays for the month average
2398	Backend . .	New shoes and dies required for the battery
2399	Backfaller .	The battery returns
2400	Backfriend .	Are losing in the battery
2401	Backgammon	There is a good battery of —— stamps
2402	Background .	There is a poor battery of —— stamps
2403	Backhanded .	A —— stamp battery
2404	Backingup .	A —— stamp battery wanted
2405	Backjoint .	A prospecting battery
2406	Backlink .	**Be**
2407	Backpiece .	To be
2408	Backraking .	Not to be
2409	Backroom .	Can be
2410	Backsight .	Cannot be
2411	Backslang .	Should be
2412	Backslider .	Should not be
2413	Backspeed .	Will be
2414	Backstream .	Will not be
2415	Backstring .	Must be
2416	Backsword .	Must not be
2417	Backwardly .	If there should be
2418	Backwards .	In order to be
2419	Backwater .	In order not to be
2420	Bacterium .	Might be
2421	Bactris . .	Might not be
2422	Baculites .	Will it be
2423	Badgeless .	Will it not be
2424	Badgered .	**Bear(s)** [soon advance
2425	Badgering .	Bear movement on —— stock; I believe it will
2426	Baffling . .	There is said to be a bear movement (on)
2427	Bafflingly .	A bear interest in
2428	Bagatelle .	A bear interest in our shares
2429	Bagfilters .	A bear movement
2430	Bagfox . .	A strong bear movement
2431	Baggagers .	The bears are active
2432	Baggy . .	The bears are endeavouring
2433	Bagnolian .	The bears have forced down
2434	Bagpiper .	Bears have succeeded in their operations
2435	Bagreefs . .	Bears have been caught short
2436	Bagwig . .	Bear operations
2437	Baikalite . .	Bear heavily

No.	Code Word.	**Bear** (*continued*)
2438	Bailbond . .	Bear interest on
2439	Bailiffs .	. **Bearer(s)**
2440	Bailiwick .	Payable to bearer
2441	Bailpieces .	The bearer of our letter to you
2442	Bailscoop .	The bearer(s) of
2443	Bairam .	. **Bearing**
2444	Baisemains .	The vein bearing about ——° East of North
2445	Baitmill . .	The vein bearing about ——° West of North
2446	Baize . . .	The vein bearing
2447	Bakedmeat .	Gold-bearing veins
2448	Bakehouse .	Are insiders bearing
2449	Bakerfoot .	Bearing the stock
2450	Bakery . .	Bearing the shares
2451	Bakester . .	Bearing interest
2452	Bakshish . .	Bearing interest at the rate of
2453	Balabeds . .	Bearing little
2454	Balance . .	Bearing much
2455	Balancing .	**Because**
2456	Balandrana .	Because of the
2457	Balaninus .	Not because of
2458	Balanoid . .	Because if there are
2459	Balaustion .	Because if not
2460	Balbutiate .	**Become(s)**
2461	Balconied .	Which will become
2462	Balcony . .	Unless it becomes
2463	Baldachin .	Do not let it become
2464	Baldeagle .	Has (have) become
2465	Balderdash .	**Bed**
2466	Baldmoney .	The main bed of ore
2467	Baldness .	Through the bed
2468	Baldrib . .	In the bed of the river
2469	Baldricks .	**Bedrock**
2470	Balefire . .	In the decomposed bedrock
2471	Balkingly .	On the bedrock
2472	Ballad . .	Close to the bedrock
2473	Balladist . .	Bedrock price
2474	Balladized .	**Been**
2475	Ballahou . .	Has (have) been
2476	Ballastage .	Has (have) been here
2477	Ballasting .	Has (have) not been
2478	Ballatoon .	Has (have) not yet been
2479	Ballcaster .	Which must have been
2480	Ballimong .	Which cannot have been
2481	Ballinggun .	Should have been
2482	Ballismus .	Should not have been
2483	Ballistic . .	If you have been
2484	Ballooning .	May have been
2485	Balloonry .	Had been
2486	Ballotant .	If it had been
2487	Ballotbox .	If it had not been

No.	Code Word.	**Been** (*continued*)
2488	Ballproof	. We have been
2489	Ballscrew	. We have not been
2490	Ballstock	. You have been
2491	Balltrain	. You have not been
2492	Balltrolly	. If he has been
2493	Ballvalve	. If he has not been
2494	Ballvein .	. If they have been
2495	Balmified	. If they have not been
2496	Balmmint	. **Before**
2497	Balneation	. Not before
2498	Balneatory	. On or before
2499	Balneology	. Before long
2500	Balsam .	. This was before
2501	Balsamic	. This was long before
2502	Balsamical	. Before I (we) can
2503	Balsamous	. Before you can
2504	Balustrade	. Before he (they) can
2505	Balzarine.	. Before the agreement can be signed
2506	Bamboo .	. Before this
2507	Bamboorat	. Before your telegram arrived
2508	Bamboozle	. Before returning home
2509	Bambusa	. Before we do
2510	Banality .	. Before you do
2511	Bandaged	. Before you set to work
2512	Bandaging	. Before you
2513	Bandagists	. Before we
2514	Bandanna	. Before he (they)
2515	Bandboxes	. If before
2516	Banddriver	. If not before
2517	Bandfish.	. Unless before
2518	Bandicoot	. Before leaving
2519	Bandoleers	. **Beforehand**
2520	Bandoline	. So as to be beforehand (with)
2521	Bandore .	. Beforehand (with)
2522	Bandpulley	. **Begin**
2523	Bandwheel	. Shall begin immediately (to)
2524	Bandyball	. Cannot begin till
2525	Bandyjig	. When can —— begin
2526	Bandyman	. When can you begin to
2527	Baneberry	. Begin work at once
2528	Bangle .	. We can begin
2529	Bangleear	. If you can begin—Can you begin
2530	Bangorian	. **Begun**
2531	Banish .	. Has (have) already begun
2532	Banisheth	. Has (have) not yet begun
2533	Banjo .	. Mill has begun to run again
2534	Bankable	. Has (have) begun to
2535	Bankagent	. Has (have) not begun to
2536	Bankbill .	. Have you begun
2537	Bankcredit	. **Behalf**

No.	Code Word.	**Behalf** (*continued*)
2538	Bankers .	On behalf of
2539	Bankfence .	Who act(s) on behalf of
2540	Banknote .	Will act on behalf of
2541	Bankpost .	Will not act on behalf of
2542	Bankrupt .	On behalf of the Vendors
2543	Bankruptcy .	On behalf of the Company
2544	Bankstock .	On behalf of the Board
2545	Banneret .	On behalf of the Shareholders
2546	Bannerless .	On behalf of all interests
2547	Banquet .	**. Believe**
2548	Banqueting .	Do you believe
2549	Banshee . .	I (we) do believe
2550	Banstickle .	You may entirely believe all that —— say(s)
2551	Bantams . .	You cannot believe what —— say(s)
2552	Bantamwork	I (we) do not believe
2553	Banyan . .	Is there any reason to believe
2554	Banyantree .	There is every reason to believe
2555	Baphia . .	There is no reason to believe
2556	Baphometic .	Do not believe
2557	Baptism . .	Believe nothing you hear from
2558	Baptismal	**. Believed**
2559	Baptistery .	(It) is generally believed (that)
2560	Baptistic .	**. Belong(s) (to)**
2561	Baptizable .	Whom does it belong to
2562	Baptized . .	Do(es) not belong to
2563	Baralipton .	Ought it (they) to belong to us
2564	Barbarian .	Claim that it belongs to
2565	Barbarisms .	Belong(s) to the same people
2566	Barbarity .	If it (they) belong(s) to
2567	Barbarized .	Which belong(s) to
2568	Barbarous .	**Belonged**
2569	Barbastel .	Formerly belonged to
2570	Barbate .	**. Belonging (to)**
2571	Barbbolt .	**. Belt**
2572	Barbecue .	A belt of hard rock
2573	Barbecuing .	A belt of
2574	Barbellate .	**Belting**
2575	Barcarole .	Leather belting
2576	Barcutter .	Rubber belting
2577	Barebones .	**Below**
2578	Barefaced .	Has (have) not been worked below
2579	Barefoot . .	Has (have) been worked below
2580	Baregnawn .	Below the lowest level
2581	Bareheaded .	From below the
2582	Barepicked .	Below its value
2583	Bareribbed .	Below the level of
2584	Baresark .	Below the water level
2585	Bargain . .	Below the surface
2586	Bargaining .	Below the point where
2587	Bargainors .	Any ore below

No.	Code Word.	
2588	Bargeboard .	**Benefit**
2589	Bargeman .	A great benefit to
2590	Bargown. .	Not any benefit to
2591	Barguest. .	Would it be any benefit to
2592	Barillets . .	**Best**
2593	Baritah . .	What do you think would be best
2594	Baritone . .	Do the best you can
2595	Barkantine .	It is the best thing to do
2596	Barkbed . .	The best plan would be
2597	Barkbound .	The best that could be done
2598	Barkchafer .	It will be best to
2599	Barkeeper .	Is this the best that can be done
2600	Barkgalled .	Doing the best we can
2601	Barklouse .	Doing the best they can
2602	Barkpaper .	Are you doing the best you can
2603	Barkpit . .	**Better**
2604	Barkstove .	Under better management
2605	Barlathe. .	With better appliances for working
2606	Barley . .	You had better
2607	Barleybird .	The better way is to
2608	Barleycorn .	What had better be done
2609	Barleymeal .	In better health
2610	Barmaids .	In no better health
2611	Barmacide .	Can do better
2612	Barmaster .	Cannot do better
2613	Barmilian .	Cannot you do better
2614	Barmkyn .	If nothing better can be done
2615	Barnabites .	Decidedly better
2616	Barnaby .	If —— can do any better
2617	Barnacle. .	Vein is looking better
2618	Barndoor .	Workings are looking much better
2619	Barnowl. .	Is (are) doing better
2620	Barnyard .	Is (are) much better
2621	Barograph .	Better do so, than
2622	Barolite . .	**Between**
2623	Barology .	The vein runs between
2624	Barometer .	Between the two
2625	Barometric .	Between this
2626	Baroncourt .	**Beware**
2627	Baroness. .	Beware of
2628	Baronetcy .	You must beware of
2629	Baronial . .	Beware of judging too hastily
2630	Baroscopes .	**Bill(s)**
2631	Barosma. .	Cannot meet the bill(s)
2632	Barouche .	Will meet the bill(s)
2633	Barpump .	You must renew the bill(s)
2634	Barraboat .	Will not renew the bill(s)
2635	Barracan .	Will renew the bill(s)
2636	Barracks. .	The bill is due on
2637	Barrator . .	In order to meet the bill(s)

No.	Code Word.	Bill(s) *(continued)*
2638	Barratrous .	In order to renew the bill(s)
2639	Barrel . .	All payments made by bills
2640	Barrelbulk .	Bill(s) provided for
2641	Barrelcurb .	Present bill(s) on
2642	Barrelled .	Do not present bill(s) (on ——)
2643	Barrelling .	Bill(s) protested
2644	Barrelloom .	Bill(s) of lading
2645	Barrelpen .	Bill(s) of lading not to hand
2646	Barrenwort .	Bill(s) of lading forwarded
2647	Barretcap .	The bill(s) has (have) been accepted
2648	Barricaded .	Bill(s) has (have) been paid
2649	Barringout .	Bill(s) has (have) not been paid
2650	Barrister . .	Will take bill(s) at —— months
2651	Barroom . .	Bill(s) not presented yet
2652	Barrowpump	What do(es) —— bill(s) amount to
2653	Barshear . .	Bill(s) amount(s) to
2654	Bartered . .	Please forward bill(s) of lading
2655	Barytic . .	Forward duplicate bill(s) of lading
2656	Basaltic . .	Heavy bill(s) falling due
2657	Basaltoid .	Bill on
2658	Basanite . .	Better take a bill of sale
2659	Baseballs .	A bill of sale
2660	Baseborn .	Bills on a first class house
2661	Basebred .	Approved bill
2662	Basebroom .	Bill payable at —— date
2663	Basecourt .	Bill payable at 3 months date, dated
2664	Baselard . .	Bill payable at 4 months date, dated
2665	Baseminded .	Bill payable at 6 months date, dated
2666	Baseness . .	Bill(s) accepted
2667	Baseplate .	Has (have) bill(s) been accepted
2668	Basering . .	Bill(s) not yet accepted
2669	Baseviol . .	**Bind**
2670	Bashaw . .	This will bind —— (to)
2671	Bashful . .	This will not bind —— (to)
2672	Bashfully .	Will this bind —— (to)
2673	Basicity . .	In order to bind —— (to)
2674	Basidium .	Will not bind ourselves to
2675	Basifugal .	If he will bind himself to
2676	Basigynium .	Be careful not to bind yourself to
2677	Basihyal . .	Get him to bind himself to
2678	Basilical . .	To bind himself to
2679	Basilisk . .	**Binding**
2680	Basilthyme .	Not sufficiently binding
2681	Basilweed .	Must make it binding (on)
2682	Basipetal .	Binding on us
2683	Basis . . .	Binding on him (them)
2684	Basisolute .	Not binding on us
2685	Basketfish .	Not binding on him (them)
2686	Baskethare .	**Bismuth**
2687	Baskethilt .	**Bitumen**

No.	Code Word.	
2688	Basketry .	. **Bituminous**
2689	Basquish .	. Bituminous coal has been found
2690	Basrelief .	. **Blame**
2691	Bassclef .	. Great blame attaches to
2692	Bassethorn .	There can be no blame attached to
2693	Bastardbar .	Can any blame be attached to
2694	Bastardice .	We blame
2695	Bastinado .	Is to blame (for)
2696	Bastionary .	Who is to blame (for)
2697	Basylous .	. **Blanket(s)**
2698	Batfowler .	Blankets for saving the gold
2699	Batfowling .	A revolving blanket
2700	Bathbrick .	. **Blasting**
2701	Bathbun .	. The character of the rock necessitates constant
2702	Bathchair .	Very little blasting is needed [blasting
2703	Bathing .	. By blasting
2704	Bathingbox .	Blasting powder
2705	Bathingtub .	After blasting
2706	Bathmetal .	Much blasting is needed
2707	Bathorse .	. **Blowpipe**
2708	Bathrooms .	Assays made by the blowpipe gave
2709	Bathstone .	Tested by the blowpipe
2710	Bathybius .	. **Blueground**
2711	Bathyergus .	Blueground hauled —— loads of 16 cubic feet
2712	Bathymetry .	Blueground washed —— loads of 16 cubic feet
2713	Batrachia .	. **Bluestone**
2714	Batrachoid .	**Board**
2715	Battalion .	. Is willing to join the Board of an English company
2716	Batteries .	. Would —— be likely to join the Board
2717	Batterrule .	Will join the Board
2718	Batterygun .	Will not join the Board
2719	Battleaxe .	Under consideration by the Board
2720	Battleclub .	Board of Directors
2721	Battledore .	The Board have resolved
2722	Battleflag .	The Board have resolved not to
2723	Battlement .	The matter was submitted to the Board
2724	Battlesong .	Before the Board can decide
2725	Baudelaire .	The Board can come to no decision till
2726	Bauxite .	. Must have —— on the Board
2727	Bayadeer .	Must not have —— on the Board
2728	Bayantler .	Will you join the Board of
2729	Bayardly .	. Nothing can be done till Board meets
2730	Baybolts .	. Has (have) joined the Board
2731	Bayice .	. The Board assent to what you propose
2732	Bayleaf .	. The Board desire you to
2733	Bayonet .	. The Board cannot agree
2734	Bayoneting .	The Board are of opinion (that)
2735	Baystalls .	. The Board considered the matter and decided
2736	Baytree .	. The Board considered the matter (referred to in)
2737	Baywindow .	Do the Board authorize me to

No.	Code Word.	Board (*continued*)
2738	Baywood .	Do the Board wish me to
2739	Bayyarn . .	Are the Board of opinion
2740	Bazaar . .	To board the men
2741	Beacon . .	Cost to board the men
2742	Beaconage .	Must board men, which will cost
2743	Beaconfire .	**Boarding-house(s)**
2744	Beadleism .	A good boarding-house at the mine
2745	Beadmould .	Good boarding-houses at the mine and mill
2746	Beadplane .	There are —— boarding-houses, but in bad state
2747	Beadproof .	There are —— boarding-houses in good repair
2748	Beadroll . .	A yearly profit of —— per cent. on the boarding-
2749	Beadsman .	Boarding-house for the men [houses)
2750	Beadsnake .	Boarding-house for the men has been built
2751	Beadswoman	**Body**
2752	Beadtools .	Which will open up a body of ore
2753	Beagle . .	Have struck a body of rich ore
2754	Beakhead .	Every indication of soon striking the body of ore
2755	Beakirons .	Prospects of large body of ore very good
2756	Beambird .	Opened up large body of ore
2757	Beamless .	Body of ore
2758	Beancaper .	**Boiler(s)**
2759	Beancod . .	Portable locomotive boiler
2760	Beanfeast .	Cornish boiler
2761	Beanflies . .	Knap's water-tube boiler
2762	Beanfly . .	Tubular boiler
2763	Beangoose .	Lancashire boiler
2764	Beanking .	Owing to boiler(s) being out of repair
2765	Beanmills .	Boiler(s) is (are) out of order
2766	Beanstalks .	Require new boiler(s) for
2767	Bearcloth .	There is (are) —— boiler(s)
2768	Bearded . .	There is (are) —— h.p. boiler(s)
2769	Beardgrass .	Send —— h.p. tubular boiler
2770	Beardleted .	Send —— h.p. portable boiler
2771	Beardmoss .	Send —— h.p. Cornish boiler
2772	Beargarden .	Send —— h.p. Lancashire boiler
2773	Bearish . .	Send —— h.p. boiler
2774	Bearleader .	Send set of boiler-tubes for —— boiler
2775	Bearlike . .	Require —— h.p. boiler for
2776	Bearpits .	**Bonanza**
2777	Bearskins .	Have struck a bonanza
2778	Bearwhelp .	It is reported you have struck a bonanza
2779	Beastly . .	A bonanza has been struck
2780	Beaterup .	**Bond**
2781	Beatific . .	The bond on the property
2782	Beatifical	Bond must be for
2783	Beatifying .	Bond can be secured for
2784	Beatitudes .	The bond will be forfeited
2785	Beaujolais .	In order to renew the bond
2786	Beauship .	Can you secure the bond
2787	Beauteous .	Cannot obtain bond

No.	Code Word.	**Bond** (*continued*)
2788	Beautiful .	Cannot renew the bond
2789	Beautify . .	Try and obtain a bond to purchase
2790	Beautiless .	Bond cannot be renewed without payment
2791	Beautyspot .	Get a working bond for —— months
2792	Beavered .	Cannot recommend bond. I would advise you to have deed in escrow [being satisfactory
2793	Beaverrat .	Get a bond subject to an examination and report
2794	Beavertree .	Bond can only be secured upon following terms
2795	Beblot . .	Will not give bond [and conditions
2796	Beblotting .	I have obtained bond for —— months
2797	Bebooted .	Bond for —— months
2798	Becafigo .	Bond for a year
2799	Becalmed .	Can you get a bond on the property
2800	Becalming .	Can get a bond on the property
2801	Beccabunga .	Cannot get a bond on the property
2802	Bechanced .	To give a bond on the property
2803	Becharm ·.	To give a bond for
2804	Bechic . .	Renew the bond
2805	Beckharman	Bond has been given
2806	Beckoned .	Must give a bond
2807	Beclipping .	There is no bond
2808	Beclipt . .	Can do without a bond
2809	Becloud . .	Do not (does not) want a bond
2810	Beclouding .	**Bonded**
2811	Become . .	Has (have) bonded the mine for
2812	Becoming .	When the mine was bonded
2813	Becomingly .	The mine was originally bonded (to)
2814	Becoronet .	To whom was the mine bonded
2815	Becripple .	To whom has the property been bonded
2816	Becuibanut .	The property has been bonded (for)
2817	Becursed .	Has (have) been bonded (for)
2818	Becursing .	**Bonus**
2819	Bedabble .	Giving a good bonus
2820	Bedabbling .	To be paid as a bonus
2821	Bedarken .	Bonus shares
2822	Bedashed .	Bonus on the sale
2823	Bedaub . .	Bonus on the transaction
2824	Bedaubing .	Will give a bonus
2825	Bedbolts . .	Have given a bonus
2826	Bedbugs . .	**Boom**
2827	Bedchair . .	A big boom in mining speculations
2828	Bedchamber	The boom is passing over
2829	Bedclothes .	In order to get it floated during the **boom**
2830	Bedecked .	There is a boom now on (in)
2831	Bedecking .	While the boom is on
2832	Bedehouse .	As soon as the boom is over
2833	Bedelry . .	**Borax**
2834	Bedevil . .	**Bored**
2835	Bedevilled .	How far have you bored
2836	Bedew . .	We have bored

No.	Code Word.	
2837	Bedfellow .	**Borehole(s)**
2838	Bedframe .	A borehole was put down
2839	Bediadem .	By a borehole(s)
2840	Bedimmed .	Borehole —— feet deep
2841	Bedismal .	**Boring(s)**
2842	Bedizened .	Boring for
2843	Bedizening .	From borings in face ot
2844	Bedkey . .	By the several borings made
2845	Bedlam . .	From the previous borings
2846	Bedlamite .	Advise boring for
2847	Bedlinen . .	In boring
2848	Bedmakers .	Do you advise boring
2849	Bedmate .	Has boring been begun
2850	Bedpheer .	Will begin boring
2851	Bedposts . .	Boring carried to
2852	Bedpresser .	Boring by air-drill
2853	Bedraggled .	Boring by diamond-drill
2854	Bedrench .	Have obtained by boring
2855	Bedridden .	Boring carried to —— feet
2856	Bedrooms .	**Both**
2857	Bedrop . .	This refers to both
2858	Bedropped .	Has (have) been to both
2859	Bedscrews .	In both the
2860	Bedstaff . .	Both are
2861	Bedsteads .	Both should be
2862	Bedstraw .	If both are
2863	Bedswerver .	When both are
2864	Bedwarf . .	Are both
2865	Bedwarfing .	In case both
2866	Beeblock .	Both of us
2867	Beebread .	**Bottom**
2868	Beechcoal .	The bottom of the winze
2869	Beechfinch .	The bottom of the shaft
2870	Beechgall .	The bottom of the mine
2871	Beechmast .	Going down into the bottom
2872	Beechnut .	The bottom seems to be out of
2873	Beechoil .	**Bought**
2874	Beechtrees .	Has (have) been bought by
2875	Beefeater .	Was (were) bought by
2876	Beefsteak .	Bought for joint account
2877	Beeftea . .	Have you bought
2878	Beefwitted .	Bought for you
2879	Beefwood .	Cannot be bought
2880	Beeglue . .	At what price have you bought
2881	Beehawk . .	Has (have) bought up
2882	Beehive . .	Why have you bought
2883	Beemaster .	Why have you not bought
2884	Beenettles .	What have you bought
2885	Beeorchis .	How much have you bought
2886	Beerengine .	Has (have) bought at

No.	Code Word.	**Bought** (*continued*)
2887	Beerhouses .	Has (have) bought for immediate delivery
2888	Beeriness .	Has (have) bought for cash
2889	Beermoney .	Has (have) bought enough for
2890	Beerpull . .	If you have bought
2891	Beershop .	If you have not bought
2892	Beerstone .	Bought at price named
2893	Beeswax . .	Bought as arranged
2894	Beeswing .	Bought the whole for
2895	Beetle . .	**Bound**
2896	Beetlebrow .	By which we are bound to
2897	Beetlehead .	By which he is (they are) bound to
2898	Beetradish .	Is (are) not bound to
2899	Beetrave . .	Has bound himself to
2900	Beetroots .	**Boundary**
2901	Beeworm .	The northern boundary (of the)
2902	Befallen . .	The southern boundary (of the)
2903	Befaria . .	The eastern boundary (of the)
2904	Befettered .	The western boundary (of the)
2905	Beflowered .	**Box**
2906	Befoam . .	Send a box of
2907	Befogged . .	Have sent a box of
2908	Befogging .	Send —— boxes
2909	Beforehand .	Have sent —— boxes
2910	Beforetime .	A box of specimens—Boxes of specimens
2911	Befreckled .	A box of samples—Boxes of samples
2912	Befriend . .	**Break**
2913	Befrill . .	A break in the
2914	Befrilling .	Breakdown in the machinery
2915	Befringe . .	Breakdown in the mill
2916	Befrizzed .	What is the cause of the break in
2917	Befuddled .	**Breast** '
2918	Begall . .	Breast of ore
2919	Beggar . .	Breast of drift in pay ore
2920	Beggarly . .	Breast promises fairly
2921	Begging . .	**Bribe**
2922	Begifted . .	As a bribe
2923	Begin . . .	Has (have) been trying to bribe
2924	Beginner .	**Broken**
2925	Beginnings .	Broken in transit
2926	Begirdle . .	The machinery is broken
2927	Begirdling .	The engine is broken
2928	Beglared . .	Broken down
2929	Beglerbeg .	Has (have) broken the
2930	Beglooming .	**Broker(s)**
2931	Begodded .	Will act as broker(s) for
2932	Begonia . .	Who will act as broker(s) for
2933	Begrease . .	The broker(s) for
2934	Begrime . .	**Brokerage**
2935	Begroaned .	The brokerage is
2936	Begrudge .	Including brokerage

No.	Code Word.	**Brokerage** (*continued*)
2937	Begrudging .	Exclusive of brokerage
2938	Beguile . .	Will you pay brokerage of
2939	Beguiling .	No brokerage can be paid
2940	Beguilty . .	Will pay brokerage
2941	Beguinage .	Vendor to pay brokerage
2942	Behalf . .	**Brought**
2943	Behappen .	Has to be brought —— miles by
2944	Behaving .	Has (have) to be brought from
2945	Behaviour .	Cannot be brought
2946	Behemoth .	Is brought to
2947	Behind . .	**Building**(s)
2948	Behindhand.	The buildings consist of
2949	Behither .	The buildings are in good condition
2950	Behold . .	The buildings are in bad condition
2951	Beholding .	Expect to finish building
2952	Behoney. .	Now building
2953	Behoneying .	**Built**
2954	Behoof . .	Have built a new
2955	Behoovable .	**Bull**
2956	Behooveth .	To bull the shares
2957	Behooving .	A bull movement in shares is likely to take place
2958	Behovèful .	The bulls have run the shares up to
2959	Behowls . .	Bull movement in the shares
2960	Beidelsar ;	There are a good many weak bulls
2961	Beingplace .	The bulls are very strong
2962	Bejade . .	In consequence of bulls operating
2963	Bejaundice .	Bull operations
2964	Bejesuit . .	The bulls are furious
2965	Bejewel . .	**Bullion** (See Bars, Ship.)
2966	Bejewelled .	What is the total amount of bullion
2967	Bejuco . .	Bullion shipped ＇
2968	Bejumble .	Shall ship bullion on
2969	Bejumbling .	Will be able to ship —— bars bullion on
2970	Bekiss . .	Ship all the bullion you possibly can before
2971	Beknaves .	No bullion on hand
2972	Beknight. .	Ship no more bullion till ordered
2973	Belabour .	Will ship bullion amounting to
2974	Belaccoyle .	Shipped bullion amounting to
2975	Beladling .	Shipment of bullion
2976	Belawgive .	What bullion are you shipping [last shipment
2977	Belaying . .	What do you estimate the bullion produced since
2978	Belch . . .	Approximate value of the bullion
2979	Belcheth . .	Actual realized value of the bullion
2980	Beleaguer .	Assay value of the bullion
2981	Belecture .	When will you ship bullion and how much
2982	Belemnites .	**Bunchy**
2983	Belepered .	The character of the ore is bunchy
2984	Belepering .	Is very bunchy
2985	Belfries . .	**Burned**
2986	Belfry. . .	The mill was burned down

No.	Code Word.	**Burned** (*continued*)
2987	Belgards . .	The —— house was burned down
2988	Belibel . .	Which has burned down
2989	Belibelled .	Is all burned down
2990	Belief. . .	Was burned down
2991	Beliefful . .	Entirely burned
2992	Believable .	Partially burned
2993	Believing .	**Burst**
2994	Belikely . .	Is (are) burst
2995	Belittle . .	**Bursting**
2996	Belittling .	Through the bursting of boiler
2997	Bellacity . .	Through the dam bursting
2998	Belladonna .	**Business**
2999	Bellatrix . .	A large business is done in
3000	Bellbird . .	No business is done
3001	Bellbuoy . .	It may lead to business
3002	Bellcrank .	Can you do the business
3003	Belled . .	Can do the business
3004	Belleric . .	Leave the business in my (our) hands
3005	Bellflower .	Rather than lose the business
3006	Bellgable .	Business in mining shares very dull
3007	Bellglass . .	Business in mining shares very good
3008	Bellhanger .	There is very little prospect of doing (the) business
3009	Bellicose . .	Is there a prospect of doing (the) business
3010	Bellique . .	Consider business safe
3011	Bellitude . .	Do not consider business safe
3012	Belljars . .	**Buy**
3013	Bellman . .	Buy —— shares for me in
3014	Bellmetal .	Buy immediately
3015	Bellowing .	Buy for next account
3016	Bellpolype .	Buy all you can (of)
3017	Bellpulls . .	Shall I buy
3018	Bellroof . .	Buy as much as you can of —— at
3019	Bellropes .	I can buy for cash
3020	Bellshaped .	I can buy cheaper here
3021	Belltowers .	Buy back if you have sold
3022	Belltraps . .	Do not buy any —— stock (or) (shares)
3023	Bellturret .	Buy for my account, and hold subject to order
3024	Bellwether .	Buy for a quick turn
3025	Bellwort . .	I advise you to buy —— (at about)
3026	Bellyache .	The market will advance, you had better buy
3027	Bellyband .	Cannot buy any
3028	Bellybrace .	Can you buy
3029	Bellychurl .	Is it safe to buy
3030	Bellyful . .	Take advantage of present weak market to buy
3031	Bellygod . .	Stocks are too high to buy at present prices
3032	Bellyguys .	Do you advise buying at present prices
3033	Bellyroll . .	Order to buy is countermanded
3034	Bellyslave .	Buy when you think it has reached the bottom
3035	Bellystay . .	Buy until the stock reaches
3036	Belomancy .	Buy if the market is strong

No.	Code Word.	**Buy** (*continued*)
3037	Belong . .	Buy if the market is weak
3038	Belongeth .	Cannot buy at limit
3039	Belongings .	It will not do to buy
3040	Belopteron .	Buy more at same price
3041	Beloved . .	Can you buy more at same price
3042	Belsire . .	**Buyers**
3043	Beltcutter .	The buyers will complete purchase
3044	Beltlacing .	The buyers will not complete purchase till
3045	Beltpipe . .	Buyers are plentiful
3046	Beltpunch .	Buyers are few
3047	Beltsaw . .	Buyers are cautious
3048	Beluga . .	No buyers for your property
3049	Bemangle .	Who are buyers
3050	Bemangling .	No buyers for —— stock (—— shares)
3051	Bemartyr .	Buyers are cautious, but the stock is held firm (the
3052	Bemaul . .	**Buying** [shares are held firm)
3053	Bemauling .	Is (are) buying largely
3054	Bemazed . .	Is (are) buying stock (shares)
3055	Bemirement	Friends are buying
3056	Bemist . .	Are insiders buying
3057	Bemitred .	Miners are buying
3058	Bemitring .	I advise buying
3059	Bemoaning .	Who is buying
3060	Bemoanable	I am (we are) buying
3061	Bemock . .	**Cable** (See Telegraph.)
3062	Bemocking .	Cable report immediately
3063	Bemonster .	By means of a cable
3064	Bemourned .	Cable me (us) —— at once; wanting money badly
3065	Bemouthing .	Cable immediately if
3066	Bemuddle .	Have received your cable (dated)
3067	Bemuddling .	As soon as you cable
3068	Bemuffled .	Cable sufficient details to enable me to prepare report
3069	Bemurmured	Cable what you know about [for the public
3070	Benchclap .	Do not send me (us) any more cables
3071	Benchers . .	Cable in plain words not code
3072	Benchhook .	Cable your next address
3073	Benchmark .	Will cable you my next address
3074	Benchreel .	First word of the cable is unintelligible, repeat it
3075	Benchstrip .	Second word of the cable is unintelligible, repeat it
3076	Benchtable .	Third word of the cable is unintelligible, repeat it
3077	Bendable .	Fourth word of the cable is unintelligible, repeat it
3078	Bendways .	Fifth word of the cable is unintelligible, repeat it
3079	Beneaped .	Sixth word of the cable is unintelligible, repeat it
3080	Beneath . .	Seventh word of the cable is unintelligible, repeat it
3081	Benedicts .	Eighth word of the cable is unintelligible, repeat it
3082	Benefactor .	Ninth word of the cable is unintelligible, repeat it
3083	Beneficial .	Tenth word of the cable is unintelligible, repeat it
3084	Benefit . .	—— word of the cable is unintelligible, repeat it
3085	Benefiting .	First word of cable should be
3086	Benevolous .	Second word of cable should be

No.	Code Word.	Cable (*continued*)
3087	Bengalroot .	Third word of cable should be
3088	Benignant .	Fourth word of cable should be
3089	Benignity .	Fifth word of cable should be
3090	Benteak . .	Sixth word of cable should be
3091	Bentgrass .	Seventh word of cable should be
3092	Benthamite .	Eighth word of cable should be
3093	Benthamism	Ninth word of cable should be
3094	Benumb . .	Tenth word of cable should be
3095	Benumbing .	—— word of cable should be
3096	Benumbment	The following words of the cable are not intelligible,
3097	Benzamide .	The words of the cable should be [repeat them
3098	Benzoic . .	**Cage(s)**
3099	Benzoline .	The cage in use is unsafe
3100	Bepaint . .	Shaft fitted with one cage
3101	Bepearl . .	Two cages
3102	Bepinched .	Shaft fitted with cages
3103	Beplaster .	**Calcite**
3104	Bepommel .	The hanging wall is calcite
3105	Bepowder .	The footwall is calcite
3106	Bepraise . .	**Calculate**
3107	Bepraising .	At what rate do you calculate
3108	Bepuckered .	How do you calculate
3109	Bepuff . .	We calculate that (there are)
3110	Bepuffing .	**Calculated**
3111	Bepurple .	Have calculated that
3112	Bepurpling .	Calculated at the rate of
3113	Bequeath .	If we had calculated
3114	Bequest . .	**Calculation(s)**
3115	Bequoted .	Calculations based upon
3116	Bequoting .	Calculation was right
3117	Berascal . .	Calculation was incorrect
3118	Berattle . .	Calculation was approximate
3119	Berberine .	An approximate calculation of
3120	Berdash . .	Upon a rough calculation
3121	Bergamot .	Must check calculations
3122	Bergander .	Calculations are correct
3123	Bergmaster .	Calculations are not correct
3124	Bergmehl .	**Calc-spar**
3125	Bergylt . .	The matrix is calc-spar
3126	Berhyme .	**California**
3127	Berhyming .	Gold mine in California
3128	Beribanded .	In California
3129	Beribbon .	In Lower California
3130	Berlinblue .	**Call**
3131	Berlinware .	Call for letters at
3132	Berlinwool .	Must make a call of —— per share
3133	Berserker .	Must make a further call (of)
3134	Berthage .	Cannot make a call
3135	Berthing .	Call on —— when you are in
3136	Beryl . . .	No further call will be made (until)

No.	Code Word.	**Call** (*continued*)
3137	Berylline . .	Call has been made (of) —— per share
3138	Berylloids .	At the time the call was made
3139	Besagne . .	When will the next call be made
3140	Besainted .	Next call is payable
3141	Besainting .	Another call
3142	Bescatter .	The last call
3143	Bescorn . .	**Called** (up)
3144	Bescoured .	Capital called up
3145	Bescouring .	Called up per share
3146	Bescratch	Have called up
3147	Bescrawl .	When —— has been called up
3148	Bescreen .	Called for
3149	Bescribble .	Has (have) called upon
3150	Bescumber .	Has (have) called upon us (to)
3151	Beseech . .	Have called upon him (to)
3152	Beseecheth .	**Came**
3153	Beseemly .	Before I came
3154	Beset . . .	Before he (they) came
3155	Besetment .	When I came
3156	Beshout . .	When he (they) came
3157	Beshrew . .	After I came
3158	Beshrewing .	After he (they) came
3159	Beshrouded .	**Can**
3160	Besidery . .	Can I (we)
3161	Besiege . .	Can you
3162	Besieging .	Can he (they)
3163	Beslabber .	I (we) can
3164	Beslave . .	You can
3165	Beslurry . .	He (they) can
3166	Besmear . .	I (we) cannot
3167	Besmearing .	You cannot
3168	Besmirch .	He (they) cannot
3169	Besmooth .	If I (we) can
3170	Besnuffed .	If you can
3171	Besotted . .	If he (they) can
3172	Besottedly .	If I (we) cannot
3173	Besought .	If you cannot
3174	Bespangle .	If he (they) cannot
3175	Bespeak . .	When you can
3176	Bespeaking .	When can you
3177	Bespeckle .	Unless you can
3178	Bespiced .	Unless he (they) can
3179	Bespoken .	What can you
3180	Bespouting .	What can he (they)
3181	Bespread .	Can you do this
3182	Besprinkle .	Can do what you want
3183	Bespurt . .	Can do nothing
3184	Bespy . .	Can send
3185	Bestiality .	Cannot you
3186	Bestialize .	Cannot he (they)

No.	Code Word.	Can (*continued*)
3187	Bestiates .	Cannot we
3188	Bestick . .	Cannot possibly
3189	Besticking .	**Cancel**
3190	Bestirred .	Can you cancel
3191	Bestorm . .	Will cancel
3192	Bestowal . .	Will not cancel
3193	Bestowing .	Cancel my order to
3194	Bestowment .	Threaten to cancel
3195	Bestraddle .	On what terms can you cancel
3196	Bestrapped .	I (we) can cancel
3197	Bestride . .	I (we) cannot cancel
3198	Bestriding .	Will cancel clause(s) No.——
3199	Beswike . .	Cancel order and substitute
3200	Betallow . .	Cancel balance of order
3201	Betelnut . .	Cancel the purchase if possible
3202	Bethlemite .	Cancel the sale if possible
3203	Bethrall . .	**Cancelled**
3204	Bethump .	Clause(s) No. —— must be cancelled
3205	Bethumping .	Has (have) cancelled
3206	Betimes . .	Has (have) not cancelled
3207	Betoken . .	The order cannot be cancelled
3208	Betokening .	The sale cannot be cancelled
3209	Betongue .	The purchase cannot be cancelled
3210	Betrayal . .	Have you cancelled
3211	Betrayment .	Will be cancelled (unless)
3212	Betroth . .	Has been cancelled
3213	Betrothals .	**Cañon**
3214	Betrothing .	The mines are situated in a cañon
3215	Betrust . .	A fine stream runs through the cañon
3216	Betterhalf .	In the cañon
3217	Bettermost .	Above the cañon
3218	Betutored .	The upper part of the cañon
3219	Between . .	**Capabilities**
3220	Betweenity .	Possessing great capabilities for
3221	Beudantite .	Possessing but small capabilities
3222	Bevelangle .	**Capable**
3223	Bevelgear .	Is (are) capable of
3224	Bevelled . .	Capable of great development
3225	Bevelling .	Capable of being worked
3226	Bevelwheel .	Capable of improvement
3227	Beverage .	Is (are) not capable (of)
3228	Bevilways .	Is (are) —— capable of
3229	Bewailable .	Not capable of judging
3230	Bewaileth .	**Capacity**
3231	Bewailing .	To the utmost of its (their) capacity
3232	Bewailment .	Is mill working to its full capacity
3233	Beware . .	The mill is working to its full capacity
3234	Beweep . .	The mill is not working to its fullest capacity
3235	Beweeping .	What is the capacity (of)
3236	Bewhisper .	Worked to its fullest capacity

D

No.	Code Word.	Capacity (*continued*)
3237	Bewilder . .	Worked to its fullest capacity; to attempt more
3238	Bewimple .	Has a capacity of [would be dangerous
3239	Bewinged .	**Capital**
3240	Bewitch . .	The capital of the company to be
3241	Bewitchful .	No capital to spend on the mine
3242	Bewitching .	A judicious investment of capital
3243	Bewondered	To bring capital into the concern
3244	Bewrayeth .	Not sufficient capital to buy the mine
3245	Bewrought .	Not sufficient capital to develop the mine
3246	Bezoardic .	To be put aside for working capital
3247	Bezoargoat .	To return —— per cent. on the capital
3248	Biacid . .	At the rate of —— per cent. on the capital of the
3249	Biangular .	On a capital of [company
3250	Biangulous .	Fully justifying the expenditure of the capital named
3251	Biassing . .	On the capital
3252	Biassness .	The capital is too large
3253	Biaxial . .	The capital is too small
3254	Bibacious .	Increase the capital to
3255	Bibasic . .	Reduce the capital to
3256	Bibcock . .	Can you get the capital underwritten
3257	Bibitory . .	Enough capital has been subscribed
3258	Bibleoath .	Not enough capital has been subscribed
3259	Biblepress .	The capital is to be reduced
3260	Biblical . .	The capital cannot be reduced
3261	Biblically .	The capital cannot be increased
3262	Biblicism .	The capital has been subscribed
3263	Bibliolite .	The capital has been underwritten
3264	Bibliomany .	What is the capital
3265	Bibliopegy .	Require you to find —— working capital
3266	Bibliopole .	Want —— working capital
3267	Biblist . .	Must provide —— working capital
3268	Bibulous . .	Capital not fully subscribed
3269	Bicallous .	If capital should not be fully subscribed
3270	Bicameral .	Particulars of capital
3271	Bicapsular .	One-fourth of the capital
3272	Bicarinate .	One-third of the capital
3273	Bicaudal . .	One-half of the capital
3274	Bicavitary .	The whole of the capital
3275	Biceps . .	Who finds the capital
3276	Bichromate .	Finds the required capital
3277	Bicipital . .	Working capital (of)
3278	Bicipitous .	Must reduce amount of working capital
3279	Bickerer . .	Must increase amount of working capital
3280	Bickerings .	The amount of nominal capital is [should be
3281	Bickerment .	What does vendor propose the nominal capital
3282	Bicoloured .	What does vendor propose the working capital to be
3283	Biconcave .	The nominal capital must not exceed
3284	Biconvex .	Vendor does not limit nominal capital
3285	Bicornous .	On capital account
3286	Bicornute .	Expenditure on capital account

No.	Code Word.	Capital (*continued*)
3287	Bicorporal .	Debited to capital
3288	Bicrural . .	Credited to capital
3289	Bicuspid . .	Chargeable to capital
3290	Bicycle . .	Should be charged to capital
3291	Bicycling .	Capital and liabilities
3292	Bicyclists .	Can you find capital for
3293	Bidders . .	Can find capital for
3294	Bident . .	Cannot find capital for
3295	Biennial .	. Capitalization
3296	Biennially .	On capitalization
3297	Bierbalk . .	Over-capitalization
3298	Bifacial .	. Capitalized (at)
3299	Biffin . . .	The concern is over-capitalized
3300	Bifidated .	. Carbonate
3301	Biflorous . .	Carbonate of copper
3302	Bifoliate . .	Carbonate of lead
3303	Biforine . .	Carbonate of lime
3304	Biforked . .	Carbonate of lime and magnesia
3305	Bifronted .	Carbonate of iron
3306	Bifurcated .	Carbonate of soda
3307	Bifurcous .	Carbonate of lead with silver
3308	Bigamy . .	Carbonate of zinc
3309	Bigaroon	. Carboniferous
3310	Bigboned .	In carboniferous rock
3311	Bigeminate .	Of the carboniferous period
3312	Bigential . .	Carboniferous strata
3313	Bighorn .	. Care
3314	Biglaurel . .	Has (have) taken care
3315	Bignamed .	Has (have) not taken care
3316	Bigot . . .	Great care will be required
3317	Bigotedly .	Take care (of)
3318	Bigotical . .	Must take care
3319	Bigotries . .	Will take the greatest care
3320	Bigotry . .	The utmost care
3321	Bigwigged .	Take care not to
3322	Bijugate . .	To the care of
3323	Bijugous . .	If you do not take care
3324	Bilander .	. Careful
3325	Bilateral . .	I will be very careful
3326	Bilberries .	Be careful not to commit
3327	Bilberry . .	Be very careful in your dealings with
3328	Bilboes . .	Make the most careful examination
3329	Bileduct . .	Has (have) made a most careful examination
3330	Bilgecoad .	Be careful how you
3331	Bilgefree . .	Be most careful (to) (in)
3332	Bilgekeel .	Be most careful not to
3333	Bilgepiece .	Be careful of
3334	Bilgeplank	. Carefully
3335	Bilgepump .	Has (have) been through everything very carefully
3336	Bilgewater .	The mines have been carefully worked

No.	Code Word.	**Carefully** (*continued*)
3337	Bilgeways .	Will have to be most carefully
3338	Biliation . .	Carefully checked
3339	Bilingual .	Carefully packed
3340	Bilinguous .	**Careless(ness)**
3341	Billbergia .	In the most careless manner
3342	Billboard .	The greatest carelessness
3343	Billbook . .	**Carriage** [costs —— per ton
3344	Billbroker .	The carriage from nearest railway station to mine
3345	Billethead .	The carriage from
3346	Billfish . .	The carriage costs —— per ton
3347	Billiard . .	The carriage on the bullion is very heavy
3348	Billicock . .	The carriage on all goods is very heavy
3349	Billposter .	By carriage —— miles from
3350	Billybiter .	By carriage over dreadful roads
3351	Billyboy . .	By carriage over —— miles, good road
3352	Bilobated .	Carriage by waggon
3353	Bimaculate .	Carriage on mule-back
3354	Bimembral .	**Carried**
3355	Bimestrial .	Has (have) been carried on
3356	Bimetallic .	Has (have) not been carried on
3357	Bimonthly .	Carried away
3358	Bimuscular .	Will have to be carried
3359	Binary . .	Will be carried on
3360	Binaural . .	Must be carried on
3361	Bindwood .	Can be carried through
3362	Binervate .	Cannot be carried through
3363	Binocular .	Can be carried (out) (or over)
3364	Binomial . .	Cannot be carried (out) (or over)
3365	Binoxyde .	**Carry**
3366	Biocellate .	To carry concern to successful issue
3367	Biodynamic .	In order to carry on the
3368	Biogenist .	Will carry on the negotiations (during)
3369	Biography .	Carry the stock
3370	Biologist . .	Can surely carry it
3371	Biolytic . .	Will carry the stock for you
3372	Biometry .	Cannot carry on
3373	Bioplasm .	Who will carry on affairs whilst you are away
3374	Bioplasmic .	To carry this out will need
3375	Biotaxy . .	Cannot carry out instructions
3376	Bipalmate .	Carry on the business
3377	Biparous . .	Carry on the negotiations
3378	Bipartible .	To enable me (us) to carry on
3379	Bipartient .	Impossible to carry on any longer
3380	Biped . . .	Carry on
3381	Bipedal . .	Carry over
3382	Bipennated .	Can you carry on
3383	Bipennis . .	Carry out our instructions
3384	Bipolar . .	Carry out —— instructions
3385	Bipolarity .	Carry over the stock (shares)
3386	Bipontine .	Cannot carry it out

No.	Code Word.	Carry (*continued*)
3387	Bipunctual .	Carry to
3388	Biquadrate .	The business will not carry
3389	Biquintile	. Carrying
3390	Biradiated .	Carrying over
3391	Birchwine .	Carrying to Capital Account
3392	Birdcages .	Carrying to Revenue Account
3393	Birdcall . .	Carrying to Stores Account
3394	Birdcherry .	Carrying over from account to account
3395	Birdeyed . .	Carrying to account of
3396	Birdfooted	. Case
3397	Birdgazer .	In case of accidents
3398	Birdlime . .	Entirely alters the case
3399	Birdorgan .	Has (have) a strong case against
3400	Birdpepper .	In case
3401	Birdseed . .	In any case
3402	Birdsmouth .	In no case
3403	Birdsnest .	In which case
3404	Birdspider .	In every case
3405	Birdstares .	In case of need
3406	Birdwitted .	In case of refusal
3407	Birken . .	It is not the case
3408	Birlawman .	The facts of the case
3409	Birthchild .	In case of delay
3410	Birthdays .	Has (have) no case against
3411	Birthdom .	In such a case
3412	Birthhour .	Has (have) —— any case against
3413	Birthmark .	The case will be heard
3414	Birthplace .	Upon the hearing of the case
3415	Birthright . .	When the case is heard
3416	Birthroot .	In that case
3417	Birthsin .	. Cash
3418	Birthsongs . .	For cash
3419	Bisaccate .	Will only accept cash payment
3420	Biscroma .	By paying down, in cash
3421	Biscuit . .	In cash
3422	Biscutate .	Bank refuses to cash your cheque
3423	Bisects . .	In cash, and the balance in shares
3424	Bisegment .	In cash, and the balance in debentures
3425	Biserial . .	No cash required
3426	Bisetous . .	Cash to be deposited with
3427	Bisexual . .	In exchange for cash
3428	Bishop . .	Will pay cash down
3429	Bishopdom .	Cash on delivery
3430	Bishopess .	Cash in —— month(s)
3431	Bishopling .	Cash in exchange for
3432	Bishoprics .	One-quarter in cash
3433	Bishopscap .	One-third in cash
3434	Bishopweed .	One-half in cash
3435	Bismuth . .	Two-thirds in cash
3436	Bismuthal	Three-fourths in cash .

No.	Code Word.	Cash (*continued*)
3437	Bispinose .	How much cash
3438	Bissextile .	How much cash, and how much **in shares**
3439	Bistipuled .	Cash must be paid within
3440	Bistoury . .	Cash must be paid on
3441	Biteth . .	We require cash
3442	Biting . .	Cash value
3443	Bitingly . .	Cash, less —— per cent.
3444	Bitless .	**. Cassiterite**
3445	Bitmouth .	**Cause**
3446	Bitstock . .	This will cause a delay of
3447	Bitterash . .	What is the cause of
3448	Bitterful . .	Cause unknown
3449	Bitterking .	This will not cause
3450	Bitterness .	This will cause
3451	Bitternut . .	Is there any cause
3452	Bitteroak .	Is there any cause for the **decline**
3453	Bittersalt .	What is the cause of the delay
3454	Bitterspar .	There is no cause for alarm
3455	Bitterweed .	There is every cause for
3456	Bitumen .	**. Caused**
3457	Bituminize .	What has caused (this)
3458	Bituminous .	Has (have) been caused by
3459	Bivalve . .	Has (have) not been caused by **any**
3460	Bivalvous .	Which would have caused
3461	Bivalvular .	Has caused great excitement
3462	Bivaulted .	Has caused the greatest alarm
3463	Biventral .	Caused a fall in the price
3464	Bivouac . .	Caused a complete panic
3465	Bivouacked .	**Caution**
3466	Biweekly .	Act with great caution
3467	Blackact . .	In spite of all the caution
3468	Blackamoor .	Has (have) acted with all due **caution**
3469	Blackball .	Caution —— against
3470	Blackbeer .	Will act with caution
3471	Blackberry	**. Cautious**
3472	Blackbirds .	Be very cautious in your dealings with
3473	Blackbooks .	Be very cautious in expressing your opinion
3474	Blackbrush .	Is (are) very cautious
3475	Blackcap .	Is (are) not very cautious
3476	Blackchalk .	You cannot be too cautious
3477	Blackcock .	We cannot be too cautious
3478	Blackdeath .	**Cave(s)**
3479	Blackdrop .	Will cave in
3480	Blackeyed .	Likely to cave in
3481	Blackflags .	Caves have occurred
3482	Blackflies .	On account of caves
3483	Blackfly .	**. Caved**
3484	Blackfoot .	Has (have) caved in [in
3485	Blackfriar .	The old workings, above the —— level, have **caved**
3486	Blackgame .	The tunnel has caved in

No.	Code Word.	**Caved** (*continued*)
3487	Blackgrass .	Caved in, in consequence of which the mines have
3488	Blackguard .	Shaft has caved in [been abandoned
3489	Blackheart .	Impossible to get into the mine, owing to the ——
3490	Blackhole .	Upper workings have caved in [having caved-in
3491	Blackiron	**. Caving**
3492	Blackjack .	From caving in
3493	Blackknot .	To prevent the tunnel(s) from caving in
3494	Blackleg . .	Every prospect of the tunnel caving in
3495	Blacklist . .	The mine is caving in
3496	Blackmatch .	Is (are) caving in
3497	Blackmonk .	All risk of caving in is now past
3498	Blackochre .	There are signs of ground caving in
3499	Blackpines .	Made all safe against caving in
3500	Blackplate .	Are there any signs of caving in
3501	Blackrod . .	No fear of caving in future
3502	Blacksalts	**. Cease**
3503	Blacksheep .	Cease buying for the present
3504	Blacksmith .	Cease selling for the present
3505	Blacksnake .	Cease forwarding for the present
3506	Blackspaul .	Cease delivering for the present
3507	Blacktail . .	Have had to cease work on account of
3508	Blackthorn .	The works practically cease through the winter
3509	Blackvomit .	You had better cease
3510	Blackwash .	Cease for the present (to)
3511	Bladder . .	Has (have) had to cease
3512	Bladdernut .	Will not cease
3513	Bladderpod .	Do not cease
3514	Bladebone	**. Ceased**
3515	Blademetal .	Opposition has now ceased
3516	Blamably	Have ceased buying
3517	Blameful	Have ceased selling
3518	Blamefully .	Have ceased operations
3519	Blameless .	Have you ceased
3520	Blancher . .	Have ceased
3521	Blanchfarm	**. Cent.** (See Per cent.)
3522	Blancmange	Cents. per ounce
3523	Blandation .	Cents. per pound
3524	Blandiment .	Has (have) advanced —— cents. per ounce
3525	Blank . .	Has (have) advanced —— cents. per pound
3526	Blankbond .	Is now —— cents. per ounce lower
3527	Blanketing .	Is now —— cents. per pound lower
3528	Blankets	**. Central**
3529	Blanktire	In a central position
3530	Blarney . .	From the central position
3531	Blarneying	**. Centre**
3532	Blasphemed .	In the centre of
3533	Blastema	**. Certain**
3534	Blasthole .	Are you certain
3535	Blastments .	I am (we are) certain
3536	Blastoderm .	I am (we are) not certain

No.	Code Word.	Certain (*continued*)
3537	Blastoidea .	It is certain (that)
3538	Blastomere .	It is not certain (that)
3539	Blastpipe .	Quite certain
3540	Blastus . .	Let me (us) hear for certain
3541	Blatant . .	It is almost certain
3542	Blatteroon .	Almost certain to float
3543	Blawort . .	If you are certain
3544	Blazers . .	Unless you are quite certain of
3545	Blazingly .	Make certain that you get
3546	Blazon . .	Must be quite certain
3547	Blazoning .	**Certainly**
3548	Blazonment .	Will certainly be
3549	Blazonries .	Will certainly not
3550	Blazonry . .	Will certainly go to
3551	Bleachery .	**Certainty**
3552	Bleachings .	With any degree of certainty
3553	Bleareyed .	There is every certainty of
3554	Blemish . .	No certainty of
3555	Blemisheth .	There is no certainty in the matter
3556	Blemishing .	**Certificate(s)**
3557	Blended . .	Have obtained a certificate
3558	Blendings .	Cannot obtain a certificate
3559	Blendous .	All the certificates
3560	Blendwater .	Share certificate(s)
3561	Blessedly .	Certificate(s) not yet received
3562	Bletonism .	Share certificate(s) will be sent on
3563	Blewits . .	Share certificate(s) will be sent when ready
3564	Blindage . .	You must get certificate
3565	Blindborn .	Certificates must be made out in name of
3566	Blindcoal .	In whose name must certificates be made out
3567	Blindfold .	We wish certificates made out in the following name
3568	Blindingly .	Certificates in name of
3569	Blindman .	Send the certificates
3570	Blindshell .	Certificates will be sent
3571	Blindside .	Certificates to bearer
3572	Blindstory .	Scrip certificates
3573	Blindworm .	Certificate number
3574	Blinkard . .	Consular certificate
3575	Bliss . . .	**Certified**
3576	Blissfully .	Are the titles certified (by)
3577	Blissless . .	The titles are certified (by)
3578	Blisterfly . .	The titles are not certified (by)
3579	Blistering .	Must be certified (by)
3580	Blitheful . .	Certified by consul
3581	Blithely . .	Have been certified
3582	Blithemeat .	**Certify**
3583	Blitheness .	Is (are) able to certify
3584	Blithesome .	Is (are) not able to certify
3585	Blobberlip .	Is (are) —— able to certify
3586	Bloblipped .	Shall certify his statement

No.	Code Word.	**Certify** (*continued*)
3587	Blobtale . .	Should —— not be able to certify
3588	Blockade .	Refuses to certify
3589	Blockading .	**Chalk**
3590	Blockhouse .	Chalk formation
3591	Blockishly .	In the chalk
3592	Blocklike .	**Chance(s)**
3593	Blocktin . .	If we now lose the chance, may not have another
3594	Blondlace .	Do not lose the chance [opportunity
3595	Blondmetal .	If by any chance
3596	Blooddrier .	Let us have the first chance
3597	Bloodheat .	There is not much chance
3598	Bloodhorse .	Lose no chance (of)
3599	Bloodhound .	We think this chance should not be lost
3600	Bloodiest .	The chances are in our favour
3601	Bloodily . .	The chances are much against
3602	Bloodiness .	The chance is too good to be lost
3603	Bloodmoney	**Change(s)**
3604	Bloodred .	A change in the character of the rock
3605	Bloodshot .	No change in the character of the rock
3606	Bloodsized .	Has any change taken place (in)
3607	Bloodstain .	There has been no change worth reporting
3608	Bloodwood .	It is necessary to change the
3609	Bloodworms	A change for the better has taken place
3610	Bloodyflux .	Change for the better
3611	Bloodyhand .	Change for the worse
3612	Bloom . .	Telegraph if any change
3613	Bloomary .	Telegraph at once if any change of importance in
3614	Bloomerism .	A great change in [the workings
3615	Bloomless .	No change in
3616	Blossom . .	Telegraph the earliest change in
3617	Blossometh .	Telegraph any material change in
3618	Blossoming .	Do not think a change advisable
3619	Blotch . .	Advise what changes can be made
3620	Blotching .	**Changed**
3621	Blottingly .	Has (have) changed
3622	Bloused . .	Has (have) not changed
3623	Blowball . .	Changed for the worse
3624	Blowerup .	Changed for the better
3625	Blowflies .	**Charge(s)**
3626	Blowfly . .	What would you charge
3627	Blowgun . .	Charge the expenses to
3628	Blowhole .	Send a competent man to take charge
3629	Blowingoff .	To whom shall I charge expenses
3630	Blowmilk .	Is in charge of the works
3631	Blowpipe .	Left in charge of the foreman
3632	Blowpoint .	Who will take charge of
3633	Blowtube .	Who will take charge while you are away
3634	Blowvalve .	Will take charge while I am away
3635	Blowzy . .	Who was in charge
3636	Blubber·d .	Has (have) been in charge of

No.	Code Word.	Charge(s) (*continued*)
3637	Blubbering .	Who will pay the extra charges
3638	Bludgeons .	Will pay all charges
3639	Bluebell . .	All charges incurred
3640	Blueberry .	After all charges are paid
3641	Bluebirds .	What charges have you incurred
3642	Blueblack .	Have incurred charges amounting to
3643	Blueblood .	On payment of all charges
3644	Bluebonnet .	All charges for freight, etc.
3645	Bluebooks .	Inclusive of all charges
3646	Bluebottle .	**Charged**
3647	Bluebreast .	Has (have) been charged (with)
3648	Bluecap . .	To be charged to Capital Account
3649	Bluecoats .	To be charged to Revenue Account
3650	Bluegowns .	Ought to be charged to
3651	Bluegrass .	Has (have) not been charged
3652	Bluehaired .	Should —— be charged with
3653	Bluejack .	**Cheap**
3654	Bluejohn .	Can be bought cheap
3655	Bluelight .	The property is cheap at the price
3656	Bluemantle .	The mine(s) can be bought cheap
3657	Bluemould .	Unless we can get it cheap, cannot entertain the
3658	Blueochre .	Think it a cheap purchase [matter
3659	Bluepeter	**Cheaper**
3660	Bluepill . .	It would be cheaper to
3661	Blueribbon .	It is cheaper to
3662	Blueruin . .	Must have it cheaper
3663	Bluespar . .	Is cheaper
3664	Bluestone	Cannot be done cheaper
3665	Bluethroat .	Will be cheaper
3666	Bluetint . .	If we cannot get the property cheaper, **must**
3667	Blueveined .	We must have cheaper labour [decline it
3668	Bluewater .	We must have cheaper
3669	Bluewings .	**Cheaply**
3670	Bluffbowed .	The mine(s) can be cheaply worked
3671	Bluffness . .	The ore can be cheaply mined and milled
3672	Bluffy . . .	The works can be cheaply carried on
3673	Bluishly . .	Can be brought to the mine very cheaply
3674	Bluishness .	Can the —— be cheaply worked
3675	Bluntly . .	Cannot be worked cheaply
3676	Bluntness .	**Check**
3677	Blushful . .	As a check on
3678	Blushing . .	What check have you upon
3679	Blushingly .	There has been no check on
3680	Bluster . .	Must check
3681	Blustereth .	Should this not check
3682	Blustrous .	It is necessary to check
3683	Blysmus . .	**Chemical**
3684	Boanerges .	The chemical works
3685	Boardable .	The erection of chemical works
3686	Boarders .	The chemical operations

No.	Code Word.	Chemical (*continued*)
3687	Boardrule .	From a chemical point of view
3688	Boardwages .	**Cheque(s)**
3689	Boarfish . .	Do not pay cheque
3690	Boarspear .	Do not pay any cheques till after
3691	Boarstag . .	Cheque has been lost
3692	Boast . . .	A cheque for
3693	Boastful . .	Send cheque for —— at once
3694	Boastfully .	Send cheque immediately to
3695	Boastingly .	——'s cheque has been dishonoured
3696	Boastless .	Cheque for —— has been presented
3697	Boatbill . .	Cheque for —— has not been presented
3698	Boathook .	I (we) send cheque
3699	Boathouse .	Have you received cheque
3700	Boatplug . .	**Chloride**
3701	Boatracing .	Chloride of gold
3702	Boatshaped .	Chloride of silver
3703	Boatshell .	Chloride of sodium
3704	Boatswain .	**Chlorination**
3705	Boattails . .	Chlorination works
3706	Boatwright .	Chlorination works will be required
3707	Bobbin . .	By chlorination process
3708	Bobbinwork .	Chlorination is necessary
3709	Bobcherry .	The Newbury Vautin chlorination process
3710	Boblincoln .	**Chloritic**
3711	Bobolinks .	Chloritic slate
3712	Bobsleigh .	The hanging wall chloritic slate
3713	Bobstay . .	The foot wall chloritic slate
3714	Bobtailed .	**Chosen**
3715	Bobtailwig .	Has (have) chosen to
3716	Bobwhite .	Has (have) chosen not to
3717	Bobwigs . .	**Chute**
3718	Bocardo . .	Chute of ore
3719	Bockbier . .	The vein outside the pay chute is barren
3720	Bockelet . .	There is a chute of pay rock about —— feet long average width
3721	Bodeful . .	What is the length and width of pay chute
3722	Bodied . .	What is average assay of pay chute
3723	Bodiless . .	Pay chute averages —— per ton
3724	Bodycloth .	From this chute we are extracting
3725	Bodycoat .	From this chute we have extracted
3726	Bodycolour .	From this chute we expect to get
3727	Bodyguard .	Has opened up a fine chute of ore
3728	Bodying . .	Cannot say yet what the character of the chute
3729	Bodyplan .	Chute above —— level [will be
3730	Bogberries .	Chute at —— level
3731	Bogberry . .	Chute below —— level
3732	Bogbumper .	This chute is now giving plenty of ore
3733	Bogearth . .	Think this chute will prove a valuable addition
3734	Bogeyism .	This chute is widening out
3735	Bogglish . .	**Cipher**

No.	Code Word.	Cipher (*continued*)
3736	Bogland . .	Cipher not intelligible
3737	Bogmoss. .	Cipher number
3738	Bogoak . .	Cipher code must be used
3739	Bogorchis .	Read this carefully : it is not all in cipher words
3740	Bogrush . .	Permit our cipher to be used by
3741	Bogspavin .	**Cinnabar**
3742	Bogtrotter .	Cinnabar lode
3743	Bogus . .	Cinnabar containing —— per cent. quicksilver
3744	Bogwhort .	Cinnabar mine
3745	Bohemians .	Cinnabar is found
3746	Bohunupas .	**Circular**
3747	Boilary . .	Official circular
3748	Boileriron .	Private circular
3749	Boilingly .	Send out circular
3750	Boisterous .	Do not send out circular
3751	Boletus . .	Our fortnightly circular
3752	Bollworm .	Our monthly circular
3753	Bolognese .	In time for our monthly circular
3754	Bolstering .	In time for our half-yearly circular
3755	Boltauger .	Half-yearly circular
3756	Boltboat . .	Quarterly circular
3757	Boltenia . .	In issuing our next circular
3758	Bolthead. .	We wish this published in your next circular
3759	Bolting . .	**Circulate**
3760	Boltingtub .	Circulate the report that
3761	Boltonite .	Do not circulate
3762	Boltrope . .	**Circulating**
3763	Boltsprit . .	Has (have) been circulating the report that
3764	Bomarea. .	Who has been circulating the report that
3765	Bombarded .	**Circulation**
3766	Bombarding	In circulation
3767	Bombardman	Not in circulation
3768	Bombast . .	**Circumstances**
3769	Bombastic .	Under the existing circumstances
3770	Bombayduck	Under no circumstances
3771	Bombazine .	Do not under any circumstances
3772	Bombchest .	If you think the circumstances justify
3773	Bombilate .	Carefully considering all the circumstances
3774	Bombilious .	Enabling me (us) fully to understand the circum-
3775	Bombketch .	Owing to the altered circumstances [stances
3776	Bombproof .	Under what circumstances
3777	Bombshells .	Entirely depends on circumstances
3778	Bombvessel.	Under these circumstances
3779	Bombycidae	According to circumstances
3780	Bombyx . .	Under any circumstances
3781	Bonasa . .	From circumstances which have come to our know-
3782	Bonassus .	Much must depend on circumstances [ledge
3783	Bonbons. .	Unforeseen circumstances which have arisen
3784	Bonchief. .	From unforeseen circumstances
3785	Bonddebt .	Are not aware of any circumstances which

No.	Code Word.	**Circumstances** (*continued*)
3786	Bondfolk .	We are justified by circumstances
3787	Bondmaid .	**Claim(s)**
3788	Bondslave .	The claim is —— feet × —— feet
3789	Bondtenant .	The adjoining claim(s)
3790	Bondtimber .	Claim is one of the best prospects in the district
3791	Bondwoman	The owner(s) of the adjoining claim(s)
3792	Boneache .	It is very important to buy the adjoining claim
3793	Bonebed . .	The property consists of one claim —— feet ×
3794	Boneblack .	Of the claim [—— feet
3795	Bonebrown .	The property consists of —— claims.
3796	Bonecave .	Of all the claims
3797	Bonedust .	There are (is) —— claim(s) —— feet × —— feet
3798	Boneglue .	It is very important to buy the adjoining claim as the vein runs into it. The owner(s) is (are)
3799	Bonemanure	Make a claim for [not aware of this
3800	Bonemill . .	Has (have) the first claim on your time
3801	Bonesetter .	Has (have) made a claim for
3802	Bonespirit .	A claim has been lodged
3803	Bonfires . .	Cannot admit the claim
3804	Boniface . .	Do (does) not claim
3805	Bonnet . .	The claim had better be paid
3806	Bonneting .	What is the amount of claim
3807	Bonnetbox .	Amount of claim is
3808	Bonniest . .	The claim had better be resisted
3809	Bonnily . .	Has (have) a claim
3810	Boobies . .	Has (have) not any claim
3811	Booby . .	Appears to be a reasonable claim
3812	Boobyhutch .	The claim is excessive
3813	Boobyism .	The claim is preposterous
3814	Bookbinder .	Claim(s) is (are) withdrawn
3815	Bookcases .	We claim that
3816	Bookdebts .	Has a legal claim
3817	Booked . .	I (we) have a claim
3818	Bookfair . .	Who makes the claim
3819	Bookformed	What is the claim for
3820	Bookholder .	The claim is for
3821	Bookhunter .	The claim is made by
3822	Bookmaker .	Dispute our claim
3823	Bookmaking	We dispute the claim
3824	Bookmonger	Send particulars of the claim
3825	Bookmuslin .	Claim the return of
3826	Bookoath .	Admit the claim
3827	Bookpost .	If the claim is pressed
3828	Bookseller .	Shall press the claim
3829	Bookslides .	Settle the claim
3830	Bookstalls .	Settle the claim on best terms possible
3831	Bookstand .	Settle the claim under advice
3832	Booktrade .	Claim under the insurance
3833	Booktrays .	Have settled the claim (for)
3834	Bookworms .	There are —— claims patented

No.	Code Word.	Claims *(continued)*
3835	Boomage .	Unpatented claims
3836	Boomerangs	Claims must be patented
3837	Boometh .	How many claims are patented and how many
3838	Boomiron .	Are all the claims patented [unpatented
3839	Boorish . .	The claim(s) will be jumped
3840	Bootcloser .	The claim(s) has (have) been jumped
3841	Bootcrimp .	Fear claim(s) will be jumped
3842	Boothalers .	Deep level claim(s)
3843	Bootikin . .	Have secured claims
3844	Bootjacks .	Spare no expense to locate claims
3845	Bootlaces .	Buy as many claims as you can
3846	Bootleg . .	Send plan of claims
3847	Bootless . .	Plan of claims has been sent
3848	Bootlessly .	**Claimed**
3849	Bootlick . .	Has (have) claimed
3850	Boottop . .	Has (have) been claimed
3851	Boottree . .	Has (have) not been claimed
3852	Bopeep . .	If not claimed
3853	Boracic . .	**Classified**
3854	Borborygm .	Have classified
3855	Borderland .	**Classify**
3856	Borderers .	Classify as follows
3857	Bordlode .	We should like you to classify
3858	Bordraging .	**Classification**
3859	Boreal . .	Classification of
3860	Boredom .	Showing classification (of)
3861	Boringbars .	Without some classification
3862	Boringbit .	**Clause**
3863	Boringmill .	Clause must be inserted
3864	Borrelist . .	Clause must be struck out
3865	Borreria . .	Insert the following clause(s)
3866	Borrowing .	Will alter the clause respecting
3867	Borsella . .	Will not alter the clause
3868	Boscage . .	Do not allow any clause to be inserted which will
3869	Bosomer . .	Clause(s) altered as follows
3870	Bosporian .	Object to —— clause
3871	Bostrichus .	A clause has been inserted by which
3872	Boswellian .	If clause approved
3873	Botanical .	If clause not approved
3874	Botanists .	Clause makes it impossible to
3875	Botanized .	Alter clause as follows
3876	Botanizing .	Insert clause
3877	Botanology .	Clause in respect to
3878	Botany . .	Strike out clause
3879	Botaurus. .	The clause is approved
3880	Botched . .	We do not approve the clause
3881	Botchedly .	The following clause
3882	Bothering .	If clause is waived
3883	Botherment .	We agree to waive the clause
3884	Bothhands .	He agrees (they agree) to waive the clause

No.	Code Word.	**Clause** (*continued*)
3885	Bothsides .	Clause in the agreement
3886	Botrychium .	**Clay**
3887	Botryogen .	In the clay
3888	Botryoid . .	Through the clay
3889	Botryoidal .	Clay iron ore
3890	Botryolite .	Clay seam
3891	Botrytis . .	Clay wall
3892	Botthammer	Clay slate
3893	Bottleboot .	**Clean-up**
3894	Bottlebump .	After —— tons were crushed, the clean-up gave
3895	Bottlefish .	This clean-up is small owing to
3896	Bottlehead .	This clean-up fairly indicates the value of the ore
3897	Bottlejack .	After our next clean-up
3898	Bottlenose .	Clean-up after —— days of 24 hours, using —— inches of water, cost of labour ——, cost of material ——, general expenses ——, profit on the run ——
3899	Bottlers . .	Clean-up after crushing —— tons of quartz, gross yield being ——, cost of labour ——, cost of material ——, general expenses ——, profit ——
3900	Bottletit . .	The result of the last clean-up was
3901	Bottom . .	Partial clean-up
3902	Bottombed .	Shall arrange to clean-up
3903	Bottomheat .	When do you intend to clean-up
3904	Bottomice .	Do not intend to clean-up until
3905	Boltomland .	Shall clean-up
3906	Bottomless .	Clean-up on —— and cable result
3907	Bottomlift .	Until clean-up cannot give actual result of
3908	Bottomry .	Until you clean-up
3909	Boudoirs. .	Days in addition to clean-up
3910	Boughpot .	Shall not clean up this month on account of
3911	Boughten .	Defer clean-up until
3912	Boulders. .	The clean-up yielded
3913	Bounced. .	Found on cleaning up-that
3914	Bouncing .	Suggest cleaning-up
3915	Bouncingly .	**Cleaned up**
3916	Bound . .	Have cleaned up after crushing —— tons of quartz
3917	Boundaries .	Have cleaned up [gross yield
3918	Boundenly .	The mill has not been cleaned up since
3919	Boundless .	**Climate**
3920	Bounties . .	Climate good all the year
3921	Bountihood .	The climate in winter very severe
3922	Bouquets .	Climate very unhealthy
3923	Bourbonism .	Climate very healthy
3924	Bourignian .	Climate fairly good
3925	Bournonite .	**Clique**
3926	Bourock . .	The clique is at work
3927	Boveycoal .	The clique has succeeded
3928	Boviform .	The clique is doing its utmost to
3929	Bovine . .	**Close**

No.	Code Word.	Close (*continued*)
3930	Bovista . .	Close with the offer
3931	Bowbacked .	Do not close until you hear from
3932	Bowbearer .	Shall I (we) close
3933	Bowbent . .	I (we) advise you to close with this offer
3934	Bowbrace .	Do not close
3935	Bowchaser .	Can you close
3936	Bowcompass	After the close (of)
3937	Bowdlerize .	We should like to close
3938	Bowdrill . .	Close at once
3939	Bowelless .	We cannot close
3940	Bowerbird .	Cannot you close
3941	Bowereaves .	Closed
3942	Bowermaid .	Has (have) closed with
3943	Bowerthane .	Has (have) not closed with
3944	Bowhand .	Have you closed
3945	Bowieknife .	Closed at
3946	Bowingly .	I (we) have closed
3947	Bowknot . .	I (we) have not closed
3948	Bowled . .	Market closed weak
3949	Bowlegged .	Market closed strong
3950	Bowpiece .	Has (have) closed
3951	Bowshot . .	If —— has (have) closed
3952	Bowtimbers .	If —— has (have) not closed
3953	Boxcrab . .	If you have closed
3954	Boxday . .	If you have not closed
3955	Boxdrain . .	If already closed
3956	Boxelder . .	The transaction should be closed
3957	Boxers . .	Have closed the accounts
3958	Boxgirder .	The shares closed at
3959	Boxhaul . .	Closed strong at
3960	Boxiana . .	Closing
3961	Boxingday .	Closing price
3962	Boxirons . .	Closing transaction
3963	Boxlobby .	Closing the business
3964	Boxmoney .	Closing the agency
3965	Boxopener .	Closing accounts
3966	Boxthorn .	Clue
3967	Boxwood .	Has (have) —— any clue to
3968	Boybishop .	Has (have) a clue to
3969	Boyblind . .	Has (have) not any clue to
3970	Boycott . .	Coal
3971	Boycotting .	Seams of coal
3972	Boyish . .	The coal lies between beds of
3973	Boyishly . .	Coal seam —— in. in thickness
3974	Boyishness .	Anthracite coal
3975	Boyqueller .	Bituminous coal
3976	Boyship . .	Brown coal
3977	Boysplay . .	Send —— tons of coal
3978	Boyuna . .	It is cheaper to burn coal than wood
3979	Bracelets .	It is cheaper to burn wood than coal

No.	Code Word.	Coal (*continued*)
3980	Brachial . .	Coal is brought from
3981	Brachionus .	Coal has been proved on the property
3982	Brachiopod .	Coal in the district
3983	Brachylogy .	The coal district
3984	Brachyura .	Coal is —— shillings per ton delivered at
3985	Bracing . .	Coal is —— per ton
3986	Bracingly .	The coal seam dips into the adjoining property
3987	Bracken . .	The coal extends over —— acres
3988	Bracteole .	Coal can be delivered at the mine for —— per ton
3989	Bradawl . .	Coal mine
3990	Bradypoda .	In the coal measures
3991	Bradypus .	The coal measures
3992	Braggart . .	The coal measures extend
3993	Braggartry .	The coal seams crop out
3994	Braggingly .	The coal is a good hard coal
3995	Brahmaic .	The coal is soft
3996	Brahmanas .	The coal is very suitable for
3997	Braidcomb .	The coal is very suitable for steam navigation
3998	Braincoral .	The coal is not at all suitable for
3999	Brainfever .	The coal is very good
4000	Brainish . .	The coal is very bad
4001	Brainless . .	The coal is of good average quality
4002	Brainpan .	The coal is of fair quality
4003	Brainthrob .	The coal is very dirty
4004	Brakebar .	No economy in use of this coal
4005	Brakebeams	Coal at the pit's mouth
4006	Brakeblock .	Coal from the deep seams
4007	Brakeshoe .	The price of coal per ton
4008	Brakesman .	What is the consumption of coal
4009	Brakewheel .	Consume about —— tons of coal per day
4010	Bramahlock .	The consumption of coal is
4011	Bramantip .	The high price of coal
4012	Bramble . .	The low price of coal
4013	Bramblenet .	The coal cannot be worked
4014	Braminical .	Coal can be produced at
4015	Braminism .	At what price can the coal be sold
4016	Branchiata .	Hauling coal
4017	Branchleaf .	Extracting coal
4018	Branchless .	Coal owners
4019	Branchline .	Has taken out —— tons of coal
4020	Branchy . .	Have produced this month —— tons of coal at a [cost of
4021	Brandgoose .	**Coarse**
4022	Brandied .	Coarse gold
4023	Brandiron .	Coarse-grained
4024	Brandisher .	The gold is very coarse
4025	Brandlings .	Coarse mesh
4026	Brandmarks .	Is very coarse
4027	Brandnew .	**Coast**
4028	Brandrith .	All along the coast
4029	Brandwine .	From here to the coast

No.	Code Word.	**Coast** (*continued*)
4030	Brantfox . .	Upwards from the coast
4031	Brassage . .	Are sent down to the coast by
4032	Brassband .	**Cobalt**
4033	Brassfoil . .	Contains —— per cent. of cobalt
4034	Brassica . .	Cobalt mine
4035	Brassleaf . .	**Code**
4036	Brasspaved .	The code now in use
4037	Brassrule .	Are sending you a private code
4038	Brattice . .	Send us a private code
4039	Bravado . .	What code is he (are they) using
4040	Bravely . .	Using the —— code
4041	Braveness .	Additions to the code
4042	Bravoes . .	**Coded**
4043	Brawlers . .	Have coded
4044	Brawney . .	**Codification**
4045	Brawniness .	Requiring codification
4046	Brazened .	**Collapse**
4047	Brazenface .	Has (have) allowed the scheme to collapse
4048	Brazilnut .	To prevent a collapse of the whole affair
4049	Braziltea . .	A total collapse of the market
4050	Brazilwood .	There has been a complete collapse
4051	Breadberry .	The whole thing will collapse
4052	Breadcorn .	**Collapsed**
4053	Breadfruit .	Has (have) collapsed
4054	Breadmeal .	The negotiations have collapsed
4055	Breadnut .	The concern has collapsed
4056	Breadroot .	**Collect**
4057	Breadsauce .	Collect all the information you can about
4058	Breadstuff .	To collect
4059	Breadth . .	Collect and send
4060	Breadtree .	Can you collect
4061	Break . . .	Collect and remit amount due
4062	Breakable .	Collect and send us a few specimens
4063	Breakdown .	**Collected**
4064	Breakfast .	Have you collected
4065	Breakjoint .	Have collected a few specimens
4066	Breakneck .	The specimens are collected from
4067	Breakshare .	Have collected
4068	Breakup . .	Have collected all the information I (we) can
4069	Breakvow .	**Colorado**
4070	Breakwater .	Gold mine in Colorado
4071	Breamflat .	Silver mine in Colorado
4072	Breastband .	Smelting works in Colorado
4073	Breastbeam .	Mine in Colorado
4074	Breastdeep .	**Combination**
4075	Breasthigh .	Unless there is a combination of
4076	Breastknot .	Owing to a combination of
4077	Breastmilk .	Forming a combination of
4078	Breastpain .	If a combination can be formed
4079	Breastrope .	A very strong combination

No.	Code Word.	Combination (*continued*)
4080	Breastwall .	Have formed a combination
4081	Breathless .	Have been unable to form a combination
4082	Brecciated .	Combination of capitalists
4083	Bredsore. .	Combination of mine owners
4084	Breechband.	Combination of workmen
4085	Breechpin .	Combination to exploit
4086	Breedbate .	In order to form a combination
4087	Breeders. .	Can you form a combination to take up this business
4088	Breezefly	. Combine
4089	Breezeless .	Will —— combine with us (to)
4090	Brevet . .	Will combine with us (to)
4091	Brevetcy. .	Will not combine with us (to)
4092	Breviature .	If —— will not combine with
4093	Brevimanu .	Offer(s) to combine with
4094	Breviped .	Do not combine
4095	Brewage. .	If necessary combine with
4096	Breweries .	Cannot combine with
4097	Brewery . .	If you will combine with
4098	Brewhouse .	If he (they) will combine with
4099	Brewster . .	Offer to combine with us in forming a company
4100	Briarean .	. Combined
4101	Briarroot. .	Has (have) combined with
4102	Bribable. .	Has (have) not combined with
4103	Bribery . .	Is (are) found combined with
4104	Bricabrac .	Chemically combined
4105	Brickbats .	Not chemically combined
4106	Brickbuilt .	Mechanically combined
4107	Brickclay	. Come
4108	Brickdust .	Come as soon as possible
4109	Brickearth .	Come home immediately
4110	Brickkiln .	Can I come
4111	Bricklayer .	When can you come
4112	Brickmaker .	I can come
4113	Brickmason.	If you can come
4114	Bricktea . .	Can you come
4115	Brickwork .	Come back
4116	Brickyard .	Come down to
4117	Bridal . .	Come up to
4118	Bridecake .	Cannot come till
4119	Brideday .	Must come back here
4120	Bridegroom.	When are you coming
4121	Brideknot .	Will come about [settled
4122	Bridesmaid . .	As soon as —— come(s), things can be definitely
4123	Bridewell .	Tell —— to come as quickly as possible
4124.	Bridgedeck .	I want your help; come here at once; leave —— in charge of everything
4125	Bridgehead .	Cannot leave till —— come(s)
4126	Bridgeless .	The matter is in abeyance till —— comes, which
4127	Bridgeward .	Until I come [will be in a few days
4128	Bridlehand .	Until he comes (they come)

No.	Code Word.	Come (*continued*)
4129	Bridlepath .	Until you come
4130	Bridleport .	When I come
4131	Bridlerein .	When he comes (they come)
4132	Bridleroad .	When you come
4133	Bridleway .	How soon can you come
4134	Bridoon . .	Why did he not come
4135	Briefest . .	As soon as you can come
4136	Briefless .	. Coming
4137	Briefly . .	Is (are) coming down
4138	Briefman .	Is (are) coming in very quickly
4139	Brigandage .	When is —— coming
4140	Brigbote . .	Telegraph if you are coming
4141	Brighten . .	Telegraph if he is (they are) coming
4142	Brightly .	. Commence
4143	Brightness .	Shall commence drifting
4144	Brightsome .	Shall commence crushing ore
4145	Brilliance .	Will commence
4146	Brimful . .	I (we) shall commence work by
4147	Brimming .	Shall commence sinking
4148	Brimmingly .	Commence as soon as possible
4149	Brimsey . .	Cannot commence till
4150	Brimstone .	When will you commence
4151	Brindled . .	When we commence
4152	Brinepump .	Until we commence
4153	Brineworm .	Until you commence
4154	Bringeth . .	Expect to commence about
4155	Briskets .	. Commenced
4156	Briskly . .	Has (have) commenced
4157	Briskness .	Has (have) not yet commenced
4158	Brittle . .	Has (have) —— commenced
4159	Brittlely . . .	Have commenced driving on the vein
4160	Broach . .	Have commenced driving the tunnel
4161	Broachers .	Work commenced (on)
4162	Broadarrow .	Commenced sinking [thoroughly overhauled
4163	Broadaxe .	Mill commenced running on ——, after being
4164	Broadbased .	The rains have commenced
4165	Broadbill	. Commencement
4166	Broadblown .	At the commencement of
4167	Broadbrim	. Commission
4168	Broadcast .	A commission has been appointed to inquire into
4169	Broadcloth .	Has (have) agreed to pay a commission of
4170	Broaden . .	What commission will have to be paid
4171	Broadening .	Must have a commission
4172	Broadeyed .	Get(s) a commission of
4173	Broadhorn .	Who will pay my (our) commission
4174	Broadly . .	Will get a commission if his report sells the mine
4175	Broadness .	Will pay a commission of
4176	Broadpiece .	Will take no commission
4177	Broadseal .	Commission from both sides
4178	Broadsides .	Commission has been paid

No.	Code Word.	Commission (*continued*)
4179	Broadsword .	What commission are you allowing
4180	Broadwise .	Will allow you a commission of
4181	Brocaded .	We accept that amount of commission
4182	Broccoli . .	Decline to accept such a small commission
4183	Brodekins .	Cannot allow higher commission
4184	Brogue . .	Will accept the commission offered
4185	Broidery . .	No commission to be paid
4186	Broiled . .	Who is to pay the commission
4187	Broiling . .	Vendor will pay commission
4188	Brokenness .	Vendor must pay commission
4189	Brokenwind	All commissions must be paid by
4190	Brokerage .	Bank commission
4191	Brokers . .	Commission on underwriting
4192	Bromegrass .	The usual commission
4193	Bromide . .	Commission to be included
4194	Bromoform .	Commission to be divided
4195	Bromyrite .	Our commission (is)
4196	Bronchi . .	Your commission (is)
4197	Bronchial .	Commission and brokerage
4198	Bronchitis .	Subject to a commission of
4199	Bronteum .	You must arrange for a commission
4200	Brontolith .	**Commissioned**
4201	Brontology .	Was commissioned to
4202	Brontozoum	Was commissioned to examine and report upon
4203	Bronzed . .	Was commissioned to negotiate
4204	Bronzified .	Was commissioned to take charge of
4205	Bronzify . .	Has been commissioned to
4206	Bronzists .	**Communicate**
4207	Broodeth .	Can you communicate with
4208	Broodmare .	I (we) can communicate with
4209	Brooklet . .	I (we) cannot communicate with
4210	Brooklime .	Communicate with
4211	Brookmint .	Communicate in future direct with
4212	Brookweed .	In order to communicate with [apply to
4213	Brooky . .	Should you not be able to communicate with ——
4214	Broomcorn .	With whom am I (are we) to communicate
4215	Broomrape .	Will communicate
4216	Broomstaff .	Communicate contents of letter to
4217	Broomstick .	Please communicate with me (us), care of
4218	Brosimum .	Communicate in confidence to
4219	Brotherly .	Communicate contents of telegram to
4220	Brougham .	Communicate with —— and report
4221	Browantler .	If you cannot communicate with us
4222	Browbands .	If you can communicate
4223	Browbeater .	Until we can communicate with
4224	Browbound . .	Have communicated with
4225	Brownblaze .	**Communication**
4226	Brownbread	Am (are) in communication with
4227	Browncoal .	Is (are) in communication with
4228	Browngull .	There are no means of communication with

No.	Code Word.	Communication (*continued*)
4229	Brownish .	No means of communication between —— and
4230	Brownpaper .	Have only postal communication with [graphic
4231	Brownrust .	No means of communication either postal or tele-
4232	Brownspar .	Telegraphic communication impeded with
4233	Brownstout ,	What communication is there between —— and
4234	Brownstudy .	All communication is cut off (from)
4235	Brownwort .	Resume communications with
4236	Browsewood	Communication is now restored
4237	Browsick .	The following communication
4238	Browsings .	In direct communication with
4239	Browsnag .	**Company**(ies) (See Registered.)
4240	Brucea . .	Dividend-paying company
4241	Bruchus . .	Non-dividend-paying company
4242	Brugmansia .	Limited liability company
4243	Bruised . .	Company in liquidation
4244	Bruisewort .	Company in difficulties
4245	Brumous . .	Can you form company
4246	Brunettes .	The existing company
4247	Brunonian .	Trying to wreck the company
4248	Brunsvigia .	Very sound company
4249	Brushburn .	Company decline to
4250	Brushiness .	I (we) advise the company to acquire
4251	Brushore .	In the hands of a company
4252	Brushwheel .	Do you advise the company to acquire
4253	Brushwood .	The mine was owned by a company which became
4254	Brusquerie .	A company is already formed [bankrupt
4255	Brutalism .	The company is now formed
4256	Brutalize . .	The company has been registered
4257	Brutally . .	The name of the company
4258	Brutish . .	What is the capital of the company
4259	Brutishly . .	Company with a capital of
4260	Bryology . .	The company is greatly over-capitalized
4261	Bryony . .	The company must go into liquidation
4262	Bryozoa . .	The company must be reconstructed
4263	Bubalus . .	If the company has to go into liquidation
4264	Bubonocele .	Reconstruction of the company
4265	Bubukles .	The company has been reconstructed
4266	Bucaneer .	The company's articles of association
4267	Buccinal . .	(To) bring out the company
4268	Buccinator .	The company's solicitors
4269	Bucco . .	The board of the company
4270	Bucconidae .	The board of the company comprises
4271	Buccula . .	The company has not enough
4272	Bucentaur .	Prior to the company taking it over
4273	Buceros . .	The company has been successfully floated
4274	Buchanite .	Transfer of the property to the company
4275	Buckbasket .	Transfer of the property to the company must be
4276	Buckbean .	The company's shares [registered
4277	Buckboard .	Upon registration of the transfer of the property to
4278	Bucketful .	The first meeting of the company [the company

No.	Code Word.	Company(ies) (*continued*)
4279	Bucketlift .	The statutory meeting of the company
4280	Bucketrod .	The company comprises —— shareholders
4281	Buckhound .	A company is being formed to
4282	Bucklandia .	The parent company
4283	Buckled . .	The affiliated companies
4284	Buckmast .	An independent company
4285	Buckram .	A subsidiary company
4286	Buckshot .	A small company
4287	Buckskin .	A very large company
4288	Buckstall	. **Comparative**
4289	Bucktooth .	The comparative cost
4290	Buckwaggon	The comparative size of
4291	Buckwheat .	Comparative economy in .
4292	Bucolic .	. **Comparatively**
4293	Bucolical .	Comparatively to
4294	Bucranium .	Comparatively worthless
4295	Buddhism .	At a comparatively small cost
4296	Buddhistic	. **Compare**
4297	Budding . .	If you compare
4298	Budeburner .	Compare my letter of —— with
4299	Budelight .	Cannot compare with
4300	Budgeness .	Compare(s) favourably with
4301	Budget . .	Compare(s) unfavourably with
4302	Buffalo . .	This —— working compares favourably with last
4303	Buffalonut	. **Compared**
4304	Buffcoat . .	Has (have) compared
4305	Buffelduck .	Has (have) not compared
4306	Bufferhead .	Has (have) —— compared
4307	Buffeted . .	As compared with
4308	Buffeting	. **Comparison**
4309	Buffjerkin .	There is no comparison (between)
4310	Buffoon . .	The comparison is unfair (to)
4311	Buffoonish .	Upon comparison with (or between)
4312	Buffoonly	. **Compartment**
4313	Buffstick. .	In one compartment
4314	Bufftip . .	In two compartments
4315	Buffwheel .	In three compartments
4316	Bufonite . .	In —— compartments
4317	Bugbear . .	How many compartments
4318	Buggyboat	. **Compel**
4319	Buglehorn .	Wants(s) to compel us to
4320	Bugleweed .	Cannot compel us to
4321	Bugloss . .	Cannot compel —— to
4322	Buhlsaw . .	Ought to compel —— to
4323	Buhlwork .	Could you not compel —— to
4324	Buhrstone .	This would compel —— to
4325	Build . . .	They can compel us to
4326	Buildings .	We can compel him (them) to
4327	Buildress	. **Compelled**
4328	Bulbaceous .	Has (have) been compelled to

No.	Code Word.	Compelled (*continued*)
4329	Bulblet . .	Has (have) compelled —— to
4330	Bulbodium .	Compelled to take legal proceedings
4331	Bulbous . .	If not compelled
4332	Bulbotuber .	**Compensated**
4333	Bulgaric . .	Has (have) been compensated
4334	Bulimus . .	Has (have) not been compensated
4335	Bulkheads .	Shall I be compensated
4336	Bulkiness .	Will be compensated
4337	Bullace . .	Will not be compensated
4338	Bullantic .	**Compensation**
4339	Bullbat . .	Demand(s) as compensation
4340	Bullbeef . .	What compensation would —— take
4341	Bullbeggar .	Claim compensation
4342	Bullcalf . .	Will require heavy compensation
4343	Bullcomber .	Will require some small compensation
4344	Bulldance .	Not content with compensation offered (of)
4345	Bulldog . .	In compensation for
4346	Bullennail .	Must have compensation for
4347	Bulletined .	Compensation for withdrawal of claims
4348	Bulletwood .	Have agreed to give compensation
4349	Bullfaced .	Have demanded compensation
4350	Bullfeast .	**Competent**
4351	Bullfight . .	A competent man to
4352	Bullfinch .	A competent expert
4353	Bullflies . .	A competent mill foreman
4354	Bullfly . .	A competent mine superintendent
4355	Bullfrog . .	A competent and reliable expert
4356	Bullidae . .	A competent and reliable manager
4357	Bullion . .	A competent and reliable engineer
4358	Bullionist .	Competent and reliable
4359	Bullseye . .	A competent and reliable expert to examine and
4360	Bullsnose .	Highly competent and trustworthy [report upon
4361	Bulltrout .	Not fully competent
4362	Bullweed . .	Not quite competent for the purpose
4363	Bullyrag .	**Competition**
4364	Bullyrook .	There is great competition
4365	Bullytree .	Little or no competition
4366	Bulrush . .	In direct competition
4367	Bultow .	**Complaint**
4368	Bumbailiff .	There is a great complaint about
4369	Bumblebee .	A complaint against
4370	Bumbledom	The cause of complaint
4371	Bumboat .	Serious cause of complaint
4372	Bumelia . .	What is the cause of complaint
4373	Bumper . .	**Complete**
4374	Bumperize .	In order to complete
4375	Bumpkin .	Ready to complete purchase
4376	Bumpkinly .	Do not complete purchase till
4377	Bumptious .	Can you complete examination by
4378	Bunchiness .	When will you complete

No.	Code Word.	**Complete** (*continued*)
4379	Bungall . .	Can complete
4380	Bungalows .	Cannot complete
4381	Bungarus .	In complete working order
4382	Bungdrawer.	Expect to complete
4383	Bunghole .	Complete arrangements
4384	Bungling .	Complete examination
4385	Bunglingly .	Complete examination must be made
4386	Bunkum . .	A complete set of
4387	Buntlines .	Unless you can complete
4388	Buoyance .	Do not think we can complete
4389	Buoyantly .	Will endeavour to complete
4390	Buoyrope .	Must complete
4391	Buphaga .	. **Completed**
4392	Burdened .	Will purchase be completed
4393	Burdening .	If purchase is not completed by
4394	Burdenous .	The purchase to be completed by
4395	Bureaucrat .	Has (have) been completed
4396	Burgess . .	Has (have) not been completed
4397	Burggrave .	Will shortly be completed
4398	Burghbote .	Cannot be completed till
4399	Burgholder	Will be completed by
4400	Burglar . .	If not completed by
4401	Burglarian .	When will —— be completed
4402	Burgward .	Provided it is completed (by)
4403	Burials . .	Must be completed (by)
4404	Buried . .	Is not yet completed
4405	Burlap . .	When —— are completed
4406	Burlesqued .	When you have completed
4407	Burnetmoth.	Have completed the arrangements
4408	Burnettize .	When arrangements are completed
4409	Burnoose .	Expect to have all completed (by)
4410	Burntear .	. **Complication(s)**
4411	Burntstone .	There is too much complication
4412	Burparsley .	Complications have arisen
4413	Burraspipe .	Should complications arise
4414	Burrelfly .	. **Composed**
4415	Burrelshot .	Is (are) composed of
4416	Burrowduck.	Of what is (are) the —— composed
4417	Bursar . .	**Comprehensive**
4418	Bursarship .	Telegraph your report, making it thoroughly com-
4419	Bursiform .	Not sufficiently comprehensive [prehensive
4420	Bursteth .	. **Compressor(s)**
4421	Burstwort .	Air-compressor
4422	Burthistle .	Air-compressor for —— drills
4423	Burying . .	Present air-compressor is too small
4424	Busbies . .	Require a new air compressor
4425	Bushbeans .	Order an air compressor
4426	Bushbuck .	**Comprise(s)**
4427	Bushcat . .	Does this comprise
4428	Bushel . .	This does not comprise

No.	Code Word.	Comprise(s) (*continued*)
4429	Bushelage .	This comprises
4430	Bushgoat .	**Comprised**
4431	Bushhammer	Comprised therein
4432	Bushharrow .	Comprised in the schedule
4433	Bushman .	Comprised in the list
4434	Bushments .	**Compromise**
4435	Bushranger .	Shall I (we) compromise
4436	Bushstrike .	Compromise if you can
4437	Bushwoman	I (we) cannot compromise
4438	Busiest . .	You must not compromise
4439	Busily . .	Try and compromise the matter
4440	Business . .	Will not make any compromise
4441	Buskined .	What will he (they) compromise for
4442	Bussupalm .	Think he (they) would compromise
4443	Bustlers . .	Refuse to compromise at all
4444	Busybodies .	If you cannot effect a compromise
4445	Busybody .	No compromise yet arranged
4446	Butcheries .	No offer to compromise will be accepted
4447	Buteagum .	Will consider any offer to compromise
4448	Butlerage .	**Compromised**
4449	Butlership .	Has (have) compromised
4450	Butment . .	Has (have) not compromised
4451	Butomus . .	Has (have) —— compromised
4452	Butshaft . .	The action has been compromised
4453	Buttchain .	All claims have been compromised
4454	Buttends . .	**Conceal**
4455	Butterbirds .	There is nothing to conceal
4456	Butterboat .	(To) conceal the fact (that)
4457	Buttercup .	(To) conceal the true state of affairs
4458	Butterfish .	**Concealed**
4459	Butterfly . .	Concealed the fact that
4460	Butterine .	Concealed the true state of affairs
4461	Butterman .	Has not concealed anything
4462	Buttermilk .	**Concentrate(s)**
4463	Butterpats .	(To) concentrate the ore
4464	Butterwife .	(To) concentrate the tailings
4465	Butterwort .	Concentrates from the ores
4466	Butterybar .	Concentrates from the tailings
4467	Butthorn . .	Concentrates from the pans
4468	Butthowel .	Concentrates from the vanners
4469	Buttocks . .	Concentrates from the
4470	Buttonbush .	Concentrates on hand
4471	Buttonhole .	Concentrates in transit
4472	Buttoning .	Concentrates on hand and in transit
4473	Buttonloom .	Concentrates assaying
4474	Buttontree .	Concentrates to the value of
4475	Buttressed .	Concentrates, granulations, etc.
4476	Buttweld .	Concentrates sent to smelters
4477	Butylamine .	Concentrates weighing —— and valued at
4478	Butyrous . .	Concentrates to be sent to the smelting works at

No.	Code Word.	**Concentrate(s)** *(continued)*
4479	Buxom	We are now sending concentrates (to)
4480	Buxomly	Returns of concentrates
4481	Buxomness	Sacks of concentrates
4482	Buyable	Sacks of concentrates weighing
4483	Buzzards	Concentrates sold to
4484	Buzzed	Moisture (per cent.) in concentrates
4485	Bybidder	Moisture (in concentrates) estimated at
4486	Byblow	Nett weight of concentrates
4487	Bybusiness	Concentrates should assay
4488	Bycocket	Refiners charges on smelting concentrates
4489	Bycorner	Loss in realization of concentrates
4490	Bydesign	Assay value of concentrates
4491	Bydrinking	Realized (or realizable) value of concentrates
4492	Byealtar	High proportion of concentrates
4493	Byeball	Low proportion of concentrates
4494	Byelaws	Loss on concentrates
4495	Byend	Gold in concentrates
4496	Byewash	Silver in concentrates
4497	Bygones	Lead in concentrates
4498	Byinterest	Percentage of gold and silver in concentrates
4499	Bymatter	**Concentration**
4500	Byname	Concentration works
4501	Byordinar	Concentration works are absolutely necessary
4502	Bypassage	In consequence of there being no concentration
4503	Bypaths	The ore requires concentration [works
4504	Byplace	Concentration before amalgamation
4505	Byproduct	Concentration after amalgamation
4506	Byraft	Concentration by
4507	Byrespect	Concentration by frue vanners
4508	Byroads	Concentration of tailings
4509	Byroom	Concentration is absolutely necessary
4510	Byrrhidae	Concentration is useless
4511	Byspeech	Concentration of the works
4512	Byspell	A proper method of concentration
4513	Byssaceous	The best method of concentration
4514	Byssoid	If concentration is employed
4515	Byssolite	The method of concentration adopted
4516	Bystander	Concentration as now carried on
4517	Bystreet	**Concession(s)**
4518	Bystroke	It is necessary to get a concession from
4519	Byttneria	Will arrange to get a concession for
4520	Byturning	Cannot get a concession from
4521	Byviews	Who is the concession from
4522	Bywalk	Who has the concession
4523	Byway	Concession of no value
4524	Bywest	Concession of great value
4525	Bywipe	Get a concession for
4526	Byzant	Have got a concession for
4527	Byzantian	Have bought concession
4528	Cabacalli	Share in the concession

CON

No.	Code Word.	Concession(s) (*continued*)
4529	Cabal . . .	Very valuable concession
4530	Cabalist . .	In exchange for the concession
4531	Cabalistic .	In purchase of the concession
4532	Cabalize. .	Concession has been sold
4533	Caballaria .	Concession has been bought
4534	Caballer . .	If we get the concession
4535	Caballing .	Has (have) given for the concession
4536	Cabaret . .	What does he (do they) get for the concession
4537	Cabassou .	Ask a heavy price for the concession
4538	Cabbagefly .	Formed to work the concession
4539	Cabbagenet .	The property covered by the concession
4540	Cabbages .	The concession includes an area of
4541	Cabbaging .	Concession has been granted
4542	Cabin . .	Land concession
4543	Cabinboy .	Mining concession
4544	Cabinetted .	Agricultural concession
4545	Cabinmate .	The concession comprises
4546	Cabirian . .	The concession estimated to be worth
4547	Cablegram .	Make no concession
4548	Cablelaid .	What concession will he (they) make
4549	Cabletier .	Cannot make any concession
4550	Cablish . .	The whole concession
4551	Cabmen . .	Part of the concession
4552	Caboceer .	Conclude
4553	Caboshed .	We conclude from this, that
4554	Cabstands .	If you conclude (from)
4555	Cacagogue .	Do not hastily conclude
4556	Cacalia . .	Concluding
4557	Cacatuinae .	The concluding remarks
4558	Cachaemia .	The concluding part of
4559	Cachectic .	The concluding part of your report
4560	Cachexy .	Conclusion(s)
4561	Cachiri . .	Upon (at) the conclusion of
4562	Cacholong .	Have arrived at the conclusion that
4563	Cacklers . .	The conclusion of the investigation
4564	Cacochymy .	The conclusion to be drawn from
4565	Cacodemon .	Conclusions have been drawn
4566	Cacodoxy .	You can draw your own conclusions
4567	Cacodyle .	The conclusions we have arrived at
4568	Cacoethes .	The conclusions you have arrived at
4569	Cacography.	The following conclusions
4570	Cacology .	In conclusion of the report
4571	Cacophony .	Please telegraph what conclusions you have arrived [at about
4572	Cacotechny .	Concur
4573	Cactaceous .	We cannot concur with your views
4574	Cadamba .	Do not concur
4575	Cadastral .	We concur with your views
4576	Cadaver . .	If you concur with our views
4577	Cadaveric .	Do you concur with
4578	Cadaverous .	Concurrence

No.	Code Word.	**Concurrence** (*continued*)
4579	Cadbait . .	With my (our) concurrence
4580	Caddice . .	Without my (our) concurrence
4581	Caddicefly .	**Concurrently**
4582	Cadenced .	Concurrently with
4583	Cadetship .	**Condemn(s)**
4584	Cadeworms .	Condemn(s) the whole concern
4585	Cadilesker .	Condemn(s) the mines as good for nothing
4586	Cadmium .	If —— condemn(s)
4587	Cadrans . .	**Condemned**
4588	Cadre. . .	The mine was condemned
4589	Caducary .	The machinery was condemned
4590	Caducean .	The mill was condemned
4591	Caducous .	The management was condemned
4592	Caecias . .	Was (were) condemned in the strongest terms
4593	Caenstone .	The Board was condemned
4594	Caesarean .	**Condition(s)**
4595	Caesarism .	In its present condition
4596	Caffeone. .	Conditions absurd
4597	Caffercorn .	In the present condition of the
4598	Caffila . .	In good condition
4599	Caged . .	In bad condition
4600	Cageling. .	Condition improved
4601	Cagmag . .	Condition worse
4602	Caimacam .	Conditions to be the same
4603	Cainozoic .	Impose(s) impossible conditions
4604	Cairned . .	Will you (they) agree to these conditions
4605	Cairngorm .	I (we) accept the conditions
4606	Caitiffly . .	Cannot agree to the conditions
4607	Caitiffs . .	Only on condition that
4608	Cajanus . .	The conditions are to be
4609	Cajeput . .	On condition that modifications are made
4610	Cajolement .	Conditions are not satisfactory
4611	Cajoleries .	The condition of the mine
4612	Cajolery . .	The condition of the property
4613	Cakebread .	The condition of the workings
4614	Cakeurchin .	Mine is in bad condition
4615	Calabash .	(In) The present condition of affairs
4616	Calaboose .	The present condition and future prospect of the
4617	Calaite . .	The condition is improving [undertaking
4618	Calamanco .	A more hopeful condition
4619	Calamary .	The condition is precarious
4620	Calamintha .	There is no change in the condition
4621	Calamist. .	The conditions meet with my (our) approval
4622	Calamites .	If the conditions meet with your approval
4623	Calamitous .	Subject to the conditions named
4624	Calangay .	**Confer**
4625	Calcaneal .	Confer with —— and report
4626	Calcareous .	Confer with —— and let us know your joint opinion
4627	Calcavella .	Confer with —— before doing anything
4628	Calceolate .	Wish to confer with

No.	Code Word.	Confer (*continued*)
4629	Calcified . .	Wish you to confer with
4630	Calciform .	**Conference**
4631	Calcify .	There will be a conference on the question
4632	Calcimine .	A conference will be held
4633	Calcinable .	The result of the conference
4634	Calcining .	**Conferring**
4635	Calcitrate .	After conferring with
4636	Calcium . .	**Confidence**
4637	Calcsinter .	I tell you this in strictest confidence; let no one [know it
4638	Calcspar . .	I (we) have every confidence in
4639	Calctuff .	I (we) have no confidence in
4640	Calculary .	In whom you may place thorough confidence
4641	Calculated .	In whom you can place no confidence
4642	Calcule . .	Is (are) unworthy of confidence
4643	Calculous .	Is (are) worthy of every confidence
4644	Caldron . .	Can I (we) have confidence in
4645	Calecannon .	Confidence has been greatly shaken
4646	Caleches . .	Confidence is not yet restored
4647	Calefying .	Confidence appears to be completely restored
4648	Calemberre .	Has (have) lost confidence in
4649	Calendar . .	Can no longer place confidence in
4650	Calendula .	Has (have) very little confidence in
4651	Calescence .	What confidence have you in
4652	Calflick . .	In the strictest confidence
4653	Calflove . .	**Confident**
4654	Calfskin . .	Can you be confident that
4655	Calfward . .	I am (we are) confident that
4656	Calibre . .	Is (are) confident that
4657	Caliduct . .	Not confident that
4658	Caligation .	**Confidential**
4659	Caliginous .	A confidential agent
4660	Caligraphy .	Private and confidential
4661	Caliphate .	Confidential reports
4662	Caliphship .	Confidential reports have reached us
4663	Calippic . .	**Confidentially**
4664	Calisayine .	We hear confidentially
4665	Calixtin . .	Has confidentially told us
4666	Callbell . .	**Confirm(s)**
4667	Callbirds .	I (we) can confirm ——'s statements
4668	Callboy . .	I (we) cannot confirm ——'s statements
4669	Callimus . .	Can you confirm ——'s statements
4670	Calliope . .	Cannot confirm ——'s statements respecting amount
4671	Callipers . .	Confirm his (their) statements [of ore in sight
4672	Callitrix . .	Telegraph whether or not you substantially confirm
4673	Callosoma .	Confirm in writing [—— reports
4674	Callous . .	Will write to confirm all his (their) statements
4675	Callously .	Confirm your statements
4676	Calluna . .	Can confirm all ——'s statements except that (those)
4677	Calmly . .	Confirm the arrangement [in regard to
4678	Calmness .	Confirm the undertaking

No.	Code Word.	Confirm(s) (*continued*)
4679	Calomel . .	Your report confirms
4680	Caloric . .	**Confirmation**
4681	Caloricity .	As soon as I (we) receive confirmation of the report
4682	Caloriduct .	Subject to the confirmation of the reports (of)
4683	Calorific. .	Confirmation of the report(s) (of)
4684	Calotropis .	Telegraph your confirmation of
4685	Calottist . .	Waiting for confirmation of ———'s report
4686	Calotype .	In confirmation of
4687	Calotypist .	In confirmation of the statements made
4688	Calpslates .	Subject to confirmation
4689	Calqued . .	Subject to confirmation by telegraph
4690	Calthrop. .	We are yet without confirmation
4691	Calumniate .	**Confirming**
4692	Calumnious.	Your telegram confirming the report received
4693	Calumny .	I (we) telegraphed confirming the report on
4694	Calvered. .	Your letter confirming
4695	Calvinize .	Your report confirming
4696	Calycoid. .	Your telegram confirming
4697	Calymene .	**Conflict(s)**
4698	Calypso . .	(To) conflict with
4699	Calyptrate .	In conflict with
4700	Calystegia .	Conflicts with what we have
4701	Calyxes . .	Conflict(s) so much with the others
4702	Calzoons .	**Conflicting**
4703	Camassia .	Reports are conflicting
4704	Camatina .	Owing to conflicting accounts
4705	Cambaye .	Very conflicting reports
4706	Camberbeam	Conflicting with each other
4707	Cambistry .	**Confusion**
4708	Cambrasine.	To prevent confusion
4709	Cambrel . .	Everything is in a state of confusion
4710	Cambrics .	The confusion caused by
4711	Camelbird .	Much confusion has been caused by
4712	Camelry . .	**Conglomerate**
4713	Camelshair .	Auriferous conglomerate
4714	Camelus . .	Conglomerate containing
4715	Cameotypes.	**Congratulate**
4716	Cameraria .	We congratulate you
4717	Camerated .	**Congratulation(s)**
4718	Camerating .	In congratulation
4719	Cameronian.	Our congratulations (to)
4720	Camestres .	Our congratulations on the success which
4721	Camletteen .	**Connect**
4722	Camletto .	In order to connect the
4723	Cammocky .	Can you connect
4724	Camomile .	**Connected**
4725	Camoys . .	Connected with
4726	Campaigner.	Connected by
4727	Campanal .	Not connected in any way with
4728	Campbeds .	We have now connected

No.	Code Word.	**Connected** (*continued*)
4729	Campestral .	(We have) now connected by a winze
4730	Campfight .	(We have) now connected by a drift
4731	Camphine .	(We have) now connected by a rise
4732	Camphogen.	The two shafts are now connected
4733	Camphor .	Are now connected
4734	Camphorate	As soon as we have connected
4735	Camphoric .	The two levels are now connected
4736	Camphoroil .	**Connecting**
4737	Camping .	Connecting levels No. —— and —— by a winze
4738	Campkettle .	Connecting the works with
4739	Campstool .	Connecting with the shafts
4740	Campylite .	Is there any way of connecting —— with
4741	Camwood .	Connecting levels —— and
4742	Canakin . .	Connecting shafts
4743	Canalboat .	**Connection**
4744	Canallift . .	In connection with
4745	Canaries . .	Must make connection with
4746	Canarybird .	Connection has been made with
4747	Canaryseed .	Have no connection with
4748	Canarywood .	Has (have) —— any connection with
4749	Canbuoys .	Avoid all connection with
4750	Cancellate .	**Consent(s)**
4751	Cancellous .	Will —— consent to
4752	Cancercell .	Will consent to
4753	Cancriform .	Will not consent to
4754	Cancrinite .	Waiting to get consent from
4755	Cancroid .	Should —— not consent to
4756	Candareen .	Let us know if you consent
4757	Candelabra .	Cannot consent to
4758	Canderos .	Decline to consent to
4759	Candicant .	If we consent (to)
4760	Candid . .	If you consent (to)
4761	Candidacy .	He consents—(They consent)
4762	Candidates .	If you do not consent
4763	Candidly .	Without our consent
4764	Candidness .	Without your consent
4765	Candified .	Without his (their) consent
4766	Candify . .	**Consented**
4767	Canditeer .	Has (have) consented
4768	Candlebark .	Has (have) not consented
4769	Candlebomb	**Consenting**
4770	Candlecase .	In consenting to
4771	Candleends .	By consenting to
4772	Candlefish .	**Consequence(s)**
4773	Candlemine	What would be the consequence (if)
4774	Candlenut .	Of great consequence (to)
4775	Candlerush .	Of little consequence
4776	Candlewood	In consequence of which
4777	Candour . .	Of no consequence
4778	Candroy . .	In consequence of

No.	Code Word.	**Consequence(s)** (*continued*)
4779	Candying .	May lead to very serious consequences
4780	Candysugar.	The most serious consequences
4781	Candytuft .	The consequences will be serious
4782	Canebrake .	The consequences that may result
4783	Canechair .	**Consider**
4784	Canegun . .	Consider and decide
4785	Canehole .	Hope the Board will consider
4786	Canemill. .	The Board consider
4787	Canephorus.	Consider the affair a good one
4788	Canesugar .	Consider the affair a bad one
4789	Canetrash .	Consider of no importance
4790	Canframe .	Consider of great importance
4791	Canhooks .	Consider it very doubtful
4792	Canina . .	Consider it expedient
4793	Canisters .	Consider it advisable
4794	Cankerbit .	Consider it bad policy
4795	Cankerfly .	**Consideration**
4796	Cankerrash .	It is a matter for grave consideration
4797	Cankerworm	Taking into consideration
4798	Cannabis .	For the consideration of
4799	Cannelcoal .	Give this your best consideration
4800	Cannibally .	Has (have) not been taken into consideration '
4801	Cannibals .	Not worth consideration
4802	Cannonade .	For your consideration
4803	Cannonball .	Consideration deferred (until)
4804	Cannoning .	Consideration of the circumstances
4805	Cannonlock.	Upon consideration (of)
4806	Cannonshot.	Upon consideration of your report
4807	Canny . .	Upon further consideration
4808	Canoebirch .	A matter for consideration
4809	Canoeclub .	**Considered**
4810	Canoeist . .	The matter has been considered
4811	Canonbit. .	The matter has been considered and deferred
4812	Canonbone .	Considered and decided
4813	Canoness. .	Was (were) considered
4814	Canonicals .	The matter was fully considered by the Board
4815	Canonistic .	The matter was fully considered and postponed until
4816	Canonlaw .	Considered satisfactory [the next meeting
4817	Canonship .	Fully considered in all its bearings
4818	Canonwise .	Fully considered and not thought advisable
4819	Canopied. .	Fully considered and was thought advisable
4820	Canopus . .	Considered objectionable
4821	Canopy . .	Considered in relation to
4822	Canopying .	Considered in conjunction with
4823	Cantabank .	Hope the Board have considered
4824	Cantaleup .	Considered doubtful
4825	Cantaliver .	**Considering**
4826	Cantatas . .	Considering the
4827	Cantatory .	Considering the course to be adopted
4828	Cantatrice .	Considering the accounts

E

No.	Code Word.	

Considering (*continued*)

4829	Canteens.	.	Considering the matter referred to
4830	Cantharis	.	Considering all the circumstances
4831	Canticles.	. **Consign**
4832	Cantonal.	.	Consign to us
4833	Cantonment.		Consign to agents
4834	Cantraip .	.	Do not consign to
4835	Cantspars	. **Consigned**
4836	Canttimber .		Consigned to you
4837	Canvas .	.	Consigned to your order
4838	Canvasback .		Consigned to agents
4839	Canvassed	.	Consigned to shipper's agents
4840	Canvassing	.	Consigned to
4841	Caoutchouc	. **Consigning**
4842	Capability	.	Who are you consigning to
4843	Capacitate	.	We are consigning
4844	Capacities	. **Consignment(s)**
4845	Capacity.	.	Upon consignment (of)
4846	Caparisons	.	After consignment (of)
4847	Capcase .	.	Before consignment (of)
4848	Capepigeon.		Against consignment (of)
4849	Caperbush	.	Consignment(s) of
4850	Caperclaw	.	Regular consignments
4851	Capering.	.	Further consignments
4852	Capersauce	. **Consist**
4853	Capertea.	.	What does the —— consist of
4854	Capillose.	.	Consist(s) of
4855	Capistrum	. **Consistent**
4856	Capital .	.	Consistent with our interests
4857	Capitalist	.	Consistent with our ideas
4858	Capitally.	.	Consistent with your interests
4859	Capitolian	.	Consistent with your ideas
4860	Capitulate	.	Consistent with his (their) interests
4861	Capivard.	.	Consistent with the truth
4862	Capmoney .		If it is not consistent
4863	Capnomancy		In what way is it not consistent
4864	Capnomor .		How far it may be (is) consistent
4865	Caponized .		So as to be consistent
4866	Caponizing	. **Consisting (of)**
4867	Capotted.	.	Consisting mostly of
4868	Capotting	.	Consisting of the following
4869	Capouched	. **Consolidate**
4870	Cappadine	.	In order to consolidate the property
4871	Cappaper	.	In order to consolidate
4872	Cappeak.	.	Consolidate all interests
4873	Cappudding.		Consolidate all claims
4874	Caprella .	.	Desirable to consolidate
4875	Capreolus	.	Consolidate the properties
4876	Caprice .	.	Agreed to consolidate
4877	Capricious	. **Consolidated**
4878	Capricorn		Have consolidated

No.	Code Word.	Consolidated (*continued*)
4879	Caprificus .	Have you consolidated
4880	Caprifole. .	If not consolidated
4881	Caprioling .	All interests should be consolidated
4882	Capriped. .	Should be consolidated
4883	Caprizant .	Cannot all interests be consolidated
4884	Capromys .	If properties are consolidated
4885	Caprovis . .	The consolidated properties
4886	Caprylene .	**Consols**
4887	Capsheaf. .	Three per cent. consols
4888	Capsicine .	Two and three quarter per cent. consols
4889	Capsicums .	New consols
4890	Capsills . .	**Consul**
4891	Capsize .	The American Consul
4892	Capsizing .	The British Consul
4893	Capsquare .	Certified by the consul
4894	Capstan . .	Is consul here
4895	Capsular . .	**Consular**
4896	Captaincy .	Consular certificate
4897	Captainess .	Consular invoice
4898	Captivated .	Consular certificate must be attached
4899	Captivity . .	Without consular certificate
4900	Capturing .	The consular authority
4901	Carabidae .	**Consulate**
4902	Carabineer .	American Consulate
4903	Carabus . .	British Consulate
4904	Caracal . .	French Consulate
4905	Caracoled .	Under the seal of the consulate
4906	Caracoling .	The invoice must be certified at the consulate.
4907	Caradoc . .	Application at the consulate
4908	Caragenine .	**Consult.** (See also Confer.)
4909	Caragheen .	You had better consult with
4910	Carapace .	Consult —— before acting
4911	Carapoil . .	Whom had I (we) better consult
4912	Caravans. .	Consult a thoroughly competent lawyer
4913	Carbamide .	Consult before telegraphing
4914	Carbazotic .	Consult counsel on the case
4915	Carbolic . .	**Consultation** [consultation with
4916	Carbonated .	Telegraph what conclusion you arrive at after your.
4917	Carbonize .	After consultation with
4918	Carbonous .	**Consulted**
4919	Carbonspar .	Have consulted (with)
4920	Carbuncle .	Have consulted the best lawyer
4921	Carburetor .	Have you consulted (with)
4922	Carcanet . .	Have not consulted
4923	Carcasses .	Has (have) been consulted
4924	Carcinoma .	Has consulted us with respect to
4925	Cardamoms .	**Consulting**
4926	Cardbasket .	Before consulting
4927	Cardboard .	After consulting
4928	Cardcases .	Do nothing without first consulting

914928

No.	Code Word.	Consulting (*continued*)
4929	Cardiacal .	After consulting, the Board decided
4930	Cardioid .	. Consumption
4931	Cardiology .	What is the consumption of coal per day
4932	Cardmatch .	Consumption of coal is —— tons per day
4933	Cardophagi .	Increasing the consumption of
4934	Cardparty .	Decreasing the consumption of
4935	Cardplayer .	What is the consumption of
4936	Cardracks .	What is the average consumption of
4937	Cardtables .	Consumption of cordwood
4938	Cardtray . .	Consumption of timber
4939	Carduelis .	Consumption of stores
4940.	Carecrazed .	Consumption of quicksilver
4941	Career . .	Consumption of salt
4942	Careering .	Consumption of coal
4943	Carefully . .	To diminish the consumption
4944	Careless . .	To increase the consumption
4945	Carelessly .	Do your best to lessen the consumption
4946	Carentane .	The consumption is excessive
4947	Caressed . .	The consumption is steady
4948	Caretaker .	The consumption is too high
4949	Caretuned .	The consumption is small
4950	Careworn .	The consumption will
4951	Careya . .	The average consumption
4952	Cargoose .	This consumption is above the average
4953	Cariboo . .	Increased consumption
4954	Caricature .	Increased consumption, owing to
4955	Carinthine .	Diminished consumption
4956	Carjacou . .	Diminished consumption, owing to
4957	Carlehemp .	If you can possibly diminish the consumption
4958	Carlism . .	The daily consumption
4959	Carlylese .	The weekly consumption
4960	Carlylian . .	The monthly consumption
4961	Carmelite .	The estimated consumption
4962	Carnaged .	. Contact
4963	Carnal . .	A contact lode
4964	Carnalism .	In contact with
4965	Carnalized .	In close contact with
4966	Carnally . .	In the contact between
4967	Carnardine .	Not exactly in the contact
4968	Carnassial .	When do you expect to come in contact with
4969	Carnations .	Think we shall soon come in contact with
4970	Carnifex . .	Think we are in close contact with
4971	Carnifying .	In contact with the footwall
4972	Carnivals .	In contact with the hanging wall
4973	Carnivora .	(To) make contact with
4974	Carolled . .	. Contain(s)
4975	Carolling .	The ore contains
4976	Carolus . .	Does it contain any
4977	Carolytic . .	It does not contain
4978	Carotid . .	It contains

No.	Code Word.	Contain(s) (*continued*)
4979	Carotidal.	It contains matters of importance
4980	Carousals	What does it contain
4981	Carouse	Whether it contains
4982	Carousing	If it contains
4983	Carpathian	The case contains
4984	Carpbream	The package contains
4985	Carpellary	The report contains
4986	Carpellum	**Contingency(ies)**
4987	Carpentry	In such a contingency
4988	Carpet	Allowing for any contingencies
4989	Carpetbags	This does not allow for any contingencies
4990	Carpeting	To provide for contingencies
4991	Carpetrod	For contingencies
4992	Carpetwalk	In view of contingencies which may arise
4993	Carpetweed	**Contingent**
4994	Carpholite	(Is) contingent upon
4995	Carphology	Must be contingent upon
4996	Carpidium	The contingent results
4997	Carpingly	Results contingent upon
4998	Carpinus	Contingent upon the course (adopted)
4999	Carpmeals	**Continuation**
5000	Carpocapsa	In continuation of
5001	Carpophaga	In continuation of our report
5002	Carpophore	In continuation of your report
5003	Carriable	In continuation of the operations
5004	Carriage	**Continue**
5005	Carrion	To continue the
5006	Carrollite	The first thing is to continue the
5007	Carronoil	Continue sinking
5008	Carrying	The stopes continue to look well
5009	Carrytale	The mine continues to look
5010	Cartbody	The vein continues to look
5011	Carterly	Shall continue (to)
5012	Carthamus	You must continue (to)
5013	Carthorse	Will still continue to
5014	Carthusian	It would be advisable to continue
5015	Cartilage	Continue to telegraph regularly
5016	Cartjade	Do not continue
5017	Cartload	Continues in ore
5018	Cartoon	Continues in pay ore
5019	Cartulary	Continues in country rock
5020	Cartway	Continues in high grade ore
5021	Cartwright	Continues in low grade ore
5022	Carucage	Continues in force
5023	Carum	Recommend you to continue
5024	Carvings	**Continued**
5025	Carwheels	Have continued
5026	Carwhichet	**Continuing**
5027	Caryatic	Continuing in (to)
5028	Caryatides	**Continuity**

No.	Code Word.	Continuity (*continued*)
5029	Caryocar. .	In continuity with
5030	Caryopsis .	Contract(s) (See also Sign.)
5031	Cascades .	Make a fresh contract
5032	Cascading .	Has (have) made a fresh contract
5033	Casebags .	No contract has been made
5034	Casebottle .	By the terms of the contract
5035	Caseharden .	The work is all being done on contract
5036	Caseknife .	Has (have) made a contract with
5037	Casemate .	Owing to the contract between —— and
5038	Caserack. .	Has (have) thrown up the contract
5039	Caseshot. .	Have you entered into any contract
5040	Cashbooks .	I (we) have made a contract in my (our) name with
5041	Cashcredit .	The contract is signed
5042	Cashewbird .	Insist on fulfilment of contract
5043	Cashewnut .	Can you make a contract in your own name with
5044	Cashewtree .	Has (have) made contract in favour of——; same
5045	Cashiered .	The contract is not yet signed [terms as last
5046	Cashiering .	Contract has fallen through
5047	Cashkeeper .	On completion of contract
5048	Cashmere .	Can you contract
5049	Cashnote .	Can contract
5050	Casketed. .	Cannot contract
5051	Cassareep .	Is the contract signed
5052	Cassia . .	Do not close the contract till
5053	Cassiabark .	If you can contract
5054	Cassiabud .	According to contract
5055	Cassiaoil. .	Upon signing contract
5056	Cassiapulp .	Upon expiration of contract
5057	Cassicus . .	To be executed by contract
5058	Cassideous .	Give the contract to
5059	Cassidony .	To whom shall we give the contract
5060	Cassiteria .	Contract has been given to
5061	Cassocked .	Contract has been cancelled
5062	Cassowary .	What work are you doing by contract
5063	Cassumunar.	What is the contract price for
5064	Cassweed .	What is the contract price for the engine and boiler
5065	Castaway .	What is the contract price for all material delivered
5066	Caste . .	Work being done by contract [at the mine
5067	Castellany .	Work being done by contract at —— per foot, con-tractors finding material
5068	Castigator .	Work being done by contract for a lump sum of
5069	Castingnet .	Machinery ordered under contract
5070	Castiron . .	Contract price to include delivery at
5071	Castknee .	Contract price to include erection
5072	Castleward .	Contract price, exclusive of cost of freight
5073	Castoff . .	Contract price for machinery
5074	Castorate .	Contract price for machinery includes cost of erec-tion, but not cost of buildings, which the company
5075	Castorbean .	Contract for machinery [will provide
5076	Castoric . .	Contract for mill

No.	Code Word.	Contract(s) *(continued)*
5077	Castoridae .	Contract for grading
5078	Castoroil . .	Contract for woodwork
5079	Castshadow .	Contract for stonework
5080	Caststeel . .	Contract for building
5081	Casual . .	Contract for shafting
5082	Casually . .	Contract for tunnelling
5083	Casualness .	Contract for drifting
5084	Casuist . .	Contract for sinking
5085	Casuistic . .	**Contracted**
5086	Catabasion .	Has (have) contracted to supply
5087	Catabrosa .	Have contracted for
5088	Cataclysm .	**Contradict**
5089	Catacomb .	Contradict the report
5090	Catadrome .	Is entirely false, and you can contradict it
5091	Catagmatic .	Wish you to enable us to contradict the rumour
5092	Catagraph .	We emphatically contradict
5093	Catalectic .	What was stated is not correct, contradict it
5094	Catalepsis .	**Contradicted**
5095	Catalogize .	Have contradicted
5096	Catalogued .	Have officially contradicted
5097	Catamaran .	Have formally contradicted the report
5098	Catamount .	Was contradicted
5099	Cataphonic .	Has been contradicted
5100	Cataphract .	**Contradiction**
5101	Cataplasm .	In contradiction of
5102	Catapult . .	What have you to say in contradiction
5103	Catapultic	Have nothing to say in contradiction
5104	Cataracts .	**Contrary**
5105	Catarrh . .	Unless you hear to the contrary
5106	Catarrhal .	Unless I (we) hear to the contrary
5107	Catarrhina .	Not having heard to the contrary
5108	Catarrhous .	Contrary to instructions
5109	Catastasis .	Contrary to our orders
5110	Catastomus .	Contrary to our ideas
5111	Catbeam . .	Contrary to our wishes
5112	Catbird . .	Contrary to our advice
5113	Catcall . .	Contrary to your
5114	Catcalling .	Contrary to his
5115	Catchable .	**Control**
5116	Catchclub .	In order to control the market
5117	Catchdrain .	In order to get control of
5118	Catcher . .	To control the stock
5119	Catchfly . .	The control of the stock is in the hands of
5120	Catching . .	Who has the control at present
5121	Catchland .	It is imperative that we should have entire control
5122	Catchmatch .	To control the
5123	Catchpenny .	Under your control
5124	Catchpoll .	Under your sole control and management
5125	Catchweed .	Under our control
5126	Catechetic .	Control the output

No.	Code Word.	Control (*continued*)
5127	Catechism .	The mine must control the mill
5128	Catechized .	**Controlled**
5129	Catechu . .	Output is controlled by
5130	Catechumen	**Copper**
5131	Categorize .	Copper mine
5132	Category . .	Very productive copper mine
5133	Catenarian .	Copper mine now paying
5134	Catenipora .	Percentage of copper
5135	Catenulate .	Copper pyrites
5136	Caterwaul .	Peacock copper ore
5137	Cateyed . .	Grey copper
5138	Catfish . .	Black oxide of copper
5139	Catfooted .	Carbonate of copper
5140	Catgold . .	Veins of copper ore
5141	Catgut . .	Copper plates
5142	Catharists .	Copper plates silvered
5143	Catharma .	Copper ore containing —— gold per ton
5144	Catharping .	Copper ore containing —— silver per ton
5145	Cathartes .	Copper stained
5146	Cathedral .	Made of copper
5147	Catheritic .	Copper now selling at
5148	Catheter . .	Copper ore now selling at —— per unit
5149	Cathode . .	Ore contains —— per cent. of copper
5150	Catholic . .	**Copperas**
5151	Catholical .	**Copy**
5152	Catholicly .	Send me (us) immediately a copy of —— by post
5153	Catilinism .	Have sent by post a copy of
5154	Catlike . .	Will send by mail of —— a copy of
5155	Catlinite . .	A copy of the agreement will be forwarded
5156	Catmint . .	A copy of the
5157	Catnip . .	Have you a copy
5158	Catoblepas .	Copy of account
5159	Catodon . .	Copy of agreement
5160	Catopsis . .	Copy of deed
5161	Catoptrics .	Copy of plan
5162	Catoptron .	Copy of drawing
5163	Catsalt . .	Copy of section
5164	Catscradle .	Copy of memorandum and articles of association
5165	Catsfoot . .	Copy of invoice
5166	Catsilver . .	Copies of plan and section
5167	Catsmilk . .	Certified copy (of)
5168	Catspaw . .	**Cords of wood**
5169	Catspurr . .	There are —— cords of wood ready stacked
5170	Catstail . .	It takes —— cords of wood per week for
5171	Catstopper .	Cord of wood costs
5172	Catthyme .	**Cordwood**
5173	Cattlepens .	Price of cordwood
5174	Cattlerun .	Cordwood consumed
5175	Caucus . .	Abundance of cordwood
5176	Caudated .	Have contracted for cordwood at —— per cord

No.	Code Word.	
5177	Caudlecup .	**Corner**
5178	Caulescent .	There is a corner in
5179	Caulicule .	Is there a corner in
5180	Cauliform .	Are trying to get a corner in
5181	Caulker .	**Correct**
5182	Caulkings .	Everything is quite correct
5183	Caumatic .	Is (are) not correct as to
5184	Cauponate .	The accounts are not correct
5185	Causable .	The account is correct
5186	Causator .	Quite correct
5187	Causeful .	Is it correct that
5188	Causeless .	Is everything correct
5189	Causeway .	The statement is correct
5190	Causidical .	The statement is not correct
5191	Causing .	Correct the mistake
5192	Caustic .	Is it not correct
5193	Causticity .	Correct your report
5194	Cauterant .	A correct report
5195	Cauterism .	Correct report of the proceedings
5196	Cauterized .	Send us a correct
5197	Cautionary .	Have sent a correct
5198	Cautioned .	**Corrected**
5199	Cautionize .	The mistake has been corrected
5200	Cautious .	Corrected the accounts
5201	Cautiously .	Corrected the mistake
5202	Cavalcaded .	**Corroborate**
5203	Cavalier .	Can you corroborate ——'s statements
5204	Cavalierly .	Money ready to purchase mine if you corroborate
5205	Cavallard .	Does it corroborate [——'s statements
5206	Caveach .	Can corroborate all ——'s statements
5207	Caveator .	Cannot corroborate ——'s statements
5208	Cavekeeper .	**Corroborated**
5209	Cavendish .	If the statements are corroborated
5210	Cavern .	The report must be corroborated
5211	Cavernal .	Will it (they) be corroborated
5212	Cavernous .	If (they) can be corroborated
5213	Cavicornia .	In corroboration of
5214	Cavillous .	**Cost (of)**
5215	Cavitary .	Cost per mile
5216	Cavities .	I estimate the cost of development at
5217	Cavolinite .	I estimate the cost of sinking at
5218	Cawquaw .	I estimate the cost of driving at
5219	Ceanothus .	I estimate the cost of milling at
5220	Cebidae .	I estimate the cost of mining at
5221	Cebipara .	I estimate the cost of prospecting at
5222	Cecidomyia .	I estimate the cost of pumping at
5223	Cecilian .	I estimate the cost of surveying at
5224	Cecity .	I estimate the cost of timbering the shaft at
5225	Cecropia .	Can be worked at small cost
5226	Cecutiency .	The cost of erecting chlorination works would be

E 2

No.	Code Word.	**Cost** (*continued*)
5227	Cedar . .	Cost of working the ore is very great, owing to
5228	Cedarbird .	Cost of erecting —— stamps mill would be
5229	Cedarlike .	Estimated cost [machinery at
5230	Cedarwood .	I estimate the cost of erecting the necessary
5231	Cedrela . .	I estimate that the additional cost would be
5232	Cedrin . .	In your estimate of cost do you include [include
5233	Ceilinged .	In your estimate of the cost of machinery do you
5234	Celadon . .	In your estimate of the cost of developments do you
5235	Celandine .	In future the cost of —— will be [include
5236	Celature . .	The difficulty of estimating the cost arises from
5237	Celebrable .	It is impossible to estimate the cost of
5238	Celebrant .	The heavy cost of shipping the ore to —— has
5239	Celeriac . .	Costs per ton [eaten up all the profits
5240	Celestial . .	The cost of shipping the ore to —— is
5241	Celestify . .	What will it cost (to)
5242	Celestine .	What is your estimate of cost
5243	Celibacy . .	Cost per foot run
5244	Celibate . .	Cost per cubic foot
5245	Celibatist .	Cost per yard
5246	Cellarage .	Cost per cord
5247	Cellarbook .	Cost per cwt.
5248	Cellarflap .	What is the cost of
5249	Cellaring .	What has been the cost of
5250	Cellarists .	What will be the cost of
5251	Cellarman .	The approximate cost
5252	Cellarous .	The average cost per ton
5253	Cellepora .	The probable cost when finished
5254	Cellular . .	The total cost
5255	Cellulated .	The total cost including
5256	Celluloid .	The total cost including freight
5257	Cellulosic .	The total cost including delivery
5258	Celosia . .	The total cost will be
5259	Celotomy .	The cost up to the present has been
5260	Celsitude .	The future cost will be
5261	Celticism .	Total cost which will be incurred by
5262	Cementeth .	Total cost, including erection
5263	Cementing .	Cost of examination and report
5264	Cemeterial .	How is the total cost made up
5265	Cenegild . .	Balance of cost required to complete
5266	Cenobite .	Balance of cost yet to be paid
5267	Cenobitism .	Have arranged for supply of —— at a cost of
5268	Cenogamy .	Have arranged for machinery at a cost of
5269	Cenotaph .	New machinery will cost
5270	Censorious .	New mill will cost .
5271	Censorship .	New mill has cost
5272	Censurable .	Estimated to cost about
5273	Centage . .	Cost incurred in respect to
5274	Centenary .	Total cost is made up as follows
5275	Centennial .	Does the cost include
5276	Centering .	Have you paid all the cost

No.	Code Word.	Cost (*continued*)
5277	Centesimal .	Who pays the cost
5278	Centigrade .	We have paid all the cost
5279	Centiped .	The greater part of the cost is already paid
5280	Centipedal .	Cost price
5281	Centonism .	Will cost
5282	Centos . .	Will cost more than
5283	Central . .	Will it cost
5284	Centralise .	Will cost too much
5285	Centrally .	Will not cover the cost
5286	Centration .	If the cost is too much
5287	Centrebit .	Cannot estimate the cost, until
5288	Centrical .	Legal costs
5289	Centricity .	Costs in the action
5290	Centriscus .	The cost of the litigation
5291	Centropus .	The cost of erecting
5292	Centumvir .	Will cover the cost
5293	Centurion .	Will more than cover the cost
5294	Cepa . . .	The cost of insurance
5295	Cepevorous .	Cost-sheet(s)
5296	Cephalalgy .	Send at once a copy of the cost-sheet
5297	Cephalata .	Have sent a copy of the cost-sheet
5298	Cephalic . .	Will send the cost-sheets
5299	Cephalicly .	Could
5300	Cephaloid .	Could you
5301	Cephalopod.	Could be
5302	Cephalotus .	Could be done
5303	Cepheus . .	Could not be
5304	Cepolidae .	Could not be done
5305	Cepphic . .	Could not
5306	Ceraceous .	If we could
5307	Cerambyx .	If you could
5308	Ceramic . .	If he (they) could
5309	Ceramidium	If this could be done
5310	Ceraphron .	If you could not
5311	Cerasinous .	If he (they) could not
5312	Cerastes . .	We could not
5313	Ceratium .	Could not be done at present
5314	Ceratocele .	Could not do this, without
5315	Ceratodus .	When could you
5316	Ceratohyal .	When could he (they)
5317	Ceratonia .	Counteract
5318	Ceraunics .	In order to counteract
5319	Cerberian .	To counteract the depression
5320	Cerberus .	To counteract the reports set afloat
5321	Cercocebus .	Will serve to counteract
5322	Cercolabes .	Counteracted
5323	Cercopidae .	Counterbalance(cing)
5324	Cerdocyon .	Will more than counterbalance
5325	Cerealia . .	(As) counterbalancing
5326	Cerebellum .	Counterbalancing the objections

No.	Code Word.	
5327	Cerebral .	. Countermand
5328	Cerebrous .	Countermand instructions
5329	Cerement .	Have decided to countermand
5330	Ceremonial .	Will countermand the order
5331	Ceremony .	Will countermand the instructions
5332	Ceriph .	. Countermanded
5333	Cerithium .	Have countermanded orders
5334	Cerograph .	Have countermanded instructions
5335	Cerosine .	. Country
5336	Cerostoma .	The country rock is
5337	Ceroxylon .	The surrounding country
5338	Cerrial . .	Through this country
5339	Certainly .	We are now in country rock
5340	Certhinae .	In country rock
5341	Certifier . .	Drifting through country rock
5342	Certify . .	The country rock is granite
5343	Certiorari .	The country rock is slate
5344	Cerulean .	The country rock is syenite
5345	Cerulific . .	The country rock is gneiss
5346	Ceruminous .	Course
5347	Cerura . .	Following the course of the vein
5348	Cerussite .	Drive in on the course of the vein
5349	Cervical . .	In the course of a few
5350	Cespitose .	In course of construction
5351	Cessant . .	In due course
5352	Cessation .	What course do you think it best to pursue
5353	Cessible . .	What course do (does) —— intend to pursue with
5354	Cessionary .	The course we advise [regard to
5355	Cesspipe . .	The course we recommend you to pursue
5356	Cesspool .	Follow your own course
5357	Cestoid . .	The course suggested
5358	Cestrum . .	If you follow the course laid down
5359	Cestvaen .	We shall pursue the course suggested
5360	Cetacea . .	The best course to adopt
5361	Cetologist .	Court
5362	Cetology . .	Pay into court
5363	Cetraria . .	Paid into court
5364	Cetylic . .	To be paid into court
5365	Ceylanite .	Court has awarded (or decreed)
5366	Ceylonmoss .	By order of the court
5367	Chabasite .	If the court should decree
5368	Chacma . .	Settled out of court
5369	Chaetodon .	Settle out of court if possible
5370	Chaetopoda	Court of first instance
5371	Chafe . . .	State court
5372	Chafery . .	Supreme Court
5373	Chafewax .	Court of Appeal
5374	Chafeweed .	Cover(s)
5375	Chaffering .	(To) cover the ground
5376	Chaffinch .	Fully covers the ground

No.	Code Word.	Cover(s) (*continued*)
5377	Chagrined .	Report covers the whole ground
5378	Chainbelts .	Cover(s) the whole ground of inquiry
5379	Chainbond .	Cover(s) all objections
5380	Chaincable .	Cover all emergencies
5381	Chaingang .	Cover all needful outlay
5382	Chainguard .	Cover(s) our share (in)
5383	Chainhook .	If you can cover
5384	Chainmail .	If you cannot cover
5385	Chainpier .	Cover our
5386	Chainplate .	Cover your
5387	Chainpump .	Cover his (their)
5388	Chainrule .	**Covered (by)**
5389	Chainshot .	Have you covered
5390	Chainwheel .	Has he covered
5391	Chainwork .	We have covered
5392	Chairbed .	You have covered
5393	Chairdays .	Has been fully covered
5394	Chairman .	The ground covered by
5395	Chairorgan .	Is completely covered by
5396	Chalaza .	**Credentials**
5397	Chalcedony .	Your credentials
5398	Chaldaism .	His (their) credentials
5399	Chaliced .	Credentials are quite satisfactory
5400	Chalkbeds .	**Credible**
5401	Chalkhill .	It is hardly credible
5402	Chalkiness .	**Credit**
5403	Chalkmark .	To be placed to credit of
5404	Chalkpit .	To your credit
5405	Chalkstone .	Has (have) good credit
5406	Chalky .	Has (have) bad credit
5407	Challenged .	Please telegraph credit
5408	Chalybean .	Credit is not good
5409	Chamaerops	Credit is good
5410	Chamber .	Has (have) telegraphed credit
5411	Chamberlin .	Letter of credit
5412	Chamecks .	Letter of credit not received
5413	Chameleon .	Cancel the credit
5414	Chamfret .	Placed to your credit on account of
5415	Chamois .	Credit opened in your favour with
5416	Champak .	Credit opened in favour of
5417	Champarty .	Please open a credit in favour of
5418	Champertor .	Open a credit with
5419	Champion .	What amount have you to my credit
5420	Chancel .	To our credit
5421	Chancellor .	To his (their) credit
5422	Chancrous .	Have placed (this) to the credit of
5423	Chandler .	Has been placed to your credit
5424	Chandlerly .	Unworthy of credit
5425	Chandoo .	Cannot credit a word of it
5426	Changeful .	Open credit

No.	Code Word.	Credit (*continued*)
5427	Changeless .	Shipping credit
5428	Chankshell .	Open credit each
5429	Channelled .	Shipping credit each
5430	Chantant .	Cannot increase your open credit
5431	Chanteth .	If necessary shall increase your open credit
5432	Chaos . .	Do you want us to increase your open credit
5433	Chaotic . .	Cannot increase shipping credit
5434	Chapbook .	If necessary shall increase shipping credit
5435	Chapelcart .	There is no need to increase shipping credit
5436	Chapelry .	Did you increase shipping credit
5437	Chapfallen .	Did you increase open credit
5438	Chaplain .	Shall we increase your shipping credit
5439	Chaplaincy .	There is no need to increase open credit
5440	Chapleted .	Increase open credit by wire
5441	Chapped . .	Increase open credit by letter
5442	Chapping .	Increase shipping credit by wire
5443	Chapteral .	Increase shipping credit by letter
5444	Charabancs .	Increased your open credit by wire
5445	Characters .	Increased your shipping credit by wire
5446	Charades .	Your letter of credit must suffice [increased
5447	Charadrius .	Telegraph whenever you want your shipping credit
5448	Charcoal .	Telegraph whenever you want your open credit
5449	Charge . .	I did not increase your credit [increased
5450	Chargeably .	I did not increase your open credit
5451	Chargeant .	Increased your open credit ——
5452	Chargeship .	Increased your open credit £1000
5453	Charioteer .	Increased your open credit £2000
5454	Chariotman .	Increased your open credit £3000
5455	Charitable .	Increased your open credit £4000
5456	Charitous .	Increased your open credit £5000
5457	Charity . .	Increased your open credit £6000
5458	Charityboy .	Increased your open credit £7000
5459	Charlatans .	Increased your open credit £8000
5460	Charlock .	Increased your open credit £9000
5461	Charmingly .	Increased your open credit £10,000
5462	Charneco .	Send further letter of credit
5463	Charpoy . .	Send further letter of credit £1000
5464	Chart . . .	Send further letter of credit £2000
5465	Charterboy .	Send further letter of credit £3000
5466	Charterist .	Send further letter of credit £4000
5467	Chartism .	Send further letter of credit £5000
5468	Charwoman .	Send further letter of credit £6000
5469	Charybdis .	Send further letter of credit £7000
5470	Chasable .	Send further letter of credit £8000
5471	Chasegun .	Send further letter of credit £9000
5472	Chasidean .	Send further letter of credit £10,000
5473	Chasms . .	Increase open credit——
5474	Chasteness .	Increase open credit £1000
5475	Chastening .	Increase open credit £2000
5476	Chastise . .	Increase open credit £3000

No.	Code Word.	Credit (*continued*)
5477	Chattelism .	Increase open credit £4000
5478	Chatterbox .	Increase open credit £5000
5479	Chattered .	Increase open credit £6000
5480	Chatwood .	Increase open credit £7000
5481	Chaukdaw .	Increase open credit £8000
5482	Chaunter .	Increase open credit £9000
5483	Chauvinist .	Increase open credit £10,000
5484	Chavica . .	We have increased shipping credit ——
5485	Chawbacon .	We have increased shipping credit £1000
5486	Cheapen. .	We have increased shipping credit £2000
5487	Cheapening.	We have increased shipping credit £3000
5488	Cheapjack .	We have increased shipping credit £4000
5489	Cheapjohn .	We have increased shipping credit £5000
5490	Cheaply . .	We have increased shipping credit £6000
5491	Cheapness .	We have increased shipping credit £7000
5492	Cheatingly .	We have increased shipping credit £8000
5493	Chebec . .	We have increased shipping credit £9000
5494	Checkbook .	We have increased shipping credit £10,000
5495	Checkclerk .	**Credited**
5496	Checking .	To what account have you credited
5497	Checkmate .	Has been credited with
5498	Checkrail .	Credited to —— account
5499	Checkrolls .	**Crediting**
5500	Checktaker .	Crediting same to —— account
5501	Cheddar . .	**Creditor(s)**
5502	Cheekbands	Creditors are pressing
5503	Cheekbone .	The principal creditor(s) is (are)
5504	Cheekpiece .	Large creditor(s)
5505	Cheekpouch	Who are the principal creditors
5506	Cheekstrap .	**Creek**
5507	Cheektooth .	In the bed of the creek
5508	Cheeky . .	The creek is never dry, even in the hottest summer
5509	Cheereth .	The creek is a very small one
5510	Cheerful. .	Plenty of water in the creek
5511	Cheerfully .	Very little water in the creek
5512	Cheerily . .	Watered by a creek
5513	Cheeriness .	**Cretaceous**
5514	Cheese . .	**Crisis**
5515	Cheesecake .	There will be a crisis
5516	Cheesefly .	Do you fear a crisis
5517	Cheeselep .	In order to weather the crisis
5518	Cheesemite .	At this crisis of affairs
5519	Cheesepress	Approaching a crisis
5520	Cheeseroom	Are now approaching a crisis in the affairs of the ⌐company
5521	Cheesevat .	To avoid a crisis
5522	Cheetah . .	The crisis is past
5523	Cheilopod .	**Croppings**
5524	Cheirology .	Croppings show the same kind of ore
5525	Cheiropter .	The croppings of the vein show for
5526	Cheirotes .	Ore taken from the croppings

No.	Code Word.	**Croppings** (*continued*)
5527	Chekoa . .	The croppings of the vein are well defined
5528	Chelicera .	The croppings are not well defined
5529	Cheliform .	**Cross-cut(s)**
5530	Chelingue .	Cross-cut from level No. ―― is driven ―― feet
5531	Chelodine .	Cross-cut to the vein
5532	Cheloid . .	The cross-cut from
5533	Chelonian .	The cross-cut to be extended
5534	Chelys . .	Cross-cut is in
5535	Chemical .	Have struck large body of ore in the cross-cut
5536	Chemically .	Have struck small body of ore in the cross-cut
5537	Chemicking .	Are now driving a cross-cut from
5538	Chemist . .	Cross-cut is getting into
5539	Chemistry .	Cross-cut is making good headway
5540	Chemitype .	How far have you to drive to cross-cut lode
5541	Chemosis .	When do you expect to cut lode in cross-cut
5542	Chequered .	Expect to cross-cut lode in ―― feet
5543	Chequering .	Expect to cross-cut lode in ―― days
5544	Chequey . .	How far has cross-cut advanced
5545	Cherimoyer .	Cross-cut has cut the lode
5546	Cherish . .	Have cut ―― lode in cross-cut
5547	Cherisheth	Drive a cross-cut (to)
5548	Cherishing .	Cross-cut to the hanging wall
5549	Cheroot . .	Cross-cut to the foot-wall
5550	Cherries . .	Drive cross-cuts to prove the walls
5551	Cherrybay .	Cross-cut to prove the walls (driven)
5552	Cherrycoal .	Cross-cuts to prove the ground !
5553	Cherrygum .	Shall drive a cross-cut (to)
5554	Cherrypit .	Have driven cross-cut
5555	Cherrytree .	Cross-cut has been driven
5556	Cherrywine .	Cross-cuts have been driven
5557	Chersonese .	Cross-cut should be driven
5558	Cherub . .	Cross-cut must be driven
5559	Cherubic .	Cross-cut to side vein
5560	Cherubical .	Cross-cut in ―― level
5561	Chervil . .	Cross-cut in ―― shoot
5562	Chessapple .	Cross-cut No. ――
5563	Chessboard .	Cross-cut No. ―― in
5564	Chessex . .	Cross-cut has advanced ―― feet
5565	Chessman .	Cross-cut ―― feet north of ―― shaft
5566	Chesstree .	Cross-cut ―― feet south of ―― shaft
5567	Chessylite .	Cross-cut ―― feet east of ―― shaft
5568	Chestnut .	Cross-cut ―― feet west of ―― shaft
5569	Chestrope .	Think you should drive cross-cuts
5570	Chestsaw .	Cross-cuts to test the lode
5571	Cheveronny .	Cross-cuts at intervals of
5572	Chevied . .	We recommend that a cross-cut should be driven
5573	Cheviots . .	We recommend that cross-cuts should be driven at
5574	Chevroned .	intervals of
5575	Chiastre . .	Position of cross-cut (or)—Give position of cross-cut
5576	Chibbal . .	Position of cross-cut shown on plan

No.	Code Word.	**Cross-cut(s)** (*continued*)
5577	Chicane . .	Want to know exact position of cross-cut
5578	Chicanery .	Where does cross-cut begin and end
5579	Chikadee .	Point of intersection of cross-cut (with)
5580	Chickenpox.	Cross-cut begins
5581	Chickpea .	Cross-cut ends
5582	Chicory . .	Cross-cut intercepts —— at
5583	Chideress .	Cross-cut driven to intercept
5584	Chidester .	Where will cross-cut intercept or cut
5585	Chidingly .	Cross-cut will intercept or cut
5586	Chiefbaron .	Cross-cut within about —— feet of point
5587	Chiefdom .	Cross-cut —— feet from shaft
5588	Chiefless. .	Cross-cut has reached the hanging-wall
5589	Chiefly . .	Cross-cut has reached the foot-wall
5590	Chiefrent .	Cross-cut has reached the side vein
5591	Chieftain .	We think we have cut —— in cross-cut
5592	Chiffchaff .	Running cross-cut to connect with
5593	Chilblain .	Cross-cut now being run
5594	Child. . .	Cross-cut now being run to intercept this lode
5595	Childage. .	Cross-cut now being run to intercept this level
5596	Childbirth .	We have cross-cut the lode at intervals
5597	Childhood .	We have cross-cut the lode at points **between**
5598	Childish . .	Of greatest importance to cross-cut
5599	Childishly .	When will cross-cut enter
5600	Childkind .	Cross-cut which has been driven
5601	Childless .	Where cross-cut will cut or intercept
5602	Childlike .	Where cross-cut should cut or intercept
5603	Children. .	Where cross-cut has cut
5604	Childwit . .	Length of cross-cut
5605	Chiliads . .	Direction of cross-cut
5606	Chiliagon .	**Cross-cutting**
5607	Chiliarch .	Constant cross-cutting should be done
5608	Chiliastic .	You must keep on cross-cutting to prove the ground
5609	Chilliness .	Cross-cutting towards
5610	Chimera. .	Cross-cutting towards the hanging wall
5611	Chimerical .	Cross-cutting towards the foot-wall
5612	Chimerize .	Cross-cutting in the —— shoot
5613	Chimney .	Cross-cutting in the —— level
5614	Chimneycap	Cross-cutting at intervals
5615	Chimneypot	Cross-cutting the lode at points **between**
5616	Chimpanzee	Cross-cutting the lode at
5617	Chinaaster .	We are at present cross-cutting
5618	Chinaclay .	Shall commence cross-cutting
5619	Chinaink .	**Crucible**
5620	Chinaroot .	Clay crucibles
5621	Chinaroses .	Plumbago crucibles
5622	Chinashop .	Forward at once —— dozen crucibles
5623	Chinaware .	**Crush**
5624	Chincapin .	How many tons can you crush per week
5625	Chincloth .	Crush only low-grade ore
5626	Chincough .	Crush only high-grade ore

No.	Code Word.	**Crush** (*continued*)
5627	Chinkbug .	Can crush —— tons per
5628	Chinscab .	Will be able to crush —— tons per
5629	Chinstraps .	The mill crushes —— tons
5630	Chintz . .	The mills crush —— tons
5631	Chiococca .	We can crush
5632	Chipaxe . .	Cannot crush more than —— tons
5633	Chipbonnet .	Cannot you crush better ore
5634	Chipchop .	Cannot crush better ore
5635	Chiphats .	. **Crushed**
5636	Chipmunk .	Total amount crushed is —— tons
5637	Chiragra . .	Total amount crushed unrecorded
5638	Chirognomy	Crushed during last seven days —— tons
5639	Chirograph .	Crushed during the last —— —— tons
5640	Chiromancy	Crushed during the week —— tons yielding
5641	Chironomus	Crushed during month —— tons yielding ——, expenses, Revenue Acccount ——, expenses, Capital Account ——, concentrates ——
5642	Chiroplast .	Let us know tonnage crushed
5643	Chirotony .	Let us know tonnage and average value of ore
5644	Chirping . .	Ore crushed in the —— mill [crushed
5645	Chirpingly .	Ore now being crushed assays
5646	Chirrup . .	During week —— mill crushed —— tons, yielded ——, expenses
5647	Chirrupeth .	During month —— mill crushed —— tons, yielded
5648	Chirruping .	Get quartz crushed [——, expenses ——
5649	Chirurgeon .	Send quartz home to be crushed
5650	Chirurgery .	We are having quartz crushed
5651	Chirurgic .	Do not have any quartz crushed
5652	Chisel .	. **Crusher(s)**
5653	Chiselled .	Require —— additional crusher(s)
5654	Chiselling .	A stone crusher
5655	Chitchat . .	A Blake's crusher
5656	Chitinous .	Are sending out a crusher
5657	Chitonidae .	Cost of a crusher
5658	Chittyface .	**Crushing**
5659	Chivalric .	Are crushing at the rate of —— tons per day
5660	Chivalrous .	Crushing in first-rate style
5661	Chlamydate .	We are now crushing ore from the
5662	Chlamys . .	About to commence crushing
5663	Chloasma .	First week's crushing
5664	Chloral . .	Second week's crushing
5665	Chloralism .	First month's crushing
5666	Chloranile .	This month's crushing
5667	Chloretic .	The ore we are now crushing
5668	Chloridize .	We are now crushing better ore
5669	Chloriodic .	We are now crushing poorer ore
5670	Chlorodyne .	We cannot get a crushing, the mills are full
5671	Chloroform .	Crushing cost —— per ton.
5672	Chloroid . .	Crushing is a failure
5673	Chloromys .	Trial crushings show —— to the ton

No.	Code Word.	Crushing (*continued*)
5674	Chlorops .	Advertise that the crushing shows —— to the ton.
5675	Chlorosis .	. Curtail
5676	Chloruret .	You must curtail the expenses
5677	Choanite .	Trying to curtail expenses
5678	Chocolate .	Cannot curtail expenses
5679	Choerogryl .	Absolutely necessary to curtail **expenses**
5680	Choiceful .	If you cannot curtail expenses
5681	Choiceless .	Have curtailed expenses
5682	Choke . .	Curtail the output
5683	Chokedamp	Have curtailed the output [practicable
5684	Chokefull .	Curtail the output and save expense by every means
5685	Chokepear .	**Custom(s)**
5686	Chokeplum .	What is the custom
5687	Chokestrap .	The custom here is to .
5688	Chokeweed .	The custom has been **to**
5689	Chokewort .	Customs dues
5690	Cholagogue.	Cannot pass the customs (except)
5691	Choleate. .	In order to pass the customs, consular invoices must
5692	Choleraic .	In order to pass the customs [be sent
5693	Cholericly .	The local customs
5694	Choliamb .	According to the custom of
5695	Choliambic .	It is not now the custom
5696	Cholic . .	The general custom (is to)
5697	Chondrify .	**Custom House**
5698	Chondroid .	The custom-house authorities
5699	Chondrus .	Custom-house authorities will permit
5700	Choosing .	Custom-house authorities will not permit
5701	Choosingly .	Will the custom-house authorities permit
5702	Chopboat .	**Cut**
5703	Chopcherry .	Hope to cut the vein
5704	Chophouse .	Have not yet cut the **vein**
5705	Choplogic .	Have cut the vein
5706	Choppy . .	How far must we drive to cut the vein
5707	Chopsticks .	We have still —— to drive to cut the vein
5708	Choragic .	Where we have cut the vein
5709	Choralists .	Have cut a vein
5710	Chorally . .	Had cut
5711	Chording .	Had not cut
5712	Choriambus .	Think we have cut the vein
5713	Choristers .	Think we shall soon cut the vein
5714	Chorobates .	Think we shall soon cut the
5715	Chorus . .	Let us know as soon as you cut the **vein**
5716	Chough . .	When you have cut the vein
5717	Choultry . .	When you have cut
5718	Chowder . .	When do you expect **to** cut
5719	Chrisinal .	**Cutting**
5720	Christdom .	Cutting into
5721	Christened .	Cutting out
5722	Christian .	Cutting into good **ore**
5723	Christless .	Good cutting ground

No.	Code Word.	**Cutting** (*continued*)
5724	Christtide .	Bad cutting ground
5725	Chromatic .	Little progress in cutting
5726	Chromealum	Making good progress in cutting
5727	Chromeiron	Cutting a vein
5728	Chromered .	Cutting into barren ground
5729	Chromides .	**Cwt(s).**
5730	Chromium .	At —— per cwt.
5731	Chromo . .	How much per cwt.
5732	Chromogen .	How many cwts.
5733	Chromolith .	**Daily**
5734	Chromotype	Daily reports
5735	Chromule. .	Must have the daily record of
5736	Chronic . .	Telegraph daily
5737	Chronicler .	What is the daily production of
5738	Chronicon .	The daily production of —— is
5739	Chronogram	What is the daily consumption of
5740	Chronology .	The daily consumption of —— is about
5741	Chrysalis .	Decreasing daily
5742	Chrysene .	Increasing daily
5743	Chrysolite .	**Dam**
5744	Chrysology .	It will be necessary to erect a dam
5745	Chrysops .	I estimate the cost of erecting dam at
5746	Chrysotype .	What would be the cost of making dam
5747	Chubbiness .	Tailing dam
5748	Chubby . .	Dam for holding back water
5749	Chubfaced .	Dam for reservoir
5750	Chuckaby .	The company's dam(s)
5751	Chuckfull .	A new dam must be erected
5752	Chuckhole .	Cost of new dam
5753	Chuckled .	Cost of new tailing dam
5754	Chuckling .	**Damage(s)**
5755	Chuffily . .	Damage caused by
5756	Chummage .	Owing to the damage done by
5757	Chumpend .	Damage must be repaired
5758	Chunam . .	Damage not yet ascertained
5759	Chupatties .	What is amount of damage
5760	Chupatty .	Great damage has occurred (to)
5761	Church . .	Slight damage has occurred (to)
5762	Churchale .	Has any damage been done
5763	Churchbred .	The amount of damage done
5764	Churchbugs .	The damage done is but trifling
5765	Churchdom .	The damage done is very serious
5766	Churchgoer .	What was the cause of the damage
5767	Churchism .	Damage has been caused by
5768	Churchland .	Has done considerable damage
5769	Churchless .	(It) will do us serious damage
5770	Churchlike .	(To) repair the damage
5771	Churchly .	Will damage your reputation
5772	Churchmode	Will damage —— reputation
5773	Churchowls .	Will damage —— credit

No.	Code Word.	Damage(s) (*continued*)
5774	Churchscot .	Cannot repair damages under
5775	Churchship .	**Damaging**
5776	Churchtown	Greatly damaging the
5777	Churchway .	Is very damaging to the affair
5778	Churchwork	**Dampness**
5779	Churchyard .	Owing to the excessive dampness of the climate
5780	Churlish . .	**Danger**
5781	Churlishly .	Is there any danger (of)
5782	Churnowl .	There is great danger (of)
5783	Churnstaff .	In danger of
5784	Churrworm .	There will be no danger of
5785	Chutney . .	There is no danger of
5786	Chylaceous .	There is great danger in delay
5787	Chylific . .	Out of danger
5788	Chylous . .	To avert the danger
5789	Chymbe . .	To avert the danger arising from
5790	Chymics . .	The danger arising from
5791	Chymified .	No danger need be feared (from)
5792	Chymify . .	The danger most to be feared
5793	Chymifying .	**Dangerous**
5794	Ciborium .	The —— is too dangerous to be continued
5795	Cicada . .	Owing to the dangerous state of the
5796	Cicadidae .	It is very dangerous (to)
5797	Cicatricle .	Dangerous to delay
5798	Cicatrized .	In a most dangerous condition
5799	Cicatrose .	**Data**
5800	Cicendia. .	There are no reliable data
5801	Ciceronian .	Are there any reliable data
5802	Cicisbeism .	From the data in our possession
5803	Cicurate . .	From the data in your possession
5804	Cicurating .	Not sufficient data
5805	Cidaris . .	We have not sufficient data to enable us
5806	Ciderists . .	Data to enable us to
5807	Ciderkin . .	**Date**
5808	Cidermill .	At an early date
5809	Ciderpress .	What date has been fixed for
5810	Cigartube .	Your telegrams received up to date are
5811	Ciliary . .	Have you received my telegrams to date
5812	Ciliata . .	The earliest date
5813	Ciliograda .	The latest date
5814	Cillo . . .	What is the date of
5815	Cillosis . .	What is the earliest date
5816	Cimbex . .	What is the latest date
5817	Cimeliarch .	You must name an earlier date
5818	Cimicidae .	You must name a later date
5819	Cimicifuga .	At what date
5820	Cimiss . .	At what date did he
5821	Cimmerian .	At what date did you
5822	Cinchona .	At what date can he
5823	Cinchonism .	At what date can you

No.	Code Word.	**Date** (*continued*)
5824	Cinclides .	At what date do you intend to
5825	Cinclosoma .	At what date shall we
5826	Cinclus . .	At what date will he
5827	Cinderbeds .	At what date may we expect to hear that
5828	Cindergirl .	What is the latest date at which we may expect
5829	Cindering .	What is the earliest date at which we may expect
5830	Cinderman .	At an earlier date
5831	Cinderpath .	At an earlier date than we were led to expect .
5832	Cindery . .	Not later than the date named
5833	Cinerary . .	The earliest date at which you can
5834	Cinereous .	The latest date you can give us
5835	Cinerulent .	What is the exact date of
5836	Cinnabar . .	At or about what date
5837	Cinnabaric .	At the date of writing
5838	Cinnamomic	At the date of receiving
5839	Cinnamon .	Cannot yet fix a date
5840	Cinnamyle .	At —— months' date
5841	Cinnyridae .	**Dated**
5842	Cinquefoil .	Letter was dated
5843	Cinquepace .	Dated respectively
5844	Cipher . .	Received your letter, dated
5845	Cipherhood .	Have you received my (our) —— dated
5846	Ciphering .	**Day(s)**
5847	Cipherkey .	How much per day
5848	Cipolin . .	How many —— per day
5849	Circean . .	From —— to —— days
5850	Circensial .	Will send report in a few days
5851	Circinal . .	During the next few days
5852	Circingle .	In the course of a few days
5853	Circle . . '	It takes —— days to get to —— from
5854	Circling . .	How many days does it take to get to
5855	Circuit . .	Causing an unusual delay of —— days
5856	Circuitous .	It will only take two or three days
5857	Circulable .	It will only take —— days
5858	Circular . .	An answer will be given in —— days
5859	Circularly .	It will take ——days to get
5860	Circulator .	The next day
5861	Circumcise .	The day after
5862	Circumduct .	From day to day
5863	Circumflex .	Per day of 24 hours [cleared of débris
5864	Circumfuse .	It will take —— days to get mine unwatered and
5865	Circumgyre .	How many days will it take to make examination
5866	Circumsail .	It will take —— days to go to ——, examine the mines, and return here
5867	Circumvent .	How many days will you be absent
5868	Circuses . .	Will be absent —— days
5869	Ciricsceat .	The mill ran —— days
5870	Cirrhopoda .	How many days
5871	Cirrhotic .	What day does —— reach
5872	Cirriped . .	How much —— can you turn out per day

No.	Code Word.	Day(s) (*continued*)
5873	Cirrostomi .	Can turn out —— per day
5874	Cirsocele .	During the past few days
5875	Cisalpine .	Day shift
5876	Ciselure . .	Days per week
5877	Cismontane.	Days per month
5878	Cispadane .	Days per year
5879	Cissoidal .	Working days
5880	Cistercian .	One day
5881	Cisterns . .	Two days
5882	Cistus . .	Three days
5883	Citadels . .	Four days
5884	Citations .	Five days
5885	Citatory . .	Six days
5886	Citizen . .	Seven days
5887	Citizeness .	Eight days
5888	Citizenize .	Nine days
5889	Citizenry .	Ten days
5890	Citrate . .	Eleven days
5891	Citric. . .	Twelve days
5892	Citrontree .	Thirteen days
5893	Citrullus . .	Fourteen days
5894	Cityward .	Fifteen days
5895	Civetcats .	Sixteen days
5896	Civil . . .	Seventeen days
5897	Civilation .	Eighteen days
5898	Civilian . .	Nineteen days
5899	Civilities. .	Twenty days
5900	Civility . .	Twenty-one days
5901	Civilized. .	Twenty-two days
5902	Civilizing .	Twenty-three days
5903	Clackbox .	Twenty-four days
5904	Clackdish .	Twenty-five days
5905	Clackmill .	Twenty-six days
5906	Clackvalve .	Twenty-seven days
5907	Cladium . .	Twenty-eight days
5908	Cladocera .	Twenty-nine days
5909	Claim . .	Thirty days
5910	Claimable .	Thirty-one days
5911	Claimants .	**Dead**
5912	Claiming .	Regret to tell you —— is dead; (died on the)
5913	Clamatores .	Have cut the vein, but it is dead
5914	Clambered .	Large amount of dead work must be done
5915	Clambering .	Has (have) done no dead work
5916	Clammy . .	The amount of dead work done
5917	Clamorous .	**Deal(s)**
5918	Clamour . .	Is (are) most honourable to deal with
5919	Clamoureth.	Is (are) not good persons (people) to deal with
5920	Clamouring .	(To) deal with
5921	Clamourist .	(To) deal in
5922	Clampirons .	We will deal with

No.	Code Word.	Deal(s) (*continued*)
5923	Clampnail .	Few deals at present in
5924	Clamshells .	**Dealers**
5925	Clanship. .	Few dealers; very little doing in the shares
5926	Clansman .	**Dealt**
5927	Clapboards .	How have you dealt with
5928	Clapbread .	If fairly dealt with, we think
5929	Clapcake .	**Dear**
5930	Clapnet . .	Too dear at the price
5931	Clapsill . .	Is (are) not dear
5932	Claptraps .	With dear money
5933	Clarence. .	Money very dear
5934	Clarenceux .	Very dear and scarce
5935	Claretcup .	**Dearness**
5936	Claretjugs .	The dearness of
5937	Claribella .	**Debenture(s)**
5938	Clarichord .	Debenture capital
5939	Clarified . .	Debenture interest
5940	Clarify . .	First debenture
5941	Clarifying .	Second debenture
5942	Clarigate. .	Debentures will be issued to amount of
5943	Clarionet .	Will you take debentures for
5944	Claritude .	Propose to issue debentures for
5945	Clashingly .	What is the amount of debentures
5946	Claspered .	Debentures bearing —— per cent.
5947	Claspknife .	Debentures being a first charge on the undertaking
5948	Clasplock .	Debenture interest will be paid
5949	Claspnail .	Debentures to amount of —— bearing interest at —— per cent.
5950	Classical . .	If debentures are issued
5951	Classicism .	Loan to be made on debentures
5952	Classifier .	Arranging loan on debentures
5953	Clatter . .	Loan on debentures
5954	Claudent .	**Debit**
5955	Claudicant .	Debit all charges to
5956	Clause . .	Do not debit
5957	Claustral .	You can debit me with
5958	Clausular .	Debit Revenue Account
5959	Clavaria . .	Debit Capital Account
5960	Claviceps .	Debit Stores Account
5961	Clavicorn .	Debit Reserve
5962	Clavicular .	To debit of what account
5963	Claviger. .	Have debited
5964	Clavipalp .	Have debited your account
5965	Clawback .	Should be debited to
5966	Clawhammer	What is the amount debited (or at debit of)
5967	Clawwrench	Amount to debit of
5968	Claybuilt .	Debit balance
5969	Claycold .	**Débris**
5970	Clayish . .	Have been filled with débris
5971	Claykilns .	Levels full of débris

No.	Code Word.	Débris *(continued)*
5972	Claymill . .	The mine, after being unwatered, was full of débris, which has to be cleared away before
5973	Claymore	Has (have) been in the habit of filling up the ——
5974	Claypit .	Cross-cuts filled with débris [with débris
5975	Clayslate .	**Debt(s)**
5976	Claystone .	Is (are) heavily in debt
5977	Claywater .	Is (are) in debt to
5978	Cleanly . .	To clear off the debt
5979	Cleansable .	Debt must be paid off
5980	Cleansed .	Debt must be paid off before we can do anything
5981	Clearance .	Debt incurred in consequence of
5982	Clearcut . .	Has (have) paid all the debt
5983	Clearsight .	If debt is not paid before
5984	Clearstory .	**Deceive**
5985	Cleavage .	Has (have) done all he (they) could to deceive
5986	Cleftgraft .	Do you think —— is (are) trying to deceive
5987	Cleftstick .	**Deceived**
5988	Clematis . .	Has (have) deceived us most shamefully (as to)
5989	Clemencies .	Has (have) deceived us with regard to
5990	Clemently .	Has (have) been deceived by
5991	Clepsammia	Has (have) —— not been deceived with regard to
5992	Clepsydra .	**Deceptive**
5993	Clergified .	It is very deceptive
5994	Clergy . .	Presented a most deceptive appearance
5995	Clergyman .	Under a deceptive aspect
5996	Clerkale . .	Appearances very deceptive
5997	Clerkless .	**Decide**
5998	Clerklike .	As soon as you decide
5999	Clerkship .	It will be necessary to decide at once
6000	Cleruchial .	Will decide
6001	Clethra . .	Will not decide
6002	Clever . .	Cannot decide until
6003	Cleverish .	To enable me (us) to decide
6004	Cleverly . .	Should —— decide not to
6005	Cleverness .	Should —— decide to
6006	Clianthus .	You must decide at once
6007	Cliency . .	Hope you will decide quickly
6008	Cliental . .	You must decide quickly ; no time is to be lost
6009	Clientship .	Decide quickly or will lose the chance
6010	Climacter .	If you do not decide promptly another party will
6011	Climatical .	Must leave it to you to decide [take it up
6012	Climatize .	**Decided**
6013	Climax . .	What has (have) —— decided
6014	Climbing .	Has (have) decided to
6015	Clinch . .	Has (have) decided not to
6016	Clinchers .	It has been decided
6017	Clingstone .	It has been decided not to
6018	Clinical . .	Not yet decided (whether)
6019	Clinkerbar ,	Telegraph as soon as it is decided
6020	Clinometer .	The Board have decided

No.	Code Word.	**Decided** *(continued)*
6021	Clionidae .	As soon as you have decided
6022	Clipfish . .	Have you yet decided to
6023	Cliquish . .	If you have decided to
6024	Clitoria . .	If he (they) has (have) decided
6025	Clivity . .	Have not decided
6026	Cloacal . .	Have decided to accept
6027	Cloakbag .	Have decided to decline
6028	Cloakroom .	Have decided to go on with the business
6029	Clobberer .	**Decipher**
6030	Clock . . .	Decipher this telegram and send it to
6031	Clockalarm .	Cannot decipher
6032	Clockcase .	**Decision**
6033	Clockstar .	As soon as we get your decision
6034	Clocktower .	The final decision is to
6035	Clockwork .	Decision in our favour
6036	Cloddish . .	The decision is against us
6037	Cloddy . .	Telegraph what decision has been arrived at
6038	Clodhopper .	What decision has been come to
6039	Clodpate . .	Awaiting your decision
6040	Clodpoll . .	Awaiting his (their) decision
6041	Clogdance .	Final decision
6042	Clogginess .	Must have your final decision
6043	Cloghead .	Pending the decision
6044	Cloister . .	The decision (is) (being) favourable
6045	Cloistral . .	The decision (is) (being) unfavourable
6046	Cloistress .	Hope you will reconsider your decision
6047	Closeness .	Hope (he) they will reconsider decision
6048	Closepent .	Upon reconsideration of former decision
6049	Closeth . .	The decision will materially affect (us)
6050	Closeting .	The decision will not seriously affect (us)
6051	Closure . .	**Declare**
6052	Clotbur . .	This will enable you to declare
6053	Clothes . .	Can we declare
6054	Clothespin .	**Decline**
6055	Clothhall .	I advise you to decline having anything further to
6056	Clothpaper .	Will decline [do with the affair
6057	Clothwheel .	Will soon decline
6058	Clothyard .	I advise you to decline
6059	Cloudage .	What is the cause of the decline in
6060	Cloudberry .	The stock is sure to decline
6061	Cloudborn .	Decline the offer
6062	Cloudbuilt .	Decline(s) to have anything further to do with
6063	Cloudcapt .	Decline(s) to go into the concern on account of
6064	Clouddrift .	Every prospect of a decline
6065	Cloudily . .	No prospect of a decline
6066	Cloudiness .	If —— decline(s)
6067	Cloudlet . .	Must decline
6068	Cloudrack .	Must decline the proposition
6069	Cloudring .	Must decline to act
6070	Clougharch .	Decline to have anything to do with

No.	Code Word.	Decline (*continued*)
6071	Clouterly .	Decline in the price of
6072	Cloutnail .	Decline in the yield
6073	Clovebark .	Decline(s) to
6074	Clovehitch .	Do you think there will be any further decline
6075	Clovehook .	There has been a rapid decline
6076	Clovepink .	There is no cause for the decline
6077	Cloversick .	The decline is said to be caused by
6078	Clovetree .	**Declined**
6079	Clown . .	Why has (have) —— declined
6080	Clownery .	Has (have) declined the offer on account of
6081	Clownishly .	Has (have) not yet positively declined
6082	Cloyless . .	Has (have) declined to act
6083	Cloyment .	Have declined
6084	Clubbable .	Why have you declined
6085	Clubbing .	Let us know if you have declined
6086	Clubfisted	Have you declined to act
6087	Clubfoot . .	(If) you have not yet declined
6088	Clubgrass .	The Board has definitely declined
6089	Clubhaul .	Definitely declined
6090	Clubheaded .	**Declining**
6091	Clubhouse .	In the event of your declining
6092	Clublaw . .	Gradually declining
6093	Clubmaster .	Has been gradually declining for some time past
6094	Clubmoss .	Declining the offer
6095	Clubrooms .	Declining the business
6096	Clubrush .	Declining the appointment
6097	Clubshaped .	**Decomposed**
6098	Cluegarnet .	Decomposed iron pyrites
6099	Clueline . .	Decomposed quartz
6100	Clumpboot .	Decomposed granite
6101	Clumsier . .	Decomposed vein matter
6102	Clumsily . .	**Decrease(ing)**
6103	Clumsiness .	How do you account for decrease
6104	Clumsy . .	There has been a great decrease in
6105	Cluniac . .	By which means we hope to decrease
6106	Clupea . .	You must try to decrease the
6107	Clupeidae .	There has been a heavy decrease all round in the
6108	Clusiaceae .	Steadily decreasing
6109	Clymenia .	Steadily decreasing returns
6110	Clypeaster .	Decreasing returns and increasing expenses
6111	Clypeate . .	There will be a slight decrease
6112	Clypeiform .	There will be a considerable decrease
6113	Clypeus . .	**Deduct**
6114	Clysmain .	To deduct
6115	Clyster . .	You must deduct from this
6116	Clysterize .	We shall have to deduct
6117	Coacervate .	Cannot deduct
6118	Coachbox .	Deduct the amount from
6119	Coached . .	Deduct the expenses
6120	Coachful . .	If he (they) will deduct

No.	Code Word.	
6121	Coachman	. **Deducted**
6122	Coachstand.	To be deducted from
6123	Coactive. .	Has been deducted from
6124	Coactively ,	Have you deducted anything (for)
6125	Coadapted .	Deducted from the amount
6126	Coadjacent .	After the amount has been deducted
6127	Coadjusted .	**Deducting**
6128	Coadjutant .	After deducting
6129	Coadjutor .	After deducting what is due
6130	Coadjutrix .	Not deducting anything for
6131	Coafforest .	**Deduction**
6132	Coagency .	Without any deduction
6133	Coagitate .	Balance after deduction of
6134	Coagulable .	(After) deduction of all expenses
6135	Coagulant .	A small deduction allowed
6136	Coagulator .	**Deed(s)**
6137	Coagulum .	All the deeds are in my (our) hands
6138	Coalbasin .	Has (have) handed over the deeds to
6139	Coalbeds .	The deeds are deposited in Bank of
6140	Coalblack .	Would recommend you to have deed(s) made out
6141	Coalbox . .	Deed(s) is (are) of no value [and placed in escrow
6142	Coalbrass .	Deed(s) has (have) been duly executed, placed on record, and forwarded you
6143	Coalbunker .	Have deeds made out and placed in escrow, subject to following conditions
6144	Coaldrop .	Deeds will be duly recorded, as requested, and sent you by mail as soon as possible afterwards
6145	Coaldust. .	Deed(s) deposited in escrow
6146	Coalesce. .	Deed(s) has (have) been executed
6147	Coalescing .	Deed(s) has (have) been duly executed and placed
6148	Coalfields .	The deeds must be registered [on record
6149	Coalfitter .	The deeds must be properly recorded
6150	Coalgas . .	The deeds must be deposited with [the hands of
6151	Coalheaver .	The deeds have been duly recorded, and placed in
6152	Coalholes .	The deeds must be attested at the consulate
6153	Coalhulk .	The deeds (are) duly recorded
6154	Coalmeter .	Deeds not yet recorded
6155	Coalmine .	Are the deeds ready
6156	Coalmining .	Are the deeds registered (or recorded)
6157	Coalmouse .	Why have the deeds not yet been executed
6158	Coaloil . .	Why are the deeds not yet recorded
6159	Coalpasser .	**Deep**
6160	Coalpits . .	Is —— feet deep
6161	Coalplant .	The deep workings
6162	Coalship. .	The shaft is —— feet deep
6163	Coalslack .	How deep have you gone
6164	Coalsmut .	Feet deep below
6165	Coalstaith .	**Deeper**
6166	Coalviewer .	Must go deeper
6167	Ccannex. .	Expect to find rich ore deeper

No.	Code Word.	Deeper (*continued*)
6168	Coannexing .	As we go deeper we are finding better ore
6169	Coarctate .	As we go deeper the ore is falling off
6170	Coarsely . .	As we go deeper
6171	Coarsened .	The deeper we go
6172	Coarseness .	Cannot go deeper until (unless)
6173	Coassessor .	Cannot go deeper on account of
6174	Coassume .	**Deepest**
6175	Coastguard .	The deepest shaft is —— feet deep
6176	Coastrat . .	The deepest workings
6177	Coastwards .	The deepest point yet reached
6178	Coastwise .	**Defeated**
6179	Coatarmour .	Has (have) been defeated
6180	Coatlink . .	Has (have) been defeated in the endeavour to
6181	Coax . . .	If we are defeated
6182	Coaxation .	If we are defeated we must be prepared to
6183	Coaxeth . .	Has (have) defeated all endeavours
6184	Coaxing . .	Has defeated all our expectations
6185	Coaxingly .	Has (have) been defeated at all points
6186	Cobaltblue .	**Defective**
6187	Cobaltic . .	Is (are) defective
6188	Cobblers .	The arrangements are very defective
6189	Cobcoal . .	The machinery is very defective
6190	Cobhorse .	Is so very defective that (unless)
6191	Cobiron . .	**Defence**
6192	Cobishops .	That we may prepare our defence
6193	Cobitis . .	There can be no defence
6194	Cobloaf . .	Are there no means of defence
6195	Cobnut . .	**Defend**
6196	Cobra . .	Defend the action
6197	Cobstones .	Defend the action at any cost
6198	Cobswan .	In order to defend ourselves
6199	Cobwall . .	We must defend ourselves
6200	Cobweb . .	Steps have been taken to defend
6201	Cobwebbed .	Have resolved to defend
6202	Cocalon . .	**Defended**
6203	Coccidae .	Defended from all attacks
6204	Coccidium .	Defended against hostile action
6205	Coccolite .	Defended from misconstruction
6206	Coccomilia .	**Defer**
6207	Coccosteus .	Defer any action
6208	Cocculus .	Defer taking proceedings
6209	Coccygeal .	Defer until you hear from us
6210	Coccyx . .	Defer going (until)
6211	Cochlearia .	Defer going for the present
6212	Cochleous .	We wish you to defer
6213	Cocinic . .	Think you had better defer
6214	Cockade . .	We defer to your wishes
6215	Cockahoop .	If you will defer
6216	Cockapert .	Cannot defer any longer
6217	Cockatoo .	**Deference**

No.	Code Word.	Deference (*continued*)
6218	Cockatrice .	In deference to
6219	Cockbill . .	In deference to instructions received
6220	Cockboat .	In deference to the directors' decision
6221	Cockchafer .	**Deferred**
6222	Cockcrow .	Has (have) deferred giving an answer till
6223	Cockerel .	Have deferred telegraphing till
6224	Cockeyed .	Have deferred sending my report till
6225	Cockfight .	Would rather have deferred giving an opinion till
6226	Cockhedge .	Have deferred writing (until)
6227	Cockhorse .	Have deferred starting
6228	Cocklaird .	Cannot be deferred any longer
6229	Cocklehat .	**Deficiency**
6230	Cockleoast .	There is a great deficiency in
6231	Cockmatch .	A heavy deficiency in his accounts
6232	Cockney . .	Is there any deficiency in
6233	Cockneydom	What is the deficiency (in)
6234	Cockneyfy .	**Deficient**
6235	Cockneyish .	Very deficient in
6236	Cockpaddle .	**Defined**
6237	Cockpit . .	In well-defined walls
6238	Cockroach .	A well-defined vein
6239	Cockscomb .	Not well defined
6240	Cocksfoot .	Is the vein well defined
6241	Cockshut .	A well-defined hanging wall
6242	Cocksorrel .	A well-defined foot-wall
6243	Cockspurs .	A well-defined lode
6244	Cocksure .	Is the —— well defined
6245	Cockswain .	Our ideas have been clearly defined
6246	Cocktail . .	Our position is clearly defined
6247	Cockwater .	**Definite**
6248	Cockweed .	Not definite enough
6249	Cocoa . .	Have you come to any definite understanding with
6250	Cocoanuts .	Has (have) come to a definite arrangement
6251	Cocoaoil. .	Has (have) not yet come to any definite under-
6252	Cocoaplum .	Very definite and clear [standing with
6253	Cocoatrees .	Arrange something definite with regard to
6254	Cocoonery .	Can get nothing definite from
6255	Coctible . .	Must have something more definite
6256	Cocuswood .	Nothing definite has yet been arranged
6257	Coddymoddy	Nothing definite has yet been done
6258	Codeine . .	Nothing definite can be done
6259	Codex . .	Nothing definite is known yet
6260	Codfishery .	Nothing can be more definite or clearer
6261	Codicils . .	Must have more definite information
6262	Codify . .	More definite information required before we can
6263	Codilla . .	A definite understanding [decide
6264	Codling . .	More definite instructions
6265	Codsound .	Your instructions were most definite; please adhere
6266	Coefficacy .	**Degrees** [to them
6267	Coelacanth .	Thermometer sometimes ——° in the shade

No.	Code Word.	**Degrees** (*continued*)
6268	Coelders . .	Degrees below zero
6269	Coelogenys .	Vein is dipping ——° to the
6270	Coelosperm .	The temperature has been down to ——°
6271	Coemption .	The temperature has been as high as ——°
6272	Coenjoyed .	By degrees
6273	Coenjoying .	We shall get on by degrees
6274	Coenoby . .	**Delay(s)**
6275	Coenoecium	Delay will be fatal
6276	Coenosarc .	Causing a delay of
6277	Coequal . .	Delay would do no harm
6278	Coequally .	Delay as long as you can
6279	Coerce . .	Avoid any delay as much as possible
6280	Coercible .	Go without delay to
6281	Coercion .	To avoid delay
6282	Coercively .	Delay is unavoidable
6283	Coeternal .	Has (have) tried to delay
6284	Coeternity .	There must be no delay
6285	Coevous . .	What caused the delay
6286	Coexecutor .	The delay was caused by
6287	Coexist . .	If there is any further delay
6288	Coexisting .	Must avoid further delay
6289	Coexpand .	These continued delays are most prejudicial to us
6290	Coextended.	Not responsible for the delay
6291	Cofactor . .	Who causes the delays
6292	Coffeecup .	Delay your departure
6293	Coffeeman .	Cannot delay any longer
6294	Coffeemill .	Delay caused by bad weather
6295	Coffeenib .	Causing further delay
6296	Coffeepots .	Delay is dangerous
6297	Coffeeroom .	Any further delay will ruin the whole affair
6298	Coffeetree .	Better not delay your return any longer
6299	Cofferdam .	Do not delay your return later than
6300	Coffership .	**Delayed**
6301	Cofferwork .	Has (have) been delayed by
6302	Coffinbone .	Unavoidably delayed by
6303	Coffinless .	Why has (have) —— delayed
6304	Coffins . .	Has (have) delayed
6305	Cogencies .	**Deliver**
6306	Cogent . .	When will —— deliver
6307	Cogently .	Will deliver
6308	Coggerie . .	Will not deliver
6309	Cogitable .	Do not deliver
6310	Cogitabund .	Can deliver
6311	Cogitateth .	Cannot deliver
6312	Cogitating .	Deliver to
6313	Cogitation .	If you can deliver
6314	Cognates .	If you cannot deliver
6315	Cognisably .	When can you deliver (up)
6316	Cognition .	We wish you to deliver
6317	Cognitive .	**Delivered**

No.	Code Word.	Delivered (*continued*)
6318	Cognizance .	To be delivered to
6319	Cognized .	Has (have) delivered to
6320	Cognizing	. **Delivery**(ies)
6321	Cognizors .	Subject to delivery (on)
6322	Cognomen .	Awaiting delivery (of)
6323	Cognominal	Against delivery
6324	Coguardian .	Will guarantee delivery (on)
6325	Cogwheels .	After delivery (of)
6326	Cohabited .	Delivery after
6327	Cohabiting .	Upon delivery (of)
6328	Coheir . .	Before delivery (of)
6329	Coheiress .	' Delivery before
6330	Coherald .	(For) immediate delivery
6331	Coherent .	Can you obtain delivery of
6332	Coherently .	Can obtain delivery of
6333	Cohésible .	Cannot obtain delivery
6334	Cohesion .	Delivery must be completed
6335	Cohesively .	Unless delivery be made at once
6336	Cohibitor .	Delivery to be made within
6337	Cohobation .	Defer delivery until
6338	Coigny . .	Deliveries to commence
6339	Coincide .	. **Demand**(s)
6340	Coinciding .	You must demand
6341	Coinhere .	There is a great demand for
6342	Cointense .	Should —— demand
6343	Cointerest .	Do not demand
6344	Cokernut .	Can meet all immediate demands
6345	Cokingkiln .	Demand immediate settlement
6346	Cokingoven	Demand the stock immediately
6347	Colander .	There is no demand for
6348	Colanut . .	What demand have you for
6349	Colaseed . .	Continue(s) in fair demand
6350	Colatitude .	There is but little demand for
6351	Colatree . .	Is there any demand for
6352	Colbertine .	The demand has fallen off (for)
6353	Coldblast .	He demands (they demand)
6354	Coldchisel .	What is the nature of the demand
6355	Coldcream .	What is the amount of the demand
6356	Coldish . .	The demand is unreasonable
6357	Coldkind .	The demand is reasonable
6358	Coldly . .	The most exaggerated demands
6359	Coldpale .	To formulate the demand
6360	Coldserved .	If his (their) demands are reasonable
6361	Coldshort .	Cannot accede to (his) demands
6362	Colegatee .	If the demands are pressed
6363	Coleophyll .	In full of all demands
6364	Coleopter .	A fair and equitable demand
6365	Coleorhiza .	**Demanded**
6366	Colerape . .	Has (have) demanded
6367	Colessor . .	Has (have) demanded immediate settlement

No.	Code Word.	**Demanded** (*continued*)
6368	Colicky . .	Have you demanded
6369	Collapse . .	Has (have) demanded immediate delivery of the
6370	Collapsing .	Has (have) not demanded
6371	Collarbeam .	If —— has (have) demanded
6372	Collarbone .	Has (have) demanded nothing but what is fair
6373	Collarday .	**Demoralized**
6374	Collatable .	Has (have) become thoroughly demoralized
6375	Collate . .	**Demurrage**
6376	Collect . .	Demurrage has been incurred
6377	Collecting .	The amount of demurrage
6378	Collection .	Demurrage has to be paid
6379	Collegial .	Have we to pay demurrage
6380	Collibert . .	Will not unload until demurrage is paid
6381	Collided . .	Pay the demurrage
6382	Colliding .	**Denounce(d)**
6383	Collieries .	You must denounce the following claims
6384	Colliery . .	Have denounced
6385	Colligated .	Have denounced the following claims
6386	Collimator .	Names of the denounced claims
6387	Collinear .	As soon as we have denounced
6388	Collingual .	**Denunciation**
6389	Colliquant .	Denunciation has been made of
6390	Collisions .	Formal denunciation of
6391	Collocate .	**Denial**
6392	Collocutor .	Have given a most emphatic denial
6393	Collodion .	Give the statement emphatic denial, there is no
6394	Collogue . .	Complete denial to the allegation [truth in it
6395	Colloguing .	**Denied**
6396	Colloidal .	If the statement is not denied
6397	Colloquial .	The statement cannot be denied
6398	Colloquize .	**Deny**
6399	Colloquy .	You can deny it emphatically
6400	Collusive .	Will you authorize us to deny it
6401	Collusory .	**Departure**
6402	Collybist . .	At the time of your departure
6403	Collyrite . .	Delay your departure
6404	Collyrium .	Departure delayed
6405	Coloboma .	Cannot you delay your departure
6406	Colocynth .	**Depend**
6407	Colonelcy .	Must depend on
6408	Colonical .	Depend upon
6409	Colonist . .	Cannot depend upon
6410	Colonizer .	You may thoroughly depend on
6411	Colonizing .	You must not depend on
6412	Colonnades .	If you can depend on
6413	Colophany .	If you cannot depend on
6414	Colophene .	Depend for supplies upon
6415	Colopholic .	Shall have to depend upon
6416	Colorature .	Can you depend upon
6417	Colossal . .	Must depend upon circumstances

F

No.	Code Word.	Depend (*continued*)
6418	Colossuses .	Think you can depend upon
6419	Colostrum .	Do not think you can depend upon
6420	Colour . .	You can depend upon him, he is most trustworthy
6421	Colourably .	**Depended**
6422	Colouring .	Can be depended upon
6423	Colourists .	Cannot be depended upon
6424	Colourless .	Has (have) depended upon
6425	Colourman .	Has (have) not depended upon
6426	Colpocele .	**Dependent(ce)**
6427	Colportage .	Dependent upon circumstances
6428	Coltevil . .	Can any dependence be placed upon
6429	Coltsfoot .	Place entire dependence upon
6430	Colubridae .	No dependence to be placed on
6431	Columbacei .	**Deposit(s)**
6432	Columbary .	A rich gold-bearing deposit
6433	Columbine .	The extent of the deposit is
6434	Columella .	Large deposits of
6435	Columnar .	Superficial deposits
6436	Columnrule .	Deposit my stock with
6437	Colymbidae .	Deposit to my (our) credit with
6438	Comatose .	Deposit with
6439	Comatula .	Deposit of —— per cent.
6440	Combacy .	Has (have) demanded deposit
6441	Combatable	Insist upon having deposit
6442	Combatants .	Must deposit sufficient to cover all expenses
6443	Combative .	Refuse(s) to make a deposit
6444	Combbroach	What deposit is required
6445	Combhoney	Require a remittance of —— for deposit to secure what I consider a very valuable property on the [following terms
6446	Combinedly	Deposit on our account
6447	Combings .	Deposit on your account
6448	Combless .	The deposits mentioned in
6449	Comboloio .	**Deposited**
6450	Combustion	Must be deposited
6451	Comedian .	This sum to be deposited with
6452	Comedietta .	Deposited to your credit with
6453	Comedies .	Has (have) deposited at —— bank the sum of
6454	Comedy .	Has (have) deposited as security
6455	Comelily .	To be deposited
6456	Comeliness .	**Depreciated**
6457	Comeoff .	Has (have) been seriously depreciated
6458	Comeouter .	Our shares are much depreciated
6459	Comephorus	**Depreciation**
6460	Comestible .	For depreciation
6461	Cometarium	Depreciation of machinery and plant
6462	Cometh . .	Depreciation fund
6463	Cometology .	Depreciation on machinery
6464	Comfit . .	Depreciation on buildings
6465	Comfiture .	Depreciation on machinery and buildings
6466	Comfortful .	Depreciation on mills

No.	Code Word.	**Depreciation** (*continued*)
6467	Comforting .	Depreciation allowed
6468	Comic . .	Depreciation for half year
6469	Comically .	Depreciation for past half year
6470	Comicry . .	Depreciation for year
6471	Comingon .	Depreciation nil
6472	Comma . .	Depreciation stands at —— in balance sheet
6473	Commandant	Depreciation will stand at —— in balance sheet
6474	Commandery	Nothing set aside or allowed for depreciation
6475	Commatical	Set aside or allowed for depreciation
6476	Commeasure	We think it advisable to allow for depreciation
6477	Commenced	Allow for depreciation
6478	Commencing	Allow for depreciation —— per cent. on
6479	Commendam	We do not think it necessary to set aside anything
6480	Comment .	**Depressed** [for depreciation
6481	Commentors	Stocks are depressed
6482	Commercial .	The market is depressed
6483	Commigrate	Owing to depressed state of
6484	Commingle .	Shareholders are much depressed
6485	Commissary	We are much depressed at having such a bad report ..
6486	Commission	**Depressing**
6487	Commissive .	Has (have) a depressing effect on
6488	Commitment	**Depth**
6489	Committals .	Depth of incline
6490	Committee .	Depth of shaft
6491	Committing	Depth of winze
6492	Commixture	At a considerable depth below the surface
6493	Commodate	At a depth of —— feet
6494	Commodious	The vein has been proved to a depth of —— feet
6495	Commonable	Winze has been sunk to a depth of —— feet
6496	Commoner .	Depth attained to date
6497	Commonly .	The mine is getting richer as depth is attained
6498	Commonness	Has reached a depth of —— feet
6499	Commonweal	Vein is improving with depth
6500	Commorancy	Vein is getting poorer with depth
6501	Commorient	The mine has been worked by shafts to a depth
6502	Commother .	To a depth of [of —— feet
6503	Commotion .	Must be sunk to a depth of
6504	Community .	They ceased work at a depth of —— feet
6505	Comocladia .	What depth is the
6506	Compact . .	A moderate depth
6507	Compactly .	**Derive**
6508	Compacture	You will be able to derive all information from
6509	Companions	Was (were) able to derive information from
6510	Companying	Was (were) not able to derive any information from
6511	Comparable	From which I (we) derive
6512	Compared .	Cannot derive
6513	Comparison	**Describe**
6514	Compass . .	Must describe fully what is wanted
6515	Compassbox	**Description**
6516	Compassing .	The description you have given

No.	Code Word.	Description (*continued*)
6517	Compassion.	A full description of the property
6518	Compatibly .	Must have a full description of the mine and property
6519	Compatriot .	If the property is at all near the description
6520	Compeer .	A very valuable mine from the description given
6521	Compelled .	Full description of the developments and workings
6522	Compelling .	Full description of the surface works
6523	Compendium	Send full description
6524	Compensate	According to the description
6525	Compesce .	Is (are) nothing like the description
6526	Competency	Full description by first post
6527	Compilator .	**Desert**
6528	Compinged .	Is (are) situated in a desert
6529	Complacent.	The country round the mine is quite a desert
6530	Complained	**Desirable**
6531	Complanate.	The Board do not consider it desirable
6532	Complease .	It is most desirable (that)
6533	Complement	It is not at all desirable (that)
6534	Completely .	Is it desirable (that)
6535	Completive .	If you consider it desirable
6536	Complexion	**Desirous**
6537	Complexity .	Extremely desirous to
6538	Complexure.	**Despatch**
6539	Compliable .	With the utmost despatch
6540	Compliant .	When will you despatch
6541	Complicacy.	Shall despatch on or about
6542	Complish .	Cannot despatch before
6543	Complotter .	When did you despatch
6544	Comply . .	Prompt despatch is essential
6545	Complying .	**Despatched**
6546	Component .	If despatched on or before
6547	Composedly	Will be despatched
6548	Compositor .	Was (were) despatched
6549	Compound .	**Destroyed**
6550	Comprehend	A fire has destroyed
6551	Compressor.	A flood has destroyed
6552	Compriest .	Has (have) been destroyed
6553	Comprisal .	How was (were) —— destroyed
6554	Comprobate	The mill was partially destroyed
6555	Compromise	The mill was wholly destroyed
6556	Comptible .	Buildings destroyed
6557	Comptrol .	The camp was nearly destroyed
6558	Compulse .	**Despair**
6559	Compulsory	Beginning to despair
6560	Compuncted	We do not despair of success
6561	Compupil .	No cause for despair
6562	Computist .	There never was less cause for despair
6563	Comradery .	**Despairing**
6564	Comrogue .	Shareholders are in a very despairing mood
6565	Comtism .	In a despairing mood and very angry
6566	Conatus . .	Despairing of success, and much disheartened

No.	Code Word.	
6567	Concavity .	**Detail(s)**
6568	Concavous .	Will furnish you with all the details
6569	Concealeth .	When can you mail details
6570	Concealing .	Details will be mailed on
6571	Concedence.	Details will be mailed in a few days
6572	Conceive .	Has (have) sent the fullest details
6573	Concentric .	Has (have) received the fullest details
6574	Concentual .	Must have fuller details of
6575	Concept . .	Send full details
6576	Conception .	Send sufficient details to enable us here to form an intelligent opinion about the matter
6577	Concertina .	We have sent you sufficient details to enable you to [judge
6578	Concettism .	**Detain**
6579	Concha . .	Will detain me (us) some days
6580	Conchacea .	Do not let anything detain you
6581	Conchifer .	This will detain
6582	Conchitic .	This will not detain
6583	Conchoidal .	Will this detain
6584	Conchology.	This need not detain
6585	Conciator .	**Detained**
6586	Conciliate .	Has (have) been detained by
6587	Concionary .	Am (are) detained here for want of money
6588	Concisely .	Detained by
6589	Concitizen .	This has detained me here
6590	Conclavist .	How long will you be detained
6591	Conclimate .	Should you be detained
6592	Concluded .	Detained in consequence of
6593	Concluding.	Will be detained —— days
6594	Conclusory .	We are detained here by business
6595	Concocteth .	(Are) detained here by bad weather
6596	Concocting .	(Are) detained by the authorities
6597	Concoction .	Detained by circumstances beyond our control
6598	Concolour .	**Determine(d)**
6599	Concord . .	Can you determine
6600	Concordly .	Cannot yet determine
6601	Concredit .	Cannot yet be determined
6602	Concrement	Has been determined (or) (have determined)
6603	Concretive .	Has not yet been determined
6604	Concubaria .	**Detrimental (to)**
6605	Conculcate .	Detrimental to the interests of the concern
6606	Concupy .	Detrimental to our interests
6607	Concurring .	Will be most detrimental (to)
6608	Concussive .	**Develop**
6609	Concutient .	In order to develop
6610	Condemneth	Will probably develop a valuable mine
6611	Condemning	To develop
6612	Condensers.	To develop the mine would require
6613	Condensity .	**Developed**
6614	Condescend	If the mine were better developed
6615	Condign . .	The mine is but little developed

No.	Code Word.	**Developed** (*continued*)
6616	Condignly .	Until better developed
6617	Condiments.	**Development(s)**
6618	Condolence.	Developments in the mine are
6619	Condoned .	Developments fully justify the advance in
6620	Conducent .	Developments do not justify advance in
6621	Conducibly .	Developments expected every day
6622	Conducive .	The only development is
6623	Conducted .	The mines are valuable prospects requiring develop-
6624	Conducting .	To continue the developments [ment
6625	Conductors .	Not sufficient development to
6626	Condurrite .	Do developments justify
6627	Condyle . .	New developments
6628	Condyloid .	Await further development
6629	Condylopod	Will probably turn out to be a good mining property : further developments necessary
6630	Condylura .	Further development necessary
6631	Conenchyma	Development tunnel completed
6632	Conepulley .	Further development of the property
6633	Coneshell .	Employ —— men in development [ment
6634	Coneyfish .	Shall I spend money in sinking shaft for develop-
6635	Confab . .	Shall I spend money in driving adit for development
6636	Confabbing .	Make a tunnel for development of the property
6637	Confabular .	Sink a shaft for development of the property
6638	Confervoid .	Employ good engineer and men to assist him in
6639	Confess . .	Total lineal development [development
6640	Confessant .	What development work has been done
6641	Confesseth .	How are the developments progressing
6642	Confestly .	What is the appearance of the recent developments
6643	Conficient .	Developments promise well
6644	Configured .	Developments in the higher levels
6645	Confinity .	Developments in the lower levels
6646	Confirmeth .	Developments opening up splendidly
6647	Confirming .	The developments and workings
6648	Confiscate .	Developments opening up ore bodies
6649	Conflated .	The ore bodies opened up by the developments
6650	Conflating .	Developments upon the —— lode
6651	Conflation .	**Diameter**
6652	Conflict . .	Having a diameter of
6653	Confluence .	Inches in diameter
6654	Conflux . .	**Diamond(s)**
6655	Confocal. .	Diamond mine
6656	Conformers .	Diamonds discovered at
6657	Conformist .	Estimated value of diamonds found
6658	Confounded.	You will have to go to the diamond mines of
6659	Confronter .	Carats of diamonds value
6660	Confused .	Diamonds found in
6661	Confusedly .	Core from the diamond drills
6662	Confutable .	By use of diamond drills
6663	Confutant .	Assays from diamond drill cores (give) [wide
6664	Congeable .	The diamond drill has searched a zone —— feet

No.	Code Word.	Diamond(s) *(continued)*
6665	Congenious .	Depth of diamond drill bore
6666	Congenital .	**Diamondiferous**
6667	Congereel .	Diamondiferous blue ground
6668	Congested .	Diamondiferous alluvial deposits
6669	Congestive .	**Diagram(s)**
6670	Congiary. .	Can do nothing without surveyor's diagrams
6671	Conglobed .	Send surveyor's diagrams at once
6672	Congopea .	Diagrams posted
6673	Congosnake	We send you a diagram, from which you will see
6674	Congreeing .	**Did**
6675	Congregate .	I (we) did
6676	Congresses .	I (we) did not
6677	Congruency	He (they) did
6678	Congruity .	He (they) did not
6679	Congruous .	Did not
6680	Conicality .	Why did you
6681	Conicity . .	Why did you not
6682	Conidium .	I (we) did not get
6683	Coniferae .	I (we) did not see
6684	Coniocyst .	Did you go to
6685	Coniotheca .	Did you
6686	Coniroster .	It did
6687	Conject . .	It did not
6688	Conjecting .	If we did
6689	Conjobble .	If you did
6690	Conjoined .	If he (they) did
6691	Conjointly .	If it did
6692	Conjugally .	If they did not
6693	Conjunctly .	What did he (they) do
6694	Conjurator .	What did you do
6695	Connascent .	Whether you did or not
6696	Connature .	Whether he did say so or not
6697	Connective .	Whether he (they) did or not
6698	Connexion .	When did you
6699	Connivancy .	When did he (they)
6700	Connubial .	Why did it
6701	Conocarp .	**Died**
6702	Conodont .	Died suddenly
6703	Conohelix .	Died of ——
6704	Conoidic .	Died of —— on the
6705	Conoidical .	**Dies**
6706	Conominee .	Shoes and dies of iron
6707	Conqueress .	Shoes and dies of steel
6708	Conquering .	Set of shoes and dies
6709	Conscience .	**Difference**
6710	Consecrate .	Is there any difference
6711	Consectary .	There is a great difference
6712	Consensus .	There is no difference
6713	Consent . .	Difference between
6714	Consenteth .	What is the difference

No.	Code Word.	**Difference** (*continued*)
6715	Consenting .	(To) pay the difference
6716	Consequent .	The difference is made up by
6717	Conserved .	The difference between our views and
6718	Consider. .	The difference has been adjusted
6719	Consigneth .	Very little difference (between)
6720	Consigning .	**Different** [led to suppose
6721	Consiliary .	I find a very different state of things to what I was
6722	Consimilar .	The reports are very different
6723	Consisted .	Has (have) been telling two different tales
6724	Consistory .	Under very different circumstances
6725	Consociate .	The mine(s) has (have) been in a number of dif-
6726	Consolator .	**Difficult** [ferent hands
6727	Console .	The mine(s) is (are) very difficult to work
6728	Consonancy	It will be a very difficult undertaking
6729	Conspectus .	Will it be difficult
6730	Conspiracy .	It will be very difficult
6731	Conspire. .	Very difficult to decide
6732	Conspiring .	Very difficult to work
6733	Constantia .	**Difficulty(ies)**
6734	Constantly .	Will there be any difficulty
6735	Constate. .	Has (have) found great difficulty
6736	Constipate .	Will have some difficulty
6737	Constrain .	Has (have) found no difficulty
6738	Constrict .	The difficulty is
6739	Constringe .	Will have no difficulty
6740	Consubsist .	What is the difficulty
6741	Consuetude.	Without difficulty
6742	Consulage .	With great difficulty
6743	Consulship .	If there is any difficulty
6744	Consultive .	There is no difficulty
6745	Consumable	We shall not have much difficulty
6746	Consumed .	The greatest difficulty we have to encounter
6747	Consumedly	Let us know if you have any (further) difficulty
6748	Consuming .	Difficulties have arisen
6749	Consummate	(To) overcome the difficulties
6750	Consumpt .	The difficulties have been overcome
6751	Consutile .	Difficulties in the way
6752	Contactual .	**Dimensions**
6753	Contagious .	The dimensions are
6754	Contagium .	What are the dimensions
6755	Contain . .	Give dimensions of the largest piece
6756	Containant .	The dimensions of the largest piece
6757	Containeth .	Dimensions of the tunnel
6758	Contango .	Dimensions of the shaft
6759	Contemper .	**Diorite**
6760	Contentful .	Dyke of diorite
6761	Context .	**Dip(s)**
6762	Continency.	The dip of the lode
6763	Continue .	Dips at an angle of ——•
6764	Continuing .	Dips to the

No.	Code Word.	
6765	Contorted .	**Direct**
6766	Contraband.	Shall go direct (to)
6767	Contracts .	Direct communication with —— is interrupted, owing to
6768	Contradict .	You had better go direct to .
6769	Contramure	There is no direct route to .
6770	Contrapose .	Direct all letters to the care of —— at
6771	Comtrarily .	Direct all letters to —— hotel at
6772	Contribute .	Direct all telegrams to the care of —— at
6773	Contrite . .	Direct all telegrams to —— hotel at
6774	Contriving .	Send direct to .
6775	Controller .	Ordered to be sent direct
6776	Controvert .	Come home by most direct route
6777	Contrusion .	**Director(s)**
6778	Contumacy .	The directors approve
6779	Conundrum.	The directors disapprove
6780	Convalesce .	The directors authorize you to
6781	Convenable.	Would the directors agree to ,
6782	Convergent.	Will join board of directors
6783	Converse .	Will you join board of directors
6784	Conversing .	(As) local director
6785	Conversion .	(As) resident director
6786	Converters .	(As) managing director
6787	Convertite .	(As) chairman of board of directors
6788	Convexedly .	Can you get me made a director
6789	Convexity .	Can you get me made managing director
6790	Convexness .	Is (are) appointed managing director(s)
6791	Convey . .	Is (are) made director
6792	Conveyance.	Would make a good director
6793	Conveying .	Join board of directors after allotment
6794	Convicious .	There are some good names among the directors
6795	Convicted .	Cannot accept the post of director
6796	Convictism .	Accept the post of director
6797	Convinces .	Good names for local directors
6798	Convivial .	Cable the names of local directors
6799	Convocated.	Have appointed local director(s)
6800	Convoked .	The directors wish you to
6801	Convolute .	The directors agree to your proposal
6802	Convulsed .	The directors decline to
6803	Convulsing .	Do the directors decline or consent
6804	Conyburrow	The directors require further information
6805	Conycatch .	The directors are of opinion that
6806	Conywool .	**Disabled**
6807	Conyza . .	Disabled by an accident
6808	Cooingly. .	Disabled by sickness
6809	Cookeries .	Is (are) disabled from
6810	Cookery . .	**Disappointed**
6811	Cookmaid .	Have been greatly disappointed
6812	Cookroom .	**Disappointment**
6813	Coolheaded .	Much disappointment is felt

No.	Code Word.	
6814	Coolness.	. **Disapproval (of)**
6815	Coolwort.	. Desire to express their entire disapproval of
6816	Cooped .	. **Disbursements**
6817	Cooperated .	Disbursements this month have been
6818	Coopering .	Disbursements last month have been
6819	Cooptate. .	What have been the disbursements for
6820	Cooptation .	The disbursements have been very heavy
6821	Coordinate .	Disbursements in respect of
6822	Copaiba . .	On account of heavy disbursements (for)
6823	Copaifera	. **Discharge(d)**
6824	Copalche . .	Discharged for
6825	Coparcener .	Has been discharged
6826	Copartment.	(Will) discharge the duties
6827	Copartners .	Who discharges the duties in your absence
6828	Coparts . .	Have had to discharge
6829	Copeck .	. **Disconnected**
6830	Copepoda .	Entirely disconnected from
6831	Copernican .	We have disconnected
6832	Copesmate	. **Discontented**
6833	Copestones .	Getting very discontented at
6834	Cophinus	. **Discontinue**
6835	Cophosis	. Will you discontinue
6836	Copied . .	Think you ought to discontinue
6837	Copiously	. You must discontinue
6838	Coplant . .	You must not discontinue
6839	Coportion .	Propose to discontinue
6840	Copper . .	Shall discontinue
6841	Copperhead.	**Discontinued**
6842	Copperish .	Has (have) discontinued driving
6843	Coppernose .	Have discontinued developments on account of
6844	Copperwork.	Why was the —— discontinued
6845	Coppice . .	Have you discontinued
6846	Coppledust .	Have discontinued
6847	Copresence	. **Discount**
6848	Copridae	. The shares are at a discount
6849	Coprolitic . .	What discount allowed for cash
6850	Coprophagi .	Discount for cash
6851	Copsewood .	Can you discount
6852	Copspinner .	Can discount
6853	Copsy . .	Cannot discount
6854	Copulate. .	At a discount of
6855	Copulation .	Rate of discount
6856	Copulatory .	Cannot allow any discount
6857	Copybook	. **Discouraging**
6858	Copyhold .	The accounts are very discouraging
6859	Copying . .	The aspect of affairs is very discouraging
6860	Copyists .	. **Discover**
6861	Copyright .	Can you discover
6862	Coquetted .	Cannot discover
6863	Coquetting .	Hope to discover

No.	Code Word.	
6864	Coquettish .	**Discovered**
6865	Coquito . .	Have you discovered
6866	Coracias . .	Has (have) discovered
6867	Coracle . .	Has (have) not discovered
6868	Coradicate .	Nothing has been discovered about
6869	Coralloid .	**Discovery**
6870	Corallum .	Important discovery (of)
6871	Coralrag .	**Discretion**
6872	Coralreefs .	You must act on your own discretion
6873	Coralwort .	At the discretion of
6874	Coranto . .	Have no discretion in the matter
6875	Corbelled .	At my (our) discretion
6876	Corbelling .	**Dishonest**
6877	Corchorus .	Is dishonest
6878	Cordately .	The concern is in dishonest hands
6879	Cordgrass .	A most dishonest report
6880	Cordial . .	A more dishonest report was never **made**
6881	Cordialize .	**Dishonestly**
6882	Cordially .	Has (have) acted dishonestly
6883	Cordiceps .	Has (have) been detected acting dishonestly
6884	Cordierite .	Is (are) acting dishonestly
6885	Cordon .	**Dishonesty**
6886	Corduroy .	Dismiss —— for his (their) dishonesty
6887	Cordwain .	Have dismissed —— for dishonesty
6888	Cordwood .	**Dishonourably**
6889	Coregonus .	Has (have) acted very dishonourably
6890	Corelation .	**Dishonoured**
6891	Corelative .	Draft(s) has (have) been dishonoured
6892	Coreopsis .	Draft will be dishonoured unless
6893	Corfhouse .	Will be dishonoured
6894	Corivalry .	Dishonoured acceptance(s)
6895	Corkcutter .	, If acceptance is dishonoured
6896	Corkfossil .	Acceptance dishonoured
6897	Corkingpin .	**Dismiss**
6898	Corkjacket .	Have been obliged to dismiss —— on account of
6899	Corkleg . .	Dismiss all the men you can possibly do without
6900	Corkscrew .	Do not dismiss
6901	Corktrees .	**Disorder**
6902	Cormogens .	The accounts are in the greatest disorder
6903	Cormophyte	There is great disorder in
6904	Cormorant .	Most disorderly
6905	Cornaceae .	**Dispensed with**
6906	Cornamute .	Must be dispensed with
6907	Cornbadger .	Can be dispensed with
6908	Cornbeef .	Cannot be dispensed with
6909	Cornbeetle .	Will be dispensed with
6910	Cornbrash .	Has (have) been dispensed with
6911	Cornbread .	**Dispute**
6912	Corncockle .	There is a dispute (about the ——)
6913	Corncrakes .	There is a dispute between

No.	Code Word.	Dispute (*continued*)
6914	Corncutter .	The dispute has been settled about the
6915	Corndrills .	The dispute is not yet settled about the
6916	Cornelian .	What is the dispute
6917	Corneous .	In the event of any dispute
6918	Cornercap .	Can you settle the dispute
6919	Cornered .	If a dispute arises, the matter to be referred to
6920	Cornering .	Refer the matter in dispute to
6921	Cornerwise .	The dispute is settled
6922	Cornetstop .	**Dissatisfaction**
6923	Cornfactor .	Very general dissatisfaction is expressed
6924	Cornfield .	There is not so much dissatisfaction
6925	Cornflag .	. **Dissatisfied**
6926	Cornflour .	Is (are) very much dissatisfied
6927	Cornice . .	Am (are) very much dissatisfied
6928	Cornished .	Shareholders are dissatisfied
6929	Cornjuice .	**Dissolution**
6930	Cornlaws .	Dissolution of partnership between —— and
6931	Cornloft .	. **Distance**
6932	Cornmeters .	The distance is
6933	Cornmoth .	A distance of
6934	Cornopeans .	Distance from the shaft to ,
6935	Cornpoppy .	What is the distance
6936	Cornrose .	**Distinct**
6937	Cornsalad .	There are —— distinct veins
6938	Cornsawfly .	Quite distinct
6939	Cornstone .	The meaning of your telegram is not distinct
6940	Cornthrips .	Is not distinct
6941	Cornucopia .	**Distinction**
6942	Cornvan . .	There is a distinction between
6943	Cornviolet .	Make a distinction between
6944	Cornweevil .	What is the distinction between
6945	Corocore .	**Distinctly**
6946	Corollary .	Must be distinctly understood
6947	Corollated .	Can be distinctly traced
6948	Corollist . .	Cannot be distinctly traced
6949	Coronamen .	Has (have) been distinctly told
6950	Coronation .	**Distinguish(ed)**
6951	Coroner . .	How shall we distinguish
6952	Coroniform .	Can be distinguished
6953	Coronilla .	Cannot be distinguished
6954	Coronoid .	**Distress(ed)**
6955	Corozonuts .	In the greatest distress
6956	Corporal . .	There is much distress
6957	Corporally .	Is (are) much distressed
6958	Corporator .	**District**
6959	Corporeity .	In the district
6960	Corporeous .	Grow(s) in the district
6961	Corporify .	In the mining district of
6962	Corpsegate .	The whole district
6963	Corpulency .	The district in which the mine is situated

No.	Code Word	
6964	Corpulent	. **Distrust**
6965	Corpuscles	. No cause for distrust
6966	Corradial	. Much distrust is felt
6967	Correctory	. Is there any cause for distrust
6968	Corregidor	. **Disturbance**
6969	Correspond	. Causes great disturbance
6970	Corridors	. Has caused great disturbance
6971	Corrigent	. The disturbance is owing to
6972	Corrigible	. Owing to the disturbance
6973	Corrigiola	. What has caused the disturbance
6974	Corrodent	. **Divide**
6975	Corrosives	. Will divide
6976	Corrugant	. Will not divide
6977	Corrugator	. Has (have) agreed to divide (the)
6978	Corrupt	. . Ought to divide
6979	Corrupteth	. Will —— divide the
6980	Corruptful	. **Divided**
6981	Corrupting	. Is apparently divided
6982	Corruption	. Divided into
6983	Corruptly	. Divided among
6984	Corsac	. . **Dividend(s)**
6985	Corsairs	. . Dividends will be larger
6986	Corticifer	. Dividend has been declared on
6987	Corticine	. Dividend is passed on
6988	Corticous	. Dividend will be
6989	Cortusa	. Cannot pay dividend [dividend-paying concern
6990	Coruscant	. Am (are) confident that the mine will prove a good
6991	Corvidae	. . Hitherto they have been unable to pay any divi-
6992	Corybant	. Will it pay a dividend [dends on account of
6993	Corybantic	. Will pay a dividend
6994	Corydalina	. Will not pay a dividend
6995	Corylaceae	. Will have to stop paying dividends (as)
6996	Corylus	. . The mines have been good dividend-paying con-
6997	Corymb	. . Dividend will be more than last [cerns
6998	Corymbiate	. Dividend of —— per cent. declared
6999	Corymbous	. Dividend will be less than last
7000	Corypha	. . Including dividend
7001	Coryphaena	. What will be the dividend
7002	Corypheus	. Will probably pay a dividend
7003	Coryphodon	. To declare a dividend of —— per share
7004	Corystes	. . Declare usual dividend
7005	Corystidae	. Dividend to be paid
7006	Cosecant	. Dividend paid
7007	Coseismal	. Dividend will be paid
7008	Cosentient	. Dividend for present quarter
7009	Cosherers	. Dividend for next quarter
7010	Cosmetic	. Dividend for half-year (to ——)
7011	Cosmetical	. Dividend for quarter ending 31st March
7012	Cosmocrat	. Dividend for quarter ending 30th June
7013	Cosmogonal	Dividend for quarter ending 30th September

No.	Code Word.	**Dividend(s)** (*continued*)
7014	Cosmogony .	Dividend for quarter ending 31st December
7015	Cosmolabe .	Dividend at the rate of ——
7016	Cosmology .	Dividend at the rate of 3 per cent.
7017	Cosmometry	Dividend at the rate of 5 per cent.
7018	Cosmorama.	Dividend at the rate of 6 per cent.
7019	Cosmos . .	Dividend at the rate of 7½ per cent.
7020	Cossonus .	Dividend at the rate of 10 per cent.
7021	Cossyphus .	Dividend at the rate of 12 per cent.
7022	Costbook .	Dividend at the rate of 12½ per cent.
7023	Costeanpit .	Dividend at the rate of 15 per cent.
7024	Costellate .	Dividend at the rate of 17½ per cent.
7025	Costfree . .	Dividend at the rate of 20 per cent.
7026	Costive . .	Dividend at the rate of 22½ per cent.
7027	Costively .	Dividend at the rate of 25 per cent.
7028	Costlier . .	Dividend at the rate of 30 per cent.
7029	Costliness .	Dividend of threepence per share
7030	Costsheet .	Dividend of sixpence per share
7031	Cosupreme .	Dividend of eightpence per share
7032	Cosureties .	Dividend of ninepence per share
7033	Cosurety. .	Dividend of tenpence per share
7034	Cotabulate .	Dividend of one shilling per share
7035	Cotangent .	Dividend of 1s. 3d. per share
7036	Cotenants .	Dividend of 1s. 4d. per share
7037	Cothurn . .	Dividend of 1s. 6d. per share
7038	Cothurnate .	Dividend of 1s. 8d. per share
7039	Cotidal . .	Dividend of 1s. 9d. per share
7040	Cotland . .	Dividend of 2s. per share
7041	Cotrustee .	Dividend of 2s. 6d. per share
7042	Cotswold .	Dividend of 3s. per share
7043	Cottagely .	We propose to pay a dividend (of ——)
7044	Cottages . .	We propose to pay an interim dividend
7045	Cottartown .	Interim dividend
7046	Cottidae . .	Interim dividend for quarter
7047	Cottierism .	Interim dividend for half-year
7048	Cottonary .	We can pay a dividend
7049	Cottongins .	We shall recommend a dividend
7050	Cottonlord .	We shall declare a dividend
7051	Cottonous .	What rate of dividend are you paying
7052	Cottonrose .	What rate of dividend are they paying
7053	Cottonweed.	What rate of dividend can you recommend
7054	Cottonwool .	Dividend will be only at the rate of
7055	Cottony . .	Cannot recommend dividend of more than
7056	Coturnix. .	In order that we may be able to determine what dividend we can with prudence recommend
7057	Cotutor . .	Declared a dividend
7058	Cotyla . .	No dividend declared
7059	Cotyliform .	Must pass dividend
7060	Cotyloid . .	If we do not declare a dividend
7061	Couchancy .	If we pay a dividend
7062	Couchgrass .	If we cannot declare more than

No.	Code Word.	**Dividend(s)** (*continued*)
7063	Couchless .	If no dividend is paid
7064	Coughing .	If no dividend is paid we can carry forward
7065	Couguar . .	In order to pay a dividend
7066	Coulter . .	In order to pay a dividend we shall want
7067	Coulterneb .	In order to pay this dividend you must remit
7068	Council . .	Balance left after paying dividend
7069	Councilist .	After dividend is paid
7070	Councillor .	When will dividend be declared
7071	Councilman.	When will dividend be paid
7072	Counited .	Dividend will not be more than
7073	Counselled .	Dividend will not be less than
7074	Countable .	Dividend is quite satisfactory
7075	Counteract .	Dividend is not satisfactory
7076	Countersea .	Much dissatisfaction expressed at small dividend
7077	Countless .	Can you pay a dividend
7078	Countrify .	Ex dividend
7079	Countryman	No dividend has been paid for
7080	Countwheel .	**Do**
7081	Couplings .	Do everything you can to
7082	Courage . .	Do what you think best
7083	Courageous .	Do as I advise
7084	Couriers . .	Do you
7085	Courtamour.	Do nothing
7086	Courtbaron .	Do it
7087	Courtbred .	Do not do it
7088	Courtcards .	I do
7089	Courtcraft .	I do not
7090	Courtdays .	What are you going to **do**
7091	Courtdress .	We do
7092	Courtesies .	We do not
7093	Courtesy. .	You do
7094	Courtesan .	You do not
7095	Courtfool .	They do—(He does)
7096	Courthand .	If we do
7097	Courthouse .	If we do not
7098	Courtledge .	If you do
7099	Courtleet .	If you do not
7100	Courtlike .	If they do—(If he does)
7101	Courtlings .	If they do not—(If he does **not**)
7102	Courtparty .	Whether he does or not
7103	Courtrolls .	Whether you do or not
7104	Courtship .	Do you not
7105	Courtsword .	Do they not
7106	Couscousou	Do all you can
7107	Cousinhood .	Will do very well
7108	Cousinly. . .	When do you
7109	Cousinship .	When do they—(When does he)
7110	Covelline .	Do not
7111	Covenantor .	Do not go
7112	Coverchief .	How do you

No.	Code Word.	**Do** (*continued*)
7113	Coveredway	How do they—(How does he) .
7114	Covereth .	Do no more
7115	Coverlid .	. **Documents**
7116	Coverpoint .	Has (have) the documents
7117	Covershame	Send me (us) the necessary documents
7118	Covertly . .	Deliver the documents
7119	Covertness .	Do not deliver the documents
7120	Covetingly .	Documents are all in order
7121	Covetise . .	Documents are not in order
7122	Covetously .	Has (have) given up the documents
7123	Coward . .	Has (have) not given up the documents
7124	Cowardeth .	Documents forwarded by mail of
7125	Cowardlike .	Documents received
7126	Cowardly .	Documents not received
7127	Cowardous .	Have you received the documents
7128	Cowardship .	Will furnish you with the necessary documents
7129	Cowbane .	Documents will be forwarded by mail of
7130	Cowberries .	The documents must be recorded
7131	Cowberry .	The documents are all in order, and you can go on
7132	Cowblakes .	Documents will be sent [with the business
7133	Cowboys. .	In exchange for the documents
7134	Cowbunting	Until you receive the documents
7135	Cowcalf . .	Until we receive the documents
7136	Cowcatcher .	Unless we get the documents
7137	Cowchervil .	**Doing**
7138	Cowdiepine .	What are you doing about
7139	Cowdoctor .	Doing everything possible with regard to
7140	Cowfeeder .	Doing nothing about
7141	Cowgrass .	By so doing
7142	Cowherb. .	In the event of his (their) doing so
7143	Cowhides .	Not doing anything
7144	Cowhiding .	Doing very well
7145	Cowhouse .	Doing better
7146	Cowkeepers .	There has been very little doing
7147	Cowleech .	Telegraph what is doing in
7148	Cowlstaff .	**Dollars**
7149	Coworkers .	Dollars in cash
7150	Cowparsley .	Dollars per ton
7151	Cowparsnip .	Dollars per cord
7152	Cowpox . .	Dollars per foot
7153	Cowquakes .	Dollars per share
7154	Cowslip . .	How many dollars
7155	Cowslipped .	The equivalent in dollars
7156	Cowwheat .	Dollars at —— pence
7157	Coxalgia. .	Pounds sterling at —— dollars
7158	Coxcombly .	**Dolomite**
7159	Coxcomical .	**Dolorite**
7160	Coxendix .	**Done**
7161	Coyness . .	Must be done
7162	Coystrel . .	Can it be done

No.	Code Word.	**Done** (*continued*)
7163	Coziest	When will it be done
7164	Cozily	What has been done
7165	Crabapple	What have you done with regard **to**
7166	Crabbedly	Will be done immediately
7167	Crabby	It can be done
7168	Crabeater	It cannot be done
7169	Crabfaced	Has (have) been done
7170	Crabgrass	Has (have) not been done
7171	Crablouse	Will have to be done
7172	Crabsidle	Nothing can be done (owing to)
7173	Crabstick	Will endeavour to get it done
7174	Crabtree	A great deal has been done (to)
7175	Crabwood	Very little has been done (to)
7176	Crabyaws	Nothing has yet been done (to)
7177	Cracidae	Ought to be done (at once)
7178	Crackhemp	Will be done
7179	Crackle	Much yet remains to be done
7180	Crackling	Something must be done immediately
7181	Cracknel	If not already done
7182	Crackrope	Nothing can be done until
7183	Crackskull	There is nothing more to be done
7184	Cracksman	Have you done
7185	Cracowes	Has he (have they) done
7186	Cradlebabe	Whether done or not
7187	Cradled	**Double**
7188	Cradlewalk	Double the quantity (or amount)
7189	Craftiest	(To) double the
7190	Craftily	(To) double the capacity
7191	Cragged	(To) double the yield
7192	Cragginess	(To) double the output
7193	Craigsman	Has been doubled
7194	Crakys	**Doubt**
7195	Crambus	Have you any doubt
7196	Crameria	Have no doubt
7197	Crammer	Have great doubt
7198	Cramming	Is there any doubt about
7199	Crampbark	There is no doubt (about) **that**
7200	Crampbone	There is a doubt about
7201	Cranberry	Beginning to doubt
7202	Craneflies	Doubt if we can (do)
7203	Cranefly	**Down**
7204	Cranesbill	The vein goes down
7205	Crangon	As we go down
7206	Craniology	Going down
7207	Crankbird	Down to
7208	Crankhook	Down to water level
7209	Crankpin	Has (have) not gone **down**
7210	Crannied	Has (have) gone down
7211	Crapaudine	**Downward**
7212	Crapefish	A downward tendency

No.	Code Word.	
7213	Crapes .	. **Draft(s)**
7214	Crapulence .	Date of draft
7215	Crapulous .	Draft payable
7216	Craspedota .	Draft dated —— at —— months
7217	Crassament .	Draft is on the way
7218	Crassitude .	Draft paid
7219	Crassness .	Will you pay draft for —— (at)
7220	Crassula . .	I will pay draft for —— (at)
7221	Crataeva . .	Draft provided for
7222	Craterous .	Send me (us) draft upon
7223	Craunch . .	Have sent you draft upon
7224	Cravatted .	Will send you draft upon
7225	Crawl . . .	Your draft will be duly met
7226	Crawling .	Have honoured your draft
7227	Crawlingly .	Your draft for —— has been dishonoured (at)
7228	Crayons . .	Draft dishonoured ; drawer had no authority to draw
7229	Crazedness .	All drafts will be duly honoured
7230	Crazemill .	Draft has been dishonoured in consequence of time
7231	Cream . .	Please pay draft [for delivery having elapsed
7232	Creamcake .	Please provide for draft
7233	Creamery .	Draft dishonoured because you departed from in-
7234	Creamfaced .	Draft dishonoured, there being no funds [structions
7235	Creamfruit .	Why was draft dishonoured
7236	Creaminess .	**Drain(age)**
7237	Creamlaid .	In order to drain the mine
7238	Creamnut .	To drain the
7239	Creampots .	Free drainage to depth of
7240	Creamslice .	**Drained**
7241	Creamwhite .	The mine has been drained
7242	Creamwove .	The mine cannot be drained
7243	Creasing . .	The mine is drained by a tunnel
7244	Creatable .	When the mine is drained
7245	Createth . .	Can you get the mine drained
7246	Creatic . .	To get the mine drained will require
7247	Creation . ,	The mine is drained by
7248	Creatural .	If the mine is to be drained
7249	Creaturely .	**Draining**
7250	Creaturize .	Effectually draining the mine
7251	Crebritude .	**Draw** (See also Sight.)
7252	Credencing .	Draw on me (us)
7253	Credendum .	Draw on me (us) for investment
7254	Credent . .	Draw on me (us) for fee
7255	Credential .	At what date may I draw
7256	Credible . .	On whom shall I (we) draw
7257	Creditably .	Arrange to draw upon
7258	Creditress .	May I (we) draw upon you (for)
7259	Creditrix . .	Have had to draw on
7260	Credulity .	Draw on —— for the amount of
7261	Credulous .	Draw at —— days' date
7262	Creeper . .	You may draw for

No.	Code Word.	**Draw** (*continued*)
7263	Creephole .	Draw at —— days' sight
7264	Creeping. .	Draw at —— date
7265	Cremate . .	You may not draw for
7266	Cremating .	Do not draw on us
7267	Cremocarp .	Draw with bill of lading attached
7268	Cremona. .	To what amount may I (we) draw
7269	Cremosine .	Not authorized to draw
7270	Crenatula .	Cannot draw out
7271	Crenelets .	Authorized to draw
7272	Crenellate .	Draw for the amount (£——) attaching certificates
7273	Crenelled .	**Drawback** [—— bank
7274	Crenulated .	The drawback is
7275	Crepitant .	A most serious drawback
7276	Crepuscle .	**Drawing**
7277	Crescentia .	Send drawing showing
7278	Crescive. .	Have sent drawing showing
7279	Cresol . .	Drawing of the engine
7280	Crestless. .	Drawing of the mill
7281	Cresttile . .	Drawing of the
7282	Crevices . .	General drawing
7283	Cribbage. .	Working drawing
7284	Cribbiter. .	Detail drawing
7285	Cribration .	Drawing showing in some detail .
7286	Cribrose . .	**Drawn**
7287	Cricket . .	Have drawn on
7288	Cricketbat .	Have drawn upon you for —— through —— at
7289	Cricketers .	Have drawn on ——, advise us by wire if draft is
7290	Cricoid . .	To what amount have you drawn [duly met
7291	Crime. . .	**Drift(s)**
7292	Crimeless .	Drift from the
7293	Criminally .	Drift from the incline is in
7294	Criminate .	Drift from the shaft is in
7295	Criminous .	How is drift looking
7296	Crimpage .	Drift is now in
7297	Crimson . .	North drift on the —— feet level (is in)
7298	Crimsoning .	South drift on the —— feet level (is in)
7299	Crinal . .	East drift on the —— feet level (is in)
7300	Crinatory . .	West drift on the —— feet level (is in)
7301	Crincun . .	The bottom drift looks well for ore
7302	Cringed . .	A sample taken from the drift assayed
7303	Cringeling .	It only appears at the bottom of the drift
7304	Cringingly .	The lower drift is looking
7305	Crinoidal .	How is the lower drift looking
7306	Crinosity. .	Drift from the —— level
7307	Criosphinx .	Drift from —— shaft
7308	Cripple . .	North drift
7309	Crippling .	South drift
7310	Crisis. . .	East drift
7311	Crispate . .	West drift
7312	Crispature .	Side drift

No.	Code Word.	**Drift(s)** (*continued*)
7313	Crisply . .	Bottom drift
7314	Crisscross .	Drift on side lode
7315	Criterion. .	Drift has advanced
7316	Crithmum .	Drift has been run in very hard ground
7317	Critically .	Drift has been run in ore
7318	Criticisms .	Drift has been run in
7319	Criticizer .	Drift is looking better
7320	Critiques. .	Drift is not looking so well
7321	Croaked . .	Drift looks well for ore
7322	Croceous .	Drift is now in a large body of ore
7323	Crockery. .	Drift still continues in barren ground
7324	Crocketed .	Drift still continues in poor ore
7325	Crocodilia .	Drift still continues in rich ore
7326	Crocoisite .	Drift passed through ore
7327	Croconate .	Drift passed through ore, giving high assays
7328	Croconic. .	Drift has been in barren ground, but has now got
7329	Crocuses. .	Shall drift on the vein [in ore
7330	Crookback .	Shall drift both ways
7331	Crookedly .	Shall drift to connect
7332	Cropeared .	Have connected by a drift
7333	Cropfull . .	Drift connecting —— and —— shaft
7334	Cropout . .	Drift connecting
7335	Cropping . .	Drift to connect
7336	Crossaisle .	We are now running a drift from —— to
7337	Crossarmed .	Shall push forward drift with all possible speed
7338	Crossarrow .	Hope to have drift completed within
7339	Crossbar. .	When do you expect drift to reach
7340	Crossbill. .	Bottom of drift
7341	Crossbirth .	Top of drift
7342	Crossbones .	Side of drift
7343	Crossbred .	Face of drift
7344	Crosschock .	Average value of ore in drift
7345	Crosscut. .	This drift does not show much promise as yet
7346	Crossdays .	Drift is showing signs of getting into
7347	Crosseyed .	Ore in this drift improving in value
7348	Crossfire. .	Ore in this drift is giving out
7349	Crossflow .	Work has been discontinued in this drift
7350	Crosshead .	**Drill(s)**
7351	Crossings .	Machine drills are doing excellent work
7352	Crossjack .	Machine drills are working badly
7353	Crosslode .	Machine rock-drill
7354	Crossly . .	Diamond drill
7355	Crosspatch .	Air drills
7356	Crosspawl .	The drills we are using
7357	Crossroads .	How many drills are you using
7358	Crosstail. .	What kind of drill are you using
7359	Crosswinds .	What drill do you consider best
7360	Crosswise .	The most suitable drill
7361	Crosswort .	Cost of the drills
7362	Crotalaria .	Estimated saving in labour by the use of drills

No.	Code Word.	**Drill(s)** (*continued*)
7363	Crotalidae .	The best drill for our purpose
7364	Crotalo . .	Repairs to drills
7365	Crotchety .	**Drive(s)**
7366	Croton . .	To drive on the vein
7367	Crotonoil .	From various drives
7368	Crotophaga .	**Driven**
7369	Crottles . .	Have driven in
7370	Crouchback .	Have driven in upon the vein for —— feet
7371	Crowbars .	Driven to connect
7372	Crowding .	How many feet have you driven (on)
7373	Crowds . .	Driven to the hanging-wall
7374	Crowflower .	Driven to the foot-wall
7375	Crowfoot . .	Driven —— feet
7376	Crownagent .	How far have you driven
7377	Crowncourt .	How far have you driven in the direction of
7378	Crowneth .	Have driven through the
7379	Crownglass .	Have driven along
7380	Crownhead .	**Driving**
7381	Crownpiece .	Have commenced driving
7382	Crownpost .	Have stopped driving in —— (on account of)
7383	Crownsaw .	Driving through good ore
7384	Crownscab .	Driving through poor ore
7385	Crownside .	Driving through country rock
7386	Crownwheel	Driving through granite
7387	Crowquill .	Driving through barren quartz
7388	Crowsbill .	Driving through slate
7389	Crowsilk . .	Driving through soft ground
7390	Crowsnest .	Driving through very hard ground
7391	Crowstone .	Driving through a dyke
7392	Croziered .	Driving through a horse
7393	Crozophora .	Driving towards
7394	Crucial . .	Driving in both directions
7395	Crucibles .	In which direction are you now **driving**
7396	Crucifer . .	Now driving on level
7397	Crucifixes .	We are now driving at the rate of
7398	Crucifying .	**Dry(ing)**
7399	Crudities . .	The mines are always dry
7400	Crudity . .	The mine is quite dry
7401	Cruel . . .	Dry weather
7402	Cruelly . .	On account of the dry weather
7403	Cruelness .	Tons of dry ore
7404	Cruentate .	Dry crushing
7405	Cruentous .	Dry or wet crushing
7406	Cruetstand .	Drying-floor
7407	Crumb . .	Revolving dryers
7408	Crumbbrush	**Dump(s)**
7409	Crumbcloth .	An abundance of ore on the dump
7410	Crumbling .	Tons of ore on the dump
7411	Crumbly . .	Tons of ore from the dump
7412	Crummable .	Assays from the dump averaged

No.	Code Word.	Dump(s) (*continued*)
7413	Crummock .	Measurement of the dump showed it to contain
7414	Crumpled .	Old waste dumps [about —— tons
7415	Crunched .	**Duplicate**
7416	Cruorin . .	Send a duplicate copy of
7417	Crusader. .	Have sent you a duplicate
7418	Crusading .	Will send you a duplicate copy (of the)
7419	Crusheth .	**Duty**
7420	Crushhat. .	Neglects his duty
7421	Crushroom .	Is unable to attend to his duty
7422	Crustacea .	Has returned to his duty
7423	Crustalogy .	What is the rate of duty on
7424	Crustation .	The duty on
7425	Crustific . .	Doing better duty
7426	Crustily . .	Can do better duty
7427	Crustiness .	Cannot do better duty
7428	Crutch . .	As good duty as
7429	Cryer. . .	The engine will do a duty of
7430	Cryolite . .	The engine is doing a duty of
7431	Cryophorus .	What duty is the engine doing
7432	Crypt. . .	Is doing his duty
7433	Cryptic . .	We are only doing our duty
7434	Cryptical. .	You will be only doing your duty
7435	Cryptogam .	Chargeable with duty
7436	Cryptology .	An *ad valorem* duty of
7437	Cryptonym .	The duty on machinery
7438	Crystallin .	**Dwelling-house**
7439	Cubation .	There is a good superintendent's dwelling-house
7440	Cubatory .	There is no manager's dwelling-house
7441	Cubature .	There is a good dwelling-house for
7442	Cubbyhole .	There are —— dwelling-houses
7443	Cubdrawn .	The dwelling-houses are built of
7444	Cubeba . .	Has a dwelling-house
7445	Cubeore . .	The dwelling-houses are in good condition
7446	Cubespar .	The dwelling-houses are in a very bad condition
7447	Cubical . .	Dwelling-house(s) for the staff
7448	Cubically .	Dwelling-house(s) for the miners
7449	Cubicular .	**Dynamite**
7450	Cubiform .	Are using dynamite
7451	Cubited . .	The dynamite used
7452	Cubocube .	Tons of dynamite
7453	Cuboidal. .	Pounds of dynamite
7454	Cuckold . .	Dynamite for blasting
7455	Cuckoldize .	Explosion of dynamite
7456	Cuckoldly .	**Dyke**
7457	Cuckoldom .	Slate dyke
7458	Cuckoo . .	Granite dyke
7459	Cuckoobud .	Porphyry dyke
7460	Cuckoospit .	Vein here intersected by a dyke
7461	Cucubalus .	**Dynamo**
7462	Cuculidae .	Dynamo for electric lighting

No.	Code Word.	Dynamo (*continued*)
7463	Cucullaris .	Want a new dynamo
7464	Cucumbers .	Put down a new dynamo
7465	Cucumiform	Alteration in the dynamos
7466	Cucurbit . .	The dynamo(s) now in use
7467	Cucurbital .	Dynamo does not work well
7468	Cudbear . .	A dynamo of greater power
7469	Cudgelled .	Dynamo requires
7470	Cudgelling .	**Dynamometer**
7471	Cudgelplay .	By testing with dynamometer
7472	Cudweed. .	Trials with the dynamometer
7473	Cueball . .	**Each**
7474	Cueing . .	In each case
7475	Culicidae .	Have visited each
7476	Culilawan .	Keep the details of each mine separate and distinct
7477	Culinary . .	Each of (them) [in your report
7478	Cullied . .	Each of the mines
7479	Cullionly. .	Each of the workings
7480	Cullizan . .	For each (of them)
7481	Cully . . .	Of each (of them)
7482	Cullyism · .	The price of each
7483	Culottic . .	Each and every one
7484	Culpable . .	If each of them
7485	Culpatory .	**Eager**
7486	Culprit . .	Is (are) very eager to
7487	Cultivator .	Is (are) not very eager to
7488	Cultrate . .	Do not appear too eager
7489	Culturable .	Is (are) —— eager to
7490	Culturist . .	**Eagerness**
7491	Cultus . .	Shows too much eagerness
7492	Culver . .	**Earliest**
7493	Culverkey .	Earliest possible
7494	Culvertail .	Shall take the earliest opportunity
7495	Cumbent .	By earliest possible mail
7496	Cumberless .	**Early**
7497	Cumbersome	Send an early answer
7498	Cumbi . .	As early as possible
7499	Cumbrance .	Hope to get early to work
7500	Cumbrous .	Not early enough
7501	Cumbrously .	It is too early yet to
7502	Cuminol . .	Much too early
7503	Cummerbund	Not so early
7504	Cumshaw .	At an early date
7505	Cumulate .	At an earlier date (than)
7506	Cumulating .	Early next month
7507	Cumulation .	Early next
7508	Cumulose . .	Take an early opportunity (to)
7509	Cumyl . .	**Earnest**
7510	Cunctation .	In earnest
7511	Cunctative .	Not in earnest
7512	Cuneal . .	As earnest of our intention

No.	Code Word.	**Earnest** (*continued*)
7513	Cuneatic . .	Thoroughly in earnest
7514	Cuniculous .	**Earnestly**
7515	Cunning . .	Impress most earnestly on
7516	Cunningman	Is (are) most earnestly
7517	Cupbearer .	**Earnings**
7518	Cupboard .	What are the net earnings (of)
7519	Cupel . . .	The net earnings are
7520	Cupeldust .	The gross earnings (are)
7521	Cupful . .	The amount of earnings
7522	Cupgall . .	Increase in earnings
7523	Cupidity . .	Decrease in earnings
7524	Cupmoss. .	The earnings last year (were)
7525	Cupressite .	Earnings during the last six months
7526	Cuproid . .	The earning capacity
7527	Cupvalve .	The total earnings during the past —— were
7528	Curability .	**Earthquake(s)**
7529	Curacy . .	The district is subject to frequent earthquakes
7530	Curassow '.	Severe shock of earthquake
7531	Curateship .	The earthquake has caused but little damage
7532	Curators . .	Suffered much damage from the earthquake which
7533	Curatrix . .	**Easier** [occurred
7534	Curba . .	Much easier than
7535	Curbless . .	We require easier terms
7536	Curbplate .	Arrange for easier terms
7537	Curbroof . .	**Easily**
7538	Curbsender .	Can you easily
7539	Curbstone .	Can easily
7540	Curculio . .	Can easily do what you want
7541	Curcuma . .	Cannot easily
7542	Curcumine .	**East** (See Table at end)
7543	Curdiness .	**Easy**
7544	Curdled . .	Easy payments
7545	Curdling . .	Arrange for easy payments
7546	Curdog . .	Balance by easy payments
7547	Curfew . .	Easy payments extending over
7548	Curiosity . ;	On easy terms
7549	Curious . .	It will be easy to
7550	Curlcloud .	It is not so easy
7551	Curledness .	Will not be so easy as you expect
7552	Curledpate .	Less easy than
7553	Curlypated .	**Economical**
7554	Curmudgeon	Not so economical as
7555	Currants . .	(Is) more economical than
7556	Currencies .	A more economical process (or method)
7557	Currency . .	Some more economical method must be adopted
7558	Curricle . .	The most economical (method)
7559	Currishly .	**Economize**
7560	Currycomb .	It is necessary to economize
7561	Cursedly . .	So as to economize
7562	Curship . .	**Economy**

No.	Code Word.	Economy (*continued*)
7563	Cursitor . .	With economy
7564	Cursive . .	Use every economy
7565	Cursively .	Every economy will be used
7566	Cursorary .	Every economy must be used
7567	Cursorial. .	Cannot you effect economies in
7568	Curstful . .	With too great economy
7569	Curstfully .	The mine has been starved from too great economy
7570	Curtail . .	Economy in
7571	Curtaildog .	Effect (of)
7572	Curtailing .	The effect will be
7573	Curtalax . .	What will be the effect of
7574	Curtana . .	To take effect on
7575	Curtly . .	This will have the effect of
7576	Curtness . .	May have some effect
7577	Curtsied . .	Do not know what the effect might be
7578	Curtsying .	To this effect
7579	Curule . .	To the same effect
7580	Curvature .	To the effect that
7581	Curvetted .	Unable to effect
7582	Curvetting .	The effect will be favourable
7583	Curvity . .	The effect will be serious
7584	Curvograph.	Cannot say what the effect will be
7585	Cuscobark .	Do you think it will have the desired effect
7586	Cushion . .	Had the desired effect
7587	Cushioning .	The effect of the arrangement
7588	Cusparine .	A better effect
7589	Cuspidate .	The desired effect
7590	Custard . .	Will probably have some effect
7591	Custodian .	The effect upon the
7592	Custom . .	The effect upon the price of the shares
7593	Customable .	A disastrous effect
7594	Customary .	Will have a good effect
7595	Custometh .	Will have a bad effect
7596	Cutaneous .	Carry this into effect
7597	Cutaway . .	Carried into effect
7598	Cutgrass . .	Hope you will give effect to this
7599	Cutlasses. .	Please to give effect to
7600	Cutlery . .	Effective
7601	Cutoff . .	The effective fall
7602	Cutpurse .	The effective horse-power
7603	Cutterbar .	If it is not effective
7604	Cutthroats .	Not so effective as
7605	Cuttlebone .	Effectual(ly)
7606	Cuttlefish .	Will this effectually (or) Will this be effectual
7607	Cuttystool .	This will effectually (or) This will be effectual
7608	Cutwater .	Fear this will not effectually
7609	Cutworm .	Effort
7610	Cyamidae .	Every effort must be made (to)
7611	Cyanamide .	Every effort will be made (to)
7612	Cyanhydric .	Every effort has been made (to)

No.	Code Word.	**Effort** (*continued*)
7613	Cyanogen .	Is (are) making no effort to
7614	Cyanometer.	Spare no effort to
7615	Cyanopathy .	**Either**
7616	Cyanosis. .	Either will do
7617	Cyanotype .	Either one or the other
7618	Cyanurate .	Either of them
7619	Cyanuric. .	Either course can be adopted
7620	Cybium . .	If either of them was (or) (should be)
7621	Cycad . .	Either that or
7622	Cycadiform .	**Elapse**
7623	Cycadite. .	The bond will elapse
7624	Cyclamen .	The mine was bonded for ——, which will elapse on
7625	Cyclantha .	Has (have) allowed the time to elapse
7626	Cyclical . .	Do not let the time elapse
7627	Cycling . .	When does time elapse
7628	Cyclograph .	**Elapsed**
7629	Cycloid . .	Has not elapsed
7630	Cycloidal .	Will have elapsed
7631	Cyclolith .	Will not have elapsed
7632	Cyclometry .	Must be done before the time has elapsed
7633	Cyclonic. .	The time has elapsed
7634	Cyclopean .	As the time has elapsed for which
7635	Cyclopedic .	**Elected**
7636	Cyclostoma .	Has (have) been elected
7637	Cyder . .	Has (have) not been elected
7638	Cydippe . .	Who has been elected (to)
7639	Cydonia . .	Has (have) elected to
7640	Cyesiology .	Has (have) elected not to
7641	Cygnet . .	What has (have) —— elected to
7642	Cygninae .	**Election**(s)
7643	Cylindric .	The election of —— as
7644	Cylindroid .	During the elections
7645	Cymbal . .	After the elections
7646	Cymbalist .	The result of the elections
7647	Cymbella .	**Electric** (See Table at end)
7648	Cymbidium .	The electric light
7649	Cymbiform .	The electric light (is) used in the mill
7650	Cymiferous .	The electric light (is) used in the mine
7651	Cyminum .	The electric light (is) generally used
7652	Cymophane .	System of electric lighting
7653	Cymous . .	By the adoption of the electric light
7654	Cynanchum .	The electric wires
7655	Cynara . .	Incandescent electric light(s)
7656	Cynegetics .	Arc electric light(s)
7657	Cynical . .	Electric transmission
7658	Cynicisms .	Electric storage
7659	Cynictis . .	Electric batteries
7660	Cynipidae .	Electric machine for blasting
7661	Cynogale .	Electric fuses
7662	Cynography .	**Electricity**

No.	Code Word.	**Electricity** (*continued*)
7663	Cynomorium	Lighting by electricity
7664	Cynorexia .	Blasting by electricity
7665	Cynosure .	Transmission by electricity
7666	Cynthia . .	Introduction of electricity
7667	Cyophoria .	Light the mine by electricity
7668	Cyphelia .	Light the mill by electricity
7669	Cyphonidae.	By electricity
7670	Cyphonism .	(To) light by electricity
7671	Cypraea . .	**Elevation** [sea
7672	Cypress . .	The elevation of the mine(s) is —— feet above the
7673	Cyprinus. .	What is the elevation of the mine(s) above the sea
7674	Cypruslawn .	At this elevation
7675	Cypselidae .	At what elevation
7676	Cypselus .	At an elevation of
7677	Cyrillic . .	Elevation of the
7678	Cyriologic .	Plan and elevation of the [sea
7679	Cyrtostyle .	At an elevation of —— feet above the level of the
7680	Cysticles. .	At an elevation of —— feet above the river (or creek)
7681	Cystidean .	At an elevation of —— feet above the valley
7682	Cystitis . .	At an elevation of —— feet above the mill
7683	Cystitome .	At what elevation is the —— above the
7684	Cystocarp .	**Elevator**
7685	Cystocele .	Elevator for
7686	Cystolith .	**Elicit**
7687	Cystose . .	Can you elicit from —— whether
7688	Cystotomy .	Trying to elicit some information
7689	Cystula . .	**Elicited**
7690	Cytinaceae .	Have you elicited anything from
7691	Cytisus . .	Has (have) elicited from —— that
7692	Cytoblast .	Has (have) not elicited anything from
7693	Cytogeny .	**Else**
7694	Czarinian .	Anything else
7695	Czarish . .	**Elsewhere**
7696	Dabblers .	Have to send elsewhere
7697	Dabchicks .	Cannot apply elsewhere
7698	Dabeocia .	Try elsewhere
7699	Dabster . .	Is it done elsewhere
7700	Dacelo . .	It has been done elsewhere
7701	Dacoity . .	**Emanated**
7702	Dacrydium .	The report emanated from [report emanated
7703	Dacryolite .	Has (have) not been able to find out from whom the
7704	Dacryoma .	From whom has this report emanated
7705	Dactyl . .	Has (have) emanated from
7706	Dactylar. .	Has (have) not emanated from
7707	Dactylion .	We hear that the rumour emanated from
7708	Dadoes . .	The report emanated from the usual clique
7709	Dadoxylon .	**Emancipated**
7710	Daemonic .	Emancipated from all control
7711	Daffodil . .	Emancipated from the control of
7712	Daggers . .	**Emancipation**

No.	Code Word.	**Emancipation** (*continued*)
7713	Daggletail .	Emancipation from control
7714	Dagswain .	**Embargo**
7715	Dagtailed .	Place an embargo upon
7716	Daguerrean .	Placed an embargo upon
7717	Dahabieh .	Embargo removed
7718	Dahlias . .	Removal of embargo
7719	Daintiest . .	**Embarrass**
7720	Daintify . .	Greatly embarrass(es) us
7721	Daintiness .	Will greatly embarrass us
7722	Daintrel . .	**Embarrassed**
7723	Dairyfarm .	Seriously embarrassed (by)
7724	Dairyhouse .	**Embarrassment**
7725	Dairying . .	Caused great embarrassment
7726	Dairymaid .	Likely to cause great embarrassment
7727	Dairyroom .	Will cause us great embarrassment if we are com-
7728	Daisied . .	Is causing great embarrassment [pelled to
7729	Daisy . . .	Financial embarrassment
7730	Dakerhen .	Embarrassments have been caused by
7731	Dalbergia .	**Embezzled**
7732	Dalesman .	Has embezzled
7733	Dallying . .	**Embezzlement**
7734	Dalmahoy .	Has committed embezzlement
7735	Dalmatica .	**Embrace**(ing)
7736	Dalriadic .	Will embrace
7737	Daltonian .	Embracing every variety of
7738	Damage . .	**Emerald**(s)
7739	Damageable	The emeralds are found in
7740	Damaging .	Emerald mines
7741	Damajavag .	You will have to go to the emerald mines of
7742	Damarresin .	Emeralds have been found
7743	Damask . .	Emeralds found, but do not pay cost of working
7744	Damaskins .	Emeralds found, but not plentiful
7745	Damaskplum	Emeralds (found) of poor quality
7746	Damaskrose	Emeralds and rubies
7747	Dambonite .	**Emergency**
7748	Damewort .	In case of emergency
7749	Damianist .	In case of emergency, we recommend you
7750	Dammara .	**Employ**
7751	Dammed .	Authorize you to employ
7752	Damnably .	It will be necessary to employ
7753	Damnatory .	The company employs —— men
7754	Damnific .	How many men does the company employ
7755	Damnifying .	You had better employ some other
7756	Damoclean .	You had better employ some one to
7757	Damoiselle .	You had better employ
7758	Dampish .	How many men is it necessary to employ
7759	Dampishly .	Must employ for this work good English miners
7760	Damplate .	Can you employ
7761	Dampness .	Can employ
7762	Dampoff . .	Cannot employ

No.	Code Word.	Employ (*continued*)
7763	Dampy . .	Do not employ
7764	Damsels . .	Do not employ any one, unless he is fully competent
7765	Damstone .	Do not employ any one, unless
7766	Dance . .	If you can employ
7767	Dancemusic	If you cannot employ
7768	Dancing . .	Can employ to better purpose
7769	Dandelion .	(To) employ the funds
7770	Dandified .	(To) employ the capital
7771	Dandiprat .	(To) employ the surplus
7772	Dandled . .	(To) employ the balance
7773	Dandruff .	Employ some one (we) you can trust
7774	Dandycock .	Employ a competent, trustworthy man
7775	Dandyhen .	Employ for the purpose named
7776	Dandyish .	Employ a first-class lawyer
7777	Dandyize .	Employ the best aid
7778	Dandylings .	Employ the best expert
7779	Danebrog .	Employ the best counsel
7780	Danegelt .	Employ on our behalf
7781	Dangerous .	Employ all possible means to
7782	Danglers .	. **Employed**
7783	Dantesque .	Has (have) employed
7784	Dapatical .	Has (have) not employed
7785	Dapedium .	Has (have) —— employed
7786	Daphne . .	Has (have) been employed
7787	Daphnidae .	Has (have) not been employed
7788	Dapifer . .	How many men are employed (in)
7789	Dapperling .	Must be employed
7790	Dapplebay .	Employed in —— interests
7791	Darapti . .	Was employed by (in)
7792	Darbyites .	(Have) employed the funds
7793	Daredevil .	(Have) employed the man (or men)
7794	Daringly . .	(Have) employed the amount
7795	Daringness .	(Have) employed the balance
7796	Darkened .	The capital employed
7797	Darkful . .	The working capital employed
7798	Darkish . .	The funds employed
7799	Darkly . .	The men employed
7800	Darkness .	Good native miners can be employed
7801	Darksome .	Natives áre employed
7802	Darnix . .	Are employed, chiefly
7803	Darootree .	If you have employed
7804	Dartles . .	If you have not employed
7805	Dartoid . .	Why have you employed
7806	Dartrous . .	Why have you not employed
7807	Dartsnake .	Can be employed to better purpose
7808	Darwinians .	Have employed all possible means (to)
7809	Darwinism .	Every one fully employed
7810	Dashboard .	**Employing**
7811	Dashism . .	We are now employing
7812	Dashwheel .	We think of employing

No.	Code Word.	Employing (*continued*)
7813	Dastardly .	In the event of employing
7814	Dasymeter .	**Employment**
7815	Dasyornis .	In whose employment was he
7816	Dasypidae .	The terms of his employment
7817	Dasyprocta .	In our employment
7818	Dasypus . .	In —— employment
7819	Dasyure . .	His employment in our interests
7820	Dateless . .	In the event of his employment
7821	Datepalm .	The employment of counsel
7822	Dateplums .	The employment of an expert
7823	Datesugar .	The employment of a first-rate lawyer
7824	Datiscine .	The employment of a surveyor
7825	Datisi . . .	The employment of a mining engineer
7826	Datura . .	**Empower**
7827	Daubery . .	Will you empower us to
7828	Daughterly .	Will you empower us to employ
7829	Daunt . .	Empower you to act
7830	Dauntless .	Empower you to
7831	Dauphiness .	Empower you to negotiate
7832	Davallia . .	**Empowered**
7833	Davenport .	You are empowered to
7834	Davidist . .	Is (are) empowered to
7835	Davina . .	Is (are) empowered to negotiate
7836	Davylamp .	Empowered by Power of Attorney
7837	Dawcock .	Fully empowered under the terms of
7838	Dawdling .	Not empowered
7839	Daybed . .	**Empowering**
7840	Daybooks .	Empowering you to
7841	Daybreak .	Empowering me (us) to
7842	Daycoal . .	Empowering him (them) to
7843	Daydreamer	Empowering —— to　　　　　[to act for us
7844	Daylabour .	(Sending out) Power of Attorney, empowering you
7845	Daylight . .	Empowering you to act on our behalf
7846	Daylilies . .	**Enable**
7847	Daylily . .	Will enable me (us) to judge
7848	Daylong . .	Will enable me (us) to
7849	Daypeep .	Will enable you to
7850	Dayroom .	Will this enable
7851	Dayrule . .	To enable —— to form an opinion
7852	Dayschool .	To enable —— to do what is wanted
7853	Dayshine .	**Enabling**
7854	Dayspring .	Enabling us to
7855	Daystar . .	Enabling you to
7856	Daytime . .	Enabling him (them) to
7857	Daywoman .	**Encouraged**
7858	Daywork . .	We are encouraged to hope
7859	Dazzle . .	Will be encouraged
7860	Dazzlement .	**Encouragement**
7861	Dazzlingly .	Do the mines give any encouragement
7862	Deaconess .	There is no encouragement

No.	Code Word.	**Encouragement** (*continued*)
7863	Deaconhood	There is every encouragement
7864	Deaconry .	Is there any encouragement
7865	Deaconship.	There is no encouragement to induce us to
7866	Deadangle .	There is but little encouragement to persevere
7867	Deadbeat .	The report holds out great encouragement
7868	Deadbell .	Without any encouragement [to expect
7869	Deadcentre .	Have not received the encouragement we were led
7870	Deadened .	If we can get sufficient encouragement
7871	Deadening .	Sufficient encouragement
7872	Deadflat . .	**Encouraging**
7873	Deadground	Report is very encouraging
7874	Deadhedge .	Report is not encouraging
7875	Deadhorse .	The tone of your letter is not encouraging
7876	Deadletter .	The purport of your letter is most encouraging
7877	Deadlift . .	Most encouraging
7878	Deadlihood .	Not at all encouraging
7879	Deadliness .	**Encroached** (on)
7880	Deadlock .	Has (have) encroached on the resources
7881	Deadneap .	Has (have) encroached on the reserve
7882	Deadnettle .	(Has) have encroached considerably (on)
7883	Deadoil . .	**Encroaching** (on)
7884	Deadpay .	Is (are) encroaching on
7885	Deadpoint .	Is (are) not encroaching on
7886	Deadripe .	Is (are) —— encroaching on
7887	Deadrising .	Encroaching on the resources
7888	Deadsheave.	Encroaching on the reserve
7889	Deadshot .	**Encroachment** (on)
7890	Deadwall .	Do not allow any encroachment (on)
7891	Deadwater .	(To) prevent any encroachment (on)
7892	Deadweight.	(To) stop at once the encroachment (on)
7893	Deadwood .	**End**
7894	Deafen . .	In the end
7895	Deafmute .	At the end of
7896	Dealbate .	An end of
7897	Dealbation .	To bring it to an end
7898	Dealfish . .	The end of
7899	Dealtree . .	To the end that
7900	Dealwine .	Bring the matter to an end
7901	Deambulate	There is an end of the matter
7902	Deaneries .	To the bitter end
7903	Deanship .	**Endeavour(s)** (to)
7904	Dearborn .	Use your best endeavours to
7905	Dearbought.	Will endeavour
7906	Dearest . .	Endeavour to ascertain
7907	Dearthful .	Endeavour to ascertain and let us know by telegraph
7908	Deathagony.	Endeavour to complete
7909	Deathbed .	**Endeavoured** (to)
7910	Deathblow .	Have endeavoured to
7911	Deathcord .	Has (have) not endeavoured (to)
7912	Deathdamp.	Have you endeavoured (to)

No.	Code Word.	
7913	Deathdance.	**Endeavouring (to)**
7914	Deathfire .	We are now endeavouring to
7915	Deathify . .	He is (they are) now endeavouring to
7916	Deathless .	Has (have) been endeavouring to
7917	Deathrate .	Has (have) been endeavouring for some time past to
7918	Deathsdoor.	Has (have) been endeavouring for some time to
7919	Deaththroe .	Endeavouring to raise capital [dispose of
7920	Deathtoken .	Endeavouring to form a company
7921	Deathward .	Endeavouring to raise sufficient funds
7922	Deathwatch.	**Ended**
7923	Deathwound	Has ended in
7924	Deathy . .	**Endorse**
7925	Deauration .	Do you endorse what —— states in his report
7926	Debacchate.	We endorse what is stated in
7927	Debarment .	Endorse what is stated
7928	Debarrass .	Refused to endorse
7929	Debarred .	**Endorsed**
7930	Debaseth .	Have endorsed
7931	Debasingly .	Have not endorsed
7932	Debated . .	Has been endorsed by
7933	Debateful .	By whom endorsed
7934	Debauchery.	Acceptance endorsed by
7935	Debellate .	**Endorsement**
7936	Debenture .	The endorsement is good
7937	Debilitant .	The endorsement is not good enough
7938	Debility . .	Endorsement guaranteed
7939	Debiting. .	Require the endorsement of
7940	Debonairly .	Endorsement irregular
7941	Debruised .	If the endorsement is
7942	Debtless . .	**Enforce**
7943	Debtors . .	You must enforce on —— the necessity of
7944	Decachord .	Will enforce
7945	Decade . .	Must enforce
7946	Decagon. .	Enforce the claim
7947	Decagonal .	Enforce the undertaking
7948	Decagynia .	Enforce the covenant
7949	Decagynous.	Take prompt steps to enforce
7950	Decahedron.	Do not enforce
7951	Decaisnea .	**Enforced**
7952	Decalcify .	Have you enforced
7953	Decalogist .	Has (have) enforced
7954	Decamped .	Must be enforced
7955	Decamping .	Cannot be enforced
7956	Decampment	The claim must be enforced
7957	Decander .	If the claim is not enforced
7958	Decandrous.	Payment must be enforced
7959	Decangular .	If payment is not enforced
7960	Decantate .	Claim enforced
7961	Decapitate .	Payment enforced
7962	Decapoda .	Have enforced a settlement

No.	Code Word.	Enforced (*continued*)
7963	Decapodous.	Have not enforced
7964	Decastyle .	Will not be enforced
7965	Decease .	.**Enforcement**
7966	Decedent .	We look to the enforcement of our claim
7967	Deceit . .	Upon the enforcement of our claim
7968	Deceitful .	The enforcement of the claim
7969	Deceitless .	**Engage(s)**
7970	Deceived .	Can you engage
7971	Decemberly	Can engage
7972	Decemfid .	Cannot engage —— for less than
7973	Decempedal	Please engage
7974	Decemviral .	Will engage
7975	Decennary .	Engage(s) freight
7976	Decennium .	Engage(s) to send out
7977	Decennoval .	Engage for the purpose
7978	Decently. .	Engage a competent
7979	Decentish .	If we engage
7980	Deceptible .	If you can engage
7981	Deception .	If you cannot engage
7982	Deceptive .	Engage a passage for
7983	Deceptory .	Engage to superintend
7984	Decerpt . .	Engage a lawyer
7985	Decidable .	Engage to report
7986	Decidedly .	Engage to audit
7987	Decidement.	Engage to examine and report upon
7988	Decidence .	Try to engage
7989	Deciduity .	We must engage
7990	Deciduous .	Engages to go out
7991	Decigram .	Engages to go out, and will examine and **report on**
7992	Decillion .	You may engage temporarily
7993	Decimal .	.**Engaged**
7994	Decimalism .	Has (have) engaged
7995	Decimally .	Has (have) not engaged
7996	Decimated .	Engaged temporarily
7997	Decimating .	Permanently engaged
7998	Decimation .	Must not be engaged
7999	Decimators .	May be engaged
8000	Deciphered .	May not be engaged
8001	Decision. .	If you have engaged
8002	Decisively .	Engaged the services of
8003	Decivilize .	**Engagement(s)**
8004	Deckbeam .	Do not make any engagements
8005	Deckcargo .	Is (are) under an engagement to
8006	Decked . .	No engagement has been made
8007	Deckhands .	What engagements have you made
8008	Deckhook .	The following engagements have been made
8009	Decking . .	As regards the engagement (of)
8010	Deckload .	Approve of the engagement
8011	Deckpipe .	Engagement binding
8012	Deckpump .	Engagement not binding

No.	Code Word.	**Engagement(s)** *(continued)*
8013	Decksheet .	More engagements than can possibly be carried out
8014	Declaimeth .	Carry out engagement(s)
8015	Declaiming .	Under engagement
8016	Declamator .	Broken the engagement
8017	Declarable .	**Engine(s)**
8018	Declarant .	The engine has been removed from
8019	Declare . .	Will replace the engine
8020	Declaredly .	The engine is out of repair
8021	Declension .	The repairs to the engine
8022	Declined .	Require —— h.p. engine
8023	Declinous .	Hoisting engine of —— h.p.
8024	Declivity .	Portable engine of —— h.p.
8025	Decoctible .	Pumping engine of —— h.p.
8026	Decocting .	Pumping and winding engine in one, of —— h.p.
8027	Decoctions .	What is h.p. of engine
8028	Decollated .	The engine is of —— h.p.
8029	Decolorant .	The engine at the mill
8030	Decolorize .	The cylinder(s) of the engine
8031	Decolour .	The condenser of the engine
8032	Decomplex .	The piston-rod of the engine
8033	Decompound	Engine(s) in good order
8034	Decoped . .	Engine(s) in bad order
8035	Decorament	Must have new engine
8036	Decoration .	Full working power of engine is
8037	Decorative .	Cost of erecting engine would be
8038	Decorous .	It would be necessary to have —— engine(s)
8039	Decorously .	What is the power and condition of engine
8040	Decorum .	The engine is in good condition
8041	Decouple .	Engine and boiler in good condition
8042	Decoy . .	Engine in good condition but boiler quite out of
8043	Decoybird .	The engine is in very bad condition [repair
8044	Decoyduck .	The engine is not sufficiently powerful
8045	Decoyman .	**Engineer(s)**
8046	Decreaseth .	A thorough mechanical engineer
8047	Decreasing .	A thoroughly efficient engineer
8048	Decree . .	Must have a competent engineer to
8049	Decreeable .	Have had to engage temporarily an engineer
8050	Decreement	Send at once an engineer, to take complete control
8051	Decrepity .	Mining engineer [of engines and machinery
8052	Decrescent .	Require an efficient engineer
8053	Decrown . .	Require an efficient mining engineer
8054	Decubitus .	A competent and reliable mining engineer
8055	Decuman .	A good engineer, who can
8056	Decumbent .	A competent and reliable mining engineer to ex-
8057	Decurions .	**Enlarge** [amine and report upon
8058	Decussate .	To enlarge the
8059	Dedal . .	Will you have to enlarge the
8060	Dedalian .	Will have to enlarge the
8061	Dedalous .	Need not at present enlarge the
8062	Dedicating .	**Enlarged**

No.	Code Word.	Enlarged (*continued*)
8063	Dedicatory .	Have enlarged the
8064	Dedimus .	Have not enlarged the
8065	Deducing .	Has been much enlargeu
8066	Deduct .	**Enlargement**
8067	Deductions .	The enlargement of
8068	Deedbox .	**Enough**
8069	Deediest . .	There is enough ore
8070	Deedless . .	There is not enough ore
8071	Deedpoll .	There will be time enough
8072	Deepbrowed	Is there enough
8073	Deepening .	There is enough
8074	Deeplaid .	There is not enough
8075	Deeply . .	Enough ore to supply the
8076	Deepsea .	The ore is rich enough to
8077	Deepsome .	Enough ore can be raised (**to**)
8078	Deepwaist .	Have you enough
8079	Deerfold . .	Have not enough of
8080	Deergrass .	Did you have enough of
8081	Deerhair . .	Shall you have enough of
8082	Deerhound .	**Enquire.** (See Inquire.)
8083	Deerneck .	**Enter**
8084	Defacement .	To enter into
8085	Defacer . .	You had better enter into
8086	Defacingly .	Do not enter into
8087	Defailure .	Will —— enter
8088	Defalcator .	Will enter
8089	Defalk . .	Will not enter
8090	Defamation .	Should —— be willing to enter
8091	Defamatory .	Should —— not be willing to enter
8092	Defamous .	(To) enter into an undertaking
8093	Defatigate .	(To) enter into an agreement
8094	Defaulter .	(To) enter the mine
8095	Defaulting .	Could not enter the mine (on account of)
8096	Defeasance .	**Entered**
8097	Defeating .	Has (have) entered
8098	Defeature .	Has (have) not entered
8099	Defectuous .	Has (have) —— entered
8100	Defences .	Has entered upon his duties
8101	Defencing .	Entered into arrangements
8102	Defensible .	**Enterprise**
8103	Defensory .	Is it a risky enterprise
8104	Defiance . .	A risky enterprise
8105	Defiantly .	A most important enterprise
8106	Deficiency .	The enterprise is one which
8107	Deficient .	Abandon the enterprise
8108	Defiers . .	After enterprise was abandoned
8109	Defiguring .	Abandonment of the enterprise
8110	Defilading .	(To) take up the enterprise
8111	Defile . .	If they take up the enterprise they will carry it
8112	Defiling . .	The results of the enterprise [through

No.	Code Word.	**Enterprise** (*continued*)
8113	Definite . .	The objects of the enterprise
8114	Definitely .	The enterprise was successful .
8115	Definitude .	The enterprise has not hitherto been successful
8116	Deflagrate .	**Enterprising**
8117	Deflected .	Most enterprising
8118	Deflexion .	**Entertain**
8119	Deflexure .	Would —— entertain
8120	Defluous . .	Would be inclined to entertain
8121	Deflux . .	Would not entertain
8122	Defoliated .	Will entertain
8123	Deforceor .	Will not entertain
8124	Deforciant .	Will you entertain a proposition
8125	Deformedly .	Will you entertain a mining property
8126	Defossion .	Entertain the proposition
8127	Defoulment .	Entertain the idea
8128	Defraud . .	Cannot entertain the proposal .
8129	Defraudeth .	Before we can entertain
8130	Defrauding .	Will entertain the proposal, subject to certain modi-
8131	Defray . .	**Entertained** [fications
8132	Defraying .	The proposition has been favourably entertained
8133	Defunct . .	Will it be entertained
8134	Defunction .	Will the proposal be entertained
8135	Defunctive .	If proposal is entertained
8136	Defying . .	**Entirely**
8137	Degarnish .	Is it entirely
8138	Degeneracy .	It is entirely
8139	Degenerate .	It is not entirely
8140	Degenerous .	Entirely out of the question
8141	Degrade . .	The matter is one entirely of
8142	Degrading .	Entirely settled
8143	Dehisce . .	**Entitle**
8144	Dehiscence .	Will this entitle
8145	Dehorter .	This will entitle
8146	Dehumanize	Will not entitle
8147	Dehusk . .	**Entitled**
8148	Deiamba .	Is (are) entitled to
8149	Deifical . .	Is (are) not entitled to
8150	Deignous .	Ought to be entitled to
8151	Deinacrida .	Fully entitled to
8152	Deinosaur .	The share to which —— is (are) entitled
8153	Deistic . .	Has entitled —— to
8154	Deities . .	We think we are fully entitled to
8155	Deity . . .	Think you are entitled to
8156	Dejected .	Not entitled to more than
8157	Dejectedly .	Entitled under the agreement
8158	Dejectory .	Entitled under the clause
8159	Dejerate . .	Entitled under
8160	Delabechea .	**Eocene**
8161	Delapse . .	In Eocene formation
8162	Delapsion .	**Epidemic**

No.	Code Word.	**Epidemic** (*continued*)
8163	Delayed . .	An epidemic of —— has broken out
8164	Delayingly .	Epidemic decreasing
8165	Delayment .	Epidemic increasing
8166	Delectable .	Owing to the prevailing epidemic
8167	Delectate .	The epidemic now prevailing
8168	Delegation .	The epidemic has subsided
8169	Delesseria .	The epidemic has broken out again
8170	Deletery . .	**Equal**
8171	Delft . . .	Is (are) equal to
8172	Delftware .	Is (are) not equal to
8173	Deliac . .	Must have equal
8174	Deliberate .	The mines are not equal to the descriptions of them
8175	Delicacies .	Quite equal to
8176	Delicacy . .	Quite equal to the requirements (of)
8177	Delicious .	Not quite equal to
8178	Delighteth .	Equal to his duties
8179	Delightful .	Must be equal to
8180	Delighting .	Of equal quality (or grade)
8181	Delimit . .	An equal division
8182	Delineator .	If fully equal to
8183	Deliniment .	**Equally**
8184	Delinquent .	About equally divided
8185	Deliquate .	Equally divided between
8186	Deliquesce .	Equally bad
8187	Deliquium .	Equally good
8188	Delirancy .	Equally necessary
8189	Deliriant . .	**Equality**
8190	Deliriums .	Upon an equality of
8191	Deliver . .	**Equitable**
8192	Deliveress .	An equitable arrangement
8193	Deliverly .	We think it a most equitable arrangement
8194	Delphic . .	The most equitable arrangement [our approval
8195	Delphinate .	Any equitable arrangement or agreement will have
8196	Delphinus .	Cannot come to an equitable arrangement
8197	Deltaic . .	Equitable terms
8198	Deltidium .	Equitable terms have been proposed
8199	Delubrum .	Equitable terms have been arranged
8200	Deluding .	**Equivalent (to)**
8201	Delusions .	At the equivalent of
8202	Delusive . .	Taken as equivalent to
8203	Delusory . .	The equivalent in dollars
8204	Delved . .	The equivalent in pounds sterling
8205	Delving . .	The equivalent in
8206	Demagogism	What equivalent has he
8207	Demagogues	What equivalent can you offer
8208	Demagogy .	This will be about equivalent to
8209	Demand . .	About equivalent to
8210	Demandable	Not equivalent to
8211	Demandeth .	Equivalent to what we are giving up
8212	Demanding .	**Equivocating**

No.	Code Word.	**Equivocating** (*continued*)
8213	Demandress	Is (are) equivocating
8214	Demarcated.	**Equivocation**
8215	Demarch .	Without any equivocation
8216	Demeanour .	Let there be no equivocation
8217	Dementate .	Without equivocation or reserve
8218	Dementia .	**Erect**
8219	Demerits .	Shall have to erect
8220	Demesne .	Must erect
8221	Demesnial .	It is positively necessary to erect
8222	Demibath .	(To) erect the mill
8223	Demicannon	(To) erect the machinery
8224	Demideify .	(To) erect the engine
8225	Demidevil .	(To) erect the
8226	Demiditone .	**Erection**
8227	Demigorge .	In course of erection
8228	Demigroat .	The erection of
8229	Demijambe .	Do you advise the erection of
8230	Demijohn .	Advise erection of
8231	Demilance .	In course of erection, will be completed
8232	Demimonde	Upon completing the erection
8233	Demiquaver	The erection of a portion of the
8234	Demirep . .	**Error**
8235	Demisable .	Owing to an error
8236	Demising .	There is an error
8237	Demisuit .	There is no error whatever
8238	Demiurgus .	Is there not some error
8239	Demivill . .	There is a great error
8240	Demiwolf .	There is an error in the accounts
8241	Demobilise .	To eliminate the error
8242	Democracy .	It was an error on our part
8243	Democrat .	As a result of the error
8244	Demogorgon	Attribute the error to
8245	Demography	An error in the telegram
8246	Demolish .	An error on the part of
8247	Demolition .	How did the error occur
8248	Demonetize.	Think you have made an error
8249	Demoniasm.	If we have made an error
8250	Demonifuge.	Admit(s) the error
8251	Demonology	We much regret the error
8252	Demonomist	**Escape**
8253	Demonry .	To escape the consequences
8254	Demonship .	To escape from
8255	Demoralize .	A narrow escape (from)
8256	Dempster .	No way of escape (from)
8257	Demulcent .	**Escaped**
8258	Demure . .	Escaped from
8259	Demurely .	Escaped with life
8260	Demureness	Escaped with life, but seriously injured
8261	Demurrable.	Has not escaped (from)
8262	Demurrage .	**Essential**

No.	Code Word.	**Essential** (*continued*)
8263	Demurrer .	It is absolutely essential
8264	Demurring .	It is not absolutely essential
8265	Denariate .	Is it absolutely essential
8266	Denaturate .	**Essentially**
8267	Dendiculus .	Report is essentially correct
8268	Dendriform .	Accounts essentially correct
8269	Dendritic .	Is essentially a question (of)
8270	Dendrobium	**Establish**
8271	Dendrodont .	(To be) able to establish
8272	Dendrodus .	Must establish
8273	Dendroid .	Cannot establish
8274	Dendroidal .	Cannot you establish
8275	Dendrolite .	Establish friendly relations with
8276	Dendrology .	Establish a trade in
8277	Dendromys .	If you could establish
8278	Dendrophis .	Establish an agency
8279	Denelage .	Establish a branch
8280	Denigrate .	**Established**
8281	Denizen . .	Was (were) established
8282	Denominate	Has (have) been established
8283	Denotable .	Must be established
8284	Denotative .	Established an agency
8285	Denounce .	Established a branch
8286	Denouncing	Has (have) established
8287	Densely . .	Has not yet been established
8288	Denseness .	**Establishing**
8289	Dentalidae .	Think of establishing
8290	Dentalium .	**Establishment**
8291	Dentary . .	The establishment of
8292	Dentately .	Advise the establishment of
8293	Denticle . .	**Estimate(s).** (See also Cost, and Time.)
8294	Dentiform .	Estimate expenses for the month at
8295	Dentifrice .	Impossible to estimate the value
8296	Dentiloquy .	Cannot estimate the quantity of ore
8297	Dentinal . .	I (we) estimate the value of the ore at
8298	Dentiscalp .	I (we) estimate the value of the machinery at
8299	Dentist . .	I (we) estimate the value of mines, plant, and
8300	Dentistry .	Do you include in the estimate [works at
8301	Dentition .	I (we) estimate the value of the machinery and
8302	Dentizing .	I estimate the expenses at [buildings at
8303	Denture . .	In your estimate of working expenses do you include
8304	Denudating .	Has (have) included in the estimate
8305	Denuded .	Has (have) not included in the estimate
8306	Denunciant .	Did not include in the estimate
8307	Denyingly .	At what do you estimate
8308	Deobstruct .	Cannot estimate
8309	Deoculate .	Impossible to estimate
8310	Deodorant .	Estimates forwarded
8311	Deodorize .	Please send estimate for
8312	Deonerated .	Estimates are exaggerated

No.	Code Word.	Estimate(s) (*continued*)
8313	Deoppilate .	Estimates for
8314	Deoxidized .	Estimate of tonnage and value of ore reserves
8315	Depainter .	Estimate of ore reserves (in)
8316	Departable .	Estimate of ore in sight
8317	Departure .	Estimates sent by last mail
8318	Depascent .	Estimates will be sent on the
8319	Dependable .	Cable what you estimate to be the amount and value of ore in sight, the rate of production, and
8320	Dependants .	Cable what you estimate [the cost of mining
8321	Depended .	Get an estimate from
8322	Dependency	Estimates given by —— are too high
8323	Depeople .	Estimates given by —— are too low
8324	Deperdit .	Upon a moderate estimate
8325	Dephal . .	Upon the basis of your estimate
8326	Dephlegm .	The estimate is based upon
8327	Depict . .	At a very low estimate
8328	Depicting .	At a very high estimate
8329	Depilation .	We think it will be safe to estimate
8330	Depilatory .	**Estimated**
8331	Depilous. .	Estimated quantity of ore in sight
8332	Deplant . .	Roughly estimated at
8333	Depleting .	Estimated cost of
8334	Depletion .	The estimated return
8335	Deplorable .	The estimated yield
8336	Deplorate .	The estimated expense
8337	Deplored .	What have you estimated —— at
8338	Deploredly .	Estimated to be worth
8339	Deploring .	Estimated to produce
8340	Deploy . .	What is the estimated
8341	Deployment	Cannot be estimated
8342	Deplume .	Estimated upon the basis of
8343	Depolarize .	**Estimation**
8344	Deponeth .	In your estimation
8345	Depopulate .	In ——'s estimation
8346	Deportment .	In the estimation of
8347	Deposit . .	**Evade**
8348	Depositing .	Is (are) trying to evade
8349	Deposition .	Will try to evade
8350	Depravedly .	Do not evade
8351	Depravity .	In order to evade
8352	Deprecable .	Cannot evade
8353	Depredator .	Cannot evade the conclusion (that)
8354	Deprehend .	Have evaded
8355	Depressant .	**Evasion**
8356	Depression .	An evasion of the difficulty
8357	Deprisure .	An evasion of the agreement
8358	Depriveth .	**Evening**
8359	Depth . .	On the evening of the
8360	Depucelate .	This evening
8361	Depulse . .	To-morrow evening

No.	Code Word.	**Evening** (*continued*)
8362	Depulsion	Last evening
8363	Depulsory	**Event(s)**
8364	Depurating	In the event of
8365	Depurator	To be prepared for any event
8366	Deputize	Recent events
8367	Deputizing	A most untoward event
8368	Deputy	A most disastrous event
8369	Dequace	In the course of events
8370	Deracinate	Events have proved
8371	Deraign	Events have falsified
8372	Derainment	**Eventually**
8373	Deranged	If it should prove eventually
8374	Deranging	Eventually proved to be
8375	Derdoing	It may eventually
8376	Derelict	**Everything**
8377	Dereworth	Everything ready
8378	Deridingly	Everything has been arranged most satisfactorily
8379	Derivably	Everything is going right
8380	Derivating	Everything is going wrong
8381	Derivation	Put everything in good order
8382	Dermalgia	Everything has been done
8383	Dermaptera	As soon as everything is going right
8384	Dermatoid	Everything now is in first-rate order
8385	Dermestes	How is everything going
8386	Dermic	Everything looks well
8387	Dermotomy	Do everything that is possible
8388	Dernful	Everything has been tried
8389	Dernly	Everything possible
8390	Derogated	**Evidence(s)**
8391	Derrick	The evidence is in favour of
8392	Derringdo	Has (have) given evidence on
8393	Derringers	There is no evidence to prove
8394	Dervish	Get all the favourable evidence you can
8395	Desatir	Is there any evidence to prove
8396	Descant	Give(s) evidence of
8397	Descanteth	There is evidence to prove
8398	Descanting	Gives no evidence of
8399	Descended	We are looking up evidence
8400	Descendant	No evidence whatever
8401	Descension	**Exact**
8402	Descensive	Is (are) exact
8403	Describe	Is (are) not exact
8404	Describing	Exact statement as to
8405	Descrier	Cannot be too exact
8406	Descry	An exact copy of—(Exact copies of)
8407	Desecrated	An exact statement of facts
8408	Desert	Your report should be very exact
8409	Deserting	Will try to exact all he can
8410	Desertions	Trying to exact
8411	Desertless	Exact a sufficient amount

No.	Code Word.	**Exact** (*continued*)
8412	Desertrix	. Is (are) sufficiently exact
8413	Deservedly	. The exact nature of
8414	Desiccant	. **Exacted**
8415	Desiderate	. Exacted as much as possible
8416	Desidiose	. **Exacting**
8417	Designable	. Exacting all he (they) can
8418	Designator	. The exacting nature of the demands
8419	Designed	. **Exactness**
8420	Designedly	. Done with the greatest exactness
8421	Designful	. The exactness of the drawings
8422	Designing	. **Exaggerated**
8423	Designment	. Exaggerated reports
8424	Desilvered	. Cannot be exaggerated
8425	Desipient	. Exaggerated statements
8426	Desireth	. . Have been greatly exaggerated
8427	Desirous	. . The most exaggerated notions have been conceived
8428	Desirously	. The statements are exaggerated [of the
8429	Desist	. . The statements (or reports) are not exaggerated
8430	Desistance	. Exaggerated and distorted
8431	Desistive	. The facts have been exaggerated and distorted to serve the purposes of
8432	Deskwork	. Do (does) not exaggerate in the slightest degree
8433	Desmidians	. **Examination.** (See also Inspection.)
8434	Desmobrya	. Subject to examination
8435	Desmodium	Must have —— before leaving for examination of
8436	Desmodus	. My fee and expenses for examination of ——
8437	Desolating	. Report of examination [mine(s) are
8438	Desolately	. As soon as you have completed examination
8439	Desolator	. Wish(es) you to make the most thorough examina-
8440	Despaired	. An examination has been made [tion of
8441	Despairful	. Have returned from making the examination of
8442	Despairing	. An immediate examination of the —— mine(s) is
8443	Despatch	. Make a searching examination [wanted by
8444	Desperados	. When making the examination of —— mines
8445	Desperate	. Examination of the —— mine finished, shall leave
8446	Despicably	. If examination is unsatisfactory [on —— for
8447	Despiteful	. Examination of the —— mine completed, report written and sent, fee and expenses amount to
8448	Despitous	. After making searching examination
8449	Despoiler	. On examination being satisfactory
8450	Despondent	Examination completed and most satisfactory
8451	Desponsate	. Upon examination, it is found
8452	Despot	. . If examination is not required now
8453	Despotical	, How soon can you make complete examination of
8454	Despotism	. Fee for examination and report inclusive of all expenses will be [will be
8455	Despumated	Fee for examination and report exclusive of expenses
8456	Desquamate	Fee for examination and report inclusive of travelling expenses [expenses but exclusive of
8457	Dessiatine	. Fee for examination and report including travelling

No.	Code Word.	Examination (*continued*)
8458	Destinable .	The result of the examination is very satisfactory
8459	Destinated .	The result of the examination is not at all satisfactory
8460	Destining .	We must rely on your examination
8461	Destinists .	Examination by competent expert
8462	Destiny . .	(Make) Searching examination of mine(s), extent of property, workings, machinery, facilities for working, labour, material, water, reserves, prospects, method and cost of working, and value of produce, and advise on improvements, and future developments, latter important
8463	Destituent .	Examination of the mining property
8464	Destroy . .	Examination of the —— mine situated at [etc.
8465	Destroying .	A thorough examination of the mine, the workings,
8466	Destruct . .	Examination will be made [property
8467	Destructor .	A thorough examination of, and report upon the
8468	Desudation .	Examination completed, shall send report
8469	Desuetude .	Have completed examination and sent report
8470	Detach . .	Subject to approval after examination
8471	Detacheth .	Cannot make examination on account of
8472	Detaching .	Examination of the accounts
8473	Detachment	**Examine.** (See also Inspect.) [mine(s) in
8474	Detailing .	Want(s) you to examine and report on the ——
8475	Detained .	What would you charge to examine and report on a mine in
8476	Deterge . .	My charge to examine a mine is ——, and expenses
8477	Determined .	Permit —— to examine the report
8478	Deterrence .	Examine —— mine, and telegraph opinion
8479	Detersive .	Examine the —— mine, and report by letter
8480	Detestable .	The mine you will have to examine is in
8481	Detestate .	Can examine both properties at the same time
8482	Detesting .	Could you examine both properties at the same time
8483	Dethrone .	When can —— start to examine the —— mine
8484	Dethroning .	Can start to examine the —— mine on
8485	Detonize .	Examine for my (our) account the —— mine(s) situated in
8486	Detonizing .	Examine for my (our) account the —— mine(s) situated in ——, and send synopsis of report by telegraph (cable), and full report by mail as soon as possible [a mine in
8487	Detort . .	Be ready to leave as soon as possible to examine
8488	Detracting .	Arrange so that you can examine the —— mine(s) at —— on your way to
8489	Detractive .	Will arrange so as to examine the —— mine, on my
8490	Detractory .	Can examine —— mine [way to
8491	Detracts . .	Examine and report upon the present condition of the —— mine(s), estimated quantity of ore in sight, developments, cost of working, estimated profit, facilities, assays, etc.
8492	Detritus . .	Has (have) arranged that you examine
8493	Detruncate .	Cannot examine

No.	Code Word.	Examine (*continued*)
8494	Deturpate .	Impossible for me to examine
8495	Deuceace .	(To) examine and report upon
8496	Deucedly .	Send a competent man to examine
8497	Deutoplasm	I cannot examine the property. until
8498	Deutoxyde .	Can examine the —— mine about
8499	Devastator .	Have sent a competent man to examine
8500	Developed .	Examine carefully into the matter
8501	Developing .	Have arranged to examine and report upon
8502	Devex . .	Examine the accounts
8503	Devexity .	Examine the statements
8504	Deviate . .	Let —— go at once to examine
8505	Deviating .	**Examined**
8506	Deviations .	I have examined the —— mine(s), and can recommend it (them) [commend it (them)
8507	Deviceful .	I have examined the —— mine(s), but cannot re-
8508	Devices . .	Who has previously examined the mine(s)
8509	Devilbirds .	I have carefully examined the property
8510	Devilfish .	When —— examined the mine(s)
8511	Devilishly .	Have examined several other mines in the district, which were more developed, so as to be able to
8512	Devilkin . .	Have you examined the [judge better
8513	Devilries . .	Have examined the
8514	Devilry . .	Have examined very carefully and find
8515	Devilsdust .	I have examined the —— mine(s), and can recommend buying for the price of —— on condition that a working capital of —— be provided
8516	Devilship .	Have examined the —— mine(s) and can recommend it (them) at the price of
8517	Devitalize .	My fee and expenses amount to —— for having examined the —— mine [reported upon
8518	Devitrify . .	The mines have been frequently examined, and
8519	Devocation .	The mines have never been examined by any re-
8520	Devolving .	Must be minutely examined [liable expert
8521	Devotary .	Appear never to have examined
8522	Devoted . .	Appears never to have been examined
8523	Devotement	Has it (have they) ever been examined
8524	Devotion .	Examined and reported upon
8525	Devotional .	Allow the mine to be examined
8526	Devourable .	Mine had been previously examined
8527	Devoureth .	Previously examined and reported upon (by)
8528	Devouring .	**Examining**
8529	Devoutful .	While examining
8530	Devoutly .	While examining the property
8531	Devoutness .	After examining
8532	Devoyre . .	Before examining
8533	Dewberries .	What is your fee for examining
8534	Dewberry .	My fee for examining the —— mine will be
8535	Dewclaw .	If, after examining, you find that
8536	Dewdrop .	**Exceed(s)**
8537	Dewfall . .	Not to exceed

No.	Code Word.	**Exceed(s)** (*continued*)
8538	Dewiness	Will exceed
8539	Dewpoint	Must not exceed
8540	Dewretting	The capital of the company ought not to **exceed**
8541	Dewstone	Ought not to exceed
8542	Dewworm	Expense(s) ought not to exceed
8543	Dexter	Greatly exceed(s) our expectations
8544	Dexterity	Does not exceed what we expected
8545	Dexterous	Probable that it will exceed
8546	Dextral	It is possible that it will not exceed
8547	Dextrality	If it does not exceed
8548	Dextrorsal	**Exceeded**
8549	Dextrously	Exceeded what we expected
8550	Dhurra	Exceeded the last
8551	Diabetes	Greatly exceeded our expections
8552	Diabetical	Exceeded the cost
8553	Diablery	The cost exceeded the value
8554	Diabolic	**Exceeding**
8555	Diabolisms	Not exceeding
8556	Diabrosis	If exceeding
8557	Diacaustic	If not exceeding
8558	Diachylon	Without exceeding
8559	Diachyma	At an expense not exceeding
8560	Diaconate	If without exceeding the cost
8561	Diacope	Is (are) rapidly exceeding
8562	Diacritic	Exceeding all previous
8563	Diadelph	Exceeding last week's
8564	Diadem	Exceeding last month's
8565	Diadexis	An outlay much exceeding
8566	Diaglyphic	**Exceedingly**
8567	Diagnose	Exceedingly rich
8568	Diagnosing	Exceedingly profitable
8569	Diagnostic	Exceedingly rich in gold
8570	Diagometer	Exceedingly rich in silver
8571	Diagonally	Exceedingly doubtful (if)
8572	Diagonial	Exceedingly difficult
8573	Diagonus	**Exception(s)**
8574	Diagram	With the exception of
8575	Diagraphs	With few exceptions
8576	Dialect	No exception can (to) be made
8577	Dialectal	Must take exception to
8578	Dialectics	Take(s) exception to the charge
8579	Diallage	With the following exception
8580	Diallelous	Without exception
8581	Dialling	Will make an exception in this **case**
8582	Diallogite	Cannot make any exception
8583	Diallyl	An exceptional case
8584	Dialogism	**Excess**
8585	Dialogize	Is (are) greatly in excess (of)
8586	Dialogues	There is no excess
8587	Dialoguing	Expenses in excess of

No.	Code Word.	Excess (*continued*)
8588	Dialplate .	Will there be any excess
8589	Dialwheel .	Owing to the excess of
8590	Dialwork .	**Excessive**
8591	Dialytic . .	An excessive strain upon
8592	Dialyzer . .	Excessive and wasteful expenditure
8593	Dialyzing .	We think the expenses are excessive, and they must
8594	Diamantine .	The charges are excessive [be reduced
8595	Diameter .	**Exchange**
8596	Diametric .	At what rate of exchange
8597	Diamonds .	If exchange is favourable
8598	Dianatic . .	What is the rate of exchange
8599	Diander . .	Rate of exchange is
8600	Diandrian .	In exchange for
8601	Diandrous .	At what rate of exchange have you calculated
8602	Dianthus .	Will you exchange
8603	Diapason .	Will —— exchange
8604	Diapente .	Will not make the exchange
8605	Diapering .	Will make the exchange
8606	Diaphanic .	At the current rate of exchange
8607	Diaphanous .	**Excitement**
8608	Diaphragm .	Great excitement in the market
8609	Diaphysis .	Great excitement prevails
8610	Diaplastic .	The excitement is subsiding
8611	Diapnoic .	To take advantage of the excitement in
8612	Diaporesis .	There is no excitement here
8613	Diapyetic .	Caused great excitement [been discounted
8614	Diarists . .	Caused but little excitement, the information had
8615	Diarrhetic .	Under the influence of the excitement
8616	Diarrhoea .	**Exclusive**
8617	Diaschisma .	Exclusive of
8618	Diaspore .	This must be exclusive of
8619	Diastaltic .	Exclusive of travelling expenses
8620	Diastasis .	Exclusive of out-of-pocket expenses
8621	Diastema .	Exclusive intelligence
8622	Diathermal .	Exclusive of all other considerations
8623	Diathesis .	**Excuse**
8624	Diatomic .	There can be no excuse for
8625	Diatomous .	Every excuse must be made
8626	Diatribe . .	Cannot excuse
8627	Diatribist .	**Execute**
8628	Diazeutic .	If you cannot execute
8629	Dibber . .	**Executed**
8630	Dibrothian .	Was (were) executed
8631	Dibstone .	Was (were) not executed
8632	Dicacity . .	The order was executed
8633	Dicaeology .	The order was not executed
8634	Dicastery .	Why was the order not executed
8635	Dicebox . .	Can the order be executed
8636	Dicecoal . .	Order has been executed
8637	Dichastic .	**Exertion**

No.	Code Word.	Exertion (*continued*)
8638	Diche. . .	Use every exertion to
8639	Dichobune .	Has (have) used every exertion to
8640	Dichodon .	**Exist(s)**
8641	Dichogamy .	Do (does) not exist
8642	Dichroism .	Exists in large quantities
8643	Dickey . .	Proving that —— exist(s)
8644	Dicksonia .	No longer exist(s)
8645	Dickybird .	**Existing**
8646	Diclesium .	Under the existing circumstances
8647	Diclinic . .	At present existing
8648	Dicoccous .	Not now existing
8649	Dicotyles .	**Exorbitant**
8650	Dicrurinae .	An exorbitant demand
8651	Dicrurus . .	The price asked is simply exorbitant
8652	Dictamnus .	Consider the demand exorbitant
8653	Dictate . .	**Expect**
8654	Dictating .	When do you expect to be able to leave for
8655	Dictation .	I fully expect to finish here by
8656	Dictatress .	I do not expect to leave before
8657	Dictatrix . .	Do not expect
8658	Dictature .	Do you expect
8659	Dictionary .	I (we) expect
8660	Dictum . .	I (we) do not expect
8661	Dictyogen .	Expect to strike the vein
8662	Dicynodon .	You may expect —— to arrive about
8663	Didactic . .	I expect —— to be here about
8664	Didactical .	You may expect
8665	Didactyl . .	I expect to be able to
8666	Didascalar .	Expect to hear
8667	Didelphian .	When do you expect
8668	Didelphys .	When may I (we) expect
8669	Dididae . .	Expect to know
8670	Didrachm .	You must not expect
8671	Didunculus .	You must not expect such a good return
8672	Didymous .	What do you expect will be the
8673	Didynamian	Expect the run will be about
8674	Dieaway . .	Expect the result will be
8675	Diegesis . .	**Expectation(s)**
8676	Diervilla . .	In the full expectation of
8677	Diesinker .	Contrary to expectations
8678	Diesinking .	Expectations have been realized
8679	Diestock . .	Expectations have not been realized
8680	Dietarian .	**Expected**
8681	Dietary . .	Expected to arrive
8682	Dietbread .	Expected to leave
8683	Dietdrink .	Not expected until
8684	Dietetic . .	Expected you were at
8685	Dietetists .	It was expected that
8686	Differeth . .	Less than was expected
8687	Differing . .	More than was expected

No.	Code Word.	Expected (*continued*)
8688	Difficile . .	As might have been expected
8689	Difficulty .	Is (are) expected
8690	Diffidence .	So different to what was expected
8691	Difflation .	Turning out as good as was expected
8692	Diffluency .	Turning out better than expected
8693	Diffluent .	. **Expecting**
8694	Difflugia . .	Has (have) been expecting to
8695	Difform .	. **Expedient(s)**
8696	Difformity .	As may be deemed expedient
8697	Diffract . .	A judicious expedient
8698	Diffused . .	The only expedient we can suggest
8699	Diffusedly .	Do not think it expedient to
8700	Diffusible .	The most expedient course
8701	Diffusion .	Act as you deem expedient
8702	Diffusive . .	All expedients have been tried but without success
8703	Difluan .	. **Expedite**
8704	Digamma .	To expedite matters
8705	Digastric .	It will expedite matters very much, if
8706	Digerent .	Can you expedite the matter
8707	Digest .	. **Expenditure**
8708	Digestedly .	A judicious expenditure of capital
8709	Digestible .	By a small expenditure
8710	Digestion .	The expenditure of capital named
8711	Digital . .	Justifying the expenditure of
8712	Digitately .	The prospects do not justify the expenditure (of)
8713	Digitiform .	The expenditure of
8714	Digitorium .	Expenditure on Capital Account
8715	Digladiate .	Expenditure on Revenue Account
8716	Diglyph . .	Expenditure on Stores Account
8717	Dignitary .	By increasing the expenditure
8718	Digress . .	In this expenditure is included
8719	Digressing .	The expenditure is rapidly increasing
8720	Digressive .	An increase in the expenditure and decrease in the
8721	Digynian .	Without any increase of expenditure　　　[yield
8722	Dihedral . .	Keep the expenditure down
8723	Dijudicant .	The expenditure on the
8724	Dilacerate .	A most lavish expenditure
8725	Dilapidate .	Extraneous expenditure
8726	Dilatable .	**Expense(s).** (See Reduce, Regulate.)
8727	Dilatation .	Telegraph me —— for travelling expenses
8728	Dilating . .	Who will pay expenses
8729	Dilative . .	Will pay expenses
8730	Dilatorily .	Will not pay expenses
8731	Dilemmatic .	Will pay any extra expense incurred
8732	Dilemmas .	The extra expense to be provided for by
8733	Diligences .	Expenses exceed receipts by
8734	Diligent . .	Expenses per ton are　　　　　[the reason that
8735	Diligently .	The expenses this month have been very heavy for
8736	Dillenia . .	Not sufficient funds to meet the additional expense(s)
8737	Dillydally .	The expenses will not be so great in future

No.	Code Word.	Expense(s) *(continued)*
8738	Dilogy	What will be the expense of
8739	Dilucid	To meet the expense of
8740	Dilucidity	Does —— include the total expenses
8741	Dilute	Expenses will be about
8742	Dilutedly	General expenses
8743	Diluteness	The expenses necessary to
8744	Dilutions	The expenses ought to be paid by
8745	Diluviated	Will pay all expenses (of)
8746	Diluvium	Expense is no object with
8747	Dimaris	Will the expenses exceed
8748	Dimensity	What will be about the month's expenses
8749	Dimeran	Expenses for the month of —— are
8750	Dimerous	Keep expenses down as much as possible
8751	Dimication	Will entail very considerable expense
8752	Dimidiated	The works have been carried on regardless of expense [expense
8753	Diminish	The works have been carried on without much
8754	Diminuent	To whom shall I (we) apply to pay expenses of
8755	Diminutely	Expense on revenue account
8756	Diminutive	Expense on capital account
8757	Dimissory	Provided that expenses do not exceed
8758	Dimities	The expense is too great
8759	Dimity	Has (have) paid —— for expenses
8760	Dimorphic	What is the amount of expense entailed
8761	Dimorphous	Expenses are guaranteed (up to)
8762	Dimpled	The expense entailed (by)
8763	Dimplement	Will pay travelling expenses
8764	Dimpling	Decline(s) to pay travelling expenses
8765	Dimpsy	The expense of working the property
8766	Dimyary	Willing to divide the expense
8767	Dinarchy	Telegraph us what the expenses are (were) of visiting
8768	Dinetical	The expenses cannot be kept below
8769	Dinginess	Mining and milling expenses
8770	Dingo	Development expenses
8771	Dingthrift	How are expenses to be paid
8772	Diningroom	Spare no expense
8773	Dinmont	What do you estimate the expense of
8774	Dinnerhour	Extraneous expenses amount to
8775	Dinnerless	Extraneous expenses include
8776	Dinnertime	Extraneous expenses for month
8777	Dinornis	To the extraneous expenses must be added
8778	Dinosauria	The expense will be very heavy
8779	Dinothere	The expense will not be so heavy
8780	Diocesans	The expense will be too heavy
8781	Diocese	Incur no further expenses on
8782	Diocesener	Expenses must be paid (by)
8783	Diodon	At any expense
8784	Dioecious	We think expenses can be reduced
8785	Dioecism	What expense has been incurred in respect of
8786	Diomedea	Great expense has been incurred by

No.	Code Word.	**Expense(s)** (*continued*)
8787	Diopsis . .	Your expenses will be paid
8788	Dioptase .	The working expenses
8789	Diopter . .	The working expenses should not exceed
8790	Dioptric . .	The expense of bringing out the company
8791	Dioptrical .	All expenses up to allotment
8792	Dioramic .	Legal expenses
8793	Diorism . .	The legal expenses will amount to
8794	Dioristic . .	The expense of advertising
8795	Diorthosis .	All expenses for brokerage, commission, etc.
8796	Dioscorea .	The expense of floating the company, including legal charges, printing, advertising, brokerage and commission
8797	Diosma . .	**Expert(s)**
8798	Diospyros .	Expert's report is
8799	Dipaschal .	Experts' reports are
8800	Dipchick .	The first expert
8801	Dipetalous .	The opinion of every expert in the country
8802	Diphda . .	Experts all speak favourably of ―― mine(s)
8803	Diphtheria .	A mining expert of great repute
8804	Diphthongs .	A mining expert of no repute
8805	Diphycerc .	A competent and reliable expert
8806	Diphyes . .	The advice of the best expert
8807	Diphydae .	Recommended as a most expert
8808	Diphyllous .	**Expiration**
8809	Diphyodont .	At the expiration of
8810	Diplogenic .	Before the expiration of
8811	Diploma . .	**Expire(s)**
8812	Diplomacy .	When will the time expire
8813	Diplomatic .	Will not expire until
8814	Diplopoda .	The bond expire(s)
8815	Diploptera .	The time expires
8816	Diplopy . .	**Expired**
8817	Diplotaxis .	The time has already expired
8818	Diplozoon .	**Explain**
8819	Dipnoi . .	Will explain everything
8820	Dipodidae .	Will explain
8821	Dipped . .	Will not explain
8822	Dippings .	Explain by telegraph
8823	Diprotodon .	Explain by letter
8824	Dipsaceae .	Cannot explain by telegraph, will write you fully
8825	Dipsas . .	Cannot explain more clearly
8826	Dipsomania	Must explain more clearly
8827	Dipsosis . .	**Explanation(s)**
8828	Diptera . .	What explanation can I give
8829	Dipterous .	Wait explanation by letter
8830	Dipteryx . .	What explanation can you give
8831	Diptych . .	Explanation satisfactory
8832	Diptychum .	Explanation not satisfactory
8833	Dipworking .	Wish to have full explanation
8834	Dipyre . .	More satisfactory explanations

No.	Code Word.	Explanation(s) *(continued)*
8835	Dipyrenous .	Send full explanation by mail
8836	Direct . .	Am writing full explanation
8837	Directeth .	Full and definite explanations
8838	Directive .	What is the explanation of
8839	Directly . .	Give no explanation
8840	Directness .	What explanation can —— give
8841	Directors .	Can —— give any explanation
8842	Directrix .	Has (have) no explanation to give
8843	Direfully .	In explanation of the plan (drawing)
8844	Dirempt . .	For full explanation await
8845	Diremption .	**Explicit(ly)**
8846	Dirge . . .	Not sufficiently explicit
8847	Dirgeful . .	Must be more explicit
8848	Dirtbed . .	Cannot be more explicit
8849	Dirteating .	Our instructions were very explicit
8850	Dirtily . .	In the most explicit term
8851	Dirtpie . .	Stated explicitly
8852	Disability .	**Exploration(s)**
8853	Disable . .	Exploration work
8854	Disabling .	Extensive explorations have been made
8855	Disabused .	More extensive explorations must be made
8856	Disaccord .	Explorations in the higher levels
8857	Disacidify .	Explorations in the lower levels
8858	Disadorned .	In carrying out the explorations
8859	Disadvance .	The recent explorations
8860	Disadvise .	The recent explorations are opening out
8861	Disaffects .	The explorations have not as yet given us much ore, but we have learnt a good deal from them of the nature of the vein
8862	Disaffirm .	The recent explorations have developed a fine body [of ore
8863	Disagree . .	The explorations promise well
8864	Disalliege .	What is the appearance and value of the explora- [tions in
8865	Disallows .	**Explosion**
8866	Disamis . .	A bad explosion has occurred
8867	Disanchor .	Boiler explosion
8868	Disangelic .	Gunpowder explosion
8869	Disanimate .	Dynamite explosion
8870	Disannex .	Colliery explosion
8871	Disannuls .	Explosion of
8872	Disapparel .	Explosion of fire-damp
8873	Disappear .	Explosion of a blasting-charge
8874	Disappoint .	The explosion destroyed
8875	Disapprove .	When the explosion occurred
8876	Disarm . .	Killed by the explosion
8877	Disarming .	Injured by the explosion
8878	Disarrange .	**Export(s)**
8879	Disassent .	Our export trade
8880	Disaster . .	The export of
8881	Disasterly .	Exports extensive
8882	Disastrous .	**Express**

No.	Code Word.	**Express** (*continued*)
8883	Disattach .	Send by express
8884	Disattune .	Sent by express
8885	Disavaunce .	Am sending by express
8886	Disavow .	. **Extend.** (See also Provided.)
8887	Disavowal .	To extend
8888	Disavowing .	Not to extend
8889	Disband . .	Will extend
8890	Disbanding .	Will not extend
8891	Disbelief .	Will —— extend
8892	Disbenched .	Extend the drift to
8893	Disblame .	Extend beyond
8894	Disblaming .	Extend as far as
8895	Disbodied .	**Extended**
8896	Disbowel .	The tunnel to be extended
8897	Disbranch .	To be extended north
8898	Disbudded .	To be extended south
8899	Disbudding .	To be extended east
8900	Disburgeon .	To be extended west
8901	Disbursers .	Was (were) extended
8902	Disburthen .	Was (were) not extended
8903	Discamp . .	Has (have) been extended
8904	Discandy .	Has (have) not been extended
8905	Discarded .	(Have) extended the time
8906	Discardure .	(Have) extended the drift
8907	Discarnate .	When the drift is extended to
8908	Discern . .	Should be further extended to
8909	Discerneth .	**Extension** (of)
8910	Discerning .	Will —— grant extension (of)
8911	Discerns . .	Extension will be granted
8912	Discharges .	Extension of time required
8913	Discharity .	Must have further extension (of)
8914	Dischidia .	Cannot grant extension (of)
8915	Dischurch .	Have asked —— days extension
8916	Disciple . .	Extension of time for which the mine is bonded
8917	Discipling .	**Extent**
8918	Disclaimed .	In extent
8919	Disclosing .	(To) the extent of
8920	Disclosure .	What is the extent of
8921	Discloud . .	To a reasonable extent
8922	Disclusion .	No extent of
8923	Discoast . .	To a great extent
8924	Discobolus .	The mine has not been worked to any great extent
8925	Discocarp .	The mine has been worked to a very great extent
8926	Discoidal .	To what extent
8927	Discolith . .	To the fullest extent
8928	Discolor . .	To the fullest extent of our resources
8929	Discomfits .	The extent is not known yet
8930	Discomfort .	The extent of the damage
8931	Discommend	Can you give the extent of
8932	Discompose	Cannot give the extent of

No.	Code Word.	
8933	Discompt	. **Extract**
8934	Disconcert	. An extract from
8935	Discophora	. In order to extract
8936	Discordant	. It is difficult to extract the ore on account of
8937	Discordful	. Cannot extract
8938	Discordous	. **Extracted**
8939	Discourage	. Is (are) easily extracted
8940	Discoursed	. Large quantities of ore have been extracted
8941	Discovery	. The means by which the ore is extracted
8942	Discradle	. What amount of ore has been extracted
8943	Discredits	. **Extraction**
8944	Discreet	. The cost of extraction
8945	Discreetly	. The labour of extraction
8946	Discrepant	. By the extraction of
8947	Discretion	. Rely on an increased extraction
8948	Discrowned	. Extraction doubled
8949	Disculpate	. Increased extraction
8950	Discumber	. **Extraordinary**
8951	Discurrent	. Most extraordinary
8952	Discuss	. Nothing extraordinary
8953	Discutient	. Has (have) taken extraordinary measures (or precautions)
8954	Disdain	. Extraordinary measures [cautions)
8955	Disdaineth	. Extraordinary precautions
8956	Disdainful	. **Extreme(ly)**
8957	Disdaining	. In the extreme
8958	Disdainous	. Extremely doubtful
8959	Diseaseful	. **Fabrication**
8960	Disedge	. The account you have heard is a complete fabrica-
8961	Disedified	. **Face** [tion
8962	Disedify	. Face is changing in appearance
8963	Disembarks	. Have had samples assayed from the face, with the
8964	Disembody	. Face of cross-cut is in [following results
8965	Disembogue	Face of cross-cut is looking
8966	Disembosom	Face of drift
8967	Disembowel	How is the face of drift looking
8968	Disembroil	. The face of drift is looking
8969	Disemploy	. From the appearance of the face
8970	Disenchant	. The face is now in
8971	Disendow	. The face is not looking so well
8972	Disengage	. The face is looking better
8973	Disennoble	. **Facilitate**
8974	Disenroll	. It will facilitate work
8975	Disenslave	. To facilitate the work
8976	Disentail	. **Facility (Facilities)**
8977	Disentitle	. Will give you every facility
8978	Disentomb	. There is every facility at hand for working the mine
8979	Disentwine	. What facilities are there [cheaply
8980	Disespouse	. There is (are) no facility(ies) for
8981	Disesteem	. Facilities for exploration
8982	Disfancies	. What facility(ies) have you

No.	Code Word.	Facility (Facilities) (*continued*)
8983	Disfancy . .	Every facility will be given
8984	Disfashion .	Give every facility
8985	Disfavour .	No working facilities whatever
8986	Disfeature .	Working facilities very inadequate
8987	Disfigured .	Working facilities fair
8988	Disflesh . .	Working facilities capable of great improvement
8989	Disfriar . .	Working facilities good
8990	Disgarland .	Working facilities specially good
8991	Disgarnish .	Have you facilities for
8992	Disgavel . .	Further facilities
8993	Disglorify .	**Fact(s)**
8994	Disglory . .	Has (have) ascertained the fact that
8995	Disgorge . .	The real facts of the case (are)
8996	Disgorging .	It is not the fact
8997	Disgospel .	Is it a fact (that)
8998	Disgrace . .	The facts are not as stated
8999	Disgregate .	What are the real facts of the case
9000	Disguised .	If the actual facts can be ascertained
9001	Disguising .	From the fact of
9002	Disgustful .	The facts, as disclosed, indicate
9003	Dishabille .	Is a very unpleasant fact
9004	Dishabit . . .	Cannot conceal the fact
9005	Disharmony	The fact must be known
9006	Dishaunted .	**Fail**
9007	Dishcatch .	Do not fail to
9008	Dishcloths .	Will not fail to
9009	Dishclout .	Must without fail
9010	Dishearten .	Must not fail to
9011	Disheir . .	Let us know without fail
9012	Disheiring .	**Failed**
9013	Dishelmed .	Has (have) failed to
9014	Disheritor .	Has (have) failed to make connection
9015	Dishevel . .	Has (have) not failed
9016	Dishful . .	The trial has failed
9017	Dishonest .	Failed in the attempt to
9018	Dishonour .	**Failure**
9019	Dishumour .	What is the failure due to
9020	Dishwasher .	Owing to the failure of
9021	Disimpark .	The failure is due to
9022	Disimprove .	The failure of
9023	Disincline .	To avoid failure
9024	Disinfects .	Is an entire failure
9025	Disinherit .	Reported to be a failure
9026	Disinhume .	Is report of ——'s failure correct
9027	Disinter . .	The report of failure is correct
9028	Disinure . .	A very heavy failure
9029	Disinvolve .	Affected by the failure (of)
9030	Disjection .	The cause of the failure (of)
9031	Disjoin . .	The process is an entire failure
9032	Disjoining .	**Fair**

No.	Code Word.	Fair (*continued*)
9033	Disjointly .	Very fair
9034	Disjunct . .	Not very fair
9035	Dislady . .	In a fair way
9036	Dislikable .	Not in a fair way
9037	Dislikeful .	Are things in a fair way (to)
9038	Disliking .	It would not be fair
9039	Dislimb . .	Think it would be more fair (if)
9040	Dislimbing .	It would be perfectly fair
9041	Dislinked .	We are in a fair way to
9042	Dislocate .	**Fairly**
9043	Dislodge . .	Has (have) acted very fairly
9044	Dislodging .	Has (have) not acted fairly
9045	Disloyal . .	Has (have) ——— acted fairly with regard to
9046	Disloyally .	**Fairness**
9047	Dismal . .	(In) fairness to both (or all) parties
9048	Dismally . .	In fairness to
9049	Dismalness .	**Faith**
9050	Dismantle .	In good faith
9051	Dismarshal .	In order to keep good faith with
9052	Dismasked .	Not in good faith
9053	Dismayful .	Good faith with all concerned
9054	Dismaying .	Am (are) beginning to lose faith in
9055	Dismember .	Has (have) lost faith in
9056	Dismettled .	There is no faith to be placed in
9057	Dismiss . .	**Fall**
9058	Dismissals .	A fall in stocks is imminent
9059	Dismissing .	A fall in our shares
9060	Dismissive .	Be prepared for a fall in
9061	Dismounted	Will cause a fall (in)
9062	Disnatured .	A fall of ——— inches per
9063	Disnest . .	A fall of ——— feet
9064	Disnesting .	**Fallen**
9065	Disobey . .	Has (have) fallen
9066	Disobeyeth .	Has (have) not yet fallen
9067	Disobeying .	Shares have fallen on the report that
9068	Disoblige .	Our shares have fallen
9069	Disomatous .	Shares have fallen from ——— to
9070	Disopinion .	**Falling**
9071	Disorder . .	Stocks are all falling
9072	Disorderly .	Stocks continue falling
9073	Disorient .	To prevent the shaft from falling in
9074	Disowneth .	Falling in
9075	Disownment	Falling off
9076	Disoxidate .	There is a great falling off in
9077	Dispansion .	**Familiar**
9078	Disparage .	Quite familiar with
9079	Disparity .	Must be familiar with
9080	Dispark . .	**Famine**
9081	Dispathy .	There is every prospect of a famine
9082	Dispauper .	Are face to face with famine

No.	Code Word.	Famine (*continued*)
9083	Dispel .	The famine is very severe
9084	Dispelled .	**Famous**
9085	Dispensary .	The —— mine(s) has (have) always been famous.
9086	Dispeopled .	**Fancy**
9087	Disperance .	Is (are) putting a fancy price upon the property
9088	Disperge .	. **Far**
9089	Dispermous .	Far down
9090	Dispersal .	Far up
9091	Dispersive .	Far from
9092	Dispirited .	Far in
9093	Displacing .	How far
9094	Displant .	Not far from
9095	Displayeth . .	As far as you can
9096	Displeased .	How far will you go
9097	Displosion .	Is it far to
9098	Displosive .	Far off
9099	Disponged .	Far from wishing —— to
9100	Disport . .	As far as can be seen
9101	Disposeth .	As far back as
9102	Disposing .	If it is not too far
9103	Dispositor .	We are still far from
9104	Dispossess .	**Farther.** (See also Further.)
9105	Disposure .	To sink farther
9106	Dispraise .	To drive farther
9107	Disprepare .	Gone up farther
9108	Disprison . .	Will have to go on farther
9109	Disprized .	Not much farther
9110	Disprizing .	A good deal farther
9111	Disprofess .	Can go no farther
9112	Disprofit . .	Can go farther in the direction of
9113	Disproof . .	If we go any farther
9114	Disproval .	How much farther
9115	Dispurpose .	The farther we go
9116	Dispurvey .	As we go farther, —— appears more promising
9117	Disputable .	To go no farther
9118	Disputed .	Farther to the
9119	Disputing .	Farther away (from)
9120	Disqualify .	Farther toward
9121	Disquiet .	Farther progress (toward)
9122	Disquietal .	On the farther side of
9123	Disquietly .	**Farthest**
9124	Disranked .	At the farthest point
9125	Disranking .	What is the farthest
9126	Disregard .	Not the farthest
9127	Disregular .	To the farthest
9128	Disrelish . .	The farthest point reached
9129	Disrepair .	The farthest point to which we have driven
9130	Disrepute .	**Fast**
9131	Disrespect .	As fast as possible
9132	Disrobe . .	Not as fast as it ought

No.	Code Word.	**Fast** (*continued*)
9133	Disrobing .	Push on as fast as you can
9134	Disrooted .	As fast as circumstances permit
9135	Disruly . .	Not so fast as we should like
9136	Disrupt .	. **Faster**
9137	Disrupting .	Cannot get on faster
9138	Disruption .	Shall be able to get on faster
9139	Disrupture .	(To) get the work done faster
9140	Dissatisfy	. **Fastest**
9141	Disscatter .	The fastest speed at which we can drive
9142	Dissect . .	The fastest speed attainable
9143	Dissecting	. **Fathom(s)**
9144	Dissection	Shaft has been sunk —— fathoms
9145	Disseize . .	Fathoms in depth
9146	Disseizors .	Fathoms apart
9147	Dissemble .	One fathom
9148	Dissented .	The —— fathom seam
9149	Dissertate .	The —— fathom level
9150	Disservice .	Fathoms below
9151	Dissevered .	How many fathoms
9152	Disshadow	. **Fault(y)**
9153	Dissheathe .	Whose fault is it
9154	Disship . .	It is the fault of
9155	Disshipped .	Through no fault of
9156	Disshiver .	Through a fault in the
9157	Dissight . .	What is the fault
9158	Dissilient .	Was (were) in fault
9159	Dissimilar .	Have come across a fault
9160	Dissipable .	Do not consider it is fault of
9161	Dissipated .	Faulty in construction
9162	Dissocial.	. **Faulted**
9163	Dissoluble .	The vein has faulted
9164	Dissolute	. **Favour**
9165	Dissolved .	Is (are) in favour of
9166	Dissolving .	Is (are) not in favour of
9167	Dissonancy .	In favour of
9168	Dissonant .	Not in favour of
9169	Disspirit . .	In my (our) favour
9170	Dissuade .	In your favour
9171	Dissuading .	In his (their) favour
9172	Dissuasion .	Can you do me (us) the favour (to)
9173	Dissuasory .	Ask(s) the favour of
9174	Disssweeten .	Think it will be in our favour
9175	Distackle .	Judgment is in our favour, with costs
9176	Distaffs . .	If in our favour
9177	Distancial .	Decision given in favour of
9178	Distancing .	Are you in favour of
9179	Distantly .	Without fear or favour
9180	Distaste . .	The utmost favour that can be
9181	Distasting .	Should be in favour of
9182	Distasture	. **Favourable**

No.	Code Word.	Favourable (*continued*)
9183	Distemper .	I consider the prospects most favourable
9184	Disthene .	Are the prospects favourable or otherwise
9185	Disthroned .	Do you consider the prospects more favourable
9186	Distichous .	It is a favourable moment to
9187	Distillate .	It is not a favourable moment to
9188	Distillery .	Is in a favourable condition
9189	Distilleth .	Is not in a favourable condition
9190	Distilment .	Is it favourable
9191	Distinct . .	Favourable news
9192	Distinctly .	Accounts are more favourable
9193	Distinctor .	Accounts are less favourable
9194	Distitle . .	Appearances are favourable
9195	Distortion .	Everything seems favourable (for)
9196	Distortive .	A very favourable report
9197	Distracted .	The report is very favourable
9198	Distrainor .	Favourable to our interests
9199	Distraught .	If the report is favourable
9200	Distream .	Of a highly favourable character for
9201	Distressed .	A favourable result may be looked for
9202	Distribute .	Very favourable results have been attained
9203	Distringas .	Under the most favourable circumstances
9204	Distrouble .	If you find the prospect favourable
9205	Disturb . .	There are favourable indications of
9206	Disturbing .	Very favourably situated (for)
9207	Distutor . .	**Fear(s)**
9208	Distyle . .	Do you think there is any fear
9209	Disulphide .	Have no fear as to the result
9210	Disuniform .	Do not have any fear
9211	Disunion. .	Our only fear is
9212	Disunited .	There is great reason to fear
9213	Disuniting .	If you have any fear
9214	Disusage .	No reason for fear or alarm
9215	Disutilize .	Has given rise to fears
9216	Disvalue . .	Our fears have been realized
9217	Disvaluing .	**Feasible**
9218	Disvelop. .	Would it be feasible
9219	Disvouched.	It would be feasible
9220	Diswitted .	It would not be feasible
9221	Disworship .	Even if feasible, we do not think it advisable
9222	Disworths .	**Fee(s)**
9223	Disyoke . .	What is your fee
9224	Disyoking .	What will be your fee
9225	Ditchdog .	My fee is —— per day and expenses
9226	Ditchwater .	My fee and expenses amount to
9227	Dithecal . .	My fee will be
9228	Ditheist . .	Do not give up report till you get the fee
9229	Ditheistic .	Who will pay fee and expenses
9230	Dithyramb .	Will pay fee and expenses
9231	Ditriglyph .	Has (have) paid —— towards the fee
9232	Ditrochean .	Has (have) paid —— towards fee and expenses

No.	Code Word.	**Fee(s)** (*continued*)
9233	Dittology	Do not go unless —— has (have) paid fee
9234	Dittybag	Have sent report, fee and expenses amount to ——, which please get before giving up the report

Feel

9235	Diuresis	**Feel**
9236	Diurnal	Feel very strongly on this matter
9237	Diurnalist	**Feeling**
9238	Diurnally	There is a strong feeling in favour of
9239	Diuturnity	There is a strong feeling against
9240	Divan	What is the present feeling
9241	Divaricate	A feeling of greater confidence exists
9242	Divedapper	A feeling of alarm has arisen
9243	Divellent	Feeling here is very good, people are buying
9244	Divergency	Feeling here is very bad, people are selling
9245	Diversify	In the present state of feeling
9246	Diversions	The prevailing feeling is one of
9247	Diverted	There is a much better feeling
9248	Diverticle	Will create a bad feeling
9249	Diverting	In the prevailing feeling of anxiety
9250	Divertised	**Feet**
9251	Divestment	Shaft is down —— feet
9252	Divide	Vein varying from —— feet to —— feet
9253	Dividends	Winze is down —— feet below
9254	Dividing	Tunnel has been extended —— feet
9255	Dividual	Tunnel is in —— feet
9256	Dividually	Vein has been prospected —— feet
9257	Dividuous	Is (are) about —— feet from
9258	Divination	How many feet
9259	Divinatory	Cubic feet
9260	Divinely	Cubic feet per second
9261	Divingbell	Cubic feet per minute
9262	Divinities	At the rate of —— feet per day
9263	Divinized	At the rate of —— feet per month
9264	Divisibly	Square feet
9265	Divisional	Feet square
9266	Divisor	Feet wide
9267	Divorce	Feet long
9268	Divorcible	Feet deep
9269	Divulgate	Feet high
9270	Divulged	Feet in diameter
9271	Divulging	Feet in circumference
9272	Divulsive	Running feet
9273	Dizened	Thousand square feet
9274	Dizzied	Average about —— feet
9275	Dizzy	Feet of water
9276	Dizzying	To a depth of —— feet
9277	Docetic	At a depth of —— feet
9278	Dochmius	Feet below
9279	Docible	We are now down —— feet
9280	Docility	From —— to —— feet
9281	Docimacy	**Felspar**

No.	Code Word.	
9282	Docimastic .	**Felspathic rock**
9283	Docimology.	The hanging wall is felspathic rock
9284	Dockage. .	The foot wall is felspathic rock
9285	Dockcress .	**Ferruginous**
9286	Docketed .	Ferruginous quartz
9287	Dockmaster.	Ferruginous matter
9288	Dockrent .	**Fever**
9289	Dockyard .	There has been an outbreak of fever
9290	Doctor . .	Malarial fever
9291	Doctorally .	Yellow fever
9292	Doctorated .	Prostrated with attack of fever
9293	Doctoress .	From fever and ague
9294	Doctorfish .	**Few**
9295	Doctorship .	Very few
9296	Doctrinal .	Only a few
9297	Document .	There are very few
9298	Documental.	To get a few
9299	Doddart . .	Send a few
9300	Doddering .	**Fiery**
9301	Dodecagon .	Fiery seams
9302	Dodecander.	Owing to the fiery nature of the seams
9303	Dodge . .	The seams are very fiery
9304	Dodging . .	A very fiery mine
9305	Dodipate .	**Figure(s)**
9306	Dodipoll. .	At this figure
9307	Dodrans . .	About the figure
9308	Doffing . .	Beyond this figure
9309	Dogapes. .	Mistake in the figures
9310	Dogbolt . .	These figures will show
9311	Dogbrier. .	Too high a figure
9312	Dogcabbage.	At a reasonable figure
9313	Dogcarts. .	Must have exact figures
9314	Dogcheap .	The exact figures are not to be had
9315	Dogdays . .	This is as close as we can figure
9316	Dogdraw. .	It figures out at .
9317	Dogfancier .	Cannot alter the figures
9318	Dogfish . .	Cannot give the exact figure
9319	Dogflies . .	Do not think figures are correct—check them
9320	Dogfly . .	The figures must be checked
9321	Dogfox . .	If you have reason to alter your figures
9322	Doggedly .	**Fill**
9323	Doggedness.	Can you fill in the time between
9324	Doggerel .	To fill in the time between
9325	Doggerman .	Mine beginning to fill
9326	Doghearted .	Fill up
9327	Dogleach .	Fill up the old workings
9328	Doglegged .	Fill up the stopes
9329	Doglichen .	Preparing to fill up
9330	Doglooked .	Do not fill
9331	Dogmatic .	If you can fill

No.	Code Word.	Fill (*continued*)
9332	Dogmatical .	Shall fill in (or up)
9333	Dogmatisms	**Filled**
9334	Dogmatized.	The stopes are filled with
9335	Dogmatory .	Filled up
9336	Dogparsley .	Filled up the old workings
9337	Dogsbane .	The shaft is filled up with débris
9338	Dogsfennel .	**Filling**
9339	Dogsick . .	Filling with water
9340	Dogskin . .	Filling up fast
9341	Dogsleep .	The filling in will take
9342	Dogsmeat .	Filling up with débris
9343	Dogsrue . .	**Final**
9344	Dogstail . .	Waiting for the final
9345	Dogstones .	Is this final
9346	Dogstongue.	This is final
9347	Dogstooth .	(To) take it as final
9348	Dogwatch .	As a final result
9349	Dogweary .	A final interview (with)
9350	Dogwhelk .	Final arrangement
9351	Doings .	**Finally**
9352	Doldrums .	We have finally determined
9353	Dolebeer .	Finally settled
9354	Dolebread .	Have finally succeeded
9355	Doleful .	**Finance(s)**
9356	Dolefully .	Can you finance
9357	Dolemeadow	The state of the finances
9358	Dolent . .	The state of the finances will not warrant
9359	Dolerite . .	In the present state of the finances
9360	Dolesome .	Finances much embarrassed
9361	Dolesomely .	Not easy to finance
9362	Dolichos. .	**Financial**
9363	Dolichurus .	Financial statement
9364	Doliolum .	The financial results
9365	Dollyshop .	From a financial point of view
9366	Dolorific. .	Financial difficulties
9367	Dolorous .	Good financial house(s)
9368	Dolorously .	Financial arrangements
9369	Dolphin . .	Financial assistance
9370	Dolphinfly .	Will give us financial aid
9371	Doltish .	**Financing**
9372	Domain . .	Will undertake the financing
9373	Domanial .	Requires for financing
9374	Dombeya .	Terms for financing
9375	Domboc . .	If we (you) can succeed in financing
9376	Domestic .	Has (have) succeeded in financing
9377	Domestical .	Aided us in financing
9378	Domiciliar .	Will assist in financing
9379	Domified .	**Find**
9380	Domify . .	I (we) find on further examination
9381	Dominancy .	I (we) find on looking into the matter

No.	Code Word.	**Find** (*continued*)
9382	Dominant .	Find(s) that
9383	Domination.	Will find (out)
9384	Dominative .	Find out about
9385	Domineer .	Cannot find —— at the address given
9386	Dominicide .	Cannot find out
9387	Dominions .	From all I (we) can find out
9388	Donatist . .	Find out from
9389	Donatistic .	Cannot find
9390	Donaught .	You must find
9391	Donax . .	We must find
9392	Donkey . .	Shall have to find
9393	Donkeyman.	If you find
9394	Donkeypump	If you do not find
9395	Donnism .	If we find that
9396	Donors . .	If he finds that (if they find that)
9397	Doodlesack.	Undertake(s) to find. (*Should be* Dudelsack)
9398	Doomful. .	Where shall we find
9399	Doomsday .	Have been unable to find any
9400	Doomsman .	You will find (that)
9401	Doomster .	As soon as you (we) find
9402	Doonga . .	**Fine**
9403	Doorcase .	Of very fine quality
9404	Doorframe .	Very fine specimens
9405	Dooring . .	Not very fine
9406	Doorkeeper.	Will be subject to a fine unless
9407	Doornail. .	What is the fine
9408	Doorplate .	The fine is (will be)
9409	Doorstead .	Upon payment of fine
9410	Doorsteps .	Fine (was) imposed upon
9411	Doorstone .	The fine has been remitted
9412	Doorway .	The fine has been paid
9413	Dorian . .	Fine reduced to
9414	Doricism .	There is fine ore in
9415	Dorippe . .	Fine ore continues to be taken (from)
9416	Dormant .	There is a fine prospect for
9417	Dormouse .	**Finish**
9418	Dorrhawk .	When can you finish
9419	Dorstenia .	I (we) hope to finish here by
9420	Doryphora .	Finish as quickly as possible
9421	Dosithean .	Shall finish
9422	Dotage . .	If you can finish by
9423	Dotardly. .	If you cannot finish (before)
9424	Doublebank.	Cannot finish till
9425	Doublebass .	**Finished.** (See also Examination.)
9426	Doublecone.	If you have finished
9427	Doubledyed.	If you have not finished [detained by
9428	Doubleface .	Expected to have finished by ——, but am (are)
9429	Doublegear .	When —— has (have) finished
9430	Doublegild .	When you have finished
9431	Doublehung.	Have finished putting up —— machinery

No.	Code Word.	Finished (*continued*)
9432	Doublelock .	Not yet finished
9433	Doubleplea .	Has (have) finished
9434	Doublestop .	Has (have) not finished
9435	Doublets . .	Could not be finished in time (to send)
9436	Doubling .	I (we) have not quite finished
9437	Doubly . .	Will be finished
9438	Doubtable .	Will be finished on or about
9439	Doubtful . .	The work is now finished
9440	Doubtfully .	Shall have finished here (by)
9441	Doubtingly .	As soon as it is finished
9442	Doubtless .	**Fire**
9443	Doubtous .	Fire has broken out in the mine
9444	Doucepere .	Fire has destroyed
9445	Doucker . .	The mill has been destroyed by fire
9446	Doughfaced .	Has (have) been destroyed by fire
9447	Doughnut .	A fire in the —— is raging
9448	Doughtiest .	In order to extinguish the fire
9449	Doughtily .	Since the fire
9450	Doutance .	On fire
9451	Dovecots .	The camp is on fire
9452	Doveeyed .	Cause of fire
9453	Dovekie . .	Timber in shaft is on fire
9454	Dovelet . .	' Has been much damaged by the fire
9455	Doveplant .	Forest fire broken out
9456	Dovesfoot .	Fire still raging, camp is in danger
9457	Doveship .	Totally destroyed by the fire
9458	Dovetail . .	Partially destroyed by the fire
9459	Dowager . .	Not much damage done by fire
9460	Doweljoint .	Estimated damage by the fire
9461	Dowelled .	Fire in wood-yard
9462	Dowelpin .	On account of the fire
9463	Downbeard .	Vigorous efforts made to extinguish the fire
9464	Downcast .	Precautions against fire
9465	Downcome .	Fire-extinctors
9466	Downfallen .	The fire has been totally extinguished
9467	Downgyved .	Fire still spreading
9468	Downhaul .	The loss by the fire will exceed
9469	Downiness .	Fire has again broken out
9470	Downline .	Could not save anything from the fire
9471	Downlooked	**Fire-damp**
9472	Downlying .	Explosion of fire-damp
9473	Downpour .	Fire-damp in the mine
9474	Downright .	**Firm(s)**
9475	Downrush .	Be firm with
9476	Downsett .	Firm at the close
9477	Downshare .	Firm at the opening, but closed weak
9478	Downstairs .	Unless you are firm
9479	Downsteepy .	Not firm enough
9480	Downstroke .	A firm offer has been made to us
9481	Downthrow .	We have made a firm offer

No.	Code Word.	**Firm(s)** (*continued*)
9482	Downtrain .	If a firm offer is made
9483	Downtrod .	Do you know the following firm(s)
9484	Downweed .	The firm consists of
9485	Downweigh .	The firm(s) you ask about is (are) highly respectable
9486	Doxologize .	Frm(s) is (are) very doubtful
9487	Doxology .	Firm(s) has (have) a very poor reputation
9488	Drabbish .	Ascertain the position of the firm
9489	Dracanth .	Have nothing to do with the firm named
9490	Drachma .	It is a first-class firm
9491	Draconic .	A well-known and most respectable firm
9492	Dracontium .	The firm is not of much account
9493	Drafthorse .	The firm is not known here
9494	Draftox . .	The firm is in difficulties
9495	Draftsman .	The firm failed some time ago
9496	Dragantine .	The firm failed recently
9497	Dragbar . .	**Firmly**
9498	Dragchain .	Firmly established
9499	Draggle . .	Not firmly established
9500	Draghook .	Am (are) firmly convinced
9501	Draglink . .	**First**
9502	Dragnets .	First opportunity
9503	Dragon . .	From first to last
9504	Dragonade .	The first thing to be done
9505	Dragonbeam	First call
9506	Dragonfish .	First class
9507	Dragonfly .	In the first place
9508	Dragontree .	First chance you have
9509	Dragooner .	For the first few days
9510	Dragooning .	The first man to
9511	Dragsheet .	Has a first-class record
9512	Drainable .	When (it) was first
9513	Drained . .	On the first of the month
9514	Draintiles .	The first tests made
9515	Draintrap .	Our first offer
9516	Drake . .	The first offer
9517	Drakeflies .	From the very first
9518	Drakestone .	For the first time
9519	Dramatical .	The first few days
9520	Dramatize .	When was the first
9521	Dramaturgy .	If not the first
9522	Dramshop .	**Fissure**
9523	Draperied .	True fissure vein
9524	Drapery . .	A contact fissure
9525	Drastic . .	A well-defined fissure vein
9526	Draughtbar .	Is it a true fissure vein
9527	Draughty .	A true and well-defined fissure vein running
9528	Dravidian .	A fissure vein with a dip of ——°
9529	Drawback .	**Fit**
9530	Drawbolt .	Not fit for anything
9531	Drawboys .	Not at all fit for our purpose(s)

No.	Code Word.	**Fit** (*continued*)
9532	Drawbridge .	Can you fit up
9533	Drawcansir .	We can fit up
9534	Drawcut . .	We cannot fit up
9535	Drawgates .	**Fitted**
9536	Drawgear .	Is (are) eminently fitted for
9537	Drawgloves .	Is (are) not fitted for
9538	Drawingawl .	Is (are) fitted with
9539	Drawingpen .	Is (are) not fitted with
9540	Drawings .	Has (have) been fitted with
9541	Drawknife .	Has (have) not been fitted with
9542	Drawlatch .	Ought to have been fitted with
9543	Drawn . .	Must be fitted with
9544	Drawplate .	Have fitted up
9545	Drawwell .	Not properly fitted
9546	Drayage . .	Fitted up in accordance with
9547	Draycarts .	Well fitted for the purpose (of)
9548	Drayhorses .	**Fix**
9549	Drayman .	In a fix about
9550	Drayplough .	If you can fix
9551	Dreadable .	If you cannot fix
9552	Dreadful .	Can you fix a time for
9553	Dreadfully .	Cannot yet fix a time for
9554	Dreadingly .	We propose to fix it
9555	Dreadless .	**Fixed**
9556	Dreadly . .	Has (have) fixed
9557	Dreameth .	Has (have) not fixed
9558	Drearihood .	Is not fixed
9559	Drearily . .	Has (have) —— fixed
9560	Dreariment .	The time fixed
9561	Dreariness .	The price fixed
9562	Drearisome .	Have you fixed the time for
9563	Dredge . .	**Flat(s)**
9564	Dredgebox .	The country around is very flat
9565	Dredgeman .	The ore occurs in flats
9566	Dredging .	The flats extend for a long distance
9567	Dreggy . .	In flats
9568	Dregs . . .	**Flaw**
9569	Drench . .	Owing to a flaw in the agreement
9570	Dresscoat .	Owing to a flaw
9571	Dressings .	There is a flaw
9572	Dressmaker .	There is a flaw in the agreement
9573	Dribblets .	The flaw in the agreement has been rectified
9574	Driftage . .	Took advantage of a flaw in
9575	Driftbolt .	**Flinty**
9576	Driftland .	Hard flinty rock
9577	Driftless . .	Owing to the flinty nature of the rock
9578	Driftnet .	**Float**
9579	Driftsail . .	Prepared to float company
9580	Driftway .	Will you endeavour to float a (the) company
9581	Driftwind .	To float the company, we require

H

No.	Code Word.	Float (*continued*)
9582	Driftwood .	Can float the company
9583	Drillbow	. **Floated**
9584	Drillpress .	To be floated
9585	Drillstock .	Company has been floated
9586	Drimys . .	Company has not been floated
9587	Drink. . .	Company floated in
9588	Drinkable .	Company cannot be floated yet
9589	Drinking .	Company cannot be floated until (or unless)
9590	Drinkless .	Can only be floated if
9591	Drinkmoney	The company can ·be floated here
9592	Dripstones .	Company cannot be floated without capital is sub-
9593	Drivelled .	When it is floated [scribed to the amount of
9594	Drivelling .	Company floated ; capital
9595	Driverant .	If the company is floated
9596	Driverboom.	After the company was floated
9597	Driveth . .	Unless the company is floated before
9598	Driving .	. **Float-gold**
9599	Drivingbox .	Cannot save the float-gold
9600	Drizzling .	Contains float-gold
9601	Drizzly . .	Percentage of gold is float
9602	Drogher . .	The gold lost occurs in the form of float
9603	Droitschka .	So much float-gold is lost
9604	Droitural .	Endeavour to save the float-gold
9605	Droll . . .	The greater part of the float-gold is lost
9606	Drolleries .	Claims to save most of the float-gold by this process
9607	Drollery .	. **Floating**
9608	Drollish . .	No chance of floating
9609	Dromedary .	Floating reef
9610	Dromond .	Property is worth floating for
9611	Dronebees .	Property is not worth floating
9612	Dronefly .	. **Flood(s)**
9613	Dronishly .	Subject to severe floods
9614	Droop . .	The flood is subsiding
9615	Droopeth .	A bad flood has destroyed
9616	Drooping .	The flood is rising steadily
9617	Droopingly .	The flood is increasing rapidly
9618	Dropax . .	Owing to the floods, we cannot
9619	Drophammer	Owing to the flood, have had to
9620	Dropletter .	The place is nearly submerged by the flood
9621	Droplight .	Heavy floods prevail
9622	Dropmeal .	After the flood had subsided
9623	Dropped . .	As soon as the flood subsides
9624	Droppress .	The floods have subsided
9625	Dropscene .	Flood subsided ; can now re-start
9626	Dropsical	. **Flooded**
9627	Dropsied .	The market is flooded with
9628	Dropsy . .	Heavy storms ; fear we shall be flooded
9629	Droptable .	The mine is flooded
9630	Droptin . .	The roads are flooded
9631	Dropwise .	The country round is flooded

No.	Code Word.	
9632	Dropwort	**Flooding**
9633	Drosera . .	Flooding the market
9634	Drosky . .	**Floor(s)**
9635	Drosometer .	Depositing floors
9636	Drosophila .	On the depositing floors
9637	Drossel . .	Dressing floors
9638	Drover . .	The dumping floor
9639	Drownage .	The drying floor
9640	Drowse . .	**Flow**
9641	Drowsily . .	Has ceased to flow
9642	Drowsiness .	To stop the flow
9643	Drowsyhead	The flow of water (in)
9644	Drudgery .	The flow of water in the mine
9645	Druggist . .	The flow of water from the
9646	Drugster . .	The flow of water is at the rate of about ——
9647	Druid . . .	In order to check the flow [gallons per minute
9648	Druidess . .	Have checked the flow
9649	Druidic . .	The flow of water is equal to —— miners' inches
9650	Druidical	**Fluctuated**
9651	Drumfish .	Fluctuated very much
9652	Drumhead .	Fluctuated between
9653	Drummajor .	The price of the shares fluctuated between
9654	Drummers .	**Fluctuation(s)**
9655	Drumroom .	On account of the fluctuations in the market
9656	Drumstick .	The fluctuations in the price [from
9657	Drunkards .	There have been rapid fluctuations ; prices ranged
9658	Drunken .	We wish these fluctuations could be avoided
9659	Drunkenly .	Fluctuations in the yield
9660	Drunkship .	The effect of these repeated fluctuations is
9661	Drupaceae .	In order to check the fluctuations [be checked
9662	Drupaceous .	It is very important that these fluctuations should
9663	Drupel . .	How can these fluctuations be avoided
9664	Druxey . .	These fluctuations have a most prejudicial effect
9665	Dryandra .	**Flume(s)**
9666	Dryas . .	Flume(s) in good condition
9667	Dryasdust .	Flume(s) is (are) out of repair
9668	Drybeat . .	The flume(s) has (have) been destroyed
9669	Drybeating .	Will have to build a flume
9670	Dryboned .	A (the) flume from the hills
9671	Drycastor .	A (the) flume from the creek
9672	Drycupping .	The company's flume
9673	Drydock .	**Fluorspar**
9674	Dryfisted	The matrix is principally fluorspar
9675	Dryfoot . .	**Flux**
9676	Dryingroom .	Used as flux for
9677	Drymeasure .	Difficulty in getting fluxes
9678	Dryness .	**Follow(s)**
9679	Drynurse .	Will follow later
9680	Drypile . .	Follow the instructions (of)
9681	Drypoint . .	Will follow in a few days

No.	Code Word.	Follow(s) (*continued*)
9682	D'ryrent . .	Follow the vein
9683	Dryrub . .	(To) follow it up
9684	Dryrubbed .	We intend to follow
9685	Dryrubbing .	Cannot follow
9686	Drysalter .	As follows
9687	Dryshod .	**Followed**
9688	Drystoves .	To be followed by
9689	Dualism . .	Have followed the course of the vein —— feet
9690	Dualistic. .	Has (have) followed
9691	Dubiate . .	Has (have) not followed
9692	Dubiating .	You ought to have followed
9693	Dubiosity .	You ought not to have followed
9694	Dubiously .	If —— had followed
9695	Dubitable .	If our instructions had been followed
9696	Dubitancy .	Has (have) followed your instructions
9697	Dubitation .	Instructions must be strictly followed
9698	Dubitative .	Followed the usual course
9699	Ducal . .	Our orders must be followed
9700	Ducally . .	Should be followed up
9701	Ducatoons .	**Following**
9702	Duchess . .	Is (are) following
9703	Duchy . .	Am (are) following
9704	Duchycourt.	Is (are) not following
9705	Duckant . .	Am (are) not following
9706	Duckbills .	Is (are) —— following
9707	Duckhawk .	In the following manner
9708	Ducklegged.	We are now following up
9709	Duckling .	Following up the clue to the
9710	Duckmole .	On the following day
9711	Ducksfoot .	This month and the following
9712	Ducksmeat .	Arrived at the following conclusions
9713	Duckweed .	Following up the lead
9714	Ductile . .	Following the course of the vein
9715	Ductilely .	Following the outcrop of the vein
9716	Duddery .	Following our letter of
9397	Dudelsack .	Undertake(s) to find (*instead of* Doodlesack, p. 190).
9717	Duebill . .	Following your letter of
9718	Dueller . .	Following the report
9719	Duellist	**Foolish**
9720	Duellum . .	It would be very foolish
9721	Duffers . .	It would be foolish to ignore the fact
9722	Dufoil . .	Behaved in a foolish manner
9723	Dufrenite .	**Foot**
9724	Dugongs. .	One foot
9725	Dugout . .	From one foot
9726	Dukedom .	From one foot to —— feet
9727	Dukeship .	Not a foot of ground
9728	Dulcamara .	Every foot of
9729	Dulcet . .	**Footing**
9730	Dulciana .	Put matters on a better footing
9731	Dulcified .	Establish a firm footing

No.	Code Word.	**Footing** (*continued*)
9732	Dulcify . .	Gain a footing in
9733	Dulcifying .	On a firmer footing
9734	Dulciloquy .	**Foot-wall**
9735	Dulcimers .	Close to the foot-wall
9736	Dulcinist .	Against the foot-wall
9737	Dulcitude .	The foot-wall is
9738	Dulcorate .	From the foot-wall to
9739	Dulcose . .	Towards the foot-wall
9740	Dulia . . .	The foot-wall is well defined
9741	Dullardism .	The foot-wall dips at an angle of ——°
9742	Dullbrowed .	A solid and good foot-wall
9743	Dulleyed . . .	Have encountered the true foot-wall
9744	Dullhead .	From foot-wall to hanging-wall
9745	Dullish . .	Have not yet got to the foot wall
9746	Dullness . .	**For**
9747	Dulocracy .	For you
9748	Dumbarge .	For me (us)
9749	Dumbbells .	For him (them)
9750	Dumbcake .	Not for
9751	Dumbcraft .	For some time
9752	Dumbfound .	For fear of
9753	Dumbledor .	Is for
9754	Dumbshow .	Is not for
9755	Dumbwaiter	It is not for us to say
9756	Dumetose .	It is for you to say whether
9757	Dumpage .	For a few days
9758	Dumpingcar	For a short time
9759	Dumpylevel	For you to return
9760	Dunciad . .	For you to remain
9761	Duncow . .	For you to send
9762	Dunderhead	For what
9763	Dunderpate .	For how long
9764	Dundivers .	If only for
9765	Dungeoned .	For the next
9766	Dungeoning	For the past
9767	Dungfork .	**Forage**
9768	Dunghill . .	It is difficult to get forage for the horses
9769	Dungiyah .	Plenty of forage can be got
9770	Dungmeer .	Forage is cheap
9771	Dungyard .	Forage is dear
9772	Duodecimal .	**Forbidden**
9773	Duodecuple	Has (have) forbidden
9774	Duodenary .	Has (have) been forbidden
9775	Duodenum .	Has (have) not forbidden
9776	Duoliteral .	Has (have) not been forbidden
9777	Duping . .	Has (have) —— been forbidden
9778	Duplex . .	Has (have) forbidden the work to go on
9779	Duplexing .	Has (have) forbidden the mine to be shown
9780	Duplicate .	**Force**
9781	Duplicity . .	With great force

No.	Code Word.	**Force** (*continued*)
9782	Durably . .	Not a sufficient force
9783	Durance . .	Wish(es) to force us to
9784	Durbar . .	Do not force
9785	Duressor . .	Force on the action
9786	Duskily . .	How long is the agreement in force
9787	Duskiness .	Will be in force until
9788	Duskishly .	There is a force of good men, but not enough
9789	Dustbrand .	The present force sufficient
9790	Dustbrush .	The present force insufficient
9791	Dustcart . .	With all the force at our command
9792	Dusted . .	Employ all the force at your command
9793	Dustpoint .	All the available force
9794	Dustyfoot .	With the full force of
9795	Duteous . .	Still in force
9796	Duteously .	Not now in force
9797	Dutifully .	. **Forced**
9798	Dutyfree . .	We have been forced to
9799	Duumvir . .	Forced to withdraw
9800	Duumviral .	Forced to conclude that
9801	Dwarf . .	The suit has been forced on us
9802	Dwarfing .	. **Forcing**
9803	Dwarfish . .	Is (are) forcing us to
9804	Dwarfishly .	Forcing on the sale
9805	Dwarfwall .	Is (are) not forcing
9806	Dwelleth .	. **Forebreast**
9807	Dwelling .	In the forebreast
9808	Dwindle . .	The forebreast of the drift
9809	Dyadic . .	Forebreast is in
9810	Dyehouse .	**Foreclose**
9811	Dyersmoss .	Foreclose the mortgage
9812	Dyersweed .	Threaten to foreclose
9813	Dyestuff . .	Do not foreclose
9814	Dyewood .	Taking steps to foreclose
9815	Dyeworks .	**Foreclosed**
9816	Dying . .	Have foreclosed the mortgage
9817	Dyingly . .	**Foreclosure**
9818	Dyingness .	Taking steps to prevent foreclosure
9819	Dynameter .	The foreclosure of the mortgage
9820	Dynametric .	**Forego**
9821	Dynamical .	Must forego
9822	Dynamism .	Not to forego
9823	Dynast . .	If —— will forego
9824	Dynastidan .	**Forehead**
9825	Dynasties .	In the forehead
9826	Dyschroa .	The vein in the forehead
9827	Dysclasite .	**Foreman**
9828	Dyscrasy .	The foreman of the
9829	Dysentery .	The foreman told me (us)
9830	Dyslogy . .	The foreman can be relied on
9831	Dysodile . .	A good mill foreman

No.	Code Word.	**Foreman** (*continued*)
9832	Dysorexia .	A good underground foreman
9833	Dyspepsia .	The mill foreman
9834	Dyspeptic .	A competent and trustworthy foreman
9835	Dysphagia .	**Foresee**
9836	Dysphony .	Foresee a difficulty
9837	Dyspnoic .	I (we) plainly foresee
9838	Dysthetic .	Do you foresee
9839	Dysthymic .	Cannot foresee
9840	Dystomous .	We plainly foresee that if
9841	Dysuria .	**Foreseen**
9842	Dytiscidae .	It has been foreseen for some time past
9843	Dytiscus .	**Forest(s).** (See also Fire.)
9844	Eachwhere .	Fine forest of
9845	Eager. . .	No forest near
9846	Eagerly . .	The mine(s) is (are) surrounded by forests (of)
9847	Eagleeyed .	The forest is almost denuded of timber
9848	Eaglehawk .	The forest can supply plenty of timber
9849	Eagleowl .	From the forest in the vicinity
9850	Eagleray .	**Forfeit(s)**
9851	Eaglestone .	Forfeits all right or claim to
9852	Eaglewood .	**Forfeited**
9853	Earache .	Has (have) forfeited
9854	Earcap . .	This sum to be forfeited if
9855	Earcockle .	Has (have) been forfeited
9856	Eardrum. .	To be forfeited
9857	Earkissing .	The concession has been forfeited
9858	Earldom. .	The concession will be forfeited (unless)
9859	Earlier . .	Has forfeited all right to
9860	Earmark .	**Forfeiture**
9861	Earnest . .	Under forfeiture clause of agreement
9862	Earnestly .	Concession is in danger of forfeiture
9863	Earpiercer .	**Forged**
9864	Earshrift. .	Forged deeds
9865	Earthapple .	Property acquired by means of forged deeds
9866	Earthbath .	**Forgery(ies)**
9867	Earthboard .	The signature is a forgery
9868	Earthborer .	The endorsement is a forgery
9869	Earthbred .	Has been guilty of forgery
9870	Earthdrake .	The forgery of the deeds
9871	Earthflax .	Forgery is suspected
9872	Earthhog .	Large forgeries have been detected in scrip
9873	Earthhouse .	**Forget**
9874	Earthiness .	Do not forget
9875	Earthlings .	Do not forget to send
9876	Earthly . .	**Forgotten**
9877	Earthmad .	Has (have) been forgotten
9878	Earthoil . .	Has (have) forgotten
9879	Earthpeas .	Do not let it be forgotten that
9880	Earthpig. .	The document forgotten to be sent with our letter
9881	Earthplate .	Has forgotten his papers

No.	Code Word.	**Forgotten** (*continued*)
9882	Earthquake .	Have you forgotten your promise
9883	Earthshine .	Had forgotten to send
9884	Earthtable .	**Form**
9885	Earthwolf .	Form a company
9886	Earthworks .	Form a just opinion
9887	Eartrumpet .	Form any
9888	Earwax . .	This will form
9889	Earwigged .	Not to form
9890	Earwigging .	Cannot form
9891	Earwigs . .	Can you form
9892	Earwitness .	If we form a company
9893	Easeful . .	Form a syndicate
9894	Easements .	In proper form
9895	Easiness . .	Have not been able to form
9896	Easterday .	**Formation** .
9897	Easterdues .	Company now in course of formation
9898	Easteregg .	Formation of company
9899	Easterling .	After the formation of the company
9900	Eastwards .	By the formation of
9901	Easychair .	To hasten the formation of
9902	Easygoing .	To prevent the formation of
9903	Eatingroom .	A water-bearing formation
9904	Eavesboard .	The vein formation
9905	Eavescatch .	From the peculiar formation of [formation
9906	Eavesdrop .	We have learnt a good deal about the geological
9907	Ebbtide . .	From further knowledge of the formation
9908	Ebeneous .	From the formation of
9909	Ebonists .	**Formed**
9910	Ebony . .	A company has been formed to acquire
9911	Ebriety . .	A company has been formed with a capital of
9912	Ebriosity .	Should be formed
9913	Ebulliated .	Has (have) been formed
9914	Ebulliency .	Has not been formed
9915	Ebullient .	Can be formed
9916	Ebullition .	Cannot be formed
9917	Eburna . .	Will be formed
9918	Eburnation .	Will not be formed
9919	Ecballium .	Have you formed
9920	Ecbasis . .	Has (have) formed
9921	Ecbolic . .	If a company can be formed
9922	Eccentric .	As soon as the company is formed
9923	Ecchymosis .	Has (have) not formed [to purchase the mine
9924	Ecclesiast .	A company should be formed with a capital of ——
9925	Eccope . .	A syndicate now being formed
9926	Eccoprotic .	Have formed a syndicate to examine, and if thought
9927	Eccrisis . .	A syndicate to be formed [fit to develop
9928	Ecderon .	**Former**
9929	Echeloned .	On my former inspection
9930	Echeneis .	On a former occasion
9931	Echeveria .	The former owner(s)

No.	Code Word.	**Former** (*continued*)
9932	Echidna . .	The former owners bought it for
9933	Echidnine .	The former owners sold it for
9934	Echimyna .	By the former
9935	Echinated .	Against the former
9936	Echinidan .	Was bought by the former
9937	Echinoderm	Sold by the former (to)
9938	Echinomys .	**Formerly**
9939	Echinops .	Formerly owned by
9940	Echinozoa .	Formerly worked by
9941	Echinulate .	The mine was formerly owned by
9942	Echium . .	Formerly occupied by
9943	Echoed . .	Formerly owned the property, but it is now in the
9944	Echoing . .	**Formidable** [hands of
9945	Echometry .	A formidable opponent
9946	Eclampsy .	Not a very formidable
9947	Eclegm . .	Formidable opposition must be expected
9948	Eclipsed .	**Fortnightly**
9949	Eclipsing .	Fortnightly report(s)
9950	Ecliptic . .	Fortnightly report(s) must be sent by
9951	Eclysis . .	Send fortnightly reports regularly
9952	Economical .	Fortnightly accounts
9953	Economized	Fortnightly statements of
9954	Economy .	Fortnightly output
9955	Ecphlysis .	**Fortunate**
9956	Ecphonema	Have been so fortunate as to
9957	Ecphonesis .	At a fortunate moment
9958	Ecphora . .	You will be fortunate (if)
9959	Ecphractic .	It was a fortunate circumstance
9960	Ecstasied	**Forward**
9961	Ecstatical .	Please forward without delay
9962	Ecthlipsis .	Pushing forward as fast as possible
9963	Ecthyma . .	Get everything forward
9964	Ectoblast .	When can you forward
9965	Ectodermal .	Push the work forward
9966	Ectopy . .	Not sufficiently forward to
9967	Ectosarc . .	Pushing forward
9968	Ectozoa . .	If you cannot forward
9969	Ectropical .	From this time forward
9970	Ectropium .	We are now well forward with
9971	Ectylotic .	**Forwarded**
9972	Ectypal . .	Has (have) been forwarded
9973	Ecumenic .	Has (have) not yet been forwarded
9974	Ecumenical .	Has (have) forwarded
9975	Eczema . .	Has (have) not forwarded
9976	Eczematous .	Has (have) —— forwarded
9977	Edacious .	Will be forwarded
9978	Edaciously .	Your telegram was forwarded
9979	Edacity . .	To be forwarded
9980	Edaphodont	Not to be forwarded
9981	Eddaic . .	Forwarded by express

H 2

No.	Code Word.	**Forwarded** (*continued*)
9982	Eddied .	Have forwarded to the care of
9983	Eddying .	Forwarded to the address of
9984	Eddywater.	Have forwarded to your address
9985	Eddywind.	Should be forwarded (by)
9986	Edenized .	When was it (were they) forwarded
9987	Edenizing .	**Found**
9988	Edentulous	Has (have) been found
9989	Edgebone.	I (we) have found on further examination
9990	Edgerail .	I (we) have found everything very satisfactory
9991	Edgetool .	Found everything to be as represented (by)
9992	Edgewise .	Found everything all right
9993	Edging .	Found nothing
9994	Edgingiron	Not yet found
9995	Edible . .	Have you found
9996	Edibleness.	Has he (have they) found
9997	Edictal. .	We have found
9998	Edificant .	Has not yet been found
9999	Edifice. .	Was (were) found to be
10,000	Edificial .	**Foundation(s)**
10,001	Edifier . .	The foundation of
10,002	Edify . .	Laying the foundation of
10,003	Edifying .	The foundation plate
10,004	Edifyingly.	The foundations are giving way
10,005	Edileship .	The foundations for the mill
10,006	Editor . .	The foundations for the engine
10,007	Editorial .	To get in the foundations will require
10,008	Editorship	Now grading for foundation
10,009	Editress .	**Foundered**
10,010	Educated .	The ship by which they were sent foundered, and
10,011	Education.	**Framing** [all were lost
10,012	Eelbasket .	Framing for mill
10,013	Eelfork. .	Framing for battery
10,014	Eelgrass .	We are now framing
10,015	Eelspout .	**Fraud**
10,016	Effable. .	It is a regular fraud
10,017	Effacement	The whole thing is a regular fraud from beginning
10,018	Effacing .	Is it a fraud [to end
10,019	Effective .	Is charged with fraud
10,020	Effectual .	We think a charge of fraud could be sustained
10,021	Effectuous	**Fraudulent**
10,022	Effeminacy	Done with a fraudulent intent
10,023	Effeminize	Fraudulent practices somewhere
10,024	Efferous .	**Free**
10,025	Effervesce.	Is (are) free to
10,026	Effete . .	Is (are) not free to
10,027	Efficacy .	Is (are) —— free to
10,028	Efficiency .	Are you free to
10,029	Efficient .	You are free to act as you think best
10,030	Effigiated .	Must be free to act as I (we) think best
10,031	Effigy . .	Free gold

No.	Code Word.	**Free** (*continued*)
10,032	Efflation .	Full of visible free gold
10,033	Effloresce .	Specks of visible free gold
10,034	Effluency .	Abound(s) in free gold
10,035	Effluvial .	A streak of the vein very rich in free gold
10,036	Efflux . .	Pieces of quartz very rich in free gold
10,037	Effluxing .	Free milling ore
10,038	Effluxions .	Not free milling ore
10,039	Effodient .	Is the mine free of water
10,040	Efforming .	Do not go unless you hear the mine is free of water
10,041	Effortless .	Shall not go unless I (we) hear the mine is free of [water
10,042	Effracture .	When the mine is free of water
10,043	Effrayable .	The mine is now free of water
10,044	Effrayed .	Could not get the mine sufficiently free of water to
10,045	Effronted .	Free of water [enable me (us) to see
10,046	Effrontery .	Should be free to
10,047	Effulcrate .	Delivery free on board
10,048	Effulge . .	Price includes delivery free on board
10,049	Effulgence	We give you a free hand
10,050	Effulging .	If you let us have a free hand
10,051	Effumable .	Free of everything
10,052	Effusions .	Let us know if you are free
10,053	Effusive .	We are free to go when you wish
10,054	Effusively .	Would like to be set free
10,055	Eggbags .	Free of duty
10,056	Eggbird .	Free of expense
10,057	Eggcup .	Free of all expense to you
10,058	Eggery. .	Free of all expense to us
10,059	Eggflip. .	Free of all charges
10,060	Eggshot .	**Freely**
10,061	Eggler . .	Not freely
10,062	Eggplant .	Buy freely
10,063	Eggsauce .	Has been freely
10,064	Eggshell .	Shares have been freely bought, and are now at
10,065	Eggspoon .	**Freight(s)**
10,066	Egilopical .	Allowing for freight and charges
10,067	Egilops .	Owing to the heavy freight rates
10,068	Eglantine .	The freight and charges are
10,069	Egoical .	As freight
10,070	Egoism .	What is rate of freight
10,071	Egoistical .	Rate of freight is
10,072	Egophonic	Who is to pay freight and carriage
10,073	Egotheism	We have paid freight on this side
10,074	Egotistic .	You have nothing to pay for freight
10,075	Egotize .	You must pay freight and charges
10,076	Egregious .	Freight, duty, and insurance
10,077	Egriot . .	Freight, duty, and insurance, to port of discharge
10,078	Egritude .	Freight, duty, and insurance right up to the mine
10,079	Egyptology	Can you get a special rate of freight
10,080	Ehlite . .	Quoted a special rate of freight
10,081	Ehretia .	Freight per ton

No.	Code Word.	Freight (*continued*)
10,082	Eiderdown	Freight per 100 lbs.
10,083	Eiderduck.	Freight per carload
10,084	Eidolon .	Freight, in carloads, per 100 lbs.
10,085	Eightday .	Freight, less than carloads, per 100 lbs.
10,086	Eightfold .	Freight from —— to
10,087	Eightscore	Have engaged freight for
10,088	Eirenarch .	Cannot engage freight on terms quoted
10,089	Eisenrahm	Your terms for freight are too high
10,090	Eisteddfod	Quote lowest terms for freight
10,091	Ejaculated	At a low freight
10,092	Ejecting .	Freight is too high
10,093	Ejectment.	Have made contract for freight and carriage, to include all charges right up to
10,094	Elaborate .	Freight and all charges will be paid here
10,095	Elaidate .	Freights will be lower (after)
10,096	Elaphine .	Freights will be higher (after)
10,097	Elaqueated	Shippers claim for freight
10,098	Elastic . .	Railroad company claims for freight
10,099	Elatchee .	**Freighters**
10,100	Elatedly .	Who are the freighters
10,101	Elatedness	The freighters are
10,102	Elateridae.	The freighters undertake to
10,103	Elaterium .	**Frequent**
10,104	Elbow . .	Is of frequent occurrence in the district
10,105	Elbowchair	Very frequent
10,106	Elbowing .	Not frequent
10,107	Elbowpiece	Frequent stoppages
10,108	Elbowroom	Frequent complaints are made
10,109	Elbuck . .	What is the cause of the frequent
10,110	Elcaja . .	Owing to the frequent
10,111	Elcesaite .	**Frequently**
10,112	Elderberry	Do not occur so frequently
10,113	Eldergun .	Frequently occur (occurring)
10,114	Elderly .	Not so frequently
10,115	Eldership .	**Fresh**
10,116	Eldertree .	Telegraph if anything fresh occurs
10,117	Elderwine .	A fresh connection
10,118	Elderwort .	Nothing fresh
10,119	Electary .	We have nothing fresh to report (since)
10,120	Electicism.	Nothing fresh since we last wrote
10,121	Electively .	**Friable**
10,122	Electorate.	Friable quartz
10,123	Electoress.	Of a friable nature
10,124	Electorial .	**Friday**
10,125	Electrical .	Every Friday
10,126	Electrify .	Every other Friday
10,127	Electrizer .	Last Friday
10,128	Electrode .	Next Friday
10,129	Elegant . .	On Friday
10,130	Elegantly .	First Friday in the month

No.	Code Word.	**Friday** (*continued*)
10,131	Elegiac .	Second Friday in the month
10,132	Elegiacal .	Third Friday in the month
10,133	Elegiambic	Fourth Friday in the month
10,134	Elegiasts .	Last Friday in the month
10,135	Elegize .	**From**
10,136	Element .	From this date
10,137	Elementary	From all I can find out
10,138	Elementoid	From whom
10,139	Elenchical	From which
10,140	Elenchize .	From the other
10,141	Elenchus .	Not from
10,142	Eleocharis.	From one (of the)
10,143	Elephants.	From what I (we) can learn, I (we) advise you to
10,144	Eleusinian.	From the nearest
10,145	Eleutheria.	From whom did you get the
10,146	Elevates .	From what part did it come
10,147	Elevating .	From all parts
10,148	Elevation .	From here
10,149	Elevatory .	From some cause or other
10,150	Elfarrow .	From us
10,151	Elfbolt. .	Not from us
10,152	Elfchild .	If not from us
10,153	Elfdart. .	From you
10,154	Elfland. .	Not from you
10,155	Elflock. .	If not from you
10,156	Elfstone .	From what we hear
10,157	Eliciting .	**Frontier**
10,158	Eliminant.	Miles from the frontier
10,159	Eliquament	Near the frontier
10,160	Elixate. .	The frontier town of
10,161	Elixirs . .	From the frontier
10,162	Elknut. .	**Frost(s)**
10,163	Ellipsis .	The severe frosts prevent
10,164	Ellipsoid .	Owing to the severe frosts
10,165	Elliptical .	As soon as the frost breaks up
10,166	Ellwands .	Severe frosts for —— months in the year
10,167	Elmidae .	Sharp frost has set in; temperature has been to
10,168	Elmosfire .	The frost is breaking up
10,169	Elmwood .	The frost is all gone
10,170	Elocution .	**Frue vanner(s)**
10,171	Eloignment	Will require —— frue vanners
10,172	Elongated.	Frue vanners working well
10,173	Elongating	Frue vanners working badly
10,174	Elongation	There are —— frue vanners
10,175	Elopement	Frue vanners are being erected
10,176	Eloping .	Frue vanners are ready to work
10,177	Eloquent .	Frue vanners are stopped
10,178	Eloquently	Frue vanners will commence work
10,179	Elsewhere.	Frue vanners commenced work
10,180	Elsewise .	Frue vanners commenced work, and are doing well

No.	Code Word.	**Frue vanner(s)** (*continued*) -
10,181	Elucidate .	Frue vanners commenced work, and are not doing
10,182	Elumbated	Frue vanners are a failure [well
10,183	Elvan . .	Frue vanners stopped for water
10,184	Elvanite .	When will vanners be ready to start
10,185	Elvish . .	How are vanners working
10,186	Elvishly .	What percentage are vanners saving out of tailings
10,187	Elydoric .	Why are vanners stopped
10,188	Elytriform	How many vanners have you at work
10,189	Elytrocele.	Working expenses of vanner per ton treated
10,190	Elytroid .	**Fuel**
10,191	Emaciate .	What fuel do you use
10,192	Emaciation	The daily consumption of fuel is
10,193	Emanated.	Fuel is very abundant
10,194	Emanating	What is price of fuel
10,195	Emancipate	The price of fuel is about
10,196	Emancipist	Fuel is scarce and expensive
10,197	Embalm .	Fuel is so abundant, that the only cost is the cutting
10,198	Embalming	No fuel to be had [and stacking
10,199	Embanked	The fuel used is wood
10,200	Embankment	The fuel used is a resinous wood
10,201	Embargo .	The fuel used is anthracite coal
10,202	Embarrass	The fuel used is coal
10,203	Embarren.	The fuel used is coal of a poor quality
10,204	Embathe .	The fuel used is coal of excellent quality
10,205	Embed. .	The fuel used is coal brought from
10,206	Embedded	The fuel used is petroleum
10,207	Embedding	Fuel and water both abundant
10,208	Embedment	Have you any fuel
10,209	Embellish .	Wood only is available for fuel
10,210	Emberdays	We have sufficient fuel on the property to last
10,211	Emberfast.	Coal is sufficiently near to be reckoned on for fuel
10,212	Embergoose	**Fulfil**
10,213	Embertide.	So as to fulfil the terms of the bond
10,214	Emberweek	To fulfil the agreement
10,215	Embezzle .	(To) fulfil my (our) part of the
10,216	Embezzling	Cannot fulfil
10,217	Embillow .	Is (are) ready to fulfil the
10,218	Embitter .	(To) enable us to fulfil our promises
10,219	Emblazing	Can you fulfil the promises you made
10,220	Emblazoned	(To) fulfil the promises made to the shareholders
10,221	Emblematic	(To) fulfil our part of the agreement
10,222	Emblemized	To fulfil the contract
10,223	Emblems .	Fully believe we can fulfil our promises
10,224	Emblica .	If we do not fulfil our promises we must expect
10,225	Embloomed	Ready to fulfil
10,226	Emblossom	If you cannot fulfil
10,227	Embodiers	If he (they) cannot fulfil
10,228	Embody .	**Fulfilled**
10,229	Embodying	Fulfilled the agreement
10,230	Embogue .	Fulfilled the contract

No.	Code Word.	**Fulfilled** (*continued*)
10,231	Emboguing	Fulfilled so much of the
10,232	Embolden	**Fulfilment**
10,233	Embolism.	In fulfilment of the agreement
10,234	Embolismal	In fulfilment of the contract
10,235	Embosom.	In fulfilment of our promises
10,236	Embossed.	**Full**
10,237	Embossing	Could see nothing, as the mine was full of water
10,238	Embossment	The —— shaft is full of water
10,239	Embottled	All the lower workings are full of water
10,240	Embowelled	All workings full of water
10,241	Embox. .	Is always full of water
10,242	Embraced.	Not full
10,243	Embracery	Quite full
10,244	Embracive	Full particulars
10,245	Embrangle	Full details
10,246	Embrasures	The shaft is full of rubbish
10,247	Embright .	For full particulars, refer to
10,248	Embrocado	Full of débris
10,249	Embrocated	Not nearly full
10,250	Embroider	As full as possible
10,251	Embroiling	**Fully**
10,252	Embronze.	Am (are) fully convinced
10,253	Embrown .	Am (are) fully satisfied with
10,254	Embrowning	Are you fully satisfied that
10,255	Embryology	Fully paid-up shares
10,256	Embryonary	**Funds**
10,257	Embryotega	Work on the mine has been stopped for want of funds
10,258	Embryotic	Not having sufficient funds in hand
10,259	Embryous.	Not having funds enough to purchase
10,260	Emendals.	Funds are running short
10,261	Emendately	Funds are exhausted
10,262	Emendation	Funds will last till
10,263	Emendicate	Funds are exhausted, please send at once
10,264	Emerald .	The available funds
10,265	Emerged .	It will require large funds to
10,266	Emergency	What funds have you in hand
10,267	Emerging .	Must provide funds for
10,268	Emerycloth	Cannot provide funds for
10,269	Emerypaper	All available funds that can be spared
10,270	Emerywheel	To provide funds (for)
10,271	Emforth .	Reserve funds
10,272	Emicant .	A reserve fund of
10,273	Emictory .	Sinking fund
10,274	Emigrating	The only available funds are those derivable from
10,275	Emilian .	Out of the fund arising from
10,276	Eminence.	The funds set apart for [how long
10,277	Eminently	Have you sufficient funds to go on with, and for
10,278	Emissaries	How do you propose to raise funds to
10,279	Emissary .	Must furnish me with necessary funds to examine
10,280	Emmantling	An accumulated fund (of)

No.	Code Word.	
10,281	Emmarble.	**Funeral**
10,282	Emmenology	Dead; funeral will take place
10,283	Emolliated	**Furnace(s)**
10,284	Emollition	Drying furnace
10,285	Emoluments	Reverberatory furnace
10,286	Emotional	Roasting furnace
10,287	Emotions .	Water-jacket furnace
10,288	Emotively .	Smelting furnace
10,289	Empaistic .	Bruckner furnace
10,290	Empanoply	Howell white furnace
10,291	Empasm .	Cupelling furnace
10,292	Empawned	Melting furnace
10,293	Empeopled	Furnaces to boilers worn out
10,294	Emperished	Must put up new furnace
10,295	Emphasized	Cost of a new furnace (will be)
10,296	Emphatic .	The best furnace
10,297	Emphatical	The best furnace to burn
10,298	Emphrensy	Furnace must be adapted to burn
10,299	Emphysema	**Furnish**
10,300	Empiric .	Will furnish full particulars of the mine(s)
10,301	Empirical .	Will furnish you with the necessary money to examine —— mine(s)
10,302	Empiricism	Will not furnish me (us) with
10,303	Emplaster.	Furnish full particulars
10,304	Emplecton	Furnish us with
10,305	Employable	Furnish and provide for
10,306	Employed .	Do not furnish
10,307	Employment	**Furnished**
10,308	Emplumed	Has not furnished
10,309	Emplunge.	Has (have) furnished us with
10,310	Emplunging	Fully furnished and equipped with all appliances
10,311	Empoison .	Has —— furnished you with full particulars
10,312	Empoldered	Has not yet furnished us with
10,313	Emporetic.	**Furnishing**
10,314	Emporium	We are now furnishing
10,315	Empowering	The cost of furnishing
10,316	Emprize .	**Furniture**
10,317	Emptier .	The furniture and equipment
10,318	Emptiness	**Further.** (See also Farther.)
10,319	Empty . .	Can do nothing further in the matter
10,320	Emptying .	After further
10,321	Empurple .	Have driven in further on the vein
10,322	Empuzzled	If you hear anything further
10,323	Empuzzling	Wait till you get further instructions
10,324	Empyocele	Am (are) waiting for further instructions
10,325	Empyreal .	Is (are) waiting for further orders
10,326	Empyreuma	Further from
10,327	Empyrosis.	Further information required
10,328	Emulations	Further down
10,329	Emulative.	Further away from

No.	Code Word.	**Further** *(continued)*
10,330	Emulatory	Further details
10,331	Emulatress	Further particulars
10,332	Emulgent .	If you do not hear further from
10,333	Emulous .	Until you hear further from
10,334	Emulously	Cannot go further
10,335	Emulsified	Cannot further your wishes
10,336	Emulsify .	Do not go any further
10,337	Emunctory	**Fuse(s)**
10,338	Emuwren .	Send —— coils safety fuse
10,339	Emydidae.	Electric fuses
10,340	Enabled .	**Future**
10,341	Enabling .	Am (are) convinced there is a brilliant future before
10,342	Enacteth .	You must in future [the Company
10,343	Enaliosaur	Future developments would determine
10,344	Enaluron .	In the future
10,345	Enambush	Some future time
10,346	Enamel .	Am (are) sure it is a mine with a great future before
10,347	Enamelled	Future developments [it
10,348	Enamelling	What is your opinion of the future prospects of the
10,349	Enamoured	The future prosperity of the mine [mine
10,350	Encalendar	Unfavourable for the future prospects of the under-
10,351	Encampeth	What are your views as to the future [taking
10,352	Encamping	The future prospects of the mine are good
10,353	Encampment	The present condition and future prospects
10,354	Encanker .	Having regard to the future and not to immediate
10,355	Encanthis .	Future runs [interests
10,356	Encardion.	Future runs should exceed
10,357	Encarpus .	Are future runs likely to be better
10,358	Encashment	We think future runs will be better
10,359	Encauma .	Consider that future runs
10,360	Encaustic .	Future yield
10,361	Encenia .	Future yield will depend upon
10,362	Encephalos	**Gain**
10,363	Enchafe .	It will be a decided gain
10,364	Enchafing.	What shall I (we) gain
10,365	Enchanters	What will be the gain
10,366	Encharging	What will he gain (by)
10,367	Encheason	By which we gain
10,368	Encheer .	By which —— will gain
10,369	Enchelya .	Do (does) not gain
10,370	Enchisel .	Will gain nothing (by)
10,371	Enchodus.	Will gain nothing by delay
10,372	Enchorial.	Shall gain very little
10,373	Enchoric .	Shall gain by the transaction
10,374	Enchymonia	**Gained**
10,375	Encincture	Has (have) —— gained
10,376	Encindered	Has (have) gained
10,377	Encircled .	Has (have) not gained
10,378	Encircling.	Has (have) gained on
10,379	Enclasp .	The water has gained on the

No.	Code Word.	Gained (*continued*)
10,380	Enclasping	Nothing to be gained by the transaction
10,381	Enclitics .	There is nothing to be gained by
10,382	Encloister .	**Gaining**
10,383	Enclosure .	We are gaining on the water daily
10,384	Encoaching	**Galena**
10,385	Encoffin .	Galena carrying —— ounces of silver per ton
10,386	Encollar .	Argentiferous galena
10,387	Encomiast	Argentiferous galena plant complete with power,
10,388	Encomion.	**Gales** · [etc., for —— tons daily
10,389	Encomiums	Heavy gales have been raging
10,390	Encompass	Damaged by the heavy gales
10,391	Encores .	The late heavy gales have prevented us
10,392	Encounter	**Gangue**
10,393	Encouraged	The gangue consists of
10,394	Encradle .	**Gas**
10,395	Encradling	The bad gas in the mine stops work
10,396	Encratites .	Carbonic acid gas
10,397	Encrimson	Water gas
10,398	Encrinal .	Coal gas
10,399	Encrinitic .	Gas explosion
10,400	Encroached	**Gather**
10,401	Encumber	Gather all the information you can
10,402	Encurtain .	From what we can gather
10,403	Encyclic .	**Gathered**
10,404	Encyclical	Gathered from various parts
10,405	Encysted .	Gathered from what was said
10,406	Endamage	**Gear**
10,407	Endamaging	Gear wants overhauling
10,408	Endangered	Pumping gear
10,409	Endarken .	Hauling gear
10,410	Endearedly	The gear will be repaired
10,411	Endeareth	Had to overhaul the gear
10,412	Endearment	Had to repair the gear
10,413	Endeavour	**General(ly)**
10,414	Endeixis .	The general feeling here is
10,415	Endemical	In general
10,416	Endemicity	General dissatisfaction (caused by)
10,417	Endermatic	General distrust of the management
10,418	Endiaper .	General distrust prevails
10,419	Endless .	General satisfaction is felt
10,420	Endlessly .	There is a general want of confidence
10,421	Endlong .	There is general dissatisfaction and disappointment
10,422	Endmost .	The general results are satisfactory
10,423	Endocarp .	The general result is disappointing
10,424	Endochrome	The general outlook is promising
10,425	Endoctrine	Matters in general are going on satisfactorily
10,426	Endocyst .	A general strike is imminent
10,427	Endoderm	Generally speaking
10,428	Endodermic	How are things looking generally
10,429	Endogamous	**Generously**

No.	Code Word.	Generously (*continued*)
10,430	Endogamy	Has (have) very generously
10,431	Endogen .	Has (have) not behaved generously
10,432	Endolymph	Genuine
10,433	Endomorph	Thoroughly genuine
10,434	Endoplast.	Not (a) genuine
10,435	Endopleura	It is a thoroughly genuine affair
10,436	Endopodite	Do not think it is a genuine affair
10,437	Endoptile .	Geological(ly)
10,438	Endorhiz .	Geological formation
10,439	Endorsable	The geological features of the district
10,440	Endorse .	From the geological formation of the lode
10,441	Endorsing	The geological formation leads us to suppose
10,442	Endosmic.	Have learned a good deal about the geological
10,443	Endosmosis	Geologically speaking [formation
10,444	Endospore	A geological map of the (district)
10,445	Endostomes	A celebrated geological expert
10,446	Endowed .	From the geological formation and character of the district we consider that there is no probability of —— being found
10,447	Endowments	A reference to the geological features [dicate
10,448	Endrudge .	There is nothing in the geological formation to in-
10,449	Endspeech	Geological conditions absolutely prohibit the existence of any metalliferous deposit
10,450	Endurably	Geological conditions are not unfavourable, but hitherto no metalliferous deposits of any value have been discovered
10,451	Endurance	Geological conditions are favourable and it is probable that metalliferous deposits of value will be found
10,452	Enduringly	Totally unprospected, and geological conditions so obscured that it is not possible to form any opinion as to metalliferous value until some work [has been done
10,453	Endwise .	Geologist
10,454	Enemies .	An experienced geologist
10,455	Energetics	Geology
10,456	Energical .	The geology of the mine
10,457	Energizer .	The geology of the district
10,458	Energizing	Geometrical
10,459	Energumen	A geometrical survey
10,460	Energy. .	Get
10,461	Enervated	Get me (us) some
10,462	Enervating	Get more of
10,463	Enervative	Get permission to examine
10,464	Enfamished	Get some of the shares at about
10,465	Enfeebled.	Get the men to wait for money till
10,466	Enfeebling	Can you get
10,467	Enfeeblish	If you cannot get
10,468	Enfeloned.	Cannot get
10,469	Enfeoff .	Can get
10,470	Enfever .	Get out of

No.	Code Word.	Get (*continued*)
10,471	Enfierced .	We shall be able to get out of the difficulty
10,472	Enfiladed .	We shall then be able to get out —— tons per day
10,473	Enfilading	What can you get out (of)
10,474	Enflesh .	Can get nothing out (of)
10,475	Enflowered	Can only get
10,476	Enfolded .	I did not get
10,477	Enfoldment	How much can you get
10,478	Enforce .	Cannot get more till
10,479	Enforcedly	To enable us to get
10,480	Enforcible	Get all you can
10,481	Enforcing .	You must get —— to
10,482	Enforested	From whom shall I (we) get
10,483	Enfortune.	Shall have to get
10,484	Enfreed .	You must get
10,485	Enfreedom	Get ready to
10,486	Enfroward	(To) get on with
10,487	Enfrozen .	Cannot get on at all with
10,488	Engaging .	To get over the difficulty
10,489	Engagingly	All that we can get out
10,490	Engaol. .	Cannot get any further
10,491	Engarboil .	Cannot get any more
10,492	Engarrison	If you cannot get more
10,493	Engender .	**Getting**
10,494	Engendrure	Getting along fast
10,495	Engineman	Getting along slowly
10,496	Enginetool	Getting ready to
10,497	Engiscope	We are getting —— tons per day out of
10,498	Engladded	Has (have) been getting
10,499	Englishry .	Is (are) getting
10,500	Englislet .	Not getting enough
10,501	Englooming	Is (are) getting all he (they) can
10,502	Englutting	Getting out
10,503	Engorged .	He is (they are) getting
10,504	Engraft .	Getting enough out to pay expenses
10,505	Engrapple	**Give**
10,506	Engrasped	Give your opinion of
10,507	Engraulis .	The last returns give an average of
10,508	Engravings	Can you give
10,509	Engreaten	Can you not give
10,510	Engross .	Can give
10,511	Engrosseth	Cannot give
10,512	Engrossing	Declines to give
10,513	Enguard .	Must give
10,514	Enguarding	Will give
10,515	Engulf. .	Will not give
10,516	Engulfing .	Will —— give
10,517	Engulfment	Will —— give up
10,518	Enhanceth	Will give up
10,519	Enhancing	Will not give up
10,520	Enharbour	What will you give

No.	Code Word.	**Give** (*continued*)
10,521	Enhardened	What will he (they) give
10,522	Enharmonic	How long can you give
10,523	Enhedge .	You must give longer time
10,524	Enhedging	Cannot give longer time
10,525	Enhunger.	If he will give
10,526	Enhydra .	He (they) will give more
10,527	Enigma .	He (they) will give longer time
10,528	Enigmatic	Give in exchange (for)
10,529	Enjoined .	Give(s) a better result
10,530	Enjoining .	If he (they) can give
10,531	Enjoinment	If we can give
10,532	Enjoyable	If you can give
10,533	Enkennel .	**Given**
10,534	Enkindle .	Has (have) given
10,535	Enkindling	Has (have) given up
10,536	Enlacement	Has (have) given out
10,537	Enlangour	Has (have) —— given up
10,538	Enlarged .	Has (have) not given
10,539	Enlargedly	Within a given time
10,540	Enleague .	Must be given
10,541	Enleaguing	Must not be given
10,542	Enlengthen	The men have given in
10,543	Enlightens	Have given better results
10,544	Enlink. .	What has been given
10,545	Enlinking.	Given in exchange for
10,546	Enlistment	**Glad**
10,547	Enlivened	You will be glad to know that
10,548	Enlivening	Am (are) very glad to be able to tell you that
10,549	Enmanche	Am (are) very glad to be able to give you good
10,550	Enmeshed	Shall be glad to hear [accounts of the mine
10,551	Enmity .	Am (are) very glad to hear
10,552	Ennobled .	I (we) shall be glad to know
10,553	Ennobling	**Globe mill**
10,554	Ennuye .	Globe mill of a capacity to crush
10,555	Enomotarch	**Gneiss**
10,556	Enormities	The hanging-wall is gneiss
10,557	Enormity .	The foot-wall is gneiss
10,558	Enormous.	The country rock is gneiss
10,559	Enormously	**Go**
10,560	Enpatron .	Be ready to go on —— to
10,561	Enquicken	When can you go to
10,562	Enquired .	Can go to —— on
10,563	Enrapture.	Cannot go till you send me
10,564	Enravish .	You must go to
10,565	Enregister	Go by way of
10,566	Enrich. .	Am (are) able to go
10,567	Enricheth.	Am (are) not able to go
10,568	Enriching.	Have to go by steamer to
10,569	Enrichment	Will have to go immediately
10,570	Enridged .	Go as quickly as possible

No.	Code Word.	**Go** (*continued*)
10,571	Enripen .	Will go with me (us)
10,572	Enrol . .	Cannot go with me (us)
10,573	Enrolled .	Am (are) ready to go
10,574	Enrounded	To go down
10,575	Ensample .	To go up
10,576	Ensampling	If —— can go
10,577	Ensanguine	If —— cannot go
10,578	Enscaled .	Can only go as far as —— by rail, and then by car-
10,579	Enscaling .	Go ahead (with the) [riage to the mine
10,580	Enschedule	Likely to go through
10,581	Ensconce .	Can you go
10,582	Ensconcing	Can —— go
10,583	Enseal . .	Must not go
10,584	Ensearch .	Can go
10,585	Ensemble .	Cannot go
10,586	Enshawl .	Will go
10,587	Enshawling	Will not go
10,588	Ensheath .	From here shall go to
10,589	Enshield .	Go at once to
10,590	Enshrine .	Can go at once to
10,591	Enshrining	We will go together if you wish
10,592	Enshroud .	It is essential that I go at once to
10,593	Ensiform .	You must go immediately to
10,594	Ensign. .	He must go immediately to
10,595	Ensigncy .	Must go with you
10,596	Ensignship	(To) go with you
10,597	Enskied .	**Goes**
10,598	Ensky . .	The ore goes down
10,599	Ensilage .	The ore goes up
10,600	Enslave .	Goes up
10,601	Enslaving .	Goes down
10,602	Ensnareth	Goes along
10,603	Ensober .	**Going**
10,604	Ensphere .	They are going to
10,605	Ensphering	Must decide quickly what he is (they are) going to do
10,606	Enstamp .	Who is going
10,607	Enstamping	Going away
10,608	Enstock .	Am (are) going (to)
10,609	Enstocking	Is (are) going (to)
10,610	Enstyle .	Am (are) not going to
10,611	Enstyling .	Is (are) not going to
10,612	Ensuable .	Are you going to
10,613	Ensued .	Is (are) —— going to
10,614	Ensuing .	If —— is (are) not going to
10,615	Enswathing	Otherwise, all is going well
10,616	Ensweep .	Telegraph when you are going (to)
10,617	Entackle .	What are you going to do (about)
10,618	Entail . .	What is he (are they) going to do
10,619	Entaileth .	Is going with me
10,620	Entailing .	Is —— going with you

No.	Code Word.	
10,621	Entailment	**Gold.** (See also Free.)
10,622	Entalent .	The mine is very rich in free gold
10,623	Entangle .	Have struck a large pocket of gold
10,624	Entangling	Have struck a small pocket of gold
10,625	Entelechy.	Showed only a trace of gold
10,626	Entellus .	A rich streak of gold ore
10,627	Entempest	Sample assayed gave no gold
10,628	Enterclose	A large quantity of gold has been taken from
10,629	Enteritis .	Pieces of rock containing visible gold
10,630	Entermewer	Is one of the richest gold-mines in the district of
10,631	Enterodela	Pieces of pure gold
10,632	Enterolith	The gold lies in the arsenical pyrites
10,633	Enterology	The gold lies in iron pyrites
10,634	Enterotome	The gold-bearing vein runs through
10,635	Enterpart .	The gold is in alluvial deposit
10,636	Enterprise	The gold-bearing veins run north and south
10,637	Entersole .	The gold-bearing veins run east and west
10,638	Entertains	The gold occurs in quartz lodes
10,639	Entheastic	Gold is found in the bed of the stream
10,640	Enthralled	Nuggets of gold have been found
10,641	Enthrone .	The gold-bearing vein can be traced for a long dis-
10,642	Enthroning	Owing to the gold being found in [tance
10,643	Enthronize	Indications show that gold exists in abundance on
10,644	Enthunder	Showing traces of gold [this property
10,645	Enthusiasm	An average assay gave —— in gold
10,646	Enthymema	Gold ore
10,647	Entice . .	Alluvial gold
10,648	Enticement	Retorted gold
10,649	Enticing .	Gold amalgam
10,650	Enticingly	Gold has been discovered
10,651	Entirely .	Ounce(s) of gold per ton
10,652	Entireness	Ounces of gold have been shipped
10,653	Entitled .	Ounces of gold will be shipped
10,654	Entomb .	How much gold has been shipped
10,655	Entombing	Crushed —— tons obtained —— ounces of gold
10,656	Entombment	The gold quartz is very free ; no base metals present
10,657	Entomoid.	The gold in quartz is not free; there are base
10,658	Entomotomy	Gold shipped to England [metals present
10,659	Entonic .	Gold shipped to
10,660	Entophytic	Proportion of gold to silver is —— per cent.
10,661	Entozoon .	This chute is very rich in gold
10,662	Entrammel	Very rich in gold
10,663	Entranced	The gold occurs in
10,664	Entrapped	How many ounces of gold
10,665	Entreasure	Ounces of gold .
10,666	Entreateth	Payable in gold
10,667	Entreating	The assay value of the gold
10,668	Entreaty .	The realized value of the gold
10,669	Entrench .	Gold mill of —— stamps
10,670	Entrochal.	Chloride of gold

No.	Code Word.	**Gold** (*continued*)
10,671	Entrust .	Gold from concentrates
10,672	Entrusting	Gold from tailings
10,673	Entrymoney	Tailings being worked for gold
10,674	Entwine .	Tons tailings produced —— ounces of gold
10,675	Entwining.	Average assay of gold from tailings
10,676	Entwisted.	Of the total amount, the value of gold is
10,677	Enucleated	Gold from the reef
10,678	Enunciable	The reef assays in gold
10,679	Enunciated	Value of gold bullion
10,680	Enuresis .	Fineness of gold
10,681	Envassal .	The gold is worth —— per ounce
10,682	Envault .	**Gone**
10,683	Envaulting	Has (have) gone (to)
10,684	Envelop .	If —— has (have) gone (to)
10,685	Envelopeth	If —— has (have) not gone (to)
10,686	Enveloping	Has (have) not gone (to)
10,687	Envenomed	Where has —— gone to
10,688	Enviably .	Think he has gone (to)
10,689	Envious .	Can you tell us where he has gone
10,690	Enviously.	**Good**
10,691	Environed	Will there be any good in
10,692	Environing	There will be no good in
10,693	Envisage .	Not any good
10,694	Envisaging	Not good enough (for)
10,695	Envolume.	As good as
10,696	Envoluming	Not so good
10,697	Envoyship	Good progress is being made
10,698	Enwallow .	To be of any good
10,699	Enwheel .	If I (we) can do any good
10,700	Enwheeling	Very good
10,701	Enwidening	Very little good
10,702	Enwoman.	Very good for the purpose
10,703	Enwombed	If equally as good as last
10,704	Enwrap .	Good to hold
10,705	Enwrapment	Good to sell
10,706	Enwreathed	Do you think it good enough **to**
10,707	Enwritten.	Good results
10,708	Enwrought	Can do good by
10,709	Enzootic .	Can you do any good
10,710	Eolipile .	Has not done much good
10,711	Epacris .	Will do a great deal of good
10,712	Epanaphora	Is a good man for the purpose
10,713	Epanados .	If you want a good man
10,714	Epaulement	In good time
10,715	Epauletted	**Goods**
10,716	Epaxial .	Goods have been shipped
10,717	Epeira . .	Goods will not be shipped
10,718	Epeiridae .	Goods not yet shipped
10,719	Epenthesy	Goods not yet to hand
10,720	Epenthetic	What has become of the **goods**

No.	Code Word.	**Goods** (*continued*)
10,721	Epergnes .	Railroad company refuse to deliver the goods
10,722	Ephah . .	Goods detained in transit
10,723	Ephelis .	Goods detained by Customs authorities
10,724	Ephemeral	Goods have been released
10,725	Ephemerist	Have got the goods at last
10,726	Ephemerous	Goods will be released upon payment of
10,727	Ephesite .	What goods were sent
10,728	Ephialtes .	The goods have been delivered
10,729	Ephippial .	**Got**
10,730	Ephods .	Not to be got
10,731	Ephoral .	Has (have) got
10,732	Ephoralty .	Has (have) not got
10,733	Epiblast .	What can —— be got for
10,734	Epiblema .	Can be got for
10,735	Epicalyx .	Have you got
10,736	Epicaridan	Has he (have they) got
10,737	Epicarp .	Got all that was to be had
10,738	Epicedial .	Got back
10,739	Epichilium	Got hold of
10,740	Epichirema	Got from
10,741	Epichorial	Got out
10,742	Epiclinal .	Not got back
10,743	Epicolic .	Not got hold of
10,744	Epicondyle	Not got from
10,745	Epicranium	Not got out
10,746	Epics . .	Got to the end of
10,747	Epictetian	Have got through
10,748	Epicure .	Have got out of
10,749	Epicurism	If you have got
10,750	Epicurized	If you have not got
10,751	Epicyclic .	When you have got
10,752	Epicycloid	**Governed**
10,753	Epidemic .	Be governed by
10,754	Epidemical	Governed by circumstances
10,755	Epidendrum	Will be materially governed by
10,756	Epiderm .	**Government**
10,757	Epidermal	Government required(s)
10,758	Epidermoid	Government requires a deposit at once of
10,759	Epididymis	The government levies a tax of
10,760	Epigastric.	The amount paid to government
10,761	Epigenesis	Concession has been granted by the government
10,762	Epigenous	Concession ratified by the government
10,763	Epigeum .	The government have cancelled the concession
10,764	Epiglot .	Have arranged with the government
10,765	Epiglottic.	Government has been overthrown
10,766	Epigonium	Can arrange with the government
10,767	Epigram .	The government will grant
10,768	Epigramist	The government will not grant
10,769	Epigraphic	**Governor**
10,770	Epilepsy .	The acting Governor

No.	Code Word.	Governor (*continued*)
10,771	Epileptic .	The Governor of the State
10,772	Epileptoid	The Governor-General
10,773	Epilogical	**Grade**
10,774	Epiloguise	High-grade ore
10,775	Epinasty .	Low-grade ore
10,776	Epinicion .	The grade is
10,777	Epinyctis .	The grade is very steep
10,778	Epiphany .	High-grade ore in abundance in sight
10,779	Epiphegus	Low-grade ore in abundance in sight
10,780	Epiphragm	The ore is high grade, but not much of it
10,781	Epiphysial	Although the ore is low grade, the mine can be worked to a good profit, on account of the abundance of ore and capabilities for cheap working
10,782	Epiphytic .	We are now milling high-grade ore
10,783	Epiplexis .	We are now milling low-grade ore
10,784	Epiploce .	Tonnage of high-grade ore
10,785	Epiploon .	Tonnage of low-grade ore
10,786	Epipolism .	Tons of high-grade ore, value
10,787	Epipolized	Tons of low-grade ore, value
10,788	Epirhizous	**Grading**
10,789	Episcopacy	We are now grading for (the)
10,790	Episcopal .	Have finished grading
10,791	Episcopize	**Gradual**
10,792	Episodes .	A gradual increase may be expected
10,793	Episodial .	A gradual decrease may be expected
10,794	Epispastic	Will be very gradual
10,795	Episperm .	**Gradually**
10,796	Epispermic	The works are getting into order gradually
10,797	Epistaxis .	Very gradually on account of
10,798	Epistle. .	Must be done gradually
10,799	Epistolary .	Gradually increasing
10,800	Epistolean	Gradually decreasing
10,801	Epistolist .	**Granite**
10,802	Epistoma .	The hanging wall is granite
10,803	Epitaph .	The foot-wall is granite
10,804	Epitaphist	In proximity to the granite
10,805	Epithalamy	A belt of granite
10,806	Epithesis .	Country rock is granite
10,807	Epitomator	A horse of granite
10,808	Epitomized	A granite dyke
10,809	Epitrite .	In granite formation
10,810	Epizeuxis .	**Grant**
10,811	Epizooty .	Will —— grant
10,812	Epochal .	Will grant
10,813	Eponymic	Will not grant
10,814	Epopt . .	If —— will not grant
10,815	Epsomite .	A grant of
10,816	Epsomsalt	To obtain a grant
10,817	Epyornis .	**Gratification**
10,818	Equability	Have much gratification in stating

No.	Code Word.	**Gratification** (*continued*)
10,819	Equably .	It was with much gratification
10,820	Equalize .	**Gravel**
10,821	Equalizing	Gravel carrying gold
10,822	Equalness.	Gravel pays —— per cubic yard
10,823	Equangular	In the gravel
10,824	Equanimity	Through the gravel
10,825	Equanimous	Auriferous gravel
10,826	Equated .	Auriferous gravel yielding —— per cubic yard
10,827	Equating .	The gravel goes down to a depth of
10,828	Equation .	Panning out the gravel
10,829	Equatorial	Gravel will pan out
10,830	Equerries .	Gravel will assay
10,831	Equerry .	Rich gravel has been found
10,832	Equestrian	There is a bed of rich gravel
10,833	Equiangled	**Great(er)**
10,834	Equicrural	Great waste of time
10,835	Equiform .	Great deal of waste
10,836	Equinoxes	Covering a great extent
10,837	Equip . .	To a great extent
10,838	Equipage .	Very great
10,839	Equipped.	Not very great
10,840	Equipping	Great many
10,841	Equiparate	As great as
10,842	Equipedal	Not so great as
10,843	Equipment	Greater than before (hitherto)
10,844	Equipoise	Greater than ever
10,845	Equirotal .	**Greatly**
10,846	Equisetum	It would be greatly to
10,847	Equison .	It would not be greatly to
10,848	Equisonant	Would be greatly to our advantage
10,849	Equitancy	Would greatly prejudice
10,850	Equitant .	Would greatly restrict our efforts
10,851	Equities .	**Gross**
10,852	Equity . .	Gross yield
10,853	Equivalent	The gross yield per ton is
10,854	Equivalved	The gross receipts
10,855	Equivocacy	The gross receipts have been
10,856	Equivocate	What are the gross receipts for
10,857	Equivorous	A gross revenue of
10,858	Eradicable	**Grossly**
10,859	Eradicated	Is (are) grossly exaggerated
10,860	Eragrostis	Grossly exaggerated accounts have been circulated
10,861	Eranthemum	Grossly insulting remarks
10,862	Erastian .	**Ground(s)**
10,863	Erectable .	There is a dispute about the ground
10,864	Erectility .	Over the whole of the ground
10,865	Erectness .	In barren ground
10,866	Erelong .	Ground much broken up
10,867	Eremitic .	Above ground
10,868	Eremitical	The ground above

No.	Code Word.	**Ground(s)** (*continued*)
10,869	Eremitish .	Below ground
10,870	Erethism .	The ground below
10,871	Erethistic .	They have no ground for complaint
10,872	Erewhiles .	On the ground of
10,873	Ergotine .	No ground for
10,874	Ergotisms	Is there any ground for
10,875	Ericaceous	The ground above level No. —— is worked out
10,876	Eridanus .	The surrounding ground
10,877	Erigeron .	The ground about here is very rich
10,878	Eriophorum	We are now in poor ground
10,879	Erlking .	The ground we are now in is poor, but improving
10,880	Erminemoth	There are grounds for complaint [as we go on
10,881	Erminites .	There are grounds for suspicion
10,882	Erminois .	There are grounds for the belief
10,883	Erodium .	There are no grounds for
10,884	Erosionist .	Upon all grounds
10,885	Erotesis .	**Group**
10,886	Erotomany	A (the) group of mines
10,887	Erpetology	The entire group
10,888	Errable .	In the group
10,889	Errabund .	Amongst a group of
10,890	Errand .	The entire group of mines offered for
10,891	Errantry .	Will accept for the entire group
10,892	Errhine .	Cannot offer for the entire group more than
10,893	Erroneous.	**Guarantee.** (See also Provided.)
10,894	Errors . .	As a guarantee
10,895	Eructate .	Will —— guarantee
10,896	Eructating	Will guarantee
10,897	Erudite .	Will not guarantee
10,898	Eruditely .	Must guarantee
10,899	Erudition .	Without a guarantee
10,900	Eruginous	Will you guarantee expense of examination
10,901	Erunda .	Will guarantee to the amount of
10,902	Eruption .	Do not go without a guarantee
10,903	Eruptional	Who will guarantee us
10,904	Eruptive .	Under the guarantee of
10,905	Ervalenta .	Must have a guarantee
10,906	Eryngium .	Joint guarantee
10,907	Eryngo .	Joint and several guarantee
10,908	Erysimum	Will not give required guarantee
10,909	Erysipelas	Against approved guarantee
10,910	Erythace .	**Guaranteed**
10,911	Erythrean .	If guaranteed (by)
10,912	Erythrosis	If not guaranteed (by)
10,913	Erythrozym	Expenses guaranteed (by)
10,914	Escalade .	Expenses guaranteed, not to exceed
10,915	Escalading	Must be guaranteed
10,916	Escallonia	**Guard**
10,917	Escape .	To guard against accidents
10,918	Escapement	Be on your guard with

No.	Code Word.	**Guard** (*continued*)
10,919	Escaping .	You must guard against
10,920	Eschara .	Guard our interests
10,921	Escharotic	**Guarded**
10,922	Escheatage	You cannot be too guarded with
10,923	Eschew .	Your interests will be guarded
10,924	Eschewers	**Guidance**
10,925	Eschewing	For my (our) guidance
10,926	Eschewment	For your guidance
10,927	Eschynite .	For my (our) guidance in preparing the **report**
10,928	Escorted .	For your guidance we forward
10,929	Escorting .	Let us have for our guidance
10,930	Esculapian	For his (their) guidance
10,931	Escutcheon	**Guide**
10,932	Esdras . .	As a guide
10,933	Eskimodog	Is no guide
10,934	Esnecy .	Must have some guide
10,935	Esophageal	**Guided**
10,936	Esopian .	Be guided by
10,937	Esoterical	Do not be guided by
10,938	Esoterism	Has (have) been guided by
10,939	Esparto .	Has (have) not been guided by
10,940	Especially	Who has (have) —— been guided by (in)
10,941	Espied . .	Guided by circumstances
10,942	Espionage	**Gulch**
10,943	Esplanades	**Gypsum**
10,944	Espousals	**Habit**
10,945	Espouser .	Is (are) in the habit of
10,946	Espousing	Is (are) not in the habit of
10,947	Espying .	The habit of working here
10,948	Essay . .	**Had**
10,949	Essayist .	Had not
10,950	Essenced .	Had you not
10,951	Essenism .	After he had (they had)
10,952	Essential .	After we had
10,953	Essonite .	After you had
10,954	Essorant .	Before he had (they had)
10,955	Esteemed .	Before we had
10,956	Esteeming	Before you had
10,957	Estiferous	If he had (they had)
10,958	Estimably	If we had
10,959	Estimate .	If you had
10,960	Estimating	If it had
10,961	Estimation	If it had not been (for)
10,962	Estimators	If there had been
10,963	Estivage .	If there had not been
10,964	Estoppel .	Must have had
10,965	Estranged	Could have had
10,966	Estranging	Could not have had
10,967	Estuaries .	You could have had
10,968	Estuary .	We could have had

No.	Code Word.	**Had** (*continued*)
10,969	Eternalist .	He (they) could have had
10,970	Eternified	He (they) had not
10,971	Eternify .	We had not
10,972	Etheling .	You had not
10,973	Ethereal .	Which had
10,974	Ethereally	What had
10,975	Etherous .	You had better
10,976	Etheriform	You had better not
10,977	Etherism .	**Half**
10,978	Etherizing	One half
10,979	Ethical .	Half the amount
10,980	Ethically .	Halfway between
10,981	Ethidene .	Not worth more than half the price
10,982	Ethiopic .	Half the year
10,983	Ethmoid .	For the half-year
10,984	Ethmoidal	Half in cash, and half in shares
10,985	Ethnarchy.	Half in cash, and half in —— days
10,986	Ethnic. .	Half next month, and half the month after
10,987	Ethnicism	Half now, and the remainder
10,988	Ethnogeny	**Handle**
10,989	Ethnologic	Difficult to handle
10,990	Ethopoetic	In order more easily to handle
10,991	Ethusa. .	To handle the stuff
10,992	Ethyl . .	So as to handle
10,993	Ethylamine	Could not handle
10,994	Etiquette .	Must handle very carefully
10,995	Etonian .	**Handling**
10,996	Ettercap .	We are now handling
10,997	Etterpyle .	**Hand(s)**
10,998	Etymic .	On hand
10,999	Etymologic	We have in hand
11,000	Eucalyn .	Short of hands
11,001	Eucalyptus	In our hands
11,002	Eucharist .	In the hands of
11,003	Euchelaion	In other hands
11,004	Euchirus .	In whose hands
11,005	Euchite .	In whose hands is the property now
11,006	Euchlore .	Property in the hands of
11,007	Euchology	Most of the hands
11,008	Euchroite.	All hands at work
11,009	Euchymy .	All hands set to work
11,010	Eucrasy .	**Hanging-wall**
11,011	Eudaemon	The hanging-wall is
11,012	Eudemonism	Against the hanging-wall
11,013	Eudiometry	Have encountered the true hanging-wall
11,014	Eudoxian .	Have not yet got to the hanging-wall
11,015	Eudyalite .	The hanging-wall is well defined
11,016	Eugenesic.	The hanging-wall is not well defined
11,017	Eugubine .	Hanging-wall solid and good
11,018	Euhemerize	**Happen(ed)**

No.	Code Word.	**Happen(ed)** (*continued*)
11,019	Eukairite .	How did it happen
11,020	Eulogical .	Happened in consequence of
11,021	Eulogist .	A serious accident has happened owing to
11,022	Eulogistic	**Happens**
11,023	Eulogiums	If it happens again
11,024	Eulogized .	It so happens that
11,025	Eulogizing	**Hard**
11,026	Eulogy .	Very hard
11,027	Eunectus .	Not very hard
11,028	Eunuch .	Very hard to
11,029	Eunuchate	It will be hard work
11,030	Eunuchism	The rock has become very hard
11,031	Eupatorine	Hard rock
11,032	Eupatrid .	We are now drifting through hard rock
11,033	Eupepsy .	**Harder**
11,034	Euphemism	If the rock becomes harder
11,035	Euphemized	If the rock does not become any harder
11,036	Euphonic .	It is a harder task than I (we) expected
11,037	Euphonical	The rock has become much harder
11,038	Euphonious	In harder rock
11,039	Euphonon	It is harder
11,040	Euphorbial	**Hardness**
11,041	Euphrasia	Owing to the hardness of the rock
11,042	Euphroe .	The hardness of the rock necessitates
11,043	Euphrosyne	The hardness of the rock retards the progress of
11,044	Euphuistic	**Harm**
11,045	Euphuize .	No harm has been done
11,046	Euplastic .	Great harm has been done by
11,047	Eupyrion .	Is (are) doing great harm by
11,048	Eurekas .	Has (have) —— done any harm by
11,049	Euripus .	To prevent —— from doing any harm to
11,050	Eurithmy .	**Harmonious(ly)**
11,051	Euroclydon	Harmonious action desirable
11,052	Europa .	Work(ing) harmoniously together
11,053	Euryale .	Work(ing) harmoniously with
11,054	Eurycerous	**Hasten(ed)**
11,055	Eusebian .	Hasten matters as much as you can
11,056	Eustachian	Cannot be hastened
11,057	Eutaxy. .	**Hasty**
11,058	Euterpe .	Is (are) too hasty
11,059	Euthanasy	Could only make a hasty inspection
11,060	Eutrophic.	After a very hasty examination
11,061	Eutychian .	Do not be too hasty
11,062	Euxanthic	**Haul**
11,063	Euxenite .	What can you haul in one day
11,064	Evacuated	Are able to haul —— tons per day
11,065	Evacuating	We hope to be able to haul —— tons per day
11,066	Evacuation	To haul
11,067	Evadable .	Cannot haul more than
11,068	Evaded .	**Haulage**

No.	Code Word.	**Haulage** (*continued*)
11,069	Evading .	Cost of haulage
11,070	Evanesced	Haulage is carried on by
11,071	Evanescing	System of haulage
11,072	Evangel .	Mechanical haulage
11,073	Evangelian	Haulage by tramway
11,074	Evangelize	Haulage by rail
11,075	Evaporated	Haulage by horse teams
11,076	Evasions .	Haulage by wire-rope tramway
11,077	Evasive .	**Hauled**
11,078	Evasively .	Hauled —— loads of 16 cubic feet
11,079	Evectics .	Water hauled —— gallons
11,080	Evenbishop	All materials have to be hauled
11,081	Evendown	Hauled —— tons
11,082	Evenfall .	**Hauling**
11,083	Evenhand	Snow has prevented hauling
11,084	Evening .	Hauling ore from —— mine at present
11,085	Eveninggun	Now hauling material
11,086	Evenkeel .	Now hauling machinery
11,087	Evenlike .	Bad state of roads impedes hauling
11,088	Evenly .	**Hauling engine.** (See Hoisting.)
11,089	Evensong .	Shall require a hauling engine of —— h.p.
11,090	Eventful .	**Have**
11,091	Eventide .	Have you
11,092	Eventerate	Have you not
11,093	Eventually	I (we) have
11,094	Everduring	I (we) have not
11,095	Everglade .	Has (have) not
11,096	Evergreen	Has (have) had
11,097	Everliving	Has (have) not had
11,098	Evernia .	Has (have) —— had
11,099	Everybody	Have you had
11,100	Everyday .	Have you not had
11,101	Everything	If you have
11,102	Everywhere	If you have not
11,103	Evestigate	If —— has (have)
11,104	Evicted .	If —— (has) have not
11,105	Evidence .	If they have not
11,106	Evidential	When you have
11,107	Evidently .	When —— has (have)
11,108	Evildoer .	When will you have
11,109	Evileyed .	When shall we have
11,110	Evilly . .	Why have you
11,111	Evilminded	Shall have
11,112	Evincement	Shall not have
11,113	Evincing .	Can have
11,114	Evincive .	Cannot have
11,115	Eviration .	Must have
11,116	Evitable .	Should have had
11,117	Eviternity .	Ought to have
11,118	Evoking .	Have not any

No.	Code Word.	**Have** (*continued*)
11,119	Evolatic .	Have you any
11,120	Evolatical	Could you not have
11,121	Evolutions	Ought not to have
11,122	Evolved .	That you might have
11,123	Evolvement	In case you have
11,124	Evomition	So that we may have
11,125	Evulgate .	Let us know if you have
11,126	Evulgation	Let us know if he (they) have
11,127	Evulsion .	**Hazardous**
11,128	Ewecheese	It would be too hazardous
11,129	Exacerbate	Would it be too hazardous
11,130	Exacinate	A very hazardous undertaking
11,131	Exactitude	Very hazardous for all concerned
11,132	Exactly .	**Heading**
11,133	Exactness	To drive a heading in
11,134	Exactor .	General appearance of heading (is)
11,135	Exalt . .	Most of the headings (are in)
11,136	Exalteth .	The heading, as far as driven
11,137	Exalting .	The heading is (headings are) being driven
11,138	Examinable	**Headway**
11,139	Examinator	Have made no headway as yet
11,140	Examiner.	We are making good headway (with)
11,141	Examining	Are you making any headway
11,142	Exampless	Hope soon to make some headway (against)
11,143	Exangia .	**Heap**
11,144	Exanguious	A great heap
11,145	Exanimated	To heap up
11,146	Exanthema	From the heap
11,147	Exanthesis	**Hear**
11,148	Exarchate	Let me (us) hear immediately
11,149	Exaristate	Let me (us) hear as soon as you can
11,150	Exasperate	When can I (we) hear (as to)
11,151	Excalceate	If you hear anything
11,152	Excalibur.	You shall hear about (the)
11,153	Excamb .	Cannot hear till
11,154	Excambion	I (we) hear on good authority that
11,155	Excavated	From what I (we) can hear
11,156	Excavating	Expect to hear
11,157	Exceedable	Can only hear
11,158	Exceeding	You ought to hear
11,159	Excellency	Should hear
11,160	Excellent .	If you do not hear from us
11,161	Excelsior .	We hear that you have
11,162	Excentral .	We hear that he has (they have)
11,163	Excepted .	Unless we hear further
11,164	Exceptant	Unless we hear to the contrary
11,165	Exceptive.	Unless they hear to the contrary
11,166	Exceptless	Unless you hear to the contrary
11,167	Excerpt .	**Heard**
11,168	Excerption	I (we) hoped to have heard (from)

I

No.	Code Word.	**Heard** (*continued*)
11,169	Excerptors	Has (have) heard (that)
11,170	Excessive	Has (have) not heard
11,171	Exchange .	Has (have) not yet heard anything from
11,172	Exchanging	Have you heard
11,173	Excheator	Has (have) heard nothing
11,174	Exchequer	Has (have) only heard
11,175	Excipients	Heard through
11,176	Excipule .	Heard from
11,177	Excise . .	We heard from
11,178	Exciseman	Have you heard anything about
11,179	Excisions .	Have you heard anything further about
11,180	Excitant .	Have heard nothing further about
11,181	Excitation	Have not heard from (of)
11,182	Exciteful .	We heard that he was
11,183	Exciteth .	Heard that you were
11,184	Excitingly	When we last heard
11,185	Exclaim .	**Heat**
11,186	Exclaimeth	The heat in summer is intense
11,187	Exclaiming	Great heat has now set in here
11,188	Exclude .	As soon as the heat abates
11,189	Excluding	Owing to the great heat
11,190	Exclusion	**Heavily**
11,191	Exclusory .	Has (have) lost heavily over the mines
11,192	Excoction	Has (have) bought heavily
11,193	Excogitate	Is (are) buying heavily
11,194	Excommune	**Height**
11,195	Excoriable	In height
11,196	Excoriated	The height of
11,197	Excrements	(At) a height of —— feet
11,198	Excrescent	**Held**
11,199	Excretory	Is (are) held by
11,200	Excruciate	Was (were) held by
11,201	Excudit .	Will be held by
11,202	Exculpable	Is (are) not held by
11,203	Exculpated	Held together by
11,204	Excurrent .	**Help**
11,205	Excursus .	Shall want help
11,206	Excusatory	Will help
11,207	Excuseless	Will not help
11,208	Excusement	To help to
11,209	Execrably .	A help in
11,210	Execrated .	No help to be had [and take the samples
11,211	Execration	Must take —— to help me make the examination
11,212	Executable	To help in making the assays
11,213	Executants	Will be a help
11,214	Executeth .	It would be no help
11,215	Executors .	Cannot help
11,216	Executrix .	To help if possible
11,217	Exegesis .	Should you want help
11,218	Exegetical	Who can help

No.	Code Word.	**Help** (*continued*)
11,219	Exemplar .	If you can possibly help it, do not
11,220	Exemplify	Take —— to help you
11,221	Exempt .	Cannot do without some (further) help
11,222	Exempteth	Can we help you in any way
11,223	Exemptible	Can be of material help
11,224	Exemption	Is a great help to me (us)
11,225	Exequatur	Help us over the
11,226	Exequial .	Unless you cannot possibly help it
11,227	Exequious	**Helpless**
11,228	Exequy .	We are quite helpless
11,229	Exercising	Too helpless to do anything
11,230	Exergue .	Is —— quite helpless
11,231	Exertment	**Here**
11,232	Exestuate .	Not here
11,233	Exfetation	I (we) must remain here till
11,234	Exfoliate .	I (we) must remain here till you send me (us) money
11,235	Exhalable.	I (we) shall remain here till
11,236	Exhalant .	Telegraph to me (us) here (till the)
11,237	Exhalation	Telegraph to me (us) here ; shall get it on my (our)
11,238	Exhaled .	Has (have) left here [return from
11,239	Exhalement	Was (were) here, but has (have) gone
11,240	Exhaust .	Write to me here [my letters are to be sent
11,241	Exhausteth	Write to me here ; I shall leave full directions where
11,242	Exhausting	Write to me here ; I shall find your letter on my
11,243	Exhaustion	Will return here [return from
11,244	Exhedra .	Shall go from here to
11,245	Exheredate	Arrived here
11,246	Exhibited .	Will be here (about) (on)
11,247	Exhibitive	Come here at once
11,248	Exhibitor .	Must be here at latest
11,249	Exhibitory	Come here as soon as
11,250	Exhumate	If you are here before
11,251	Exhuming	Cannot remain here longer than
11,252	Exiccating	When will you be here
11,253	Exigencies	When will he (they) be here
11,254	Exigency .	If not here before (or by)
11,255	Exigendary	Passed through here
11,256	Exigible .	Now here, on way to
11,257	Exiguity .	**Hesitate**
11,258	Exintine .	Do not hesitate
11,259	Existence .	Hesitate(s) to
11,260	Exocetus .	Do (does) not hesitate to declare
11,261	Exofficial .	Hesitate before taking action
11,262	Exomphalos	**Hesitation**
11,263	Exonerator	After much hesitation
11,264	Exonship .	**Hide**
11,265	Exopodite	To hide the fact
11,266	Exorbitant	Is (are) trying to hide
11,267	Exorciseth	Do (does) not try to hide the fact
11,268	Exorcism .	It is no use trying to hide the fact

No.	Code Word.	Hide (*continued*)
11,269	Exordiums	Even if we wished to hide the fact
11,270	Exorganic.	**High**
11,271	Exorhizal .	High enough
11,272	Exorhizous	Too high
11,273	Exortive .	Too high a price is asked for the
11,274	Exosculate	Very high
11,275	Exosmose.	Not very high
11,276	Exosmotic	The shares keep high
11,277	Exosporous	We think the shares are too high
11,278	Exossated	Expenses too high ; must be kept down
11,279	Exossation	This seems very high ; is it right
11,280	Exosseous	The mines stand high
11,281	Exostemma	The mines are surrounded by high hills
11,282	Exostosis .	On the side of a high hill
11,283	Exotery .	**Higher**
11,284	Exothecium	Will they go any higher
11,285	Exòtic . .	They will go higher
11,286	Exotical .	They are not likely to go any higher
11,287	Expand .	Is (are) fetching a higher price
11,288	Expansible	Is (are) going higher
11,289	Expansions	May possibly go higher
11,290	Expatiated	Will undoubtedly go considerably higher
11,291	Expectancy	Telegraph if you think they will go higher
11,292	Expectedly	Unless you think they will go higher
11,293	Expediency	The higher levels
11,294	Expedient	The higher workings
11,295	Expedite .	In the higher districts
11,296	Expeditely	At a higher altitude
11,297	Expedition	**Highest**
11,298	Expelled .	Has (have) reached the highest figure
11,299	Expenditor	The highest point
11,300	Expenseful	Not yet at the highest point
11,301	Expensive	**Hill(s)**
11,302	Experience	The mine(s) is (are) at the top of a very steep hill
11,303	Experiment	Down the hill
11,304	Expertly .	Up the hill
11,305	Expiating .	On the top of an almost perpendicular hill
11,306	Expiatists.	To carry the ore down the hill
11,307	Expiatory.	The hills are impassable in consequence of
11,308	Expirable .	On the top of the hill
11,309	Expirant .	On the slope of the hill
11,310	Expiry . .	The hill slopes steeply
11,311	Expiscate .	Halfway up the hill or thereabouts
11,312	Explanate.	Over the hill
11,313	Expletives	Along the side of the hill
11,314	Expletory .	At the foot of the hill
11,315	Explicable	To get it up the hill
11,316	Explicator	**Hindered**
11,317	Explicit .	Have been greatly hindered by
11,318	Explicitly .	We are no longer hindered by

No.	Code Word.	
11,319	Explode .	**Hindrance**
11,320	Exploding	Has been a great hindrance
11,321	Exploiture	The want of —— has proved a great hindrance
11,322	Explosion	A hindrance caused by
11,323	Explosive .	Would not be any hindrance to
11,324	Expolish .	**Hire**
11,325	Exponent .	Shall have to hire
11,326	Exposed .	Can you hire
11,327	Exposing .	Try to hire
11,328	Exposition	If you cannot hire
11,329	Expositive	Cannot hire
11,330	Expository	Has (have) been able to hire
11,331	Exposures	**Hoist**
11,332	Expounder	To hoist
11,333	Express .	Can hoist —— tons per day
11,334	Expressage	Cannot hoist
11,335	Expresseth	Can you hoist
11,336	Expression	So as to be able to hoist
11,337	Expressly .	To hoist —— tons per day
11,338	Exprime .	To hoist —— tons per day from a depth of
11,339	Exprobrate	**Hoisted**
11,340	Expugn .	Hoisted last week —— tons
11,341	Expugnable	Has (have) hoisted
11,342	Expunction	**Hoisting**
11,343	Expunged	Hoisting engine
11,344	Expunging	Hoisting machinery
11,345	Expurgate	Engine capable of hoisting —— tons
11,346	Exquisite .	A hoisting engine with gear to lift —— tons per day
11,347	Exscind .	We are hoisting daily
11,348	Exscribed	Placing hoisting engine in position
11,349	Exscript .	Have started hoisting engine
11,350	Exsertile .	When will hoisting engine be ready to start
11,351	Exsiccant .	We propose to place hoisting engine in
11,352	Exsiccator	**Hold(s)**
11,353	Exsputory	To hold
11,354	Exsuccous	Hold(s) all the stock
11,355	Exsufflate .	Hold(s) balance
11,356	Extancy .	Hold your stock
11,357	Extant . .	Hold on
11,358	Extemporal	Hold yourself in readiness to
11,359	Extendant	Hold back
11,360	Extendedly	Try to get hold of
11,361	Extendeth	It is desirable to get hold of
11,362	Extendible	Has (have) hold of
11,363	Extendless	Do you think it safe to hold
11,364	Extension	We hold —— shares
11,365	Extenuator	He holds —— shares
11,366	Exteriorly .	We recommend you to hold on
11,367	Exteriors .	How many do you hold
11,368	Extermine	Can you hold on

No.	Code Word.	Hold(s) *(continued)*
11,369	Extern . .	Think it safe to hold
11,370	External .	Hold largely in
11,371	Externally	**Holders**
11,372	Extimulate	Holders have great confidence in
11,373	Extincted .	Holders will not sell at present prices
11,374	Extinguish	The holders of
11,375	Extirpable	Is the holder of
11,376	Extirpator	Who are the principal holders
11,377	Extirped .	Many holders would be glad to
11,378	Extolleth .	The principal holders are
11,379	Extolling .	A few large holders
11,380	Extolment	A few large holders control the market
11,381	Extorters .	**Holding**
11,382	Extortious	What is ——'s holding in
11,383	Extra . .	Is (are) holding back
11,384	Extracted .	Is (are) holding the stock
11,385	Extracting	**Hole(s)**
11,386	Extraction	Only a hole in the ground
11,387	Extractors	A few holes have been made
11,388	Extradite .	**Holed**
11,389	Extradosed	Holed the rise
11,390	Extradotal	Holed the winze
11,391	Extramural	Holed through to the old workings
11,392	Extraneity	Have holed
11,393	Extraneous	**Holidays**
11,394	Extrasolar	Nothing now will be done till after the holidays
11,395	Extraught	Before the holidays
11,396	Extreme .	During the holidays
11,397	Extremely	After the holidays
11,398	Extremists	It must be put before the public before the holidays
11,399	Extricated	The Company must be brought out as soon as the
11,400	Extrinsic	Directly after the holidays [holidays are over
11,401	Exuberancy	The banks here are closed on account of a national holiday, which will detain me —— days
11,402	Exuberant	Yesterday was Bank Holiday
11,403	Exudating	To-day is Bank Holiday
11,404	Exulcerate	To-morrow will be Bank Holiday
11,405	Exultancy.	On account of the holidays
11,406	Exultant .	Is taking a holiday ; will return
11,407	Exultingly	**Home**
11,408	Exundate .	Shall be home about the
11,409	Exundation	Am (are) anxious to return home
11,410	Exungulate	You must return home at once ; —— is very ill
11,411	Exuperable	You must return home at once ; —— died on the
11,412	Exurgent .	Before returning home
11,413	Exuvial .	Am (are) on my (our) road home
11,414	Exvotos .	Shall return home by way of
11,415	Eyasmusket	Can you arrange to return home by way of
11,416	Eydent .	Is (are) away from home
11,417	Eyeballs .	Impossible to return home till

No.	Code Word.	**Home** (*continued*)
11,418	Eyebeams	Come home immediately
11,419	Eyebolt .	Will come home as fast as possible
11,420	Eyebrow .	How are all at home
11,421	Eyedoctor	All at home are well
11,422	Eyedrops .	Will come home as soon as the business is settled
11,423	Eyeflaps .	**Honest**
11,424	Eyeful . .	Has (have) not the reputation of being too honest
11,425	Eyeglance	Is (are) thoroughly honest and upright
11,426	Eyeglasses	**Honeycombed**
11,427	Eyeing. .	In honeycombed quartz
11,428	Eyelash .	In honeycombed rock
11,429	Eyeleteer .	**Honour**
11,430	Eyelethole	Of unimpeachable honour
11,431	Eyelid . .	Please to honour our draft
11,432	Eyepieces .	Will honour your draft
11,433	Eyeservant	**Honourable**
11,434	Eyeservice	Is a most honourable man
11,435	Eyeshot .	Are most honourable men
11,436	Eyesight .	Is (are) not considered to be very honourable
11,437	Eyesores .	Has (have) behaved in a most honourable way about
11,438	Eyesplice .	Not a very honourable proceeding
11,439	Eyespotted	Anything but honourable
11,440	Eyestrings	Strictly honourable
11,441	Eyetooth .	**Honourably**
11,442	Eyewater .	Has not behaved honourably
11,443	Eyewink .	We do not consider that he acted at all honourably
11,444	Eyewitness	**Honoured** [in the matter
11,445	Eyliads .	Has (have) honoured the draft
11,446	Fabaceae .	**Hope**
11,447	Fabaceous	Hope you will be able
11,448	Fabian. .	Hope I (we) shall be able to
11,449	Fabric . .	Hope you will
11,450	Fabricant .	Hope you will not
11,451	Fabricator	Has (have) no hope of
11,452	Fabricked .	Has (have) great hope
11,453	Fabulist .	Is there any hope that (of)
11,454	Fabulized .	There is every hope that
11,455	Fabulizing	Hope you will send
11,456	Fabulosity	Must hope for the best
11,457	Fabulous .	There is little hope of
11,458	Faburthen	There is very little hope of ——'s recovery
11,459	Façades .	Hope to inform you soon
11,460	Faceache .	We are still in hope of doing better
11,461	Facecloth .	Have lost all hope and confidence
11,462	Faceguard	Our only hope now lies in
11,463	Facehammer	**Hopeful**
11,464	Facemould	Am (are) very hopeful of
11,465	Facepiece .	Not hopeful of
11,466	Faceplates	Are you hopeful of
11,467	Faceted .	**Hopeless**

No.	Code Word.	**Hopeless** (*continued*)
11,468	Facetious .	Not hopeless
11,469	Facewheel	Is it hopeless
11,470	Facially .	It is quite hopeless
11,471	Facileness	Do not think the case quite hopeless
11,472	Facilitate .	It is a hopeless matter
11,473	Facility .	**Hopelessly**
11,474	Facingly .	Hopelessly lost
11,475	Facingsand	Hopelessly involved
11,476	Facsimile .	**Horizontally**
11,477	Factionary	Worked horizontally
11,478	Factiously .	The vein is pinching out horizontally
11,479	Factitive .	**Hornblende**
11,480	Factorage .	**Horse(s)**
11,481	Factoress .	Cutting a horse
11,482	Factorized	A horse of granite
11,483	Factorship	A horse of slate
11,484	Factóry .	The horse met with
11,485	Factotums	There are —— horses
11,486	Factum .	Horse-whim
11,487	Faculties .	**Horse-power**
11,488	Facundious	Engine of —— horse-power
11,489	Faded . .	Nominal horse-power
11,490	Fadedly .	Effective horse-power
11,491	Fadeless .	Indicated horse-power
11,492	Fading. .	Effective horse-power required for
11,493	Fadingness	**Hospital**
11,494	Fagend .	Is now in hospital
11,495	Faggotvote	Had to leave him in hospital
11,496	Fagopyrum	Now convalescent, and has left hospital
11,497	Fagots . .	Hospital fund
11,498	Fahamtea .	**Hostile**
11,499	Fahlerz .	Decidedly hostile to
11,500	Fahlore .	Was hostile, but now supports
11,501	Fahlunite .	If he takes a hostile course of action
11,502	Fahrenheit	Hostile proceedings threatened
11,503	Failance .	Hostile proceedings commenced
11,504	Failingly .	**Hostility**
11,505	Failures .	We have to face his (their) hostility
11,506	Faintdraw.	In the event of his (their) hostility
11,507	Faintish .	Displays the greatest hostility to
11,508	Faintling .	Have to meet bitter hostility from
11,509	Faintness .	**Hour(s)**
11,510	Fairfaced .	In —— hour(s)
11,511	Fairhood .	In the twenty-four hours
11,512	Fairies . .	How many hours does it take to get to —— from
11,513	Fairishly .	It takes —— hours to get from here to
11,514	Fairleader	It takes —— hours to get from —— to
11,515	Fairminded	A journey of —— hours
11,516	Fairplay .	Must have a reply in —— hours
11,517	Fairspoken	A delay of —— hours has occurred

No.	Code Word.	Hour(s) *(continued)*
11,518	Fairtold .	After a delay of —— hours
11,519	Fairworld .	Will take —— hours to
11,520	Fairy . .	Every hour
11,521	Fairyism .	Hours' shift
11,522	Fairyking .	During —— hours
11,523	Fairyland .	Within —— hours
11,524	Fairylike .	More than —— hours
11,525	Fairymoney	Less than —— hours
11,526	Fairyqueen	How many hours
11,527	Fairytales .	One hour
11,528	Faithful .	Two hours
11,529	Faithfully .	Three hours
11,530	Faithless .	Four hours
11,531	Falcated .	Five hours
11,532	Falchion .	Six hours
11,533	Falcon . .	Seven hours
11,534	Falconers .	Eight hours
11,535	Falconidae	Nine hours
11,536	Falconry .	Ten hours
11,537	Falderall .	Eleven hours
11,538	Faldfee .	Twelve hours
11,539	Faldistory .	Thirteen hours
11,540	Faldstool .	Fourteen hours
11,541	Faldworth	Fifteen hours
11,542	Falernian .	Sixteen hours
11,543	Fallacies .	Seventeen hours
11,544	Fallacy .	Eighteen hours
11,545	Fallals . .	Nineteen hours
11,546	Fallboard .	Twenty hours
11,547	Fallibly .	Twenty-one hours
11,548	Fallopian .	Twenty-two hours
11,549	Fallowchat	Twenty-three hours
11,550	Fallowcrop	Twenty-four hours
11.551	Fallowdeer	**Housing.** (See Dwelling-House.)
11,552	Fallowists .	Housing for battery
11,553	Fallowness	Housing for rolls
11,554	Falltrank .	Housing for the men
11,555	False . .	**How**
11,556	Falsefaced	How is (are)
11,557	Falseheart	How long
11,558	Falsehood	How shall I (we)
11,559	Falsely . .	How has (have)
11,560	Falseness .	How many
11,561	Falsifying .	How soon
11,562	Falsism .	How far
11,563	Falsities .	How much is there
11,564	Faltering .	How is it that
11,565	Famacide .	How much
11,566	Famblecrop	How large is
11,567	Familiar .	How can you best

No.	Code Word.	**How** (*continued*)
11,568	Familiarly	How can
11,569	Familistic .	How high is
11,570	Family . .	(And) it not, how
11,571	Familyhead	**However**
11,572	Familyman	However the case may go
11,573	Famished .	However you may
11,574	Famishment	We think, however, that
11,575	Famosity .	**Hundredweight.** (See Cwt.)
11,576	Famously .	About one cwt.
11,577	Famousness	Send us about —— cwt.
11,578	Fanatic .	Weight about —— cwt.
11,579	Fanatical .	**Hunt**
11,580	Fanaticism	To hunt up
11,581	Fanaticize	Must hunt up
11,582	Fanblast .	**Huntington mill(s)**
11,583	Fanblower	Huntington mill(s), with engine and boiler
11,584	Fancied .	Huntington mill(s), with engine and boiler, tables
11,585	Fanciful .	Huntington mill(s), with turbine [and vanners
11,586	Fancoral .	Huntington mill(s), with turbine, tables, and vanners
11,587	Fancricket	Huntington mill(s), with tables and vanners
11,588	Fancyball .	Arrangement of Huntington mill(s)
11,589	Fancyfair .	**Hurry**
11,590	Fancyfree	Too great a hurry
11,591	Fancygoods	Is (are) in too great a hurry to get the matter settled
11,592	Fancying .	Must hurry on the sale
11,593	Fancysick .	Hurry back to
11,594	Fancywork	Hurry over
11,595	Fancywoven	Impossible to hurry
11,596	Fangleness	It will be impossible to get it done in such a hurry
11,597	Fanlight .	Must hurry on the works
11,598	Fannerved	In the hurry of
11,599	Fanshaped	Do not be in too great a hurry
11,600	Fantastic .	There is no need to hurry
11,601	Fantomcorn	By undue hurry all may be spoiled (or lost)
11,602	Fantracery	No use in trying to hurry
11,603	Fanveined	Owing to the hurry of business
11,604	Fanwheels	In the hurry of leaving, forgot to
11,605	Fanwindow	Must hurry up in order to
11,606	Farabouts .	Hurry shipment ; very important
11,607	Farad . .	**Hurt**
11,608	Faradism .	Has any one been hurt
11,609	Farandams	Is (are) hurt
11,610	Farbrought	No one was hurt
11,611	Farcically .	Was (were) not hurt
11,612	Farcybud .	Seriously hurt by a railway accident
11,613	Fardingale	Seriously hurt by fall from a horse
11,614	Fardingbag	Much hurt ; compelled to lie over
11,615	Farewells .	Hurt by accident ; fear seriously
11,616	Farfetched	Much hurt by what was said
11,617	Farforth .	**Hydraulic**

No.	Code Word.	**Hydraulic** (*continued*)
11,618	Farinose .	Hydraulic machinery
11,619	Farinosely	Hydraulic mining
11,620	Farmable .	Hydraulicked during past week
11,621	Farmership	**Hydraulicking**
11,622	Farmhouse	Want man experienced in hydraulicking
11,623	Farmoffice	A competent hydraulicking engineer
11,624	Farmstead	Hydraulicking operations
11,625	Farmstock	**Hypothecated**
11,626	Farmyard .	Hypothecated to cover advance
11,627	Farobank .	Hypothecated to
11,628	Faroff . .	Property hypothecated
11,629	Farseeing .	**Hypothecation**
11,630	Farsighted	**Ice**
11,631	Farsought .	Blocked with ice
11,632	Fascialis .	Blocked for many months by the ice
11,633	Fascicle .	Over the ice
11,634	Fasciculus .	**Idea**
11,635	Fascinate .	Can form no idea
11,636	Fascinous .	Can you form any idea (whether)
11,637	Fashioneth	Have no idea that
11,638	Fastday .	Abandoned the idea
11,639	Fastenings	What are your ideas on the matter
11,640	Fasthanded	We have an idea that
11,641	Fastidious	Above our ideas of its value
11,642	Fastigium .	The idea appears to be a good one
11,643	Fastingday	**Identical**
11,644	Fastingman	Identical with
11,645	Fastnesses	**Identification**
11,646	Fastuosity .	Upon identification
11,647	Fatalistic .	Identification of the properties (named)
11,648	Fatality .	**Identified**
11,649	Fatalness .	Must be identified
11,650	Fatbrained	Until identified
11,651	Fatefully .	**Identify**
11,652	Fatherhood	In order to identify the
11,653	Fatherland	Can you identify the
11,654	Fatherless	Will serve to identify
11,655	Fatherly .	**Idle**
11,656	Fathership	Idle from want of
11,657	Fathom .	Idle from want of funds
11,658	Fathomable	Idle from want of water
11,659	Fathoming	The mills are standing idle for want of
11,660	Fathomless	Idle on account of
11,661	Fathomwood	Again started after standing idle
11,662	Fatigable .	**Igneous**
11,663	Fatigation	Dyke of igneous rock
11,664	Fatigued .	In igneous rock
11,665	Fatiguing .	Igneous formation
11,666	Fatiscence	**Ignorance**
11,667	Fatlute .	Astonished at the apparent ignorance (of)

No.	Code Word.	Ignorance (*continued*)
11,668	Fatness .	Complain of being kept in ignorance
11,669	Fattened .	Still in ignorance of
11,670	Fatty . .	Have been kept in ignorance (of)
11,671	Fatuitous .	**Ignorant**
11,672	Fatwitted .	Being ignorant of
11,673	Faultful .	Kept ignorant about
11,674	Faultily .	Quite ignorant with regard to
11,675	Faultless .	Could not have been ignorant of
11,676	Fauna . .	Ignorant of what was going on (or being done)
11,677	Faunist .	Why were we kept ignorant of
11,678	Fausens .	**Ill**
11,679	Favaginous	I am very ill, and must go for advice to
11,680	Favella .	Has (have) been dangerously ill, but is (are) now
11,681	Favonian .	Is (are) dangerously ill with [better
11,682	Favourably	Has (have) been ill with
11,683	Favouress.	I am very ill (with)
11,684	Favourite .	Is very ill (with)
11,685	Favularia .	**Illness**
11,686	Fawned .	Leaving owing to illness
11,687	Fawningly	Illness appears to be serious
11,688	Fazzolets .	Illness is not serious
11,689	Fearbabe .	**Illustration(s)**
11,690	Fearful. .	In illustration of
11,691	Fearfully .	**Imagine**
11,692	Fearless .	Do not imagine that
11,693	Fearlessly .	Seem(s) to imagine that
11,694	Fearnaught	Cannot imagine why
11,695	Fearsome .	**Immaterial**
11,696	Feastrite .	It is quite immaterial (whether)
11,697	Feastwon .	**Immediate**
11,698	Featbodied	Immediate attention is requested
11,699	Feateously	Take immediate action
11,700	Featherbed	Immediate measures must be taken
11,701	Feathering	**Immediately**
11,702	Feathertop	See —— immediately
11,703	Featured .	Must be done immediately
11,704	Febricula .	Must have immediately
11,705	Febrific .	I acted immediately
11,706	Febrifugal	Telegraph immediately
11,707	Februation	Send immediately
11,708	Fecifork .	Cannot go immediately on account of
11,709	Feculency	Come immediately
11,710	Feculent .	Wanted immediately
11,711	Fecund .	Come or send some one immediately
11,712	Fecundate	Immediately after
11,713	Fecundify.	Immediately before
11,714	Federacy .	**Immense**
11,715	Federalist.	An immense expense
11,716	Feebleness.	On an immense scale
11,717	Feebly .	Covering an immense tract of country

No.	Code Word.	**Immense** (*continued*)
11,718	Feedhead .	Worked to an immense extent
11,719	Feedmotion	Immense profits have been made
11,720	Feedpipe .	**Imparted**
11,721	Feedpump	Imparted to me in confidence
11,722	Feedwater	Imparted the information (to)
11,723	Feefarm .	**Impartial**
11,724	Feelingly .	An impartial report
11,725	Feesimple	From an impartial point of view
11,726	Feigned .	**Impartially**
11,727	Feignedly.	Considered impartially
11,728	Feigning .	**Impassable**
11,729	Feigningly	The roads are impassable
11,730	Felapton .	Impassable in winter
11,731	Feldspar .	**Impatient**
11,732	Felicified .	Is (are) very impatient
11,733	Felicify .	Getting very impatient
11,734	Felicitate .	Impatient for the report
11,735	Felicitous .	Too impatient
11,736	Felixian .	Impatient to get
11,737	Fellable .	Getting very impatient at the delay
11,738	Fellinic .	Very impatient at the want of further information
11,739	Fellmonger	To allay the impatience
11,740	Fellowfeel	**Impediment**
11,741	Fellowlike	An impediment in the way
11,742	Fellowly .	Until this impediment is removed
11,743	Fellowship	**Impending**
11,744	Felonious .	The impending trial
11,745	Felonwort .	The impending lawsuit
11,746	Felony . .	**Imperative**
11,747	Felspath .	It is most imperative that you should
11,748	Felspathic	It is most imperative that we should
11,749	Felstone .	This is most imperative
11,750	Feltcloth .	It is imperative upon us
11,751	Feltgrain .	It is imperative upon him (them) to
11,752	Felthat .	Our instructions are imperative and leave no alter- [native
11,753	Feltmaker	It is an imperative necessity
11,754	Femalist .	**Imperceptible**
11,755	Femalized	Almost imperceptible
11,756	Feminancy	**Imperceptibly**
11,757	Feminine .	**Imperfect**
11,758	Femininely	The imperfect means of working
11,759	Fenberries	Is (are) imperfect
11,760	Fenberry .	Imperfect in many ways
11,761	Fenboat .	Very imperfect, but can be improved
11,762	Fenceless .	The arrangements are very imperfect
11,763	Fencemonth	Imperfect, wants renewing
11,764	Fenceroof.	So imperfect, that we recommend you to replace by [new
11,765	Fencible .	The accounts are imperfect
11,766	Fender .	The description is imperfect
11,767	Fenderbolt	The method employed is very imperfect

No.	Code Word.	Imperfect (*continued*)
11,768	Fenderpile	The machine (or machinery) is very imperfect
11,769	Fenduck .	**Imperfection(s)**
11,770	Fenestella	To remedy the imperfections
11,771	Fenestral .	Can the imperfections be remedied
11,772	Fenfowl .	With all imperfections
11,773	Fengeld .	Owing to the imperfections
11,774	Fengoose .	Free from imperfection
11,775	Fenianism	Is there any imperfection in
11,776	Fenland .	Imperfection can be remedied
11,777	Fenugreek	**Imperil**
11,778	Feodality .	Will imperil our interests
11,779	Feodaries .	Will imperil your interests
11,780	Feodary .	Imperil the interests of
11,781	Feoffment .	Will not imperil in any way
11,782	Ferdigrew .	Does not imperil our interests
11,783	Ferdwit .	**Imperilled**
11,784	Ferineness	Will our interests be imperilled by
11,785	Feringhees	Interests are imperilled by
11,786	Ferment .	**Implicated**
11,787	Fermental	Is (are) implicated in the matter
11,788	Fermenting	Is (are) not implicated in the matter
11,789	Fermillet .	Is (are) —— implicated in the
11,790	Fernandina	Implicated in
11,791	Fernowls .	Implicated in the swindle
11,792	Fernseed .	If —— is implicated in
11,793	Fernshaw .	**Implicit**
11,794	Fernticle .	Place(s) implicit faith in
11,795	Ferocious .	Have had implicit confidence in
11,796	Ferocity .	Implicit obedience to orders
11,797	Ferraria .	**Implicitly**
11,798	Ferried .	Has (have) trusted implicitly
11,799	Ferrilite .	Was implicitly trusted
11,800	Ferrotype .	Your instructions will be implicitly obeyed
11,801	Ferrugo .	Implicitly rely on
11,802	Ferrules .	**Imply**
11,803	Ferryboat .	Do (does) —— mean to imply
11,804	Ferrying .	What do you imply by
11,805	Ferryman .	Mean(s) to imply
11,806	Fertilizer .	Do (does) not mean to imply
11,807	Fervence .	**Impolitic**
11,808	Fervently .	Think it would be very impolitic
11,809	Fervescent	Think it was very impolitic
11,810	Fervidity .	The course taken very impolitic
11,811	Fervidness	If such an impolitic course were pursued
11,812	Fervour .	**Import**
11,813	Fesapo .	Will have to import
11,814	Fessitude .	Need no longer import
11,815	Festal . .	When they had to import
11,816	Festally .	**Importance**
11,817	Festino .	It is of the greatest importance

No.	Code Word.	**Importance** (*continued*)
11,818	Festivals .	It is of great importance to myself and friends
11,819	Festivity .	It is not of any importance
11,820	Festoon .	It is of great importance to secure
11,821	Festooning	Not of sufficient importance
11,822	Festuca .	Think it of sufficient importance
11,823	Festucous .	Cannot too much urge the importance of
11,824	Fetichism .	Over-estimate the importance of
11,825	Fetisely .	**Important**
11,826	Fetlocked .	Important to know
11,827	Fetlow . .	Too important to be overlooked
11,828	Fetterlock .	A most important point
11,829	Feudal . .	Has (have) gone to —— on important business
11,830	Feudalism	Important business [connected with
11,831	Feudalized	Sufficiently important to warrant
11,832	Feudally .	Very important that you should
11,833	Feudatory .	Very important that we should be kept well posted
11,834	Feudbote .	Important that we have the earliest information as to
11,835	Feuillans .	If it is important
11,836	Feverbush	Important to get it out quickly
11,837	Feverfew .	It is very important (to)
11,838	Feverish .	It is the more important (to)
11,839	Feverishly	Which is the more important
11,840	Feverly .	**Imposing**
11,841	Feverous .	Is (are) imposing absurd conditions
11,842	Feverously	Is (are) imposing too many restrictions
11,843	Fevertree .	Is (are) imposing on
11,844	Feverweed	Is (are) —— not imposing on
11,845	Feverwort .	**Impossible**
11,846	Fewtrils .	It is impossible
11,847	Fibreless .	It will be impossible
11,848	Fibres . .	It is not impossible
11,849	Fibriform .	It is impossible to get any information about
11,850	Fibrilla .	Quite impossible at present
11,851	Fibrillous .	We found it impossible to do as you wish
11,852	Fibroin .	It was quite impossible
11,853	Fibrolite .	Is not impossible, but very difficult
11,854	Ficaria . .	Impossible to answer by telegraph [write fully
11,855	Fichtelite .	Impossible to answer all your questions now—will
11,856	Fickle . .	**Impracticable**
11,857	Fickleness	Is quite impracticable
11,858	Fictional .	The idea is quite impracticable
11,859	Fictionist .	Do you consider the idea impracticable
11,860	Fictitious .	If impracticable, then try
11,861	Fiddlebow	**Impress**
11,862	Fiddledock	You must impress upon
11,863	Fiddlers .	Has (have) tried to impress upon
11,864	Fiddlewood	Cannot too strongly impress upon
11,865	Fidejussor	Will impress upon
11,866	Fidelity .	**Impressed**
11,867	Fidgety .	Favourably impressed by what was stated

No.	Code Word.	**Impressed** (*continued*)
11,868	Fidhammers	Unfavourably impressed by what was stated
11,869	Fidonia .	Most favourably impressed by (with)
11,870	Fiducial .	Much impressed by the proposal
11,871	Fiducially.	Favourably impressed by, but cannot at present
11,872	Fieldale .	**Impression(s)** [entertain, the proposal
11,873	Fieldbasil.	Has (have) formed a most favourable impression of
11,874	Fieldbeds .	Has (have) formed an unfavourable impression of
11,875	Fieldbook.	What is your impression as to
11,876	Fieldday .	(Our) impression of the matter is decidedly favourable
11,877	Fieldduck.	Acting upon my (our) impressions, have decided to
11,878	Fieldfares .	The impression which he formed from the examina-
11,879	Fieldglass.	My (our) impression is [tion was
11,880	Fieldgun .	Our impressions after reading your letter were
11,881	Fieldhand.	The impression which remains in our minds is that
11,882	Fieldhouse	We still retain the impression (formed after)
11,883	Fieldnotes	We cannot divest ourselves of the impression that
11,884	Fieldpiece	We hope you will be able to remove these im-
11,885	Fieldroom	A most disagreeable impression [pressions
11,886	Fieldsman	The shareholders have formed an impression
11,887	Fieldstaff .	An impression founded on the perusal of the report
11,888	Fieldtrain .	There is an impression of insecurity
11,889	Fieldvole .	An impression of pending disaster [which prevails
11,890	Fieldwork .	We shall do our best to remove the bad impression
11,891	Fiendful .	In order to remove the bad impression
11,892	Fiendfully.	**Improbability**
11,893	Fiendish .	The evident improbability of the statement
11,894	Fiendishly	Its improbability must be evident
11,895	Fiendlike .	There is an air of improbability about it
11,896	Fierce . .	**Improbable**
11,897	Fiercely .	It is very improbable
11,898	Fierceness	Not at all improbable (that)
11,899	Fierycross	**Improper**
11,900	Fieryhot .	In a very improper manner
11,901	Fierynew .	It was very improper of him (them)
11,902	Fieryshort	Strongly disapprove such improper conduct
11,903	Fifemajor .	**Improperly**
11,904	Fiferail .	Has (have) acted very improperly
11,905	Figapple .	Has (have) very improperly allowed
11,906	Figaro . .	Has been improperly worked
11,907	Figcake .	The property has been improperly managed
11,908	Figeater .	Improperly and without system
11,909	Figgnats .	Unfairly and improperly taken advantage of
11,910	Figgum .	**Improve(s)**
11,911	Fight . .	Is it likely to improve
11,912	Fighting .	The mine continues to improve
11,913	Figleaf. .	The mine does not improve
11,914	Figments .	Does the mine improve in depth
11,915	Figpecker.	Is likely to improve
11,916	Figshells .	Not likely to improve
11,917	Figtrees .	Do you think the runs will improve

No.	Code Word.	**Improve(s)** (*continued*)
11,918	Figurable .	We think the runs will improve
11,919	Figural .	Think it (they) will improve
11,920	Figurately .	Ought to improve
11,921	Figurative .	The vein improves as it goes down
11,922	Figurehead	The face improves as we go on
11,923	Figureth .	The quality of ore improves
11,924	Figwort .	**Improved**
11,925	Filament .	Has improved
11,926	Filatories .	Has not improved
11,927	Filatory .	Has improved since we last wrote
11,928	Filbert . .	Has not improved since we last wrote
11,929	Filcher .	**Improvement(s)**
11,930	Filchingly .	Improvements consist of
11,931	Filecutter .	There is a great improvement in
11,932	Filefish .	Do you see any improvement in
11,933	Fileleader .	There are indications of improvement (in)
11,934	Filial . .	There are no indications of improvement (in)
11,935	Filibuster .	No improvements of any kind
11,936	Filiciform .	There is no improvement (in)
11,937	Filicoid .	Owing to the improvements in [velopments
11,938	Filicology .	Is there any improvement in the workings and de-
11,939	Filiformia .	The workings show considerable improvement
11,940	Fillagree .	The developments show some improvement
11,941	Fillhorse .	We think the improvement will continue
11,942	Fillibegs .	Does the improvement continue
11,943	Filliped .	The improvement continues
11,944	Filliping .	**Imprudent**
11,945	Fillister .	We think it was very imprudent
11,946	Fillyfoal .	Think it will be most imprudent
11,947	Filmy . .	A most imprudent statement
11,948	Filoplume	**Inaccessible**
11,949	Filtered .	Very inaccessible
11,950	Filth . .	The mines are situated in a most inaccessible place
11,951	Filthiest .	Quite inaccessible except by
11,952	Filthily .	Owing to the inaccessible position of the property
11,953	Filthiness .	**Inaccuracy**
11,954	Filtrate .	There is some inaccuracy in
11,955	Filtrating .	**Inaccurate**
11,956	Filtration .	The reports are most inaccurate
11,957	Fimblehemp	Do you find the reports inaccurate
11,958	Fimbriate .	The figures are inaccurate
11,959	Finance .	The statement sent is inaccurate, must ask you to
11,960	Financial .	**Inadequate** [rectify it
11,961	Financing .	Totally inadequate
11,962	Findfault .	Quite inadequate to the
11,963	Finedrawn	Remuneration is inadequate
11,964	Finery . .	**Inadequately**
11,965	Finespoken	Inadequately paid for the risk run
11,966	Finespun .	Very inadequately provided for
11,967	Finestuff .	**Inadmissible**

No.	Code Word.	
11,968	Finger . .	**Inadvertency**
11,969	Fingerbowl	Owing to inadvertency
11,970	Fingerfern	**Inapplicable**
11,971	Fingerpost	Inapplicable under the circumstances
11,972	Fingrigo .	**Inattention**
11,973	Finical. .	Inattention to our interests
11,974	Finically .	Owing to inattention
11,975	Finicking .	The least inattention will
11,976	Finified .	**Incalculable**
11,977	Finingpot .	Will do incalculable injury
11,978	Finish . .	**Incapable**
11,979	Finisheth .	I consider —— quite incapable of
11,980	Finpike .	Is (are) incapable of
11,981	Finscales .	**Incautious**
11,982	Finspined .	The incautious use of
11,983	Fintoed .	If —— is so incautious as to
11,984	Firealarm .	**Inch, Inches**
11,985	Firearms .	The vein is several inches in width
11,986	Firearrow .	Using —— inches of water
11,987	Fireballs .	Per miner's inch
11,988	Firebarrel .	Inches wide
11,989	Firebasket	Inches deep
11,990	Firebavin .	Inches thick
11,991	Fireblast .	Inches in diameter
11,992	Fireboard .	Inches in circumference
11,993	Firebox .	A stringer from —— to —— inches thick
11,994	Firebrand .	One quarter of an inch
11,995	Firebrick .	Half-an-inch
11,996	Firebridge	Three-quarters of an inch
11,997	Firebrief .	One inch
11,998	Firebrush .	Two inches
11,999	Firebucket	Three inches
12,000	Fireclay .	Four inches
12,001	Firecock .	Five inches
12,002	Firedamp .	Six inches
12,003	Fireengine	Seven inches
12,004	Fireescape	Eight inches
12,005	Firefanged	Nine inches
12,006	Fireflags .	Ten inches
12,007	Fireflaire .	Eleven inches
12,008	Fireflies .	Twelve inches
12,009	Firefly . .	Thirteen inches
12,010	Firehooks .	Fourteen inches
12,011	Firehouse .	Fifteen inches
12,012	Firekiln .	Sixteen inches
12,013	Fireladder .	Seventeen inches
12,014	Firelight .	Eighteen inches
12,015	Firemains .	**Incidental(ly)**
12,016	Firemarble	Incidental expenses
12,017	Firemaster	Incidental to the main question

No.	Code Word.	**Incidental(ly)** (*continued*)
12,018	Fireopal .	Incidentally stated
12,019	Fireordeal.	**Inclination**
12,020	Firepan .	At an inclination of ——° from the perpendicular
12,021	Fireplaces.	At an inclination of ——° from the horizontal
12,022	Fireplug .	There is an inclination on the part of
12,023	Firepolicy.	What is the inclination of
12,024	Fireproof .	Have no inclination to ,
12,025	Fireraft. .	**Incline**
12,026	Fireroll. .	Incline is down
12,027	Firescreen.	Incline shaft
12,028	Fireships .	On the incline of the lode
12,029	Fireshovel.	North of the incline shaft
12,030	Fireside .	South of the incline shaft
12,031	Firesteel .	East of the incline shaft
12,032	Firestop .	West of the incline shaft
12,033	Fireswab .	There is an incline shaft started
12,034	Firetower .	The shaft is on an incline of ——° from the perpen-
12,035	Firetube .	At an incline of ——° [dicular
12,036	Firewarden	At an incline of one in
12,037	Firewater .	**Inclined**
12,038	Fireworks .	Is (are) —— inclined to
12,039	Firinbond .	Is (are) inclined to
12,040	Firingiron .	Is (are) not inclined to
12,041	Firkins . .	Not inclined to act as suggested
12,042	Firlot . .	Whether —— would be inclined
12,043	Firmaments	If they do not feel inclined
12,044	Firmfooted	The shareholders (may) feel inclined
12,045	Firmity .	Not inclined to go into
12,046	Firmlier .	**Include(s)**
12,047	Firmness .	To include
12,048	Firrape .	Do (does) not include
12,049	Firstbegot .	Does this include
12,050	Firstborn .	You must include
12,051	Firstclass .	Will include
12,052	Firstfloor .	Will not include
12,053	Firstfoot .	This includes
12,054	Firstfruit .	**Included(ing)**
12,055	Firsthand .	Was this included (in the)
12,056	Firstlings .	Must be included (in the)
12,057	Firstly . .	Must not be included (in the)
12,058	Firstmate .	Was included (in the)
12,059	Firstmover	Was not included (in the)
12,060	Fiscal . .	Everything included
12,061	Fishable .	Which is not included
12,062	Fishbacked	Without including
12,063	Fishbeam .	**Inclusive (of)**
12,064	Fishblock .	Is —— inclusive of
12,065	Fishcarver.	Is inclusive of all expenses
12,066	Fishdavit .	Is inclusive (of)
12,067	Fisherboat	Not inclusive (of)

No.	Code Word.	Inclusive (of) (continued)
12,068	Fisheries .	Inclusive of everything except
12,069	Fisherman.	**Income**
12,070	Fishertown	Is (are) receiving a large income from the mines
12,071	Fishfags . .	The income from these mines is —— per annum
12,072	Fishflour .	The income from these works is —— per annum
12,073	Fishgarth .	Increasing the income
12,074	Fishglue .	At what do you estimate the income from the
12,075	Fishguano.	The income of the property for the past —— years
12,076	Fishhawks.	Income tax [has been —— per ann.
12,077	Fishhook .	**Incommensurate** (with)
12,078	Fishified .	**Incompetent**
12,079	Fishiness .	Incompetent for the duties
12,080	Fishing. .	Quite incompetent; cannot recommend him
12,081	Fishingfly .	**Incomplete**
12,082	Fishingnet.	The arrangements are as yet incomplete
12,083	Fishingrod	Still incomplete
12,084	Fishjoint .	**Incompletely**
12,085	Fishkettle .	Incompletely equipped
12,086	Fishknife .	Incompletely organized
12,087	Fishlike .	**Inconsiderable**
12,088	Fishhouse.	An inconsiderable portion of the
12,089	Fishmarket	**Inconsistent** (with)
12,090	Fishmonger	Inconsistent with what was stated
12,091	Fishoil . .	Inconsistent with the report
12,092	Fishplates.	Inconsistent with the preceding
12,093	Fishponds.	They are inconsistent
12,094	Fishpool .	Inconsistent with the facts
12,095	Fishrooms.	**Incontestible**
12,096	Fishsauce .	It is incontestible that
12,097	Fishskin .	An incontestible reply
12,098	Fishsound.	It is an incontestible fact that
12,099	Fishspear .	**Incontrovertible**
12,100	Fishtackle.	We have incontrovertible proof
12,101	Fishtongue	There is incontrovertible evidence
12,102	Fishtrowel	**Inconvenience**
12,103	Fishway .	The inconvenience resulting from
12,104	Fishweirs .	Cause (causing) great inconvenience
12,105	Fishwife .	Great inconvenience (from)
12,106	Fishwoman	Without inconvenience
12,107	Fishy . .	To counterbalance the inconvenience
12,108	Fissipara .	**Inconvenient**
12,109	Fissipeds .	It will be most inconvenient to
12,110	Fissuring .	Will it be inconvenient to
12,111	Fistiana .	It is very inconvenient
12,112	Fistic . .	If not inconvenient to you
12,113	Fisticuffs .	Would it be inconvenient for you to
12,114	Fistulary .	**Incorporate**
12,115	Fistulous .	To incorporate the mine (or property)
12,116	Fitchbrush	Wish(es) to incorporate the mine (or property)
12,117	Fitful . .	Will agree to incorporate

No.	Code Word.	**Incorporate** (*continued*)
12,118	Fitfulness .	Will not agree to incorporate
12,119	Fittest . .	Will —— agree to incorporate
12,120	Fittingly .	I (we) shall not agree to incorporate, unless they
12,121	Fittingout.	In order to incorporate [will pool the stock
12,122	Fittings .	Do not consent to incorporate the mine unless they
12,123	Fittingup .	**Incorporation** [will pool the stock
12,124	Fitweed .	Before incorporation
12,125	Fivebar .	After incorporation
12,126	Fivebarred	As a condition of incorporation
12,127	Fivecleft .	Deed of incorporation
12,128	Fivefold .	Incorporation decided upon
12,129	Fiveparted	The parties to the incorporation
12,130	Fivescourt.	The scheme of incorporation
12,131	Fixative .	**Incorrect**
12,132	Fixedly .	The accounts of the mine are incorrect
12,133	Fixedness .	The statement is incorrect, that
12,134	Fixity . .	Telegram is incorrect, repeat from —— to
12,135	Fixtures .	The drawings appear to be incorrect, send revised
12,136	Fizgig . .	We have only an incorrect [ones
12,137	Flabbily .	**Increase**
12,138	Flabbiness	To increase the
12,139	Flabby . .	An increase of
12,140	Flabel . .	Can you increase the
12,141	Flabellate .	Cannot increase
12,142	Flabergast.	Will increase
12,143	Flaccidity .	Will not increase
12,144	Flagbearer	Must increase
12,145	Flagellant .	Do not increase
12,146	Flageolets .	Increase the output
12,147	Flaggingly.	Shall have to increase
12,148	Flagitate .	Intend(s) to increase
12,149	Flagitious .	Can you increase the —— to
12,150	Flagman .	There will be an increase in
12,151	Flagrantly.	We think there will be a considerable increase in the
12,152	Flagshare .	No increase can be looked for
12,153	Flagship .	Any increase, this month or next [without
12,154	Flagside .	Must not increase the yield if it cannot be done
12,155	Flagstaff .	To increase the yield we should have to
12,156	Flagstones	We expect an increase of
12,157	Flakewhite	(To) increase the monthly returns
12,158	Flameeyed	Increase in the production
12,159	Flamelet .	Increase in the expenses
12,160	Flamingoes	Material increase
12,161	Flammation	A slight increase
12,162	Flange . .	**Increased**
12,163	Flangerail .	Can be increased
12,164	Flankard .	Cannot be increased
12,165	Flankfile .	Must be increased
12,166	Flapdoodle	An increased output
12,167	Flapdragon	An increased yield

No.	Code Word.	Increased (*continued*)
12,168	Flapeared .	Output has increased from
12,169	Flapjack .	Yield has increased
12,170	Flappers .	Must not be increased
12,171	Flareup .	Materially increased
12,172	Flashed .	Not likely to be increased
12,173	Flashily .	**Increasing**
12,174	Flataft . .	Thereby largely increasing
12,175	Flatbill. .	By increasing the output
12,176	Flatcap. .	By increasing the number of
12,177	Flatfooted .	Developments justify increasing the
12,178	Flatheaded	Increasing the crushing capabilities of the mill
12,179	Flatiron .	Increasing the shipment of bullion
12,180	Flatlings .	Increasing the quantity of ore
12,181	Flatness .	Increasing the working expenses
12,182	Flatotchil .	Increasing the capacity of
12,183	Flatrace .	**Incur**
12,184	Flatrods .	Incur further expense
12,185	Flatterers .	Incur further liability
12,186	Flattery .	Incur responsibility
12,187	Flattish .	**Indebted**
12,188	Flatulency .	Have been much indebted to —— for
12,189	Flatulent .	Greatly indebted to
12,190	Flatuosity .	Am (are) not at all indebted to
12,191	Flatwise .	Indebted to our bankers for overdraft
12,192	Flatworm .	**Indefatigable**
12,193	Flaunted .	Has been indefatigable in his exertions to
12,194	Flavedo .	**Indefinite**
12,195	Flaveria .	Is (are) too indefinite
12,196	Flavescent	Your report is too indefinite with regard to
12,197	Flavindin .	My (our) report is indefinite on account of
12,198	Flavorous .	It is impossible to act on such indefinite instructions
12,199	Flavoured .	Your instructions are too indefinite
12,200	Flaxbush .	**Indemnify**
12,201	Flaxcomb .	Indemnify against all risk
12,202	Flaxen . .	Who will indemnify us
12,203	Flaxlily. .	Will he (they) indemnify
12,204	Flaxmill .	Will you indemnify
12,205	Flaxplant .	Will indemnify
12,206	Flaxseed .	Refuse to indemnify
12,207	Flaxstar .	**Indemnity**
12,208	Flaxwench	If an indemnity is given
12,209	Flayflint .	Must give an indemnity
12,210	Fleabeetle .	**Independent**
12,211	Fleabiting .	Is (are) quite independent of
12,212	Fleabitten .	You must be independent of
12,213	Flection .	An independent opinion
12,214	Fledgeling .	We prefer to be independent
12,215	Fledgy . .	**Independently**
12,216	Fledwite .	Shall act independently
12,217	Fleeced .	Better to work independently

No.	Code Word.	
12,218	Fleecewool	**Indication(s)**
12,219	Fleeringly .	Surface indications
12,220	Fleetfoot .	Very good indications
12,221	Fleetly . .	There are strong indications (that) or (of)
12,222	Fleshbrush	There is every indication to show we are close to
12,223	Fleshflies .	By all the indications [the vein
12,224	Fleshfly .	No indications of any pay ore
12,225	Fleshful .	Indications are favourable for
12,226	Fleshhood .	Indications are unfavourable for
12,227	Fleshjuice .	Are the indications favourable
12,228	Fleshling .	Cable if indications are favourable or otherwise
12,229	Fleshmeat .	Certain indications lead us to believe
12,230	Fleshpot .	There are slight indications
12,231	Fleshquake	Indications of improvement
12,232	Fleshtint .	There are indications of improvement in the lower
12,233	Fleshworm	**Indiscreetly** [workings
12,234	Fleshwound	Has (have) acted very indiscreetly
12,235	Flexibly .	**Indiscretion**
12,236	Flexile . .	By indiscretion
12,237	Flexor . .	**Indorse**
12,238	Flexuous .	Can indorse
12,239	Flibbergib .	Cannot indorse
12,240	Flicflac . .	Can you indorse
12,241	Flicker . .	Can —— indorse
12,242	Flickereth .	Will —— indorse
12,243	Flickering .	Will indorse
12,244	Flightiest .	Will not indorse
12,245	Flightshot .	I (we) can entirely indorse ——'s statements
12,246	Flightwite .	I (we) cannot indorse ——'s statements
12,247	Flimflam .	Can you indorse ——'s report
12,248	Flimsily .	Can entirely indorse ——'s report
12,249	Flinched .	Unless you can indorse the report (of)
12,250	Flinching .	**Indorsed**
12,251	Flindersia .	Who has (have) indorsed
12,252	Flingdust .	Has been indorsed by
12,253	Flinters .	Has not been indorsed (by)
12,254	Flintglass .	Ought to be indorsed (by)
12,255	Flintheart .	We have indorsed the
12,256	Flintlock .	Will have to be indorsed by
12,257	Flintstone .	**Induce**
12,258	Flinty . .	You must try to induce
12,259	Flipdog .	Has (have) tried to induce
12,260	Flippancy .	Cannot induce
12,261	Flippant .	Must induce —— to join the Board
12,262	Flirtation .	If you can induce
12,263	Flirtigig .	If you cannot induce
12,264	Fliskmahoy	**Inducement**
12,265	Floatboard	Offers great inducement
12,266	Floatcase .	Do (does) not offer sufficient inducemen'
12,267	Floating .	What inducement can you offer

No.	Code Word.	**Inducement** (*continued*)
12,268	Floatingly .	If sufficient inducement is offered
12,269	Floccose .	Unless better inducement is offered
12,270	Floccosely	**Inevitable**
12,271	Flocculent	The inevitable result will be
12,272	Flockbeds.	Inevitable loss
12,273	Flockling .	Almost inevitable
12,274	Flockly .	**Inexcusable**
12,275	Flockmel .	Inexcusable indifference
12,276	Flockpaper	Inexcusable neglect
12,277	Flogged .	Inexcusable and culpable
12,278	Flood .	**Infer**
12,279	Floodgates	Are we to infer from
12,280	Floodmark	Do you infer from
12,281	Floodtide .	What do you infer by
12,282	Floorcloth.	What do (does) —— infer by
12,283	Floorguide	I (we) infer from what —— said
12,284	Floorhead .	Do (does) not mean to infer
12,285	Floorless .	Mean(s) to infer
12,286	Florally .	Infer from this
12,287	Floramour.	**Inferred**
12,288	Florascope	It must not be inferred from
12,289	Floreated .	It will be inferred from
12,290	Florentine.	**Influence**
12,291	Florid . .	Is a man of great influence
12,292	Floridity .	Has (have) no influence here
12,293	Floriform .	Is (are) —— considered to have any influence
12,294	Florinean .	To influence
12,295	Floryboat .	Not to influence
12,296	Floscular .	Cannot influence
12,297	Flosculous.	Would have great influence
12,298	Flosferri .	Has (have) great influence in mining circles
12,299	Flosssilk .	Is (are) using all his (their) influence
12,300	Flossy . .	Use all the influence you possess
12,301	Flossyarn .	Using powerful influence
12,302	Flotant. .	Bring strong influence to bear
12,303	Flotation .	**Inform**
12,304	Flotsam .	Please inform
12,305	Flouncing.	Do not inform
12,306	Floundered	Inform me (us) of
12,307	Flourbox .	Inform me (us) of any change
12,308	Flourish .	To inform —— of
12,309	Flourmill .	Inform at once
12,310	Flowbog .	Will inform
12,311	Flowerage.	Inform me (us) by telegraph at once
12,312	Flowerbud	Will inform you
12,313	Flowerful .	Am (are) glad to inform you
12,314	Flowering.	Am (are) sorry to inform you
12,315	Flowerleaf.	Have to inform you that
12,316	Flowerpot.	Wish (wishes) you to inform —— that
12,317	Flowers .	Inform our agent (at)

No.	Code Word.	**Inform** (*continued*)
12,318	Flowerwork	Our agent informs us (that)
12,319	Flowmoss .	Our solicitors inform us (that)
12,320	Flowretry .	**Informant**
12,321	Floxedsilk.	Our informant is
12,322	Flucan . .	Who is your informant
12,323	Fluctuable.	**Information**
12,324	Fluellin .	Have you any reliable information with regard to
12,325	Fluency .	Have no reliable information concerning
12,326	Fluent . .	I cannot get any reliable information about
12,327	Fluentness	Do not place any reliance in the information (as it
12,328	Flueplate .	Give no information whatever (to) [came from)
12,329	Fluff . .	Give earliest possible information
12,330	Fluffgib .	I (we) place great reliance on the information
12,331	Fluidize .	Get all the information you can (about)
12,332	Fluidizing.	I (we) learned from various sources of information
12,333	Flukeworms	I (we) gather from the information I (we) have
12,334	Flummery.	Send us information as to [collected
12,335	Flunkey .	From whom I (we) got a great deal of information
12,336	Flunkeydom	What information have you of
12,337	Flunkeyism	Full information can be obtained of
12,338	Fluoberate	Has (have) had private information that
12,339	Fluorated .	This is for your private information only
12,340	Fluoric. .	No information can be obtained from
12,341	Fluoroid .	No information worth sending
12,342	Fluorotype	From whom did you get this information
12,343	Fluorous .	From private information
12,344	Fluorspar .	Has (have) received information that
12,345	Flurried .	The information you ask for goes by first post
12,346	Flurry . .	Wire at once fullest information about
12,347	Flurrying .	Have reliable information that
12,348	Flushness .	The information obtained from —— is reliable
12,349	Flustered .	The information obtained from —— is not reliable
12,350	Flustering.	According to the best information we can get
12,351	Flustra. .	**Informed**
12,352	Flustradae.	We are informed by (that)
12,353	Flutebit .	We have informed
12,354	Fluted . .	Has (have) been informed by
12,355	Flutelike .	Was (were) not informed of it (until)
12,356	Flutenist .	Keep us fully informed as to
12,357	Flutestop .	**Infringement**
12,358	Flutework.	Infringement of the patent
12,359	Fluvial. .	An infringement of rights
12,360	Fluvialist .	**Injunction**
12,361	Fluviatic .	The injunction has been granted
12,362	Fluxation .	The injunction has been refused
12,363	Fluxility .	Get an injunction to restrain
12,364	Fluxionary	We can get an injunction
12,365	Fluxionist.	He (they) can get an injunction
12,366	Fluxive. .	An injunction to restrain —— from
12,367	Flyagaric .	**Injure**

No.	Code Word.	**Injure** (*continued*)
12,368	Flybane .	Will injure us
12,369	Flyblock .	To injure
12,370	Flyblown .	It will not injure
12,371	Flyboards .	It will injure the company
12,372	Flyboat .	It will injure the concern if ——'s name appears in
12,373	Flycatcher	Cannot injure us in any way [connection with it
12,374	Flydrill .	Will try to injure
12,375	Flyfish . .	To prevent —— from being able to injure us in
12,376	Flyfishing .	To injure ——('s) reputation [any way
12,377	Flyflap . .	**Injured** [injured
12,378	Flyflapper .	A bad accident in the mine; —— men fatally
12,379	Flyingarmy	A bad accident in the mine; —— men badly
12,380	Flyingcamp	Man (men) injured [injured
12,381	Flyingfish .	An accident in the mine; —— men injured, but
12,382	Flyingfox .	No one was injured [not seriously
12,383	Flyingjib .	No one was seriously injured
12,384	Flyingsap .	**Injurious**
12,385	Flyingshot	We think it will be injurious to
12,386	Flyleaf . .	Do you think it likely to be injurious
12,387	Flyleaves .	The most injurious statements have been made
12,388	Flymaggot	Very injurious to the best interests of the concern
12,389	Flynet . .	**Injury**
12,390	Flyorches .	Great injury has occurred (from)
12,391	Flypapers .	Slight injury has occurred (from)
12,392	Flypenning	No injury has been done (by)
12,393	Flypowder	What injury has been done
12,394	Flypress .	No injury to
12,395	Flyrail . .	Has done considerable injury (to)
12,396	Flyshuttle .	**Inquire**
12,397	Flyspeck .	Please inquire about
12,398	Flytraps .	Please inquire and let us know by telegraph
12,399	Flywater .	If he (they) inquire(s)
12,400	Flywheel .	**Inquiry(ies)**
12,401	Flywort .	The result of the inquiry
12,402	Foalfoot .	The inquiry failed to elicit anything
12,403	Foalteeth .	A committee of inquiry
12,404	Foamcock .	Make inquiries about —— of
12,405	Foamy . .	Will answer all inquiries
12,406	Focillate .	On making inquiries in reliable quarters
12,407	Focimeter .	In answer to my (our) inquiries
12,408	Fodderers	We have made inquiries
12,409	Fodient .	From inquiries we have made
12,410	Foehood .	Made inquiries, but cannot obtain any reliable in
12,411	Foeman .	Most careful inquiries were made [formation
12,412	Foeniculum	**Insert**
12,413	Fogbanks .	Insert in the newspapers
12,414	Fogbell .	Insert in the
12,415	Foggiest .	Insert in your report
12,416	Fogginess .	Do not insert that part relating to
12,417	Foggy . .	**Inserted**

No.	Code Word.	**Inserted** (*continued*)
12,418	Foghorn .	A clause must be inserted
12,419	Fogring .	The clause need not be inserted
12,420	Fogsignal .	Will be sure to have a clause inserted
12,421	Fogsmoke	Has (have) had inserted
12,422	Fogwhistle	Will not have inserted
12,423	Foistied .	It is necessary to have inserted
12,424	Foisty . .	The particulars we are sending should be inserted
12,425	Foldless .	To be inserted in the next issue [in your report
12,426	Foldnet .	To be inserted in the next circular
12,427	Foldyard .	Will be inserted in the report
12,428	Folelarge .	Have not inserted
12,429	Foliaceous	Not to be inserted
12,430	Foliage. .	Should not be inserted
12,431	Foliating .	A paragraph should be inserted **in**
12,432	Foliature .	Do not want (it) inserted
12,433	Foliosity .	**Inside**
12,434	Folkland .	Inside the
12,435	Folklore .	The inside of the
12,436	Folkmoter	On the inside
12,437	Folkright .	**Insiders**
12,438	Folkspeech	Insiders are
12,439	Follicle .	**Insist**
12,440	Follicular.	Will not insist upon
12,441	Follies . .	You must insist upon
12,442	Folliful .	I (we) insist upon
12,443	Follower .	He (they) insist(s) upon
12,444	Followings	If —— insist(s) upon
12,445	Fomalhaut	Do (does) not insist upon
12,446	Fomenters	You need not insist upon
12,447	Fondest .	Shall I (we) insist upon
12,448	Fondly .	**Insisted**
12,449	Fontanalis	If insisted upon
12,450	Foodful .	If not insisted upon
12,451	Foolbegged	Must be insisted upon
12,452	Fooleries .	Unless insisted upon
12,453	Foolery .	**Insolvent**
12,454	Foolfish .	Is (are) insolvent
12,455	Foolhardy	Owing to —— being insolvent
12,456	Foolishly .	Is (are) likely to become insolvent
12,457	Foolscap .	**Inspect.** (See also Examine.)
12,458	Footballs .	Inspect and report on
12,459	Footbank .	You will shortly have to inspect
12,460	Footboy .	Be ready shortly to inspect
12,461	Footbridge	To inspect works at
12,462	Footcloth .	Shall be ready by —— to inspect
12,463	Footfast .	Wish(es) you to inspect —— mine, but do not go unless you hear for certain [at
12,464	Footgeld .	Wish(es) you to inspect another mine for him (them)
12,465	Footglove .	Do not arrange to inspect any mines without telling
12,466	Footguards	Who will go with you to inspect [us (as

No.	Code Word.	Inspect (*continued*)
12,467	Foothold .	Cannot inspect and report on —— property
12,468	Footiron .	Can you go to inspect and report on
12,469	Footjaw .	Can inspect and report on —— property
12,470	Footlicker	Am I to inspect and report on —— mines
12,471	Footlights	Telegraph when you can inspect
12,472	Footmantle	Have arranged for a first-class man to inspect
12,473	Footmen .	A first-class man to go and inspect
12,474	Footmuff .	**Inspected**
12,475	Footnotes.	Mine has been inspected
12,476	Footpads .	Mine has not been inspected
12,477	Footplate.	Mine is being inspected
12,478	Footplough	Have you inspected the —— property
12,479	Footpost .	Mine cannot be inspected owing to
12,480	Footpound	Mine will be inspected
12,481	Footprints	**Inspection.** (See also Examination.)
12,482	Footrules .	When you have finished your inspection of the mine telegraph your report
12,483	Footsore .	Since my previous inspection
12,484	Footstalk .	With regard to other inspections
12,485	Footstep .	Who is to pay for inspection of —— mine
12,486	Footstool .	Will pay for inspection of —— mine
12,487	Footvalve .	Could make inspection of —— mine on my way to
12,488	Footwaling	Immediate inspection urgent
12,489	Footwarmer	Inspection not yet arranged
12,490	Footways .	Inspection must be made immediately
12,491	Fopdoodles	Fee for inspection and report
12,492	Foplings .	Subject to inspection by
12,493	Foppish .	Previous to inspection
12,494	Foppishly.	After inspection
12,495	Foragecap	Inspection arranged
12,496	Foraging .	Inspection by first expert
12,497	Foralite .	Inspection by second expert
12,498	Foraminous	**Instance**
12,499	Foraminule	In the first instance
12,500	Forasmuch	There has never been any instance of
12,501	Forbade .	At the instance of
12,502	Forbear .	At whose instance
12,503	Forbearant	**Instead**
12,504	Forbidden	Instead of
12,505	Forbreak .	Not instead of
12,506	Forbruised	**Instructions**
12,507	Forcefully.	Give instructions to —— (to)
12,508	Forceless .	Awaiting instructions
12,509	Forcemeat	Has (have) given instructions to
12,510	Forcepiece	Wait for instructions
12,511	Forcepump	Had not given any instructions as to
12,512	Forcibly .	Act only upon written instructions
12,513	Forcingpit	Send instructions at once
12,514	Forcipated	Telegraph me (us) instructions
12,515	Forclosure	Am (are) waiting here for instructions

No.	Code Word.	Instructions (*continued*)
12,516	Fordable .	Telegraph instructions direct to
12,517	Fordrive .	Your instructions are to proceed at once (to)
12,518	Foreadvise	Require more explicit instructions
12,519	Foreallege	Instructions have been given to
12,520	Foreanent	Letter of instructions
12,521	Forebelief.	Have instructions been given to [day or two
12,522	Foreboded	Await full instructions, which will be sent you in a
12,523	Foreboding	Has (have) received instructions to
12,524	Forecabin.	Has (have) not received any instructions as to
12,525	Forecasts .	Full instructions have been sent by post
12,526	Forechosen	Send full instructions to me (at)
12,527	Forecovert	Instructions are being carried out
12,528	Foredated	As soon as you get instructions from
12,529	Foredeck .	Your instructions shall be carried out at once
12,530	Foredeem.	Final instructions from [and meet you at
12,531	Foredesign	Full instructions have been sent to —— to telegraph
12,532	Foredoomed	Please carry out the instructions of the Board
12,533	Foreelders	Please carry out instructions of
12,534	Forefather	According to instructions
12,535	Forefeels .	Instructions arrived too late
12,536	Forefences	Full instructions given
12,537	Forefinger	If no instructions to the contrary
12,538	Foreflow .	You have not followed instructions
12,539	Forefront .	To whom shall I apply for instructions
12,540	Foregame.	Have you any instructions from
12,541	Foregirth .	Instructions could not be carried out
12,542	Forego. .	**Insurance**
12,543	Foregoing.	Policy of insurance
12,544	Foreground	The insurance of
12,545	Foreguess.	To effect an insurance on
12,546	Forehead .	The rate of insurance (upon)
12,547	Foreign .	What is the rate of insurance (upon)
12,548	Foreignism	Amount paid for insurance
12,549	Forejudge	Insurance company(ies)
12,550	Foreknown	Rate of insurance —— per cent.
12,551	Foreleg .	Have you any insurance upon
12,552	Forelift .	Better renew insurance (to cover)
12,553	Foremost .	Claim on the insurance company for
12,554	Foremostly	Premium on insurance
12,555	Forenoon .	Premium on insurance is payable
12,556	Forenotice	**Insure**
12,557	Forensic .	Shall I insure
12,558	Forensical	How much shall I insure
12,559	Foreordain	You need not insure
12,560	Forepart .	At what rate can you insure
12,561	Forepassed	Cannot insure on lower terms
12,562	Forepeak .	Can you not insure at a lower rate
12,563	Foreplane.	Cannot insure
12,564	Foreprized	**Insured**
12,565	Forequoted	None of the buildings are insured

No.	Code Word.	**Insured** (*continued*)
12,566	Forereach .	All the buildings are fully insured [for
12,567	Forerun .	Have insured the works in the —— Insurance Office
12,568	Forerunner	Are all the buildings insured? if not, insure them
12,569	Foresaying	Have you insured against fire [at once
12,570	Forescent .	If you have not insured, do so at once
12,571	Foreseized	Mill is insured for
12,572	Foresettle .	Everything else insured for
12,573	Foreshadow	Have insured the —— at —— per cent.
12,574	Foreshamed	Has (have) been insured for
12,575	Foreshore .	**Insurrection**
12,576	Foreskirt .	Is in a state of insurrection
12,577	Foresight .	The insurrection has been crushed
12,578	Foreslack .	**Intelligent**
12,579	Foresleeve	Is very intelligent
12,580	Forespeech	Not very intelligent
12,581	Forespend	**Intend**
12,582	Forespoken	We intend to
12,583	Forest . .	Do not intend to
12,584	Forestaff .	Do you intend to
12,585	Forestage .	Does he (do they) intend to
12,586	Forestfly .	Did you intend us to
12,587	Forestoak .	If you do not intend to
12,588	Foresttree	Do you not intend to
12,589	Foretackle	Provided you intend to
12,590	Foreteller .	Intend later on to
12,591	Forethink .	Intend in the meantime to
12,592	Foretop .	Did not seriously intend to
12,593	Foretopman	**Intention(s)**
12,594	Forevouch	As earnest of our intention
12,595	Foreweary	To prove our intention
12,596	Forewinds	My intention, unless I hear to the contrary, is to
12,597	Forewoman	What is ——'s intention
12,598	Foreyards .	——'s intention is to
12,599	Forfeited .	The intention was
12,600	Forfeiting .	Has (have) abandoned their intention of
12,601	Forfeiture .	Has (have) —— any intention of
12,602	Forficula .	Has (have) no intention of
12,603	Forgeman	Has (have) not the least intention to
12,604	Forgeries .	That is our intention
12,605	Forgetable	Have changed our intentions
12,606	Forgetful .	**Intercepted**
12,607	Forgetteth	**Intercepting**
12,608	Forgewater	**Interception**
12,609	Forgiving .	The interception of
12,610	Forgotten .	**Intercession**
12,611	Forjudging	By the friendly intercession of
12,612	Forkchuck	**Interest(s)**
12,613	Forky . .	Will protect your interest(s)
12,614	Forlays .	Has (have) sold his (their) interest in
12,615	Forlorn ,	Interest at the rate of

No.	Code Word.	Interest(s) (*continued*)
12,616	Forlornly .	To pay interest at the rate of
12,617	Formalisms	What interest do you think can be paid
12,618	Formality.	Have you any interest in
12,619	Formalizer	I have an interest in
12,620	Formative	I have no interest in
12,621	Formedon	Has no interest in
12,622	Formerly .	A quarter's interest at the rate of —— per cent. per
12,623	Formful .	Interest for the half-year [annum.
12,624	Formica .	Will you sell your interest in
12,625	Formicidae	A small interest
12,626	Formidably	What interest
12,627	Formulary	A good interest on the money
12,628	Formulated	Has (have) a large interest in
12,629	Forrayers .	Take an interest in
12,630	Forsaketh .	Would take an interest in
12,631	Forsongen	Do (does) not care to take an interest in
12,632	Forsooth .	One-eighth interest in
12,633	Forsterite .	One-sixth interest in
12,634	Forsworn .	One-fifth interest in
12,635	Fortalice .	One-fourth interest in
12,636	Forthby .	One-third interest in
12,637	Forthright	One-half interest in
12,638	Forthward	Two-fifths interest in
12,639	Forthwith .	Three-fifths interest in
12,640	Fortified .	Four-fifths interest in
12,641	Fortify. .	Two-thirds interest in
12,642	Fortifying.	Three-quarters interest in
12,643	Fortitude .	Five-sixths interest in
12,644	Fortmajor.	Three-eighths interest in
12,645	Fortressed	Five eighths interest in
12,646	Fortrodden	Seven-eighths interest in
12,647	Fortuity .	The remaining interest in
12,648	Fortunate.	Not entitled to interest
12,649	Fortunes .	What is the rate of interest
12,650	Forward .	Interest will be allowed at the rate of
12,651	Forwarding	Entitled to interest
12,652	Forwardly	Must charge interest upon
12,653	Forwearied	A strong bear interest at work
12,654	Forweep .	A strong bull interest at work
12,655	Forwounded	The largest interest (in)
12,656	Forwrapped	Outstanding interest
12,657	Fossick .	Try and purchase as many interests as you can
12,658	Fossicking	Have bought all his (their) interests in
12,659	Fossil . .	Interests in —— are valued at
12,660	Fossilcork	All interests of every description
12,661	Fossilflax .	Our interests
12,662	Fossilify .	His (their) interests
12,663	Fossilism .	Your interests
12,664	Fossilized .	Watch our interests
12,665	Fossilogy .	(To) protect —— interests

No.	Code Word.	Interest(s) *(continued)*
12,666	Fossilwood	We look to you to protect our interests
12,667	Fossores .	The interests of the other parties
12,668	Fossorial .	The interests of the respective parties
12,669	Fossulate .	——'s interests will be looked after by
12,670	Fosterbabe	Has (have) bought up the other interests in the mine
12,671	Fosterdam	**Interested**
12,672	Fosterling .	Owing to the large number interested
12,673	Fosterment	Is (are) —— interested in
12,674	Fostership	Is (are) not interested (in)
12,675	Fosterson .	Interested to a large extent (in)
12,676	Foully . .	Interested to a small extent only (in)
12,677	Foulness .	The parties interested (in)
12,678	Foulspoken	Has (have) interested himself (themselves)
12,679	Foundation	Have interested ourselves
12,680	Foundering	Much interested in the
12,681	Foundress	How are we interested
12,682	Foundry .	We cannot see how we are interested
12,683	Fountain .	Chiefly interested
12,684	Fountful .	**Interesting**
12,685	Fouquiera .	A very interesting report
12,686	Fouredged	Some interesting facts in relation to
12,687	Fourfold .	(The) statement is very interesting
12,688	Fourfooted	Most interesting and valuable
12,689	Fourhanded	**Interfere**
12,690	Fourhorse .	To interfere with
12,691	Fourinhand	Not to interfere with
12,692	Fourposter	Will this interfere with
12,693	Fourscore .	This will interfere with
12,694	Foursquare	This will not interfere with
12,695	Foveolated	If it will not interfere with
12,696	Foxbat . .	If it will interfere with
12,697	Foxbrush .	Do not interfere
12,698	Foxchase .	Do not let it interfere with
12,699	Foxearth .	Must interfere with
12,700	Foxery . .	It would interfere with our plans
12,701	Foxfish .	I think you had better interfere
12,702	Foxglove .	I think you had better not interfere
12,703	Foxgrape .	Has interfered with
12,704	Foxhounds	**Interference**
12,705	Foxhunt .	The interference of
12,706	Foxhunting	**Interfering**
12,707	Foxlike .	If it can be done without interfering (with)
12,708	Foxshark .	It can be done without interfering (with)
12,709	Foxship .	It cannot be done without interfering (with)
12,710	Foxtailed .	If it cannot be done without interfering (with)
12,711	Foxtraps .	Without interfering (with)
12,712	Foxtrot .	Is (are) interfering (with)
12,713	Fractious .	Has (have) been interfering (with)
12,714	Fracture .	**Intermediate**
12,715	Fracturing	The intermediate drift

No.	Code Word.	
12,716	Fragaria	**Interrupt**
12,717	Fragilely	To interrupt the work
12,718	Fragmental	**Interrupted**
12,719	Fragments	Has much interrupted the work
12,720	Fragrancy	The business has been interrupted by
12,721	Fragrant	Progress interrupted by
12,722	Frailness	Has been greatly interrupted by
12,723	Frailty	**Interruption**
12,724	Framable	The interruption to the work (owing to)
12,725	Framboesia	The interruption to the negotiations
12,726	Framesaw	**Intersect**
12,727	Frampold	To intersect
12,728	Franchised	Hope to intersect
12,729	Franciscan	The veins intersect
12,730	Francisque	Think we shall soon intersect
12,731	Frangent	If we do not soon intersect
12,732	Frangible	When do you expect to intersect
12,733	Frangipane	**Intersected**
12,734	Frangulin	The vein was intersected
12,735	Frankbank	Have intersected
12,736	Frankchase	Intersected at a distance of —— feet
12,737	Frankenia	Intersected by a cross-cut
12,738	Frankferm	**Intersection**
12,739	Frankfold	At the point of intersection (of the)
12,740	Franking	(At) the intersection of the two veins
12,741	Frankish	At the intersection of the tunnel with the vein
12,742	Franklinic	**Interval**
12,743	Frankly	The interval between
12,744	Frankness	The interval between is filled with
12,745	Frantic	Can you fill up interval
12,746	Franticly	Fill up interval
12,747	Fratercaul	**Interview**
12,748	Fraternism	Interview —— with regard to
12,749	Fraternity	Has (have) had an interview with
12,750	Fratriage	Cannot have any interview with —— (till)
12,751	Fratricide	Can you arrange an interview between —— and
12,752	Fraud	To interview —— with the object of
12,753	Fraudful	Am (are) to have an interview with —— on
12,754	Fraudfully	It was arranged in an interview with
12,755	Fraudless	Nothing was arranged in the interview
12,756	Fraudulent	Interview with —— most satisfactory
12,757	Fraughtage	Interview with —— very unsatisfactory
12,758	Fraughting	Nothing settled definitely in the interview with ——;
12,759	Fraxinella	After an interview [we meet again on
12,760	Fraxinus	Cannot arrange without interview
12,761	Freakishly	Have not yet been able to interview
12,762	Fredstole	Our interview was most friendly
12,763	Freebench	**Intimation**
12,764	Freebooty	The first intimation we received (that)
12,765	Freeborn	Upon the first intimation (of) (that)

K

No.	Code Word.	Intimation (*continued*)
12,766	Freechapel	Have no intimation (of) (that)
12,767	Freecharge	Have had intimation (of) (that)
12,768	Freecity .	Have given intimation (that)
12,769	Freedman	**Into**
12,770	Freedom .	Go(es) into
12,771	Freefisher	Run(s) into
12,772	Freefooted	Are into
12,773	Freegrace.	Not yet into
12,774	Freeholder	How deep does it go into
12,775	Freelance.	How far into
12,776 ·	Freeliving.	Into the side of the hill
12,777	Freelove .	As soon as we get into
12,778	Freely . .	**Intolerable**
12,779	Freemartin	Has become intolerable
12,780	Freemason	Threatens to become intolerable
12,781	Freeminded	**Intractable**
12,782	Freepass .	Intractable to ordinary treatment
12,783	Freeschool	Intractable without treatment by
12,784	Freesocage	The ore is very intractable owing to
12,785	Freesoil .	Very intractable
12,786	Freespoken	**Introduce**
12,787	Freestate .	· Can you introduce
12,788	Freestuff .	Can introduce
12,789	Freetrader	Wish to introduce
12,790	Freewarren	**Introduced**
12,791	Freewill .	**Introducing**
12,792	Freeze. .	**Introduction**
12,793	Freezing .	Intended as an introduction to
12,794	Freightcar	By the introduction of
12,795	Freisleben	**Introductory**
12,796	Frenchbean	The introductory remarks
12,797	Frenchfake	**Intrust**
12,798	Frenchhorn	Whom will you intrust with
12,799	Frenchify.	Will intrust the care of —— to
12,800	Frenchpie	Do not like to intrust it to
12,801	Frenchplum	Can intrust it to
12,802	Frenchroof	Must not intrust it to
12,803	Frenchtub	**Intrusted**
12,804	Frenzical .	Have intrusted —— (with)
12,805	Frenzied .	Has (have) been intrusted (to)
12,806	Frenziedly	Intrusted to carry out
12,807	Frenzying.	Intrusted with full powers
12,808	Frequence	Intrusted as our agent(s)
12,809	Frequently	Intrusted with the charge of the mine
12,810	Frescade .	Will be intrusted to (with)
12,811	Fresco. .	Will not be intrusted to (with)
12,812	Frescoing .	**Inundation**
12,813	Freshblown	The inundation of the district around
12,814	Freshened.	**Invalid**
12,815	Freshening	Is (are) quite invalid

No.	Code Word.	**Invalid** (*continued*)
12,816	Freshforce	The concession is invalid
12,817	Freshman .	**Invalidate**
12,818	Freshshot .	Goes far to invalidate
12,819	Freshwater	Quite sufficient to invalidate
12,820	Fresison .	**Invalidated**
12,821	Fretful . .	Invalidated the concession
12,822	Fretfully .	Has invalidated the title
12,823	Fretting .	Rights have become invalidated by
12,824	Friability .	**Invalidation (of the)**
12,825	Friarbird .	By the invalidation of the concession
12,826	Friaries .	**Invalidity (of the)**
12,827	Friarlike .	The invalidity of the title
12,828	Friarly . .	The invalidity of the claims
-12,829	Friarscowl	**Invariable**
12,830	Friarskate .	The invariable custom
12,831	Fribble .	As an almost invariable rule we find
12,832	Fricando .	**Invariably**
12,833	Fricative .	Invariably found in
12,834	Fricatrice .	Invariably find in connection with
12,835	Friending .	Almost invariably
12,836	Friendlily .	Not invariably, though very often
12,837	Friendship	**Invention**
12,838	Friezerail .	A new invention for
12,839	Frigatoons	By this invention
12,840	Frigerate .	The saving effected by this invention
12,841	Frightened	This invention is expected to
12,842	Frightful .	**Inventory**
12,843	Frightless .	An inventory of all stores, loose plant, etc.
12,844	Frightment	An inventory of the machinery, etc.
12,845	Frigorific .	A complete inventory of everything on the place
12,846	Frilled . .	According to inventory
12,847	Fringeless	A monied inventory of
12,848	Fringelike	**Invest**
12,849	Fringetree	Invest immediately on my account
12,850	Fringilla .	Will invest on your account
12,851	Frippery .	Do not invest in
12,852	Friskful .	I advise you to invest in
12,853	Friskily .	Do you advise me to invest in
12,854	Frislet . .	Intend(s) to invest in
12,855	Frithsplot .	Will do well to invest
12,856	Frithstool .	**Invested**
12,857	Fritillary .	Has (have) invested in
12,858	Fritinancy	Has (have) not invested in
12,859	Frittering .	Has (have) —— invested in
12,860	Frivolism .	Has (have) not invested very much
12,861	Frivolity .	Invested on your account
12,862	Frivolous .	To be invested in
12,863	Frizzy . .	How much is invested (in)
12,864	Frockcoat .	Has been invested in
12,865	Frocked .	Invested in the undertaking

No.	Code Word.	Invested (*continued*)
12,866	Frockless .	Invested in machinery
12,867	Frogbit .	Invested in the mine
12,868	Frogcheese	Invested in stores
12,869	Frogeater .	Reserve fund (is) invested in
12,870	Frogfish .	**Investigate**
12,871	Frogflies .	Investigate, and let me know
12,872	Frogfly. .	Will investigate and let you know the result
12,873	Froggrass .	**Investigated**
12,874	Froghopper	Have investigated the matter
12,875	Frogorchis	**Investigation**
12,876	Frogshell .	Committee of investigation
12,877	Frogspit .	Will make a thorough investigation
12,878	Frolic . .	Has (have) made a thorough investigation
12,879	Frolicful .	Could not make a thorough investigation
12,880	Frolicked .	Make a thorough investigation
12,881	Flolicking .	The investigation shows
12,882	Frolicly .	After a close investigation, I am of opinion that
12,883	Frolicsome	Will bear the closest investigation
12,884	Frondesce	(After) a thorough investigation by a disinterested
12,885	Frondous .	Desire a thorough investigation [party
12,886	Front . .	A thorough investigation of all the circumstances
12,887	Frontage .	The result of the investigation is favourable
12,888	Frontdoor.	The result of the investigation is unfavourable
12,889	Frontiers .	Will it bear investigation
12,890	Fronting .	**Investment**
12,891	Frontlets .	Capital investment (of)
12,892	Frostbite .	I strongly recommend the investment
12,893	Frostbound	I do not recommend the investment
12,894	Frostiness.	A good investment for a capital of
12,895	Frostlamp.	I consider the mine a good investment
12,896	Frostmist .	It is a sufficiently good investment
12,897	Frostnail .	I am sure it would prove a bad investment
12,898	Frostsmoke	Do you recommend investment in
12,899	Frostweed	By an investment of
12,900	Frostwork.	I consider expenditure stated a judicious investment
12,901	Frosty . .	This investment will [of capital
12,902	Frothily .	It would necessitate an investment of
12,903	Frothworm	It is an investment bringing in —— per cent.
12,904	Frouzy. .	**Invisible**
12,905	Frownful .	The rock is rich in gold, but it is invisible to the
12,906	Frowningly	Invisible to the eye [eye
12,907	Frozen. .	**Invoice**
12,908	Frozenness	Invoice in duplicate
12,909	Fructified.	Invoice in triplicate
12,910	Fructist .	Must send invoice in duplicate
12,911	Fructose .	Consular invoice
12,912	Fructuary .	Invoice must be certified by consul
12,913	Frugal . .	Invoice and bill of lading
12,914	Frugality .	Invoice and list of contents of packages
12,915	Frugalness	Charged at invoice price

No.	Code Word.	Invoice (*continued*)
12,916	Frugivora .	The invoice price is
12,917	Fruitbud .	Invoice price free on board
12,918	Fruitcrow .	**Involve**
12,919	Fruiterers .	This will involve
12,920	Fruitery .	This will not involve
12,921	Fruitflies .	Will this involve
12,922	Fruitful .	Involve an additional expense of
12,923	Fruitfully .	Not to involve
12,924	Fruitive .	Involve us in an expense of
12,925	Fruitknife.	Involve us in litigation
12,926	Fruitloft .	Involve us in disputes with
12,927	Fruitshows	Involve alterations in
12,928	Fruitstall .	**Involved**
12,929	Fruitsugar	We are involved in
12,930	Fruittree .	Heavily involved　　　　　[with him (them)
12,931	Frumgyld .	Is (are) heavily involved—be careful what you do
12,932	Frumpish .	Our reputation is involved
12,933	Frumpy .	Your reputation is involved
12,934	Frustrated	His reputation is involved
12,935	Frustulent	Reputation and honour involved
12,936	Frustulose	Seriously involved by failure
12,937	Frutescent	Not at all involved
12,938	Frutical .	**Iron**
12,939	Fruticous .	Iron ore
12,940	Fryingpan	Iron pyrites
12,941	Fubsy . .	Found in connection with iron pyrites
12,942	Fuchsia .	Oxide of iron
12,943	Fucivorous	Stained with iron oxide
12,944	Fucoid. .	Brown iron ore
12,945	Fuelfeeder	Hæmatite iron ore
12,946	Fuelled .	Magnetic iron ore
12,947	Fuelling .	Bog iron ore
12,948	Fugacity .	The iron occurs
12,949	Fugitively.	A rich deposit of iron ore
12,950	Fugitives .	To divide the —— from the iron
12,951	Fugleman .	Veins of iron ore
12,952	Fuguist .	Traces of iron are found
12,953	Fulciment.	Cast iron
12,954	Fulcrate .	Wrought iron
12,955	Fulcrums .	Corrugated iron
12,956	Fulfilleth .	**Irregular**
12,957	Fulfilling .	The vein is very irregular
12,958	Fulfilment	The veins being so irregular makes it difficult to
12,959	Fulgency .	Very irregular
12,960	Fulgent .	In a most irregular manner
12,961	Fulgidity .	The proceedings were very irregular
12,962	Fulgoridae	A very irregular course
12,963	Fulgurated	**Irregularity(ies)**
12,964	Fulgurous .	Characterized by great irregularity
12,965	Fulgury . .	There must be great irregularity somewhere

No.	Code Word.	**Irregularity(ies)** (*continued*)
12,966	Fuliginose.	Irregularities in the accounts
12,967	Fullborn .	Irregularities in the management
12,968	Fullbottom	Such irregularities deserve most severe censure
12,969	Fullbutt .	**Irrespective (of)**
12,970	Fullcentre	Is this irrespective of
12,971	Fullchisel .	This is irrespective of
12,972	Fulldress .	This is not irrespective of
12,973	Fulldrive .	**Irritated**
12,974	Fulleared .	Much irritated by '
12,975	Fullfed. .	**Irritation**
12,976	Fullformed	Much irritation exists
12,977	Fullgrown.	To allay the irritation
12,978	Fullhot .	**Is**
12,979	Fulllength	Is it
12,980	Fullmanned	Is it not
12,981	Fullmoon .	Is now being
12,982	Fullorbed .	If it is
12,983	Fullsailed .	If it is not
12,984	Fullsplit .	If he is
12,985	Fullsummed	Is it so
12,986	Fullswing .	It is so
12,987	Fullvoiced	Why is it (that)
12,988	Fullwinged	When it is
12,989	Fulmar. .	When is it
12,990	Fulmineous	**Issue**
12,991	Fulmining	In order to bring matters to a successful issue
12,992	Fulsamic .	Have brought the affair to a successful issue
12,993	Fulsome .	Have failed to bring matters to a successful issue
12,994	Fulsomely	Will —— issue
12,995	Fulvid . .	Will issue
12,996	Fulvous .	Will not issue
12,997	Fumaramide	How much is the issue
12,998	Fumarate .	Before we can issue
12,999	Fumblingly	The first issue of shares
13,000	Fumeless .	The second issue
13,001	Fumifugist	We propose to issue
13,002	Fumigated	We can issue shares
13,003	Fumigating	Cannot issue shares until
13,004	Fumigatory	As soon as we issue
13,005	Fumingly .	The first issue will not exceed
13,006	Fumishness	The issue of the shares
13,007	Funambulus	The certificates are ready to issue
13,008	Function .	Before issue we require
13,009	Functional	Issue of debentures
13,010	Fundament	Issue of preference shares
13,011	Fundholder	Before we can issue certificates
13,012	Fundungi .	Before we can issue certificates, please give us
13,013	Fundus .	**Issued** [names of your nominees
13,014	Funebrial .	Have any shares been issued
13,015	Funebrious	Has (have) issued

No.	Code Word.	**Issued** (*continued*)
13,016	Funeral .	Has (have) not issued
13,017	Funeralale	Certificates will be issued
13,018	Funerally .	Certificates must be issued
13,019	Funeration	Certificates issued
13,020	Fungi . .	Certificates will not be issued until completion of
13,021	Fungidae .	Against —— shares issued [transfer of property
13,022	Fungiform	Before any shares can be issued
13,023	Funginous	Before any certificates can be issued
13,024	Fungology	As soon as they are issued
13,025	Fungosity .	The shares were taken up as soon as issued
13,026	Funguspit.	Issued at
13,027	Funicle .	Issued at a premium of
13,028	Funicular .	Issued at a discount of
13,029	Funiliform	If shares are not issued
13,030	Funnelnets	Shares cannot be issued
13,031	Funnyman	Shares can be issued
13,032	Furbelow .	Issued to you or your nominees
13,033	Furbish .	**Issuing**
13,034	Furbishers	A favourable time for issuing
13,035	Furfurine .	Are issuing
13,036	Furfurous .	Propose issuing
13,037	Furibund .	Do not think of issuing at present
13,038	Furibundal	In the event of our issuing
13,039	Furiosant .	Are not issuing
13,040	Furiously .	**Join**
13,041	Furlough .	I wish —— to join me as soon as possible
13,042	Furnace .	To join with
13,043	Furnacebar	Do not join
13,044	Furnishing	Will not join
13,045	Furriery .	If —— will join
13,046	Furrowweed	If you will join
13,047	Furtherer .	Is (are) willing to join with us in
13,048	Furthest .	Cannot join on account of
13,049	Furtive .	Will you join to the extent of
13,050	Furze . .	Will join to the extent of
13,051	Furzechat.	Join in the expense of
13,052	Furzeling .	If you will join us in the expense
13,053	Furzewren	If —— will not join
13,054	Fuscation .	Join in the expedition
13,055	Fuscite .	Join in the purchase of
13,056	Fuseloil .	Join —— interests with ours
13,057	Fusibility .	**Joined**
13,058	Fusible .	Has (have) joined
13,059	Fusilier .	Has (have) not joined
13,060	Fusilladed	Has (have) —— joined
13,061	Fusion. .	Joined the Board
13,062	Fusionless	**Joint**
13,063	Fussiest .	On joint account
13,064	Fussily. .	By joint action
13,065	Fusteric .	A joint survey

No.	Code Word.	Joint (*continued*)
13,066	Fustianist .	A joint expedition to survey
13,067	Fustiness .	Joint expense
13,068	Fusulina .	The joint properties
13,069	Futchell .	A joint-stock company
13,070	Futilous .	Our joint interests
13,071	Futtock .	A joint committee
13,072	Futurable .	A joint interest in
13,073	Futurists .	**Journey**
13,074	Futurities .	On the journey [by coach
13,075	Futurition .	The first part of the journey is by rail, the last part
13,076	Fuzees . .	The first part of the journey is by ——, the last by
13,077	Fuzzball .	A long journey from here to
13,078	Fylfot . .	A difficult journey to make, on account of
13,079	Fyrdung .	Can you undertake the journey
13,080	Gabarage .	Too ill to undertake the journey
13,081	Gabbronite	Cannot possibly undertake the journey to
13,082	Gabellman	Had to make the journey by
13,083	Gabilla. .	**Judge**
13,084	Gabionnade	Could you judge
13,085	Gableends	I could not judge
13,086	Gableroof .	In order to judge
13,087	Gablever .	To judge by
13,088	Gablifters .	It is difficult to judge on account of
13,089	Gablock .	As far as I can judge
13,090	Gabrielite .	To judge of
13,091	Gadabout .	The judge was very strong in his remarks
13,092	Gaddingly	The judge was evidently in favour of
13,093	Gaddish .	The judge ruled that
13,094	Gadflies .	The judge declined to
13,095	Gadhelic .	The judge condemned
13,096	Gadolinite	You can best judge (whether)
13,097	Gadsteel .	**Judging (by)**
13,098	Gaelic . .	No means of judging
13,099	Gaff. . .	Judging by what has been done before
13,100	Gaffhook .	**Judgment**
13,101	Gafflocks .	Use your own judgment
13,102	Gaffsman .	On mature consideration my judgment is
13,103	Gafol . .	Judgment has been given for us
13,104	Gafolgild .	Judgment has been given against us
13,105	Gafolland .	Judgment in favour of
13,106	Gagrein .	Judgment in our favour with costs
13,107	Gagrunner	Used my own judgment in
13,108	Gagtoothed	Judgment will be given on
13,109	Gahnite .	When will judgment be given
13,110	Gaidheal .	In whose favour is judgment given
13,111	Gaiety . .	Cannot form a judgment
13,112	Gainable .	Rely on our judgment
13,113	Gained .	We rely on your judgment
13,114	Gainfully .	Your judgment must be confirmed by expert's report
13,115	Gaingiving	Our judgment has been confirmed by

No.	Code Word.	**Judgment** (*continued*)
13,116	Gaining .	Shows a great want of judgment
13,117	Gainsaid .	Judgment in favour of plaintiff(s)
13,118	Gainsayer.	Judgment for the defendant(s)
13,119	Gainsaying	Judgment in favour of the company
13,120	Gainsome.	If judgment is given against us, we must appeal
13,121	Gainstrive	Judgment deferred
13,122	Gairfowl .	The judgment carries with it
13,123	Gaitered .	According to the best of our judgment
13,124	Galactia .	Cannot say when judgment will be given
13,125	Galadress .	**Judicious**
13,126	Galanthus	Would it be judicious (to)
13,127	Galantine.	It would be judicious (to)
13,128	Galavance	It would not be judicious (to)
13,129	Galaxies .	Think(s) it a judicious step to take
13,130	Galaxy . .	Do(es) not think it a judicious step to take
13,131	Galbanum	**Judiciously**
13,132	Galbula .	We think you acted very judiciously in the matter
13,133	Galeas . .	Acted judiciously
13,134	Galeated .	A little money judiciously spent may prove the
13,135	Galecynus	**Jumped** [property to be of considerable value
13,136	Galemys .	Have not been jumped
13,137	Galenical .	May be jumped
13,138	Galenism .	Otherwise our claims could be jumped, according
13,139	Galeocerdo	Has (have) jumped [to the custom here
13,140	Galeodes .	To prevent our claims being jumped
13,141	Galeodidac	Claims have been jumped
13,142	Galeopsis .	Rumoured that claims have been jumped
13,143	Galimatias	Jumped because not worked for thirty consecutive
13,144	Galingale .	Nothing to prevent property being jumped [days
13,145	Galipots .	Claims jumped ; non-payment rental on due date
13,146	Gallant .	Jumped because licences not paid
13,147	Gallantise.	**Jumpers**
13,148	Gallantly .	Lost case against jumpers
13,149	Gallaox .	Won case against jumpers
13,150	Gallature .	Lost case against jumpers, and are appealing
13,151	Gallduct .	Jumpers are appealing against decision
13,152	Galleria .	We consider jumpers have a good case
13,153	Galletyle .	We consider jumpers have a poor case
13,154	Galleyfire.	Jumpers will consent to draw their pegs
13,155	Gallfly . .	Do not think jumpers will obtain judgment in their
13,156	Galliambic	Have compromised with jumpers for [favour
13,157	Galliard .	Jumpers decline to compromise
13,158	Gallicised.	Suggest you compromise with jumpers
13,159	Gallicolae.	Jumpers accede to proposal
13,160	Gallinha .	Jumpers decline proposal
13,161	Gallinsect	**Jumping**
13,162	Gallinule .	Carrying jumping case into court
13,163	Gallium .	Believe rumour as to jumping is without foundation
13,164	Gallivant .	Are inquiring as to alleged jumping
13,165	Galliwasp .	**Junction**

No.	Code Word.	**Junction** (*continued*)
13,166	Gallnut .	Effect a junction with
13,167	Gallomania	Can you effect a junction with
13,168	Gallooned.	Cannot effect a junction with
13,169	Gallopaded	At the junction of
13,170	Galloping .	Have effected a junction with
13,171	Gallowglas	**Juncture**
13,172	Gallows .	At this juncture
13,173	Gallowstop	**Jurisdiction**
13,174	Gallpipe .	Under the jurisdiction of
13,175	Gallstone .	Beyond the jurisdiction of
13,176	Gallyworm	Has no jurisdiction
13,177	Galuncha .	(If) the court has no jurisdiction
13,178	Galvanical	**Jury**
13,179	Galvanism	The case to be tried by special jury
13,180	Gamagrass	As the jury cannot agree, there is to be a fresh trial
13,181	Gamass .	The jury have given a verdict of
13,182	Gambado .	**Just**
13,183	Gambist .	Do you consider it just (that)
13,184	Gambol .	I (we) consider it quite just (that)
13,185	Gambolled	I (we) do not think it just (that)
13,186	Gambolling	Not just yet
13,187	Gambroon	Has (have) just
13,188	Gamebag .	Has (have) just had
13,189	Gamecocks	It is not just
13,190	Gameegg .	Just come out of the —— mine
13,191	Gamefowl.	**Justice**
13,192	Gamekeeper	In justice to
13,193	Gamelaws.	To do justice to
13,194	Gameless .	Do not see the justice of
13,195	Gamely .	**Justified**
13,196	Gamesomely	Is (are) —— justified in
13,197	Gammarus	Is (are) justified in
13,198	Gammoned	Is (are) not justified in
13,199	Gamut . .	Are we justified in
13,200	Gangboard	I (we) think we are justified in
13,201	Gangcask .	I (we) do not think we are justified in
13,202	Gangetic .	Should we not be justified in
13,203	Gangliated	**Justify**
13,204	Gangliform	Does not justify
13,205	Ganglionic	Does this justify
13,206	Gangmaster	Nothing can justify
13,207	Gangplough	Do not think will justify the expense
13,208	Gangpunch	**Keep**
13,209	Gangrenate	Keep me (us) regularly informed
13,210	Gangrening	Keep me (us) well posted up
13,211	Gangrenous	Keep the accounts separate
13,212	Gangtide .	Keep this to yourself
13,213	Gangways.	Shall keep the men working on
13,214	Gangweek	Keep us informed of any change in
13,215	Ganoid .	We keep on the old hands

No.	Code Word.	**Keep** (*continued*)
13,216	Ganoidal .	Keep a look-out for
13,217	Gantlope .	Keep it down—(or) keep down the
13,218	Gaolbird .	Keep it quiet
13,219	Gaolfever .	Keep it up—(or) keep up the
13,220	Garanceux	Keep expenses down as much as possible
13,221	Garangan .	Keep a sharp watch upon
13,222	Garbled .	Keep a record of
13,223	Garbling .	Keep in reserve
13,224	Garcinia .	Keep in bounds
13,225	Gardener .	If we keep on at the rate we have been doing
13,226	Gardenless	We cannot keep on at the
13,227	Gardenplot	**Keeping**
13,228	Gardenship	In safe keeping
13,229	Gardenware	Keeping back
13,230	Garfish .	Not keeping any
13,231	Gargalized	**Kept**
13,232	Garganey .	Has (have) been kept
13,233	Gargoyle .	Has (have) not been kept
13,234	Garibaldi .	Has (have) —— been kept
13,235	Garishness	The accounts have not been kept posted
13,236	Garlicky .	Not well kept
13,237	Garlicpear	Kept in reserve
13,238	Garmenture	Kept in bounds
13,239	Garnished	Has been kept down as much as possible
13,240	Garookah .	**Killed**
13,241	Garpike .	Has (have) been killed
13,242	Garreteers	Reported to be killed
13,243	Garrisons .	No one killed
13,244	Garrulity .	Persons killed
13,245	Garrulous .	Killed, and —— injured
13,246	Garterfish .	The names of the killed
13,247	Garterking	Killed in a fight
13,248	Garthman.	Killed by an accident
13,249	Garvie . .	Several killed and hurt
13,250	Gasalier .	**Kind**
13,251	Gasbath .	There is every kind of
13,252	Gasbracket	There is no kind of
13,253	Gasburner	A kind of
13,254	Gascheck .	Be so kind as to
13,255	Gascoal .	What kind of
13,256	Gascoynes	What kind of ore have you
13,257	Gasefy . .	The kind of ore
13,258	Gasengine	Very kind of
13,259	Gaseous .	Not the kind of man we want
13,260	Gasfitter .	The kind you want
13,261	Gasfixture.	**Kindness**
13,262	Gasfurnace	Ask —— to have the kindness to
13,263	Gasgauge .	Through the kindness of —— I was (we were) able to
13,264	Gashliness	**Knew**
13,265	Gasholders	Knew nothing about it

No.	Code Word	Knew (*continued*)
13,266	Gasified .	Knew some time ago
13,267	Gasjet . .	Knew all the time that
13,268	Gaslamp .	Knew everything that was being done
13,269	Gaslantern	Knew everything that took place
13,270	Gaslight .	If you knew anything
13,271	Gasmain .	If he (they) knew
13,272	Gasmeter .	**Know(s)**
13,273	Gasometric	Do you know anything about
13,274	Gasoscope	Do you know if
13,275	Gaspereaux	Know(s) all about
13,276	Gaspipes .	Know(s) nothing about
13,277	Gasretort .	Do (does) not know
13,278	Gasservice	Did not know at the time
13,279	Gasstove .	Do (does) not know whether
13,280	Gastanks .	Will know soon if
13,281	Gasteropod	Must know by
13,282	Gastornis .	Will not know before
13,283	Gastralgia	Let me (us) know by telegraph
13,284	Gastric. .	See if —— know(s) anything about
13,285	Gastridium	Know nothing of the suitability of
13,286	Gastritis .	Want to know what we are to do
13,287	Gastrocele	Wants to know what he is to do
13,288	Gastrology	Do you know the party named
13,289	Gastromyth	What do you know about
13,290	Gastronome	Know nothing of the man
13,291	Gastrula .	Do you know when
13,292	Gaswork .	Do not know what to do
13,293	Gatevein .	Let us know the worst
13,294	Gateward .	If you know
13,295	Gatewise .	If you do not know
13,296	Gather. .	What they want to know (is)
13,297	Gatherable	**Knowledge**
13,298	Gathereth .	It has come to my knowledge that
13,299	Gatherings	Has (have) great knowledge of the mines
13,300	Gatlinggun	Has (have) great knowledge of the mining laws of
13,301	Gaudiest .	Owing to the knowledge of [this country
13,302	Gaudiness	Done without our knowledge
13,303	Gaudyday.	Without my (our) knowledge or consent
13,304	Gauffered :	If done without your knowledge
13,305	Gaugeable	Without his knowledge
13,306	Gaugecock	Without (——'s) knowledge or sanction
13,307	Gaugeglass	Without your knowledge
13,308	Gaugelamp	The knowledge we have gained
13,309	Gaugepoint	It has come to the knowledge of
13,310	Gaugingrod	**Known**
13,311	Gaultheria	Is it known that
13,312	Gauntlets .	It is not generally known that
13,313	Gausabey .	It is well known that
13,314	Gavelet .	Nothing is known of it here
13,315	Gavelkind	We fear the worst is not yet known

No.	Code Word.	**Known** (*continued*)
13,316	Gawky . .	As soon as it is known (that)
13,317	Gawntree .	If it is known (that)
13,318	Gaybine .	Nothing can be known until
13,319	Gaylussite	Nothing more is known (of or about)
13,320	Gazeebo .	Is anything more known (of or about)
13,321	Gazeful .	If we had known
13,322	Gazehound	If we had known in time
13,323	Gazement .	Has been known to
13,324	Gazetted .	Has it yet been made known
13,325	Gazetting .	**Laborious**
13,326	Gazogene .	It will be a most laborious task to
13,327	Gazolytes .	It has proved a most laborious task to
13,328	Gazzatum .	A most laborious journey
13,329	Gearcutter	Too laborious
13,330	Gearwheel	Very laborious
13,331	Gecarcinus	**Labour**
13,332	Geckotidae	Cost of labour to produce
13,333	Gehlenite .	Labour is scarce
13,334	Gelasimus .	Labour is expensive
13,335	Gelatinize .	Labour is scarce and dear
13,336	Gelatinous	Labour is plentiful
13,337	Geldable .	Labour is cheap
13,338	Gelderrose	Labour is plentiful and cheap
13,339	Geldings .	What is the cost of labour
13,340	Gelidity .	The labour question
13,341	Geloscopy	The labour question is giving much trouble
13,342	Gelsemium	How does the labour question affect
13,343	Geminate .	Cannot get sufficient labour
13,344	Gemination	Owing to the lack of labour
13,345	Geminy .	Skilled labour
13,346	Gemmaceous	Unskilled labour
13,347	Gemmary .	**Labourers**
13,348	Gemmels .	Labourers earn
13,349	Gemmeous	**Lack**
13,350	Gemmipara	For lack of
13,351	Gendarmery	No lack of
13,352	Genealogy	Developing works were stopped for lack of funds
13,353	Genearch .	A lack of funds
13,354	Generality	Owing to lack of funds we have had to
13,355	Generalize	**Ladder(s)**
13,356	Generated	The ladders in the shaft are unsafe
13,357	Generating	Put ladders down
13,358	Generative	**Laid**
13,359	Generatrix	Has been laid down
13,360	Generic .	Not yet laid down
13,361	Generical .	Cannot be laid down
13,362	Generous .	Laid under
13,363	Generously	**Lain**
13,364	Genethliac	Has (have) lain idle for want of
13,365	Genevanism	Has (have) lain

No.	Code Word.	Lain (*continued*)
13,366	Genially .	Has (have) not lain
13,367	Genialness	From having lain
13,368	Geniculate	Lain under water
13,369	Genipap .	**Land**
13,370	Genista .	The value of the land
13,371	Genitival .	The land comprises —— acres
13,372	Geniture .	(Of) land suitable for
13,373	Genius. .	Good agricultural land
13,374	Genteel .	Land of no value
13,375	Genteelish	Land is worth —— per acre
13,376	Genteelly .	The land is valued at
13,377	Gentian .	The land about here
13,378	Gentilized	The land on which —— is situate
13,379	Gentlefolk	The land required
13,380	Gentleman	The land has been bought for
13,381	Gentleness	The land has been secured for
13,382	Gentleship	Land for tailings
13,383	Genuflect .	Land for mill site
13,384	Genuinely.	(To) secure the land—(or) securing the land
13,385	Geocentric	If you can buy the land at a reasonable price
13,386	Geocorisae	Can you buy the land at
13,387	Geocronite	Do not want the land
13,388	Geocyclic .	(To) buy the land to protect
13,389	Geodephaga	Must acquire the land
13,390	Geodesian	The adjoining land
13,391	Geodesy .	The land on both sides
13,392	Geodetic .	The land to the
13,393	Geodetical	Their land adjoins
13,394	Geoffroyia	The land in question
13,395	Geoglossum	**Landed**
13,396	Geognost .	The goods have been landed
13,397	Geognostic	The goods not yet landed
13,398	Geogonical	Have had the —— landed and stored
13,399	Geography	Landed and sent up country
13,400	Geologian.	Landed and sent off by rail
13,401	Geology .	As soon as they are landed
13,402	Geometral	As soon as landed, will have them forwarded
13,403	Geometrize	Have them landed and forwarded
13,404	Geometry.	Hope to have them landed
13,405	Geophagism	As soon as you have landed
13,406	Geophagous	**Language**
13,407	Geophila .	Use plain language
13,408	Georgos .	Used very plain language
13,409	Geosaurus.	The language is too strong
13,410	Geoselenic	The language is not at all too strong
13,411	Geostatic .	**Lapse**(d)
13,412	Geoteuthis	Will lapse
13,413	Geothermic	Otherwise it will lapse
13,414	Geotropism	The concession has lapsed
13,415	Geraniums	Lapsed through effluxion of time

No.	Code Word.	
13,416	Gerbil . .	**Large**
13,417	Germander	The vein is large and well defined
13,418	Germanisms	Covering a large extent
13,419	Germcell .	A large percentage of
13,420	Germtheory	It will require a large outlay
13,421	Gerocomia	Not very large
13,422	Gerontes .	Large enough to
13,423	Geropigia .	Hardly large enough **to**
13,424	Gerund .	Not large enough
13,425	Gerundive	If not too large
13,426	Gerusia ·..	How large is
13,427	Gervas. .	Too large
13,428	Gervillia .	That size is too large
13,429	Gesnera .	Must not be too large
13,430	Gestural .	If too large, will not pass through
13,431	Gesturing.	The pieces are too large
13,432	Getpenny .	**Largely**
13,433	Gewgaws .	The mine has been largely worked
13,434	Geyser. .	The mine has not been very largely worked
13,435	Ghainorik	The mine has been largely and badly worked, and there is every likelihood of its caving
13,436	Gharry .	Very largely
13,437	Ghastful .	Must increase very largely
13,438	Ghastfully.	This will add largely to
13,439	Ghastliest.	Largely encroached upon
13,440	Ghastly .	Largely availed of
13,441	Gherkin .	Added largely to our reserves
13,442	Ghibelline	Has (have) been largely drawn upon
13,443	Ghittern .	**Larger**
13,444	Ghost . .	Must not be larger than
13,445	Ghostlike .	Cannot be made larger
13,446	Ghostmoth	If larger than
13,447	Ghostseer.	If not larger than
13,448	Ghoststory	**Largest**
13,449	Ghyll . .	The largest piece must not exceed
13,450	Giantize .	The largest will measure
13,451	Giantly .	The largest will measure —— feet by —— feet
13,452	Giantship.	What is the largest-sized piece that will pass through
13,453	Giaours .	The largest will weigh
13,454	Gibberish.	The largest we can
13,455	Gibbose .	**Last** [prospects of any more
13,456	Gibbosity.	The ore will not last much longer, and there are no
13,457	Gibbsite .	The ore will not last more than —— months, but we hope soon to cut another body of ore
13,458	Gibcat. .	The stopes will not last much longer
13,459	Gibeonite.	The last shipment of bullion
13,460	Gibingly .	Cannot last
13,461	Gibship .	Cannot last much longer
13,462	Gibstaff .	Will last some time longer
13,463	Giddily .	From the last

No.	Code Word.	**Last** (*continued*)
13,464	Giddyhead	I estimate the ore will last
13,465	Giddying .	To last (about)
13,466	Giddypaced	By last mail
13,467	Giereagle .	The last telegram
13,468	Gieseckite	If you refer to my last letter you will see
13,469	Giftedness	If you refer to my last telegram you will see
13,470	Giftrope .	The last time
13,471	Gigantean	Last chance
13,472	Gigantic .	Will last
13,473	Gigantical	Will not last
13,474	Giggling .	Will not last more than
13,475	Gighorse .	The last report on the mines
13,476	Gigmachine	The last carload
13,477	Gigster .	The last few
13,478	Gilbertine.	We have at last
13,479	Gilded . .	We sent the last
13,480	Gilhooter .	The last which we sent
13,481	Gillaroo .	Send the last
13,482	Gillcover .	From first to last
13,483	Gillflap .	The last remaining
13,484	Gillyvor .	Keeping to the last
13,485	Gilpy . . **Late**	
13,486	Gilthead .	Will it be too late
13,487	Gilttail . .	Is it too late
13,488	Gimcracks	It is too late now
13,489	Gimlet . .	It is not too late
13,490	Gimleteye	If —— put(s) off the examination till so late
13,491	Gimmalbit	Shall be late
13,492	Gingerade	Too late to
13,493	Gingerbeer	Came too late
13,494	Gingerly .	Too late to be of any use
13,495	Gingerness **Lately**	
13,496	Gingerpop	Has (have) been lately begun
13,497	Gingerwine	Has (have) been lately stopped
13,498	Gingham .	Lately they have been working
13,499	Gingival .	Lately have only been working low-grade ore
13,500	Ginglymoid	Not lately
13,501	Ginglymus	The mine was lately examined by
13,502	Ginhouses	The bunch of ore lately struck averages
13,503	Ginpalace.	Have been running the mill lately on
13,504	Gipsies .	Have been getting lately our supply of ore from
13,505	Gipsy . .	Was (were) lately at
13,506	Giraffe . .	Has (have) not been very lately to
13,507	Giraffina .	See ——, who was lately at ——, and get from him
13,508	Girdlebelt.	There was lately. [all particulars about
13,509	Girlhood .	(Has) have fallen off lately
13,510	Girlish . .	(Has) have lately begun to
13,511	Girlishly .	(Has) have only lately
13,512	Gironde . **Later**	
13,513	Girondists.	Can you give us any later news (or particulars of)

No.	Code Word.	**Later** (*continued*)
13,514	Gironny .	Cannot give you later than
13,515	Girtline .	Want something later
13,516	Gismondine	Have nothing later
13,517	Glabrity .	Will be later than expected
13,518	Glabrous .	Later in the year
13,519	Glacialist .	Not later than
13,520	Glaciarium	Later on [wait for assay of samples
13,521	Glaciers .	The report will follow by post later on, as I have to
13,522	Gladdened	Later on will have to
13,523	Gladdening	Later on in the month
13,524	Gladiolus .	Send us later on
13,525	Gladship . **Latest**	
13,526	Gladsome .	By the latest
13,527	Gladsomely	By the latest mill returns, I find that
13,528	Gladwyn .	The latest returns show
13,529	Glagolitic .	The latest accounts show that
13,530	Glairine .	You shall have the report by —— at the latest
13,531	Glairy . .	At the latest
13,532	Glama . .	The latest possible time (for)
13,533	Glamoury .	What is the latest possible time you can give
13,534	Glancecoal	Must be ready at the latest
13,535	Glanced .	From the latest advices
13,536	Glandage .	We want this at the latest by
13,537	Glandiform **Lavished**	
13,538	Glandular .	Have been lavished on
13,539	Glandulous	So much has been lavished on
13,540	Glaringly . **Law(s)**	
13,541	Glasscase .	According to the mining laws of
13,542	Glasschord	According to the law
13,543	Glasscoach	As the law now stands
13,544	Glasscrab .	By law we are forbidden to
13,545	Glasseyes .	The laws are powerless
13,546	Glassfaced	(To) go to law
13,547	Glassful .	In defiance of the law
13,548	Glassgall . **Lawsuit**	
13,549	Glasshive .	The lawsuit in which we are engaged
13,550	Glasshouse	We are liable to get into a lawsuit about
13,551	Glassmen .	Threaten(s) us with a lawsuit about
13,552	Glassmetal	Must try and prevent getting into a lawsuit (over)
13,553	Glasspaper	Has (have) instituted a lawsuit
13,554	Glassrope .	In the event of a lawsuit being brought
13,555	Glassshade	(In) the recent lawsuit
13,556	Glasstears .	(In) the lawsuit now pending between —— and
13,557	Glauberite	Involved in a lawsuit
13,558	Glaucium . **Lawyer(s)** [if they are all right	
13,559	Glaucolite	Get a reliable lawyer to examine the titles, and see
13,560	Glaucoma .	Consult a thoroughly competent lawyer
13,561	Glaucopis .	Employ a lawyer to
13,562	Glaudkin .	It is not necessary to employ a lawyer
13,563	Glebeland	I am (we are) employing a lawyer to

No.	Code Word.	**Lawyer(s)** (*continued*)
13,564	Glebeless .	See lawyer, and arrange as deemed best
13,565	Glechoma .	What does your lawyer advise
13,566	Gleeclub .	Have consulted a first-rate lawyer on the subject
13,567	Gleeful .	Have put the affair into a lawyer's hands
13,568	Gleemaiden	The lawyer(s) advise(s)
13,569	Gleichenia	The lawyer(s) does (do) not advise
13,570	Glenoid .	The lawyer's fee will be
13,571	Gliadine .	Have engaged the best lawyer we can get
13,572	Glibbery .	A lawyer of great experience (in)
13,573	Glibly . .	A lawyer of the highest standing
13,574	Glimmer .	Have got lawyer's opinion
13,575	Glimmering	Have consulted ——, as he is considered the lawyer
13,576	Glimpse . **Lax**	[of highest standing here
13,577	Glimpsing	Too lax
13,578	Glissade .	Has (have) been very lax in the way of
13,579	Glisten . . **Lay**	
13,580	Gloaming .	To lay bare the vein
13,581	Globard .	To lay before the Board of Directors
13,582	Globe . .	Lay ourselves open to
13,583	Globedaisy	You lay yourselves open to
13,584	Globefish .	Laid open to
13,585	Globeglass **Layers**	
13,586	Globosity .	In layers (of)
13,587	Globular .	Between layers (of)
13,588	Globularly	Superimposed layers (of)
13,589	Globulism	Alternate layers (of)
13,590	Glochidate **Laying**	
13,591	Gloiocarp .	Is (are) laying down
13,592	Glomerate	By laying down
13,593	Glomerous	By laying out
13,594	Gloomiest .	By laying down a tram line
13,595	Gloomily .	Is (are) laying out
13,596	Gloominess	Is (are) not laying out
13,597	Gloriable . **Lead**	
13,598	Gloriation	Lead ore
13,599	Glorified .	Argentiferous lead ore
13,600	Glorify . .	Veins of lead ore
13,601	Glorifying .	The ore contains a large percentage of lead
13,602	Gloriosa .	How many tons of lead ore to the fathom
13,603	Gloriously .	Carbonate of lead carrying —— ounces of silver
13,604	Gloss . .	The ore assays —— per cent. of lead and ——
13,605	Glossarial .	Tons of lead ore to the fathom [ozs. of silver
13,606	Glossist .	Lead to no result
13,607	Glossocele	A strong lead
13,608	Glossotomy	The lead continues strong
13,609	Glottal . .	Upon the lead
13,610	Glottalite .	Main lead
13,611	Glottology **Leader(s)**	
13,612	Gloveband	There are many leaders
13,613	Gloveclasp	There are no reefs, only leaders

No.	Code Word.	Leader(s) *(continued)*
13,614	Glovemoney	The leaders appear to be
13,615	Glowingly	**Leading**
13,616	Glowworm	Leading to (towards)
13,617	Gloxinia .	Not leading to
13,618	Glucose .	Appears to be leading towards
13,619	Glucoside .	**Leak**
13,620	Glucosuria	Owing to a leak (in)
13,621	Glueboiler	Sprung a leak in
13,622	Gluepot .	The leak has been stopped
13,623	Glueyness .	The leak cannot be stopped
13,624	Gluish . .	Is (are) very leaky
13,625	Glumaceous	The leak is increasing
13,626	Glumales .	If leak is not stopped
13,627	Glumella .	**Learn**
13,628	Glumiferae	I (we) learn on good authority that
13,629	Glumous .	Will have to learn
13,630	Glumpy .	**Lease(s)**
13,631	Glutinate .	We can get a lease of the property for —— years
13,632	Glutman .	——'s lease expires [for
13,633	Glutted .	Lease will expire
13,634	Glutting .	To grant leases for
13,635	Gluttonish	Lease will be renewed
13,636	Gluttonize	Lease will not be renewed
13,637	Glyceria .	We can take over ——'s lease for
13,638	Glycerule .	The lease expires ——; it can be renewed for
13,639	Glycocoll .	The lease expires ——; it can be renewed on the
13,640	Glycogen .	Can you get a lease of [same terms
13,641	Glycogenic	Can get lease for —— at
13,642	Glycol . .	Taken a lease of the property for
13,643	Glyconic .	Lease will run until
13,644	Glyph . .	How long has the lease to run
13,645	Glyphideae	Can you renew the lease
13,646	Glyptodon	Cannot obtain renewal of lease
13,647	Glyptothek	Offers to renew the lease
13,648	Gnaphalium	On lease for —— years
13,649	Gnarly . .	**Leased**
13,650	Gnasheth .	Has been leased to us
13,651	Gnashing .	Have leased to them
13,652	Gnashingly	Leased for —— years
13,653	Gnatflower	Leased at a rental of
13,654	Gnathitis .	At what rental is it leased
13,655	Gnathodon	**Leasehold**
13,656	Gnathonic	Leasehold property
13,657	Gnatworm	Is leasehold
13,658	Gnawing .	Part leasehold, part freehold
13,659	Gneiss . .	**Leave(s).** (See also Telegraph.)
13,660	Gneissic .	Cannot leave
13,661	Gneissoid .	Must leave
13,662	Gnetum .	Will leave
13,663	Gnomic .	I leave on —— for

No.	Code Word.	**Leave(s)** (*continued*)
13,664	Gnomical .	He (they) will leave on —— for
13,665	Gnomologic	When do you leave
13,666	Gnomon .	Will not leave (until)
13,667	Gnomonist	You must not leave (until)
13,668	Gnosticism	Will leave as soon as possible for
13,669	Goading .	Will leave immediately for
13,670	Goatbeard	Before you leave
13,671	Goatchafer	Expect to leave about
13,672	Goatfish .	I (we) leave it to you
13,673	Goatishly .	You had better leave it to me (us) to
13,674	Goatmilker	Now ready to leave
13,675	Goatmoth .	Get leave from —— to examine the
13,676	Goatpepper	Leave instructions with
13,677	Goatroot .	Unless you wire, will leave immediately for
13,678	Goatsbane	Before —— leave(s)
13,679	Goatsrue .	Shall leave for England
13,680	Goatsthorn	Shall leave —— in charge
13,681	Goatsucker	When does —— leave
13,682	Goatswheat	Leave here to-day, and shall arrive at —— about
13,683	Goatweed .	Leave it to the decision of
13,684	Gobbetly .	Want you to grant leave of absence on account of
13,685	Gobbling .	Granted leave of absence (to)
13,686	Gobetween	You can take leave of absence
13,687	Gocart. .	Obtain leave from
13,688	Godbote .	Nothing can leave
13,689	Godchild .	Leave(s) by the next steamer
13,690	Goddesses	**Leaving**
13,691	Godenda .	Is (are) leaving on —— for
13,692	Godfathers	Is (are) not leaving for —— till
13,693	Godfearing	Am not leaving for —— till
13,694	Godhood .	When you are leaving ——, telegraph
13,695	Godlessly .	Telegraph when you intend leaving for
13,696	Godlike .	Shall be leaving
13,697	Godly . .	Will not be leaving
13,698	Godlyhead	Am (are) leaving to-day for
13,699	Godmother	Am (are) leaving to-morrow for
13,700	Godown .	Before leaving
13,701	Godsend .	After leaving
13,702	Godship .	Leaving no doubt that
13,703	Godshouse	**Ledge(s)**
13,704	Godspeed	Taken from one ledge
13,705	Godspenny	Every ledge struck shows
13,706	Godwards.	The ledge(s) is (are) situated
13,707	Godwinia .	**Left**
13,708	Godwit .	Left here
13,709	Goffering .	Left over
13,710	Gofnick .	Left on the
13,711	Goggleeyed	Left on hand
13,712	Goggles .	Left —— in charge
13,713	Goitrous .	Has (have) left

No.	Code Word.	Left (*continued*)
13,714	Golaba. .	Has (have) not left
13,715	Golandause	Has (have) been left
13,716	Goldbeater	Left to —— to decide
13,717	Goldbound	Left here on his way to
13,718	Goldcloth .	Is anything left
13,719	Goldcradle	There is nothing left
13,720	Goldcup .	How much is left
13,721	Goldcutter	**Legal**
13,722	Golddigger	Get the best legal opinion on
13,723	Golddust .	The best legal opinion on
13,724	Goldenbug	Legal proceedings will be taken
13,725	Goldencarp	Acting under legal advice
13,726	Goldenclub	It is perfectly legal
13,727	Goldeneye	Would it be legal
13,728	Goldenfish	It would not be legal
13,729	Goldenhair	A legal opinion in a similar case
13,730	Goldenknop	Have (you) legal proof
13,731	Goldenrod	**Legalized**
13,732	Goldenwasp	Must be legalized
13,733	Goldfever .	Send all documents properly legalized
13,734	Goldfield .	**Legally**
13,735	Goldfinch .	Cannot legally
13,736	Goldfinder	If it can be legally done
13,737	Goldfinny .	**Legitimate**
13,738	Goldfoil .	Is it a legitimate transaction
13,739	Goldhammer	It is not a legitimate transaction
13,740	Goldleaf .	It is a perfectly legitimate transaction
13,741	Goldlily .	Not legitimate
13,742	Goldney .	Quite legitimate
13,743	Goldplate .	**Lend**
13,744	Goldproof	Will lend
13,745	Goldsize .	Can you lend
13,746	Goldsmith	Will not lend
13,747	Goldstick .	If —— will lend
13,748	Goldthread	If —— will not lend
13,749	Goldwasher	Would —— lend
13,750	Goldylocks	Bank will lend us
13,751	Golfclub .	Bank declines to lend us
13,752	Golgotha .	Will not lend without good security
13,753	Goliardery	Will not lend more than
13,754	Goliathus .	Will lend himself (themselves) to anything for a [commission
13,755	Goloeshoe	**Length** (of the)
13,756	Gomarist .	The vein has been proved for a length of —— feet
13,757	Gomelin .	What will be the length
13,758	Gomphiasis	What length of time
13,759	Gompholite	The entire length
13,760	Gomphonema	The whole length of the
13,761	Gonangium	In length
13,762	Gondolier.	The total length is
13,763	Gonggong	For a length of (—— feet)

No.	Code Word.	Length *(continued)*
13,764	Gongmetal	With a combined length of
13,765	Gongylus .	What is the length of
13,766	Goniaster .	The vein has been prospected for the length of ——.
13,767	Goniatites.	What is the greatest length [feet
13,768	Goniometer	The greatest length that will **pass**
13,769	Gonocalyx	The greatest length
13,770	Gonophore	The length of the tunnel
13,771	Gonoplax .	**Lengthen**
13,772	Gonopteryx	In order to lengthen
13,773	Goodbye .	Will lengthen
13,774	Gooddeed	Will not lengthen
13,775	Goodeven	Do not lengthen
13,776	Goodfaced	**Lent**
13,777	Goodfellow	Has (have) lent
13,778	Goodfolk .	Has (have) not lent
13,779	Goodhumour	We have lent to
13,780	Goodliest .	How much have you lent to
13,781	Goodluck .	**Less**
13,782	Goodnature	Less than was expected
13,783	Goodness .	Less than
13,784	Goodnight	Not less than
13,785	Goodnow .	Not to pay less than —— per cent.
13,786	Goodsense	The mine has never paid less than —— per cent.
13,787	Goodsshed	The profits this half-year will be less on account of
13,788	Goodstrain	The expenses in future will be less
13,789	Goodstruck	The cost of —— will be less than at first estimated
13,790	Goodswagon	The mines must be worked at less cost, otherwise
13,791	Goodwill .	Less 5 per cent. [there will be no profits
13,792	Goodwoman	Less 10 per cent.
13,793	Goodyera .	Less —— per cent.
13,794	Goodygood	Will take less
13,795	Goodyship	Will not take less (than)
13,796	Goosander	The amount of ore hauled has been less for the
13,797	Goose . .	Less will not do [—— because of
13,798	Gooseberry	Would, I feel sure, take a less price for the mine,
13,799	Goosecap .	Can do with less [if the offer was firm
13,800	Goosecorn	How much less
13,801	Gooseflesh	The amount of ore in sight is much less than ——
13,802	Goosefoot .	Will be less on account of [estimated
13,803	Goosegrass	Think —— will be less
13,804	Gooseherd	Less for the past week
13,805	Gooseneck	Less for the past month
13,806	Goosepie .	Less for the next week
13,807	Goosequill	Less for the next month
13,808	Goosery .	No less will be taken
13,809	Gooseskin	Less than was estimated
13,810	Goosestep	Proved to be less
13,811	Goosetansy	**Lessee(s)**
13,812	Goosewing	The previous lessee
13,813	Gopher .	The present lessee(s) of the mine is (are)

No.	Code Word.	**Lessee(s)** (*continued*)
13,814	Gopherwood	The present lessee(s) of the mine state(s) that
13,815	Goracco .	Can you ascertain from the lessee(s)
13,816	Gorbelly .	The present lessee(s) of the mine has (have) been
13,817	Gorcrow .	**Lessen** [making a profit
13,818	Gordiacea	(To) lessen the expenses
13,819	Gordian .	(To) lessen the cost of
13,820	Gorebill .	**Lesser**
13,821	Gorfly . .	In a lesser degree
13,822	Gorgeous .	On a lesser capital
13,823	Gorgeously	**Let**
13,824	Gorgonean	Let it (them)
13,825	Gorgonized	Let it (them) alone
13,826	Gormagon	Let me (us)
13,827	Gormandize	Let me (us) have
13,828	Goshawk .	Let me (us) know
13,829	Gospel . .	Let him know
13,830	Gospelize .	Let it remain
13,831	Gospellary	Let it stand (in)
13,832	Gossamery	Do not let —— know
13,833	Gossip . .	Let —— know
13,834	Gossiping .	Do not let the
13,835	Gossipred .	Do not let anything keep you from
13,836	Gossoon .	Will let you
13,837	Gossypium	Will you let us
13,838	Gothamist .	If you let
13,839	Gothic . .	If we let
13,840	Gothical .	Cannot let the
13,841	Gothicism	Can you let us have
13,842	Gothicized	Have let the
13,843	Gougebit .	**Letter(s).** (See Registered, and Table at end.)
13,844	Gougeslip .	Letter is on the way about
13,845	Goura . .	Letter received about
13,846	Gourdtrees	Letter sent relating to
13,847	Gourdworm	Letter sent explaining
13,848	Gourdy .	Letter will give you all particulars about (the)
13,849	Gourmand	Did not get your letter
13,850	Goutily .	Did not get your letter in time
13,851	Goutwort .	Your letter received ; will answer as soon as possible
13,852	Govern .	Letter missing
13,853	Governable	Letter of advice
13,854	Governess	Letter of guarantee
13,855	Governing	Letter of introduction
13,856	Government	Please address your letter(s) to
13,857	Governor .	Please address, till further notice, my letters to
13,858	Gowdnook	Please send letter of introduction to
13,859	Gownpiece	Have you received my (our) letter of —— date
13,860	Gownsman	Have not received your letter of
13,861	Graafian .	Have not received the letter referred to
13,862	Grabgame .	I (we) have received your letter dated
13,863	Gracecups	If you have received letter from

No.	Code Word.	**Letter(s)** (*continued*)
13,864	Graceful .	Have you received a letter from
13,865	Gracenote	If you have not received my (our) letter of
13,866	Gracilent .	Has (have) received a letter from
13,867	Gracility .	Has (have) not received a letter from
13,868	Graciously	No letter to hand yet [about
13,869	Gracy . .	I am (we are) anxiously expecting letters from you
13,870	Gradations	My (our) last letter to you was dated
13,871	Gradatory .	Letter of credit
13,872	Gradient . ·	Letter(s) now in the post for you to care of
13,873	Gradually .	My (our) letter of —— was registered (and contained)
13,874	Graduating	Since my (our) last letter
13,875	Gradus . .	Letter containing full instructions
13,876	Graffages .	Mail arrived, and no letter from you
13,877	Grafter . .	Letter forwarded by mistake to
13,878	Grain . .	Letter(s) for you, was (were) posted (on)
13,879	Grainage .	Your letter of —— gone astray ; if contents important, telegraph [contents important
13,880	Grainmill .	Your usual letter not to hand; please telegraph if
13,881	Grainmoth	Your letter dated —— to hand to-day
13,882	Grainstaff .	Upon receipt of letter from
13,883	Gramashes	Please refer to my (our) letter
13,884	Gramercy .	Waiting for letter from —— before acting
13,885	Gramineal	Referring to your letter of
13,886	Gramineous	On referring to ——'s letter
13,887	Grammarian	Wait for my letter of —— date
13,888	Grammatic	Your letter of —— explaining about —— to hand ; I (we) consider it very satisfactory
13,889	Granaries .	My (our) letter of —— explains everything
13,890	Granary .	Your letter of —— explaining about —— to hand ; I (we) consider it very unsatisfactory
13,891	Grandaunt	You do not explain in your letter
13,892	Grandchild	Telegraph in answer to my (our) letter of the
13,893	Grandcross	Tell us by letter
13,894	Grandduke	Have given a letter of introduction to you, in favour of ——, who is (are) I (we) believe worthy of every confidence
13,895	Grandest .	Have been obliged in some measure to give a letter of introduction to you in favour of —— ; be very cautious in your dealings
13,896	Grandevity	When was your letter posted
13,897	Grandguard	Referred to in your letter
13,898	Grandific .	Send letter of introduction
13,899	Grandinous	Letters waiting for you at
13,900	Grandjuror	Address letters to —— till —— after that to
13,901	Grandmamma	Before receipt of your letter
13,902	Grandness	Letter sent to our agents
13,903	Grandniece	Inquire for letter(s) at
13,904	Grandpiano	As per your letter of
13,905	Grandsire .	As per our letter of
13,906	Grandsons	The enclosure referred to in my letter

No.	Code Word.	Letter(s) (*continued*)
13,907	Grandstand	The enclosure referred to in my letter (of ——) was omitted you will receive it by next mail
13,908	Granduncle	Referred to in his (their) letter of
13,909	Grandvicar	As per his (their) letter of
13,910	Granite .	**Level(s)**
13,911	Granitical .	The level is driven in —— feet
13,912	Granitoid .	From the level mouth
13,913	Granted .	Above the level of the sea
13,914	Granulary .	The mines are situated —— feet above the level of
13,915	Granulated	Proposed level [the sea
13,916	Grapeshot	On the same level
13,917	Grapesugar	Below this level
13,918	Grapevines	Near the level mouth
13,919	Grapewort	Above the water level
13,920	Graphic .	Below the water level
13,921	Graphical .	Down to the water level
13,922	Graphitoid	Above this level a stope has been started
13,923	Grapholite	The —— ft. level has been extended
13,924	Graphotype	The level should be continued —— feet
13,925	Grappled .	The lowest level
13,926	Grapsus .	Next to the lowest level
13,927	Graptopora	On the —— ft. level
13,928	Grasp . .	Above the —— ft. level
13,929	Graspable .	To the —— ft. level
13,930	Grasping .	The —— ft. level is in
13,931	Graspingly	From the —— ft. level
13,932	Graspless .	From the —— ft. level to —— ft. level
13,933	Grassant .	On the No. —— level
13,934	Grassblade	Above the No. —— level
13,935	Grasscloth	To the No. —— level
13,936	Grassfinch	The No. —— level is in
13,937	Grassgrown	From the No. —— level
13,938	Grassland .	From the No. —— level to the No. —— level
13,939	Grassmoth	To drive a level (levels)
13,940	Grassoil .	From the surface to the —— level
13,941	Grassplot .	Driving a level
13,942	Grasspoly .	Driving levels
13,943	Grassquit .	Driving north level
13,944	Grasstree .	Driving south level
13,945	Grassvetch	Driving levels north and south
13,946	Grassweek	Driving levels east and west
13,947	Grasswidow	The levels, both north and south
13,948	Grasswrack	The levels, both east and west
13,949	Grassy . .	The distances between the levels
13,950	Gratefully .	North level
13,951	Gratelupia	South level
13,952	Gratified .	East level
13,953	Gratify . .	West level
13,954	Gratifying .	No. —— level
13,955	Gratiola .	Have driven level(s)

No.	Code Word.	**Level(s)** (*continued*)
13,956	Gratis . .	Are you driving level(s)
13,957	Gratitude .	Have you driven level(s)
13,958	Gratuitous	Level should be driven
13,959	Gratuity .	Levels should be driven north and south
13,960	Gratulant .	Levels should be driven east and west
13,961	Gravamen	Levels at every —— feet
13,962	Gravel . .	How many feet have you driven level
13,963	Graveless .	Level has been discontinued
13,964	Gravelly .	Level does not show any promise of ore
13,965	Gravelpit .	Level has been run through quartz
13,966	Gravelwalk	Level has been run through granite
13,967	Gravemaker	Level has been run through country-rock
13,968	Gravestone	Level has been run through very hard ground
13,969	Graveyard	Rate of progress in the level has been slow
13,970	Gravigrade	We have made good progress with this level
13,971	Gravimeter	Level has been advanced —— feet since
13,972	Gravitate .	It is my intention to discontinue work in this level
13,973	Gravities .	It is my intention to push on this level as quickly
13,974	Gravity .	Drive levels at every —— feet [as possible
13,975	Graybird .	Drive level(s) to intercept
13,976	Grayfalcon	Drive level(s) to make connection with
13,977	Grayflies .	Drive level(s) as quickly as possible
13,978	Grayfly .	There is little change to report in this level
13,979	Grayish .	There is not any improvement to report in this level
13,980	Grayle . .	The appearance of the ore-body in this level
13,981	Graylings .	The appearance ot the ore-body in this level is improving [so good
13,982	Graymalkin	The appearance of the ore-body in this level is not
13,983	Graymare .	The ore-body in this level is somewhat larger
13,984	Graymillet	The ore-body in this level is somewhat larger and well defined [its contents are valueless
13,985	Grayowl .	The ore-body in this level is somewhat larger, but
13,986	Graypease	Have struck a large ore-body in the —— level
13,987	Graywacke	Have struck a large ore-body in the —— level ; ore
13,988	Grazierly .	Expect this level will strike [assayed
13,989	Greasebox	Expect this level will strike —— in —— days
13,990	Greasecock	Expect this level will reach the —— in about
13,991	Greatborn	When do you expect level will reach (or strike)
13,992	Greatcoats	How far have you driven level in the direction of
13,993	Greatest .	How many feet have the following levels been driven
13,994	Greatly .	Level has again entered a body of
13,995	Grecianize	The level is now —— feet from the boundary line
13,996	Grecism .	**Levelling**
13,997	Grecized .	Levelling over
13,998	Grecizing .	By levelling it was found to be
13,999	Greedy .	**Levied**
14,000	Greedygut	Has (have) —— levied
14,001	Greenback	Has (have) levied
14,002	Greenbird .	Has (have) not levied
14,003	Greenbone	Not yet levied

No.	Code Word.	Levied (*continued*)
14,004	Greenbrier	The government have levied a tax upon
14,005	Greenbroom	The taxes are levied upon the income
14,006	Greencloth	The taxes are levied upon the rateable value
14,007	Greencrop	Upon which the taxes are levied
14,008	Greenebony	Income tax levied
14,009	Greeneyed	**Levy**
14,010	Greenfinch	Levy a tax of —— upon
14,011	Greengage	Do (does) —— intend to levy
14,012	Greenheart	Will levy
14,013	Greenhouse	Will not levy
14,014	Greenish .	Intend(s) next month to levy
14,015	Greenlaver	**Liability (Liabilities)**
14,016	Greenshank	Incurring heavy liabilities
14,017	Greensnake	We shall thus incur a liability of
14,018	Greenstall .	No liability
14,019	Greensward	What is (are) the liability(ies)
14,020	Greentea .	The liability(ies) is (are) heavy
14,021	Greenwax.	The liability(ies) is (are) slight
14,022	Greenwood	No further liability
14,023	Greetings .	Liability(ies) probably about
14,024	Gregarian .	Cannot accept the liability
14,025	Gremial .	Will you accept the liability
14,026	Grenadiers	Repudiate(s) the liability
14,027	Grenadillo	Incur no further liability
14,028	Grenatite .	Cannot undertake any further liability
14,029	Gressorial.	No liability can be incurred
14,030	Grewia. .	No further liability will be incurred
14,031	Greybeards	The liabilities incurred amount to
14,032	Greyhound	A Limited Liability Company
14,033	Greylag .	Failed with liabilities amounting to
14,034	Gridelin .	Diminish the liability to
14,035	Gridirons .	Increase the liability to
14,036	Griefshot .	The liabilities have been lessened to
14,037	Griego . .	The liabilities are so heavy
14,038	Grievable .	Admit any liability for
14,039	Grievances	**Liable**
14,040	Grieved .	Is (are) liable (for)
14,041	Grievingly.	Is (are) not liable (for)
14,042	Grievous .	Will be liable to
14,043	Grievously	Will not be liable to
14,044	Griffin . .	Liable to cave in at any moment
14,045	Griffinism .	**Liberal(ly)**
14,046	Grillroom .	Is offered on very liberal terms
14,047	Grilse . .	Has (have) behaved in a very liberal manner
14,048	Grimaced .	Has (have) not behaved in a liberal manner
14,049	Grimacing.	Most liberal terms
14,050	Grimalkins	Anything but liberal
14,051	Grimily .	A very liberal offer
14,052	Griminess.	(Has been) treated very liberally
14,053	Grindery .	**Liberty**

No.	Code Word.	**Liberty** (*continued*)
14,054	Grindlet .	Have perfect liberty to
14,055	Grindstone	Must have liberty to
14,056	Gripepenny	Liberty to act as —— think(s) proper
14,057	Gripesegg.	Are we at liberty to
14,058	Gripleness.	You are at liberty to
14,059	Grisamber.	**License(s)**
14,060	Griskin. .	Prospecting license(s)
14,061	Gristmill .	Diggers' license(s)
14,062	Gritrock .	Paying licenses on
14,063	Groanful .	Licenses are issued
14,064	Grobman .	Have received license to
14,065	Groceries .	Licenses have been paid
14,066	Grocery .	Licenses are in arrears
14,067	Grogginess	Pay licenses at once
14,068	Groggy. .	Take out licenses
14,069	Grogram .	Licenses quote wrong numbers
14,070	Grogshop .	Licenses have been amended
14,071	Groomlet .	Cannot get licenses amended
14,072	Grooving .	Licenses granted
14,073	Groroilite .	**Lie(s)**
14,074	Grossfed .	Report is a tissue of lies
14,075	Grossular .	The statement is a lie
14,076	Grotesque.	The mine(s) lie(s)
14,077	Grotto . .	The lode(s) lie(s)
14,078	Grottowork	The strata lie
14,079	Groundash	Lies in
14,080	Groundbait	Lie(s) under
14,081	Groundedly	Lie(s) over
14,082	Groundform	Lie(s) at the side
14,083	Groundgame	Lie(s) on either side
14,084	Groundhold	Lie(s) over against
14,085	Groundivy	Lie(s) conformably to
14,086	Groundlaw	How do they (does it) lie
14,087	Groundless	**Lien**
14,088	Groundling	Have (has) a lien upon
14,089	Groundnut	Have (has) no lien upon
14,090	Groundoaks	Has he (have they) a lien upon
14,091	Groundpigs	Retaining a lien upon
14,092	Groundplum	A lien to the amount of
14,093	Groundrent	Free from any lien whatever
14,094	Groundroom	No further lien upon
14,095	Groundrope	**Lighter**
14,096	Groundsill	The work is much lighter on account of
14,097	Groundtier	The work is now lighter as we are
14,098	Groundwork	**Lighterage**
14,099	Grouped .	Lighterage incurred upon
14,100	Grouping .	Charges for lighterage
14,101	Grouty . .	**Like**
14,102	Groveller .	If you like
14,103	Grovelling.	Would like to have

No.	Code Word.	Like (*continued*)
14,104	Growth .	Is (are) like
14,105	Growthead	Is (are) not like
14,106	Grubaxe .	What would you like
14,107	Grubworm	We do not like
14,108	Grudged .	**Likely**
14,109	Grudgeful .	(It) is very likely (to)
14,110	Grudgekin.	(It) is not likely (to)
14,111	Grudgingly	Is it likely (that)
14,112	Grudgings .	Is there likely to be
14,113	Grudgment	Would —— be likely to
14,114	Gruesome .	If —— were likely to
14,115	Gruffish .	Would most likely accept if it were offered to him
14,116	Gruffly . .	Likely to occur
14,117	Grugeons .	Would most likely open up ore bodies
14,118	Grugru . .	**Limestone**
14,119	Grumblers	Magnesian limestone
14,120	Grumpily .	Hard blue limestone
14,121	Grunts . .	In the limestone
14,122	Gryde . .	In carboniferous limestone
14,123	Gryfon . .	Consists of limestone
14,124	Gryllidae .	Argillaceous limestone
14,125	Gryphaea .	**Limit(s)**
14,126	Gryphosis .	Reduce limits
14,127	Grysboc .	Increase limits
14,128	Guaranteed	All limits removed
14,129	Guarantors	All limits removed, but act with great caution
14,130	Guaranty .	Cannot get on at the limit
14,131	Guarapo .	Could not act on account of your limit
14,132	Guardable.	Limit too high
14,133	Guardage .	Limit too low
14,134	Guardboat	At or about the limit of
14,135	Guarded .	I (we) cannot increase my (our) limits
14,136	Guardedly.	Beyond my (our) limit(s)
14,137	Guardful .	Is this beyond your limits
14,138	Guardfully	My limit is
14,139	Guardian .	Within the prescribed limits
14,140	Guardirons	Not to exeeed my (our) last limits
14,141	Guardless .	At the extreme limits of
14,142	Guardroom	May we increase limits
14,143	Guardship.	May we reduce limits
14,144	Guardsman	Can do nothing at present limits
14,145	Guavajelly	Do not limit you with regard to
14,146	Guaza . .	**Limited (to)**
14,147	Guazuma .	The mines are too limited in extent
14 148	Guelphic .	Limited liability
14,149	Guerillist .	Of limited capacity
14,150	Guesser .	The output has been limited
14,151	Guessingly	Must be limited to
14,152	Guesswork	My (our) liability must be limited to
14,153	Guest . .	Limited demand

No.	Code Word.	**Limited** (*continued*)
14,154	Guestrope .	Too limited
14,155	Guesttaker	Within the limited time
14,156	Guestwise.	In the limited space
14,157	Guffaw. .	Space is very limited
14,158	Guffawing.	We are limited by
14,159	Guianabark	Not limited with regard to
14,160	Guibas. .	The length must be limited (to)
14,161	Guidance .	The workings are too limited in extent
14,162	Guide . .	On a very limited scale
14,163	Guidebar .	**Line(s)**.
14,164	Guideblock	Up to the boundary line
14,165	Guidebooks	On the line
14,166	Guidepost.	Beyond the line
14,167	Guiderail .	The line of cleavage
14,168	Guidetubes	The line of demarcation
14,169	Guidonian	Within the line (of)
14,170	Guildhall .	Through one side line
14,171	Guildrent .	Through one end line
14,172	Guildry .	The line extends to
14,173	Guileful .	Define side and end lines on plan
14,174	Guillevat .	Lode passes through side line
14,175	Guillotine .	Lode passes through end line
14,176	Guiltilike .	Does the lode pass through or cut side or end line
14,177	Guiltily .	The line follows
14,178	Guiltiness .	In a straight line towards
14,179	Guiltsick .	A straight line from —— to
14,180	Guineacorn	The boundary line(s) of the claim
14,181	Guineafowl	The line shown on plan (marked)
14,182	Guineahen	Our line of action
14,183	Guineapig.	The plan shows the line (of)
14,184	Guineaplum	The line beyond which it is not prudent to go
14,185	Guineas .	The lines laid down (by)
14,186	Guitar . .	The side lines
14,187	Gulaund .	The end lines
14,188	Gulfstream	Through both lines
14,189	Gulielma .	The side and end lines
14,190	Gullible .	**Liquidated**
14,191	Gullygut .	Being officially liquidated
14,192	Gullyhole .	Being voluntarily liquidated
14,193	Gulosity .	**Liquidation**
14,194	Gulped .	The company is in liquidation
14 195	Gulping .	Will be put in liquidation
14,196	Gumanimal	To put into liquidation
14,197	Gumarabic	Endeavour to avoid liquidation
14,198	Gumboils .	Cannot avoid liquidation
14,199	Gumcistus	Liquidation will now be avoided
14,200	Gumdragon	Has (have) filed a petition of liquidation
14,201	Gumelastic	Resolved upon voluntary liquidation
14,202	Gumelemi.	(In) liquidation for purposes of reconstruction
14,203	Gumjuniper	Has application been made for liquidation (for)

No.	Code Word.	Liquidation (*continued*)
14,204	Gumlac .	No application has been made for liquidation (for)
14,205	Gumminess	Going into liquidation
14,206	Gumption .	Will have to go into liquidation
14,207	Gumrash .	May have to go into liquidation
14,208	Gumresin .	**Liquidator**
14,209	Gumsenegal	Who is the liquidator
14,210	Gumstick .	The liquidator is
14,211	Gumtrees .	The official liquidator
14,212	Gumwater .	**Little**
14,213	Gunbarrel .	Very little
14,214	Guncotton	Will be very little
14,215	Gundecks .	Too little
14,216	Gundelet .	Too little to be of any use
14,217	Gunfire .	Of very little use
14,218	Gunflint .	**Load(s)**
14,219	Gunlocks .	Loads of 16 cubic feet
14,220	Gunmetal .	Loads of —— cubic feet
14,221	Gunnel. .	To load
14,222	Gunport .	Loads on the floors
14,223	Gunpowder	Not to load
14,224	Gunreach .	Must not load with too much capital
14,225	Gunroom .	Will load it with
14,226	Gunshots .	Can you load
14,227	Gunsmith .	How many car-loads
14,228	Gunstone .	Freighted —— car-loads
14,229	Guntackle .	Car-loads of
14,230	Gunwadding	Freight per load
14,231	Gurgle . .	We have now ready to load
14,232	Gurhofite .	**Loaded**
14,233	Gustable .	Has (have) loaded it too heavily
14,234	Gustation .	Has (have) loaded
14,235	Gustatory .	Has (have) not loaded
14,236	Gustful. .	Has (have) —— loaded
14,237	Gustless .	If not too heavily loaded
14,238	Guttatrap .	Has been too heavily loaded
14,239	Guttering .	Loaded up
14,240	Guttifers .	Loaded with
14 241	Gutturally .	Heavily loaded (with)
14,242	Gutturized	Loaded on board
14,243	Gwyniad .	Loaded in the cars
14,244	Gymnasium	**Loading**
14,245	Gymnast .	Loading up with
14,246	Gymnastics	Have commenced loading
14,247	Gymnic .	Have finished loading
14,248	Gymnical .	Loading instructions
14,249	Gymnodont	The loading is progressing
14,250	Gymnogen	Now loading for
14,251	Gymnosophy	Expect to finish loading
14,252	Gymnosperm	**Loan**
14,253	Gymnotidae	Is (are) trying to raise a loan

No.	Code Word.	**Loan** (*continued*)
14,254	Gymnotus .	Can you raise a loan
14,255	Gymnura .	Cannot raise a loan
14,256	Gynandria	Want a loan of
14,257	Gynandrous	A loan of —— has been
14,258	Gynarchy .	(To) raise a loan on terms to be arranged
14,259	Gyneceum	Can you obtain a loan on the terms named
14,260	Gyneocracy	Obtain loan from company's bankers
14,261	Gynethusia	Do not think it prudent to ask for a loan
14,262	Gypaetus .	Negotiating a loan
14,263	Gypseous .	Loan against satisfactory security
14,264	Gypsology.	Will not loan more than
14,265	Gypsoplast	**Locality**
14,266	Gypsum .	I visited other mines in the locality to help me to
14,267	Gypsyhat .	In the same locality [form an opinion
14,268	Gypsyism .	In the locality of
14,269	Gypsywort	The locality of the property
14,270	Gyration .	Full particulars as to the locality of the mine or
14,271	Gyrational	The locality is very healthy [property
14,272	Gyratory .	The locality is unhealthy
14,273	Gyreful .	The mine is situated in a very healthy locality
14,274	Gyrfalcon .	(To) choose a suitable locality
14,275	Gyrinidae .	The locality is very suitable for the purpose
14,276	Gyrocarpus	The locality is excellent; there is plenty of wood
14,277	Gyrogonite	The locality referred to [and water
14,278	Gyroidal .	**Localize**
14,279	Gyrolepis .	To localize
14,280	Gyroma .	(To) enable you to localize
14,281	Gyromancy	So that we may localize
14,282	Gyrophora	The localization of
14,283	Gyroscope	**Locate**
14,284	Gyrostat .	To locate the claim
14,285	Gyved . .	Will —— locate
14,286	Gyving. .	Will locate
14,287	Haarkies .	Will not locate
14,288	Habenaria	Should be located
14,289	Habendum	Have located
14,290	Habenry .	**Lock-out**
14,291	Haberdash	Lock-out is expected
14,292	Haberdine	Lock-out is not expected to last long
14,293	Habergeon	Think we must have a lock-out
14,294	Habilatory	Lock-out has begun
14,295	Habilitate.	Lock-out has ended
14,296	Habitance	Prospect of a lock-out
14,297	Habitation	Have been locked out
14,298	Habitmaker	**Lock(ed) up**
14,299	Habits . .	A lock-up of the capital
14,300	Habitshirt.	The amount of capital locked up
14,301	Habitually	Amount locked up in
14,302	Habituated	So much has been locked up
14,303	Habitudes.	Too much locked up

No.	Code Word.	**Lock(ed) up** (*continued*)
14,304	Habranthus	Has been locked up
14,305	Habrocoma	**Lode**
14,306	Habromania	Outcrop of the lode
14,307	Habroneme	A well-defined lode
14,308	Habzelia .	The lode runs north and south
14,309	Hachure .	The lode runs east and west
14,310	Hackery .	The lode runs ——° east of north
14,311	Hacklog .	The lode runs ——° west of north
14,312	Hackmatack	The lode runs between walls of
14,313	Hackneyed	The lode runs
14,314	Hackneyman	Average width of lode is —— feet
14,315	Hackster .	Average assays from the lode give
14,316	Hackwatch	The lode carries gold in paying quantities
14,317	Haddock .	The lode does not carry gold in paying quantities
14,318	Haecceity.	Prospecting for a long time in same lode
14 319	Haemagogue	Lode opening up well
14,320	Haemal .	Lode is getting narrower
14,321	Haemalopia	Lode is getting wider
14,322	Haematinic	The lode has been located by different persons
14,323	Haematodes	An adit driven —— feet has cut the lode
14,324	Haematoid	The lode has been proved to a depth of —— feet
14,325	Haematopus	An inclined shaft has been sunk on the lode
14,326	Haematozoa	The metal-bearing part of the lode lies against the
14,327	Haemoptoe	A continuation of the same lode [hanging wall
14,328	Haemulon	The metal-bearing part of the lode lies against the
14,329	Hagada .	The lode is dipping [foot-wall
14,330	Hagberries	The only development is an adit driven in ——
14,331	Hagberry .	The lode contains [feet on the lode
14,332	Hagbut .	A compact, well-defined lode between walls of
14,333	Haggard .	How far does the lode extend
14,334	Haggardly	The lode extends
14,335	Haggis. .	The lode extends (or extending) eastward
14,336	Haggled .	The lode extends (or extending) westward
14,337	Haggling .	The lode extends (or extending) northward
14,338	Hagiarchy	The lode extends (or extending) southward
14,339	Hagiocracy	The lode can be traced
14,340	Hagiology	In direct extension of the ——- lode
14,341	Hagioscope	Parallel to the —— lode
14,342	Hagridden	Contiguous to the —— lode
14,343	Haiduck .	In conjunction with the —— lode
14,344	Hailed. .	Intend to follow the lode
14,345	Hailfellow.	The lode appears to be lost
14,346	Hailmixed	The lode is split up
14,347	Hailshot .	The full breadth of the lode
14,348	Hailstones	A branch of the main lode
14,349	Hailstorm.	At the junction of the —— lodes
14,350	Haimura .	The lode is barren
14,351	Hairbell .	The lode is very rich
14,352	Hairbroom	A very rich lode
14,353	Hairbrush.	Lode not well defined

L

No.	Code Word.	**Lode** (*continued*)
14,354	Hairdye .	Has (have) struck the lode
14,355	Hairglove .	Struck another lode
14,356	Hairhung .	The lode has pinched out
14,357	Hairlichen	The lode is pinching out
14,358	Hairlike .	The lode is looking promising .
14,359	Hairneedle	The lode gets poorer as we go
14,360	Hairnets .	Plenty of ore to be got from this lode
14,361	Hairoil. .	We think this lode will yield an ore-body of con-
14,362	Hairpencil	The lode, so far, has yielded [siderable value
14,363	Hairpins .	The lode, so far, has not yielded
14,364	Hairpowder	The surface indications of the lode
14,365	Hairsalt .	The lode, in the upper level(s) .
14,366	Hairshaped	The lode, in the lower level(s)
14,367	Hairshirt .	Prospected on the lode
14,368	Hairsieve .	The lode has been worked by
14,369	Hairspace .	**Log(s)**
14,370	Hairspring	The buildings are made of rough logs
14,371	Hairworker	A log building
14,372	Hairworms	**Long**
14,373	Hakesdame	For a long time
14,374	Haladroma	The mines are situated a long distance from
14,375	Halation .	How long will it take to go to —— and examine
14,376	Halberd .	How long [the —— mine(s)
14,377	Halberdier	How long will it be
14,378	Halcyon .	How long will it (they) take
14,379	Halcyonian	It (they) will not take long
14,380	Halcyornis	It is (they are) too long
14,381	Haldanite .	It is (they are) not long enough
14,382	Halecret .	There is not long enough time to
14,383	Halfbatta .	Cannot give as long as —— wish(es)
14,384	Halfblood .	How long will —— give
14,385	Halfbred .	You must not be long
14,386	Halfcaste .	The mines have not been worked for a long time
14,387	Halfcheek .	Some long time ago
14,388	Halfcocked	For a long distance
14,389	Halffaced .	Before long
14,390	Halfheader	We hope before long to
14,391	Halfhourly	As long as
14,392	Halfkirtle .	As long as you can
14,393	Halfflap. .	As long as we can
14,394	Halflength	How long can you
14,395	Halfmerlon	How long is (are)
14,396	Halfmoon .	How long will —— last
14,397	Halfpike .	How long is it since
14,398	Halfport .	If you are too long about it
14,399	Halfpress .	**Longer**
14,400	Halfshift .	Cannot give longer time
14,401	Halfsister .	Cannot remain longer
14,402	Halfsword	Not much longer
14,403	Halftimber	A little longer

No.	Code Word.	Longer (*continued*)
14,404	Halftongue	Not any longer
14,405	Halftruth .	Do not be longer than
14,406	Halfway .	How much longer shall you
14,407	Halfwitted	Shall not remain much longer [the inspection
14,408	Halfyear .	Will take —— longer than I (we) expected to make .
14,409	Halfyearly	I (we) hope to be in —— by the ——, but may be
14,410	Halicore .	Longer than necessary [a day or two longer
14,411	Halidom .	Longer than was expected
14,412	Halieutics.	Longer than
14,413	Halimass .	A little longer than
14,414	Haliotidae	Must have longer
14,415	Haliotoid .	If it is longer, it will not pass
14,416	Halituous .	We must have a little longer time
14,417	Halldinner	About a week longer
14,418	Hallelujah	About —— days longer
14,419	Halllamp .	About —— weeks longer
14,420	Hallowfair	About a month longer
14,421	Hallowmas	**Longest**
14,422	Hallowtide	The longest (of the)
14,423	Hallux . .	At the longest
14,424	Halmalille	The longest piece must not exceed —— in length
14,425	Halmaturus	**Look(s)**
14,426	Halmote .	How does it look
14,427	Halogenous	Look(s) encouraging
14,428	Haloid . .	Looks like a failure
14,429	Halophytes	Looks better than expected
14,430	Halorageae	Looks very favourable
14,431	Haloscope	Look out for
14,432	Halosel .	Will look out for
14,433	Halsening.	Look into the matter immediately
14,434	Halterman	Look after
14,435	Haltersack	Look into
14,436	Halteth .	Look into the matter the first opportunity
14,437	Halticidae	Have had to look into
14,438	Halvanner	Have not had time to look into the matter
14,439	Halvans .	Look to
14,440	Halvenet .	Look up
14,441	Halyard .	Telegraph when you can look to
14,442	Halysites .	Am very pleased with the look of the mine
14,443	Hamcurer.	Am very disappointed with the look of the mine
14,444	Hamiform.	You must not look for
14,445	Hamiltonia	Look for an increase in
14,446	Hamitic .	Look for a decease in
14,447	Hamlets .	Look for increased monthly runs
14,448	Hammer .	Look for decreased monthly runs
14,449	Hammerable	If you look for
14,450	Hammeraxes	We certainly look for
14,451	Hammerbeam	The shareholders look for
14,452	Hammerfish	We look to you for
14,453	Hammering	Must look to him (them) for

No.	Code Word.	**Look(s)** (*continued*)
14,454	Hammerman	Look to me (us) for
14,455	Hampered	**Looked** [he is right in every respect
14,456	Hamshackle	I have looked though ——'s report on the mine;
14,457	Hamstring	I have looked through ——'s report; his statements
14,458	Hamulus .	The mine looked well [are wrong
14,459	Hanaper .	The mine looked very bad
14,460	Hanchinol	Looked promising
14,461	Handball .	**Looking**
14,462	Handbarrow	Looking bad
14,463	Handbasket	Looking well
14,464	Handblows	Looking worse
14,465	Handbook	Looking better
14,466	Handbrace	Looking first-rate
14,467	Handcloth	Not looking so well
14,468	Handcuff .	Is looking the same
14,469	Handdrop	Where the vein is looking
14,470	Handfast .	Is (are) looking ahead
14,471	Handfastly	Has (have) not been looking ahead, and consequently
14,472	Handfetter	Looking well after
14,473	Handflower	Looking out for
14,474	Handfooted	This will require looking after at once
14,475	Handgallop	Is (are) looking out (for)
14,476	Handglass	I am (we are) looking after it (them)
14,477	Handgripe	How is the vein looking
14,478	Handgrith	How is the —— looking
14,479	Handgun .	**Lose**
14,480	Handicap .	Will be likely to lose
14,481	Handicraft	Will probably lose all that has been put into it
14,482	Handiest .	How much will —— lose by
14,483	Handinhand	Do(es) not lose
14,484	Handiworks	Do not lose any time in
14,485	Handjar .	Shall lose
14,486	Handleable	Shall not lose
14,487	Handling .	Do not lose sight of
14,488	Handmaiden	Lose(s) sight of
14,489	Handmaking	They lose at present
14,490	Handmallet	They lose —— per cent. of the gold from not being
14,491	Handmills	May lose by it [able to work the sulphurets
14,492	Handorgan	Will lose by it
14,493	Handplant	Do not lose any chance of
14,494	Handpress	If we lose
14,495	Handpump	If you lose
14,496	Handsaws.	Not likely to lose
14,497	Handscrew	Lose more than
14,498	Handshoes	If you now lose the chance you may not have
14,499	Handsmooth	(To) lose the chance of [another
14,500	Handsome	How much shall we lose
14,501	Handsomely	Not lose any time in making the necessary arrange-
14,502	Handspike	Not lose so much [ments
14,503	Handstaff.	Lose too much

No.	Code Word.	Loss (*continued*)
14,504	Handstaves	Lose at present
14,505	Handstroke	Lose by the present method
14,506	Handtight	**Loser(s)**
14,507	Handtimber	Will be the loser(s)
14,508	Handvice .	Will not be any loser(s) by
14,509	Handweapon	I (we) shall be the loser(s)
14,510	Handwinged	Loser(s) by this transaction
14,511	Handwrite	**Losing** [working
14,512	Handy . .	Are losing —— per cent. of the —— through bad
14,513	Handybilly	Are losing —— per cent. of the
14,514	Handyblow	We are losing
14,515	Handydandy	We are not losing
14,516	Handyfight	Are we losing
14,517	Hangbies .	We are losing weekly
14,518	Hangdog .	We are losing monthly
14,519	Hanged .	Losing —— per cent. of the ore in the tailings
14,520	Hangerson	Losing —— per cent. of the gold and —— per cent.
14,521	Hangings .	Losing more than [of the silver
14,522	Hangman .	Losing in the concentrates
14,523	Hangnest .	Now losing at the rate of
14,524	Hankering	Losing by the present system
14,525	Hanselines	Losing by the new system
14,526	Hansomcab	**Loss(es)**
14,527	Hapalidae	Loss of
14,528	Hapharlot	What is the estimated loss
14,529	Haphazard	Cannot estimate the loss
14,530	Hapless .	Do not think there will be any loss
14,531	Happening	Great loss of —— through imperfect working
14,532	Happiest .	Detained here by loss of luggage
14,533	Happiness	There has been a loss of —— days owing to
14,534	Harangue .	Through the loss of
14,535	Haranguing	Estimate the loss of
14,536	Harass . .	Loss of time
14,537	Harassing .	A total loss of
14,538	Harassment	To prevent any loss
14,539	Harbinger	Loss will be about
14,540	Harbour .	Has (have) incurred severe loss
14,541	Harbourage	Has (have) sustained loss of
14,542	Harbourlog	At a loss of
14,543	Hardbake	The loss will be great
14,544	Hardbilled	The loss will be small
14,545	Hardbound	The loss is heavy
14,546	Hardearned	The loss is trifling
14,547	Hardens .	Loss by depreciation of
14,548	Hardfern .	Loss by depreciation of silver
14,549	Hardfisted	The loss caused by
14,550	Hardfought	The average loss
14,551	Hardgot .	The loss in tailings
14,552	Hardgotten	The loss in working
14,553	Hardhanded	The loss is greatest in

No.	Code Word.	**Loss(es)** (*continued*)
14,554	Hardhead .	The loss is least in
14,555	Hardihood	Assays from —— show a loss of
14,556	Hardruled	Show a loss of
14,557	Hardtack .	Loss of life
14,558	Hardware .	Loss of life and property
14,559	Hardyshrew	How is it there is so much loss in
14,560	Harebrain	Will have to stand the loss
14,561	Harehounds	The loss will be counterbalanced by
14,562	Harelipped	Hope still to come out without loss
14,563	Haremint .	The loss per cent.
14,564	Harepipe .	The loss per ton
14,565	Haresear .	In estimating the loss
14,566	Haresform	Provided the loss does not exceed
14,567	Harfang .	The loss will not exceed
14,568	Haricot .	There is a heavy loss (in or on)
14,569	Harlequins	The loss of the ship conveying
14,570	Harmaline	The loss in weight
14,571	Harmanbeck	After such a heavy loss
14,572	Harmattan	The loss by the present system
14,573	Harmfully	The loss by
14,574	Harmless .	Loss for the quarter
14,575	Harmonical	Loss for the half-year
14,576	Harmonious	Will entail a loss of
14,577	Harmonist	**Lost**
14,578	Harmonized	Too much is lost in the tailings
14,579	Harmony .	Lost in the working
14,580	Harmotome	Per cent. is lost through thieving
14,581	Harnessed	Is (are) lost
14,582	Harnesstub	Is (are) not lost
14,583	Harpax .	Per cent. of the —— is lost
14,584	Harpoon .	The rest is lost
14,585	Harpoongun	Telegraph how much is (are) lost in
14,586	Harpooning	Has (have) lost
14,587	Harpseal .	Has (have) not lost
14,588	Harpsichon	Has (have) been lost
14,589	Harpster .	Have lost time through
14,590	Harpyeagle	Lost by
14,591	Harquebuse	Will be lost
14,592	Harrateen	Lost in the tailings
14,593	Harridan .	Lost in concentrates
14,594	Harriers .	Lost from want of
14,595	Harrower .	Lost heavily (on)
14,596	Harshening	Must have been lost
14,597	Harshly .	Could not have been lost
14,598	Harshness	We have lost during the past
14,599	Hartall .	Lost in transit
14,600	Hartcrop .	Much of this has been lost by
14,601	Hartroyal .	Lost more than
14,602	Hartshorn	Lost nearly as much **as**
14,603	Haruspex .	**Lot**

No.	Code Word.	**Lot** (*continued*)
14,604	Haruspicy	A lot of
14,605	Harvest .	**Low.** (See also Grade.)
14,606	Harvestbug	Very low
14,607	Harvesters	Low as you can
14,608	Harvestfly	The water is very low in
14,609	Harvestman	Low price
14,610	Haschish .	Of a very low grade
14,611	Hashedmeat	How low will it go
14,612	Haspicoll .	The —— stock is very low
14,613	Hastated .	**Lower**
14,614	Hastive .	I (we) think the —— will go lower
14,615	Hatband .	I (we) think the —— will not go lower
14,616	Hatblocks	We must sink —— feet lower
14,617	Hatbox .	Will have to sink the —— lower
14,618	Hatbrush .	Must have lower terms
14,619	Hatchboat	At the lower level
14,620	Hatcheller	Cannot lower
14,621	Hatchetine	Cannot go lower without
14,622	Hatchment	Can go lower
14,623	Hatchways	As we go lower
14,624	Hatchy .	Think we must go lower (before)
14,625	Hateful .	Lower than the —— level
14,626	Hatmould	Will be lower this week
14,627	Hatracks .	Will be lower this month
14,628	Hatred . .	Will be lower next month
14,629	Hattemist.	Will not be lower (than)
14,630	Hatteria .	Lower for the next few weeks
14,631	Hatworship	Lower for the next month or two
14,632	Hauberk .	A little lower
14,633	Hauerite .	If the price is not lower
14,634	Haughtiest	A lower temperature
14,635	Haughty .	Must not be lower (than)
14,636	Haulse . .	Must be lower
14,637	Haunched	Lower than it has ever been
14,638	Haustellum	Must have lower rates
14,639	Haustorium	Quote lower rates
14,640	Haustus .	The lower run is owing to
14,641	Hautboy .	Lower quotations
14,642	Hautboyist	Cannot quote lower than we have already done
14,643	Hautepace	Owing to the lower
14,644	Hauteur .	**Lowest**
14,645	Hautgout .	On the lowest terms
14,646	Hauyne .	The lowest level
14,647	Havenage.	Telegraph lowest price
14,648	Haverbread	The lowest price
14,649	Havermeal	——'s lowest price is
14,650	Haversack	What is the lowest price
14,651	Haversian	What is the lowest sum you will take (for)
14,652	Haverstraw	Is the lowest price; no reduction can be made
14,653	Havoc . .	The mine has been worked at the lowest cost

No.	Code Word.	**Lowest** (*continued*)
14,654	Hawbucks	The cost of working must be brought down to the
14,655	Hawcubite	Is this the lowest you can [lowest figure
14,656	Hawfinch .	Insured at lowest possible rate
14,657	Hawhaw .	At the lowest possible estimate
14,658	Hawhawing	Carried at lowest possible rate
14,659	Hawk . .	The lowest possible cost
14,660	Hawkbit .	The lowest point (reached)
14,661	Hawkeyed	The lowest workings
14,662	Hawking .	**Luck**
14,663	Hawkmoth	Through a piece of luck
14,664	Hawknosed	**Lucky**
14,665	Hawkowl .	Has (have) been so lucky as to
14,666	Hawksbeard	Has (have) not been lucky in
14,667	Hawksbill	**Lucrative**
14,668	Hawkweed	I feel sure it will be a lucrative venture
14,669	Hawsebag	Not sufficiently lucrative
14,670	Hawseblock	Has (have) been very lucrative
14,671	Hawsehook	Has (have) never been lucrative
14,672	Hawsepiece	Will it be sufficiently lucrative
14,673	Hawsepipe	**Luggage**
14,674	Hawseplug	As soon as my (our) luggage arrives
14,675	Hawserlaid	Our luggage has not yet come to hand
14,676	Hawsewood	**Lumber**
14,677	Hawthorns	Waiting for lumber
14,678	Hayasthma	Plenty of lumber
14,679	Haybird .	Built of lumber
14,680	Haycock .	The houses are built of lumber
14,681	Haydenite	A scarcity of lumber
14,682	Hayfever .	Lumber for the —— mill
14,683	Hayfield .	Lumber for housing
14,684	Hayingtime	Specification of lumber, length and size
14,685	Hayknife .	Cost of lumber —— per 1000 superficial feet
14,686	Hayloft .	Can get lumber from
14,687	Haymaids .	Cannot get lumber nearer than
14,688	Haymaker	Superficial feet of lumber
14,689	Haymow .	**Machine**
14,690	Hayricks .	Will make a trial of the machine
14,691	Haystalk .	A very useful machine
14,692	Haytea .	Machine not suitable for
14,693	Haytedder	Machine such as we require
14,694	Hazardable	**Machinery**
14,695	Hazardize .	Machinery broke down ; had to stop hoisting
14,696	Hazardous	Machinery broke down ; had to stop pumping
14,697	Hazardry .	The only machinery is
14,698	Hazelearth	There is no machinery whatever at
14,699	Hazelly .	What hoisting machinery is there at the mine
14,700	Hazelnut .	Machinery (at ——) complete in every way
14,701	Headbands	What machinery will be required
14,702	Headblock	Machinery received
14,703	Headcheese	Have commenced erecting machinery

No.	Code Word.	**Machinery** (*continued*)
14,704	Headcourt	I estimate it will cost —— to put up the necessary
14,705	Headdress	Machinery out of repair [machinery
14,706	Headfast .	I estimate the value of the machinery at
14,707	Headgargle	The machinery that will have to be erected is
14,708	Headgear .	The machinery works badly
14,709	Heading .	The machinery will be up by
14,710	Headknot .	The machinery has been sent
14,711	Headlights	The machinery cannot be sent out (until)
14,712	Headlongly	The repairs to the machinery will take
14,713	Headlugged	The best way to send the machinery
14,714	Headmark	Am (are) sending machinery
14,715	Headmaster	What kind of machinery do you want
14,716	Headmoney	Have not got the machinery to do it
14,717	Headmost	There is no machinery at all
14,718	Headpiece	Have tested the machinery
14,719	Headrail .	The machinery is in good condition
14,720	Headranger	The new machinery will cost
14,721	Headrope .	The new machinery has cost
14,722	Headsea .	The new machinery is working well
14,723	Headshake	The new machinery not yet arrived
14,724	Headsilver	The new machinery not yet finished
14,725	Headspring	The machinery for the
14,726	Headstall .	Put machinery into thorough repair
14,727	Headstrong	Put the machinery into good working order
14,728	Headsword	**Made**
14,729	Headwind	Made money over
14,730	Headyard	Made money out of
14,731	Healable .	Will have to have the —— made
14,732	Healingbox	I am (we are) getting it (them) made
14,733	Healthful .	Shall I (we) get it (them) made here
14,734	Healthier .	Can be made here
14,735	Healthless	Cannot be made here
14,736	Healthsome	When will it (they) be made
14,737	Healthy .	When was it (were they) made
14,738	Heapkeeper	It (they) will be made about
14,739	Hearkener	Must be made immediately
14,740	Hearsay .	You had better have —— made
14,741	Hearselike	If you can get it (them) made
14,742	Heartache	If it was (they were) made
14,743	Heartblood	It was (they were) made
14,744	Heartbreak	If you cannot get it (them) made
14,745	Heartburn	Could not be made out
14,746	Heartfelt .	Made the best terms I could
14,747	Heartfree .	Can —— be made at
14,748	Heartgrief	Made by
14,749	Heartheavy	Made up to
14,750	Hearthrug	Cannot be made up
14,751	Heartily .	Has (have) been made
14,752	Heartleaf .	Has (have) not been made
14,753	Heartlings	Trial made; result satisfactory

L 2

No.	Code Word.	**Made** (*continued*)
14,754	Heartquake	Trial made; result unsatisfactory
14,755	Heartsease	Not made by
14,756	Heartsick .	Made up of
14,757	Heartsore .	Made too heavy
14,758	Heartwhole	Made too light
14,759	Heartwood	Will have to be made
14,760	Heartyhale	If you have not made
14,761	Heatengine	If he has (they have) made
14,762	Heathbell .	Must be made in accordance with specification
14,763	Heathberry	Made to drawing
14,764	Heathclad	Made to specification
14,765	Heathcock	Get it made at once
14,766	Heathen .	Made in order to
14,767	Heathendom	Must be made (up) not later than
14,768	Heathenish	**Magnitude**
14,769	Heathenize	The magnitude of the concern
14,770	Heathgame	Of what magnitude is
14,771	Heathgrass	Of great magnitude
14,772	Heathpea .	Of no great magnitude
14,773	Heathpoult	**Mail(s)**
14,774	Heavenborn	Sending report by to-day's mail
14,775	Heavenbred	Sending report by to-morrow's mail
14,776	Heavenly .	Sending report by mail of
14,777	Heavenward	Report sent by mail of
14,778	Heaviest .	Will write by next mail
14,779	Heaviness	Will send full particulars by next mail
14,780	Heavyarmed	Wait for the mail of
14,781	Heavyladen	By the first mail
14,782	Heavyspar	By last mail
14,783	Heavystone	By last mail which left here on
14,784	Hebdomad	By next mail
14,785	Hebdomadal	By the next mail which leaves on
14,786	Hebevase .	No dependence can be placed on the mails
14,787	Hebraic .	Mail arrived too late for me (us) to
14,788	Hebraical .	Mail delayed through accident to
14,789	Hebraicize	Mail delayed owing to bad weather
14,790	Hebraisms	Leaving by mail steamer of —— for
14,791	Hebraized	Leaving by mail train (on) —— for
14,792	Hebraizing	Write by next mail about
14,793	Hebrew .	There is a weekly mail here from
14,794	Hebrewess	There is a monthly mail here from
14,795	Hecatombs	There is a fortnightly mail here from
14,796	Heckle .	Not hearing from you by last mail
14,797	Heckling .	Have sent you by the mail of
14,798	Hectastyle	Have sent you the accounts for the year ending ——
14,799	Hectic . .	We are sending by this mail [by mail of
14,800	Hectical .	You will receive it by the mail of
14,801	Hectogram	Letters, etc., sent by mail of —— lost; we send
14,802	Hectorian.	Mail of —— lost in steamer [copies
14,803	Hectorly .	Please send by first mail

No.	Code Word.	Mail(s) (continued)
14,804	Heddleeye	Mail steamer broken down
14,805	Hedeoma . **Mailed**	
14,806	Hedgeborn	Was mailed in the
14,807	Hedgehogs	If not already mailed, please send immediately
14,808	Hedgeknife	Mailed in time for
14,809	Hedgenote **Main**	
14,810	Hedgepig .	Main drift
14,811	Hedgepress	Main shaft
14,812	Hedgerhyme	In the main drift
14,813	Hedgestake	In the main shaft
14,814	Hedonics .	The main shaft is down —— ft
14,815	Hedyphane	The main tunnel
14,816	Hedysarum	From the main shaft to
14,817	Heedfully .	Main drift has been advanced
14,818	Heedily .	Our main object is to
14,819	Heedlessly	The main object in view
14,820	Heelknee . **Make**	
14,821	Heelpost .	Prepared to make over
14,822	Heeltap .	Not prepared to make over
14,823	Heeltool .	Willing to make
14,824	Hegelian .	Not willing to make
14,825	Hegemonic	Must make some alteration
14,826	Heifer . .	Cannot make any alteration
14,827	Heighho .	Do not make any alteration
14,828	Heightener	Make a good run
14,829	Heinous .	Make a better run
14,830	Heinously	Make a better
14,831	Heirlooms	Make (any) alteration (in)
14,832	Heirs . .	Please make out and send at once
14,833	Heirship .	Make arrangements to go to —— to examine ——
14,834	Hejalap .	To make money [mine
14,835	Helamys .	When can you mak
14,836	Helarctos .	Can make
14,837	Helbeh .	Can you make
14,838	Helicoidal	Cannot make
14,839	Heliconian	Cannot you make
14,840	Helicteres.	If you cannot make
14,841	Helictis .	Can make use of it
14,842	Heliograph	Make the most of it
14,843	Heliolites .	Make terms with
14,844	Heliometer	Make the best of the matter
14,845	Heliostat .	Make up the accounts
14,846	Heliotype.	Do not know what to make of it
14,847	Hellanodic	We think you ought to make
14,848	Hellbender	Make a difference
14,849	Hellblack .	Do not make it the same as you
14,850	Hellbrewed	How much do you make it
14,851	Hellbroth . **Making**	
14,852	Helldoomed	Am (are) now making up the accounts
14,853	Helleborus	Making no profits, only paying expenses

No.	Code Word.	**Making** (*continued*)
14,854	Hellenic .	Now making sufficient to
14,855	Hellfire .	Not making enough to
14,856	Hellgates .	Am (are) not making more than
14,857	Hellhag .	Are you making
14,858	Hellhound	Hope soon to be making
14,859	Hellish .	Making good progress
14,860	Hellishly .	**Man**
14,861	Hellkite .	A man of no reputation
14,862	Hellward .	Is a man notorious for writing exaggerated reports
14,863	Helmage .	Must have a reliable man to
14,864	Helmets .	Is a man who gets paid a commission only if his re-
14,865	Helminth .	A man of high standing [port sells the mine
14,866	Helminthic	**Manage**
14,867	Helmport .	Can you manage
14,868	Helmsman	Can —— manage (to)
14,869	Helmwind	Can manage (to)
14,870	Heloderma	Cannot manage (to)
14,871	Helodus .	If you can manage (to)
14,872	Helopidae	Has (have) not been able to manage (to)
14,873	Helotism .	If you cannot manage (to)
14,874	Helotry .	Very difficult to manage
14,875	Helpfellow	Shall be able to manage (to)
14,876	Helpful .	Has (have) not known how to manage the
14,877	Helping .	Will you manage (to)
14,878	Helpless .	Has (have) been able to manage
14,879	Helplessly	Will undertake to manage
14,880	Helpmeet .	**Managed**
14,881	Helvella .	Has been badly managed
14,882	Helvetic .	Can be easily managed
14,883	Hemachate	The property has been badly managed
14,884	Hemachrome	The property has been well managed
14,885	Hemanthus	It is a well-managed mine
14,886	Hemastatic	If not well managed, heavy loss will result
14,887	Hematherm	Must be better managed henceforth
14,888	Hematocele	Managed it badly
14,889	Hematology	**Management**
14,890	Hematosine	Will —— take the management
14,891	Hematuria	Owing to bad management
14,892	Hemelytron	Will you take the management
14,893	Hemerobian	With proper management
14,894	Hemicarp .	Will not take the management
14,895	Hemicrany	Without proper management
14,896	Hemicycle	Will take the management
14,897	Hemidactyl	I (we) will leave the management of —— to
14,898	Hemidesmus	It will be best to leave the management of —— to
14,899	Hemigamous	Will require some little management
14,900	Hemiglyph	To whom can you leave the management of
14,901	Hemihedron	Will leave the management to
14,902	Hemiope .	The sole management (or) [settled
14,903	Hemiopsy	With careful management the affair may be easily

No.	Code Word.	**Management** (*continued*)
14,904	Hemiplegia	Management left to
14,905	Hemiplexy	Change in the management
14,906	Hemipode	The previous management
14,907	Hemipodius	Your predecessor in the management
14,908	Hemiprism	(Under) the present management
14,909	Hemipter .	**Manager (of)**
14,910	Hemipteral	Who is the manager of
14,911	Hemitropy	To appoint as manager
14,912	Hemlock .	Appoint —— as manager
14,913	Hemoptysis	Manager of —— mine is leaving
14,914	Hemorrhage	Will not do as manager
14,915	Hemorrhoid	Has been appointed manager
14,916	Hempnettle	Must have a competent man as **manager**
14,917	Hemppalm	Resident manager (of)
14,918	Hempseed	Assistant manager (of)
14,919	Hemstitch	The resident manager's report
14,920	Henbanes .	The assistant manager's report
14,921	Henceforth	The manager's report
14,922	Henchboy	**Manganese**
14,923	Henchman	Black oxide of manganese
14,924	Hencoop .	The manganese deposit
14,925	Hendecagon	**Manifest**
14,926	Hendiadys	It is manifest that
14,927	Hendriver	**Manipulated**
14,928	Henfish .	The stock has been manipulated
14,929	Henharrier	Is being manipulated
14,930	Henhearted	Is the stock being manipulated
14,931	Henhouses	Manipulated for their own purposes
14,932	Henhussy.	Must not be manipulated
14,933	Hennery .	Has been manipulated in some way
14,934	Henotheism	**Manipulating**
14,935	Henpeck .	**Manipulation**
14,936	Henpeckery	If it cannot be done without manipulation
14,937	Henrician	With any manipulation
14,938	Henroost .	Great waste in manipulation
14,939	Hensfoot .	**Manner**
14,940	Henwife .	In what manner
14,941	Hepatalgia	In the same manner as
14,942	Hepatic .	In like manner
14,943	Hepatitis .	In some manner
14,944	Hepatize .	In no other manner
14,945	Hepatizing	In the manner explained
14,946	Hepbramble	The only manner in which the ore can be **worked**
14,947	Hepbriar .	Cannot in any manner [to any profit is by
14,948	Hephaestos	In a very proper manner
14,949	Hepialidae	This is the only manner
14,950	Heptachord	**Manufactory(ies)**
14,951	Heptad .	From the manufactory (of)
14,952	Heptaglot .	Several manufactories
14,953	Heptagonal	**Manufacture(s)**

No.	Code Word.	Manufacture(s) (*continued*)
14,954	Heptagynia	In the manufacture of
14,955	Heptameron	Manufactures very limited
14,956	Heptander	Process of manufacture
14,957	Heptaphony	(Who) are the manufacturers of
14,958	Heptarchic	**Many**
14,959	Heptateuch	How many do you want
14,960	Heptylene	Do you want many
14,961	Heraclidan	I (we) do not want many
14,962	Herald . .	I (we) shall want a great **many**
14,963	Heraldcrab	As many as you can get
14,964	Heralding	Not so many as
14,965	Heraldship	As many as possible
14,966	Herbaceous	Not very many
14,967	Herbage .	Is found in many of the mines in the district
14,968	Herbalists	In many instances
14,969	Herbarium	Many miners
14,970	Herbelet .	Many of the mines about here
14,971	Herbescent	How many
14,972	Herbgerard	In many places
14,973	Herbgrace	In many cases
14,974	Herbist .	Many of the men
14,975	Herbivora	**Map(s)**
14,976	Herborized	Send me (us), if you can, a map of the country
14,977	Herbparis .	Map of the country in which the —— mines are
14,978	Herbrobert	No reliable map has ever been made of [situated
14,979	Herbulent	A government map of the
14,980	Herbwoman	You will see by the map that
14,981	Herculean	Map of the district
14,982	Hercynian	Map of the state
14,983	Herderite .	Map of the country
14,984	Herdewich	Map of the concession
14,985	Herdgroom	Map of the property
14,986	Herdsgrass	Geological map
14,987	Hereabouts	A contour map
14,988	Hereafter .	Map showing the mines
14,989	Herebote .	Map showing the concession(s)
14,990	Hereby .	Map showing the position of the property
14,991	Heredipety	Map showing the reefs
14,992	Hereditary	Map showing all the locations (or claims)
14,993	Heregild .	Map to accompany (accompanying) the report
14,994	Herehence	Sketch map
14,995	Herein . .	Map showing the roads, rivers, etc.
14,996	Heretoch .	Map showing the geological features
14,997	Heretofore	Can you send a good map of
14,998	Hereupon	Have sent you a map of
14,999	Herewith .	The map is reliable
15,000	Heriot . .	The map is unreliable
15,001	Heriotable	Map published by
15,002	Herisson .	A good map of —— published by
15,003	Heritably .	Map and report(s)

No.	Code Word.	
15,004	Heritance .	**Margin(s)**
15,005	Heritrix .	Must have a margin for contingencies
15,006	Hermaic .	A further margin of
15,007	Hermaical	Will not give any margin
15,008	Hermannia	Will give a margin of
15,009	Herminium	Unless it (they) leave a margin
15,010	Hermitage	Leaving no margin
15,011	Hermitcrab	Leaving a good margin for
15,012	Hermitess	The requisite margin
15,013	Heroerrant	Quite sufficient margin
15,014	Heroically	Margin not sufficient
15,015	Heroicness	Margin must be made good on
15,016	Heroified .	Margin for working
15,017	Heroify .	Margin of —— for contingencies
15,018	Heronsbill	Margin absorbed by
15,019	Heronshaws	The margin of profit very small
15,020	Heroship .	No margin necessary
15,021	Herpestes .	The margin will be absorbed by
15,022	Herpeton .	**Mark**
15,023	Herringbus	There is a mark
15,024	Herrnhuter	There is no mark
15,025	Hersillon .	Left a mark
15,026	Herstpan .	The mark is
15,027	Herygoud .	The distinguishing mark
15,028	Hesitative.	Mark cannot be found
15,029	Hesperian	Mark the place (in the)
15,030	Hessianbit	**Marked**
15,031	Hessianfly	None seem to have been marked
15,032	Hesychast	Have marked
15,033	Hetaristic .	Have marked the places (in the)
15,034	Heterarchy	**Market**
15,035	Heterocera	Market advancing
15,036	Heterodoxy	Market broke
15,037	Heterogeny	Market broke so fast, could not sell more than
15,038	Heterogyna	Market closed firmer with more inquiry
15,039	Heteromys	Market closed dull with less inquiry
15,040	Heterophyl	Market declining
15,041	Heterosis .	Market dull
15,042	Heuchera .	Market dull, but prices are firm
15,043	Heulandite	Market has an upward tendency
15,044	Hewhole .	Market has a downward tendency
15,045	Hexade .	Market has indications of a panic
15,046	Hexagyn .	Market is dull and unsettled
15,047	Hexagynian	Market is dull owing to
15,048	Hexagynous	Market is strong on
15,049	Hexahedral	Market is weak on
15,050	Hexamerous	Market opened weak, but closed strong
15,051	Hexameter	Market opened strong, but closed weak
15,052	Hexametric	Market quiet, but firm
15,053	Hexandrian	Market is improving

No.	Code Word.	Market (*continued*)
15,054	Hexangular	Market appears to have reached top
15,055	Hexapla .	Market appears to have reached the bottom
15,056	Hexapods	Market much affected by the report
15,057	Hexastich.	Market firming rapidly
15,058	Hexastylar	Market weakening rapidly
15,059	Hexyl . .	Market very sensitive [output this month
15,060	Heyday .	Market will probably rise in view of expected large
15,061	Heydeguy	Market will probably firm in view of expected large
15,062	Heypass .	Market still falling [output next month
15,063	Hiatus . .	Market opened and closed firm
15,064	Hibernacle	Market opened and closed weak
15,065	Hibiscus .	Market slightly better
15,066	Hiccough .	Market void of buying orders
15,067	Hiccuping	Market spirited, owing to receipt of foreign orders
15,068	Hickjoint .	Market affected by
15,069	Hickwall .	All stocks are advancing; the market is very strong
15,070	Hidden .	All stocks are declining; the market indicates a further downward tendency
15,071	Hiddenly .	All stocks are declining; market very weak
15,072	Hiddenness	How is the market? Please telegraph
15,073	Hideous .	If the market is strong
15,074	Hideously .	Telegraph if there is any change in the market
15,075	Hiderope .	As fast as the market will stand it
15,076	Hidrotic .	From appearance of the market
15,077	Hieracian .	From appearance of the market I think I had better buy. What is your advice
15,078	Hierapicra	From appearance of the market I think I had better sell. What is your advice
15,079	Hierarch .	From appearance of the market, stock will soon
15,080	Hierarchal	If the market is weak [decline
15,081	Hieratic .	Think there is a prospect for a good market
15,082	Hieratical.	The report has had a depressing effect on the
15,083	Hierochloa	To excite the market [market
15,084	Hieroglyph	The report has had a favourable effect on the
15,085	Hierolatry	I think the market will keep up [market
15,086	Hierologic	I think the market will not keep up
15,087	Hieromancy	To excite the market as little as possible
15,088	Hierophant	What do you think of the market [bottom
15,089	Hieroscopy	Watch the market, and buy quickly if it shows
15,090	Hierourgy	The circular depressed the market
15,091	Highaimed	Depressed the market considerably
15,092	Highaltar .	Raised the tone of the market
15,093	Highblest .	Panic in the market all round
15,094	Highbound	Had no effect on the market
15,095	Highbuilt .	Impossible to say whether market will rise or fall
15,096	Highcaste	There has been a complete panic in the market;
15,097	Highchurch	London market [our shares declined to
15,098	Highest .	Paris market
15,099	Highfed .	Berlin market
15,100	Highflier .	Continental markets

No.	Code Word.	**Market** (*continued*)
15,101	Highflown	Do not expect any improvement in market before
15,102	Highflying	Do not expect much improvement in market before
15,103	Highgoing	Banks pressing customers, so forcing market
15,104	Highheeled	No dependence to be placed on market
15,105	Highhorse	How is your market
15,106	Highlife	What are prospects of your market
15,107	Highlows .	Is your market to be relied on
15,108	Highmass .	What has deranged market
15,109	Highmen .	There is no market here for
15,110	Highminded	**Material(s)**
15,111	Highpalmed	Materials can be easily obtained
15,112	Highplaced	Shall have to send to —— for necessary materials to
15,113	Highpriest	No materials at hand for
15,114	Highproof	The necessary material for
15,115	Highraised	For the materials to
15,116	Highroad .	Is it material
15,117	Highseas .	Material to —— interests
15,118	Highsouled	Very material to the purpose
15,119	Highstrung	It is material that we should
15,120	Highswoln	What are the materials used
15,121	Hightaper .	Materials used are principally
15,122	Hightasted	Cost of materials and labour
15,123	Hightide .	**Matrix**
15,124	Hightoned	The matrix being
15,125	Highvoiced	The matrix is
15,126	Highway .	**Matter(s).** (See also Settle.)
15,127	Highwayman	Go into the matter carefully
15,128	Hilarate .	Have referred the matter to
15,129	Hilarious .	What is the matter with
15,130	Hilary . .	The matter has been settled by
15,131	Hillfever .	The matter is not yet settled
15,132	Hillfolk .	When the matter is settled
15,133	Hillfoot .	Till the matter is settled
15,134	Hillocky .	I (we) leave the matter to —— to arrange
15,135	Hillside .	The matter is now in hand
15,136	Hilltop .	(Will) have nothing to do with the matter
15,137	Hilsah . .	The matter will soon be settled
15,138	Himantopus	Matter(s) appear(s) very unsettled
15,139	Himself .	It will not matter
15,140	Hindberry	The matter is in the hands of
15,141	Hindbow .	The matter is no longer in the hands of
15,142	Hindcalf .	There is nothing the matter with
15,143	Hinderling	It does not matter
15,144	Hindleg .	It is a matter of the greatest importance
15,145	Hindooism	It is a matter of no importance
15,146	Hindrance	How do matters stand
15,147	Hingejoint	It is a matter of
15,148	Hinoideus	Settle the matter as we best can
15,149	Hipbath .	Settle the matter as you best can
15,150	Hipgout .	The matter is now out of our hands

No.	Code Word.	**Matter(s)** *(continued)*
15,151	Hipjoint .	Bring the matter forward (at the)
15,152	Hipknob .	Cannot matters be settled without litigation
15,153	Hipmould	Taken action in the matter
15,154	Hipparchia	The matter has been adjourned
15,155	Hipparion	The matter will be reconsidered
15,156	Hippelaph	Cannot you reconsider the matter
15,157	Hippides .	Expedite matters
15,158	Hippobosca	Is anything the matter (with)
15,159	Hippobroma	Considering the matter
15,160	Hippocamp	The matter must be settled (by)
15,161	Hippocras	Nothing to do with the matter
15,162	Hippogryph	We leave the matter to your discretion
15,163	Hippolith .	The matter has been left
15,164	Hipponyx .	**May**
15,165	Hippophagi	How soon may I (we)
15,166	Hippopus .	How much may I (we)
15,167	Hippuric .	How long may I (we)
15,168	Hippurites	When may I (we)
15,169	Hiprafter .	May I (we)
15,170	Hipshot .	May you
15,171	Hiptile .	I (we) may
15,172	Hircinous .	I (we) may not
15,173	Hirsute .	You may
15,174	Hirudinea	You may not
15,175	Hirudo .	You may not have
15,176	Hispid . .	May he (they)
15,177	Hispidity .	He (they) may
15,178	Hissingly .	If we may
15,179	Histeridae	If you may
15,180	Histiology	If he (they) may
15,181	Histogeny	If —— may
15,182	Histologic	If —— may not
15,183	Histolysis .	May I (we) do so
15,184	Historical .	If I (we) may do so
15,185	Historify .	May go up again
15,186	Histrionic	May go down
15,187	Hithermost	May decline
15,188	Hitherto .	In case it may
15,189	Hitherward	**Meagre**
15,190	Hoarding .	We have very meagre details
15,191	Hoarfrost .	**Mean(s)**
15,192	Hoariness	Do not understand what you mean in your letter of
15,193	Hoarstone	Do not understand what you mean in your telegram
15,194	Hoatzin .	By what means [of
15,195	Hobbadehoy	By which means
15,196	Hobbism .	What do (does) —— mean
15,197	Hobblebush	By all means
15,198	Hobbled .	By no means
15,199	Hobbleshow	Has (have) the means
15,200	Hobblingly	Has (have) not the means

No.	Code Word.	Mean(s) (*continued*)
15,201	Hobbyhorse	Take such means as you think best
15,202	Hobgoblin	Shall take other means (to)
15,203	Hobnail .	Take every possible means (to)
15,204	Hobnob .	What does it mean
15,205	Hobnobbing	If he means
15,206	Hockamore	If you mean
15,207	Hockday .	The means to be taken
15,208	Hockherb.	Has (have) taken means to
15,209	Hockleaf .	Some effectual means must be taken
15,210	Hocktide .	Must take other means
15,211	Hocus . .	Has (have) ample means at
15,212	Hocuspocus	Has (have) unlimited means
15,213	Hocussed .	The best means of knowing
15,214	Hocussing.	Means have been provided for
15,215	Hoddengray	Provide means for
15,216	Hoddypeak	What do the following words mean
15,217	Hodgepodge	Must take prompt means
15,218	Hodiern .	If we do not take prompt means
15,219	Hodiernal.	Cannot think that —— mean(s)
15,220	Hodman .	Tried all possible means
15,221	Hodmandod	Have you tried all possible means
15,222	Hoecake .	Means will not permit
15,223	Hoemother	Take energetic means
15,224	Hoffmanist	He means (they mean) to
15,225	Hogbacked	We mean to
15,226	Hogcote .	Do you mean to
15,227	Hogframes	The only means available
15,228	Hoggerpump	The only means possible
15,229	Hoggishly.	The most effectual means
15,230	Hogpeanut	Prompt and energetic means
15,231	Hogplums.	Means (are) adequate to
15,232	Hogreeve .	(If) no other means can be taken
15,233	Hogringer.	**Meaning**
15,234	Hogrubber	What is the meaning of
15,235	Hogsback .	The meaning is clear
15,236	Hogscore .	Our meaning cannot be clearer
15,237	Hogsheads	(To) make the meaning clear
15,238	Hogslard .	**Meant**
15,239	Hogsteer .	I (we) meant to say
15,240	Hogwallow	What was meant
15,241	Hogwash .	Was not meant
15,242	Hogweed .	What is meant (by)
15,243	Hohlspath	Who is meant
15,244	Hoiden .	Meant to convey our
15,245	Hoidenhood	**Meantime**
15,246	Hoidenish	In the meantime
15,247	Hoistway .	If in the meantime
15,248	Hoitytoity.	Shall be glad if, in the meantime
15,249	Hokerly .	**Measure**
15,250	Holaster .	Measure carefully

No.	Code Word.	**Measure** (*continued*)
15,251	Holdback .	Measure from top to
15,252	Holdbeam	Measure from bottom to
15,253	Holdfast .	Measure from centre to
15,254	Holding .	Measure ——— inches from top to
15,255	Holectypus	Measure ——— inches from bottom to
15,256	Holiday .	Measure ——— inches from centre to
15,257	Holingaxe	Measure ——— feet from top to
15,258	Hollandish	Measure ——— feet from bottom to
15,259	Hollowest.	Measure ——— feet from centre to
15,260	Holloweyed	Measure ——— yards from top to
15,261	Hollowly .	Measure ——— yards from bottom to
15,262	Hollowrail	Measure ——— yards from centre to
15,263	Hollowroot	**Measured**
15,264	Hollowspar	Has the ——— been measured
15,265	Hollyhocks	Cannot be measured exactly
15,266	Hollyoak .	Has (have) been measured up
15,267	Hollytrees	**Measurement(s)**
15,268	Holmoak .	Careful measurements have been taken
15,269	Holocaust.	Take accurate measurements
15,270	Holohedral	We find after taking accurate measurements
15,271	Holophotal	Careful measurements prove that
15,272	Holosteric.	**Measure(s).** (See also Means.)
15,273	Holosteum	Take strong measures
15,274	Holothuria	Such measures as may be necessary
15,275	Holstered .	What measures have been taken
15,276	Holusbolus	The measures we have taken
15,277	Holycross.	The measure was strongly supported
15,278	Holycruel.	The measures (by which) you propose
15,279	Holyfire .	Measures are being taken to
15,280	Holystone	Do not wish to take harsh measures
15,281	Holywater	**Mechanic(al)**
15,282	Holyweek.	A skilled mechanic
15,283	Homagejury	The master mechanic
15,284	Homagium	A competent master mechanic
15,285	Homarus .	A mechanic for general use in mines and mills
15,286	Homeblow	Suitable mechanics
15,287	Homeborn	(The) mechanical means
15,288	Homebrewed	(The) mechanical action
15,289	Homebuilt	Is perfect in its mechanical action
15,290	Homecircle	The action is purely mechanical
15,291	Homefelt .	From the mechanical point of view
15,292	Homegrown	**Mechanism**
15,293	Homeless .	The mechanism is very simple
15,294	Homelike.	The mechanism can be considerably improved
15,295	Homemade	**Medium**
15,296	Homeopathy	The ore is of medium quality
15,297	Homesick .	Only of medium quality
15,298	Homespun	A medium grade of ore, assaying
15,299	Homestall	Affording a medium for
15,300	Homethrust	The only medium available

No.	Code Word.	
15,301	Homeward	**Meet**
15,302	Homiletics	Meet me (us) at
15,303	Homilist .	Meet me (us) on arrival, at the —— hotel
15,304	Hominy .	Will meet you at
15,305	Homocerc	Will meet me (us) at
15,306	Homogamy	Will not be able to meet you at
15,307	Homogene	Will not be able to meet you at——, but will follow
15,308	Homogeneal	If —— do (does) not meet you
15,309	Homography	If you cannot meet
15,310	Homologate	Must have —— to meet current expenses
15,311	Homologous	Telegraph to —— to meet me at
15,312	Homomalous	Telegraph —— to the —— bank to meet me on my
15,313	Homonym	Can you meet —— at [arrival at
15,314	Homonymic	We have to meet
15,315	Homonymous	Have to meet all charges
15,316	Homophone	The charges we have to meet
15,317	Homoplasmy	When we next meet
15,318	Homopteran	Tell him to meet you
15,319	Homorgana	Has been told to meet
15,320	Homostyled	Cannot meet
15,321	Homotaxis	Was unable to meet
15,322	Homotony	Have to meet the bill
15,323	Homotropal	Meet our engagements
15,324	Homunculus	Meet his engagements
15,325	Honestate.	**Meeting**
15,326	Honestly .	Call a meeting of
15,327	Honey . .	Statutory meeting
15,328	Honeyant .	A public meeting
15,329	Honeybag .	Has (have) called a meeting of
15,330	Honeybees	Annual general meeting will be held
15,331	Honeyberry	Extraordinary general meeting
15,332	Honeycomb	The meeting was highly satisfactory
15,333	Honeycrock	At the meeting [telegraph
15,334	Honeydew	Please attend meeting of —— and report result by
15,335	Honeyeater	A meeting will be held on —— (to)
15,336	Honeygnats	A Board meeting
15,337	Honeyguide	The last Board meeting
15,338	Honeyless	When will the next Board meeting take place
15,339	Honeymonth	The next Board meeting will take place
15,340	Honeymoon	At the next Board meeting
15,341	Honeystalk	At a meeting held —— it was decided to
15,342	Honeysugar	Nothing was decided at the meeting
15,343	Honeysweet	A meeting will shortly be held to decide
15,344	Honorarium	Please attend the meeting, and watch our interests
15,345	Honorary .	A meeting of creditors
15,346	Honorific .	At a meeting of the creditors of —— payment of —— in the pound was offered
15,347	Honour .	A Committee meeting
15,348	Honourable	At the last meeting
15,349	Honoureth	The annual general meeting

No.	Code Word.	**Meeting** (*continued*)
15,350	Honouring	The half-yearly general meeting
15,351	Honourless	The half-yearly general meeting will be held
15,352	Hoodcap .	General meeting of shareholders
15,353	Hoodless .	Meeting of bondholders
15,354	Hoodlum .	Must be ready for the general meeting
15,355	Hoodmould	In time for the general meeting
15,356	Hoodsheaf	A meeting of the principal shareholders
15,357	Hoodwink	A private meeting [holders
15,358	Hoofbound	At a private meeting of a few of the largest share-
15,359	Hoofed .	General meeting held ; passed off satisfactorily
15,360	Hoofmark.	At the general meeting, which passed off very well, a vote of thanks was passed to yourself
15,361	Hookah .	At the meeting much dissatisfaction was expressed
15,362	Hookbeaked	The meeting passed off very well
15,363	Hookbilled	The meeting passed off better than we expected
15,364	Hookbone	It was resolved at the meeting
15,365	Hookedback	The meeting was stormy
15,366	Hookedness	The meeting was adjourned until
15,367	Hookladder	At the meeting, the Directors had to resign
15,368	Hookland .	At the meeting some uneasiness was felt
15,369	Hookmotion	Report was read at the meeting
15,370	Hookpin .	When will meeting be of
15,371	Hookrope.	Attend the meeting on my behalf
15,372	Hooksquid	At a meeting of the —— company
15,373	Hoopash .	Convene meeting of shareholders by advertisement
15,374	Hoopskirt.	(——) company meeting ; statements made thereat in favour of company highly misleading
15,375	Hopbinds.	(——) company meeting; statements made by chairman and others in favour of company are
15,376	Hopeful .	Meeting was not held [fair and true
15,377	Hopefully.	There was a great difference of opinion at the
15,378	Hopelessly	**Member(s)** [meeting
15,379	Hopfactor.	Is (are) member(s) of
15,380	Hopflea .	Is (are) not member(s) of
15,381	Hopgarden	Is (are) —— member(s) of
15,382	Hopingly .	By a member of
15,383	Hopkinsian	Member of Trades Union
15,384	Hoplotheke	Member of Parliament
15,385	Hopped .	Member of Council
15,386	Hopperboy	Member of Congress
15,387	Hoppickers	Member of the Committee
15,388	Hopping .	Member of the Legislature
15,389	Hoppocket	Member of the Government
15,390	Hoppoles .	Member of the Stock Exchange
15,391	Hopscotch	**Men**
15,392	Hoptrefoil	Short of men
15,393	Hopvine .	Require —— men to
15,394	Hopyards .	Require —— experienced men to
15,395	Horatian .	There are plenty of men
15,396	Hordeine .	How many men will be required

No.	Code Word.	**Men** (*continued*)
15,397	Hordeolum	Sufficient men will be sent
15,398	Horizon .	Send some good men
15,399	Horizontal	A great scarcity of good men
15,400	Hornbeak.	Can get plenty of good men
15,401	Hornbill .	Men are beginning to
15,402	Hornblende	Has (have) got a very good lot of **men**
15,403	Hornbugs.	Waiting for more men
15,404	Horned .	Stopped for want of men
15,405	Hornedowl	Will you send men, or shall we get them
15,406	Hornedpout	Men down with sickness
15,407	Hornfish .	Several men have died
15,408	Hornfoot .	We shall have trouble with the men
15,409	Hornify .	Are having trouble with the men
15,410	Hornmaker	Several of the leading men
15,411	Hornpipe .	**Mention**
15,412	Hornpock.	Do not mention to any one
15,413	Hornsilver	You had better mention (this) **to**
15,414	Hornslate.	Will mention the matter to
15,415	Hornspoon	Will make no mention
15,416	Hornthumb	You do not mention
15,417	Hornwrack	Did not mention
15,418	Hornyfrog	**Mentioned**
15,419	Hornywink	Has not been mentioned.
15,420	Horologist	Was mentioned to me privately
15,421	Horometry	Is not to be mentioned (at)
15,422	Horopter .	Was mentioned as a positive fact
15,423	Horoscopic	As mentioned in last
15,424	Horrendous	Think we heard it mentioned
15,425	Horrent .	The matter was never mentioned
15,426	Horrible .	Has been mentioned more than once
15,427	Horrified .	**Merchant(s)**
15,428	Horrifying	Is (are) the leading merchant(s) of this place
15,429	Horseboat	Is (are) leading merchant(s) of
15,430	Horseboxes	Is (are) merchant(s) in ——, but in a small **way**
15,431	Horsecloth	On behalf of ——, who is (are) merchant(s) **in**
15,432	Horsecrab	The vendors are merchants in
15,433	Horsefair .	Several merchants here
15,434	Horsefinch	**Merchantable**
15,435	Horseflesh	Not merchantable
15,436	Horsefly .	**Mercury.** (See Quicksilver.)
15,437	Horsegin .	Losing —— mercury per ton of ore (owing to)
15,438	Horseiron.	**Merit(s)**
15,439	Horseknave	The great merit is
15,440	Horseknop	On its merits
15,441	Horselaugh	Its chief merit is
15,442	Horseload	**Message**
15,443	Horsemeat	I (we) do not think you understand the message
15,444	Horsenails	Send following message to
15,445	Horsepath	Repeat the message
15,446	Horsepicks	Last message received from you was dated

No.	Code Word.	**Message** (*continued*)
15,447	Horseplay	My (our) last message
15,448	Horsepond	My (our) last message to you was dated
15,449	Horsepower	Have not received the message referred to (of)
15,450	Horserace	Has (have) received a message from —— saying
15,451	Horserugs	From message received
15,452	Horseshoe	Has (have) sent no message
15,453	Horsesugar	I (we) do not understand the message
15,454	Horsethief	A message was sent
15,455	Horsevetch	Message was received
15,456	Horseweed	Received no message from you
15,457	Horsewhips	The message was forwarded on the
15,458	Horsewoman	The message was delayed
15,459	Hostation .	When did you dispatch the message
15,460	Hortatory .	In reply to your message
15,461	Hortensial	**Messenger**
15,462	Hosanna .	Have sent a messenger
15,463	Hoseheeler	By a special messenger, if necessary
15,464	Hosiery .	**Met**
15,465	Hospitably	Was (were) met with
15,466	Hospital .	Met me (us) at
15,467	Hospitious	Has (have) not met
15,468	Hostages .	Met with
15,469	Hostelry .	Met by
15,470	Hostess .	We met, but could not agree
15,471	Hostillar .	**Metal(s)**
15,472	Hotbed .	The precious metals
15,473	Hotblast .	The precious metals are found here
15,474	Hotblooded	The baser metals
15,475	Hotbrained	The only metals which are found
15,476	Hotchpotch	**Metallic**
15,477	Hotcockles	Of a metallic nature
15,478	Hotflue .	Metallic alloy(s)
15,479	Hotheaded	**Metalliferous**
15,480	Hothouse .	Metalliferous deposits
15,481	Hotlivered	Metalliferous veins
15,482	Hotmouthed	The metalliferous district includes
15,483	Hotpress .	**Metamorphic**
15,484	Hotshort .	Metamorphic slate
15,485	Hotspurred	The foot-wall is metamorphic slate
15,486	Hotwalls .	The hanging-wall is metamorphic slate
15,487	Hotwater .	The formation is metamorphic slate
15,488	Hounded .	**Method**
15,489	Hourangle	The method by which the gold is extracted
15,490	Hourglass.	I want to go to —— to see a new method for
15,491	Hourhands	A new method of extracting the —— from the
15,492	Hourline .	A great saving is accomplished by the method which
15,493	Hourly . .	By this method [has been adopted
15,494	Hourplate	(By) the present method
15,495	Houseagent	The method now used is
15,496	Housedog .	(By) the new method of treatment

No.	Code Word.	**Method** (*continued*)
15,497	Households	What about the new method
15,498	Houselamb	Discarded the old method
15,499	Houseleeks	This method will not suit
15,500	Houseling .	We think the method will suit
15,501	Housemaid	There is no other method
15,502	Houseroom	We are now trying the —— method
15,503	Housewarm	We recommend you to try this method
15,504	Housewife .	**Micaceous**
15,505	Hovel . .	Micaceous schist
15,506	Hovelhouse	Micaceous rock
15,507	Hovelled .	**Michaelmas** (day)
15,508	Howdie .	Up to Michaelmas
15,509	However .	Next Michaelmas
15,510	Howitzer .	**Middle**
15,511	Howsoever	In the middle (of)
15,512	Hubbub .	From the middle (of)
15,513	Hubbubboo	Through the middle (of)
15,514	Huckaback	At the middle (of)
15,515	Hucklebone	**Midst**
15,516	Huckstress	In the midst of
15,517	Huffcap .	Now in the midst of preparations for
15,518	Huffish .	**Midsummer** (day)
15,519	Huffishly .	Up to midsummer day
15,520	Hugeness .	Next midsummer
15,521	Hullabaloo	**Midwinter**
15,522	Humane .	**Might**
15,523	Humanely	Might I (we)
15,524	Humanics .	Might he (they)
15,525	Humanified	You might
15,526	Humanistic	He (they) might
15,527	Humanized	Might I (we) not
15,528	Humanizing	It might
15,529	Humankind	It might not
15,530	Humation .	Might be able to
15,531	Humbird .	If it might not be
15,532	Humbleness	If it might be
15,533	Humblepie	Might be
15,534	Humbleth .	Might it be
15,535	Humbling .	It might be done
15,536	Humbly .	It might lead to
15,537	Humbugged	Might not be
15,538	Humbugging	Might it not be
15,539	Humectant	What might
15,540	Humective	Which might
15,541	Humefied .	Which might be
15,542	Humefying	Which might not
15,543	Humhum .	**Mile(s)**
15,544	Humidity .	Is about —— miles from the mine(s)
15,545	Humidness	Miles from
15,546	Humifuse .	The —— mine(s) is (are) —— miles from

No.	Code Word.	**Mile(s)** (*continued*)
15,547	Humiliated	The mill is —— miles from the mine(s)
15,548	Hummeller	How many miles is (are) the mine's) from
15,549	Hummingtop	The last —— miles of the journey, the road is very
15,550	Hummocky	Can be traced for many miles [bad
15,551	Humoral .	The first —— miles
15,552	Humoralism	The last —— miles are over the mountains
15,553	Humoristic	Less than a mile
15,554	Humorous	More than a mile
15,555	Humorously	More than —— miles (long)
15,556	Humorsome	Square miles
15,557	Humpbacked	The property has an area of —— square miles
15,558	Humph .	Is —— miles from the nearest railroad
15,559	Hunchback	From thence —— miles on horseback
15,560	Hungbeef .	Miles over bad roads
15,561	Hungerbit.	Miles of fairly good road
15,562	Hungerly .	The area is —— miles long by —— miles wide
15,563	Hunkerism	How many miles is it from
15,564	Huntingbox	Has to be hauled —— miles
15,565	Huntress .	A journey of —— miles
15,566	Huntsman	There are —— miles of very good road
15,567	Huontree .	There are —— miles of very bad road
15,568	Huraulite .	**Mill(s).** (See also Stamps.)
15,569	Hurdlerace	5 stamp mill —— (or) 5 stamps
15,570	Hurlbat .	10 stamp mill —— (or) 10 stamps
15,571	Hurlwind .	15 stamp mill —— (or) 15 stamps
15,572	Hurlyburly	20 stamp mill —— (or) 20 stamps
15,573	Huronian .	25 stamp mill —— (or) 25 stamps
15,574	Hurrah .	30 stamp mill —— (or) 30 stamps
15,575	Hurricane.	35 stamp mill —— (or) 35 stamps
15,576	Hurtlessly .	40 stamp mill —— (or) 40 stamps
15,577	Hurtsickle	50 stamp mill —— (or) 50 stamps
15,578	Husband .	60 stamp mill —— (or) 60 stamps
15,579	Husbandage	70 stamp mill —— (or) 70 stamps
15,580	Husbandman	80 stamp mill —— (or) 80 stamps
15,581	Husbandry	90 stamp mill —— (or) 90 stamps
15,582	Hushaby .	100 stamp mill —— (or) 100 stamps
15,583	Hushing .	120 stamp mill —— (or) 120 stamps
15,584	Hushmoney	150 stamp mill —— (or) 150 stamps
15,585	Husker	Mill is (mills are) running on ore from
15,586	Huskingbee	Mill returns show
15,587	Hustled .	Have based my calculation on the mill run for the
15,588	Huswifely .	There is a —— stamp mill [last year
15,589	Hutchinsia	A new —— stamp mill required
15,590	Huzzaing .	The mill can crush —— tons in twenty-four hours
15,591	Hyaenodon	Mill stopped for repairs on the ——, will start work
15,592	Hyaline .	Mill starts again on [again on
15,593	Hyalomelan	Mill stopped crushing to clean up
15,594	Hyalotype	Has (have) used his (their) mill to crush ore from
15,595	Hybernate	The present mill is [neighbouring mines
15,596	Hyblaean .	There must be a new mill

No.	Code Word.	**Mill(s)** (*continued*)
15,597	Hybodont	Mill working splendidly
15,598	Hybodus .	Mills are idle from want of water
15,599	Hybrid. .	Mill stopped for boiler to be cleaned and repaired
15,600	Hybridisms	The present mill is of no use
15,601	Hybridity .	With the present incomplete mill
15,602	Hybridizer	Mill has been working day and night for —— days, and has crushed —— tons in that time
15,603	Hycsos .	Mill working —— hours per day
15,604	Hydage .	It will be necessary to erect a —— stamp mill
15,605	Hydatism .	Mill closed in consequence of
15,606	Hydracid .	Mill arrived, will start at once erecting it
15,607	Hydraform	Chilian mill
15,608	Hydragogue	Globe mill
15,609	Hydrangea	Huntington centrifugal roller quartz mill
15,610	Hydrastis .	Wiswell mill
15,611	Hydration	Gold mill with plates and vanners
15,612	Hydratuba	Wet stamp mill
15,613	Hydraulics	Dry stamp mill
15,614	Hydriodic.	Cannot work the mill full time owing to
15,615	Hydrocele	Mill working day and night
15,616	Hydracyst	Mill crushed in that time
15,617	Hydroecium	Mill working again, plenty of water
15,618	Hydrogen.	Mill working again, think it will be all right
15,619	Hydrology	Working mill to its utmost capacity
15,620	Hydromania	The mill will work night and day
15,621	Hydromancy	The present mill-returns satisfactory
15,622	Hydromys	The present mill-returns unsatisfactory
15,623	Hydropathy	During the week mill worked —— days, crushed —— tons, yielded —— ozs. gold, and —— ozs. silver; expenses
15,624	Hydrophoby	During the month mill worked —— days, crushed —— tons, yielded —— ozs. gold, and —— ozs. silver; expenses
15,625	Hydrophyte	Value of the production of the mill
15,626	Hydropical	Mill running on high-grade ore
15,627	Hydropsy .	Mill running on low-grade ore
15,628	Hydropult	Mill returns will improve
15,629	Hydrorhiza	Sufficient ore for the mill
15,630	Hydroscope	If the mill is to be kept going
15,631	Hydrosoma	The mill(s) cannot be kept going
15,632	Hydrostat.	In order to keep the mill going
15,633	Hydrotheca	Have ordered new mill
15,634	Hydrous .	Have begun to erect new mill
15,635	Hydroxyde	Grading for the new mill
15,636	Hydrozoal	Think we shall complete new mill
15,637	Hyemation	Contract for the new mill
15,638	Hyenadog	Estimate for the new mill
15,639	Hyetometer	Mill broken down owing to accident to machinery
15,640	Hygeian .	Mill broken down; repairs now being done
15,641	Hygienic .	**Mill site**

No.	Code Word.	**Mill site** (*continued*)
15,642	Hygrometry	Mill site is in good situation ; plenty of water
15,643	Hyleosaur	Mill site is in a poor situation and water is scarce
15,644	Hylism .	Mill site very bad ; inaccessible
15,645	Hylogeny .	Mill site good ; with existing mill
15,646	Hylonomus	There is a good situation for a mill site
15,647	Hylozoical	The mill site is close to the river
15,648	Hylozoism	Water can be conveyed to the mill site
15,649	Hymenium	**Milling**
15,650	Hymenopter	Milling delayed owing to
15,651	Hymenotomy	Milling ore extracted during the past
15,652	Hymnals .	What will be cost of mining and milling the ore with the present appliances
15,653	Hymnbook	I estimate the cost of mining and milling the ore with the present appliances at
15,654	Hymnody .	Always sufficient water for milling purposes
15,655	Hyoideal .	Not sufficient water for milling purposes
15,656	Hyopotamus	Milling has turned out well
15,657	Hypallage .	Milling has turned out badly
15,658	Hypanthium	Cost of milling per ton
15,659	Hypaspist .	Cannot you reduce the cost of milling
15,660	Hyperaemia	Cost of milling has been reduced to —— per ton
15,661	Hyperbaton	Cost of milling has increased to —— per ton
15,662	Hyperbolic	We have been milling ore (from)
15,663	Hyperdulia	We are now milling ore from
15,664	Hyperite .	Suitable for milling
15,665	Hypermeter	Not worth milling
15,666	Hyperstene	What is the total yield and cost of milling
15,667	Hypethral	The ore we are now milling
15,668	Hyphasma	**Milled**
15,669	Hyphened	Amount of ore milled
15,670	Hyphening	Grade of ore milled
15,671	Hypnotism	Value of ore milled
15,672	Hypnotized	Tonnage and value of ore milled
15,673	Hypnum .	Milled —— ; value
15,674	Hypoblast	Milled during
15,675	Hypobole .	**Mill test**
15,676	Hypochil .	First mill test
15,677	Hypocotyl	Second mill test
15,678	Hypocrisy	Last mill test
15,679	Hypocritic	Mill test on ores from
15,680	Hypodermal	Mill test will be made
15,681	Hypogeum	Shall make a mill test
15,682	Hypogynous	Shall make another mill test
15,683	Hyponasty	After a prolonged mill test
15,684	Hypophet .	Several mill tests have been made
15,685	Hypophysis	Result of mill tests satisfactory
15,686	Hypopyon	Result of mill tests unsatisfactory
15,687	Hypostatic	Result of mill tests disappointing
15,688	Hypostroma	Result of mill test disappointing but think it is
15,689	Hypostyle	Result of first mill test　　　　　　[unreliable

No.	Code Word.	**Mill test** (*continued*)
15,690	Hypotenuse	Result of second mill test
15,691	Hypothec .	Mill test was not reliable
15,692	Hypsodon	What was the result of the mill test
15,693	Hyracium .	The value of the ore according to the mill test
15,694	Hyracoidea	**Mind** [was
15,695	Hyrax . .	You must bear in mind
15,696	Hyssop .	Do not mind
15,697	Hyssopus .	If you would not mind
15,698	Hysteric .	Bearing in mind
15,699	Hysterical	Bearing in mind how important it is
15,700	Hysteroid .	**Mine(s)**
15,701	Ibigau . .	The mine(s) is (are) [ore is discovered
15,702	Iceanchor.	The mine is exhausted unless another streak of pay-
15,703	Icebergs .	The mine(s) is (are) well worthy of investigation
15,704	Iceblink .	The mine(s) is (are) not worthy of investigation
15,705	Iceboat .	The mine(s) look(s) splendid
15,706	Icebound .	The mine(s) is (are) situated in a district having a
15,707	Icebrook .	On the mine [reputation for good properties
15,708	Icebuilt .	The mine(s) is (are) not situated in a district cele-
15,709	Icechisel .	In the mine [brated for good properties
15,710	Icecold .	I consider the —— mine(s) very valuable
15,711	Icedcream	The mine(s) is (are) worked out
15,712	Icedrops .	I strongly recommend the purchase of the ——
15,713	Icefender .	At the mine(s) [mine(s) for
15,714	Icefern. .	I do not consider the —— mine(s) worth more than ——; a sum I feel sure the vendors will
15,715	Icefield .	How is (are) the mine(s) looking [accept
15,716	Icefloe. .	Will be necessary to thoroughly equip the mines
15,717	Icefoot. .	The mine is not worth
15,718	Iceglazed .	Mine(s) is (are) full of water
15,719	Icehooks .	From all parts of the mine
15,720	Icehouse .	I cannot advise the purchase of the mine(s)
15,721	Iceisland .	I advise purchase of the mine(s) for
15,722	Icemaster .	The mine(s) in its (their) present condition
15,723	Icepails .	The mine(s) fall(s) far short of the report by
15,724	Iceplant .	The mine has been so badly worked that there is great danger of its caving
15,725	Iceplough	The mine(s) is (are) within easy access of
15,726	Icequake .	The mine(s) is (are) worth the price
15,727	Icesafe. .	The mine(s) is (are) not worth the price
15,728	Icespar .	The mine has been relocated
15,729	Icetable .	Have a mine of undoubted value offered me for ——; will pay a dividend of —— per cent. per annum. Could you do anything with it
15,730	Icetongs .	Look about quietly, and if you can get hold of a thoroughly good —— mine for about —— let me know. I would like to have a really good
15,731	Icewater .	Mine has improved [property to offer
15,732	Ichneumon	The mine does not look as well as I expected
15,733	Ichnolite .	Mine does not look encouraging

No.	Code Word.	Mine(s) (*continued*)
15,734	Ichnology.	Mine is being badly managed
15,735	Ichorous .	Mine(s) is (are) being worked all right
15,736	Ichthine .	Mine(s) is (are) better than I expected
15,737	Ichthulin .	Mine(s) is (are) closed
15,738	Ichthyic .	Mine(s) is (are) getting into rich ore
15,739	Ichthyocol	Mine(s) is (are) improving
15,740	Ichthyodea	Mine(s) is (are) looking poor
15,741	Ichthyosis.	Mine(s) is (are) looking better
15,742	Ichthys .	Mine(s) is (are) looking worse
15,743	Icicle . .	Mine(s) is (are) looking well
15,744	Iconisms .	Mine(s) is (are) almost exhausted
15,745	Iconize .	Must shut down the mine
15,746	Iconoclasm	It is reported that the mine has been shut down
15,747	Iconolatry	Admit no one into the mine(s)
15,748	Icosander .	Have examined the —— mine(s)
15,749	Icosandria	Nothing new in the mine
15,750	Icterical .	Keep me (us) posted of every change in the mine(s)
15,751	Icteritous .	No truth in the reported developments in the mine(s)
15,752	Icteroid .	From other parts of the mine samples taken assay
15,753	Ictides . .	Nothing new in any of the mines
15,754	Icypearled	Average samples from different parts of the ——
15,755	Idealist .	Keep the mine(s) closed [mine(s) assay
15,756	Idealistic .	The present condition of the mine would not justify
15,757	Ideality .	I consider the developments in the mine(s) fully justify the erection of —— stamps, in addition to those already in use
15,758	Idealogue.	A better mine than
15,759	Ideation .	Indications in the mine show that we must be near
15,760	Ideational.	There is nothing in the mine to justify the advance
15,761	Identic .	In the neighbourhood of the mine(s)
15,762	Identical .	This mine (these mines)
15,763	Identified.	Mine(s) not good enough
15,764	Identity .	In different parts of the mine
15,765	Ideogram .	If the —— mine(s) is (are) all right
15,766	Ideography	Nothing wrong in the mine(s)
15,767	Ideologist.	No merit in the mine(s)
15,768	Ideology .	Cannot get into the mine(s), but from reliable in-
15,769	Ideomotion	I was in the mine(s) [formation I learn that
15,770	Idiocrasy .	Cannot get into —— mine(s)
15,771	Idiocratic .	The mine(s) has (have) very good indications of
15,772	Idiocy . .	Have stopped work on the —— in the mine
15,773	Idiopathy .	Have struck good ore in the —— mine(s)
15,774	Idiotish .	Have struck it rich in the —— mine(s)
15,775	Idiotype .	Have struck quartz in the —— mine(s)
15,776	Idleheaded	Have struck the vein in the —— mine(s)
15,777	Idlely . .	Have been in the —— mine to-day, and think the recent developments show
15,778	Idleness .	Has been in the —— mine(s). See him and learn all you can about it (them) [on it (them)
15,779	Idlewheel .	All who have visited mine(s) report very favourably

No.	Code Word.	**Mine(s)** (*continued*)
15,780	Idling	Go into the —— mine(s) at once, examine the reported developments, and telegraph me (us) promptly [mine(s) situated at
15,781	Idolatrize	I have arranged for your inspection of the ——
15,782	Idolatrous	Estimate value of —— mine in its present state at
15,783	Idolatry	Containing rich silver mines
15,784	Idolfires	Mine is not what it was represented to be
15,785	Idolify	It is a very poor mine; won't pay to work
15,786	Idolized	All who have visited the mine report
15,787	Idolizing	Examined mine; find statements grossly exaggerated [stantially correct
15,788	Idolous	Have examined mine, and find representations sub-
15,789	Idolshell	What do you think of the —— mine
15,790	Idrialine	Will examine the —— mine and report at once
15,791	Idyll	A valuable gold mine
15,792	Idyllic	A valuable silver mine
15,793	Igneous	A valuable copper mine
15,794	Ignifluous	A valuable silver-lead mine
15,795	Ignigenous	The mines included in the concession
15,796	Ignite	Running expenses of the mine
15,797	Ignitible	In the lower workings of the mine
15,798	Ignivomous	Reports of the mine being exhausted
15,799	Ignobility	Picking out the eyes of the mine
15,800	Ignominy	Have picked out the eyes of the mine
15,801	Ignoramus	The mine is best reached viâ
15,802	Ignorance	No other mine in the vicinity worthy of trial
15,803	Ignorantly	Mine is being worked by
15,804	Ignored	Mine has been worked by —— and paid well
15,805	Ignorement	Mine has produced in the past
15,806	Ignoring	The operations in the mine
15,807	Ignoscible	The developments of the mine fully justify the expectations which have been formed
15,808	Iguanidae	State if mine from miner's standpoint can fairly be
15,809	Iguanodon	The condition of the mine [worked
15,810	Ilicine	Mine must be fairly worked in a miner-like manner
15,811	Ilixanthin	Has mine (mine has) been fairly and properly
15,812	Illadvised	Telegraph opinion and value of the mine [worked
15,813	Illapsable	The mine has been salted
15,814	Illapse	Was the mine salted
15,815	Illaqueate	The mine must have been salted
15,816	Illation	Could they have salted the mine
15,817	Illative	Send reliable miner to ascertain the intrinsic value
15,818	Illatively	Mine equipments [of the mine
15,819	Illaudable	The mine is equipped with
15,820	Illblood	**Miner(s)**
15,821	Illbred	Miners are
15,822	Illcontent	Miners demand
15,823	Illecebrum	Are good miners
15,824	Illegalize	Are bad miners [for the work
15,825	Illegibly	It will be necessary to engage good English miners

No.	Code Word.	**Miner(s)** (*continued*)
15,826	Illerected .	Much sickness among the miners
15,827	Illeviable .	Miners have struck ; they demand —— extra
15,828	Illfated .	In a miner-like manner [wages per day
15,829	Illgot . .	The miners here are a lawless set, and likely to give
15,830	Illhumour .	Impossible to get good miners [much trouble
15,831	Illiberal .	The miners prefer working on contract
15,832	Illicit . .	Miners will not work on contract
15,833	Illicitly .	We are likely to have trouble with the miners
15,834	Illicitous .	The miners are giving a good deal of trouble
15,835	Illicium .	Miners are paid —— per day
15,836	Illighten .	From a miner's standpoint
15,837	Illimited .	The Miners' Union
15,838	Illiquid .	**Mineral**
15,839	Illiteracy .	The mineral belt of
15,840	Illjudged .	The mineral district of
15,841	Illlived . .	The mineral zone
15,842	Illlooked .	A very rich mineral district
15,843	Illlooking .	Mineral oil
15,844	Illluck . .	**Minimum**
15,845	Illmatched	The minimum price
15,846	Illmeaning	The minimum production
15,847	Illnature .	As a minimum
15,848	Illocable .	The minimum cost
15,849	Illocality .	Has reached the minimum since
15,850	Illogical .	**Mining** [—— per ton
15,851	Illomened .	The cost of mining and milling the ore has been
15,852	Illset . .	Estimated cost of mining and milling
15,853	Illstarred .	Placer mining [primitive fashion
15,854	Illtimed .	The mining in this district is carried on in a very
15,855	Illtreat .	Mining interests are looking up
15,856	Illuminary	The cost of mining and milling, with the improvements I have specified, would be, I estimate, ——
15,857	Illuminate	Have resumed mining operations [per ton
15,858	Illuming .	Facilities for mining exceptionally good
15,859	Illuminism	The mines are situated in the mining district of
15,860	Illured . .	Mining operations suspended
15,861	Illusory .	Mining operations not yet commenced
15,862	Illustrate .	Mining stocks generally strong
15,863	Illwill . .	Mining stocks generally weak
15,864	Illwisher .	Mining can only be carried on for —— months in
15,865	Ilmenite .	Is a fair mining venture [the year
15,866	Imageable	What is the cost of mining and milling ore
15,867	Imageless .	Mining properties
15,868	Imagemaker	The mining rights
15,869	Imageman	A good mining captain
15,870	Imagery .	**Mining engineer(s)**
15,871	Imagined .	Send mining engineer
15,872	Imaginous	Mining engineer will be sent to examine property
15,873	Imbanking	Is (are) mining engineer(s) of great repute
15,874	Imbannered	Is a (are) mining engineer(s) of no repute

No.	Code Word.	**Mining engineer(s)** (*continued*)
15,875	Imbecility	Has (have) the reputation of being mining engineer(s) of ability and integrity
15,876	Imbellic .	Has (have) the reputation of being mining engineer(s) of ability, but unscrupulous
15,877	Imbenching	Mining engineer has examined, and confirms reports
15,878	Imbibe .	Mining engineer does not confirm reports
15,879	Imbibing .	**Minister**
15,880	Imblaze .	The British Minister at
15,881	Imboil . .	The Minister for Foreign Affairs
15,882	Imbonity .	The Minister of Public Works
15,883	Imbordered	The Minister of Commerce
15,884	Imbruement	The United States Minister at
15,885	Imbruing .	The Minister for
15,886	Imburse .	**Minority**
15,887	Imbution .	The Board were in a minority
15,888	Imitancy .	Our opponents were in a minority
15,889	Imitated .	A discontented minority
15,890	Imitations	A small, but influential, minority
15,891	Imitative .	**Mint**
15,892	Imitatress.	To the mint
15,893	Imitatrix .	All our bullion is sold to the mint
15,894	Immaculate	At the mint of
15,895	Immailed .	The bullion is shipped to the mint at
15,896	Immanifest	To what mint is the bullion shipped
15,897	Immaturely	At the mint price
15,898	Immensely	At the mint price of gold
15,899	Immersed.	The English mint price of
15,900	Immersible	The United States mint price of
15,901	Immeshing	**Minute(s)**
15,902	Immethoded	How many minutes
15,903	Immingle .	How many strokes per minute
15,904	Immingling	How many revolutions per minute
15,905	Immix . .	How many feet per minute
15,906	Immixable	Is making —— strokes per minute
15,907	Immobile .	Is making —— revolutions per minute
15,908	Immobility	Makes —— revolutions per minute and throws
15,909	Immoderacy	Feet per minute
15,910	Immoderate	Number of gallons per minute
15,911	Immodest.	Speed of engine in revolutions per minute
15,912	Immodestly	Speed of piston in feet per minute
15,913	Immolating	**Minutely**
15,914	Immolation	Must be minutely examined
15,915	Immortally	Have gone most minutely into all the details of
15,916	Immould .	You must go very minutely into all the details
15,917	Immunities	Had not time to go very minutely into all the details
15,918	Immunity .	**Misapprehension**
15,919	Immutably	There has been some misapprehension
15,920	Impackment	There has been no misapprehension
15,921	Impact .	Has there not been some misapprehension
15,922	Impaireth .	Labouring under a misapprehension

M

No.	Code Word.	
15,923	Impairing .	**Miscellaneous**
15,924	Impalsied .	Miscellaneous expenses
15,925	Impalsy .	A miscellaneous collection of
15,926	Impannel .	**Misconception**
15,927	Imparlance	Owing to a misconception on the part of
15,928	Impartial .	It was a misconception on ——'s part
15,929	Impassably	**Misconduct**
15,930	Impatience	Dismissed for gross misconduct
15,931	Impavid .	Through his gross misconduct
15,932	Impavidly	**Misconstruction**
15,933	Impeach .	By a misconstruction of what was said
15,934	Impeacheth	Owing to a misconstruction
15,935	Impeaching	**Misconstrue(d)**
15,936	Impeccancy	Misconstrue the meaning
15,937	Impeccant	You have misconstrued what we
15,938	Impede .	**Misconstruing**
15,939	Impedible	**Misfortune**
15,940	Impeding .	Has (have) had the misfortune to
15,941	Impellent .	A great misfortune to have
15,942	Impendence	It would be a great misfortune to
15,943	Imperatory	**Mishap**
15,944	Imperfect .	Mishap to battery
15,945	Imperially	Mishap to pump
15,946	Imperilled	Mishap to engine
15,947	Impersonal	**Misinformed**
15,948	Impervious	You have been totally misinformed (as to)
15,949	Impester .	I (we) have been totally misinformed (as to)
15,950	Impetigo .	Has (have) been misinformed (as to)
15,951	Impetrable	**Mislead**
15,952	Impetus .	Intended to mislead
15,953	Impeyan .	**Misleading**
15,954	Impicture .	The accounts are very misleading
15,955	Impiety .	The accounts of the mine(s) are very misleading, although not absolutely untrue
15,956	Impinguate	Some very misleading statements have appeared in
15,957	Impiously .	**Misled** [the papers
15,958	Impish . .	We have been misled (by)
15,959	Impishly .	Appears to have been misled (by)
15,960	Implements	Do not be misled (by)
15,961	Implexous	Do not allow shareholders to be misled (by)
15,962	Implicit .	**Mismanaged**
15,963	Implicitly .	Was (has been) mismanaged in every way
15,964	Impliedly .	Has been shamefully mismanaged
15,965	Implorator	**Mismanagement**
15,966	Imploring .	Through gross mismanagement
15,967	Impocket .	The mismanagement has been extreme
15,968	Impolarily	Mismanagement has been stopped
15,969	Impolicy .	**Misrepresent(ed)**
15,970	Impolite .	Do not misrepresent
15,971	Imporosity	Have been greatly misrepresented

No.	Code Word.	**Misrepresent(ed)** (*continued*)
15,972	Imporous .	Do not allow the case to be misrepresented
15,973	Importless	**Misrepresentation(s)**
15,974	Impose .	Misrepresentation of the facts of the case
15,975	Imposement	**Miss**
15,976	Imposingly	Do not miss
15,977	Impositor .	Shall miss
15,978	Impossible	Will not miss
15,979	Imposthume	If you miss
15,980	Impostrix .	You should not miss
15,981	Impostured	**Missed**
15,982	Impotently	Have you missed
15,983	Impound .	Missed the steamer (owing to)
15,984	Impoundage	Missed the mail (owing to)
15,985	Impounding	Missed hearing from
15,986	Impoverish	Missed seeing
15,987	Imprecated	Has (have) missed
15,988	Impregnant	**Missing**
15,989	Impress .	The —— is (are) missing
15,990	Impresseth	How much is missing
15,991	Impressive	The inclosure in your last letter, relating to —— is missing; telegraph synopsis
15,992	Imprimatur	Has arrived; parts missing are
15,993	Imprimery	Send us missing parts at once
15,994	Imprisoned	Have received the missing
15,995	Improbity .	Send copy of the missing
15,996	Improduced	The missing parts
15,997	Improperly	Missing documents
15,998	Improve .	Missing report
15,999	Improving	**Miss-statements**
16,000	Improviser	Is full of miss-statements
16,001	Imprudence	Such miss-statements do a great deal of harm
16,002	Impuberal	Through the miss-statements made (by)
16,003	Impudent .	**Mistake(s)**
16,004	Impugn .	You have made a mistake
16,005	Impugnable	Is not —— a mistake
16,006	Impugneth	Is a mistake; it ought to have been
16,007	Impugning	Please correct mistake immediately
16,008	Impugnment	There is some mistake (in)
16,009	Impulsion .	There is some mistake in the figures; send corrected
16,010	Impulsive .	Has the mistake been found out [statement
16,011	Impunctate	There is not any mistake
16,012	Impurely .	Has (have) made a mistake [read
16,013	Impureness	Have made a mistake in telegram of ——; for ——
16,014	Imputative	Is not —— word in your telegram of —— a mistake? Telegraph correctly at once
16,015	Imputing .	A mistake has been made in
16,016	Inability .	There must be a mistake in your telegram of ——; are the following words correct? if not, telegraph
16,017	Inaccuracy	We made a mistake [the right words at once
16,018	Inaccurate	If you find any mistake in

No.	Code Word.	**Mistake(s)** (*continued*)
16,019	Inactivity .	It would be a great mistake to
16,020	Inadequacy	It was a great mistake to
16,021	Inadequate	Was sent by mistake by
16,022	Inadherent	Is there not some mistake (in)
16,023	Inadhesion	Made a mistake in the figures
16,024	Inanimate .	There is a mistake in description
16,025	Inanition .	There is no mistake in description
16,026	Inapathy .	There was a mistake in description
16,027	Inapertous	There was no mistake in description
16,028	Inapposite	There was a mistake in
16,029	Inapt . .	**Misunderstand**
16,030	Inaptness .	You misunderstand our views
16,031	Inaquation	Misunderstand(s) the instructions given
16,032	Inasmuch .	**Misunderstanding**
16,033	Inaudibly .	There is evidently a misunderstanding about
16,034	Inaugur .	Misunderstanding between
16,035	Inbeaming	No misunderstanding at all
16,036	Inbeing .	Let there be no misunderstanding
16,037	Inblown .	Will lead to misunderstanding and cause confusion
16,038	Inboard .	**Misunderstood**
16,039	Inbreeding	You must have misunderstood
16,040	Incapably .	Has (have) misunderstood
16,041	Incarnate .	Misunderstood —— instructions
16,042	Incask . .	Misunderstood our meaning
16,043	Incautious	**Mix**
16,044	Incavated .	Do not mix up the two matters
16,045	Incaverned	Do not mix up
16,046	Incendiary .	**Mixed**
16,047	Incendious	The ore is largely mixed with
16,048	Incensed .	Mixed up with
16,049	Incensory .	So as not to be mixed up with
16,050	Inception .	The accounts are so mixed up, you can tell nothing
16,051	Inceptive .	**Mode**
16,052	Incertum .	The mode in which they work the
16,053	Incessancy	Go to —— and see their mode of working the
16,054	Inchamber	The mode in which they have hitherto worked is
16,055	Inchested .	The mode of working
16,056	Inchmeal .	The mode in which the accounts are (were) kept
16,057	Inchoately	**Model**
16,058	Inchoation	A model of the mine
16,059	Inchoative	A model of the mill
16,060	Inchpin .	A model of the machine
16,061	Incident .	A model to the scale of —— inch(es) to the foot
16,062	Incidently .	**Moderate(s)**
16,063	Incipiency	At a moderate computation
16,064	Incipient .	At a moderate expense
16,065	Incisely . .	Must be very moderate
16,066	Incitative .	A very moderate
16,067	Incitement	A moderate output
16,068	Incivility .	A moderate increase in the

No.	Code Word.	**Moderate(s)** *(continued)*
16,069	Incivism .	A moderate profit
16,070	Inclemency	When the weather moderates
16,071	Inclement.	**Moderated**
16,072	Inclineth .	Has been moderated
16,073	Inclip . .	The weather has moderated
16,074	Inclipped .	**Moderation**
16,075	Inclipping	By proceeding with moderation
16,076	Include .	**Modification(s)**
16,077	Includible	What modification would be required
16,078	Including .	Not much modification
16,079	Incoact .	With but little modification
16,080	Incocted .	Some modification will be necessary
16,081	Incogitant.	Have you any modifications to make
16,082	Incolumity	**Modify**
16,083	Income .	Does it modify your views
16,084	Incometax	Will you modify your
16,085	Incoming .	See no cause to modify our
16,086	Incompact	Modify so far as
16,087	Incomplete	Intended to modify
16,088	Incomposed	**Modified**
16,089	Inconcoct	Our views are not modified
16,090	Incorpse .	Will be modified in respect to
16,091	Incorpsing	Cannot be modified
16,092	Incorrupt .	Considerably modified
16,093	Increase .	The whole arrangement must be modified
16,094	Increasing	Must be modified
16,095	Incredibly	Modified with a view to
16,096	Increscent	**Moment**
16,097	Incruental	A favourable moment
16,098	Incrustate.	An unfavourable moment
16,099	Incubate .	It is of no moment
16,100	Incubation	Do not lose a moment
16,101	Incubatory	Every moment is of importance
16,102	Incubuses.	**Monday**
16,103	Inculcator.	On Monday
16,104	Inculpated	Last Monday
16,105	Incult . .	Next Monday
16,106	Incumbency	Every Monday
16,107	Incumbent	Every other Monday
16,108	Incumbrous	The first thing Monday morning
16,109	Incurable .	Every second Monday
16,110	Incurred .	Every third Monday
16,111	Incurrence	First Monday in the month
16,112	Incurring .	Second Monday in the month
16,113	Incurvity .	Third Monday in the month
16,114	Indagate .	Fourth Monday in the month
16,115	Indebt . .	Last Monday in the month
16,116	Indebtment	**Money**
16,117	Indecent .	Am (are) badly in want of money; you must send me (us) at once

No.	Code Word.	**Money** (*continued*)
16,118	Indecisive	Shall want money when I get to ——; telegraph
16,119	Indecorous	Do you want any money [—— to
16,120	Indecorum	Have sent you —— thinking you will be in want
16,121	Indefinite .	The money is already subscribed [of money for
16,122	Indelibly .	Can pay the purchase money as soon as your report
16,123	Indelicacy.	Have the money ready [reaches us
16,124	Indemnify	The money to be paid on the
16,125	Indented .	There will be a difficulty in raising the money
16,126	Indentedly	There will be no difficulty in raising the money
16,127	Indentures	Require an advance of money
16,128	Indesert .	How much money do you want
16,129	Indevoutly	Demand(s) a deposit of money as a guarantee
16,130	Indexerror	You must demand a deposit of money as a guarantee
16,131	Indexglass	Have demanded a deposit of money as a guarantee
16,232	Indexing .	We are prepared to pay down the purchase money as soon as we receive your report advising pur-
16,133	Indiaman .	Will want money by [chase of
16,134	Indianeer .	It will require a large outlay of money to
16,135	Indianlike.	Will cause a forfeiture of all money paid thereon
16,136	Indiapaper	Shall not want any money till
16,137	Indicated .	Will remit the money to
16,138	Indication	There is no money in it
16,139	Indicative .	There is money in it
16,140	Indicatory	We are short of money
16,141	Indicolite .	Shall want money before going to ——; send me
16,142	Indictable	Who will find the money for [—— at once
16,143	Indicteth .	Will find the money for
16,144	Indicting .	Will not find the money for
16,145	Indigence.	Has (have) provided the money for
16,146	Indigenous	Money is easy
16,147	Indigently	Money is tight
16,148	Indign . .	Money is plentiful
16,149	Indignant .	Money is on the way
16,150	Indigobird	Must have the money by
16,151	Indigoblue	Must have the money for men's wages by
16,152	Indigofera	Money to be paid down
16,153	Indigogen .	Must first have the money
16,154	Indigotic .	The money will be paid down
16,155	Indiligent .	The money will not be paid (without)
16,156	Indiscreet.	Do you think the money is all right
16,157	Indisposed	I (we) think the money is all right
16,158	Indistancy	I (we) do not think the money is all right
16,159	Inditched .	What amount of purchase money is required
16,160	Inditer. .	Purchase money required is
16,161	Individual	The supply of money has increased
16,162	Indivinity .	Stop the monthly money of
16,163	Indocility .	From the want of money
16,164	Indolence.	Cannot send money
16,165	Indolently	If the money is not paid on or before
16,166	Indomable	Money required will be supplied by

No.	Code Word.	**Money** (*continued*)
16,167	Indoors .	From whom shall I draw money
16,168	Indraught .	Draw what money you require from
16,169	Indrawn .	Supply of money very scarce
16,170	Indubious .	No money received yet, please send me (us) at [once
16,171	Inducement	Money is no object
16,172	Inducing .	Money down is an important object in the case
16,173	Inductric .	Will send money
16,174	Indulgeth .	I (we) have not sufficient money in hand for
16,175	Indulging .	I (we) have sufficient money in hand for
16,176	Indumentum	I am (we are) entirely without money
16,177	Industrial .	Will keep you supplied with money
16,178	Industry .	May calculate on having money by
16,179	Indwell .	Will send money very soon
16,180	Indwelling	Send money at once to pay
16,181	Inearthed .	Has (have) enough money for the present
16,182	Inebriated .	Has (have) only enough money to
16,183	Inebriety .	Money for travelling expenses
16,184	Inebrious .	Without any money in hand
16,185	Inefficacy .	**Monopolize(s)**
16,186	Inelastic .	**Monopolized**
16,187	Inelegancy	Entirely monopolized (by)
16,188	Ineloquent	**Month(s)**
16,189	Ineptitude	During last month
16,190	Inequable .	During next month
16,191	Inequality .	The month after next
16,192	Inequitate .	This month
16,193	Inerrably .	Last month
16,194	Inerratic .	Next month
16,195	Inerudite .	One month
16,196	Inevasible .	Two months
16,197	Inevidence	Three months
16,198	Inexact .	Four months
16,199	Inexertion	Five months
16,200	Inexistent .	Six months
16,201	Inexorably	Seven months
16,202	Inexpert .	Eight months
16,203	Inexpiable	Nine months
16,204	Inexplicit .	Ten months
16,205	Inexposure	Eleven months
16,206	Inextended	Twelve months
16,207	Inextinct .	By the month (of)
16,208	Infall . .	For the month (of)
16,209	Infallible .	In the month (of)
16,210	Infamized .	During the month (of)
16,211	Infamous .	After the month (of)
16,212	Infamously	Since the month (of)
16,213	Infamy. .	Before the month (of)
16,214	Infanthood	The beginning of the month
16,215	Infantile .	At the beginning of each month
16,216	Infantlike .	The end of the month

No.	Code Word.	**Month(s)** (*continued*)
16,217	Infatuated.	At the end of each month
16,218	Infausting.	In a month from this date
16,219	Infectious.	In —— month(s) from
16,220	Infective .	Per month
16,221	Infecund .	In —— month(s)
16,222	Infelicity .	Will leave early this month for
16,223	Inferable .	Will leave towards the end of the month for
16,224	Inference .	Will leave early next month for
16,225	Inferior .	Will leave towards the end of next month for
16,226	Infidelity .	Every alternate month
16,227	Infiltrate .	About the middle of the month
16,228	Infinitely .	Not later than the —— of the month
16,229	Infinitude.	On or before the —— of the month
16,230	Infirmary .	(For) this present month
16,231	Infirmness	(For) this month and next
16,232	Infixed .	(For) the next two months
16,233	Inflame .	(For) the next three months
16,234	Inflaming .	The return for last month
16,235	Inflatable .	The return for this month
16,236	Inflatus .	The return for next month
16,237	Inflected .	For the past month
16,238	Inflex . .	For the past three months
16,239	Inflexure .	For the past —— months
16,240	Inflicting .	For the first half of the month
16,241	Infliction .	For the second half of the month
16,242	Influenced	The accounts for the month (of)
16,243	Influxious.	Half this month, and half next
16,244	Informal .	Has been made (or done) during the month
16,245	Informally	The work done during the month
16,246	Informants	**Monthly**
16,247	Informed .	What are the monthly expenses
16,248	Informing.	At what do you estimate monthly profits
16,249	Infossous .	Monthly payments are
16,250	Infragrant.	Monthly shipment of bars averages
16,251	Infrapose .	Monthly shipment(s) of ore
16,252	Infrequent	Monthly profits estimated at
16,253	Infringe .	(Send) monthly statement of
16,254	Infringing.	(Send) monthly statement of expenses and returns
16,255	Infrugal .	(Send) monthly statements so that they arrive here
16,256	Infuriate .	Monthly report [by the —— of each month
16,257	Infuse . .	Regular monthly report
16,258	Infusing .	The monthly accounts
16,259	Infusorial .	Monthly production
16,260	Infusory .	Have not yet received the monthly
16,261	Ingeminate	The usual monthly
16,262	Ingendered	Our monthly circular
16,263	Ingirt . .	The monthly return
16,264	Inglenook.	Total monthly payments
16,265	Inglobate .	**More**
16,266	Inglorious.	Will want more

No.	Code Word.	More (*continued*)
16,267	Ingluvial .	No more than
16,268	Ingrained .	But no more
16,269	Ingraining	A great deal more than
16,270	Ingrate . .	No more can be got
16,271	Ingrately .	From a great many more
16,272	Ingratiate .	No more wanted
16,273	Inguilty .	More than
16,274	Ingustable	More than was expected
16,275	Ingwort .	If no more than
16,276	Inhabit .	If there is no more
16,277	Inhabiting	Will probably want more
16,278	Inharmony	More development work must be done
16,279	Inherently	Will not in all probability want more
16,280	Inheritors .	Will not cost more than
16,281	Inheritrix .	Will cost more than
16,282	Inhibitory .	Will be a little more
16,283	Inholdeth .	Will be a great deal more
16,284	Inhoop . .	Will not be any more
16,285	Inhuman .	Do you think it will be more
16,286	Inhumanly	Prove(s) to be more than I first stated
16,287	Inimicous .	When we want more
16,288	Inimitable	If more wanted, telegraph
16,289	Iniquitous .	Do you want more
16,290	Iniquity .	How much more
16,291	Initialled .	Can —— have more if required
16,292	Initiatory .	Have more
16,293	Inject . .	Have you any more
16,294	Injecting .	We can do more if necessary
16,295	Injellied .	Can you do any more
16,296	Injelly . .	We cannot do more
16,297	Injudicial .	Will take more than
16,298	Injurious .	**Morning**
16,299	Injustice .	This morning
16,300	Inkbag . .	Any morning
16,301	Inkberries	To-morrow morning
16,302	Inkberry .	Yesterday morning
16,303	Inkblurred	At —— o'clock this morning
16,304	Inkbottle .	Early in the morning
16,305	Inkfish . .	Morning shift
16,306	Inkglass .	**Mortars**
16,307	Inkholder .	Wet crushing mortars
16,308	Inkhorn .	Dry crushing mortars
16,309	Inkiness .	**Mortgage(s)** [per cent. interest
16,310	Inkling . .	Will be allowed to remain on mortgage at ——
16,311	Inkmaker .	Mortgage must be either paid or arranged (for)
16,312	Inkpot . .	With mortgage on record nothing can be done
16,313	Inkstand .	Has (have) a mortgage on the property
16,314	Inlaid . .	Ascertain if there is any mortgage on the
16,315	Inlandish .	There is no mortgage on
16,316	Inlapidate	There is (are) mortgage(s) on the property as follows

No.	Code Word.	**Mortgage(s)** (*continued*)
16,317	Inlaying .	The only mortgage on the property is for —— to
16,318	Inmates .	Mortgage on the land
16,319	Innately .	Mortgage must be paid off
16,320	Innermost	Mortgage will be foreclosed (if)
16,321	Innerplate	Warranted free from mortgage
16,322	Innkeeper .	Will give a mortgage on the property
16,323	Innocuity .	Will take a mortgage on the property
16,324	Innominate	The mortgage deeds
16,325	Innovated .	Foreclosure of the mortgage
16,326	Innovating	Has foreclosed the mortgage
16,327	Innovative	**Mortgaged**
16,328	Innoxious .	The property is mortgaged for
16,329	Innumerous	The property is mortgaged up to the hilt
16,330	Inobedient	**Mortgagee(s)**
16,331	Inocarpin .	Who is (are) the mortgagee(s)
16,332	Inoceramus	Telegraph name(s) of mortgagee(s)
16,333	Inoculable	Mortgagee threatens to foreclose
16,334	Inoculator	If the mortgagee forecloses
16,335	Inodorous .	**Most**
16,336	Inopinate .	The most of the
16,337	Inopulent .	Most of the work
16,338	Inordinacy	Most of the ore now being extracted
16,339	Inordinate	Most of the ore now being milled
16,340	Inorganic .	By far the most
16,341	Inosculate	The most that can be done
16,342	Inosic . .	What is the most that can be
16,343	Inpenny .	The most that was ever done (made)
16,344	Inquest .	What is the most
16,345	Inquinate .	Have made the most
16,346	Inquirable	**Mould(s)**
16,347	Inquiring .	Ingot mould(s)
16,348	Inquiry .	Bullion mould(s)
16,349	Inquisitor .	Lead mould(s)
16,350	Inroads .	Silver mould(s)
16,351	Insalutary .	Gold mould(s)
16,352	Insanably .	**Mountain(s)**
16,353	Insaneness	The side of the mountain
16,354	Insanified .	The summit of the mountain
16,355	Insanify .	The foot of the mountain
16,356	Insapory .	Driving into the mountain
16,357	Insatiety .	The other side of the mountain
16,358	Inscribe .	Driven in the side of the mountain
16,359	Inscribing .	The mountain is mostly
16,360	Inscroll .	These mountains
16,361	Insculp .	Situated in —— mountains
16,362	Insculping	A mountain road
16,363	Insectator	On the precipitous side of a mountain
16,364	Insectile .	Halfway up the mountain
16,365	Insects .	**Mountainous**
16,366	Insecure .	A very mountainous district

No.	Code Word.	**Mountainous** (*continued*)
16,367	Insecurely.	A very mountainous road
16,368	Insensibly.	**Move**
16,369	Insensuous	To move in the matter
16,370	Inseparate	Do not move in the matter
16,371	Inserting .	To move away
16,372	Inservient.	The first thing is to move the
16,373	Insessores	We shall have to move
16,374	Inshelled .	Will move
16,375	Inshore .	Will not move
16,376	Inside . .	Try to move
16,377	Insidiated .	Not to move
16,378	Insight. .	Has (have) tried to move
16,379	Insincere .	(To) move the court
16,380	Insinewed .	(To) move that the action be
16,381	Insinuator.	Watch every move (of)
16,382	Insipid. .	Cannot move without
16,383	Insipidly .	**Moved**
16,384	Insipience	Moved the court
16,385	Insisted .	Moved to set aside
16,386	Insisture .	**Movement(s)**
16,387	Insnaring .	A movement in
16,388	Insolvable	There is a considerable movement in
16,389	Insolvency	There is but little movement in
16,390	Insolvent .	No movement in
16,391	Insomnia .	A movement is being made to
16,392	Insomnious	Every movement
16,393	Insooth .	What are your movements
16,394	Insoul . .	Please wire at once your movements
16,395	Inspected .	Watch his (their) movements
16,396	Insperse .	**Moving**
16,397	Inspeximus	Is (are) —— moving in the matter
16,398	Inspirable.	Is (are) moving in the matter
16,399	Inspissate	Is (are) not moving in the matter
16,400	Install . .	Will soon be moving
16,401	Installing .	Now moving to upset the
16,402	Instalment	**Much**
16,403	Instancing	Not much
16,404	Instantly .	Very much
16,405	Instaurate.	As much as
16,406	Insteeped .	As much as you can
16,407	Insteeping	If we do as much as
16,408	Instigator .	If you do as much as
16,409	Instinctly .	Cannot do as much
16,410	Instituted .	Much less (than)
16,411	Instruct .	Much more (than)
16,412	Instructor .	How much (shall)
16,413	Instrument	Without much more
16,414	Insuavity .	Will not be much more
16,415	Insularity .	Shall get as much
16,416	Insulate .	Shall not get as much

No.	Code Word.	**Much** (*continued*)
16,417	Insulation.	Without much
16,418	Insulous .	Quite as much as we expected
16,419	Insulted .	Much more than expected
16,420	Insulting .	How much more do you expect to get
16,421	Insultment	We expect to get much more
16,422	Insurance .	Have you much further to go
16,423	Insurgent .	We have not much further to go
16,424	Inswathed.	Much further
16,425	Intangible	Not very much (of)
16,426	Integrally .	**Multiplication**
16,427	Integrity .	**Multiplying**
16,428	Integument	**Murdered**
16,429	Intendancy	Has been murdered
16,430	Intendant .	Several murders have been committed
16,431	Intendedly	**Must**
16,432	Intendment	Must be
16,433	Intensify .	Must not be
16,434	Intent . .	Must this be
16,435	Interact .	You must
16,436	Interagent	You must not
16,437	Interaulic .	I (we) must
16,438	Interaxal .	I (we) must not
16,439	Interblend	He (they) must
16,440	Intercalar .	He (they) must not
16,441	Intercedes	Must I (we)
16,442	Intercept .	Must I (we) not
16,443	Interchain	Must proceed
16,444	Intercloud	Must not proceed
16,445	Intercome	If I (we) must
16,446	Intercross.	If you must
16,447	Interested.	When you must
16,448	Interfold .	When must you
16,449	Interfuse .	Why must
16,450	Interhemal	(It) must be done
16,451	Interim .	(It) must not be done
16,452	Interimist.	Must do it
16,453	Interjects .	Must not do it
16,454	Interknit .	You must come at once
16,455	Interknow.	You must go at once
16,456	Interlaced.	**Mutilated**
16,457	Interlapse.	The message was mutilated in being cabled
16,458	Interleaf .	The telegram was evidently mutilated
16,459	Interlibel .	A mutilated and garbled report
16,460	Interlink .	In a mutilated form
16,461	Interloper.	**Mutual(ly)**
16,462	Interlunar.	To our mutual advantage
16,463	Intermarry	A mutual arrangement
16,464	Intermixed	A mutual exchange
16,465	Intermural	It is mutually agreed
16,466	Internodal	On mutually advantageous terms

No.	Code Word.	
16,467	Interpale .	**Myself**
16,468	Interplay .	By myself
16,469	Interplead	For myself and friends
16,470	Interposit.	Cannot do the work by myself
16,471	Interreign.	Reserve for myself
16,472	Interrupt .	To be allotted to myself
16,473	Interscind	**Name(s) (of)**
16,474	Intershock .	Name of the company
16,475	Intersour .	Telegraph the right name (of)
16,476	Intertalk .	What is (are) the name(s) of
16,477	Intertwist .	The name of a good maker of
16,478	Intervened	In my (our) name
16,479	Interview .	In your name
16,480	Intervisit .	In the name of
16,481	Intervital .	Name the date
16,482	Intervolve.	Do you know the name(s) of (the)
16,483	Interwish .	Name(s) not known
16,484	Interwork .	Name(s) not much liked
16,485	Interwound	Name(s) stand(s) well
16,486	Interwoven	Cannot name a day
16,487	Intestable.	The name of the mine is
16,488	Intestacy .	Do not name
16,489	Intexture .	Cannot ascertain name
16,490	'Inthirsted.	Name has a bad reputation
16,491	Inthrall .	What name will you give the company
16,492	Intimacy .	Sign your name to
16,493	Intimating	First-class names
16,494	Intimidate	Names of firms
16,495	Intolerant.	Names of firms from whom we are to obtain
16,496	Intoxicant.	Names of firms who supply
16,497	Intreatful .	Names of some of the best firms
16,498	Intrepid .	Names of some of the best makers
16;499	Intrepidly.	**Narrow**
16,500	Intricable .	A narrow strip of
16,501	Intricacy .	The vein is narrow
16,502	Intriguing.	Becomes narrow
16,503	Intrinsic .	Is too narrow to work profitably
16,504	Introduced	In a narrow
16,505	Introspect	Is found in narrow
16,506	Introsume.	In the —— level the vein becomes narrow
16,507	Introvert .	**Narrowing**
16,508	Intruder .	The vein is narrowing
16,509	Intrudress.	The vein is narrowing as we get deeper
16,510	Intrusive .	**Native(s)**
16,511	Intrusted .	Native silver
16,512	Intumesce	Native copper
16,513	Inundating	In a native state
16,514	Inurbane .	The natives in the district are very quiet
16,515	Inurbanely	The natives in the district are very troublesome
16,516	Invalidism	Plenty of native labour to be got

No.	Code Word.	**Native(s)** (*continued*)
16,517	Invalidity .	Native workmen
16,518	Invalorous	Native miners
16,519	Invaluably	Have procured some natives from
16,520	Invariable.	Natives have been substituted
16,521	Inveigh .	We have brought up some native miners
16,522	Inveigheth	Are chiefly natives
16,523	Inveighing	Both natives and foreigners
16,524	Invenoming	Native labour is cheap and abundant
16,525	Inventory .	Native labour is scarce
16,526	Inventress	Natives earn from —— to —— a day
16,527	Inversely .	Native labour difficulties
16,528	Investment	We have a good deal of difficulty with the **natives**
16,529	Inveteracy	**Nature**
16,530	Inveterate.	The nature of the
16,531	Invidious .	What is the nature of the
16,532	Invigorate	Of a dangerous nature
16,533	Invillaged.	Owing to the dangerous nature of the
16,534	Inviolacy .	Owing to the peculiar nature of
16,535	Inviolated	Of a temporary nature
16,536	Invisibly .	The inflammable nature of the
16,537	Invitingly .	The nature of the vein system
16,538	Invocate .	From the nature of the
16,539	Invocation	**Navigable**
16,540	Invocatory	A navigable river
16,541	Invoice .	Not navigable for
16,542	Invoked .	**Navigation**
16,543	Involucral	Navigation closed
16,544	Involuted .	Navigation opened
16,545	Involutina	Navigation will not be open until
16,546	Involving .	Navigation expected to
16,547	Inweave .	**Near (the)**
16,548	Inweaving	The mines are situated near (the)
16,549	Inwheeled	The mines are not near any
16,550	Inwoven .	Near the vein
16,551	Iodized .	Near the hanging wall
16,552	Iodizing .	Near the foot-wall
16,553	Iodoform .	As near as possible
16,554	Iracund .	Near the bottom (of)
16,555	Irascible .	Quite near enough
16,556	Ireful . .	Not near enough
16,557	Irenicon .	Very near
16,558	Iricism. .	We should be very near (the)
16,559	Iridectomy	Not near
16,560	Iridescent.	Near the surface
16,561	Iridosmium	We are getting near
16,562	Irisated .	As we get near the
16,563	Iriscope .	Near the shaft
16,564	Irishmoss .	Near the boundary
16,565	Iritis . .	**Nearer**
16,566	Irksome .	Nearer the

No.	Code Word.	**Nearer** (*continued*)
16,567	Irksomely .	Nearer —— than
16,568	Ironbark .	Will be nearer —— than
16,569	Ironbound	Getting nearer the
16,570	Ironcased .	Not nearer than
16,571	Ironclad .	When nearer
16,572	Ironcrown	Is it (are they) any nearer to
16,573	Ironfisted .	No nearer
16,574	Ironflint .	**Nearly**
16,575	Irongray .	Has (have) nearly reached the
16,576	Ironhat .	Has (have) nearly completed (the)
16,577	Ironical .	The stopes above level No. —— are nearly done
16,578	Ironically .	The ore is nearly worked out
16,579	Ironingbox	Not nearly done
16,580	Ironish. .	Not nearly finished
16,581	Ironliquor.	Not nearly
16,582	Ironlord .	Very nearly
16,583	Ironmaster	The works are nearly completed
16,584	Ironmonger	Erection of —— nearly completed
16,585	Ironmould	Examination nearly completed
16,586	Ironsand .	**Necessary**
16,587	Ironsick .	The first thing necessary is to
16,588	Ironsides .	To do what is absolutely necessary
16,589	Ironstone .	Do (does) not think it will be necessary
16,590	Ironware .	It will be necessary (to)
16,591	Ironwood .	It will not be necessary (to)
16,592	Ironworks	Under these circumstances it will be necessary
16,593	Irradiancy.	But, if necessary, you can
16,594	Irradiate .	If necessary
16,595	Irrational .	If not necessary
16,596	Irregular .	If necessary to give
16,597	Irregulous	If necessary to take
16,598	Irrelative .	If not necessary to give
16,599	Irrelevant.	If not necessary to take
16,600	Irremoval .	It is necessary (to)
16,601	Irrenowned	It is not necessary (to)
16,602	Irresolute .	If absolutely necessary
16,603	Irrigate .	It is absolutely necessary
16,604	Irrigating .	It is not absolutely necessary
16,605	Irritation .	Will it be necessary (to)
16,606	Irritatory .	If you consider it necessary (to)
16,607	Irrubrical .	I (we) do not consider it necessary (to)
16,608	Irrupted .	More than is necessary
16,609	Irvingite .	Provide all that is necessary
16,610	Isabelline .	I am (we are) taking the necessary steps (to)
16,611	Isagogic .	Take (taking) the necessary means (to)
16,612	Isagogical .	**Necessitate**
16,613	Ischiadic .	(This) will necessitate
16,614	Ischiagra .	(This) need not necessitate
16,615	Ischiocele.	(This) will necessitate renewing the bond
16,616	Ischium .	(This) will necessitate his (their)

No.	Code Word.	**Necessitate** (*continued*)
16,617	Ischuretic.	(This) will necessitate my (our)
16,618	Ischury .	(This) will necessitate your
16,619	Ischyodon	Will this necessitate
16,620	Isidoid . .	It will not necessitate
16,621	Isinglass .	Provided it does not necessitate
16,622	Islamism .	**Necessity**
16,623	Islamitic .	Is there any necessity to (for)
16,624	Islamize .	I (we) do not see any necessity to (for)
16,625	Islamizing	There is great necessity
16,626	Islanders .	In case of necessity
16,627	Islandy. .	Unless compelled by absolute necessity
16,628	Isnardia .	**Need**
16,629	Isobar . .	Do (does) not need
16,630	Isobaric .	There is no need
16,631	Isobrious .	Shall not need
16,632	Isochimal .	In case of need
16,633	Isochronon	In need of
16,634	Isochrous .	I am (we are) much in need of
16,635	Isoclinic .	In case of need, apply to
16,636	Isocryme .	In case of need, to whom shall I (we) apply
16,637	Isodomum	Has (have) no need of
16,638	Isodynamic	Is (are) in great need of
16,639	Isogonic .	You need not
16,640	Isography .	Procure whatever you need (from)
16,641	Isohyetose	**Negative**
16,642	Isolable .	In the negative
16,643	Isolated .	Telegraph at once if negative
16,644	Isolatedly .	If the reply is in the negative
16,645	Isolating .	**Neglect**
16,646	Isomerical	There is great neglect
16,647	Isomerism	Very great neglect
16,648	Isomerous	Do not neglect
16,649	Isonandra .	I (we) will not neglect
16,650	Isonomy .	Do not think there has been any neglect
16,651	Isopathy .	There has been no neglect on my (our) part
16,652	Isopoda .	There has been no neglect on —— part
16,653	Isopodous	The result of neglect
16,654	Isopolity .	Owing to the neglect of
16,655	Isopyre .	Owing to the neglect of proper precautions
16,656	Isosceles .	**Neglected**
16,657	Isotherm .	Has (have) neglected to
16,658	Isothermal	Has been very much neglected
16,659	Isotropic .	Was (has been) entirely neglected
16,660	Israelitic .	You have neglected to
16,661	Issuably .	I (we) neglected to
16,662	Issuance .	I (we) have not neglected to
16,663	Issueless .	Why have you neglected to
16,664	Issuing .	Why has (have) —— neglected to
16,665	Isthmian .	**Neglecting**
16,666	Isthmitis .	In consequence of neglecting to

No.	Code Word.	
16,667	Itaberite .	**Negotiate**
16,668	Italic . .	Can you negotiate
16,669	Italicism .	Can you negotiate bill (on)
16,670	Italicized .	I (we) can negotiate
16,671	Itinerancy	I (we) cannot negotiate
16,672	Itinerant .	Hope to be able to negotiate
16,673	Ittrium . .	Shall be able to negotiate
16,674	Ivories . .	Cannot negotiate
16,675	Ivory . .	Cannot negotiate bill (on)
16,676	Ivoryblack	Could not negotiate
16,677	Ivorynut .	Do not negotiate
16,678	Ivorypalm.	**Negotiated**
16,679	Ivoryshell.	Have negotiated a bill (on)
16,680	Ivygum .	Has (have) negotiated terms of agreement
16,681	Ivymantled	**Negotiating**
16,682	Ixodes . .	We are now negotiating for
16,683	Ixolyte . .	Negotiating terms
16,684	Jaalgoat .	Negotiating the purchase of
16,685	Jabber . .	Negotiating the sale of
16,686	Jabbering .	**Negotiation(s)**
16,687	Jabberment	Negotiations still pending
16,688	Jabbernowl	Negotiations are not satisfactory
16,689	Jabiru . .	Negotiations are very satisfactory
16,690	Jaborandi.	Am (are) carrying on negotiations
16,691	Jacamars .	Is (are) carrying on negotiations
16,692	Jacaranda.	Negotiations stopped for the present
16,693	Jacare . .	Negotiations have failed
16,694	Jacinth .	Negotiations satisfactorily carried through
16,695	Jackadandy	Negotiations are going on for the acquisition of
16,696	Jackalent .	Stop all negotiations (concerning)
16,697	Jackanapes	Have stopped all negotiations (concerning)
16,698	Jackarch .	The result of the negotiations (with or about)
16,699	Jackbacks .	To facilitate negotiations
16,700	Jackblock.	Negotiations will have the result of
16,701	Jackboot .	Negotiations broken off
16,702	Jackchain.	Negotiations have been resumed
16,703	Jackdaws .	Negotiations are concluded
16,704	Jackflag .	The matter is ripe for negotiation
16,705	Jackfruit .	No time for negotiation
16,706	Jackknife .	The time for negotiation is past
16,707	Jackrafter .	Negotiation for the purchase of
16,708	Jackrib .	**Neighbourhood**
16,709	Jacksauce.	In the neighbourhood
16,710	Jackscrew.	When you are in the neighbourhood
16,711	Jackslave .	As I (we) shall be in the neighbourhood
16,712	Jacksnipe .	Being in the neighbourhood
16,713	Jackstaff .	Not in the neighbourhood
16,714	Jackstays .	**Neither**
16,715	Jacktimber	In neither
16,716	Jacktowel .	From neither

No.	Code Word.	**Neither** (*continued*)
16,717	Jacktree .	Neither of the
16,718	Jackwood .	Neither one nor the other
16,719	Jacobinism	Neither seen nor heard of
16,720	Jacobinly .	Neither of the —— has
16,721	Jacobus .	In neither of the
16,722	Jactancy .	**Net**
16,723	Jactation .	What has been the net yield per ton
16,724	Jaculating.	What is the net yield
16,725	Jaculator .	What is the net value
16,726	Jadery . .	The net yield per ton is
16,727	Jaggedness	The net value of the mine(s) is
16,728	Jagghery .	The net value of the bullion is
16,729	Jaghirdar .	The net value (of)
16,730	Jaghire .	The net weight
16,731	Jaguars .	The net profits
16,732	Jailbird .	The net proceeds
16,733	Jailfever .	Has (have) netted
16,734	Jailkeeper.	**Neutral**
16,735	Jailors . .	Will remain neutral
16,736	Jalap . .	Advise you to remain neutral
16,737	Jalapic .	Is (are) perfectly neutral
16,738	Jaloose .	Will remain neutral unless our interests are in danger
16,739	Jamdari .	**Never**
16,740	Jamesonite	It will never do to
16,741	Jamming .	It has never been known
16,742	Jamnut .	It has never been mentioned
16,743	Jampanees	It can never be done unless
16,744	Jamrosade	You must never
16,745	Jangleress.	Never mind
16,746	Jangling .	Never before
16,747	Janglour .	Has (have) never known a similar case
16,748	Janissary .	**Nevertheless**
16,749	Janitrix .	We think, nevertheless
16,750	Janizar .	**New**
16,751	Janizarian.	Is there anything new
16,752	Jansenism.	Is (are) quite new
16,753	Janthina .	Will require new
16,754	Janus . .	Nothing new
16,755	Janusfaced	Everything new
16,756	Japanblack	Send immediately new
16,757	Japanearth	Have made a new shaft
16,758	Japanner .	The new developments
16,759	Japanning	The new workings
16,760	Japannish .	The new machinery
16,761	Jargoned .	The new shaft
16,762	Jargonists.	It will be better to have new
16,763	Jargonized	Must have new
16,764	Jarool . .	Make a new
16,765	Jarringly .	Need not at present have new
16,766	Jashawk .	Everything new and in good order

No.	Code Word.	New (*continued*)
16,767	Jaspachate	This is perfectly new to us
16,768	Jasperated	New and improved plant
16,769	Jaspideous	For new machinery, building, etc.
16,770	Jaspoid	. **News**
16,771	Jasponyx .	Waiting anxiously for news of
16,772	Jatamansi.	What is the latest news
16,773	Jateorhiza.	What news have you of
16,774	Jatropha .	Have you any news
16,775	Jaundice .	I (we) have no news
16,776	Jaundicing	The latest news is
16,777	Jauntily .	Very important news
16,778	Jauntiness	No news of importance
16,779	Javelining.	Have no further news (from)
16,780	Javelins .	Send all the news you can about
16,781	Jawfallen .	Unless you have favourable news do not telegraph
16,782	Jawfoot .	Unless you have later news of a favourable character
16,783	Jawlever .	News not reliable
16,784	Jawrope .	News quite correct
16,785	Jawteeth .	Owing to the news that
16,786	Jawwedge.	Very glad to receive news of
16,787	Jazerant .	Very sorry to receive news of
16,788	Jealous .	Alarming news (from)
16,789	Jealously .	Favourable news
16,790	Jeddingaxe	The only news we could get
16,791	Jehovistic .	The latest news is very good
16,792	Jejunely .	The latest news is unfavourable
16,793	Jejuneness	What further news concerning
16,794	Jellied . .	The news upset everything
16,795	Jellybag	. **Newspapers**
16,796	Jellyfish .	Send newspapers
16,797	Jemidar .	Send all newspapers which refer to
16,798	Jemminess	A paragraph in the newspapers
16,799	Jenneting .	A newspaper notice
16,800	Jennyass .	The newspapers have reported
16,801	Jeopardize	The newspaper report
16,802	Jeopardous	Most of the newspapers
16,803	Jeopardy .	The best newspaper
16,804	Jerfalcon .	(By) advertisement in the newspapers
16,805	Jerguer .	Has appeared in the newspapers
16,806	Jerked . .	You will have seen by the newspapers
16,807	Jerking	. **Next**
16,808	Jerkinhead	Leaving by next steamer
16,809	Jeronymite	Will send by next steamer
16,810	Jessamine.	In your next letter
16,811	Jesseraunt	In your next telegram
16,812	Jestbooks .	**Nickel**
16,813	Jesters . .	Nickel ore
16,814	Jestful . .	A nickel mine
16,815	Jestmonger	Nickel has been found
16,816	Jestword	. **Night(s)**

No.	Code Word.	**Night(s)** (*continued*)
16,817	Jesuit . .	During the night
16,818	Jesuitess .	Night shift
16,819	Jesuitical .	Last night
16,820	Jesuitry .	Received last night
16,821	Jesuitsnut.	Left last night
16,822	Jetblack .	To-morrow night
16,823	Jetsam . .	Will leave to-morrow night for
16,824	Jetties . .	Last night. after office had closed
16,825	Jettisoned.	Too late last night for reply
16,826	Jetty . .	**No**
16,827	Jettyhead .	No more
16,828	Jewbush .	There is (are) no
16,829	Jewel . .	There can be no
16,830	Jewelblock	There has been no
16,831	Jewelcase .	We have no
16,832	Jewelhouse	We say at once, no
16,833	Jewellery .	Decidedly no
16,834	Jewellike .	If no, then
16,835	Jewelweed	Is (are) there no
16,836	Jewesseye.	No more to be had
16,837	Jewish . .	No less
16,838	Jewishly .	**Nominal**
16,839	Jewishness	For a nominal consideration
16,840	Jewsapple.	At a nominal rate (or rent)
16,841	Jewsear .	A nominal payment
16,842	Jewsharp .	Quotations nominal
16,843	Jewsmallow	**Nominally**
16,844	Jewspitch .	Nominally quoted at
16,845	Jewstrump	**Nominate**
16,846	Jezid . .	We must nominate some one to
16,847	Jibbing .	Whom can you nominate to
16,848	Jibboom .	Is (are) the best man (men) to nominate
16,849	Jiboya . .	Do not nominate
16,850	Jickajog .	Cannot nominate
16,851	Jiffy . .	You must nominate some one
16,852	Jiggumbob	Nominate some one to act for you
16,853	Jigjog . .	Nominate as trustee
16,854	Jigmaker .	Nominate as —— agent
16,855	Jigpin . .	**Nominated**
16,856	Jigsaw . .	Have nominated
16,857	Jillflirt . .	Have nominated the following
16,858	Jilt . . .	Whom have you nominated
16,859	Jimps . .	Nominated as director
16,860	Jippo . .	Nominated by the Board
16,861	Jirkinet .	Nominated by the vendor
16,862	Jobbery .	**Nominee(s)**
16,863	Jobmaster.	Our nominee(s)
16,864	Jobprinter	His (their) nominee(s)
16,865	Jobspost .	Your nominee(s)
16,866	Jobstears .	The vendor's nominee

No.	Code Word.	Nominee(s) (*continued*)
16,867	Jobwatch .	(In) the names of my (our) nominees
16,868	Jockey .	. Non-
16,869	Jockeyclub	Non-acceptance (of)
16,870	Jockeyism.	Non-appearance
16,871	Jockeyship	Non-committal
16,872	Jockied .	Non-compliance (with)
16,873	Jocose . .	Non-concurrence (with)
16,874	Jocosely .	Non-delivery (of)
16,875	Jocoseness	Non-fulfilment (of)
16,876	Jocular .	Non-payment (of)
16,877	Jocularity.	Non-performance (of)
16,878	Jocund .	Non-presentation (of)
16,879	Jocundly .	Non-receipt (of)
16,880	Jocundness	**None**
16,881	Jogtrot. .	None of the
16,882	Johannite .	Are there none to be had
16,883	Johnapple.	None have been sent
16,884	Johncrow .	There are none to be had
16,885	Johndory .	**Nonsense**
16,886	Johnnycake	It is utter nonsense
16,887	Johnnyraw	A great deal of nonsense was talked
16,888	Johnsonese	Treat it as utter nonsense
16,889	Johnsonian	Stand no nonsense
16,890	Johnswort.	**North.** (See Table at end.)
16,891	Joinhand .	**Not**
16,892	Joining .	Not any
16,893	Jointchair.	Not at present
16,894	Jointevil .	Not before
16,895	Jointfiat .	Not after
16,896	Jointheir .	Not to be
16,897	Jointless .	Not until
16,898	Jointly . .	Not unless
16,899	Jointstock.	Not much (not many)
16,900	Jointstool .	If not so
16,901	Jointures .	If I am (we are) not to
16,902	Jointuring.	If it is not
16,903	Jointworm	Is it not
16,904	Jokingly .	Is it not so
16,905	Jokish . .	Are we not
16,906	Jollier . .	Are you not
16,907	Jollily . .	Are (is) there not
16,908	Jolliment .	Is he (are they) not
16,909	Jolliness .	Must not
16,910	Jollyboat .	Must not be
16,911	Jollyhead .	Not more (than)
16,912	Jolterhead	Not less (than)
16,913	Jorum . .	**Notarial(ly)**
16,914	Joss . .	Notarial attestation required
16,915	Josshouse .	Notarial charges
16,916	Jossstick .	Must be notarially attested and certified by consul

No.	Code Word.	**Notarial(ly)** (*continued*)
16,917	Jostle . .	Notarially attested
16,918	Jostling .	**Note(s)**
16,919	Jotted . .	Be very careful to note
16,920	Jougs . .	Have taken great care to note
16,921	Journalary	Note the
16,922	Journalbox	We have taken note of
16,923	Journalism	(To) take notes
16,924	Journalize .	Will take notes
16,925	Journey .	**Nothing**
16,926	Journeying	If there is nothing suitable
16,927	Journeyman	Nothing can be done
16,928	Jouster. .	Can nothing be done
16,929	Jovial . .	Nothing doing
16,930	Jovialist .	Nothing to prevent
16,931	Jovialized .	Nothing yet
16,932	Jovially .	Nothing has been done
16,933	Jovialness .	Nothing new worth reporting
16,934	Jowler . .	Do nothing without first consulting
16,935	Jowlopped	Do nothing until you hear
16,936	Joyance .	Can do nothing till I hear from
16,937	Joybells .	Is (are) doing nothing to
16,938	Joyful . .	There is nothing in it
16,939	Joyfully .	There is nothing in the mine
16,940	Joyfulness .	Nothing in sight
16,941	Joyless. .	Nothing left
16,942	Joylessly .	It is worth nothing
16,943	Joyous . .	Can see nothing owing to
16,944	Joyously .	Can hear nothing of
16,945	Joyousness	Nothing at all
16,946	Jubilant .	Nothing more can be done
16,947	Jubilation .	Have nothing to do with it (him)
16,948	Judaical .	Nothing to do with
16,949	Judaism .	Nothing more to do with
16,950	Judaistic .	Nothing can be further from
16,951	Judaize .	Nothing was further from our intention (or wish)
16,952	Judaizing .	Thought nothing of it
16,953	Judashole .	That nothing be done
16,954	Judasly .	Nothing later than
16,955	Judastree .	Nothing prevents
16,956	Judcock .	Nothing will be done until we hear from you
16,957	Judged .	**Notice**
16,958	Judgeship .	Has (have) given notice
16,959	Judging .	Has (have) received no notice
16,960	Judgingly .	Has (have) ―― sent any notice
16,961	Judgments	Due notice will be sent
16,962	Judica . .	Give notice that
16,963	Judicable .	Give ―― notice
16,964	Judicatory	Until further notice
16,965	Judicature	Notice to terminate the agreement
16,966	Judicially .	What notice is necessary

No.	Code Word.	Notice (*continued*)
16,967	Judicious .	Has (have) taken no notice
16,968	Jugglery .	Take no notice of
16,969	Juglandine	Have you given notice
16,970	Juglans .	Shall we give notice
16,971	Jugulate .	Has he given notice
16,972	Juiceful .	If notice has not been given
16,973	Juiceless .	No notice was given
16,974	Juices . .	A notice of —— days was given
16,975	Julianist .	Due notice has been given
16,976	Juliform .	The proper notice must be given
16,977	Julyflower.	Final notice
16,978	Jumbled .	Formal notice
16,979	Jumblement	If —— do not give notice
16,980	Jumpingrat	Not worthy of notice
16,981	Jumpseat .	This is to give due notice
16,982	Jumpweld.	Public notice must be given
16,983	Juncaceous	Without further notice
16,984	Juncous .	Notices have been posted
16,985	Juncture .	Notice is published (that)
16,986	Juneating .	**Notification**
16,987	Juneberry.	Official notification
16,988	Junglefowl	Upon notification of
16,989	Jungly . .	Notification of the numbers
16,990	Junglygau.	Without proper notification
16,991	Junior . .	Received notification
16,992	Juniority .	**Notified**
16,993	Juniorship	Has (have) notified me (us) that
16,994	Juniper .	Has (have) not yet been notified
16,995	Junkbottle	Have you been notified of
16,996	Junkerite .	Unless I am (we are) notified
16,997	Junket . .	Has (have) been notified
16,998	Junketings	Unless you are notified
16,999	Jupatipalm	Will be notified in due time
17,000	Jurassic .	Must be notified
17,001	Juration .	**Notify**
17,002	Juratory .	Failed to notify
17,003	Juridic . .	Notify all parties concerned
17,004	Juridical .	**Notorious**
17,005	Jurybox .	Notorious for
17,006	Juryman .	Is notorious as having been the
17,007	Jurymasts .	It is notorious as a regular swindle
17,008	Juryrigged	**Notwithstanding**
17,009	Juryrudder	Notwithstanding this
17,010	Jussel . .	**Now**
17,011	Justice . .	Now is a favourable time to
17,012	Justiciary .	It must be done now
17,013	Justified .	It is now the time to
17,014	Justifying .	Now is the time to act
17,015	Justinian .	Not now
17,016	Justly . .	Now on the way

No.	Code Word.	**Now** (*continued*)
17,017	Justness .	Now ready to leave
17,018	Jutwindow	Now or never
17,019	Juveniles .	If not done now
17,020	Juvenility .	We are now
17,021	Juventate .	We are now in a position to
17,022	Juvia . .	We are not now
17,023	Juwanza .	Are you now
17,024	Juxtapose.	Is (are) now
17,025	Juzail . .	Now and then
17,026	Jymold .	From now until
17,027	Jysse . .	We should now
17,028	Kabani .	We must now
17,029	Kadiaster .	**Nowhere**
17,030	Kafilah .	Go nowhere without first telegraphing me (us)
17,031	Kafirbread	Nowhere else
17,032	Kainsi . .	Nowhere to be found
17,033	Kajugaru .	Nowhere in this neighbourhood
17,034	Kakaralli .	**Nugget(s)**
17,035	Kakoxene	The nugget was found
17,036	Kalaf . .	Have found a nugget weighing —— ozs.
17,037	Kalends .	Nuggets have frequently been found in the
17,038	Kalmia .	Have found —— nuggets weighing altogether [—— ozs.
17,039	Kaloyer .	**Nuisance**
17,040	Kamala .	It will be a great nuisance
17,041	Kamarband	The nuisance is almost unbearable
17,042	Kamichi .	**Number(s)**
17,043	Kanchil .	A large number
17,044	Kangaroo .	A small number
17,045	Kantists .	Has (have) obtained a number of
17,046	Kaoline .	Number of certificates
17,047	Karaskier .	What is the number of the certificates
17,048	Karatas .	Hold(s) a number of
17,049	Karengia .	(A) number of shares
17,050	Karstenite	The number of men employed
17,051	Katydid .	The number of —— employed
17,052	Kauripine .	What is the exact number (of)
17,053	Keblah .	Numbers and marks (of)
17,054	Kecklish .	Numbers and dates (of)
17,055	Kedgerope	Numbers and dates of the respective
17,056	Keelage .	Number of square on plan
17,057	Keelblock	See square No. ——, letter
17,058	Keelboat .	Number and letter of square on plan
17,059	Keelfat .	Number and letter of square on section
17,060	Keelhaul .	What is the number of
17,061	Keelson .	What are the numbers of
17,062	Keelstaple	The numbers of the shares are
17,063	Keenest .	**Numeral(s)**
17,064	Keeneyed .	The following numerals
17,065	Keenwitted	Indicated by the numerals
17,066	Keeperless	In numerical order

No.	Code Word.	**Numeral(s)** (*continued*)
17,067	Keepers .	Numeral —— and letter —— indicate
17,068	Keepership	**Numerous**
17,069	Keepsake .	Numerous complaints
17,070	Keepworthy	From numerous sources
17,071	Keffekil .	Numerous inquiries
17,072	Kehul .	**Oak**
17,073	Kelpies .	Oak timber
17,074	Kempt. .	Oak guides
17,075	Kennel .	Framing of oak
17,076	Kennelled	We are using oak
17,077	Kentledge	**Obedience**
17,078	Keplerian .	In obedience to your instructions
17,079	Keratome .	Will have ready and willing obedience
17,080	Kerchief .	**Obey**
17,081	Kernbaby.	You must obey instructions
17,082	Kernelwort	Do not obey
17,083	Kerodon .	Will obey
17,084	Kerosolene	Will not obey
17,085	Kersey. .	If —— will not obey orders
17,086	Kerseymere	**Obeyed**
17,087	Keslop. .	We expect our orders will be obeyed
17,088	Kettledrum	**Object**
17,089	Kettlehat .	What is the object
17,090	Kettlepins	What is ——'s object in
17,091	Kevelhead	Their only object is to
17,092	Kexy . .	Unless you object
17,093	Keyboard .	I (we) object very strongly to
17,094	Keybugle .	You must object to
17,095	Keycold .	Why do (does) —— object to
17,096	Keycolour	We do not object to what you propose
17,097	Keyscrew .	Cannot see the object of
17,098	Keyseat .	The object is evidently to
17,099	Keystone .	What object have you in view
17,100	Keyway .	Have no other object than
17,101	Khamsin .	Our object has been, and still is, to
17,102	Khansamah	So long as this object is attained
17,103	Khawass .	It must be our object to
17,104	Khaya . .	The object for which we are striving
17,105	Khenna .	**Objected**
17,106	Khuskhus.	Objected to
17,107	Kibbal . .	**Objection(s)**
17,108	Kick . .	Has (have) raised an objection to
17,109	Kickable .	Make no objection
17,110	Kickshaws	What is the objection
17,111	Kickup .	No objection to your
17,112	Kiddows .	There would be no objection to
17,113	Kiddypie .	I (we) have no objection
17,114	Kidfox. .	Have you any objection to
17,115	Kidlings .	We shall raise objection
17,116	Kidnap .	Will raise objection

No.	Code Word.	**Objection(s)** *(continued)*
17,117	Kidnapped	The chief objection
17,118	Kidnapping	**Obligation(s)**
17,119	Kidneybean	Avoid putting us under any obligation to
17,120	Kidneylipt	Thereby preventing our being under any obligation
17,121	Kidneywort	Putting us under an obligation to [to
17,122	Kilderkin .	Is (are) under a great obligation to
17,123	Killadar .	Under obligations (to)
17,124	Killas . .	Fulfil —— obligations
17,125	Killcow .	Can obligations be fulfilled
17,126	Killeth .	. **Oblige**
17,127	Killigrew .	This will oblige us to
17,128	Killingly .	This will not oblige us to
17,129	Killinite .	To oblige
17,130	Kiln . .	**Obliged**
17,131	Kilndried .	Shall be obliged (to)
17,132	Kilndry .	Will not be obliged (to)
17,133	Kilndrying	We have been obliged to
17,134	Kilnhole .	Are we obliged (to)
17,135	Kilodyne .	Obliged to be very cautious
17,136	Kilogramme	Are you obliged to
17,137	Kilolitre .	**Observation**
17,138	Kilostere .	A close observation leads us to think
17,139	Kimcoal .	After the closest observation
17,140	Kincob .	The closest observation failed to show
17,141	Kindlecoal	**Observe**
17,142	Kindlefire .	Did not observe
17,143	Kindliness	**Observed**
17,144	Kindly . .	Has (have) been observed
17,145	Kindspoken	Has (have) not been observed
17,146	Kinesodic .	Have you ever observed
17,147	Kinetic .	Has (have) been recently observed
17,148	Kingapple	Have observed that
17,149	Kingatarms	**Obstacle**
17,150	Kingbird .	A great obstacle
17,151	Kingcrabs .	This obstacle must be overcome at once
17,152	Kingcrow .	An unforeseen obstacle
17,153	Kingcup .	Place (placing) every obstacle in our way
17,154	Kingdom .	(To) remove the obstacle
17,155	Kingfisher	This obstacle will be the cause of much delay
17,156	Kinggeld .	Expense must be no obstacle
17,157	Kingkiller .	The greatest obstacle is the
17,158	Kinglihood	**Obtain**
17,159	Kinglypoor	Can you obtain
17,160	Kingmullet	You must obtain
17,161	Kingpiece	You must obtain longer time
17,162	Kingpost .	Try and obtain easier terms
17,163	Kingsevil .	Endeavour to obtain
17,164	Kingspear .	I (we) can easily obtain
17,165	Kingstones	I (we) cannot obtain
17,166	Kingtable .	Difficult to obtain

No.	Code Word.	Obtain (*continued*)
17,167	Kingtruss .	I (we) hope to obtain
17,168	Kinkajou .	Likely to obtain
17,169	Kinked .	Important to obtain
17,170	Kinkhaust	Think we can obtain better terms
17,171	Kinking .	Think we can obtain better results
17,172	Kinless .	How much can you obtain
17,173	Kinone .	If you cannot obtain all
17,174	Kinsfolk .	**Obtainable**
17,175	Kinship .	Must be obtainable
17,176	Kinsman .	Obtainable under the deed
17,177	Kinswoman	The property not now obtainable
17,178	Kippage .	May be obtainable later on
17,179	Kippered .	**Obtained**
17,180	Kippernut	Have obtained all we want
17,181	Kipskin .	Have obtained our ends
17,182	Kirkyard .	Have obtained
17,183	Kirtle . .	Has (have) not obtained
17,184	Kirwanite.	Obtained from this test
17,185	Kismet .	Can be obtained from
17,186	Kiss . .	Cannot be obtained from
17,187	Kissing .	Can be obtained locally
17,188	Kissmiss .	Cannot be obtained locally
17,189	Kitcat . .	The results obtained
17,190	Kitcatroll .	The results obtained would justify
17,191	Kitchen .	**Obtaining**
17,192	Kitchendom	Did not succeed in obtaining
17,193	Kitchenry.	Succeeded in obtaining
17,194	Kiteflier .	**Obviate**
17,195	Kiteflying.	To obviate this
17,196	Kitefoot .	To obviate the necessity
17,197	Kithara .	Thereby obviating the necessity of
17,198	Kitten . .	**Occasion**
17,199	Kittenhood	Is there any occasion for
17,200	Kittenish .	There is no occasion for
17,201	Kittiwake .	If the occasion occurs
17,202	Kleeneboc	On the next occasion
17,203	Klopemania	May have occasion to (for)
17,204	Knabbing.	**Occasional(ly)**
17,205	Knabble .	Except for an occasional
17,206	Knack . .	Occasionally occur
17,207	Knackish .	**Occupation**
17,208	Knapbottle	Can you obtain occupation for
17,209	Knappeth.	Occupation has been to
17,210	Knappia .	There is plenty of occupation for
17,211	Knapsack .	**Occupied**
17,212	Knapweed	At present occupied in
17,213	Knautia .	Occupied in preparing statement
17,214	Knave . .	If not already occupied
17,215	Knavery .	How long shall you be occupied
17,216	Knaveship	The ground is already occupied

No.	Code Word.	
17,217	Knavish	. **Occupy**
17,218	Knavishly.	This will occupy
17,219	Kneaded .	This will occupy me (us) —— days
17,220	Knebelite.	This will not occupy
17,221	Kneebrush	Will this occupy
17,222	Kneecords	Occupy yourself with other inspections till
17,223	Kneedeep.	**Occurred**
17,224	Kneehigh .	Nothing of importance has yet occurred
17,225	Kneeholly	Telegraph if anything of importance has occurred
17,226	Kneejoint.	What has occurred that we do not hear
17,227	Kneeling .	A disagreement has occurred between
17,228	Kneepan .	Has occurred
17,229	Kneepiece	Has anything occurred
17,230	Kneerafter	Has anything occurred to lead you to suppose
17,231	Kneestop .	Nothing has occurred to lead us to believe
17,232	Kneestring	Have from time to time occurred
17,233	Kneeswell	What has occurred makes us anxious
17,234	Kneetimber	After what has occurred, we think it will be best
17,235	Knells . .	**Occurrence**
17,236	Knickknack	Is (are) of frequent occurrence
17,237	Knifeblade	Is (are) not of frequent occurrence
17,238	Knifeboard	Is (are) —— of frequent occurrence
17,239	Knifebox .	**Offer(s)**
17,240	Knifeedge.	You had better make —— an offer
17,241	Kniferest .	Has (have) made an offer [through,
17,242	Knifetray .	With firm offer I am (we are) certain to put it.
17,243	Knight .	I (we) would advise you to withdraw the offer
17,244	Knightage	Cannot make better offer
17,245	Knighthood	Offer accepted
17,246	Knightless	Offer declined
17,247	Knightlike	Cannot obtain better offer
17,248	Knightly .	Has (have) withdrawn his (their) offer
17,249	Knightsfee	If the offer is firm
17,250	Knightship	With a firm offer
17,251	Knitster .	Has (have) made no offer as yet
17,252	Knittable .	Has (have) declined the offer
17,253	Knitting .	Is (are) —— likely to make any offer
17,254	Knobbiness	A reasonable offer
17,255	Knobby .	An unreasonable offer
17,256	Knobkerrie	Do you accept the offer
17,257	Knobstick	Does he (do they) accept our offer
17,258	Knockdown	Have submitted your offer to
17,259	Knocking.	If the offer is not accepted
17,260	Knockkneed	Shall we accept the offer
17,261	Knockstone	Accept the offer
17,262	Knoppern	If we accept the offer
17,263	Knorria .	Renewal of offer
17,264	Knosp . .	Accept offer if it is the best you can get
17,265	Knotberry	Offer to do it for
17,266	Knotgrass.	Wire the best offer you can get (for)

No.	Code Word.	Offer(s) (*continued*)
17,267	Knotless .	Offer will be kept open for
17,268	Knotted .	Cannot keep offer open any longer
17,269	Knottiest .	Best offer that can be got (is)
17,270	Knotwort .	(Is) now under offer
17,271	Knout . .	Consider offer withdrawn
17,272	Knowable.	Cannot get an offer
17,273	Knoweth .	Made an offer
17,274	Knowing .	Keep the offer open
17,275	Knowingly	Offers the property for
17,276	Knowledge	Offer(s) that part of the property, known as
17,277	Knowltonia	Do not offer more than
17,278	Knubs . .	At the time the offer·was made
17,279	Knuckles .	A *bonâ-fide* offer
17,280	Knurled .	Accept the offer, provided that
17,281	Knurry .	Offer accepted ; come over at once
17,282	Kobellite .	Offer is too late
17,283	Kobold .	Have refused offer of
17,284	Koeleria .	Offer withdrawn
17,285	Koff . .	Received better offer [take the matter up
17,286	Kohlrabi .	Offer made is not sufficient inducement to us to
17,287	Kokob . .	Offer is under consideration
17,288	Kokrawood	Have submitted your offer
17,289	Konigite .	Offer subject to immediate reply
17,290	Koninckia	Offer subject to reply within
17,291	Koordish .	**Offered**
17,292	Korkalett .	Has (have) offered to
17,293	Kousso .	I (we) have offered to
17,294	Kraals .	Has (have) —— offered to
17,295	Kriegspiel.	The property was offered
17,296	Kritarchy .	The mine has been offered
17,297	Kruller .	I am offered
17,298	Krummhorn	I have offered
17,299	Kufic . .	The property is being offered at
17,300	Kuhhorn .	Offered and refused
17,301	Kuhnia .	Offered and accepted
17,302	Kuichua .	Offered the mine for a sum of
17,303	Kumquat .	If the mine (or property) is offered
17,304	Kundahoil	**Offering**
17,305	Kyanize .	Is (are) offering the mine for
17,306	Kyanizing.	What is (are) —— offering the mine for
17,307	Kyanol .	Has (have) been offering (to)
17,308	Kymograph	**Office**
17,309	Kyrie . .	In the office
17,310	Kyriolexy.	Has (have) an office in
17,311	Kyrsin . .	A good man to manage the office and keep the
17,312	Labadist .	For record in the office [accounts
17,313	Labarri .	**Official(ly)**
17,314	Labefied .	An official statement
17,315	Labefy . .	An official declaration
17,316	Labefying .	(The) official authorization

No.	Code Word.	**Official(ly)** (*continued*)
17,317	Labelled .	(The) official notification
17,318	Labellum .	Official map (of)
17,319	Labially .	Official map, showing
17,320	Labiated .	Officially recorded
17,321	Labipalpi .	Officially reported
17,322	Laboratory	**Often**
17,323	Laborious .	Which often occurs
17,324	Labour. .	Has often been found
17,325	Labouring	Not very often
17,326	Labourless	Cannot often be seen
17,327	Laboursome	The mine(s) has (have) often been
17,328	Labrax . .	How often do you
17,329	Labridan .	It often occurs (that)
17,330	Laburnum	**Old**
17,331	Labyrinth .	Old workings full of water
17,332	Laccic . .	In the old workings
17,333	Lacdye .	In the old shaft
17,334	Lacebark .	The buildings are old and dilapidated
17,335	Laceboot .	Too old for
17,336	Laceframe	Very old
17,337	Laceleaf .	Below the old workings
17,338	Lacemakers	Extensive old workings
17,339	Laceman .	The old workings appear to have been
17,340	Lacepaper	Old workings quite exhausted
17,341	Lacepillow	Old workings caved in
17,342	Lacerable .	Cannot yet get into the old workings
17,343	Lacerating	Old workings extend to a depth of
17,344	Lacerative	Under the old management
17,345	Lacertian .	The old company
17,346	Lacertilia .	The old board
17,347	Lacewoman	The mine is a very old one, and has been worked
17,348	Lachesis .	The machinery is very old and nearly worn out
17,349	Lachesness	**Omission**
17,350	Lachrymal	By the omission of
17,351	Lachrymose	The omission may involve
17,352	Lacings .	The omission of all reference to
17,353	Laciniform	**Omit(s)**
17,354	Lackadaisy	Do not omit to
17,355	Lackalls .	Do not omit to mention
17,356	Lackbrain .	Do not omit to see
17,357	Lackeymoth	Do not omit to send
17,358	Lackeys .	You must not omit
17,359	Lacklinen .	If you omit to do so
17,360	Lacklustre	If he omits—If they omit
17,361	Laconic .	Will not omit (to)
17,362	Laconical .	We think you should omit
17,363	Laconicism	Omit that part of
17,364	Laconized	**Omitted**
17,365	Lacquered	Has (have) omitted to
17,366	Lacquering	The —— has (have) been omitted

No.	Code Word.	**Omitted** (*continued*)
17,367	Lacrymable	It has been omitted
17,368	Lactamide	Has (have) omitted to mention
17,369	Lacteal .	Has (have) omitted to send
17,370	Lacteally .	Omitted in my (our) last
17,371	Lacteously	**On**
17,372	Lactific .	On and after the
17,373	Lactifuge .	Has (have) now on hand
17,374	Lactometer	On hand and in transit
17,375	Lactoscope	On the whole
17,376	Lactose .	**Once**
17,377	Lactuca .	Go at once to
17,378	Lactumen .	You must see —— at once
17,379	Lacunaria .	Can you at once
17,380	Lacustrine	Cannot be done at once owing to
17,381	Ladderwork	This must be seen to at once
17,382	Ladiesman	We want at once
17,383	Ladkin .	**Only**
17,384	Ladleful .	I (we) could only judge by
17,385	Ladling .	I (we) could only
17,386	Ladybird .	Can only be done at
17,387	Ladybrach	Appears the only thing to be done
17,388	Ladyclock	The mill is only running —— days per month
17,389	Ladycourt	Is (are) only working —— days per month
17,390	Ladycow .	Only one sale at that price
17,391	Ladyfern	There has been only one
17,392	Ladyhood .	Can only be done by
17,393	Ladylike .	There is (are) only
17,394	Ladylove .	Only a few
17,395	Ladysbower	Only enough
17,396	Ladyscomb	If only a little
17,397	Ladysgown	Only once
17,398	Ladyshair .	Can only do
17,399	Ladyship .	Can only be
17,400	Ladysmaid	If you can only
17,401	Ladysseal .	Had only
17,402	Ladyssmock	Had only just enough
17,403	Lagbellied	Have only just learnt that
17,404	Lagomys .	We can only
17,405	Lagoon .	We can only suppose that
17,406	Lagopus .	**Onyx**
17,407	Lagostoma	Mexican onyx
17,408	Lagostomys	Onyx and other minerals
17,409	Lagothrix .	**Open**
17,410	Lagurus .	Open to an offer [valuable mine
17,411	Laic . .	I am (we are) convinced that this will open up a
17,412	Laical . .	Will this open up a desirable property
17,413	Laicality .	In order to open up
17,414	Laird . .	Open an account at
17,415	Lairdship .	Open to consider
17,416	Lakebasin .	Am (are) open to

No.	Code Word.	Open (*continued*)
17,417	Lakelet .	Are you open to
17,418	Lakepoet .	Is he (are they) open to
17,419	Lakewake.	Not now open to offer
17,420	Lamaism .	Open as long as possible
17,421	Lamarckism	When do you expect to open up
17,422	Lamasery .	Before we can open up
17,423	Lambative	**Open-cuts**
17,424	Lambdacism	The open-cuts have fallen in [abandoned
17,425	Lambdoidal	When the open-cuts became dangerous the mine was
17,426	Lambkins .	The mine has been extensively worked by open-cuts
17,427	Lamblike .	The open-cuts have been filled in
17,428	Lambswool	(By) open-cuts and trial shafts
17,429	Lameduck	(By) open-cuts and prospect tunnels
17,430	Lamella .	**Opened.** (See also Account and Market.)
17,431	Lamellarly	At what bank have you opened an account
17,432	Lamellated	Have opened up a fine body of ore
17,433	Lamellose.	The mine has not been opened up
17,434	Lament .	The mine has been opened up well
17,435	Lamentable	Have you yet opened up
17,436	Lamenteth	**Opening**
17,437	Lamenting	What will the cost be of opening up the mine
17,438	Lamination	By opening up
17,439	Lammas .	Must be set aside for opening up the mine
17,440	Lammasday	To make an opening
17,441	Lammastide	We are now opening up
17,442	Lampblack	Opening communication with
17,443	Lampereel	**Operated**
17,444	Lamperns .	The quantity operated upon
17,445	Lampetians	Have operated upon
17,446	Lampglass	Is being operated on
17,447	Lampooned	Operated on by the new process
17,448	Lampoonry	**Operation**(s)
17,449	Lamppost .	In full operation
17,450	Lamprey .	Not in operation
17,451	Lampshade	The mill was not in operation when I was there
17,452	Lampshell	Before commencing operations
17,453	Lampyrine	Commence operations
17,454	Lancegay .	Put into operation
17,455	Lancehead	Limit your operations to
17,456	Lanceolar .	Are limiting our operations to
17,457	Lancetfish	Limit operation to those workings (only)
17,458	Lancewood	By limiting the operations
17,459	Lancinated	Large " bull " operations (in)
17,460	Landagent	Large " bear " operations (in)
17,461	Landaulet	Operations have resulted in
17,462	Landblink	No longer in operation
17,463	Landbreeze	Repeat the operation
17,464	Landbug .	Continue operations
17,465	Landcrab .	Undergone an operation
17,466	Landdamn	Had to undergo an operation

No.	Code Word.	**Operation(s)** (*continued*)
17,467	Landfall .	Operations are most satisfactory
17,468	Landflood.	Operations are not satisfactory
17,469	Landforce.	**Opinion**
17,470	Landfowls	Telegraph sufficient details to enable us to form here an intelligent opinion of the value of the [mine
17,471	Landgabel	My opinion is very unfavourable
17,472	Landholder	In my opinion
17,473	Landice .	Is (are) of opinion that
17,474	Landladies	Ascertain ——'s opinion of
17,475	Landlady .	Have changed my opinion since
17,476	Landleaper	Telegraph your opinion (of)
17,477	Landless .	What is your opinion with regard to
17,478	Landlocked	The opinion of all who have visited the mine is
17,479	Landloping	Has (have) formed a very poor opinion of
17,480	Landlords.	Has (have) formed a very high opinion of
17,481	Landlubber	The general opinion is
17,482	Landlurch	What is the legal opinion
17,483	Landman .	What is the general opinion
17,484	Landmarks	Expert's opinion (is)
17,485	Landoffice	Counsel's opinion (taken)
17,486	Landowner	Take counsel's opinion and act as he advises
17,487	Landpilot.	Let us have your candid opinion
17,488	Landreeve	Opinion not worth anything
17,489	Landrents.	In the opinion of
17,490	Landroll .	If your opinion is unchanged
17,491	Landscapes	In my opinion mine will develop into a fine property
17,492	Landscurvy	In my opinion mine is worthless
17,493	Landshark	Opinion based upon
17,494	Landside .	Too soon to form an opinion
17,495	Landslips .	Has (have) formed a decided opinion
17,496	Landspout	Our opinion is unchanged
17,497	Landspring	Is (are) of the same opinion
17,498	Landstrait	Write fully your opinion
17,499	Landtaxes	Take the opinion of——, and cable
17,500	Landturtle	Have taken the opinion of
17,501	Landurchin	A very unfavourable opinion
17,502	Landwaiter	Cannot express opinion without further information
17,503	Landworker	**Opponent(s)**
17,504	Langaha .	Is (are) our opponent(s) as to
17,505	Langsettle	Is (are) very formidable opponent(s)
17,506	Langsyne .	Are there any opponents to fear
17,507	Languages	Is a (are) opponent(s) not to be much feared
17,508	Languid .	Opponents who can be squared
17,509	Languidly.	Our opponents are
17,510	Languor .	Our opponents have
17,511	Languorous	Has (have) many opponents
17,512	Laniary .	Our opponents intend to
17,513	Laniferous	Our opponents did not venture to
17,514	Lanifical .	Our opponents were successful
17,515	Lanius. .	Give opponents no opportunity

N

No.	Code Word.	
17,516	Lansium .	**Opportunity**
17,517	Lansquenet	When will there be a good opportunity
17,518	Lantcha .	Should the opportunity occur
17,519	Lanterloo .	At the earliest opportunity
17,520	Lanternfly	If I can get the opportunity I will
17,521	Lanthorn .	Have not had an opportunity of
17,522	Laophis .	Have lost the opportunity (to)
17,523	Laparocele	If favourable opportunity occurs
17,524	Lapdogs .	Presents no opportunity
17,525	Lapicide .	Will lose the opportunity if you do not
17,526	Lapidarian	There is a favourable opportunity
17,527	Lapidator .	If you lose the opportunity
17,528	Lapideous	**Oppose(s)**
17,529	Lapidified .	Intend(s) to oppose
17,530	Lapidify .	It would not be wise at first to oppose
17,531	Lapidose .	Oppose(s) us in every way
17,532	Lapilli . .	Do not oppose
17,533	Lapjointed	Will not oppose
17,534	Laplap. . .	**Opposed**
17,535	Lappaceous	The directors are opposed to
17,536	Lappior .	Am (are) opposed to
17,537	Lappish .	Is (are) opposed (to)
17,538	Lapponian	Is (are) not opposed (to)
17,539	Lapsable .	We were opposed by
17,540	Lapsana .	We shall be opposed by
17,541	Lapstones .	In the event of our being opposed
17,542	Lapstreak .	**Opposition**
17,543	Laputan .	Have you met with any opposition
17,544	Lapwings .	If you meet with any opposition
17,545	Lapwork .	Has (have) met with great opposition
17,546	Laquear .	Has (have) not met with any opposition
17,547	Larboard .	There was considerable opposition
17,548	Larcenist .	After considerable opposition we succeeded in
17,549	Larcenous	Offering great opposition
17,550	Larceny .	Opposition from a large section of the shareholders
17,551	Lardaceous	Opposition being got up by
17,552	Lardoil .	Expect opposition from
17,553	Large . .	Meets with great opposition
17,554	Largeacred	Little or no opposition
17,555	Largifical .	The opposition was very feeble
17,556	Largition .	The opposition has collapsed
17,557	Larksheel .	You are acting in opposition to your own interests
17,558	Larkspur .	The opposition comes from a few large shareholders
17,559	Larrup . .	The opposition comes from
17,560	Larruping .	In the event of opposition to the proposals
17,561	Larvated .	In opposition to the wishes of the Board
17,562	Larviform .	In opposition to our wishes and instructions
17,563	Larvipara .	**Option**
17,564	Laryngeal .	At my option
17,565	Laryngitis .	At your option

No.	Code Word.	**Option** (*continued*)
17,566	Lascivient	At his option [option
17,567	Lascivious	Do you authorize my (our) paying —— to secure
17,568	Lashfree .	We must have the option of
17,569	Lasionite .	Will give us no option in the matter
17,570	Lassitude .	Give(s) us the option of
17,571	Lasslorn .	To have the option of
17,572	Lasso . .	An option on the property
17,573	Lassoing .	Can have option by giving bonds for
17,574	Lastcourt .	Option for —— months
17,575	Lastheir .	Can we have the option of
17,576	Lastingly .	If you can get an option on the property
17,577	Lastrea .	Try and get the option of
17,578	Latakia .	There is no option in the matter
17,579	Latchets .	This company has the option on the property
17,580	Latchkey .	**Order(s)**
17,581	Latency .	Act strictly according to order(s) from
17,582	Laterality .	Order at once
17,583	Latescence	Do not order
17,584	Lathebed .	Out of order
17,585	Lathraea .	When will everything be in working order
17,586	Lathreeve .	To your order
17,587	Lathwork .	To my (our) order
17,588	Lathyrus .	Is (are) in good working order
17,589	Latialite .	Have placed order with
17,590	Latibulize .	Have not yet placed order
17,591	Laticlave .	Before placing order
17,592	Latinism .	Get estimate and tender before placing order
17,593	Latinistic .	From whom shall we order
17,594	Latinity .	Think you had better place order with
17,595	Latiseptae .	Have executed the order(s)
17,596	Latitancy .	Work on order
17,597	Latrine .	Now under order from
17,598	Latrociny .	We authorize you to order
17,599	Latterly .	Decline the order
17,600	Lattermath	Have written to decline the order
17,601	Lattices .	Declining to place the order with
17,602	Latticing .	If —— accept(s) the order
17,603	Laudably .	Order still in force
17,604	Laudanum	Order cancelled
17,605	Laudation	Cannot cancel order
17,606	Laudative .	Waiting for orders
17,607	Lauder . .	If all is in order
17,608	Laughable	Everything is now in order
17,609	Laughingly	Consigned to your orders
17,610	Laughs . .	Until receipt of orders
17,611	Laughsome	Until further orders
17,612	Laughter .	Can you execute the order
17,613	Laumontite	Cannot carry out the orders
17,614	Launch .	(Upon) completion of the order
17,615	Launching	Received an order (from)

No.	Code Word.	Order(s) *(continued)*
17,616	Launderer	In order to prevent
17,617	Laundress	In order to allow
17,618	Laundry .	**Ordered**
17,619	Laureate .	Have already ordered
17,620	Laureating	If it has (they have) not been already ordered
17,621	Laurel . .	Has been ordered to
17,622	Laurelled .	As ordered in the first instance
17,623	Laurencia .	When was —— ordered
17,624	Laurestine	Too late ; had already ordered
17,625	Lava . .	**Ordinary**
17,626	Lavandula	Is it an ordinary thing
17,627	Lavatera .	Nothing but the ordinary
17,628	Lavatories	In the ordinary manner
17,629	Laveeared	The ordinary amalgamation process
17,630	Lavender .	By the ordinary process
17,631	Lavish . .	**Ore.** (See also Estimate, Free, Grade.)
17,632	Lavishly .	Ore chamber [the week was —— tons
17,633	Lavishment	Ore extracted and forwarded to the mill(s) during
17,634	Lavishness	All ready to ship ore to the extent of —— tons
17,635	Lawabiding	Have discovered a good body of pay ore
17,636	Lawbooks .	Assorted ore [tons
17,637	Lawbreach	I (we) estimate the amount of ore in sight at ——
17,638	Lawbreaker	I (we) estimate the value of the ore at
17,639	Lawburrows	The reserve of ore
17,640	Lawcalf .	Good ore is still being found
17,641	Lawday .	Ore of fair grade
17,642	Lawful . .	The bunch of ore is pinching out [ever
17,643	Lawfully .	Ore is —— inches thick, and continues as rich as
17,644	Lawfulness	We are now putting high-grade ore through the mill
17,645	Lawgiver .	We are now putting low-grade ore through the mill
17,646	Lawgiving	You must put high-grade ore through the mill in order to have sufficient money for the quarter's
17,647	Lawlessly .	The ore from these workings [dividend
17,648	Lawlists .	A streak of ore in this mine gives assays of
17,649	Lawlore .	Now producing large quantities of ore
17,650	Lawmaker	Now producing large quantities of rich ore
17,651	Lawmaking	There is no pay ore below the —— level
17,652	Lawmonger	There is no pay ore above —— level
17,653	Lawn . .	Owing to the ore being very refractory
17,654	Lawnmower	We are taking out —— tons of ore in the twenty-
17,655	Lawnsleeve	The ore is very rich [four hours
17,656	Lawntennis	The amount of ore in sight is —— tons, yielding a
17,657	Lawofficer	Ore pays [net profit of —— per ton
17,658	Lawsonia .	The ore is getting poor (in)
17,659	Lawsuit .	The ore is giving out (in)
17,660	Lawwriter.	Ore in shaft is improving as it goes down
17,661	Lawyer .	The ore in sight will last for
17,662	Lawyerlike	Ore in stope is improving
17,663	Lawyerly .	Ore in winze is improving
17,664	Laxities .	Ore is good and will pay

No.	Code Word.	Ore (*continued*)
17,665	Laxness .	Ore is improving in the drift as we advance
17,666	Laydown .	The ore is milling better than was expected
17,667	Layerboard	Ore in the —— level is very rich
17,668	Layering .	Ore struck, assays show —— per ton
17,669	Layerout .	Are out of ore
17,670	Layfigure .	The ore is found in
17,671	Layrace .	Is there any ore in sight
17,672	Laysermon	Have found ore
17,673	Lazarhouse	Number of tons of first-class ore is
17,674	Lazarists .	Number of tons of second-class ore is
17,675	Lazarlike .	Number of tons of third-class ore is
17,676	Lazarwort .	Number of tons of ore on the dump is
17,677	Laziest . .	Shipping —— tons of ore to the mill(s) per day
17,678	Lazuli . .	Take out all the ore you can
17,679	Lazy . .	The ore cannot be worked except by
17,680	Lazybed .	The body of ore is narrowing
17,681	Lazybones	The body of ore is increasing in width
17,682	Lazytongs .	So as to open up fresh bodies of ore
17,683	Leadarming	I (we) cannot procure ore on account of
17,684	Leadash .	We get our principal supply of ore out of
17,685	Leaderette	Ore improving in quality as developed
17,686	Leadership	First-class ore
17,687	Leadglance	Drift now in ore, assaying —— per ton
17,688	Leadgray .	Fine body of ore
17,689	Leadmill .	Has the appearance of good body of ore
17,690	Leadmine .	No pay ore found yet
17,691	Leadplants	Tons of ore in sight in mine
17,692	Leadscrew	Free milling ore
17,693	Leadspar .	Refractory ore
17,694	Leadworks	The body of ore is getting poorer
17,695	Leafbridge	Teams cannot haul ore at present on account of
17,696	Leafbud .	Ore contains no visible gold
17,697	Leafcutter .	Selected ore
17,698	Leaffat . .	Rich ore
17,699	Leafgold .	Very rich ore ; averaging —— per ton
17,700	Leafhopper	Poor ore
17,701	Leafinsect	Poor ore ; does not average more than
17,702	Leaflard .	Have struck pay ore
17,703	Leafless .	Have struck very rich ore
17,704	Leaflouse .	Have struck very rich ore, which promises to yield
17,705	Leafmetal .	Have struck very rich ore, and expect it to continue
17,706	Leafmould	Have struck ore of fair quality
17,707	Leafstalk .	Have struck high-grade ore
17,708	Leafy . .	Are in good ore
17,709	League .	Ore continues so far very good
17,710	Leaguelong	Ore continues to improve
17,711	Leaguerer .	Coming into good ore
17,712	Leakage .	Shows a body of ore —— wide
17,713	Leakiness .	Has developed a large body of high-grade ore
17,714	Leanfaced .	Has developed a large body of low-grade ore

No.	Code Word.	Ore (*continued*)
17,715	Leanto .	Will place at our command a valuable body (or stope) of high-grade ore
17,716	Leapeth .	Has placed in our possession a large body of ore
17,717	Leapfrog .	There is a very good prospect of finding ore
17,718	Leapingly .	No prospect of finding ore in this
17,719	Leapweel .	A rich streak of ore has been found
17,720	Leapyear .	A rich streak of ore, assaying about
17,721	Learn . .	Ore that will pay expenses of mining and milling
17,722	Learnable .	Ore that will assay —— per ton
17,723	Learnedish	Ore that will not pay expenses
17,724	Learnedly .	Ore blocked out
17,725	Leasable .	Ore in sight averages about —— per ton
17,726	Leasehold	How many tons of ore are you getting daily
17,727	Leashing .	How many tons of ore have you open as reserve
17,728	Leastways.	How many tons of ore are stored on dump
17,729	Leastwise .	How is the ore looking in the
17,730	Leathery .	A daily supply of ore sufficient for the mill(s)
17,731	Leaveneth	Cannot mine sufficient ore to keep —— mill(s) going
17,732	Leavening	Sufficient quantity of high-grade ore
17,733	Leavenous	Sufficient quantity of low-grade ore
17,734	Lecanora .	Bearing strictly in view average tonnage value of ore
17,735	Lecher . .	Tonnage of ore [reserves
17,736	Lecherous	Tonnage and gross value of ore
17,737	Leckstone	Tonnage and value of ore reserves
17,738	Lecterns .	Tonnage of ore stored on dump
17,739	Lecturing .	Tonnage and value of ore milled
17,740	Lecythis .	Tonnage and value of ore shipped
17,741	Ledgerbook	Ore shipped to smelters [value of
17,742	Ledgerline	Tons of ore shipped during —— of an estimated
17,743	Leeboards.	Shipped —— tons of ore to smelters
17,744	Leech . .	Ore reserves
17,745	Leechcraft	Ore reserves, high grade
17,746	Leechline .	Ore reserves, low grade
17,747	Leechrope	Ore reserves, estimated average —— per ton
17,748	Leefange .	Approximate tonnage and value of ore reserves
17,749	Leekgreen	Ore reserves now in sight ; estimated tonnage ——;
17,750	Leelite. .	The grade of ore is very good [estimated value
17,751	Leelurch .	The grade of ore is not up to my expectations
17,752	Leetides .	Ore now being extracted from
17,753	Leetman .	First-class ore from —— to ——.per ton
17,754	Leewardly	Second-class ore from —— to —— per ton
17,755	Lefthanded	Third-class ore from —— to —— per ton
17,756	Leftoff . .	Ore suitable for the —— mill
17,757	Leftward .	Ore assays : gold, —— ozs. ; silver, —— ozs.
17,758	Leftwitted	Ore averages —— ozs. silver ; —— per cent. copper ; —— per cent. lead
17,759	Legacies .	Ore averages —— per cent. lead
17,760	Legacy .	Ore averages —— per cent. copper
17,761	Legal . .	Ore assays per ton —— ozs. of silver ; —— per
17,762	Legalities .	Ore from face of drift [cent. of lead

No.	Code Word.	Ore (*continued*)
17,763	Legalism .	Ore from winze
17,764	Legalize .	Ore from stope
17,765	Legalizing	Had not any better ore to crush
17,766	Legalness .	The ore is in thin stringers
17,767	Legantine. .	Interspersed with the ore
17,768	Legateship	The ore shows a considerable quantity of
17,769	Legbail .	A good breast of ore
17,770	Leggism .	Rich ore going down in
17,771	Leggy . .	Feet of rich ore has been struck
17,772	Legibility .	Ore is high in sulphurets
17,773	Legibly .	Ore is rich in gold
17,774	Legionary .	Ore shows free gold
17,775	Legions .	Ore is copper stained
17,776	Legislator .	Ore shows ruby silver
17,777	Legitim .	Have forwarded samples of ore
17,778	Legitimacy	Samples of ore
17,779	Legitimate	Ore is largely charged with
17,780	Legitimism	The character of the ore
17,781	Legless .	The ore mills tolerably free
17,782	Leglock .	The ore is somewhat refractory
17,783	Leguleian .	The ore wants roasting
17,784	Leguminous	Ore only suitable for milling
17,785	Leiodon .	Ore only suitable for smelting
17,786	Leiothrix .	Is there any ore to be seen
17,787	Leiotrichi .	The ore comes mostly from
17,788	Leipoa. .	Ore delivered at mill this week
17,789	Leipothymy	Ore reduced this week
17,790	Leisurable	Ore smelted this week
17,791	Leisure .	Ore shipped this week
17,792	Leisurely .	Ore delivered at mill this month
17,793	Lemniscata	Ore reduced this month
17,794	Lemodipoda	Ore smelted this month
17,795	Lemonade	Ore shipped this month
17,796	Lemongrass	Tons of ore in the twenty-four hours
17,797	Lemonjuice	Ore breakers
17,798	Lemonkali	Ore feeders
17,799	Lemonpeel	**Orebody (Orebodies)**
17,800	Lemuridae	Orebody to the north
17,801	Lengthened	Orebody to the south
17,802	Lengthful .	Orebody to the east
17,803	Lengthier .	Orebody to the west
17,804	Lengthways	Orebody —— feet long and —— feet wide
17,805	Lengthwise	Orebody —— feet, width not proved
17,806	Lengthy .	Orebody now being developed
17,807	Lenience .	Developed a fine orebody
17,808	Leniently .	A fine orebody of high-grade ore
17,809	Lenitive .	A large orebody of low-grade ore
17,810	Lenocinant	A large orebody, but very low grade
17,811	Lenocinium	(From) the orebodies now being worked
17,812	Lenses . .	The orebody in the —— chute

No.	Code Word.	**Orebody** (*continued*)
17,813	Lenten. .	A very large orebody
17,814	Lenticular.	The orebody in the —— level
17,815	Lentigo .	Large orebodies have been worked
17,816	Lentiscus .	Orebody not so large as was supposed
17,817	Lentitude .	Struck a very valuable orebody
17,818	Lentoid .	Hope to find this orebody in the level below
17,819	Lenzinite .	This orebody continues in
17,820	Leonides .	**Organization**
17,821	Leoninely .	The organization of the concern.
17,822	Leonurus .	Perfect organization
17,823	Lepadite .	**Organize**
17,824	Lepidoid .	It will be necessary to organize
17,825	Lepidolite	You must organize
17,826	Lepidotus .	**Organized**
17,827	Lepismidae	Not fully organized
17,828	Leporine .	Have organized a
17,829	Lepraria .	When the arrangements are fully organized
17,830	Leprosy .	**Organizing**
17,831	Leprous .	We are now organizing
17,832	Leptolepis	**Origin**
17,833	Leptology .	The origin of the whole thing
17,834	Lernean .	From its origin
17,835	Lessening .	What was the origin of
17,836	Lessons .	**Original**
17,837	Letchtub .	According to the original agreement
17,838	Letgame .	The original document(s)
17,839	Lethality .	(Copy) copies of the original (documents)
17,840	Lethargize	Send a copy of the original
17,841	Lethargy .	**Originally**
17,842	Lethean .	The mines were originally worked by
17,843	Letheonize	Who originally worked the
17,844	Letterbox .	**Originated**
17,845	Lettercase	Originated the scheme
17,846	Letterclip .	The fire originated in
17,847	Letterize .	The report has originated from
17,848	Letterling .	The report has not originated from
17,849	Letterwood	From whom has the report originated
17,850	Letticecap	**Originating**
17,851	Leucine .	Originating in
17,852	Leucitic .	**Other(s)**
17,853	Leucitoid .	In no other way
17,854	Leucojum .	From other
17,855	Leucol. ·	Other parts of the mine
17,856	Leucopathy	By other
17,857	Leucophane	By no other
17,858	Leucosis .	Other parties
17,859	Leucostine	If by any other means
17,860	Levanters .	Is there any other
17,861	Levelcoil .	Others have said
17,862	Levelisin .	If, in any other way

No.	Code Word.	**Other(s)** *(continued)*
17,863	Levelness .	Other than
17,864	Leverboard	Or other
17,865	Levervalve	In other respects
17,866	Levesell .	On the other hand
17,867	Leviable .	**Otherwise**
17,868	Leviathan .	Otherwise it will be
17,869	Levied . .	Otherwise you will have to
17,870	Levigate .	Otherwise we run the chance of
17,871	Levigating	Otherwise it cannot be done
17,872	Levigation	Otherwise all going well
17,873	Levinbrand	But if otherwise
17,874	Levitic. .	**Ought**
17,875	Levitical .	It ought to be
17,876	Levogyrate	It ought not to be
17,877	Levulose .	Ought not to
17,878	Levying .	Which ought
17,879	Lewdly .	If we ought
17,880	Lewdness .	But which ought
17,881	Lewdster .	Do not think we ought to
17,882	Lexical .	If you think —— ought to
17,883	Lexically .	**Ounce(s)**
17,884	Lexicology	Ounce(s) of gold per ton
17,885	Lexiconist	Ounce(s) of silver per ton
17,886	Lexigraphy	How many ounces to the ton
17,887	Lexiphanic	Assuming the ounce to be worth
17,888	Leydenjar	(At) the United States standard price of silver of
17,889	Lherzolite	Weighing —— ounce(s) [1·2929 dollar per ounce
17,890	Liability .	(At) the United States standard price of gold of
17,891	Liableness	Less than half an ounce [20·67 dollars per ounce
17,892	Liafail . .	(At) the English standard of gold per ounce
17,893	Liassic . .	(At) the English standard of silver per ounce
17,894	Libations .	Taking the value of the ounce of gold at
17,895	Libatory .	Taking the value of the ounce of silver at
17,896	Libel . .	The present price of gold per ounce
17,897	Libelling .	The present price of silver per ounce
17,898	Libellist .	Less than an ounce
17,899	Libellous .	More than an ounce
17,900	Libellula .	More than —— ounces
17,901	Liberal .	Ore assaying less than —— ounce will not pay
17,902	Liberalism	The ore assays more than —— ounces
17,903	Liberality .	**Our(selves)**
17,904	Liberation	If our
17,905	Liberties .	Is (are) our
17,906	Liberty .	Can our
17,907	Libidinist .	And if our
17,908	Libidinous	Our own
17,909	Librarian .	We can reserve for ourselves
17,910	Library .	Ourselves and friends
17,911	Licensable	**Out (of)**
17,912	License .	We are quite out of

N 2

No.	Code Word.	Out (of) (*continued*)
17,913	Licensing .	Left out
17,914	Licensure .	Out of time
17,915	Licentiate	Out and home
17,916	Licentious	(To) carry out
17,917	Lichened .	**Outcrop**
17,918	Lichfowl .	The outcrop extends for many miles
17,919	Lichgate .	Judging from the outcrop
17,920	Lichway .	Having an outcrop of considerable width
17,921	Lickerish .	Samples from the outcrop in various places assayed
17,922	Lickspigot	Samples from this outcrop assayed [—— per ton
17,923	Lieabeds .	Having a bold outcrop
17,924	Lieberkuhn	The outcrop of the vein
17,925	Liegeman .	The outcrop can be traced
17,926	Lienteric .	The outcrop runs north and south
17,927	Lievrite .	The outcrop runs east and west
17,928	Lifearrow .	The outcrop is visible in places
17,929	Lifebelts .	The outcrop is strong
17,930	Lifeblood .	The outcrop yielded
17,931	Lifeboat .	To judge from the outcrop
17,932	Lifebuoys .	The outcrop runs diagonally across the claim
17,933	Lifedrop .	The outcrop runs straight through the end lines
17,934	Lifeful . .	There is an outcrop
17,935	Lifegiving .	Outcrop in several of the claims
17,936	Lifeguard .	Outcrop is plain on adjoining property, and the strike of formation is towards and through the claims
17,937	Lifehold .	There is no outcrop, but reef is probably present
17,938	Lifeless .	There is no outcrop, doubtful if reef is in the ground
17,939	Lifelessly .	**Outfit**
17,940	Lifelike .	A complete outfit (of)
17,941	Lifelong .	**Outlay**
17,942	Lifemortar	The mine(s) can be worked without a great outlay
17,943	Lifeoffice .	The state of the mine(s) necessitates a large outlay
17,944	Liferate .	An outlay of [in order to
17,945	Liferenter .	What outlay will be necessary at first (to)
17,946	Liferocket	Will this outlay be sufficient for
17,947	Lifeshot .	I (we) recommend an outlay of
17,948	Lifesome .	The first outlay need not be more than
17,949	Lifespring .	An outlay of —— will be sufficient
17,950	Lifetime .	A considerable outlay
17,951	Lifeweary .	The outlay will be soon repaid
17,952	Lifthammer	(Will) repay the outlay
17,953	Liftingrod .	**Outlying**
17,954	Liftlock .	Some outlying claims
17,955	Liftpump .	In the outlying
17,956	Ligamental	If we obtain these outlying claims
17,957	Ligaments	The outlying claims should be bought
17,958	Lightable .	Outlying properties
17,959	Lightballs .	**Output** [last
17,960	Lightboat .	What has been the output from the mine(s) in the

No.	Code Word.	Output (*continued*)
17,961	Lightbrain	The output has been —— tons in —— months
17,962	Lightdue .	The output has been —— tons in —— weeks
17,963	Lightened	Assuming an output of
17,964	Lighterage	In order to increase the output
17,965	Lighterman	To prevent any decrease in the output
17,966	Lightfoot .	Output was —— tons
17,667	Lighthouse	Output is satisfactory
17,968	Lightly .	Output is unsatisfactory
17,969	Lightmaker	Output very poor
17,970	Lightnings	Probable output
17,971	Lightship .	Output is more than
17,972	Lightsome	Output is less than
17,973	Lightwood	Output steadily increasing
17,974	Lignaloes .	Output decreasing
17,975	Lignitic .	What was the output
17,976	Ligusticum	What is the present output
17,977	Likeable .	What is the output likely to be
17,978	Likelihood	An output of over [of ore
17,979	Likelier .	Expect largely increased output from better quality
17,980	Likely . .	Expect a largely increased output from increased number of stamps now in operation
17,981	Likeminded	Expect largely increased output from exhaustive
17,982	Likeness .	Last month's poor output attributable to [effort
17,983	Lilac . .	Last month's output of these fields was
17,984	Lilacine .	Outside
17,985	Liliaceous.	On the outside
17,986	Liliputian	From the outside
17,987	Lilliput .	Are now outside
17,988	Lillypilly .	Outside of the boundaries of
17,989	Lilybeetle.	Outside our
17,990	Lilyfaced .	Outstanding
17,991	Lilyhanded	To get in all outstanding claims
17,992	Lilypad .	To pay all outstanding accounts
17,993	Lilywhite .	Are there any outstanding accounts
17,994	Limawood	No outstanding accounts
17,995	Limbeck .	Outstanding accounts
17,996	Limberhole	Outstanding liabilities (to the amount of)
17,997	Limberness	Outstanding since (last)
17,998	Limbmeal.	Outstanding debts amount to
17,999	Limbo . .	There are no outstanding claims
18,000	Limeburner	All outstanding claims
18,001	Limejuice.	Over
18,002	Limekilns.	Anything over —— will do
18,003	Limelight .	Not over
18,004	Limenean.	Nothing (none) over
18,005	Limepit .	Over and above
18,006	Limerod .	Come over
18,007	Limesink .	Send over
18,008	Limestone	There will be over
18,009	Limetwig .	We have over

No.	Code Word.	**Over** (*continued*)
18,010	Limewash .	We expect —— will soon be over
18,011	Limewater	**Overboard**
18,012	Limitarian	Were (was) thrown overboard
18,013	Limitary .	**Overcharge(s)**
18,014	Limited .	Is an overcharge ; should be deducted
18,015	Limitedly .	Overcharge will be credited next account
18,016	Limitless .	Debit —— with the overcharge
18,017	Limnoria .	All overcharges
18,018	Limonite .	**Overcome**
18,019	Limosa .	There will be great difficulties to overcome
18,020	Limosella .	In order to overcome
18,021	Limp . .	Can you overcome
18,022	Limpidity .	**Overdraft**
18,023	Limpidness	Overdraft at bankers
18,024	Limpingly .	Overdraft at bank has been reduced
18,025	Linaloa .	Overdraft has been paid off
18,026	Linctus .	Have asked bank to allow an overdraft
18,027	Linear . .	Bankers want overdraft reduced
18,028	Linearly .	Overdraft not to exceed
18,029	Lingence .	Ask bank to allow overdraft (of)
18,030	Lingered .	Will not allow further overdraft (until) or (unless)
18,031	Lingering .	**Overdrawn**
18,032	Linguiform	Account is overdrawn by
18,033	Linguist .	Has (have) overdrawn
18,034	Linguistic .	Has (have) —— overdrawn
18,035	Lingulate .	Cannot allow account to be overdrawn
18,036	Link . .	**Overdue**
18,037	Linkboy .	Is (are) overdue
18,038	Linkman .	Overdue acceptance(s)
18,039	Linkmotion	Mail is overdue
18,040	Linkwork .	Accounts overdue
18,041	Linoleum .	Account is overdue and must be paid
18,042	Linsang .	**Overhaul**
18,043	Linseed .	Must overhaul thoroughly
18,044	Linseedoil	**Overhauled**
18,045	Linstock .	Has (have) been thoroughly overhauled
18,046	Lintwhite .	**Overlap(s)**
18,047	Lionant .	Claims overlap
18,048	Lionced .	Overlap(s) the
18,049	Liondog .	A question of overlap
18,050	Lionesses .	**Overlapped**
18,051	Lionheart .	Is overlapped by
18,052	Lionism .	Overlapped by a prior location
18,053	Lionize .	**Overlapping**
18,054	Lionizing .	The overlapping claims
18,055	Lionlizard .	Shown on the plan as overlapping
18,056	Lionly . .	**Overlook**
18,057	Lionsfoot .	Do not overlook
18,058	Lionship .	You must on no account overlook (the)
18,059	Lionsleaf .	Cannot overlook

No.	Code Word.	Overlook (*continued*)
18,060	Lionsmouth	Will overlook
18,061	Liontiger .	**Overlooked**
18,062	Lipborn .	Was (were) overlooked
18,063	Lipcomfort	Has (have) been overlooked
18,064	Lipgood .	**Overlying**
18,065	Liplabour .	The overlying beds
18,066	Liplet . .	The overlying strata
18,067	Lipogram .	Overlying the reef
18,068	Lipothymia	**Overpowered (by)**
18,069	Lippia . .	Has overpowered the
18,070	Lippitude .	Has been overpowered by
18,071	Lipreading	**Overpowering**
18,072	Lipservice	Overpowering all efforts made to check it
18,073	Lipwisdom	**Overrated**
18,074	Liquable .	Much overrated
18,075	Liquate .	The value has been much overrated
18,076	Liquating .	**Overreached**
18,077	Liquation .	Overreached themselves (himself)
18,078	Liquefying	**Overreaching**
18,079	Liquescent	**Overrun (with)**
18,080	Liquidable	**Oversight**
18,081	Liquidator	By an oversight
18,082	Liquidize .	It has been a gross oversight on the part of
18,083	Liquidness	By an oversight on our part
18,084	Liquids .	By an oversight on your part
18,085	Liquorice .	**Owing**
18,086	Liricone .	Owing to the
18,087	Liripoop .	We are now owing
18,088	Lisping .	He is (they are) now owing
18,089	Lissome .	How much are we owing
18,090	Listeners .	Not owing (to)
18,091	Listeth. .	Nothing owing to them
18,092	Listful . .	Nothing owing by us
18,093	Listless .	Nothing owing by him (them)
18,094	Listlessly .	Owing nothing to
18,095	Litanies .	**Own(s)**
18,096	Litany . .	He (they) own(s)
18,097	Literalize .	Own(s) one-half of the property
18,098	Literally .	Own(s) one-quarter of the property
18,099	Literature.	Own(s) one-third of the property
18,100	Lithagogue	Own(s) two-thirds of the property
18,101	Litherlie .	Own(s) the largest share in the
18,102	Litherness	Own(s) the greatest numbers of shares in the
18,103	Lithiasis .	Own(s) the adjoining claim(s) [company
18,104	Lithiate .	Own(s) the entire property
18,105	Lithic . .	Own(s) the mine
18,106	Lithocarp .	Own(s) the concession
18,107	Lithocyst .	Own(s) a —— part of the concession
18,108	Lithodome	Who owns
18,109	Lithodomus	One company should own both mines

No.	Code Word.	**Own(s)** *(continued)*
18,110	Lithograph	By this means we shall own
18,111	Lithoid .	Do (does) not own
18,112	Lithoidal .	Do you know who owns
18,113	Litholabe .	Will own
18,114	Litholatry .	We now own
18,115	Lithology .	Who now own the
18,116	Lithomancy	Unless we can own the largest part
18,117	Lithomarge	To own the controlling interests in the mine(s)
18,118	Lithophyl .	It is necessary you should own the
18,119	Lithornis .	**Owned**
18,120	Lithotint .	The mine(s) is (are) owned by
18,121	Lithotomic	The mine(s) has (have) been owned by
18,122	Lithotrite .	Ascertain who are the different parties who have owned the
18,123	Lithotypy .	When the mine(s) was (were) owned by
18,124	Lithoxyle .	Owned by
18,125	Litigator .	Never owned by
18,126	Litigious .	Is jointly owned by —— owners
18,127	Litmus .	**Owner(s)**
18,128	Litotes . .	The owner(s) of the mine(s) will meet you at
18,129	Littleease .	See —— the owner(s) of the —— mine(s)
18,130	Littlego .	The mines have frequently changed owners
18,131	Littleness .	The owner(s) wish(es) to retain an interest in the
18,132	Littoral .	The owner(s) represent(s) [mine(s)
18,133	Littorella .	The owner(s) will accept —— in cash, and ——
18,134	Littorina .	On behalf of the owner(s) [in shares
18,135	Lituolida .	Is (are) the largest owner(s) in the mine(s)
18,136	Liturate .	The owner(s) of the adjacent claim(s)
18,137	Liturgical .	Have seen the owner(s) of the mine(s) (who)
18,138	Liturgist .	The owner(s) of the property
18,139	Liveliest .	The owner(s) of —— of the property
18,140	Livelily .	The previous owner(s)
18,141	Liveoak .	The present owner(s)
18,142	Liverfluke .	Stated by the owner(s) to be
18,143	Liveried .	Who are the owners of the property
18,144	Liverspots	The owner(s) of the
18,145	Liverstone	The owner offers
18,146	Liverwort .	Owner offers the mine for
18,147	Liverycoat	What does the owner say
18,148	Liverygown	The owner's share is
18,149	Liveryman	The owner(s) will take
18,150	Lividness .	**Package(s)**
18,151	Livonian .	Package contains (containing)
18,152	Lixiviated .	Every package
18,153	Lixivium .	Package not received
18,154	Lizards .	Each package not to weigh more than
18,155	Lizardtail .	Package of samples
18,156	Loadmanage	In packages not over —— lbs. in weight
18,157	Loadsman	In packages for mule transport
18,158	Loadstones	Packages have been opened

No.	Code Word.	**Package(s)** (*continued*)
18,159	Loafers .	Packages will be examined
18,160	Loafsugar .	Weight of packages
18,161	Loanmonger	**Packed**
18,162	Loanoffice	Packed in bags
18,163	Loathful .	Packed in case(s)
18,164	Loathing .	Sent, packed in bags
18,165	Loathingly	Securely packed
18,166	Loathly .	Must be carefully packed
18,167	Loathness .	**Paid**
18,168	Loathsome	Has (have) paid
18,169	Lobated .	This sum to be paid on
18,170	Lobbies .	To be paid
18,171	Lobby . .	Cannot be paid till
18,172	Lobbyist .	Has draft been paid
18,173	Lobcock .	Has not been paid
18,174	Lobefoot .	Has been paid
18,175	Lobelias .	Has (have) hitherto paid
18,176	Lobiole .	What has been so far paid
18,177	Lobiped .	To be paid immediately, otherwise
18,178	Loblolly .	Will have to be paid
18,179	Lobscouse	Have paid to your account
18,180	Lobspound	Bill has been paid
18,181	Lobulated .	Bill has not been paid
18,182	Lobworm .	Bill will be paid
18,183	Localisms .	Bill will not be paid
18,184	Localize .	Has (have) not paid
18,185	Localizing	To be paid as follows
18,186	Locating .	Unless paid by
18,187	Locator .	Have paid to your credit at the —— bank the sum
18,188	Lochial .	To be paid in advance [of
18,189	Lockband .	Have you paid
18,190	Lockers .	Must be paid
18,191	Lockerup .	Should be paid
18,192	Lockfast .	Should not be paid
18,193	Lockgate .	Will be paid
18,194	Lockjaw .	Will not be paid until
18,195	Lockkeeper	To be paid on delivery
18,196	Lockout .	To be paid on completion
18,197	Lockpiece .	To be paid in cash ; —— in shares
18,198	Locksill .	Remainder to be paid in instalments (of)
18,199	Locksmiths	Paid on your account
18,200	Lockspit .	Freight paid here
18,201	Lockstep .	Freight, duty, and transport paid here
18,202	Lockstitch	Freight and duty paid here
18,203	Lockweir .	Freight to be paid at destination
18,204	Locofoco .	Freight to be paid in advance
18,205	Locomotion	Have paid into court
18,206	Locomotive	Duty paid
18,207	Loculament	Paid a profit last year of
18,208	Loculous .	Paid a profit for last half-year

No.	Code Word.	
18,209	Locustbean	**Pan (out)**
18,210	Locustelle	Pan amalgamation
18,211	Locustidae	Pan gearing
18,212	Locusts .	Pan shoes
18,213	Locusttree	Pan mullers
18,214	Locutory .	Pan bottoms
18,215	Lodam. .	Pans and settlers
18,216	Lodgeable	Can pan out gold
18,217	Lodgegate	Have panned out
18,218	Lodgment	**Panning(s)**
18,219	Loftiest .	In panning out
18,220	Loftily . .	In panning out, visible gold is seen
18,221	Loftiness .	Pannings give excellent results
18,222	Lofty . .	Pannings give good results
18,223	Logarithm	Pannings give fair results
18,224	Logbook .	Pannings give encouraging results
18,225	Logcabin .	**Panic**
18,226	Logcanoe .	There has been a complete panic
18,227	Loggerhead	Things settling down after the panic
18,228	Logglass .	**Papers**
18,229	Logheaps .	Papers have been sent
18,230	Loghouse .	Duplicates of papers wanted
18,231	Loghut .	All needful papers
18,232	Logic . .	Send all papers relating to
18,233	Logical .	Have sent all papers relating to
18,234	Logically .	All the papers are ready
18,235	Logicising	All the papers are ready, and will be sent as soon as
18,236	Logistic .	**Par**
18,237	Logistical .	At par
18,238	Logman .	The shares are now at par
18,239	Logocracy	When the shares are at par, sell
18,240	Logography	Are to be had at par
18,241	Logogryph	Has (have) bought at par
18,242	Logomachy	To get the shares above par
18,243	Logomania	**Paragraph**
18,244	Logometer	The paragraph in the —— about
18,245	Logometric	The paragraph in the —— is absurd
18,246	Logothete .	The paragraph is a tissue of lies
18,247	Logotype .	Paragraph in the report (about)
18,248	Logreel .	Let us have a paragraph about
18,249	Logrolls .	**Part(s)**
18,250	Logslate .	A part of
18,251	Logwood .	Do not part with
18,252	Lohock .	On the part of
18,253	Loiterer .	Divided into —— parts
18,254	Loitering .	Each part
18,255	Loligidae .	The lesser part (of)
18,256	Loligo . .	The greater part (of)
18,257	Lollard .	On our part
18,258	Lollingly .	On your part

No.	Code Word.	Part(s) (*continued*)
18,259	Lollipops .	On his (their) part
18,260	Lollop . .	Only take a —— part
18,261	Londonize	**Participate**
18,262	Loneness .	Who participate in
18,263	Lonesome	Do you want to participate
18,264	Lonesomely	Should like to participate
18,265	Longago .	**Participating**
18,266	Longbows.	Participating equally
18,267	Longdozen	**Participation**
18,268	Longest .	In participation with
18,269	Longeval .	**Particularly**
18,270	Longevity.	You must particularly
18,271	Longevous	Will particularly
18,272	Longhand	Has (have) particularly
18,273	Longheaded	**Particulars**
18,274	Longhorned	Get full particulars about
18,275	Longicorn	Telegraph full particulars
18,276	Longimetry	Has (have) obtained full particulars as to
18,277	Longingly .	Send the fullest particulars
18,278	Longipalp	Have sent full particulars by letter
18,279	Longitude	Furnish him with full particulars
18,280	Longlegs .	Will give particulars in letter
18,281	Longlived.	Could get no particulars as to
18,282	Longmynd	Let us have further particulars by post
18,283	Longnecked	Have sent further particulars
18,284	Longnose .	**Parties.** (See Party.)
18,285	Longprimer	You must see all the parties interested
18,286	Longshore	Can you send me the names of the parties interested
18,287	Longspun.	Parties here impatient at delay
18,288	Longstop .	Parties have money ready
18,289	Longtailed	The names of the parties
18,290	Longtom .	The parties who circulate these reports
18,291	Longtongue	Circulated by the usual parties
18,292	Longways.	The parties are well known
18,293	Longwinded	There are no other parties
18,294	Longwise .	The interested parties
18,295	Lonicera .	The parties chiefly interested
18,296	Looking .	Other parties
18,297	Lookoutman	Other parties have also to be considered
18,298	Loomed .	Parties to the transaction
18,299	Loomgale.	Opposing parties
18,300	Loopholed	**Partly**
18,301	Looplight .	The mine(s) has (have) only been partly worked
18,302	Loopline .	Could only partly examine the mine(s) owing to
18,303	Loosebox .	Is (are) partly
18,304	Loosehouse	Was (were) partly
18,305	Loosely .	**Partner(s)**
18,306	Looseneth	Are partners in the concern
18,307	Loosening	Are no longer partners
18,308	Lootable .	Partner objects to

No.	Code Word.	**Partner(s)** (*continued*)
18,309	Lopestaff .	One of the partners is now in
18,310	Lophiodon	One of the partners is on his way to
18,311	Lophius .	As soon as the other partner(s) arrive(s)
18,312	Lophophore	Has bought out his partner(s)
18,313	Lophopoda	Could only see ——'s partner
18,314	Lophyrus .	**Partnership**
18,315	Lopsided .	Has gone into partnership with
18,316	Loquacious	Have formed partnership
18,317	Loquacity .	Have dissolved partnership
18,318	Loquela .	**Party.** (See Parties.)
18,319	Loranthus	Party to the transaction
18,320	Lorcha . .	The other interested party
18,321	Lorddom .	Another party
18,322	Lordeth .	The party referred to
18,323	Lordlike .	Is there any other party
18,324	Lordliness	Who is the party
18,325	Lordmayor	No other party
18,326	Lordolatry	Party interested
18,327	Lordsday .	Party well known
18,328	Lorettine .	Party not known
18,329	Loricated .	**Pass**
18,330	Loricating	To pass over
18,331	Lorication	As you will pass —— on your way to
18,332	Lorikeet .	Will you pass —— on your way to
18,333	Lorimer .	Do (does) not pass
18,334	Losengeour	Shall pass
18,335	Losing . .	Shall you pass .
18,336	Lossful .	Must pass dividend
18,337	Lotion . .	(To) pass a dividend
18,338	Lotteries .	Through having to pass
18,339	Lottery .	**Passage**
18,340	Lotus . .	Made a quick passage
18,341	Lotuseater	Made a slow passage
18,342	Lotusland .	Passage-money to be paid here
18,343	Loudful .	Agreed to pay his passage (to)
18,344	Loudlunged	First-class free passage to
18,345	Loudvoiced	Second-class free passage to
18,346	Lounge .	Pay his passage back
18,347	Lousewort	Wants his passage paid
18,348	Loutishly .	Passage money
18,349	Loveapple	Has (have) taken his passage ; to leave on the
18,350	Lovebird .	Taken passage for him by the
18,351	Lovebroker	**Passed**
18,352	Lovecause	Have passed dividend
18,353	Lovecharm	Have passed the
18,354	Lovechild .	Have you passed
18,355	Lovedrink	**Passenger**
18,356	Lovefavour	Passenger on board
18,357	Lovefeast .	Passengers all saved
18,358	Lovegrass .	**Past**

No.	Code Word.	**Past** (*continued*)
18,359	Lovejuice .	In the past
18,360	Loveknot .	The past records
18,361	Loveless .	Judging from the work done in the past
18,362	Loveletter .	**Patent(s)**
18,363	Loveling .	Patents all in order
18,364	Lovely . .	Number and date of patent
18,365	Lovemaking	Number and dates of respective patents
18,366	Lovematch	Unless you can get patents
18,367	Lovemonger	Insist on the patents being obtained
18,368	Lovepined	Title to property is United States patent
18,369	Lovescene	Have the patents examined by an efficient lawyer
18,370	Loveshaft .	Patent has been applied for
18,371	Lovesick .	Patent has not yet been granted, but all is in order
18,372	Lovespell .	**Patented**
18,373	Lovesuit .	Is the property patented
18,374	Lovetoys .	The property is patented
18,375	Lovingcup	The property is partly patented and partly not
18,376	Lovingness	Have nothing to do with the mine(s) unless patented
18,377	Lowbell .	Each claim is patented
18,378	Lowborn .	This claim is patented
18,379	Lowcaste .	Is the claim (are the claims) patented
18,380	Lowercase	See that the claim(s) is (are) patented
18,381	Lowerclass	The mine is patented
18,382	Loweringly	Patented mill site
18,383	Lowermost	Patented mining claim
18,384	Lowlands .	If not patented cannot do anything
18,385	Lowlife .	The mine(s) must be patented
18,386	Lowlihead	Not yet patented, but provisionally protected
18,387	Lowlily .	**Patience**
18,388	Lowminded	If you will have a little patience
18,389	Lownecked	If shareholders will only exercise a little patience
18,390	Lowpitched	Have a little patience
18,391	Lowstudded	Think your patience will soon be rewarded
18,392	Lowwater .	Have shown a good deal of patience
18,393	Loxabark .	**Pay(s)**
18,394	Loxarthrus	Pay to the credit of
18,395	Loxia . .	Do not pay
18,396	Loxodon .	Who will pay
18,397	Loxodromic	Agree(s) to pay
18,398	Loxosoma	Do not make —— pay till
18,399	Loyalness .	Will have to pay
18,400	Lozengy .	Will not pay
18,401	Lubberly .	Will not have to pay
18,402	Lubricated	Pay to the order of
18,403	Lubricity .	(To) pay me (us)
18,404	Lucanus .	(To) pay you
18,405	Lucent. .	You must not pay more than
18,406	Lucidly .	Can you pay
18,407	Lucidness .	Can he (they) pay
18,408	Luciferian	If you can pay

No.	Code Word.	**Pay(s)** (*continued*)
18,409	Lucimeter	If you cannot pay
18,410	Luckiest .	Cannot pay so much
18,411	Luckily .	Please pay for my account
18,412	Lucklessly	(To) pay him (them)
18,413	Luckpenny	Pay by telegraph
18,414	Lucrative .	Will it pay (to)
18,415	Lucrous .	It will pay to
18,416	Luctation .	Pay assessment on
18,417	Lucubrator	Is (are) unable to pay
18,418	Luculently	To whom shall we pay your dividend
18,419	Lucullite .	To whom shall we pay the amount due
18,420	Lucuma .	Shall we pay for your account
18,421	Luddite .	Shall I pay
18,422	Ludicrous.	You have nothing to pay; all charges were paid here
18,423	Luffed . .	No instructions to pay
18,424	Luffhook .	Will pay you
18,425	Lufftackle.	Will you pay
18,426	Luggage .	How much will —— pay
18,427	Luggagevan	Will pay him
18,428	Lugmark .	Not pay more than
18,429	Lugsail .	Pay as little as you can in cash
18,430	Lugubrious	Pay by instalments
18,431	Lukewarm	Pay all charges
18,432	Lukewarmly	The mine pays annually
18,433	Lullabies .	The mine will pay
18,434	Lullaby .	In what way is it proposed to pay
18,435	Lumachella	Pay off all hands
18,436	Lumbago .	**Payable**
18,437	Lumber .	Payable in —— month(s)
18,438	Lumberman	Payable in instalments of
18,439	Lumberroom	Payable at the bank of
18,440	Lumberyard	Payable to the credit of
18,441	Lumbric .	Payable to the order of
18,442	Lumbrical	Have sent —— payable to your order
18,443	Luminance	Payable to you at
18,444	Luminaries	Payable quarterly
18,445	Luminary .	Payable half-yearly
18,446	Luminosity	Payable in cash
18,447	Luminous.	Payable —— in cash, and the balance in shares
18,448	Lumpfish .	Payable —— in cash, and the balance in debentures
18,449	Lumpsucker	Duty payable [bearing —— per cent. interest
18,450	Lumpsugar	Payable entirely in fully paid shares
18,451	Lunacy .	How is it payable
18,452	Lunarian .	It is (they are) payable to bearer
18,453	Lunatics .	Shares payable to bearer
18,454	Lungeous .	Payable on delivery
18,455	Lunggrown	Payable on shipment, the balance
18,456	Lungwort .	Payable monthly
18,457	Lunisolar .	Payable against bills of lading
18,458	Lunistice .	The first call is payable

No.	Code Word.	**Payable** (*continued*)
18,459	Lunitidal .	The second call is payable
18,460	Lunular .	The balance payable
18,461	Lupercal .	**Paying**
18,462	Lupiform .	The mine(s) is a (are) dividend-paying property
18,463	Lupinus .	A non-paying concern
18,464	Lupuline .	Has (have) been paying
18,465	Lurcheth .	Has (have) not been paying
18,466	Lurching .	Is (are) the —— mine(s) a paying concern
18,467	Lurdane .	By paying —— in cash you would secure a most
18,468	Luscinia .	By paying [valuable property
18,469	Luscious .	Are not paying
18,470	Lusciously	Paying at the rate of
18,471	Lusiad . .	Are not paying sufficient attention to
18,472	Lustdieted	The mine is a paying concern
18,473	Lustfully .	Before paying the balance
18,474	Lustihood	Before paying his account
18,475	Lustreless .	The mine has been paying for the last
18,476	Lustrical .	Paying through the nose for
18,477	Lustrums .	We are now paying
18,478	Lusty . .	Is now paying
18,479	Lutebacked	**Payment(s)**
18,480	Luteoline .	In part payment
18,481	Lutheran .	Payment in full
18,482	Lutherism .	Future payments
18,483	Lutraria .	Payment(s) to be made to
18,484	Luxating .	Make no payment(s) till after
18,485	Luxuriancy	Demand payment
18,486	Luxuriant .	Will guarantee payment
18,487	Luxurious .	No arrangement made as to payment
18,488	Luxurist .	Stop the payment of
18,489	Luxury .	Payment on delivery at
18,490	Lycaena .	Payment in full of all demands
18,491	Lyceum .	Guaranteed payment
18,492	Lychnis .	What are the arrangements for payment
18,493	Lychnobite	Payment to be made as follows :
18,494	Lycodon .	**Peace**
18,495	Lycoperdon	Prospect of peace
18,496	Lycopodite	There will be no peace
18,497	Lycopodium	**Peaceable**
18,498	Lycopus .	Peaceable possession
18,499	Lycotropal	Have taken peaceable possession of the property
18,500	Lygodium .	Peaceable intentions
18,501	Lymegrass	**Peculation**
18,502	Lymexylon	Peculation has been going on
18,503	Lymhound	**Pending**
18,504	Lymnite .	Pending further instructions
18,505	Lymphatic	Pending the settlement of affairs
18,506	Lymphoduct	Pending the decision of the court
18,507	Lynch . .	Whilst this is pending
18,508	Lynching .	While the action is pending

No.	Code Word.	**Pending** (*continued*)
18,509	Lynchlaw .	Pending the hearing of the case
18,510	Lyndentree	While negotiations are pending
18,511	Lynxeyed .	Pending the result of
18,512	Lyoncourt	**Penetrated (through)**
18,513	Lyre . .	**Penetrating**
18,514	Lyrebird .	**Penetration**
18,515	Lyrical. .	Through the penetration of
18,516	Lyrichord .	By constant penetration
18,517	Lyricisms .	**Pennyweight(s)**
18,518	Lyrist . .	Less than one pennyweight
18,519	Lysimachia	Not more than one pennyweight
18,520	Lyssa . .	About one pennyweight
18,521	Lyterian .	Pennyweights per ton
18,522	Lythraceae	**Per**
18,523	Lythrum .	Yielding —— per cent. on capital
18,524	Maasha .	Per annum
18,525	Macacus .	Per cent.
18,526	Macadamize	Per cent. per annum
18,527	Macarized	Per diem
18,528	Macaronian	Per mensem
18,529	Macaroon.	How much per cent.
18,530	Macaws .	Per cent. is lost by
18,531	Macawtree	A profit of —— per cent.
18,532	Maccoboy	Will yield a profit of —— per cent.
18,533	Maccouba	There is a loss of —— per cent.
18,534	Maceale .	There is a gain of —— per cent.
18,535	Macebearer	**Percentage**
18,536	Maceproof	What percentage do you estimate
18,537	Maceration	A large percentage
18,538	Macereed .	A small percentage
18,539	Machinal .	**Perceptible**
18,540	Machinator	Very perceptible
18,541	Machinery	Hardly perceptible
18,542	Machinists	Perceptible to the naked eye
18,543	Macilency	**Peremptory**
18,544	Mackerel .	Peremptory orders have been given
18,545	Mackintosh	Under peremptory orders to
18,546	Maclurite .	**Perfect**
18,547	Macrodome	Perfect in every respect
18,548	Macrology	Quite perfect, except
18,549	Macrometer	Very far from perfect
18,550	Macropiper	The method is as nearly perfect as can be
18,551	Macropod.	Has been started, and found to work perfectly
18,552	Macropodal	**Perform(s)**
18,553	Macropus .	Can you perform your promise
18,554	Macrural .	Cannot perform our promise
18,555	Macrurous	Will he perform what he promised
18,556	Mactator .	Will perform
18,557	Mactra .	Performs all that was promised
18,558	Maculating	Does not perform all that was promised

No.	Code Word.	
18,559	Maculature	**Performance**
18,560	Madapple .	Performance not satisfactory
18,561	Madarosis	A very satisfactory performance
18,562	Madbrained	Performance might be better
18,563	Madbred .	Performance not as good as
18,564	Madcaps .	**Performed**
18,565	Madden .	Has (have) performed
18,566	Maddening	Has (have) not performed
18,567	Madeiranut	Must be performed
18,568	Madheaded	Has not been performed
18,569	Madhouses	We have performed our part of the contract
18,570	Madjoun .	**Perhaps**
18,571	Madman .	Perhaps you can
18,572	Madness .	Perhaps we might
18,573	Madoqua .	Perhaps you might
18,574	Madreporal	Perhaps might be
18,575	Madrigals .	It may perhaps be advisable to
18,576	Madroma .	**Period**
18,577	Maelstrom	At this period
18,578	Magazines	At what period
18,579	Magazining	At a later period
18,580	Magbote .	At an earlier period
18,581	Magdaleon	A period of great anxiety
18,582	Magellanic	**Permission.** (See also Get.)
18,583	Maggotish	Do not give permission
18,584	Maggoty .	Has (have) given permission
18,585	Magianism	Will not give permission
18,586	Magic . .	Has (have) refused permission
18,587	Magically .	Has (have) obtained permission to
18,588	Magicians.	With the full permission and sanction of
18,589	Magilph .	Try and get permission
18,590	Magistery .	Will you give permission to
18,591	Magistracy	Permission to inspect the mine
18,592	Magistral .	You have full permission (to)
18,593	Magnality.	With our permission
18,594	Magnates .	Without our permission
18,595	Magnesite.	Without any permission
18,596	Magnesium	**Permit(s)**
18,597	Magnetical	Permit our cipher to be used by
18,598	Magnetized	Permit —— to examine the
18,599	Magnific .	Will you permit
18,600	Magnifiers	Will you permit me (us) to
18,601	Magnifying	Will permit
18,602	Magnitude	To get a permit
18,603	Magnolia .	Do (does) not permit
18,604	Magotpie .	Cannot permit
18,605	Magpiemoth	It is advisable to permit
18,606	Magpies .	Permit(s) us to
18,607	Magydare .	**Permitted**
18,608	Mahaleb .	Was (were) permitted to

No.	Code Word.	**Permitted** (*continued*)
18,609	Maharmah	Was (were) not permitted
18,610	Mahoganize	Have permitted
18,611	Mahomedan	If it was not permitted
18,612	Mahometism	**Permitting**
18,613	Mahometry	Permitting us to
18,614	Mahonia .	**Perseverance**
18,615	Maidchild.	With patience and perseverance
18,616	Maidenhair	**Persevere**
18,617	Maidenhood	You must persevere
18,618	Maidenish	We intend to persevere
18,619	Maidenlike	If we persevere
18,620	Maidenly .	**Persist(s)**
18,621	Maidenmeek	Persist in doing (proceeding)
18,622	Maidenpink	Persist(s) in the determination (to)
18,623	Maidenplum	Persist in the course of action
18,624	Maidenship	If he (they) persist
18,625	Maidmarian	Persisted in
18,626	Maidpale .	**Persistence**
18,627	Maigrefood	Steady persistence in
18,628	Mailable .	**Person(s)**
18,629	Mailbags .	A most unsuitable person
18,630	Mailboat .	The person(s) concerned
18,631	Mailclad .	Who is the person referred to
18,632	Mailcoach	A suitable person
18,633	Mailguard.	**Personal**
18,634	Mailing .	Personal expenses
18,635	Mailmaster	Personal guarantee
18,636	Mailroom .	Personal liability
18,637	Mailroute .	Personal and joint undertaking (or guarantee)
18,638	Mailstage .	**Personally**
18,639	Mailtrain .	You must see —— personally
18,640	Maimedness	Has (have) not personally any knowledge of
18,641	Mainboom	Has (have) seen personally
18,642	Maincouple	Has (have) —— personally seen the mine(s)
18,643	Maindecks	Have you personally examined
18,644	Maineport	Personally responsible (for)
18,645	Mainhamper	Personally and collectively
18,646	Mainhatch	**Persuaded**
18,647	Mainhold .	Am (are) persuaded that
18,648	Mainkeel .	Cannot be persuaded
18,649	Mainland .	Can he (they) not be persuaded
18,650	Mainly . .	**Petition**
18,651	Mainmasts	To petition against
18,652	Mainpernor	Have presented a petition
18,653	Mainpost .	Petition is refused
18,654	Mainprize.	Petition is granted
18,655	Mainsail .	You had better petition against
18,656	Mainsheet	The hearing of the petition
18,657	Mainspring	Now preparing petition
18,658	Mainstay .	**Petroleum**

No.	Code Word.	**Petroleum** (*continued*)
18,659	Mainsworn	The petroleum occurs in strata of
18,660	Maintackle	The petroleum is found at —— feet
18,661	Maintains .	Borehole(s) struck petroleum at —— feet
18,662	Maintop .	Well(s) produce(s) —— gallons petroleum per hour
18,663	Mainyard .	The supply of petroleum is inexhaustible
18,664	Majestical.	Petroleum is found largely in the district
18,665	Majesties .	For burning petroleum
18,666	Majesty .	Petroleum is used
18,667	Majolica . **Pick**	
18,668	Majoration	If you have the chance to pick up a good mine for about ——, do so. This is a good moment to [sell
18,669	Majorities.	To pick up
18,670	Majority .	To pick over
18,671	Majuscula	Have been able to pick up a good —— mine, time
18,672	Makebate . **Picked**	[—— months, titles all right
18,673	Makeless .	Picked ore
18,674	Makepeace	Picked specimens
18,675	Makeshift .	Has (have) picked up
18,676	Makeup .	Picked up a very promising
18,677	Makeweight	Picked over
18,678	Makingiron	Has (have) been picked over
18,679	Malacatune	Has (have) not been picked over
18,680	Malacoderm	Hand-picked samples
18,681	Malacology	Can be picked up
18,682	Maladies .	Some promising mines can be picked up cheap
18,683	Malakanes **Pinched**	
18,684	Malapert .	The vein is very much pinched
18,685	Malapertly	Pinched out
18,686	Malapropos	The vein has pinched out
18,687	Malarious.	Much pinched for want of
18,688	Malax . . **Pinching**	
18,689	Malaxate .	The vein is pinching out laterally
18,690	Malaxation	The vein is pinching out vertically
18,691	Malbrouk .	The veins are pinching out
18,692	Maledict .	The vein is showing signs of pinching out
18,693	Malefactor **Pipe(s)**	
18,694	Malefern .	Lay down pipes
18,695	Maleficent	By laying down pipes
18,696	Malengine	Miles of pipes
18,697	Maletolt .	In (a) pipe(s)
18,698	Malevolent	The water has to be brought in pipes
18,699	Malevolous	Cast-iron pipes
18,700	Maliced .	Copper pipes
18,701	Malign. .	Gun-metal pipes
18,702	Malignancy	Wrought-iron pipes
18,703	Malignity .	Steel pipes
18,704	Malingery.	Pipes —— inches diameter [feet ; thickness
18,705	Malison .	Cast-iron pipes : diameter —— inches ; length ——
18,706	Malleate .	Wrought-iron pipes : diameter —— inches ; length
18,707	Malleating	Steam pipe(s) [—— feet ; thickness

No.	Code Word.	**Pipe(s)** (*continued*)
18,708	Malleation	State length and diameter of pipes required
18,709	Mallemock	Length, diameter, and thickness of pipes
18,710	Mallenders	Pipes have been laid
18,711	Malleolar .	Pipes will have to be renewed
18,712	Malleus .	The cost of pipes will be
18,713	Mallophaga	Pipes to discharge —— gallons
18,714	Mallotus .	Pipes to convey the water (from)
18,715	Malmbricks	Pipes with flanges
18,716	Malmrock	Flanges of pipes
18,717	Malmsey .	Flanges of pipes to be —— inches diameter
18,718	Malodorous	Pipes with socket joints
18,719	Malodour .	Through the bursting of a pipe
18,720	Malpighia . **Piping**	
18,721	Maltbarn .	Feet of wrought-iron piping —— inch diameter
18,722	Maltdrink .	Feet of cast-iron piping —— inch diameter
18,723	Maltdust . **Place(s)**	
18,724	Maltfloor .	Have no one to take the place of
18,725	Malthouse	In no place are the conditions so favourable for
18,726	Malthusian	In place of [cheap working
18,727	Maltkiln .	At what place
18,728	Maltliquor	At this place
18,729	Maltmill .	Can you place
18,730	Maltose .	Can place
18,731	Maltreat .	If we can place
18,732	Maltster .	Cannot place
18,733	Maltworm .	In several places
18,734	Malurus .	In no place
18,735	Mamma .	The place at which
18,736	Mammalian	The place where we are
18,737	Mammalogy	The place where we found
18,738	Mammary	The place where it was
18,739	Mammeated	Who can take your place
18,740	Mammifer	Will take my place
18,741	Mammillate	Goes in place of
18,742	Mammilloid	Can you place any shares of
18,743	Mammock	What amount of shares can you place
18,744	Mammonists	Have been able to place
18,745	Mammonize	Have not been able to place
18,746	Mammose	Want to know the exact place
18,747	Mammoths	Place in the hands of
18,748	Manacle .	Places me (us) in an awkward position
18,749	Manacling **Placed**	
18,750	Manageably	The matter is placed before
18,751	Managed .	Has the matter been placed before
18,752	Manageless	Has (have) placed
18,753	Management	Has (have) not yet placed
18,754	Managerial	Placed in position
18,755	Manatus .	Have been placed
18,756	Manbound	The shares have all been placed
18,757	Manchineel	Has been placed in the hands of

No.	Code Word.	**Placed** (*continued*)
18,758	Manciple .	Have placed the matter in ——'s hands
18,759	Mancusa .	In whose hands have you placed the business
18,760	Mandarinic	**Placing**
18,761	Mandatory	Am (are) now placing
18,762	Mandibula	While we are placing
18,763	Mandil . .	After placing
18,764	Mandilion	**Plant**
18,765	Mandioc .	What plant is there
18,766	Mandragora	Does this include all the plant
18,767	Mandrakes	Not including the plant
18,768	Manducable	Including all the plant
18,769	Manducated	The necessary plant
18,770	Manducus	All the plant
18,771	Manefaire.	The plant is not worth anything
18,772	Manettia .	The plant is old-fashioned and almost worn out
18,773	Manfully .	The plant is almost new
18,774	Manfulness	The plant is in first-rate condition
18,775	Manganate	Plant and machinery
18,776	Manganesic	Plant, machinery, and buildings
18,777	Manganium	Inventory of the plant, etc.
18,778	Mangcorn.	The plant, at a fair valuation, is worth
18,779	Mangifera.	The machinery and plant cost [and stores
18,780	Mangily .	Send a complete inventory of the plant, buildings,
18,781	Manginess	**Plan(s)**
18,782	Manglers .	When will you send plan
18,783	Mango . .	The plans of the (——) mine(s) were sent you on
18,784	Mangobird	The plans are now in the office
18,785	Mangofish	Send as soon as possible plans of
18,786	Mangonel.	Has (have) sent the plans of
18,787	Mangonism	You will see by the plans that
18,788	Mangostan	There are no good plans of the
18,789	Mangotree	Have plans made of the (——) mine(s)
18,790	Mangrove.	From the plans
18,791	Manhater .	It will be seen upon the plan sent
18,792	Manholes .	Plan and section of the mine
18,793	Maniacal .	Plan of the different locations
18,794	Manicate .	Complete plan, showing all surface features
18,795	Manichean	Plan showing each level
18,796	Manicheism	Sketch plan
18,797	Manicordon	Progress plan(s) and section(s)
18,798	Manifest .	Plan to accompany the report
18,799	Manifestly	Plan of the mills and buildings
18,800	Manifoldly	Plan and specification
18,801	Maniform .	Letter and number of square on plan
18,802	Manilahemp	See square on plan : letter ——, No. ——
18,803	Manipular	Square on plan indicating position of
18,804	Manitrunk	**Platinum**
18,805	Mankind .	The platinum is found in
18,806	Manlessly.	Large quantities of platinum
18,807	Manmercer	Small quantities of platinum

No.	Code Word.	
18,808	Manminded	**Pleased**
18,809	Mannacroup	You will be pleased to learn
18,810	Mannerism	Are pleased to hear
18,811	Mannerly .	Much pleased with
18,812	Mannishly	Much pleased with the appearance of
18,813	Manorchis	**Pledged**
18,814	Manorhouse	Is (are) distinctly pledged (to)
18,815	Manorial .	**Plentiful**
18,816	Manorseat	Plentiful supply of
18,817	Manoscopy	A plentiful supply of water at all times
18,818	Manovery .	Is there a plentiful supply of
18,819	Manpleaser	Is very plentiful in the district
18,820	Manqueller	Is not very plentiful in the district
18,821	Manrent .	**Plenty**
18,822	Manrope .	Can you get plenty of
18,823	Manservant	There is plenty of
18,824	Manslayer	To be found in plenty
18,825	Manstealer	Plenty coming forward
18,826	Mansties .	Plenty of room in
18,827	Mansty .	**Plunder**
18,828	Mansuete .	His share of the plunder
18,829	Mansuetude	(To) disgorge some of the plunder
18,830	Mantellia .	**Plundered**
18,831	Manteltree	Has plundered and robbed
18,832	Mantichor	Has (have) been plundered every way
18,833	Mantiger .	Much of —— was plundered
18,834	Mantiscrab	**Pocket(s)**
18,835	Mantispa .	The ore is mostly found in pockets
18,836	Mantled .	Have struck a large pocket of ore
18,837	Mantraps .	Have struck a small pocket of ore
18,838	Manualist .	Several pockets of very rich ore have been found
18,839	Manually .	Owing to the ore being mostly in pockets
18,840	Manubrium	**Pockety**
18,841	Manucaptor	Pay ore very pockety
18,842	Manuducent	The mine is said to be very pockety
18,843	Manuductor	A very pockety mine
18,844	Manumise	**Policy**
18,845	Manumitted	The best policy would be
18,846	Manumotive	The line of policy we recommend you to adopt
18,847	Manumotor	What policy do you recommend
18,848	Manurance	Think your best policy is to
18,849	Manure .	Think it will be good policy
18,850	Manuscript	Think it will not be good policy
18,851	Manworship	A very judicious policy
18,852	Manworthy	Cannot alter our policy
18,853	Manyheaded	Must adhere to the policy laid down
18,854	Manyplies .	Policy based upon
18,855	Manysided	Would it be good policy to
18,856	Manyways	See no reason to alter our policy
18,857	Manywise .	The policy to be adopted

No.	Code Word.	
18,858	Maple . .	**Politic(al)**
18,859	Mapmounter	A very politic course
18,860	Mappery .	Most politic measures
18,861	Maracan .	In the present state of politics
18,862	Maranatha	Political disturbances affect us
18,863	Marasmus	Political questions
18,864	Maraud .	Great political uneasiness
18,865	Marauding	**Poor.** (See also Quality.)
18,866	Marbleize .	Poor returns
18,867	Marbly .	Too poor to pay .
18,868	Marcasitic	Too poor for milling purposes
18,869	Marceline .	Not quite so poor
18,870	Marcescent	Very poor
18,871	Marchest .	The vein looks poor
18,872	Marching .	The ore is too poor to pay for extraction
18,873	Marchpane	Getting only poor ore
18,874	Marchward	**Poorer**
18,875	Marcidity .	Poorer and poorer
18,876	Marcionite	If the vein gets poorer
18,877	Marcosian	The vein is getting poorer
18,878	Mareca .	The ore is getting poorer in depth
18,879	Marekanite	Ore gets poorer as we advance
18,880	Marestail .	**Porphyry**
18,881	Margaric .	The hanging wall is porphyry
18,882	Margarous	The foot-wall is porphyry
18,883	Marginal .	The country rock is porphyry
18,884	Marginally	**Porphyritic**
18,885	Margining	Porphyritic rock
18,886	Marginline	Porphyritic dyke
18,887	Margosa .	**Port**
18,888	Margravate	Port of loading
18,889	Marguerite	Port of discharge
18,890	Marigold .	Port charges
18,891	Marigraph	Bound for the port of
18,892	Marikina .	Now in port waiting for
18,893	Marineglue	**Portion**
18,894	Marinorama	A large portion of which
18,895	Mariolater	A small portion of which
18,896	Mariput .	If only a small portion
18,897	Markab .	In a portion of
18,898	Marketable	As our portion
18,899	Marketbell	The greater portion
18,900	Marketday	To retain the greater portion
18,901	Marketers .	A considerable portion of (the)
18,902	Marketrate	The largest portion of the
18,903	Markettown	What portion (of the)
18,904	Markingink	**Position**
18,905	Markingnut	Telegraph position and standing of
18,906	Marksman	What is the position of affairs with regard to
18,907	Marlaceous	To strengthen our position

No.	Code Word.	Position (*continued*)
18,908	Marlitic .	To improve the position
18,909	Marlpit .	What is the position of
18,910	Marlstones	Position very serious
18,911	Marmalade	Is (are) in a wrong position .
18,912	Marmatite	In a very dangerous position
18,913	Marmorated	The position of
18,914	Maronite .	Are now in a position to
18,915	Maroon .	Not in a position to
18,916	Marooning	Send plan showing position of
18,917	Marplot .	Present position is such that .
18,918	Marquees .	Having regard to the present position
18,919	Marquisate	Present position and future prospects
18,920	Marquisdom	Position is not encouraging
18,921	Marrowbone	Our position will be serious if you cannot
18,922	Marrowfat	What is the position likely to be, in the event **of**
18,923	Marrowish	The position is unchanged
18,924	Marrowless	In our present position it will be best to
18,925	Marrymuffe	In a very questionable position
18,926	Marsdenia	Position has improved
18,927	Marshalled	Position improved, but time is needed to
18,928	Marshalsea	What is our position
18,929	Marshelder	Cable your opinion of the position and prospects
18,930	Marshgas .	Has (have) taken up a strong position
18,931	Marshiest .	Position taken by —— is very weak
18,932	Marshiness	Are you in a position to
18,933	Marshnut .	**Positive**
18,934	Marshy .	Nothing positive as yet arranged
18,935	Marsupial .	Are you positive
18,936	Martello .	Cannot be positive till
18,937	Martext .	As soon as anything positive is arranged
18,938	Martial .	Something positive must be settled at **once**
18,939	Martialism	Am positive that
18,940	Martialize .	Is (are) positive that
18,941	Martingale	**Possession**
18,942	Martinmas	In possession of
18,943	Martyrdom	Possession will be given
18,944	Martyrest .	(To) put us in possession of
18,945	Martyrized	Cannot get possession
18,946	Martyrly .	(To) gain possession of
18,947	Marvel. .	Is (are) in possession
18,948	Marvelling	Is (are) not in possession of the **property**
18,949	Marvellous	Am in possession (of)
18,950	Marybud .	Not in possession (of)
18,951	Maryolatry	When shall we be in possession (of)
18,952	Masahib .	Have retaken possession
18,953	Mascagnin	Forcible possession
18,954	Masculated	Gave up possession
18,955	Masculy .	Refuse(s) to give up possession
18,956	Mashingtub	Can give possession, upon
18,957	Mashvat .	Have you got possesion (of)

No.	Code Word.	Possession (*continued*)
18,958	Masked .	Have taken possession (of) [session
18,959	Maslach .	The money is ready—are you prepared to give pos-
18,960	Masonbee.	Will give possession upon receipt of
18,961	Masonry .	Must be in our possession
18,962	Masonwasp	If possession is not given [trouble
18,963	Masoretic .	Possession of these claims will, relieve us of all
18,964	Masorite .	**Possible**
18,965	Masquerade	As soon as possible
18,966	Massacring	As much as possible
18,967	Massday .	Is it possible
18,968	Masseter .	It is possible
18,969	Masshouse	(It) is not possible
18,970	Massilia .	Would it be possible
18,971	Massively .	If possible
18,972	Massoybark	Would not be possible
18,973	Masspriest	Would it not be possible to
18,974	Massuelle .	Do what you possibly can (to)
18,975	Mastax .	It is hardly possible (to)
18,976	Mastcoat .	If not possible
18,977	Masterdom	It is just possible that
18,978	Masterful .	**Possibility**
18,979	Masterhood	The mere possibility of
18,980	Mastering .	Is there any possibility of
18,981	Masterjest .	Not the slightest possibility of
18,982	Masterless	If there is the slightest possibility of
18,983	Masterlode	No possibility of doing anything
18,984	Mastermind	**Post.** (See also Mail.)
18,985	Masterous	A very responsible post
18,986	Mastership	Is the best man for the post
18,987	Masterwork	Will —— be fit for the post
18,988	Masthead .	Is not fit for the post
18,989	Masthoop.	By last post
18,990	Masticable	By this evening's post
18,991	Masticator	By the next post
18,992	Masticine .	Do not post
18,993	Mastiffbat.	Do not post any letters to me
18,994	Mastiffs .	Shall not post any more letters
18,995	Mastigopod	**Postpone**
18,996	Mastitis .	You must postpone
18,997	Mastodon .	Will postpone
18,998	Mastoid .	Will not postpone
18,999	Mastoideal	Will —— postpone
19,000	Masttree .	Cannot postpone
19,001	Masulaboat	Postpone going to —— till
19,002	Mataco .	Postpone all operations
19,003	Matafund .	**Postponed**
19,004	Matamata.	Postponed on account of
19,005	Matchcloth	Have postponed my departure
19,006	Matchcoat	**Pound(s)**
19,007	Matchless .	Pounds weight

No.	Code Word.	**Pound(s)** (*continued*)
19,008	Matchlock	How many pounds
19,009	Matchmaker	We require —— pounds
19,010	Matchplane	Pounds sterling
19,011	Matchtub .	Pounds sterling equivalent to —— dollars
19,012	Materially .	Pounds sterling equivalent to —— rupees
19,013	Materiated	**Powder**
19,014	Materious .	Waiting for powder in order to
19,015	Maternity .	Send —— lbs. of giant powder
19,016	Matfelon .	No powder to be had here
19,017	Mathematic	Have only —— lbs. of powder in store
19,018	Mathemeg	Powder in store
19,019	Mathesis .	Mammoth powder
19,020	Matindog .	Blasting powder
19,021	Matins.	. **Power**
19,022	Matricaria	Has (have) great power
19,023	Matricide .	Has (have) no power
19,024	Matrimony	Of what power
19,025	Matronal .	It is in the power of
19,026	Matronhood	It is not in the power of
19,027	Matronize .	If it is in the power of
19,028	Matronlike	Is it in ——'s power
19,029	Matronly .	It is in your power (to)
19,030	Matthiola .	Is it in your power (to)
19,031	Mattresses	It is in (my) our power (to)
19,032	Mattulla .	It is not in my (our) power (to)
19,033	Maturative	Power of attorney to be made in name of
19,034	Maturely .	In whose name is power of attorney to be made
19,035	Matureness	Power of attorney will be required
19,036	Matwork .	Who holds power of attorney
19,037	Maudlin .	Give power of attorney to
19,038	Maudlinism	Send power of attorney
19,039	Maulstick .	Send power of attorney to sue
19,040	Maumetrie	Cannot send power of attorney
19,041	Maumletdar	Has (have) sent power of attorney; see it is duly
19,042	Maund . .	Will send power of attorney [recorded
19,043	Maundering	Have received power of attorney
19,044	Mauresque	Power of attorney notarially attested
19,045	Mausoleum	Power of attorney in favour of
19,046	Mauveine .	Power of attorney must be recorded
19,047	Mawkish .	Power of attorney recorded
19,048	Mawseed .	Hereby withdraw and cancel power of attorney
19,049	Maxilla .	Power of attorney cancelled
19,050	Maxillary .	Date when power was given
19,051	Maxilliped	Date when power was revoked
19,052	Maximist .	Power of attorney not on record
19,053	Maximized	Special power of attorney
19,054	Maximum	Sufficient power to
19,055	Maybeetle	Not enough power to
19,056	Maybloom	More power required
19,057	Maybug .	Adding more power

No.	Code Word.	**Power** (*continued*)
19,058	Maybush .	More power can be obtained by increasing **fall**
19,059	Maydew .	Of greater power
19,060	Mayflower	An engine of greater power required
19,061	Mayflies .	Engine-power
19,062	Mayfly .	Water-power
19,063	Maygame .	Steam-power
19,064	Mayhap .	Steam power necessary to drive
19,065	Maymorn .	Electric-power
19,066	Mayonnaise	Is electric power obtainable
19,067	Mayor . .	Power obtained is sufficient (to)
19,068	Mayoralty	**Powerful**
19,069	Mayoress .	Is (are) all-powerful in the district
19,070	Mayorship	Not sufficiently powerful
19,071	Maypole .	Is —— sufficiently powerful
19,072	Mayqueen	Will require more powerful machinery
19,073	Mazdean .	Very powerful
19,074	Mazdeism .	**Practicable**
19,075	Maze . .	Is it practicable
19,076	Mazologist	It is quite practicable
19,077	Mazurka .	It is not practicable
19,078	Mazzard .	If it is not quite practicable
19,079	Meacock .	Has been found practicable
19,080	Meadow .	**Practically**
19,081	Meadowlark	**Practice**
19,082	Meadowpink	Has been found in practice
19,083	Meadowrue	**Precaution(s)**
19,084	Meadowsage	Take every precaution
19,085	Meadowwort	Has (have) taken every precaution
19,086	Meagrely .	As a measure of precaution
19,087	Meagreness	Has (have) not taken any precautions
19,088	Meagrim .	What precautions have you taken against
19,089	Mealbeetle	**Precedent(s)**
19,090	Mealies .	Is there any precedent for
19,091	Mealiness .	The case is without precedent
19,092	Mealmonger	There is no well-established precedent
19,093	Mealmoth .	If a precedent can be found
19,094	Mealsmeat	Precedents have been looked up
19,095	Mealtime .	**Preceding**
19,096	Mealtub .	The preceding points
19,097	Mealworm	To explain the preceding
19,098	Mealy . .	During the preceding
19,099	Meander .	**Precisely**
19,100	Meandrian	Want to know precisely
19,101	Meanest .	**Preclude(s)**
19,102	Meanwhile	Would preclude us from
19,103	Measelry .	Preclude(s) us from
19,104	Measles .	Precludes them (him) from
19,105	Measurable	To preclude them (him) from
19,106	Measured .	Does not preclude us from
19,107	Measuring	Preclude(s) the possibility

o

No.	Code Word.	
19,108	Meatflies .	**Prefer(s)**
19,109	Meatfly .	Which do you prefer
19,110	Meatpie .	Would you prefer
19,111	Meatscreen	Would prefer
19,112	Mechanic .	Would prefer not to
19,113	Mechanical	If you prefer
19,114	Mechanurgy	We should prefer to
19,115	Mechlin .	**Preferable**
19,116	Mecometer	Would it not be preferable
19,117	Medalet .	Would be preferable
19,118	Medallic .	Would not be preferable
19,119	Medallions	**Preference**
19,120	Mediacy .	By preference
19,121	Mediaeval	We give the preference to
19,122	Mediatize .	We have the preference
19,123	Mediator .	You will have the preference
19,124	Medicable	Should give the preference to
19,125	Medicago .	**Prejudice(s)**
19,126	Medicean .	Without prejudice (to)
19,127	Medicinal .	Offer made without prejudice
19,128	Medick .	We consent, but without prejudice to our rights
19,129	Mediocral .	A strong prejudice against [and claims
19,130	Mediocrist	Will prejudice our case
19,131	Meditate .	If it will not prejudice our case or claim
19,132	Meditating	Must seriously prejudice us
19,133	Meditation	Seriously prejudices our efforts to obtain
19,134	Medjidie .	**Prejudiced**
19,135	Medlars .	Is (are) prejudiced against
19,136	Medullary	Is (are) prejudiced in favour of
19,137	Medullated	Without being prejudiced
19,138	Medullose	A prejudiced opinion
19,139	Medusoid .	**Preliminaries**
19,140	Meekeyed	Preliminaries arranged
19,141	Meekly .	Preliminaries not yet arranged
19,142	Meerkat .	**Preliminary**
19,143	Meerschaum	As a preliminary step
19,144	Megaceros	A preliminary agreement
19,145	Megachile .	**Premature**
19,146	Megacosm	Premature disclosure
19,147	Megafarad	Would be at present premature
19,148	Megalanea	Do nothing premature
19,149	Megalithic	Is at present premature
19,150	Megalodon	The statement (or report) is premature
19,151	Megalonyx	**Premium**
19,152	Megalosaur	What is the premium (on)
19,153	Megalotis .	The premium is too heavy
19,154	Megaptera.	At a premium of
19,155	Megarian .	The shares are now at a premium
19,156	Megascopes	The shares, after touching a premium, fell off
19,157	Megaspore	Is a premium upon dishonesty

No.	Code Word.	
19,158	Megasthene	**Prepaid**
19,159	Megaweber	Must be prepaid
19,160	Meggelup..	The freight was prepaid
19,161	Megohm .	All charges have been prepaid
19,162	Meibomian	We are prepaying
19,163	Meiosis .	**Preparation(s)**
19,164	Meith . .	Preparations being made
19,165	Meiwell .	Preparations have been made
19,166	Melaconite	Preparations delayed owing to
19,167	Melada .	Time to make preparations
19,168	Melaleuca.	Making preparations (to)
19,169	Melampode	Make all necessary preparations
19,170	Melampyrum	All the necessary preparations are now made
19,171	Melancholy	**Prepare**
19,172	Melaniline	Prepare as quickly as possible
19,173	Melanism .	Prepare everything for
19,174	Melanoma	Will prepare
19,175	Melanopsis	Cannot prepare
19,176	Melanotic .	Is there time to prepare
19,177	Melanotype	**Prepared**
19,178	Melanure .	Has (have) prepared
19,179	Melaphyre.	Has (have) not yet prepared
19,180	Melarosa .	I am (we are) prepared for
19,181	Melasmic .	Not prepared for
19,182	Melchite .	Have you prepared
19,183	Melder .	Cannot be prepared
19,184	Meleagrina	You must be prepared for
19,185	Melicerous	Prepared to accept an offer
19,186	Melicgrass	Is (are) not prepared to
19,187	Melidae .	Shall you be prepared
19,188	Melilotus .	Shall be fully prepared when the time arrives
19,189	Meliorate .	**Preparing**
19,190	Meliphaga	When preparing for
19,191	Melissa .	Now preparing for
19,192	Melitose .	We are preparing to
19,193	Melittis .	**Prepayment**
19,194	Mellific .	Upon prepayment of
19,195	Melligo .	**Presence**
19,196	Mellilite .	In the presence of
19,197	Mellivora .	Your presence needed here
19,198	Melloca .	Will my presence be needed
19,199	Melocactus	**Present(s)**
19,200	Melocoton	At the present time
19,201	Melodious	Up to the present
19,202	Melodist .	For the present
19,203	Melodize .	Not at present
19,204	Melodizing	Do not present
19,205	Melodrama	That may present itself
19,206	Melody .	Now presents a better appearance
19,207	Melolontha	**Presented**

No.	Code Word.	**Presented** (*continued*)
19,208	Melopiano	Has (have) been presented
19,209	Meltingly .	Has (have) not been presented
19,210	Meltingpot	When presented
19,211	Melyridae .	Has (have) the —— been presented
19,212	Membership	The bill has been presented for payment but
19,213	Membrane **Preservation**	[returned dishonoured
19,214	Membranous	In a good state of preservation
19,215	Memento . **Preserve**	
19,216	Memoir .	In order to preserve our rights
19,217	Memoirist	To preserve what is left
19,218	Memorable	Preserve all rights intact
19,219	Memorandum**Preserved**	
19,220	Memorative	All rights preserved intact
19,221	Memoriter	Our interests have been preserved
19,222	Memorize . **President**	
19,223	Memorizing	The president of the
19,224	Memphian	Is president of
19,225	Menace .	Who is the president of
19,226	Menacingly	Have laid the matter before the president
19,227	Menagerie	The president has promised
19,228	Menagogue **Press**	
19,229	Mendacious	Press for an answer
19,230	Mendicancy	Press for payment
19,231	Mendicity . **Pressed**	
19,232	Mendose .	Has (have) pressed
19,233	Mengite .	We have been pressed to
19,234	Menhaden	Pressed for time
19,235	Meningeal **Pressing**	
19,236	Meningitis	Are now pressing
19,237	Meniscoid .	Is (are) now pressing his (their) claims against
19,238	Meniscuses **Pressure**	
19,239	Meniver .	Under a heavy pressure
19,240	Mennonite	The heavy pressure
19,241	Menologium	Owing to the pressure put upon
19,242	Menology .	Bring all the pressure you can get to bear (upon)
19,243	Menopome **Prevail**	
19,244	Menostasis	Hope to prevail on —— to
19,245	Mensa . .	Cannot prevail on —— to
19,246	Mensurate	Can you prevail on —— to
19,247	Mentagra .	We have prevailed upon
19,248	Mentally . **Prevalence**	
19,249	Menthene .	Owing to the prevalence of
19,250	Mentoniere **Prevent**	
19,251	Mentum .	Cannot prevent
19,252	Menuridae	Try and prevent
19,253	Menyanthes	Will prevent
19,254	Menyie .	To prevent expense and delay
19,255	Menziesia .	You must prevent him (them) from
19,256	Mephitic .	There was nothing to prevent
19,257	Mephitical	Is there anything to prevent

No.	Code Word.	**Prevent** (*continued*)
19,258	Mercantile	Do all you can to prevent
19,259	Mercaptan	Did all we could to prevent
19,260	Mercatante	To prevent others from
19,261	Mercature	To prevent others from stepping in
19,262	Mercenary	Impossible to prevent
19,263	Mercership	Do you know of anything likely to prevent
19,264	Merchand	Can prevent
19,265	Merchandry	If anything happens to prevent
19,266	Merciable.	**Prevented**
19,267	Merciament	Was (were) prevented from
19,268	Merciful .	Could not have been prevented
19,269	Mercifully.	Which has prevented
19,270	Merciless .	Been prevented by circumstances
19,271	Mercurial.	Cannot be prevented
19,272	Mercurify.	If we had not been prevented (by)
19,273	Mercurism	We have been prevented from the want of
19,274	Mercurous	**Preventing**
19,275	Mercyseat	Thereby preventing —— from
19,276	Meregoutte	**Previous**
19,277	Merenchyma	On my previous visit
19,278	Merestead	Since the previous
19,279	Merganser	In your previous letter
19,280	Mergus .	Previous to
19,281	Meridian .	Had been done previous to
19,282	Meridional	In its (their) previous state
19,283	Merismatic	Previous orders withdrawn
19,284	Merit . .	Since your previous instructions
19,285	Meritedly .	**Price(s).** (See also Shares.)
19,286	Merithal .	Is the price asked
19,287	Meriting .	At what price can-you
19,288	Merlangus	At what price did you
19,289	Merlucius .	At present price(s)
19,290	Mermaids.	At what price can you bond
19,291	Merman .	What is the lowest price you will take
19,292	Merocele .	Wire present price of
19,293	Meropidae	Is the price asked, but vendors would take
19,294	Merosome	Is (are) getting good prices for
19,295	Merrily .	Is (are) worth the price
19,296	Merrimake	The price asked is reasonable
19,297	Merrynight	The price asked is unreasonable
19,298	Merryquilt	Has (have) asked too high a price
19,299	Merula. .	This price can probably be reduced
19,300	Mervaille .	What is the price of
19,301	Mesartin .	Is the price of
19,302	Mesdames	Would get a high price for
19,303	Meseems .	With the high price of
19,304	Mesenteric	Can get a high price for
19,305	Meshwork	Reserve price
19,306	Mesitule .	The reserve price set upon the property is
19,307	Mesitylene	At a reasonable price

No.	Code Word.	**Price(s)** (*continued*)
19,308	Mesmerical	The price is too high
19,309	Mesmerism	If —— can reduce the price to
19,310	Mesmerized	If the price does not exceed
19,311	Mesnality .	Cannot touch it at the price
19,312	Mesne . .	Price should be reduced to at least
19,313	Mesoblast	Original invoice price
19,314	Mesocaecum	Prices rising
19,315	Mesocarp .	Prices have risen
19,316	Mesoderm	Prices steady
19,317	Mesoleucos	Prices declining
19,318	Mesolite .	Prices will go lower
19,319	Mesolobar	Prices nominal
19,320	Mesomelas	Prices continue to increase
19,321	Mesopodium	Prices continue to fall
19,322	Mesorectum	The price to include
19,323	Mesosperm	At what price was it offered
19,324	Mesothesis	What price does he expect to get
19,325	Mesothorax	Much higher prices
19,326	Mesotype .	Much lower prices
19,327	Mesoxalic .	At an increased price
19,328	Mesozoic .	At a reduced price
19,329	Mespilus .	Price does not include
19,330	Mesprise .	Does the price include
19,331	Messalian .	**Principal**
19,332	Messdeck .	Our principal object (is)
19,333	Messenger	The principal thing is
19,334	Messianic .	In the absence of our principal
19,335	Messidor .	Principal will not be here
19,336	Messmates	Our principal
19,337	Messtable .	Their principal
19,338	Mesteque .	**Principally**
19,339	Mesymnicum	**Principle**
19,340	Metabasis .	Object upon principle
19,341	Metabolic .	The principle is quite wrong
19,342	Metacarpal	The principle upon which
19,343	Metacentre	Cannot approve of the principle
19,344	Metacetone	**Private**
19,345	Metacism .	This is for your private information only
19,346	Metacresol	This is to be kept quite private
19,347	Metagallic	Tell —— in private
19,348	Metaleptic	Told me in private that
19,349	Metallical .	From private information
19,350	Metalling .	Private and confidential
19,351	Metallists .	Has been kept private
19,352	Metallize .	By private contract
19,353	Metalloid .	**Privilege(s)**
19,354	Metallurgy	(To) have the privilege of
19,355	Metalman .	Our privileges and rights
19,356	Metameric	Their privileges
19,357	Metaphor .	Cannot have the privileges

No.	Code Word.	**Privilege(s)** (*continued*)
19,358	Metaphrase	The privileges now enjoyed
19,359	Metaphysic	**Probability**
19,360	Metaplasm	In all probability
19,361	Metaptosis	Is there any probability of
19,362	Metastatic	There seems a probability of
19,363	Metastoma	There is no probability of
19,364	Metatarsal	If there is no probability of
19,365	Metathetic	**Probable (Probably)**
19,366	Metazoa .	Is it probable (that)
19,367	Metecorn .	It is very probable (that)
19,368	Metegavel.	Not very probab'e (that)
19,369	Metempiric	Will probably be able to
19,370	Meteor .	Will probably
19,371	Meteorism	**Procedure**
19,372	Meteorites	By this procedure
19,373	Meteoroid	This method of procedure
19,374	Metestick	**Proceed**
19,375	Meteyard .	Proceed at once to ——, make a thorough examination of the property, telegraph opinion, and
19,376	Metheglin.	Will proceed as requested [send by post full report
19,377	Methinks .	You need not now proceed to
19,378	Methodical	Instruct —— to proceed to —— (to)
19,379	Methodists	Cannot proceed at once; can do so on
19,380	Methodized	Can you proceed at once to
19,381	Methought	Unable to proceed to
19,382	Methule .	Must proceed at once to
19,383	Methylated	Now ready to proceed
19,384	Methylic .	Proceed against
19,385	Meticulous	Cannot proceed further (without)
19,386	Metonymy	**Proceeding(s)**
19,387	Metrical .	Am (are) now proceeding against
19,388	Metrically	Commence proceedings
19,389	Metrifier .	Proceedings have begun
19,390	Metrists .	Legal proceedings
19,391	Metromania	Proceedings are illegal
19,392	Metronymic	Proceedings have been instituted to
19,393	Metrotome	**Proceeds**
19,394	Metroxylon	Remit proceeds
19,395	Mewl . .	Telegraph proceeds of
19,396	Mewling .	Proceeds of the —— amount to
19,397	Meynt . .	Proceeds of the sale
19,398	Mezuzoth .	Proceeds of the shipment(s)
19,399	Mezzo . .	The proceeds to be devoted to
19,400	Mezzotint.	Gross proceeds
19,401	Miargyrite	Net proceeds
19,402	Miaskite .	The net proceeds resulting from
19,403	Miasm . .	The proceeds will be more than sufficient to
19,404	Miasmal .	**Process**
19,405	Miasmatic	A new process
19,406	Miasmology	I am going to —— to see a new process of

No.	Code Word.	**Process** (*continued*)
19,407	Michaelite	By which process
19,408	Michaelmas	By an improved process
19,409	Michelia .	The new process is now at work
19,410	Micraster .	Saving by the new process
19,411	Microcosm	What process is most suitable
19,412	Microdon .	This process is well adapted for
19,413	Microfarad	**Procured**
19,414	Micrograph	Can be procured
19,415	Microhm .	Cannot be procured
19,416	Microlite .	Must be procured
19,417	Micrometry	**Produce**
19,418	Microphone	Hope to produce
19,419	Micropyle.	Cannot produce
19,420	Microscope	What do the mines produce per week
19,421	Microzoa .	For some time past the produce has been
19,422	Midchannel	The annual produce
19,423	Midcouples	The monthly produce
19,424	Midcourse	The weekly produce
19,425	Midday .	The entire produce from
19,426	Middle. .	Will produce on an average
19,427	Middleage	Cannot produce more than
19,428	Middleman	**Produced**
19,429	Middlemost	The mine(s) has (have) produced
19,430	Middletint	Have produced per week
19,431	Middlingly	Have produced per month for the last —— months
19,432	Midearth .	**Producing**
19,433	Midfeather	The mine(s) is (are) producing —— tons
19,434	Midheaven	The mine(s) is (are) producing —— tons per week
19,435	Midhour .	The mine(s) is (are) producing very fine ore
19,436	Midleg. .	Hope shortly to be producing —— tons per week
19,437	Midmost .	The ore chute is now producing
19,438	Midnight .	Stope is now producing
19,439	Midnoon .	**Production**
19,440	Midrash .	To increase the production
19,441	Midrib . .	To maintain this production
19,442	Midsea. .	Production will increase
19,443	Midship .	What is the average production
19,444	Midshipman	What is the production likely to be
19,445	Midsky .	The production has increased
19,446	Midstream	The production has fallen off (to)
19,447	Midsummer	**Profit(s)**
19,448	Midwicket	Profit in sight
19,449	Midwife .	Last month's profit was
19,450	Midwifery	Profit this month will be
19,451	Midwifish .	Profit will be
19,452	Midwinter	Likely to give large profits
19,453	Miemite .	An annual profit of
19,454	Miff. . .	The net profit for the year is
19,455	Mightful .	What are the profits upon [been paid
19,456	Mightier .	What profit will there be after all expenses have

No.	Code Word.	**Profit(s)** (*continued*)
19,457	Mightily .	Will net profits for month be
19,458	Mightiness	A monthly profit (of)
19,459	Mighty . .	(At) a profit of
19,460	Migrant .	The profits have been very large
19,461	Migration .	The profits at first will not be large
19,462	Migratory .	Telegraph the probable net profit for the month of
19,463	Mikania .	Estimated profit
19,464	Milch . .	Leaves too small a profit
19,465	Mildest .	I estimate we shall make a profit of
19,466	Mildewy .	Profit on the run
19,467	Mildness .	Have been making small profits all the time
19,468	Mildspoken	Has been worked at a profit (of)
19,469	Mileage .	Profits may be expected inside of
19,470	Milepost .	Profit may be expected to be not less than
19,471	Milestone .	There will be no profit
19,472	Milfoil . .	How much profit do we make
19,473	Miliaria .	Actual realized profit
19,474	Miliola .	Impossible to estimate profit
19,475	Miliolitic .	A good margin of profit
19,476	Militancy .	Profit for the half-year
19,477	Militiaman	Profit and loss account
19,478	Milium .	Think the mine will make large profits
19,479	Milkenway	Afraid there is very little profit to be made
19,480	Milkfever .	Profits to be divided
19,481	Milkglass .	Out of the profit for
19,482	Milkhedge	Out of the profits made we can declare a dividend
19,483	Milkily . .	Leaves no profit [at the rate of
19,484	Milkmaids	Leaves a large profit
19,485	Milkman .	Leaves a small profit
19,486	Milkmolar	**Profitable**
19,487	Milkpunch	A very profitable undertaking
19,488	Milkquartz	Am (are) sure it will be a very profitable investment
19,489	Milkrack .	Not sufficiently profitable
19,490	Milksnake	**Progress**
19,491	Milksop .	What progress is being made
19,492	Milksopism	Good progress is being made
19,493	Milksugar .	Cannot you make better progress
19,494	Milkthrush	Not making much progress at present
19,495	Milktree .	Not making much progress, owing to
19,496	Milkvats .	**Promise(s)**
19,497	Milkvessel	Promise has not been fulfilled
19,498	Milkvetch .	Will —— promise to
19,499	Milkwarm .	Cannot promise to
19,500	Milkwhite	Would not promise
19,501	Milkwood .	Can you promise
19,502	Milky . .	Can you fulfil your promise
19,503	Millbar .	Rely on the fulfilment of your promise
19,504	Millcog .	Can promise faithfully
19,505	Millennial .	Promises to do all he can
19,506	Milleped .	The mines promise well

No.	Code Word.	Promise(s) *(continued)*
19,507	Millepora .	The shareholders do not want promises but divi-
19,508	Millerite .	**Promised** [dends
19,509	Millesimal	Has (have) promised to .
19,510	Milletbeer.	Has (have) promised not to
19,511	Milleyes .	I (we) have promised
19,512	Millgang .	Promised to give a definite answer
19,513	Millhands.	If you had not promised
19,514	Millholm .	If —— had not promised
19,515	Milligram .	Shareholders were promised
19,516	Millilitre .	**Promissory note**
19,517	Milliner .	Has (have) given a promissory note (for)
19,518	Millionist .	Promissory note presented and honoured
19,519	Millocrat .	Promissory note presented and returned unpaid
19,520	Millpick .	**Promote**
19,521	Millpond .	(To) promote a company
19,522	Millrace .	Cannot promote
19,523	Millrea .	(To) promote a good feeling
19,524	Milltail .	(To) promote a better feeling
19,525	Milltooth .	Will not tend to promote
19,526	Millwheel.	Will promote
19,527	Millworks.	(To) promote your interests
19,528	Millwright	(To) promote the interests of all concerned
19,529	Miltonic .	**Promoted**
19,530	Miltwaste .	Company has been promoted
19,531	Milvinae .	Company now being promoted
19,532	Milvus . .	Promoted the suit
19,533	Mimetene.	Promoted a better feeling among all concerned
19,534	Mimetism .	**Promoter(s)**
19,535	Mimical .	Are strong promoters of the
19,536	Mimically.	The promoters (to) get
19,537	Mimicker.	Cannot allow the promoters
19,538	Mimicking	Promoters and underwriters get
19,539	Mimicry .	The promoters of the undertaking
19,540	Mimulus .	**Promoting**
19,541	Mimusops	Are promoting a company
19,542	Minaccioso	Are promoting a movement for
19,543	Minarets .	**Promotion**
19,544	Minargent	Expenses of promotion
19,545	Minatorily	Promotion expenses to be borne by
19,546	Minaul .	All promotion expenses, including legal charges, printing, brokerage, and commission
19,547	Mincemeat	The promotion of the company
19,548	Mincepie .	The promotion of the lawsuit
19,549	Minded .	**Prompt**
19,550	Mindful .	Has (have) been as prompt as possible
19,551	Mineon .	Be as prompt as you can
19,552	Mineral .	Prompt action is required
19,553	Mineralist.	Prompt action will be taken
19,554	Mingleable	**Promptly**
19,555	Mingledly	This must be done promptly

No.	Code Word.	**Promptly** (*continued*)
19,556	Minglement	Promptly on account of
19,557	Mingling .	Act very promptly, if you wish to
19,558	Mingrelian	Must act promptly in order to prevent
19,559	Miniard .	**Proof**
19,560	Miniardize	As a proof
19,561	Miniatures	There is no proof
19,562	Minieball .	The proof is
19,563	Minierifle .	What proof is there that
19,564	Minimizing	Can give sufficient proof to convince you that
19,565	Minioning	Can show no proof .
19,566	Minionlike	Could give no proof
19,567	Minionship	There is every proof that
19,568	Ministered	If you want proof
19,569	Ministrant	**Properly**
19,570	Ministry .	Fairly and properly worked
19,571	Minnows .	Accounts were properly made out
19,572	Minoration	Not properly done
19,573	Minoress .	**Property**
19,574	Minorities	I consider it a most valuable property
19,575	Minority .	The property is worth
19,576	Minorship	I do not consider the property worth having .
19,577	Minstrel .	It is not as valuable a property as represented
19,578	Minstrelsy	The property is situated
19,579	Mintage .	The situation of the property is bad [desired
19,580	Mintjulep .	The situation of the property is all that can be
19,581	Mintmark .	The property comprises
19,582	Mintmaster	Owing to the large extent of the property
19,583	Mintsauce	How is the property situated
19,584	Mintwarden	The position of the property
19,585	Minuend .	In the same locality as the —— company's property
19,586	Minuscula	The property is in the —— district
19,587	Minutebell	What is the property worth
19,588	Minutebook	What will they take for the property
19,589	Minutegun	Is the property worth having
19,590	Minutehand	What does the property consist of
19,591	Minutejack	The property is in the hands of
19,592	Minutely .	The code word of the property to be
19,593	Minuteness	Property code word altered to
19,594	Minutia .	The property is situated near the —— Co.'s property
19,595	Minutiose.	The property is situated on —— side of
19,596	Minxotter.	In my opinion, the mine will develop into a fine
19,597	Miocene .	Has (have) examined the property [property
19,598	Miohippus	Can get lease of property for —— years at
19,599	Miquelet .	Can get lease of property for —— months at
19,600	Mirabilary	The property is offered to others. If we can conclude at once, believe we can secure it over their
19,601	Mirabilis .	A most promising property [heads
19,602	Miraculize	Property in an excellent district
19,603	Miraculous	Is the property auriferous
19,604	Mirador .	Do not vouch for the property being auriferous

No.	Code Word.	**Property** (*continued*)
19,605	Mirbane .	Property represented as auriferous
19,606	Mirecrow .	Property probably is auriferous
19,607	Miredrum .	Property is auriferous
19,608	Mirfack .	A —— interest is offered to us in a property for —— . Shall we accept on your account
19,609	Mirific . .	Have personally inspected property, and recommend
19,610	Mirificent .	The property is a good one [it on terms stated
19,611	Mirligoes .	What work has been done on the property
19,612	Mirthful .	Please visit property again and confirm your report
19,613	Mirthless .	Thoroughly good property [by wire
19,614	Misaccompt	Please give your opinion of it as a payable gold
19,615	Misadjust .	Fairly good property [property
19,616	Misaimed .	The property may be auriferous, but not in our opinion payably so
19,617	Misapplied	The property is a fair one, and at a moderate figure a good investment
19,618	Misapply .	What price do you intend giving for the property
19,619	Misascribe	We know nothing of the property
19,620	Misassign .	Property is not known of here
19,621	Misattend .	The property is practically a —— dwt. property
19,622	Misbecome	The property is a —— oz. property
19,623	Misbeget .	We have closed all connection with the property
19,624	Misbehave	The property has been specially registered
19,625	Misbeseem	Adjoining property has been abandoned
19,626	Misbestow	Refused access to property
19,627	Misboden .	Transfer of the property
19,628	Miscalling .	**Proportion(s)**
19,629	Miscarried	Contain(s) a proportion of
19,630	Miscarry .	Contains a large proportion of
19,631	Miscast .	A large proportion of which
19,632	Miscellany	In proportion to
19,633	Miscentre .	In what proportion
19,634	Mischance	Is not in proportion to
19,635	Mischief .	Our proportion will be
19,636	Mischna .	Your proportion will be
19,637	Mischoose	Divide(d) in equal proportions
19,638	Misclaim .	**Proportionally**
19,639	Miscognize	Proportionally divided among
19,640	Miscollect	**Proposal(s)**
19,641	Miscompute	Has (have) made a proposal to
19,642	Misconduct	Have asked —— to make a proposal
19,643	Misconster	Proposal cannot be entertained
19,644	Miscovet .	A very fair proposal
19,645	Miscreants	Must have a more definite proposal
19,646	Misdated .	Accept the proposal
19,647	Misdaub .	Decline the proposal
19,648	Misdealing	Proposal accepted, subject to modification
19,649	Misdeeds .	Proposal accepted, provided that
19,650	Misdemean	Must modifiy proposal
19,651	Misderive .	Any further proposals

No.	Code Word.	
19,652	Misdesert .	**Propose(s)**
19,653	Misdiet .	Would propose to
19,654	Misdirects	Do (does) not propose to
19,655	Misdivide .	What do (does) —— propose doing
19,656	Misdo . .	Propose(s) the following
19,657	Misdoings	Have you anything to propose
19,658	Misdoubt .	I have nothing to propose
19,659	Misdread .	**Proposed**
19,660	Miseasy .	It was proposed at the meeting **to**
19,661	Misedition	It has been proposed
19,662	Miseducate	**Proposition.** (See Proposal.)
19,663	Misemoney	Wire if proposition is accepted
19,664	Misentries	A very fair proposition
19,665	Misentry .	Proposition has been altered
19,666	Miser . .	**Prosecute**
19,667	Miserable .	Will prosecute
19,668	Miserect .	Will not prosecute
19,669	Miserly .	Will —— prosecute
19,670	Misexpound	In order to prosecute
19,671	Misfaith .	Not to prosecute
19,672	Misfare .	**Prospect(s)**
19,673	Misfeasor .	Prospect thoroughly
19,674	Misfeign .	What is the prospect of
19,675	Misfit . .	If any immediate prospect
19,676	Misfitting .	The mine is little more than a prospect
19,677	Misformed	A valuable prospect
19,678	Misfortune	There seems every prospect of
19,679	Misframe .	There seems no prospect of
19,680	Misframing	No prospect of our being able to
19,681	Misgivings	A poor prospect
19,682	Misgotten	A fine prospect
19,683	Misgovern	Think the mine is a good prospect
19,684	Misgraff .	With this prospect in view
19,685	Misground	Unless there be some prospect of
19,686	Misgrowth	Think there is little prospect
19,687	Misguess .	If no prospect of
19,688	Misguide .	Prospect very doubtful
19,689	Mishandle	Can see no immediate prospect **of**
19,690	Mishappen	There are good prospects
19,691	Mishaps .	Prospects are discouraging
19,692	Mishmash .	Prospects are encouraging
19,693	Misinform	Prospects not encouraging enough to warrant
19,694	Misintend .	The future prospects of the
19,695	Misintreat .	Prospects do not seem to improve
19,696	Misjoinder	As prospects are improving
19,697	Misjudge .	As prospects do not improve
19,698	Misjudging	**Prospecting**
19,699	Miskenning	Have sunk a prospecting shaft
19,700	Miskindled	Prospecting a long time on the same vein
19,701	Misknow .	Must at once begin prospecting

No.	Code Word.	**Prospecting** (*continued*)
19,702	Misknowing	No prospecting has been done
19,703	Mislaid .	Shall continue prospecting in
19,704	Misleadeth	Shall stop prospecting in
19,705	Mislearned	Has (have) stopped prospecting
19,706	Mislight .	Stop prospecting
19,707	Misluck .	Am (are) prospecting
19,708	Mismake .	Has (have) been prospecting for
19,709	Mismanage	Are prospecting in every direction
19,710	Mismanners	What prospecting has been done
19,711	Mismatch .	Prospecting adit(s)
19,712	Mismeasure	Prospecting shaft
19,713	Misnomers	Prospecting vigorously carried on
19,714	Misnumber	You can spend in prospecting
19,715	Misnurture	How much can we spend in prospecting
19,716	Misobserve	Desirable to go on prospecting
19,717	Misogamist	Examining and prospecting the
19,718	Misogamy.	Syndicate for prospecting and developing
19,719	Mispense .	Prospecting over the entire property
19,720	Mispikel .	I recommend your prospecting (it)
19,721	Misplace .	Do you recommend us to accept it on prospecting
19,722	Mispoint .	Set men on prospecting [terms
19,723	Mispolicy .	**Prospectus**
19,724	Misprinted	Send me prospectus
19,725	Misproud .	Sending prospectus
19,726	Misquoted	Send your prospectus
19,727	Misquoting	Prospectus will be issued
19,728	Misraise .	Prospectus issued
19,729	Misreceive	Prospectus not issued
19,730	Misrecital.	Shall we issue prospectus
19,731	Misreckon	We advise issue of prospectus
19,732	Misrelate .	Prospectus withdrawn
19,733	Misrender.	Withdraw prospectus until
19,734	Misreport .	Defer issuing prospectus
19,735	Missal .	**Protect(s)**
19,736	Missayer .	It will be necessary to protect
19,737	Misseek .	(To) protect yourself
19,738	Misseldine	(To) protect ourselves against
19,739	Misshaping	Will protect
19,740	Missionary	Will not protect
19,741	Missives .	Will this protect
19,742	Missound .	Protect our rights and interests
19,743	Misspeech	Protect all interests
19,744	Misspell .	Protect us from molestation
19,745	Misstep .	We undertake to protect
19,746	Missuccess	**Protected**
19,747	Misswear .	Protected by
19,748	Mistakenly	Not protected by
19,749	Mistaking .	Must be fully protected
19,750	Misteach .	**Protection**
19,751	Mistflower	What protection is there

PRO

No.	Code Word.	**Protection** (*continued*)
19,752	Misthrive .	There is no protection against
19 753	Misthrow .	Sufficient protection against
19,754	Mistico .	Every protection
19,755	Mistide .	Every protection has been afforded us by the
19,756	Mistihead .	For our protection
19,757	Mistily. .	For whose protection
19,758	Mistletoe .	For the protection of our interests
19,759	Mistook .	The proper protection of our rights
19,760	Mistrals .	Under the protection of the court
19,761	Mistruster.	**Protest**(s)
19,762	Mistryst .	Protest against
19,763	Mistuned .	Pay under protest
19,764	Mistutored	Has (have) paid under protest
19,765	Misty . .	No use to protest
19,766	Misurato .	Protest against —— action in the matter
19,767	Misuse. .	A formal protest against
19,768	Misusement	Must protest strongly against
19,769	Misusing .	Under protest
19,770	Misvouch .	**Prove**(s)
19,771	Miswed .	Can you prove
19,772	Miswedding	Cannot prove
19,773	Miswrite .	We can prove
19,774	Miswriting	Incontestably prove(s)
19,775	Miswrought	If we can only prove
19,776	Miszealous	Evidence to prove
19,777	Mitaine .	Prove(s) the truth of
19,778	Mithras .	Can —— prove the existence of
19,779	Mithridate	To prove the existence of
19,780	Mitigable .	To prove the vein in depth
19,781	Mitigant .	Likely to prove a valuable ore body
19,782	Mitigation	**Proved**
19,783	Mitigatory	Has (have) been proved by
19,784	Mitisgreen	Has (have) not been proved (by)
19,785	Mitrailled.	This has proved
19,786	Mitrebox .	Unless it has been proved that
19,787	Mitredrain	Width has not yet been proved
19,788	Mitrejoint.	Length not proved
19,789	Mitresill .	Length —— feet; width not yet proved
19,790	Mitrewheel	This has proved to be
19,791	Mitriform .	Proved to be in excess of
19,792	Mittimus .	Proved less valuable than
19,793	Mixable .	**Provide**(s)
19,794	Mixtion .	To provide against all emergencies
19,795	Mizmaze .	It is necessary to provide
19,796	Mizzled .	To provide against
19,797	Moachibo.	Will —— provide (for)
19,798	Mobcaps .	Will provide
19,799	Mobile. .	Will not provide
19,800	Mobilised.	Must provide for
19,801	Mobility .	Will this provide

No.	Code Word.	Provide (*continued*)
19,802	Moblaw .	Provide what is necessary
19,803	Mobocracy	Provide funds for
19,804	Mobocratic	Provide a reserve for
19,805	Mobreader	Provide against all contingencies
19,806	Mobsman .	Provide a sufficient stock of
19,807	Mobstory .	To provide against possible loss
19,808	Moccasin .	Provide for the proper working of
19,809	Mochastone	**Provided**
19,810	Mockable .	Provided —— will extend the time
19,811	Mockadour	Provided —— will give a guarantee
19,812	Mockbird .	Not provided in any way for
19,813	Mockeries	Has (have) not provided against
19,814	Mockery .	Be provided for
19,815	Mockheroic	Has been provided for
19,816	Mockingly	Funds provided for
19,817	Mockish .	**Providing**
19,818	Mocklead .	While providing for
19,819	Mockorange	Providing for future use
19,820	Mockore .	**Provision(s)**
19,821	Mocksun .	Provision must be made for
19,822	Mockturtle	Has (have) made provision for
19,823	Mockvelvet	Has (have not) made provision for
19,824	Modalist .	Sufficient provision for
19,825	Modelize .	Insufficient provision for
19,826	Modelled .	What provision has there been made for
19,827	Modelling.	Owing to the difficulty of getting provisions
19,828	Moderable	**Provisional**
19,829	Moderance	Provisional agreement
19,830	Moderately	Provisional arrangement
19,831	Moderating	Provisional certificates
19,832	Moderation	Provisional accommodation
19,833	Moderators	Provisional measures
19,834	Moderatrix	**Prudent**
19,835	Modern .	It will be more prudent to
19,836	Modernisms	Be very prudent in
19,837	Modernized	Must be more prudent
19,838	Modesties.	Not sufficiently prudent
19,839	Modicity .	Will it be prudent (to)
19,840	Modicum .	We think it would be prudent
19,841	Modifiable	**Public**
19,842	Modifier .	For the public
19,843	Modiola .	To make public
19,844	Moduleth .	Has not yet been made public
19,845	Moehringia	The public are eager for mining speculations
19,846	Moenchia .	The public will not at present touch any mining
19,847	Moggan .	Before the public [speculation
19,848	Mograbian	You can make this public
19,849	Mohair .	Do not make this public til.
19,850	Moholi .	Stated in public that
19,851	Mohsite .	Before it is made public

No.	Code Word.	**Public** (*continued*)
19,852	Mohwatree	This will not be made public till
19,853	Moidore .	The public are investing largely in
19,854	Moieties .	The public took
19,855	Moiety .	Did not go down with the public
19,856	Moineau .	We shall not have to go to the public
19,857	Moist . .	We shall have to go to the public for the money
19,858	Moistener .	**Publication**
19,859	Moistening	For publication in
19,860	Moistless .	Intended for publication
19,861	Moisture .	**Publish**
19,862	Molasses .	Intend(s) to publish
19,863	Molecast .	Will not publish
19,864	Molech .	Do not publish
19,865	Molecular .	Do not think it advisable to publish
19,866	Molehill .	Think it desirable to publish
19,867	Molerat .	We must publish
19,868	Moleskin .	Be careful what you publish
19,869	Molest . .	Must be careful what we publish
19,870	Molestful .	Publish in extenso
19,871	Molesting .	Shall only publish
19,872	Moletrack	Publish only that part of the report
19,873	Moletree .	You can publish full report
19,874	Molewarp .	Can we publish
19,875	Moliminous	When shall you publish
19,876	Molinism .	**Published** [be published
19,877	Mollebart .	Be very guarded in your report, as all you say will
19,878	Mollemoke	This ought, I think, to be published
19,879	Molleton .	This had better not be published
19,880	Molliently	Should this be published
19,881	Mollify .	Has (have) been published
19,882	Mollities .	Has (have) not yet been published
19,883	Mollitude .	Has this been published
19,884	Mollusc .	When this (it) is published
19,885	Molluscoid	As soon as published
19,886	Molluskite	Telegraph fully all the information which you con-
19,887	Molochize	**Pulp** [sider should be published
19,888	Molokan .	Pulp assays average
19,889	Molopes .	Pulp assays for one week average
19,890	Molothrus	Pulp assays for one month average
19,891	Molten .	What is the weekly average of the pulp assay
19,892	Molunghee	**Pump(s)**
19,893	Molybdate	Will erect the necessary pumps
19,894	Molybdic .	Purchasers to take over pumps at valuation
19,895	Molybdous	The pumps are out of repair
19,896	Momentary	The pumps cannot cope with the water
19,897	Momentous	Must at once erect pumps
19,898	Momentum	Pump capable of raising
19,899	Momier .	Send —— inch bucket-lift pump —— feet stroke
19,900	Mommery	Send —— inch plunger pump —— feet stroke
19,901	Momordica	Send —— feet of —— inch pumps

No.	Code Word.	Pump(s) (*continued*)
19,902	Momotinae	Will this sum include the pumps
19,903	Monadaria	Pumps will have to be sent from
19,904	Monadelph	Pumps now working
19,905	Monadic .	Will require new pumps
19,906	Monadology	Delay of —— days owing to pump(s) being out of
19,907	Monandry	Windbore for —— inch pump [order
19,908	Monanthous	Working barrel —— inch pump
19,909	Monarch .	Pump is not able to keep the water down
19,910	Monarchal	Pump is gaining rapidly on the water
19,911	Monarchism	What pumps exist at the mine
19,912	Monarchize	Can do nothing till pumps arrive [—— feet
19,913	Monarda .	Require pumps to raise —— gallons per hour
19,914	Monastery	Pumps will raise —— gallons per minute [hours
19,915	Monastical	Pumps are now lifting —— gallons every twelve
19,916	Monaulos .	Pump that will lift —— gallons per minute
19,917	Mondjourou	To pump —— gallons per minute from a depth of
19,918	Monecian .	Pump to raise —— gallons per hour [—— feet
19,919	Monetizing	Have you started pump(s)
19,920	Moneyage	Have started pump(s)
19,921	Moneybill .	Have started; pumps working well
19,922	Moneyland	Cannot do anything without steam pump
19,923	Moneymaker	Pump continually going
19,924	Moneyorder	Steam pump will be sent
19,925	Mongolidae	Send steam pump at once
19,926	Mongolioid	Shall start steam pump
19,927	Mongoose	What size pumps do you require
19,928	Mongrel .	Pumps are inadequate
19,929	Mongrelize	Pump capacity
19,930	Monilifer .	Pumps working day and night
19,931	Moniliform	Pumps working —— hours
19,932	Moniours .	How many hours in twenty-four are pumps at work
19,933	Monisher .	Are pumps at work
19,934	Monishment	Pumps making —— strokes per minute
19,935	Monition .	Pumps can make —— strokes per minute
19,936	Monitive .	Pump used whilst sinking
19,937	Monitorial	This pump handles all the water coming from
19,938	Monitory .	Total cost of engine and pump-work
19,939	Monitress .	Pumps and ironwork complete
19,940	Monkbat .	Pumps and ironwork complete for a —— feet
19,941	Monkeyboat	Pumps in shaft [standing lift
19,942	Monkeycup	Suction pump(s)
19,943	Monkeyism	Plunger pump(s)
19,944	Monkeypot	Pump throws the water to the surface
19,945	Monkeyrail	Pump delivers the water at the —— level
19,946	Monkfish .	Water thrown by these pumps
19,947	Monkhood	Pump-work in shaft
19,948	Monkseal .	Wood-work in pump-rods, guides, bearers, etc.
19,949	Monobasic	Pump is working —— hours out of twenty-four
19,950	Monocarpic	Pump is able to cope with all the water we have
19,951	Monoceros	Stroke of pumps —— feet

No.	Code Word.	Pump(s) *(continued)*
19,952	Monochord	Pump is capable of raising —— gallons per hour
19,953	Monochrome	Pump is not able to do more than
19,954	Monoclinal	Pump is working at its full capacity
19,955	Monocotyle	Pump keeps the water down easily
19,956	Monocrat .	Pump(s) —— inch diameter, and —— feet stroke
19,957	Monocular	Particulars of pumps [cation
19,958	Monodical	Contract price for engine and pumps as per specifi-
19,959	Monodist .	Contract price for engine and pump-work
19,960	Monodrama	Contract price for pumps and pump-work
19,961	Monoecious	Steam pump
19,962	Monogamist	Pulsometer pump
19,963	Monogamous	Centrifugal pump
19,964	Monogamy	Pump buckets
19,965	Monograph	Pump bob(s)
19,966	Monogyn .	Pump plunger(s)
19,967	Monolith .	Pump rod(s)
19,968	Monolithal	Length of pump rods
19,969	Monologist	Wooden pump rods —— inches square
19,970	Monologue	Wrought-iron pump rods —— inches diameter
19,971	Monomachy	Six-inch pumps
19,972	Monomane	Seven-inch pumps
19,973	Monomaniac	Eight-inch pumps
19,974	Monometric	Nine-inch pumps
19,975	Monomial .	Ten-inch pumps
19,976	Monomyaria	Eleven-inch pumps
19,977	Monopathic	Twelve-inch pumps
19,978	Monophonic	Thirteen-inch pumps
19,979	Monopnoa	Fourteen-inch pumps
19,980	Monopoly .	Fifteen-inch pumps
19,981	Monopteral	Sixteen-inch pumps
19,982	Monoptote	Seventeen-inch pumps
19,983	Monorganic	Lift of pumps —— feet
19,984	Monorhyme	**Pumped**
19,985	Monosperm	Water has been pumped out
19,986	Monostich	Cannot be pumped out, without
19,987	Monostyle	As soon as the water is pumped out
19,988	Monothecal	**Pumping**
19,989	Monotheism	Commence pumping at once
19,990	Monotonic	Have to be constantly pumping
19,991	Monotonous	Continue pumping
19,992	Monotreme	Cease pumping
19,993	Monotropa	Pumping the water out of the mine
19,994	Monotype.	Are pumping —— gallons in twenty-four hours
19,995	Monovalent	Are not pumping
19,996	Monoxylous	Have had to stop pumping
19,997	Monsoon .	Recommence pumping
19,998	Monstrance	Pumping engine
19,999	Monstrator	Pumping engine of —— h.p.
20,000	Monstrous	Pumping engine and machinery
20,001	Montanism	Cornish pumping engine (single acting)

No.	Code Word.	**Pumping** (*continued*)
20,002	Montanize	Cornish pumping engine (double acting)
20,003	Montant .	Differential compound condensing pumping engine
20,004	Monterocap	Direct-acting pumping engine
20,005	Montezuma	Must have pumping power sufficient
20,006	Monthling	Pumping machinery in shaft
20,007	Monticle .	Pumping machinery in shaft now being erected
20,008	Montross .	Pumping machinery in shaft will be completed
20,009	Monument	Pumping machinery in shaft now at work
20,010	Monumental	Particulars of pumping engine and machinery
20,011	Moodily .	**Purchase**
20,012	Moodishly	With the option of purchase
20,013	Moodymad	Tried to purchase the mine, but failed
20,014	Moonbeam	Will take purchase money in shares
20,015	Moonblind	The purchase completed
20,016	Mooncalf .	The purchase not yet completed
20,017	Moondial .	The purchase must be completed by
20,018	Mooneye .	I (we) advise purchase of mine(s) for .
20,019	Moonface .	I (we) cannot advise purchase
20,020	Moonfern .	Do you advise purchase
20,021	Moonflower	We authorize the purchase (of)
20,022	Moongus .	Money ready if you advise purchase
20,023	Moonlight	Will interfere with the purchase
20,024	Moonlit .	It is a very good purchase
20,025	Moonloved	It is a very bad purchase
20,026	Moonmilk	To complete the purchase
20,027	Moonmonth	Completion of the purchase
20,028	Moonraker	Purchase was made with the clear understanding that
20,029	Moonraking	Purchase made under following conditions
20,030	Moonrise .	Ready to complete purchase [purchase
20,031	Moonseed	Telegraph as soon as you are ready to complete
20,032	Moonshaped	Purchase can be completed as soon as deeds are
20,033	Moonshine	The purchase of [ready
20,034	Moonstones	Delay the purchase
20,035	Moonstruck	To complete purchase, money must be paid by
20,036	Moonyear.	In order to complete purchase, please have deeds recorded and hand them to
20,037	Moorage .	**Purchaser(s)**
20,038	Moorbred.	Purchaser(s) to deposit as guarantee
20,039	Moorcha .	Purchaser(s) has (have) money ready
20,040	Moorfowl .	Purchaser(s) to take over plant at a valuation
20,041	Moorgrass	Can you find a purchaser for
20,042	Moorhen .	Have found a purchaser for
20,043	Moorlands	Cannot get a purchaser for
20,044	Moory . .	The purchaser now here
20,045	Moosedeer	**Purchasing**
20,046	Moosewood	Think of purchasing
20,047	Mootcourt	On the point of purchasing
20,048	Moothall .	**Purport**
20,049	Moothouse	The purport of the report
20,050	Mootpoint	The purport of the telegram

No.	Code Word.	**Purport** (*continued*)
20,051	Mopboard	The purport of our remarks
20,052	Mopeeyed	Misunderstood the purport of
20,053	Mopfair .	What is the purport of
20,054	Mopish .	Do not understand the purport **of**
20,055	Mopishly .	**Purported**
20,056	Mopsical .	Purported to be
20,057	Moraceae .	Purporting to be
20,058	Moraine .	Purporting to come from
20,059	Moralized.	**Purpose**
20,060	Moralizing	For what purpose
20,061	Morassy .	For the purpose of
20,062	Morbid .	On purpose
20,063	Morbidezza	For no purpose
20,064	Morbidly .	Will answer our purpose very well
20,065	Morbidness	Does it answer the purpose
20,066	Morbific .	Does (——) not answer the purpose
20,067	Morbifical.	Will answer the purpose, if we
20,068	Morbillous	Too large for the purpose
20,069	Morbosity.	Not enough for the purpose
20,070	Morceau .	Too little (too small) for the purpose
20,071	Morchella	More than needed for the purpose
20,072	Mordente .	We purpose doing so
20,073	Mordicant	**Push**
20,074	Morehough	Push the sale of
20,075	Moreover .	Do not push the
20,076	Morganatic	Push ahead with
20,077	Morgay .	You must push on the
20,078	Moribund.	To push it through
20,079	Morigerate	Hope(s) to push through
20,080	Morigerous	Push the matter forward with all speed
20,081	Moringa .	Push the matter forward or it will be too late
20,082	Morisonian	**Pushing**
20,083	Mormonism	Pushing the matter forward with all possible speed
20,084	Mormonites	**Put**
20,085	Mormyridae	Do not put
20,086	Mormyrus	Not to be put in
20,087	Morose .	Put in your next
20,088	Morosely .	To be put in
20,089	Moroseness	We have not put
20,090	Morosoph.	Why did you put
20,091	Morosous .	What has been put down for
20,092	Moroxite .	Nothing put down for
20,093	Morphia .	Put everything in good order
20,094	Morphology.	Put off payment of all the bills you can, and only
20,095	Morphosis	Put off [draw for what you are compelled to pay
20,096	Morpion .	Put on all the men you can to work
20,097	Morpunkee	**Putting**
20,098	Morrhua .	We are now putting up
20,099	Morriced .	Now putting up the ——; and expect it will be
20,100	Morrimal .	Recommend putting off [ready by

No.	Code Word.	
20,101	Morrispike	**Pyrites**
20,102	Morrow .	A sample with pyrites assayed —— per ton
20,103	Morsel . .	A sample rejecting pyrites assayed —— per ton
20,104	Mortality .	The pyrites assay —— per ton
20,105	Mortalness	Arsenical pyrites
20,106	Mortarbed	Magnetic pyrites
20,107	Mortcloth .	If the pyrites were concentrated
20,108	Mortgage .	A sample of pyrites assayed
20,109	Mortgageor	The principal part of the gold is in the pyrites
20,110	Mortgaging	**Quadrant**
20,111	Mortising .	Wrought-iron quadrant
20,112	Mortstone	Quadrant and fittings complete
20,113	Mortuaries	Connection with quadrant and spears
20,114	Mortuary .	Main caps for connecting quadrant and spears
20,115	Mosaic .	**Qualification(s)**
20,116	Mosaical .	Possessing the necessary qualifications
20,117	Mosaically	**Qualified**
20,118	Mosaicist .	Qualified surveyor
20,119	Mosaicwork	A fully qualified
20,120	Mosasaurus	**Quality.** (See also Grade.)
20,121	Moschatel	Medium quality
20,122	Moschidae	Good quality
20,123	Moschine .	Fair quality
20,124	Moslem .	Usual quality
20,125	Mosquito .	Inferior quality
20,126	Mossagate	What is the quality of the ore
20,127	Mossbunker	Quality of ore is improving
20,128	Mosscapped	Quality of ore is not so good
20,129	Mossclad .	Same quality as last
20,130	Mossgrown	Better quality than last
20,131	Mossiness.	Quality of ore fully equal to
20,132	Mossland .	Quality of ore equal to what it was in the
20,133	Mosspink .	Plenty of ore, but the quality is not very good
20,134	Mossrose .	Plenty of ore of good high-grade quality
20,135	Mostahiba	**Quantity(ies)**
20,136	Mostick .	The quantity and quality of the ore are all that can
20,137	Mostra . .	A large quantity has been taken [be desired
20,138	Motacil .	What quantity of —— is there
20,139	Motazilite	A large quantity of
20,140	Motella .	A small quantity of
20,141	Mothblight	What quantity have you
20,142	Motheaten	A much larger quantity
20,143	Mothercell	A less quantity
20,144	Mothercoal	From the quantity of
20,145	Motherhood	A large quantity of ore is now being extracted from
20,146	Motherland	Increase the quantity
20,147	Motherless	If the quantity is increased
20,148	Motherly .	The quantity can be increased
20,149	Motherspot	A considerable quantity still on hand
20,150	Motherwit	In large quantities

No.	Code Word.	**Quantity(ies)** (*continued*)
20,151	Motherwort	In small quantities
20,152	Mothgnat .	(In) quantity and quality
20,153	Mothhunter	**Quarrel(s)**
20,154	Motific. .	There has been a quarrel between
20,155	Motioned .	Owing to the quarrel (between)
20,156	Motioning	Avoid any quarrel with
20,157	Motionless	To prevent any quarrel
20,158	Motives .	**Quarry(ies)**
20,159	Motley .	Large quarries of —— exist
20,160	Motliest .	Limestone quarry (quarries)
20,161	Motmot .	Marble quarry (quarries)
20,162	Motorpathy	Granite quarry (quarries)
20,163	Mottoed .	You can quarry it
20,164	Mould . .	Can get quarry for
20,165	Mouldable	Quarry producing
20,166	Mouldboard	Quarry has been worked for
20,167	Mouldery .	**Quarrying**
20,168	Mouldeth .	Cost of quarrying
20,169	Mouldiness	In quarrying for
20,170	Mouldloft .	**Quarter(s)**
20,171	Mouldstone	A quarter of
20,172	Mouldwarp	One quarter
20,173	Mountance	Three quarters
20,174	Mountebank	Less than a quarter
20,175	Mounted .	**Quartz**
20,176	Mountingly	Quartz coming in (in face of)
20,177	Mournfully	Veins of gold-bearing quartz
20,178	Mourning .	The average thickness of the quartz is
20,179	Mournival .	Rejecting the barren quartz
20,180	Mournsome	No available quartz
20,181	Mousebird	Contains no quartz
20,182	Mouseear .	Composed of decomposed quartz
20,183	Mousefall .	Rich quartz
20,184	Mousehawk	Barren quartz
20,185	Mousehole	The quartz being very white
20,186	Mousehunt	Quartz vein
20,187	Mousekin .	From specimens of the quartz
20,188	Mousesight	The matrix is quartz
20,189	Mousetail .	Quartz assayed badly
20,190	Mousetrap	Quartz assayed
20,191	Mousing .	Composed of quartz and
20,192	Mouthglass	Gold-bearing quartz
20,193	Mouthmade	Quartz is charged with
20,194	Mouthpiece	Quartz carries traces of gold
20,195	Mouthorgan	Samples of quartz (from the)
20,196	Movement	Very promising quartz
20,197	Mowburn .	**Quartzite**
20,198	Mowyer .	Quartzite formation
20,199	Mozarabic	The hanging-wall is quartzite
20,200	Mucate .	The foot-wall is quartzite

No.	Code Word.	
20,201	Mucedineae	**Question(s)**
20,202	Mucedinous	It is a question whether
20,203	Mucilage .	There is no question as to
20,204	Muciparous	What is the question
20,205	Mucivora .	In answer to your question
20,206	Muckender	Answer this question at once
20,207	Muckforks	This is a question which will have to be answered
20,208	Muckheap	Out of the question
20,209	Muckiness	It is a question of
20,210	Muckrakes	It is a very important question
20,211	Mucksweat	Must give plain answer to the question
20,212	Mucksy .	Questions relating to title
20,213	Muckthrift	Questions relating to
20,214	Muckworm	A question of the greatest importance for the future
20,215	Mucocele .	The question is sure to be put [of the concern
20,216	Mucoraceae	Cannot answer the question
20,217	Mucosity .	**Quick.** (See also Prompt.)
20,218	Mucous .	Be as quick as you can
20,219	Mucousness	Not as quick as it should be
20,220	Mucronate	Will be as quick as possible
20,221	Muculent .	**Quickly**
20,222	Mucusine .	As quickly as
20,223	Mudbath .	Do it quickly
20,224	Muddevil .	Must be done quickly
20,225	Muddiest .	Go as quickly as you can
20,226	Muddrag .	Will be sent as quickly as possible
20,227	Muddredger	Send as quickly as you can
20,228	Mudfish .	Has been done as quickly as possible
20,229	Mudflat .	Was not done as quickly as it might have been
20,230	Mudhole .	Cannot get things done quickly here
20,231	Mudlark .	Very quickly
20,232	Mudplug .	Not very quickly
20,233	Mudsill .	Working quickly
20,234	Mudstone.	**Quicksilver**
20,235	Mudsucker	A quicksilver mine
20,236	Mudturtle .	The quicksilver is found
20,237	Mudvalves	Causing the quicksilver to sicken
20,238	Mudwalled	Great difficulty in getting quicksilver for
20,239	Mudworms	What is present price of quicksilver in London
20,240	Muffettees	What is the present price of quicksilver in San Fran-
20,241	Muffin . .	Quicksilver is quoted —— per flask [cisco
20,242	Muffincap.	Can get quicksilver delivered at mine at —— per
20,243	Mufflers .	Have ordered —— flasks of quicksilver [flask
20,244	Mufti . .	You can order —— flasks of quicksilver
20,245	Mugiency .	Quicksilver has advanced
20,246	Mugiloid .	Quicksilver has fallen considerably; think it a good
20,247	Mugweed .	Quicksilver in store [time to buy
20,248	Mulatto .	Freight of quicksilver per car-load
20,249	Mulattress	Quicksilver consumed
20,250	Mulberry .	The loss of quicksilver

No.	Code Word.	Quicksilver (*continued*)
20,251	Mulctuary .	What is the consumption of quicksilver
20,252	Muledriver	Quicksilver saved
20,253	Mulejenny	Offer quicksilver at
20,254	Muleyhead	Quicksilver in great demand
20,255	Muleysaw .	Little demand at present for quicksilver
20,256	Mulgedium	The production of quicksilver is
20,257	Muliebrity	Average consumption of quicksilver per ton of ore
20,258	Mulierly .	**Quiet** [milled
20,259	Mulishness	Everything is now quiet
20,260	Mullet . .	**Quietly**
20,261	Mullidae .	Look about quietly, and tell me (us)
20,262	Mulligrubs	Go to work quietly
20,263	Multeity .	We are quietly working
20,264	Multifaced	Quietly and steadily
20,265	Multifid .	**Quit**
20,266	Multiflue .	Will quit work
20,267	Multiform .	Has (have) not yet quit
20,268	Multipedes	Is (are) likely to quit
20,269	Multiplied	**Quite**
20,270	Multisect .	Is (are) quite
20,271	Multitude .	Is (are) not quite
20,272	Multivalve	Quite right
20,273	Multivious	Quite sufficient
20,274	Multivocal	Quite the same
20,275	Multoca .	Not quite the same
20,276	Multocular	Are you quite sure
20,277	Mumblenews	Must be quite sure
20,278	Mumbudget	Is (are) quite correct
20,279	Mumchance	Quite in order
20,280	Mummachog	**Quotation(s)**
20,281	Mummeries	The quotations are
20,282	Mummycloth	Want lowest possible quotation for
20,283	Mummywheat	Telegraph quotation(s) of
20,284	Mumpsimus	The following are the latest quotations
20,285	Mundanely	Quotations the last week have been
20,286	Mundation	Want quotation for freight
20,287	Mundify .	Want quotation for freight, insurance, and all
20,288	Munerary .	Not exceeding your latest quotation [charges
20,289	Municipal .	The quotation includes freight and insurance
20,290	Munificate	Quotation is for delivery at mine
20,291	Muniments	Quotation is F. O. B.
20,292	Munjah .	Quotation is loaded in cars at
20,293	Munjistin .	At the present quotation(s)
20,294	Munsiff .	**Quoted**
20,295	Muntin .	Quoted at —— premium
20,296	Muraenoid	Quoted at ——discount
20,297	Murdering	Quoted at par
20,298	Murderment	Telegraph at what —— shares are being quoted
20,299	Murderous	Shares are being quoted
20,300	Murex . .	Shares are not quoted

No.	Code Word.	Quoted (*continued*)
20,301	Murexide .	Shares now quoted at —— *cum div.*
20,302	Murgeon .	Shares now quoted at —— *ex div.*
20,303	Muriatic .	Shares now quoted at ——; but on a weak market
20,304	Muricinae.	Shares now quoted at ——; think they will go
20,305	Murkily .	**Rail(s)** [better
20,306	Murmured	By rail to ——; thence by
20,307	Murmurings	Hours journey by rail
20,308	Murrain .	By rail all the way
20,309	Musaceous	Can get the goods by rail
20,310	Musalchees	Will go forward by rail
20,311	Musaph .	Iron rails —— lbs. per yard
20,312	Muscales .	Steel rails —— lbs. per yard
20,313	Muscardine	Bridge rails
20,314	Muscicapa	Flat-bottomed rails
20,315	Muscle .	**Railway(s)**
20,316	Muscling .	Railway washed out by floods
20,317	Muscoid .	Railway from —— to —— snowed up
20,318	Muscology	The railway passes within —— mile(s) of the mine
20,319	Muscovado	A railway is projected which will pass within ——
20,320	Muscular .	A branch railway to [miles of the mines
20,321	Muscularly	The nearest railway station is —— (distant ——
20,322	Musculous	A proposed line of railway [mile(s)
20,323	Museful .	There is no railway within —— miles of the mine
20,324	Museless .	Railway from —— to
20,325	Museum .	Railway blocked by landslip
20,326	Mushrooms	Railway blocked by floods
20,327	Musicai .	Railway blocked; bridge broken
20,328	Musicalbox	If a railway is constructed
20,329	Musically .	Until the railway is completed
20,330	Musicbook	Delivered on the railway at
20,331	Musicfolio	**Rain(s)**
20,332	Musicloft .	The mine was flooded owing to heavy rains
20,333	Musicpaper	As soon as the rains set in
20,334	Musicshell	Only after the rains
20,335	Musicsmith	Till the rains
20,336	Musicstand	During the rains
20,337	Musicstool	No rain yet
20,338	Musictype	Heavy rains
20,339	Musimon .	Was (were) washed away by heavy rains
20,340	Musingly .	Has (have) been delayed by the rain
20,341	Muskbag .	Great damage has been done by the late rains
20,342	Muskball .	Owing to the heavy rains
20,343	Muskbeaver	The rainy season has now begun
20,344	Muskbeetle	The rainy season will last until
20,345	Muskcake.	The rainy season is over
20,346	Muskcavy.	Inaccessible during the rainy season
20,347	Muskduck	**Raise**
20,348	Musketeer	Raise to meet the
20,349	Musketoon	To raise
20,350	Musketrest	Can you raise

No.	Code Word.	**Raise** (*continued*)
20,351	Musketry .	Will raise
20,352	Muskmallow	Cannot raise
20,353	Muskmelon	If you can raise
20,354	Muskorchis	If you cannot raise
20,355	Muskox .	Try and raise
20,356	Muskpear .	Trying to raise
20,357	Muskplants	How much can you raise
20,358	Muskplum	Can you raise enough money **to**
20,359	Muskrat .	Can raise enough
20,360	Muskrose .	Cannot raise enough
20,361	Muskseed .	Will raise the price (of)
20,362	Muskwood	**Raised**
20,363	Muslinet .	Has been raised
20,364	Muslinkail	Has not been raised
20,365	Musnud .	Have you raised
20,366	Musomania	Will be raised
20,367	Musquash	Have raised enough ore to
20,368	Musquaw .	Raised enough ore to keep us going **until**
20,369	Musselbed	Raised enough capital
20,370	Mussulman	Have raised on the
20,371	Mustang .	**Raising**
20,372	Mustardpot	Raising per day
20,373	Musteline .	Raising up on the
20,374	Musterbook	Are now raising on the
20,375	Musterroll	**Rate(s).** (See Freight.)
20,376	Mutability	The rates are very high
20,377	Mutatory .	The freight rates are very **high**
20,378	Mutilated .	To increase the rates
20,379	Mutilating	To lessen the rates
20,380	Mutineers.	The high rate of
20,381	Mutinied .	The low rate of
20,382	Mutinous .	At the rate of
20,383	Mutinously	At what rate
20,384	Mutiny .	What are the rates for
20,385	Mutinying.	At any rate
20,386	Mutism .	At what rate can you
20,387	Muttonchop	Rates are falling
20,388	Muttonfist	Rates moderate
20,389	Muttonham	Rates are rising
20,390	Muttonpie	Can you maintain the present rate **of**
20,391	Mutual .	Can maintain the present rate of
20,392	Mutually .	What is the lowest possible rate
20,393	Muzzy . .	Special rate fixed
20,394	Myadae .	Can you get a special rate
20,395	Myalgia .	Cannot get a special rate (for)
20,396	Myallwood	At the present rate
20,397	Mycelium .	At the present rate of increase
20,398	Mycetes .	Rates are likely to be advanced
20,399	Mycina .	Rates will probably be lowered
20,400	Mycoderm	**Rather**

No.	Code Word.	**Rather** (*continued*)
20,401	Mycodermic	Rather than
20,402	Mycologic	Would rather not
20,403	Mycose .	Would rather
20,404	Mydriasis .	It would be rather
20,405	Myelitis .	Rather venturesome to predict
20,406	Mygale .	Rather too much
20,407	Mylabris .	Rather too little
20,408	Mylocarium	**Reach**
20,409	Mylodon .	Hope(s) to reach —— on
20,410	Mylohyoid	Telegraph when you reach
20,411	Mynchery.	To reach me (us) at
20,412	Myocaris .	Will not reach in time
20,413	Myographic	To reach you at
20,414	Myolemma	When do you expect to reach
20,415	Myological	Will reach you at
20,416	Myologist.	I shall reach —— on
20,417	Myomancy	Ought to reach you by
20,418	Myonicity.	May probably reach you by
20,419	Myonosus.	As soon as —— reaches
20,420	Myopathia	**Reached**
20,421	Myopic .	This is the highest yet reached
20,422	Myosis. .	The news reached us
20,423	Myositic .	Has now been reached
20,424	Myosurus .	Have now reached
20,425	Myotomy .	Have you reached
20,426	Myoxidae.	**Readiness**
20,427	Myriad .	In readiness for
20,428	Myriagram	**Ready**
20,429	Myrialitre .	Everything ready to begin work
20,430	Myriapod .	Are you ready to
20,431	Myriarch .	Be all ready to
20,432	Myricin .	Must be ready by
20,433	Myriologue	Have everything ready
20,434	Myriorama	Not yet ready
20,435	Myristica .	All will be ready by
20,436	Myrmidons	When will you be ready to
20,437	Myrobalan	Is everything ready for
20,438	Myronic .	Shall be ready on
20,439	Myropolist	Cannot be ready before
20,440	Myroxylic.	As soon as everything is ready
20,441	Myrrh . .	Not ready until
20,442	Myrrhic .	Hold yourself ready to go to
20,443	Myrtaceae	All ready and awaiting shipment
20,444	Myrtaceous	Will be ready in time
20,445	Myrtiform.	Has (have) been ready for some time
20,446	Myrtle. .	Has (have) been ready since
20,447	Myrtlewax	Nothing ready yet
20,448	Myself. .	Everything is ready (for)
20,449	Mysisstage	Will not be ready for some time
20,450	Mysorine .	The mill will be ready to start

No.	Code Word.	**Ready** (*continued*)
20,451	Mystagogic	The engine will be ready to start
20,452	Mysterial .	When will you be ready to start
20,453	Mysterious	We are now getting ready to
20,454	Mystic. .	**Realizable**
20,455	Mysticism	The amount realizable (upon)
20,456	Mystified .	**Realization**
20,457	Mystify .	Net value of bullion on realization
20,458	Mystifying	Net value of concentrates on realization
20,459	Myth . .	Net value of gold on realization
20,460	Mythical .	Net value of silver on realization
20,461	Mythically	Net value of —— on realization
20,462	Mythologue	Loss on realization
20,463	Mythoplasm	**Realize**
20,464	Mythopoeic	Ought to realize
20,465	Mytiloid .	What ought the —— to realize
20,466	Mytilus .	Will realize about
20,467	Myxinidae	Had better realize at once
20,468	Myxopoda	**Realized**
20,469	Nabobs .	Net realized or realizable value of bullion
20,470	Nachlaut .	Net realized or realizable value of concentrates
20,471	Nacodar .	Has realized
20,472	Nacreous .	After everything is realized
20,473	Nacrite .	If now realized will fetch about
20,474	Naevose .	**Reason**
20,475	Nagelfluh .	In reason
20,476	Nagyagite.	The reason is
20,477	Naiades .	Do not know the reason
20,478	Naididae .	Out of all reason
20,479	Nailball .	Anything —— ask(s) in reason
20,480	Nailbrush .	There is every reason to believe
20,481	Naileress .	There is no reason for
20,482	Nailfile .	Is there any reason for
20,483	Nailhead .	Do everything in reason to help
20,484	Nailwort .	For reasons already stated
20,485	Nainsook .	The reason given is insufficient
20,486	Naique .	What is the reason for
20,487	Naissant .	**Reasonable**
20,488	Naively .	What —— ask(s) is fair and reasonable
20,489	Naked . .	Quite reasonable
20,490	Nakedly .	Is it reasonable
20,491	Nakedness	It is not reasonable
20,492	Namaycush	Will do anything reasonable
20,493	Namaz. .	Will accept any reasonable offer
20,494	Nambypamby	Will take a reasonable price
20,495	Nameable.	**Rebate**
20,496	Namelessly	Rebate allowed of
20,497	Nameplate	A rebate of
20,498	Namesake.	**Receipt(s)**
20,499	Nankeen .	Expenses exceed receipts by
20,500	Nanosaur .	Am (are) in receipt of

No.	Code Word	Receipt(s) (*continued*)
20,501	Nanosaurus	Is (are) in receipt of
20,502	Napecrest.	Receipts exceed expenses by
20,503	Naphawater	What are the receipts for the month of
20,504	Naphtha .	When you are in receipt of
20,505	Napiform .	Telegraph each month's receipts
20,506	Napkinring	Receipts from the sale of
20,507	Napkins .	**Receive**
20,508	Napped .	To receive
20,509	Naptakings	Did not receive
20,510	Napwarp .	Will receive
20,511	Narceine .	Cannot receive before
20,512	Narcissus .	As soon as you receive
20,513	Narcotical.	You will receive full instructions at
20,514	Narcotism.	As I (we) did not receive any
20,515	Narcotized	When shall I (we) receive
20,516	Nardus .	When did you receive
20,517	Narghile .	When do you expect to receive
20,518	Narica. .	If you receive
20,519	Narrating .	If you do not receive
20,520	Narrations	·Will not receive
20,521	Narratory .	You will receive
20,522	Narrowest	Receive it on account of
20,523	Narrowly .	Receive it on our account
20,524	Narthecium	**Received**
20,525	Narwhal .	Received —— from
20,526	Nasal . .	Received —— on accoun;
20,527	Nasalizing.	Has (have) received
20,528	Nasally .	Has (have) not received
20,529	Nascency .	Have you received
20,530	Nascent .	Has (have) —— received
20,531	Naseberry.	Wait till you have received
20,532	Nasicornia	If you have not received
20,533	Nasolabial	Has (have) received your orders to
20,534	Nasturtion	Received your orders; will attend to them at once
20,535	Nastyman	Will by this time have received
20,536	Natalitial .	Have received no such message; repeat at once
20,537	Natation .	Has (have) not been received
20,538	Natatores .	Was (were) received by
20,539	Nathmore.	You have probably by this time received
20,540	National .	From the statement received
20,541	Nativeness	No return has been received
20,542	Nativity .	**Recent**
20,543	Natrolite .	Recent developments show
20,544	Natterjack	Quite recent
20,545	Nattiest .	Has this been recent
20,546	Nattily. .	The recent discoveries
20,547	Nattiness .	**Recently**
20,548	Naturalism	Has (have) quite recently
20,549	Naturality.	Has (have) been recently
20,550	Naturalize	Has (have) not been very recently

No.	Code Word.	**Recently** (*continued*)
20,551	Naturemyth	Has (have) —— been recently
20,552	Naturists .	Has (have) been recently begun
20,553	Nauclea. .	Has (have) been recently discovered
20,554	Naufrages.	**Reckon.** (See Rely.)
20,555	Naulage .	To reckon on
20,556	Naumannite	You must reckon on
20,557	Nauscopy .	You must not reckon on
20,558	Nauseant .	You may reckon on
20,559	Nauseating	Can I (we) reckon on
20,560	Nauseation	**Reckoned**
20,561	Nauseous .	Reckoned on the basis of
20,562	Nauseously	**Recognition**
20,563	Nausity .	In recognition of
20,564	Nautch .	**Recognize**
20,565	Nautchgirl	Will not recognize
20,566	Nautical .	Will not in any way recognize —— claims or title
20,567	Nautically.	To recognize
20,568	Nautiloid .	Not to recognize
20,569	Navagium.	Must recognize
20,570	Navarchy .	If he (they) will not recognize
20,571	Navehole .	**Recognized**
20,572	Navelgall .	Been fully recognized
20,573	Navelled .	Must be fully recognized
20,574	Navelwort.	Our claim has been recognized
20,575	Navicular .	**Recommence**
20,576	Navigably.	When do you recommence
20,577	Navigate .	Not able to recommence before
20,578	Navigating	We are about to recommence
20,579	Navigation	**Recommenced**
20,580	Navigators	Recommenced operations
20,581	Navigerous	Recommenced sinking
20,582	Navvies .	Recommenced drifting
20,583	Navvy .	**Recommencing**
20,584	Navybill .	**Recommend**
20,585	Nazaritic .	I (we) strongly recommend
20,586	Neaptide .	I (we) would recommend you
20,587	Nearlegged	I (we) cannot recommend you
20,588	Nearness .	Do you recommend
20,589	Neathanded	What would you recommend
20,590	Neatherd .	Telegraph if you can recommend
20,591	Neathouse	Ask —— if he (they) can recommend
20,592	Nebalia .	Have you any one to recommend
20,593	Nebneb .	Have no one I (we) can recommend
20,594	Nebulists .	**Recommendation**
20,595	Nebulized.	Acting upon his recommendation
20,596	Nebulose .	We think the recommendation a very judicious one
20,597	Nebulosity	If the recommendation is not acted upon
20,598	Necessary .	**Recommended**
20,599	Necessism	Has (have) been recommended
20,600	Neckbands	Has (have) been recommended not to

No.	Code Word.	Recommended (*continued*)
20,601	Neckbeef .	Was (were) not recommended
20,602	Neckcloth	Strongly recommended us to
20,603	Necklaces	Recommended by
20,604	Necklet .	**Reconsider**
20,605	Neckmould	To reconsider the matter and decided
20,606	Neckpiece	It would be advisable to reconsider
20,607	Neckties .	Would —— reconsider
20,608	Neckverse	It is useless to ask —— to reconsider
20,609	Neckweed	Open to reconsider
20,610	Necrolatry	Will you reconsider the matter
20,611	Necrologic	**Reconsidered**
20,612	Necronite.	Have reconsidered the matter and decided
20,613	Necrophaga	After having reconsidered the
20,614	Necrophoby	**Reconstruct**
20,615	Necropsy .	(To) reconstruct the company
20,616	Necrotomy	**Reconstructed**
20,617	Necrosed .	Company has been reconstructed
20,618	Nectandra	Is now reconstructed
20,619	Nectareal .	Must be reconstructed
20,620	Nectareous	**Reconstruction**
20,621	Nectarine .	Scheme of reconstruction
20,622	Nextocalyx	No course open but reconstruction
20,623	Nectosac .	If reconstruction carried out
20,624	Needfire .	Advise reconstruction of company
20,625	Needlebook	**Record(s)**
20,626	Needlecase	The only records
20,627	Needlefish	There is no record to show
20,628	Needleful .	Before any record was kept
20,629	Needlegun	Is there any record of
20,630	Needlework	(To) keep a strict record of
20,631	Needment	According to the record
20,632	Neemtree .	I (we) gathered from the record(s) that
20,633	Neesewort	On record in the record office (of)
20,634	Nefand .	Must have deed to record in
20,635	Negations.	Place on record
20,636	Negative .	In the record office
20,637	Negatively	The official record of
20,638	Negativing	**Recorded**
20,639	Negatory .	It is recorded that
20,640	Neglect .	Has (have) been recorded
20,641	Neglecteth	Has (have) not been recorded
20,642	Neglectful	To be recorded
20,643	Neglection	Deed(s) must be recorded
20,644	Negligence	If deed is not recorded (before)
20,645	Negligible	All papers must be recorded
20,646	Negoce .	All papers have been recorded
20,647	Negotiable	Transfer of the property recorded
20,648	Negotiant.	As soon as the transfer is recorded
20,649	Negotiator	Deeds recorded
20,650	Negotious.	When deed(s) has (have) been recorded

No.	Code Word.	Recorded (*continued*)
20,651	Negress .	If the —— is (are) not recorded
20,652	Negro . .	Let us know if —— is (are) recorded
20,653	Negrocorn	(To) be recorded not later than
20,654	Negroflies.	That it should be recorded
20,655	Negrofly .	Was (were) recorded on the
20,656	Negrohead	Recorded in the official registry
20,657	Negrooid .	**Recover**
20,658	Negundo .	What do you expect to recover
20,659	Neighbour	Ought to recover
20,660	Neighing .	Will try to recover
20,661	Nelumbium	Can you recover
20,662	Nelumbo .	Cannot recover
20,663	Nemalite .	Expect to recover
20,664	Nematelmia	Do not think we can recover more
20,665	Nematocyst	Can recover
20,666	Nematoda	If we recover
20,667	Nemausa .	If you can recover
20,668	Nemertean	Recover all you can
20,669	Nemertid .	Cannot recover more than
20,670	Nemocera	Not likely to recover
20,671	Nemophila	Doubtful if we shall ever recover
20,672	Nemorose.	Think we shall recover
20,673	Nempne .	Think he will recover
20,674	Nenuphar.	Recover what has been spent
20,675	Neoarctic .	The amount which we have been able to recover
20,676	Neocomian	**Recovered**
20,677	Neocosmic	Have recovered from
20,678	Neodamode	Recovered from the tailings
20,679	Neogamist	How much —— have you recovered
20,680	Neolatin .	Have you recovered
20,681	Neolithic .	Have recovered some of the
20,682	Neological	Recovered a considerable portion of the
20,683	Neologism	Very little has been recovered
20,684	Neologized	If no more can be recovered
20,685	Neomorpha	**Recovering**
20,686	Neophron.	Now recovering from
20,687	Neophytes	Succeeded in recovering
20,688	Neoplastic	**Recovery**
20,689	Neoteric .	Recovery very doubtful
20,690	Neoterized	The recovery of the
20,691	Neozoic .	No hope of recovery
20,692	Nephalist .	**Rectification**
20,693	Nepheloid	Rectification of the boundary line
20,694	Nephew .	**Rectified**
20,695	Nephrite .	The mistake cannot be rectified
20,696	Nephrodium	Cannot be rectified
20,697	Nephrops .	Will be rectified
20,698	Nephrotomy	Must have it (them) rectified
20,699	Nepotal .	Must be rectified
20,700	Neptunian	Have been rectified

P

No.	Code Word.	**Rectified** (*continued*)
20,701	Nereids .	How can —— be rectified
20,702	Neritacea .	**Rectify**
20,703	Nerium .	Rectify the mistake
20,704	Neroli . .	Rectify the accounts
20,705	Nervation .	Rectify the boundaries
20,706	Nervecell .	**Recurrence**
20,707	Nerved .	To prevent any recurrence (of)
20,708	Nervefibre	Has there been any recurrence
20,709	Nerveless .	There has been no recurrence (of)
20,710	Nervetube	**Redress**
20,711	Nervimotor	What redress can we get
20,712	Nervosity .	The only redress will be
20,713	Nervous .	Can get no redress
20,714	Nervously .	If they refuse redress, no other course is open but to
20,715	Nervure .	Court will give redress
20,716	Nescience	Court refuses to give redress
20,717	Nesodon .	**Reduce**
20,718	Nestegg .	You must reduce
20,719	Nestle . .	It is necessary to reduce
20,720	Nestling .	Can reduce
20,721	Nestor . .	Cannot reduce
20,722	Nestorian .	Will reduce
20,723	Nethermost	Will not reduce
20,724	Nethinim .	In order to reduce
20,725	Netloom .	Reduce the staff
20,726	Netmaking	Reduce expenses
20,727	Netmasonry	Cannot reduce the amount
20,728	Nettapus .	If you can reduce
20,729	Netted .	If we can reduce
20,730	Nettlerash	**Reduced**
20,731	Nettletree .	Reduced by this means (to)
20,732	Nettlewort	Has (have) been reduced
20,733	Neuralgia .	Can be reduced to
20,734	Neurectomy	Cannot be reduced
20,735	Neurilemma	Have reduced expenses as far as possible
20,736	Neuritis .	Expenses must be reduced as much as possible
20,737	Neurocity .	**Reducing**
20,738	Neurology	Instead of reducing
20,739	Neuroma .	By reducing the
20,740	Neuropathy	**Reduction**
20,741	Neuropter	No reduction can be made
20,742	Neurospast	Reduction works
20,743	Neurotome	By a reduction of
20,744	Neutralist .	Endeavour to make (get) a reduction
20,745	Neutrally .	The reduction of the ores
20,746	Nevermore	The reduction of the ores by this process
20,747	Newborn .	Process of reduction
20,748	Newcomer	Smelting and reduction works
20,749	Newcreate	The reduction works are situated
20,750	Newfangle	Reduction by the —— process

No.	Code Word.	**Reduction** (*continued*)
20,751	Newfashion	The reduction of the ores can only be effected by
20,752	Newfledged	This new process of reduction
20,753	Newlaid .	What method of reduction would you recommend
20,754	Newly . .	The effect of the reduction
20,755	Newmodel	Owing to the reduction we have made in
20,756	Newsagent	A large part of the reduction arises from
20,757	Newsbooks	Some reduction must be made
20,758	Newsboy .	Some reduction can no doubt be made
20,759	Newsletter	Machinery used in reduction of ores
20,760	Newsmen .	**Reef(s)**
20,761	Newsmonger	Fall of reef has occurred which will stop hauling
20,762	Newspaper	The reef is dangerous [blueground for
20,763	Newsroom	In order to handle the reef
20,764	Newsvendor	Heavy falls of reef are constantly occurring
20,765	Newswriter	The expenses of hauling the reef are —— per load
20,766	Newyear .	What are the expenses of hauling the reef per load
20,767	Nexible .	Have discovered another reef on the property
20,768	Nexus . .	No other reef of importance on property
20,769	Nias . .	Several small reefs on property
20,770	Nibbling .	No reef of any importance on property
20,771	Nibblingly	Every probability that reefs pass through property
20,772	Niblick .	Now prospecting reefs
20,773	Niccolite .	Reef proved to a depth of —— feet
20,774	Nicelings .	Reef proved for a length of —— feet
20,775	Nicest . .	Reef has a dip of ——° from the horizontal
20,776	Niceties .	Reef has an average assay value of —— per ton
20,777	Nichar. .	Reef gives a mill result of
20,778	Nickartree	Reef has an average assay value of $\frac{1}{2}$ oz. per ton
20,779	Nickeared	Reef has an average assay value of 1 oz. per ton
20,780	Nickelic .	Reef has an average assay value of $1\frac{1}{2}$ oz. per ton
20,781	Nickname.	Reef has an average assay value of 2 ozs. per ton
20,782	Nicknaming	Reef has an average assay value of $2\frac{1}{2}$ ozs. per ton
20,783	Nicksticks	Reef has an average assay value of 3 ozs. per ton
20,784	Nicolaitan	Reef has an average assay value of $3\frac{1}{2}$ ozs. per ton
20,785	Nicotylia .	Reef has an average assay value of 4 ozs. per ton
20,786	Nidamental	Reef —— feet wide, and assays [property
20,787	Niddicock	Reef is a splendid one, and extends full length of
20,788	Niddui. .	Reef visible at surface, —— wide
20,789	Nidering .	Reef not visible at surface
20,790	Nidgery .	Reef visible at surface, but much broken and dis-
20,791	Nidificate .	The width of reef is —— feet [turbed
20,792	Nidnod .	The reefs on the property amount in number to
20,793	Nidnodding	The reefs extend about —— miles
20,794	Nidor . .	The reefs are exposed
20,795	Nidorose .	The reefs show visible gold
20,796	Nidorosity	Expect to strike reef
20,797	Nidulant .	Expect to strike reef at a depth of —— feet
20,798	Nidularium	Have struck reef at a depth of —— feet
20,799	Nidulation	Have claims on reef and dip [feet
20,800	Nigella .	Have claims on reef, but will probably lose at ——

No.	Code Word.	Reef(s) (*continued*)
20,801	Niggard .	Reef outcrops through claims
20,802	Niggardise	There is no outcrop; reef is probably present
20,803	Niggardly .	There is no outcrop; doubtful if reef is in the ground
20,804	Nigglers .	**Re-examination**
20,805	Nightbell .	After a re-examination of the
20,806	Nightcaps.	A thorough re-examination made
20,807	Nightcharm	**Re-examine**
20,808	Nightcrow	When I (we) go to re-examine the mine
20,809	Nightdew .	When you re-examine the mine
20,810	Nightdogs	I (we) shall go about —— to re-examine the mine
20,811	Nightdress	Cannot re-examine the mine till
20,812	Nighteyed	When can you re-examine the mine
20,813	Nightfire .	**Refer(s).** (See also Tables.)
20,814	Nightflies .	If you will refer to
20,815	Nightfly .	You had better refer to
20,816	Nightglass	To whom can I (we) refer (about)
20,817	Nightgowns	Refer him to me (us)
20,818	Nightheron	Refer to my (our) letter dated
20,819	Nightish .	Refer to my (our) telegram of
20,820	Nightjar .	Refer to your letter of
20,821	Nightlamp	Refer to your telegram of
20,822	Nightlight.	Refer to your report (No. ——) (dated)
20,823	Nightlong.	**Reference(s)**
20,824	Nightmares	With reference to
20,825	Nightpiece	Has this any reference to
20,826	Nightrail .	This has no reference (to)
20,827	Nightraven	Without any reference to
20,828	Nightshade	Get reference in regard to
20,829	Nightshirt .	Reference is satisfactory
20,830	Nightsnap	Reference is not at all good
20,831	Nightspell	By a reference to
20,832	Nightsteed	With reference to what occurred
20,833	Nighttaper	Has very good references
20,834	Nightwalk	**Referred (to)**
20,835	Nightwatch	Has referred him (them) to you
20,836	Nigrescent	Referred to in our letter of
20,837	Nigrine .	Referred to in our cable of
20,838	Nigritude .	Referred to in your letter of
20,839	Nihil . .	Referred to in your cable of
20,840	Nihilism .	The matter has been referred to
20,841	Nihilistic .	**Referring**
20,842	Nihility .	Upon referring to
20,843	Nilometer.	**Refined**
20,844	Niloscope.	To be refined
20,845	Nilotic. .	Refined bullion
20,846	Nimbly .	The bullion is sent to be refined
20,847	Nimbose .	Sent to —— to be refined
20,848	Nincompoop	After having been refined
20,849	Ninekiller.	**Refining**
20,850	Ninepins .	Refining charges

No.	Code Word.	Refining (*continued*)
20,851	Ninetyknot	Smelters' refining charges
20,852	Ninny . .	The expenses of refining
20,853	Niobium .	Refining works
20,854	Nipadites .	Refining works at —— shut down
20,855	Nipperkins	Smelting and refining plant would have to be erected
20,856	Nippermen	Would be advisable to erect smelting and refining
20,857	Nippingly .	Refining furnaces [works
20,858	Nippitate .	Smelting and refining works would cost
20,859	Nipplewort	Smelting and refining plant to treat
20,860	Nipter . .	**Refractory.** (See Ore.)
20,861	Nirles . .	The ore is too refractory to be treated in the usual
20,862	Nirvana .	Ore very refractory; requires roasting [way
20,863	Nitency .	Ore is very refractory; must be treated by the ——
20,864	Nitidous .	The ore is very refractory [process
20,865	Nitramidin	**Refund**
20,866	Nitranisic .	Will —— refund
20,867	Nitraria .	Will refund
20,868	Nitrates .	Refuse(s) to refund
20,869	Nitriary .	Must refund the amount
20,870	Nitrogen .	Refund the amount spent in
20,871	Nitroleum .	Will refund amount, as soon as
20,872	Nitrous .	Who will refund the
20,873	Nitryl . .	**Refunded**
20,874	Noachian .	The amount has been refunded
20,875	Nobbily .	Has not been refunded
20,876	Nobiliary .	To be (must be) refunded as soon as
20,877	Nobilities .	On the amount being refunded
20,878	Nobleman	If the amount is refunded
20,879	Nobleness .	On condition that it is refunded
20,880	Nobodies .	**Refusal**
20,881	Nobody .	Get the refusal of
20,882	Nocent .	Has (have) the refusal of
20,883	Nocently .	Since the refusal by
20,884	Nocive .	Will you give refusal till
20,885	Noctambulo	Give you the refusal until
20,886	Noctidial .	Try and get the refusal extended to
20,887	Noctiluca .	Have got the refusal extended to
20,888	Noctograph	Cannot get refusal extended
20,889	Noctuary .	**Refuse**
20,890	Nocturnal .	I (we) would advise you to refuse
20,891	Nocuous .	To refuse
20,892	Nocuously	Do not refuse
20,893	Noddingly	You must refuse
20,894	Nodosaria .	Positively refuse(s) (to)
20,895	Nodose .	The directors refuse to
20,896	Nodosities	We must refuse to
20,897	Nodosity .	Refuse to be a party to
20,898	Nodulous .	If we refuse
20,899	Noematical	If you refuse
20,900	Noemics .	If he refuses (if they refuse)

No.	Code Word.	**Refuse** (*continued*)
20,901	Noetian .	Reasons why we refuse
20,902	Nohow .	Reason why you should refuse
20,903	Noiseful .	Will not refuse
20,904	Noiseless .	Refuse(s) to take the matter up
20,905	Noisily .	Refuse(s) to have anything to do with
20,906	Noisome .	**Refused**
20,907	Noisomely	Have you refused
20,908	Nomada .	Has (have) refused
20,909	Nomadism	Has (have) been refused
20,910	Nomadizing	Sorry you have refused
20,911	Nomarchy	**Regard**
20,912	Nominalism	Having regard to the future of the concern
20,913	Nominalize	Having regard to the fact that we cannot
20,914	Nominally	Having regard to the falling off in
20,915	Nominating	Having regard to these points, do you consider we
20,916	Nominor .	Pay no regard to [are justified in
20,917	Nomocanon	Having regard to the very low grade of the ore re-
20,918	Nomothesy	With regard to [serves
20,919	Nomothetic	Having regard to the low grade of ore now being
20,920	Nonaccess	Without regard to [worked
20,921	Nonacid .	Having regard to the large quantity and high value
20,922	Nonadult .	Do not regard it as hopeless [of the ore
20,923	Nonagons .	Regard it as almost hopeless
20,924	Nonarrival	Regard our position with great anxiety
20,925	Nonce . .	Do not regard it as
20,926	Nonchalant	Having no regard to
20,927	Nonclaim .	Regard it as lost
20,928	Nonconcur	**Regarded**
20,929	Noncontent	Is it regarded
20,930	Nonego .	Regarded as
20,931	Nonelastic	Regarded as lost
20,932	Nonelect .	Regarded as of the highest importance
20,933	Nonentity	Regarded in the light of past experience
20,934	Nonesuch .	Should be regarded as
20,935	Nonillion .	**Regardless (of)**
20,936	Nonionina	Regardless of consequences
20,937	Nonius .	Regardless of the past
20,938	Nonjoinder	**Register**
20,939	Nonjurable	Register in the name of
20,940	Nonjurant	Register the deed at once
20,941	Nonjuror .	Register the transfer
20,942	Nonmember	**Registered**
20,943	Nonpayment	To be legally registered
20,944	Nonplus .	Registered in the name of
20,945	Nonplussed	What is your registered telegraphic address
20,946	Nonregent	What is registered telegraphic address of
20,947	Nonreturn	Registered address is
20,948	Nonsense .	Has (have) no registered address
20,949	Nonsociety	The company is registered under the name of
20,950	Nonsolvent	Has (have) been registered

No.	Code Word.	**Registered** (*continued*)
20,951	Nonsparing	Has not been registered
20,952	Nonsuit .	Must be registered
20,953	Nonsuiting	As soon as the company is registered
20,954	Nonsurety	Send by registered letter
20,955	Nontenure	Has (have) sent in registered letter
20,956	Nonterm .	Was the letter registered
20,957	Nontronite	Has the deed been registered
20,958	Nonuplet .	Deeds must be registered and deposited with
20,959	Nonusance	Has transfer been registered
20,960	Nonuser .	Transfer must be registered
20,961	Noodle .	Transfer was registered
20,962	Noodledom	The deeds were registered
20,963	Nook . .	(Has) have been registered, and (is) are now in the
20,964	Noonday .	Registered in the office of [hands of
20,965	Nooning .	As soon as —— is (are) registered
20,966	Noonstead	The registered owners
20,967	Noontide .	**Registration**
20,968	Nootkadog	Fees to be paid upon registration
20,969	Nopopery.	Stamp duty upon registration
20,970	Nopster .	Stamp and fees payable on registration
20,971	Noraghi .	Stamp duty of 1 per mille on the capital to be paid
20,972	Norbertine	**Regret(s)** [on registration
20,973	Norimon .	We cannot but regret that
20,974	Normalcy .	**Regular(ly)**
20,975	Norroys .	All in regular order
20,976	Northerner	A regular system of accounts
20,977	Northernly	In the regular course
20,978	Northmost	Regularly sent
20,979	Northness.	Take care that —— are regularly sent
20,980	Northpolar	**Regulate**
20,981	Northstar .	In order to regulate
20,982	Northwind	You must regulate
20,983	Nosebands	To regulate the
20,984	Nosebit .	**Regulated**
20,985	Nosebleed	**Regulating**
20,986	Nosegay .	**Regulation(s)**
20,987	Noseherb .	By careful regulation of the
20,988	Nosepiece.	Under new regulations
20,989	Noserings.	The regulations now in force
20,990	Nosesmart	**Reject**
20,991	Nosethrill.	Have had to reject
20,992	Nosocomial	Could not reject
20,993	Nosography	You must reject
20,994	Nosologist	**Rejected**
20,995	Nosology .	We have rejected all, as useless
20,996	Nosopoetic	Petition was rejected
20,997	Nosotaxy .	The motion was rejected
20,998	Nostalgia .	Has been rejected
20,999	Nostoc .	**Rejecting**
21,000	Nostomania	Now rejecting all

No.	Code Word.	**Rejecting** (*continued*)
21,001	Nostril . .	Rejecting all ore that does not assay
21,002	Nostrums .	**Relation(s)**
21,003	Notabilia .	Our relations with
21,004	Notandum	His (their) relations with
21,005	Notarial .	Has (have) no relations with
21,006	Notarially .	Your relations with
21,007	Notchblock	**Relative**
21,008	Notchboard	Everything relative to
21,009	Notchweed	Is this relative to
21,010	Notchwing	This is relative to
21,011	Notebook .	This is not relative to
21,012	Notedly .	Telegraph everything of interest relative to the
21,013	Notedness	**Relax** [workings
21,014	Notepaper	Do not relax your vigilance
21,015	Noteworthy	Do not relax your efforts
21,016	Nothingism	Will not relax my efforts
21,017	Noticeable	**Release(s)**
21,018	Notices .	Will release us from our agreement
21,019	Noticing .	Will not release us from our agreement
21,020	Notidanus	Will —— release us from our agreement
21,021	Notified .	Will not release you
21,022	Notify . .	Release, under the terms of the
21,023	Notifying .	Release(s) us from
21,024	Notionate .	Release(s) you from
21,025	Notionists	Release him (them) from
21,026	Notist . .	Releasing us from
21,027	Notochord	**Reliable**
21,028	Notonecta	Is (are) —— reliable
21,029	Notopodium	Is (are) perfectly reliable
21,030	Notorhizal	Is (are) not reliable
21,031	Notoriety .	Reliable statement
21,032	Notorious .	**Reliance**
21,033	Notornis .	Can any reliance be placed on (in)
21,034	Nottpated	Reliance can be placed on (in)
21,035	Notturno .	Place full reliance on (in)
21,036	Notwheat .	Place very little reliance on (in)
21,037	Nougat .	**Relied**
21,038	Noumenal	Relied entirely upon
21,039	Nounize .	Upon which we relied
21,040	Nounverb .	**Relieve(s)**
21,041	Nourisher .	Relieve us from great anxiety
21,042	Nourishing	Relieve the strain upon
21,043	Novaculite	Relieves us from responsibility
21,044	Novargent	**Relieved**
21,045	Novator .	Wish (wishes) to be relieved of
21,046	Novelette .	Before being relieved of
21,047	Novelism .	I had hoped to be relieved of the strain upon me
21,048	Novelize .	To be relieved from
21,049	Novelizing	Relieved from
21,050	Novelties .	**Relinquish(ed)**

No.	Code Word.	Relinquish(ed) (*continued*)
21,051	Novelty .	Relinquish all rights to
21,052	Novem .	Relinquished all rights
21,053	Novenary .	Relinquished in —— favour
21,054	Novennial	**Rely.** (See Reckon.)
21,055	Novercal .	You must rely on
21,056	Noviceship	I (we) rely on you
21,057	Novilunar.	Rely on me
21,058	Novitiate .	You may rely on
21,059	Novitious .	Obliged to rely on
21,060	Novodamus	Do not rely on
21,061	Nowadays	Can you rely safely on
21,062	Nowhere .	**Relying**
21,063	Nowhither	Relying upon
21,064	Noxious .	**Remain(s)**
21,065	Noxiously.	Cannot remain any longer
21,066	Noyade .	Remain where you are (till)
21,067	Nozzle , .	How long am I to remain
21,068	Nuance .	I (we) must remain here till
21,069	Nubecula .	I (we) shall remain —— days at
21,070	Nubigenous	How long do you remain at
21,071	Nubilated.	Remain at —— till you hear from
21,072	Nubilous .	Will remain at
21,073	Nucifraga.	How long will —— remain at
21,074	Nucleiform	Remain till all is settled
21,075	Nucleoid .	Remain till my successor comes
21,076	Nucleolite	All that remains of
21,077	Nucleus .	Very little remains of the
21,078	Nucula .	Several of the —— still remain
21,079	Nuculanium	Much of the ore still remains
21,080	Nudibranch	Still remain to be accounted for
21,081	Nudicaul .	There still remain on the dump
21,082	Nudifidian	**Remaining**
21,083	Nudities .	Remaining on hand
21,084	Nudity . .	Remaining until
21,085	Nuggets .	There are points still remaining from which
21,086	Nuisances.	**Remainder**
21,087	Nulledwork	Send remainder to
21,088	Nullibiety.	Remainder has been sent to
21,089	Nullifier .	Dispose of the remainder
21,090	Nullifying.	Remainder awaits
21,091	Numberous	The remainder of the
21,092	Numbfish .	**Remedied**
21,093	Numenius.	Can only be remedied by
21,094	Numerally	Can be remedied (by)
21,095	Numerals .	Must be remedied
21,096	Numeration	These faults must be remedied before we can
21,097	Numerical	**Remedy**
21,098	Numerosity	The only remedy is
21,099	Numerously	The only remedy we can suggest is
21,100	Numidian.	What remedy can you suggest

P 2

No.	Code Word.	**Remedy** (*continued*)
21,101	Numismatic	Take steps to remedy the evil
21,102	Nummular	There is no other remedy
21,103	Nummulitic	**Remember**
21,104	Numskull .	Do not remember
21,105	Nunbuoy .	Remember having
21,106	Nunciature	You will remember that
21,107	Nuncupate	**Remind**
21,108	Nundinal .	We beg to remind you that
21,109	Nuptials .	**Remit**
21,110	Nursechild	Remit immediately if disaster is to be prevented
21,111	Nursemaid	Remit by cheque
21,112	Nursename	Will remit in a few days
21,113	Nursepond	Will remit by
21,114	Nurseries .	Are prepared to remit
21,115	Nursery .	Will remit as soon as we are advised that
21,116	Nurseryman	Why do you not remit
21,117	Nurture .	Remit immediately
21,118	Nurturing.	When will you remit
21,119	Nussierite.	When you remit
21,120	Nutbone .	Remit by telegraph
21,121	Nutbrown.	Unless you remit
21,122	Nutcracker	Will remit you in —— days
21,123	Nutgall .	Cannot remit any
21,124	Nuthatch .	What can you remit
21,125	Nuthetes .	Remit by draft upon
21,126	Nuthooks .	Remit in bullion
21,127	Nutjobber	How much do you think you will be able to remit
21,128	Nutlock .	How much can you remit next
21,129	Nutmeg .	Can you remit —— by telegraph not later than
21,130	Nutmeggy	Will remit —— by telegraph not later than
21,131	Nutmegtree	How much can you conveniently remit, and when
21,132	Nutoil . .	How much can you remit on account
21,133	Nutpecker	To do what is proposed we should require you to
21,134	Nutpines .	Will certainly remit not later than [remit
21,135	Nutriment	Remit by telegraph next week
21,136	Nutritial .	Will telegraph on —— how much we can remit
21,137	Nutritious.	Will be able to remit
21,138	Nutritive .	Shall not be able to remit
21,139	Nutshell .	Not able to remit more than
21,140	Nuttrees .	Do your best to remit
21,141	Nutty . .	(To) remit what you want
21,142	Nutweevil.	To remit what you want, we shall have to
21,143	Nutwrench	If you cannot remit
21,144	Nuxvomica	Will remit to-morrow
21,145	Nuzzerana	Will remit next week
21,146	Nyctalopia	Cannot you remit
21,147	Nyctibius .	Do you want to remit
21,148	Nycticebus	Will remit —— by telegraph as soon as
21,149	Nycticorax	Remit on account of
21,150	Nyctinomus	Remit —— now, and balance on

No.	Code Word.	Remit (*continued*)
21,151	Nyctisaura	Please remit through
21,152	Nylgau .	Remit what you can now, the rest can follow later on
21,153	Nymphets	How much do you want us to remit
21,154	Nymphical	Can you remit bullion
21,155	Nymphish	Remit on account of profits
21,156	Nymphlike	Remit for credit of
21,157	Nymphly .	Remit in gold
21,158	Nymphotomy	Remit in dollars
21,159	Nyroca .	Have instructed —— to remit you
21,160	Nystagmus	Have paid —— to —— to remit you through their
21,161	Oafish . .	Let us have all you can possibly remit [agents
21,162	Oakapple .	Remittance(s)
21,163	Oakbeauty	Require a remittance of —— on
21,164	Oakenpin .	Have sent remittance by
21,165	Oakgall .	What remittance do you require
21,166	Oakleather	Will send the remittance to-day
21,167	Oaklungs .	You must not rely upon receiving remittance
21,168	Oakpaper .	Must have remittance at once
21,169	Oakspangle	No remittance received yet
21,170	Oaktree .	Remittances to cease after
21,171	Oakum .	Remittance of —— received
21,172	Oakwart .	In order to do this, we require a remittance of
21,173	Oarfooted.	A remittance of —— would meet all requirements
21,174	Oaritis . .	Have received remittance as per your telegram of
21,175	Oarlocks .	Have received remittance
21,176	Oarsman .	How shall I make remittance
21,177	Oarswivel.	Remittance will be sent on the
21,178	Oarweed .	Daily expecting remittance
21,179	Oasthouse	**Remitted**
21,180	Oatcake .	Has (have) remitted you —— to
21,181	Oatgrass .	Remitted through Messrs. —— to be paid you with
21,182	Oathable .	Must be remitted at once [the sanction of
21,183	Oathrite .	Must be remitted by telegraph
21,184	Oatmalt .	Has been remitted to
21,185	Obambulate	Have remitted —— through
21,186	Obbligato.	Have remitted by cable this day
21,187	Obclavate.	Remitted in cash
21,188	Obconic. .	Remitted in bullion
21,189	Obconical.	Remitted by telegraph
21,190	Obcordate	Remitted by draft upon
21,191	Obduracy .	Remitted in gold
21,192	Obdurate .	Remitted in dollars
21,193	Obduration	Remitted in bars
21,194	Obdureness	Remitted through our bankers
21,195	Obeah . .	Remitted for credit of
21,196	Obedible .	Remitted on account of profits
21,197	Obedience	Remitted proceeds of
21,198	Obediently	**Remonstrate**
21,199	Obeisancy	Remonstrate strongly against
21,200	Obeisant .	You must remonstrate as strongly as possible

No.	Code Word.	
21,201	Obeliscal .	**Remonstrated**
21,202	Obelisks .	Have remonstrated strongly
21,203	Obequitate	**Removal**
21,204	Obesity .	The removal of
21,205	Obeyer .	The removal of the case to the Superior Court
21,206	Obeying .	**Remove**
21,207	Obeyingly	Propose to remove the
21,208	Obfirmate.	If we remove the
21,209	Obfuscated	Can we remove the
21,210	Obfusque .	Remove the injunction
21,211	Obitual .	**Removed**
21,212	Obituaries	Have now removed
21,213	Obituarily.	Has (have) been removed
21,214	Object . .	Has (have) not been removed
21,215	Objectable	Was (were) removed by
21,216	Objecteth.	The action has been removed to
21,217	Objectify .	Injunction has been removed
21,218	Objecting .	If you can get it removed
21,219	Objectize .	Can get it removed
21,220	Objectless	Removed all difficulty
21,221	Objectors .	**Removing**
21,222	Objicient .	**Remunerate**
21,223	Objurgate.	Will remunerate you
21,224	Oblate . .	**Remunerated**
21,225	Oblateness	Will be remunerated accordingly
21,226	Oblations .	Will be remunerated by
21,227	Oblectated	**Remuneration**
21,228	Obligant .	What remuneration is offered
21,229	Obligation	Our remuneration
21,230	Obligatory	Your remuneration
21,231	Obliged .	The remuneration is not sufficient
21,232	Obligement	Will increase your remuneration
21,233	Obliging .	In remuneration for
21,234	Obligingly	Unless further remuneration be given
21,235	Obligulate	It is a question of remuneration
21,236	Oblique .	Remuneration for services rendered
21,237	Obliquely .	Very adequate remuneration
21,238	Obliterate.	The remuneration offered is not enough
21,239	Oblivial .	**Remunerative**
21,240	Oblivious .	Is not remunerative
21,241	Oblocutor.	Is not sufficiently remunerative
21,242	Oblong .	Must make it more remunerative
21,243	Oblongish.	Can be made remunerative
21,244	Oblongly .	Cannot be made remunerative
21,245	Oblongness	**Renew**
21,246	Obloquy .	Will you renew
21,247	Obnoxious	Wish(es) to renew
21,248	Obnubilate	Renew the bond
21,249	Oboist . .	Will renew the bond for a further period of
21,250	Obolary .	Think it desirable to renew

No.	Code Word.	Renew (*continued*)
21,251	Obole . .	To renew on favourable terms
21,252	Obolize .	Will not renew
21,253	Obolizing .	**Renewed**
21,254	Oboval .	We have renewed
21,255	Obreption .	The whole of the —— has been renewed
21,256	Obrotund .	Has not been renewed since
21,257	Obscene .	**Reorganization**
21,258	Obscenely	The reorganization of the company
21,259	Obscenous	Reorganization now completed
21,260	Obscurant	Reorganization will be effected as soon as
21,261	Obscurer .	Reorganization upon an entirely new basis
21,262	Obscurity .	Reorganization will effect a saving of
21,263	Obsecrate .	**Reorganize**
21,264	Obsequent	You must reorganize
21,265	Obsequious	It will be necessary to reorganize
21,266	Obsequies .	Reorganize the undertaking
21,267	Obsequy .	Reorganize the entire system
21,268	Observably	As soon as we are able to reorganize
21,269	Observanda	**Reorganized**
21,270	Observator	Reorganized on the basis of
21,271	Observe .	Has been completely reorganized
21,272	Observing .	**Reorganizing**
21,273	Obsidian .	We are now reorganizing
21,274	Obsidional	**Repaid**
21,275	Obsignate .	Must be repaid by
21,276	Obsolete .	Has been repaid
21,277	Obsoletism	We shall be repaid
21,278	Obstacle .	**Repair(s)**
21,279	Obstetric .	Repairs to mill
21,280	Obstinacy .	Repairs to engine
21,281	Obstructed	Repairs to boiler
21,282	Obstupefy .	Repairs to machinery
21,283	Obtain . .	Repairs to shaft
21,284	Obtainment	Repairs to pumps
21,285	Obtected .	The —— is (are) out of repair
21,286	Obtemper .	To repair the damage done
21,287	Obtend .	Is (are) the —— in good repair
21,288	Obtension	The repairs are nearly finished
21,289	Obtrition .	Estimate cost of repairs to machinery
21,290	Obtrude .	The repairs are not completed owing to
21,291	Obtruding .	What do you estimate will be the cost of repairs
21,292	Obtruncate	Cost of repairs will be about [(to)
21,293	Obtrusive .	Had to shut down on account of repairs
21,294	Obtundent	Repairs will take about
21,295	Obturator .	Owing to repairs, could not work the
21,296	Obtuse . .	Repairs can be done here
21,297	Obtusely .	Repairs have been well done
21,298	Obtuseness	**Repaired**
21,299	Obuncous .	Cannot be repaired
21,300	Obversant	Will be repaired at once

No.	Code Word.	Repaired (*continued*)
21,301	Obverse .	Damage has been repaired
21,302	Obversely .	Repaired, and will restart work
21,303	Obverting .	Have repaired all damage caused by the
21,304	Obviated .	**Repairing**
21,305	Obvious .	Repairing damages to
21,306	Obviously .	Repairing —— as fast as we can
21,307	Obvoluted	Commenced repairing
21,308	Occasional	Finished repairing
21,309	Occasive .	**Repay**
21,310	Occiduous	This will repay
21,311	Occipital .	Will not repay
21,312	Occiput .	Will this repay
21,313	Occlude .	To enable us to repay
21,314	Occluding .	Can repay the loan out of (the)
21,315	Occlusion .	**Repaying**
21,316	Occrustate	Repaying us for
21,317	Occult .	**Repayment**
21,318	Occultly .	In repayment of advance
21,319	Occultness	Repayment of charges
21,320	Occupancy	Repayment of the loan
21,321	Occupant .	In repayment of all demands
21,322	Occupation	**Repeat**
21,323	Occupier .	You must repeat
21,324	Occupy .	Repeat your telegram [in plain words
21,325	Occupying	Do not understand your telegram ; please repeat
21,326	Occurred .	Do not understand following words ; please repeat
21,327	Occurrence	Repeat first part of your telegram [them
21,328	Ocean . .	Repeat latter part of your telegram
21,329	Oceanology	If you can repeat
21,330	Ocellaria .	Repeat the offer
21,331	Ocellus .	Can only repeat what we previously stated
21,332	Ocelot . .	Must not expect us to repeat
21,333	Ochletic .	**Repeated**
21,334	Ochlocracy	Will not be repeated
21,335	Ochraceous	Cannot be repeated
21,336	Ochre . .	If repeated, will do much harm
21,337	Ochreate .	**Repeatedly**
21,338	Ochroma .	It has been repeatedly done
21,339	Ochymy .	It has been repeatedly stated
21,340	Ocimum .	**Repetition**
21,341	Octachord	A repetition of
21,342	Octagon .	A repetition of such a run
21,343	Octagonal	**Replace**
21,344	Octagynous	Replace him by a better man
21,345	Octahedral	To replace it by
21,346	Octameter	Will replace
21,347	Octandrian	Will not replace
21,348	Octandrous	Must replace
21,349	Octangular	**Replaced**
21,350	Octans . .	Has been replaced by

No.	Code Word.	**Replaced** (*continued*)
21,351	Octapla .	If not soon replaced
21,352	Octarchy .	**Replied**
21,353	Octaroon .	Have replied seriatim to the inquiries made
21,354	Octastyle .	Fully replied to in our last
21,355	Octateuch .	Has replied that he will
21,356	Octennial .	Replied in the affirmative
21,357	Octillion . .	Replied in the negative
21,358	Octobass .	Replied, assenting to
21,359	Octodecimo	If you have not already replied
21,360	Octoedrite	**Reply**
21,361	Octofid .	Whilst waiting for your reply
21,362	Octogamy .	Will give a reply after
21,363	Octogenary	Cannot reply for a few days
21,364	Octogynia .	Has (have) received no reply from
21,365	Octopedes	Awaiting a reply from
21,366	Octopus .	In reply to
21,367	Octroi . .	Subject to reply by
21,368	Octuple .	You must reply at once
21,369	Octyl . .	Cannot have a reply before
21,370	Octylamine	As soon as a reply is received from
21,371	Octylic .	Reply by wire at once
21,372	Ocubawax	Reply by letter
21,373	Oculary .	Must have a reply in time for
21,374	Oculina .	Can get no definite reply from
21,375	Oculist . .	Have you any reply from
21,376	Ocypoda .	Reply to care of
21,377	Ocypodian	What reply shall I make
21,378	Odalisque .	Telegraph reply to our letter of
21,379	Oddfellow	Wire reply to our telegram of
21,380	Oddities .	In reply to your inquiry
21,381	Oddity . .	If we have no reply to
21,382	Oddlooking	Want reply to our inquiry
21,383	Oddness .	Subject to immediate reply
21,384	Odds . .	Subject to reply within —— days
21,385	Odemaker	Cannot reply by wire ; will write you
21,386	Odically .	Cannot reply yet ; awaiting
21,387	Odiousness	Why do you not reply to
21,388	Odium . .	Awaiting your reply
21,389	Odize . .	Must have definite reply
21,390	Odometry .	Reply to hand ; it is quite satisfactory
21,391	Odontagra	A very unsatisfactory reply
21,392	Odontalgic	Your reply will be considered confidential
21,393	Odontaspis	Whatever may be said in reply
21,394	Odontitis .	**Replying**
21,395	Odonto .	Replying fully to
21,396	Odontogeny	**Report(s).** (See also Telegraph, Proceed, **and**
21,397	Odontoid .	Send copy of report to [Examine.)
21,398	Odontolite	Is your report favourable or otherwise
21,399	Odontology	Report is favourable
21,400	Odontrypy	Report is unfavourable

No.	Code Word.	Report(s) (*continued*)
21,401	Odorament ·	Report is utterly false
21,402	Odorate .	Telegraph me permission to report on
21,403	Odorating.	You are permitted to report on
21,404	Odorous .	Cannot permit you to report on
21,405	Odour . .	No change to report
21,406	Odourless.	Telegraph your report in condensed form
21,407	Odynerus .	How long will it take you to report in detail on
21,408	Odyssey .	Prepare your report, and telegraph from first avail-
21,409	Oeillade .	Separate reports [able office
21,410	Oenanthic	On your confirming ——'s guaranteed report
21,411	Oenanthyl	Telegraph your report at the earliest opportunity
21,412	Oenothera	Can you, and if so, when, report fully on the
21,413	Oesophagus	Since last report [following property(ies)
21,414	Ofcome .	(In) the first report
21,415	Offal . .	(In) the second report
21,416	Offcap . .	Cannot wait your written report; telegraph
21,417	Offcapped	When can you report on
21,418	Offcapping	Can you report for us, on
21,419	Offcasts .	Has (have) unfavourable report(s) of
21,420	Offcolour .	Has (have) received unfavourable reports from
21,421	Offcorn .	To report as soon as possible [regard to
21,422	Offcut . .	——'s report conveys an erroneous impression with
21,423	Offday . .	When will you be able to send report
21,424	Offenceful	Reports are favourable from
21,425	Offending .	There is a report that
21,426	Offendress	Is notorious here for writing reports on mines for
21,427	Offensive .	Report by letter [a commission out of the sale
21,428	Offerable .	Report by telegraph
21,429	Offered .	Adverse reports current here relative to
21,430	Offertory .	Examine and report fully on present condition of the —— mine, and what additional capital will be necessary to put all in good condition
21,431	Offerture .	Be very careful in your report (to) (on)
21,432	Offhand .	Be very guarded in your report (on)
21,433	Official. .	Shall start on —— for —— to report on (a) mine(s)
21,434	Officially .	Report for the week ending [for
21,435	Officiated .	Report immediately as to
21,436	Officious .	Weekly report
21,437	Offlets . .	Fortnightly report
21,438	Offscum .	Monthly report
21,439	Offseason .	Reports are untrue in every essential particular
21,440	Offsetting .	Concise report (upon)
21,441	Offshoots .	Detailed report upon
21,442	Offside . .	Engineer's report (upon)
21,443	Offskip .	First expert's report (upon)
21,444	Offspring .	Mine superintendent's report (upon)
21,445	Offstreet .	Report and plan
21,446	Oftencomer	Report with plan and section
21,447	Oftentimes	Report received, but not plan
21,448	Ogdoad .	Wire cost of report

No.	Code Word.	Report(s) *(continued)*
21,449	Ogdoastich	Wire preliminary report
21,450	Ogganition	Can we accept the report of ——; is he reliable and respectable; and does he carry any weight in the place [of here
21,451	Ogham .	You can accept ——'s report; he is thought well
21,452	Ogival . .	No good, and his report is not worth having
21,453	Ogle . .	——'s reports are not thought much of here
21,454	Ogling . .	——'s report cannot be relied on
21,455	Ogrillon .	Report, owing to press of business, not despatched
21,456	Ogynian .	There is no truth in the report [until
21,457	Ohmad .	Can confirm first expert's report
21,458	Oidemia .	Can confirm ——'s report
21,459	Oilbag . .	Have examined the mine and workings; can fully confirm ——'s report in all essential particulars
21,460	Oilbeetle .	There is a report current here that
21,461	Oilbird . .	Can we contradict the report
21,462	Oilboxes .	You can contradict the report; it is quite unfounded
21,463	Oilcake .	If the report is confirmed
21,464	Oilcans .	We have contradicted the report
21,465	Oilcloth .	The report has been traced to its source (from)
21,466	Oilcoal .	Cannot ascertain the origin of the report
21,467	Oilcolour .	Treat this report as confidential
21,468	Oilcup . .	Copy of report has been sent
21,469	Oiliness .	Report received too late
21,470	Oilmills .	Too late to be inserted with report
21,471	Oilnuts .	This report compares favourably with the last
21,472	Oilpalm .	Reports must be regularly kept up
21,473	Oilpress .	The expert reports favourably
21,474	Oilpumps .	The expert reports unfavourably
21,475	Oilseed .	Your report by letter is required on —— situate
21,476	Oilshale .	Get an expert's report on [—— agent —— at
21,477	Oilskin . .	Wire report briefly as to —— situate —— agent
21,478	Oilspring .	What are your terms for report on [—— at
21,479	Oilstone .	Report briefly by wire and fully by letter on —— situate —— agent —— at
21,480	Oiltree . .	Points chiefly requiring report (are)
21,481	Oily. ·. .	Our report is adverse to any speculation in the
21,482	Oilygrain .	Full report is sent by mail [property
21,483	Ointment .	**Reported**
21,484	Oisanite .	The mine has been favourably reported on
21,485	Okro . .	The mine has been unfavourably reported on
21,486	Oldened .	Do not know who reported upon (it) (the)
21,487	Oldhamia .	Do you know who reported upon (it) (the)
21,488	Oldmaidish	When previously reported on
21,489	Oldoil . .	Has (have) reported upon
21,490	Oldsaid .	The mine was reported on (by)
21,491	Oldster .	**Reporting**
21,492	Oldworld .	Since reporting
21,493	Oleaginous	Charge for reporting
21,494	Oleamide .	When reporting upon

No.	Code Word.	
21,495	Oleander .	**Represent(s)**
21,496	Oleasters .	Who will represent us
21,497	Olecranon	Will represent us
21,498	Olefiant .	Cannot represent us on account of
21,499	Olefine .	Does it represent
21,500	Oleic . .	Does it represent the actual state of
21,501	Oleographs	Do they represent
21,502	Oleoptene	Do (does) not represent
21,503	Oleoresin .	**Representation(s)**
21,504	Olfact . .	A representation has been made
21,505	Olfaction .	Made a representation (to)
21,506	Olfactive .	Representation of —— interests
21,507	Olfactory .	The representations made were
21,508	Olibanum .	The representations made by —— are substantially
21,509	Olidous .	**Representative(s)** [correct
21,510	Oligaemia .	My (our) representative
21,511	Oligarch .	His (their) representative
21,512	Oligarchal .	See —— or his representative
21,513	Oligist . .	**Represented**
21,514	Oligistic .	It has been represented to us
21,515	Oligocene .	If this is properly represented to
21,516	Oligoclase .	Not represented
21,517	Oligodon .	Is not at all what it was represented to be
21,518	Oligonite .	**Representing**
21,519	Oligonspar	As representing all interests
21,520	Olitory . .	Representing the vendor
21,521	Oliva . .	Representing the company
21,522	Olivaceous	Representing the present state of
21,523	Olivegreen	**Repudiate(s)**
21,524	Olivenite .	Entirely repudiate
21,525	Oliveoil .	Do you repudiate
21,526	Olivewood	**Repudiated**
21,527	Oliveyard .	Has (have) repudiated
21,528	Olivinoid .	If not repudiated
21,529	Olympiad .	Repudiated all their undertakings
21,530	Olympionic	Repudiated their indebtedness
21,531	Omagra .	**Repudiation**
21,532	Omasum .	Repudiation of the statement
21,533	Ombrometer	Repudiation of the agreement
21,534	Omenings .	**Reputation**
21,535	Omiletical	Is a man of good reputation
21,536	Ominously	An ignorant man and of no reputation
21,537	Omissible .	What sort of reputation has (have)
21,538	Omission .	Has (have) the reputation of being
21,539	Omissive .	Known to me (us) by reputation only
21,540	Omissively	**Reputed**
21,541	Omittance	Is reputed to be
21,542	Omitted .	**Request(s)**
21,543	Omitting .	Your request will be attended to
21,544	Omneity .	Your request cannot be allowed

No.	Code Word.	**Request(s)** (*continued*)
21,545	Omnibus .	At your request
21,546	Omniform.	At (the) request of
21,547	Omnifying	Your request is under consideration
21,548	Omnigraph	Request considered and approved (with regard to)
21,549	Omniparity	Refused the request
21,550	Omnipotent	We must request you to
21,551	Omniscient	**Requested**
21,552	Omniscious	Has (have) requested
21,553	Omnivalent	Has (have) been requested
21,554	Omnivora .	**Require(s)**
21,555	Omnivorous	Will require
21,556	Omohyoid	Will not require
21,557	Omoplate .	If you require
21,558	Omphacine	When will you require
21,559	Omphalic .	How much will you require
21,560	Omphalode	Require immediately
21,561	Onager .	Do you require
21,562	Onagraceae	I (we) shall require
21,563	Oncidium .	Require on or before
21,564	Ondatra .	What do you require to enable you to
21,565	Oneberries	Requires us to
21,566	Oneberry .	**Required**
21,567	Onehorse .	As required by the contract
21,568	Oneirology	Will be required
21,569	Onerary .	Will not be required
21,570	Onesided .	If it is required
21,571	Onesidedly	When will it be required
21,572	Onhanger.	You know best what will be required
21,573	Onicolo .	Will not be required for some time
21,574	Onioneyed	Will be required not later than
21,575	Onionshell	How much money will be required (to)
21,576	Oniscidae.	How long a time will be required (to)
21,577	Oniscus .	**Requirements**
21,578	Onlooker .	Send particulars of all your requirements
21,579	Onlooking	(To) meet all requirements
21,580	Onobrychis	Not equal to our requirements
21,581	Onocentaur	Equal to the requirements of
21,582	Onomantia	Does not come up to the requirements
21,583	Onomastic	Cannot comply with your requirements
21,584	Onomatope	**Requisite**
21,585	Ononis .	The machinery requisite for
21,586	Onopordum	To get only what is requisite
21,587	Onosma .	Cannot get the requisite
21,588	Onrush .	To get the requisite
21,589	Onset . .	(With) the requisite pumps and machinery
21,590	Onslaught.	The requisite notice
21,591	Onstead .	(To) give the requisite notice
21,592	Ontogeny .	Requisite notice has been given
21,593	Ontologic .	The requisite time
21,594	Onward .	The requisite funds for

No.	Code Word.	**Requisite** (*continued*)
21,595	Onycha .	The requisite output
21,596	Onychite .	The requisite milling power
21,597	Onyx . .	The requisite developments
21,598	Oocyst .	. **Requisition**
21,599	Oogonium	A requisition has been made
21,600	Oolak . .	An official requisition
21,601	Oolitic .	. **Reservation**
21,602	Oolysis .	Without any reservation
21,603	Oomiac .	With considerable reservation
21,604	Oosphere .	Reservation would seem to imply
21,605	Oosporange	**Reserve(s)**
21,606	Oospore .	A reserve of at least
21,607	Oostegite .	There are no reserves of ore in sight
21,608	Ooticoid .	I (we) estimate reserves of ore at
21,609	Oozed . .	Impossible to estimate reserve of ore
21,610	Opacity .	Telegraph what reserves of ore are available
21,611	Opacous .	Telegraph reserve of ore
21,612	Opalesce .	Not a great reserve of ore
21,613	Opalescing	The reserves of ore are very large
21,614	Opaljasper	In his report —— estimates reserves of ore at ——
21,615	Opaquely .	Prospective reserves [tons ; is this correct
21,616	Opaqueness	Estimate of ore reserves —— tons is correct
21,617	Openbill .	Guarantee(s) the ore reserve at —— tons
21,618	Opencast .	Think this chute will place us in possession of
21,619	Opendoored	Must reserve [valuable reserves
21,620	Openeyed .	The ore body should give us good reserves
21,621	Openhanded	Likely to develop large reserves
21,622	Openly. .	Large reserves may be expected from
21,623	Openness .	Large reserves of ore, but of low grade
21,624	Opentide .	Large reserves of ore, of fair grade
21,625	Openwork	Large reserves of ore, which may be reckoned on
21,626	Opera . .	Large reserves of [to average
21,627	Operacloak	Reserves of ore between —— and —— levels
21,628	Operaglass	Reserves of ore above the
21,629	Operahat .	Reserves of ore below the
21,630	Operahouse	Can you reserve
21,631	Operameter	Will reserve
21,632	Operancy .	If you do not reserve
21,633	Operatical	Can reserves be relied upon
21,634	Operating .	Reserves practically all low grade
21,635	Operations	Sufficient high-grade reserves
21,636	Operative .	**Reserve Fund**
21,637	Opercular .	Reserve fund
21,638	Operose .	Reserve fund should be formed
21,639	Operosity .	Reserve fund now stands at
21,640	Ophiasis .	Carried to reserve fund
21,641	Ophicalcic	Debit reserve fund
21,642	Ophicleide	Credit reserve fund
21,643	Ophidia .	**Reserved**
21,644	Ophidious	Has (have) reserved

No.	Code Word.	**Reserved** (*continued*)
21,645	Ophiolatry	Has (have) not reserved
21,646	Ophiolite .	Have reserved to themselves the right of
21,647	Ophiologic	Have reserved to ourselves the right of
21,648	Ophiomancy	**Reserving**
21,649	Ophiops .	Reserving to ourselves
21,650	Ophiorhiza	Reserving to himself (themselves)
21,651	Ophioxylon	Reserving the right to
21,652	Ophite. .	Reserving all claims
21,653	Ophiuchus	Reserving intact all rights and privileges
21,654	Ophiuridae	Reserving the full use of
21,655	Ophrys .	Reserving full freedom to
21,656	Ophthalmia	Reserving only
21,657	Opianic .	But not reserving
21,658	Opiferous .	**Resign**
21,659	Opinants .	Must resign
21,660	Opinator .	Must not resign
21,661	Opined .	Wish(es) to resign
21,662	Opining .	Will not resign
21,663	Opiniastre	Wishes to resign on account of ill-health
21,664	Opiniated .	Wishes to resign in order to
21,665	Opiniatry .	Has had to resign
21,666	Opinicus .	Take his place when he resigns
21,667	Opinionate	Can allow —— to resign
21,668	Opiumeater	Must ask —— to resign
21,669	Opletree .	Have asked him to resign
21,670	Opobalsam	If he does not resign
21,671	Opodeldoc	**Resignation**
21,672	Opoponax	Sent in his resignation
21,673	Oporice .	Resignation sent in ; have accepted it
21,674	Opossum .	Resignation sent in ; but not accepted
21,675	Oppidan .	Wish —— to withdraw resignation
21,676	Oppilation	**Resigned**
21,677	Oppilative	Has (have) resigned
21,678	Opponency	Has (have) not resigned
21,679	Opponent .	The position which he resigned
21,680	Opportune	**Resigning**
21,681	Opposable	Is about resigning on account of
21,682	Opposed .	By resigning at the present time
21,683	Opposeless	**Resolution(s)**
21,684	Opposing .	The following resolution was passed.
21,685	Oppositely	The resolution was not carried
21,686	Opposition	Passed a resolution to
21,687	Oppositive	Resolution of the Board
21,688	Oppress .	Resolution of shareholders
21,689	Oppresseth	Resolution for winding up
21,690	Oppressing	Resolution appointing a committee of inquiry
21,691	Oppressive	Resolution calling upon the Board to resign
21,692	Oppressor .	Resolution proposed, but fell through for want of
21,693	Opprobium	Supported the resolution [support
21,694	Opprobry .	Opposed the resolution

No.	Code Word.	**Resolution** (*continued*)
21,695	Oppugn .	If the resolution is carried
21,696	Oppugnancy	A resolution will be proposed [lution to
21,697	Oppugning	A section of the shareholders intend to move a reso-
21,698	Opsimathy	Will vigorously oppose the resolution
21,699	Opsiometer	There was some opposition, but the resolution was
21,700	Opsomania	Will support the resolution [ultimately carried
21,701	Optated .	Resolution will strengthen our hands
21,702	Optatively	Resolution of want of confidence in the Board
21,703	Optically .	Resolution is one involving grave censure on
21,704	Opticians .	Resolution will much hamper us
21,705	Optics . .	The resolution was carried by a majority of
21,706	Optigraph	**Resolved**
21,707	Optimacy .	Has (have) resolved to
21,708	Optimates	Has (have) resolved not to
21,709	Optimism	What have you resolved with regard to
21,710	Optimistic	Resolved to carry out
21,711	Optimized	Resolved to wind up voluntarily
21,712	Optimizing	Resolved to raise more capital
21,713	Option . .	Resolved to go into liquidation
21,714	Optional .	Have —— resolved
21,715	Optogram .	Hope you have resolved
21,716	Opulency .	If you have not resolved
21,717	Opulent .	Resolved what action to take ,
21,718	Opunctly .	**Resource(s)**
21,719	Opuntia .	Exhausting too much the resources
21,720	Opuscle .	Not to exhaust the resources of
21,721	Opusculum	The resources of the mine
21,722	Oracular .	The resources of the mine are very great
21,723	Oracularly	What are the resources of the mine
21,724	Oragious .	The resources at hand
21,725	Orangeade	No resources at hand
21,726	Orangebird	Must not count on our resources being large enough
21,727	Orangelily	The resources of the mine are not sufficient to
21,728	Orangeism	At the end of our resources
21,729	Orangemen	Our resources will only hold out
21,730	Orangemusk	How long will your resources hold out
21,731	Orangepeel	(With) your present resources
21,732	Orangery .	(With) our present resources
21,733	Orangeskin	(With) his (their) present resources
21,734	Orangetip .	The only resources we have now to depend on are
21,735	Orangewife	The present available resources
21,736	Orangutan	As a last resource
21,737	Orator . .	Resources we can depend on may be taken at
21,738	Oratorial .	**Respect**
21,739	Oratorious	With respect to
21,740	Oratorize .	Is (are) held in great respect
21,741	Orb. . .	In every respect
21,742	Orbfish .	In this respect
21,743	Orbical .	In what respect
21,744	Orbiculina	If, in any respect

No.	Code Word.	Respect (*continued*)
21,745	Orbitelae .	With respect to the future
21,746	Orbitude .	With respect to what was done
21,747	Orbity .	. **Respectable**
21,748	Orblike .	Is (are) the most respectable here in his (their)
21,749	Orchal . .	Are most respectable people [profession
21,750	Orcharding	Not respectable people
21,751	Orchards .	Very respectable
21,752	Orchestral	Is (are) —— respectable
21,753	Orchideous	**Responsibility(ies)**
21,754	Orchiocele	Take full responsibility (for)
21,755	Orchotomy	Will accept the entire responsibility
21,756	Ordain . .	The full responsibility devolves on
21,757	Ordainable	The responsibilities entailed upon
21,758	Ordaineth .	Acting on our own responsibility
21,759	Ordaining .	Responsibility ceases
21,760	Ordainment	Responsibility now at an end
21,761	Ordalian .	No further responsibility
21,762	Ordeal . .	No responsibility incurred
21,763	Ordealbean	Will incur no further responsibility
21,764	Ordealnut.	Incur no responsibility
21,765	Ordealroot	Not prepared to undertake the responsibility
21,766	Ordealtree	Do not care to incur such onerous responsibilities
21,767	Orderable .	Undertaken all responsibilities
21,768	Orderbook	**Responsible**
21,769	Orderings .	Responsible parties
21,770	Orderly .	Is (are) not responsible for
21,771	Ordinalism	Will not be responsible for
21,772	Ordinary .	Is (are) responsible for
21,773	Ordinative	Who is responsible for
21,774	Ordonnant	Hold you responsible
21,775	Oreala . .	Hold him (them) responsible
21,776	Orehearth .	We shall be responsible for
21,777	Oreillet .	Be held responsible for
21,778	Organdy .	Make us responsible for
21,779	Organfish .	Make you responsible for
21,780	Organic .	Make him responsible for
21,781	Organical .	Unless —— are held responsible
21,782	Organicism	If —— are responsible
21,783	Organista .	**Rest(s)**
21,784	Organized	The rest (of the)
21,785	Organizing	Of the rest
21,786	Organloft .	To rest upon
21,787	Organogen	Rests upon
21,788	Organology	It rests with
21,789	Organpipe	Take the rest
21,790	Organpoint	You can send the rest
21,791	Organstop	**Restricted**
21,792	Organzine	Are we restricted to
21,793	Orgasm .	(Is) are restricted to
21,794	Orgiastic .	(Is) are not restricted to

No.	Code Word.	
21,795	Orgyia . .	**Restriction(s)**
21,796	Oribatidae	All restrictions are now removed
21,797	Orichalcum	The restrictions have been withdrawn
21,798	Oriel . .	Without any restriction
21,799	Oriency .	Proposes to withdraw restrictions
21,800	Orientals .	Impose (imposed) the following restrictions
21,801	Orientness	With such restrictions as may be needful to
21,802	Orifex . .	With no restriction as to cost or time
21,803	Orifices .	The restrictions imposed upon us
21,804	Oriflamb .	Much fettered by the restrictions
21,805	Origanum .	**Result(s)**
21,806	Origin . .	The result of which has been to
21,807	Original .	The result will be
21,808	Originally .	Is the result satisfactory
21,809	Originator	Result is satisfactory
21,810	Oriolinae .	Result is unsatisfactory
21,811	Oriolus .	With no great result at present
21,812	Orion . .	What is the result of
21,813	Orismology	Telegraph the result (of)
21,814	Orlop . .	Result encouraging
21,815	Ormolu	Result disappointing
21,816	Ornament .	With no better result than the former
21,817	Ornamental	Give (gives) a better result than we expected
21,818	Ornately .	Result worse than we expected
21,819	Ornateness	The result was not as good as we should have liked
21,820	Ornature .	It will result in
21,821	Ornithon .	What do you think the result will be
21,822	Ornithopus	The result will probably be
21,823	Orobanche	Think the result will be satisfactory
21,824	Orographic	The result will be most prejudicial to our interests
21,825	Orological	The result of the experiment
21,826	Orology .	The result of the trial
21,827	Orontiad .	Gave certain results, which
21,828	Orphalines	The result is very uncertain
21,829	Orphan .	If you cannot get a good result
21,830	Orphanage	Good results can only be attained by
21,831	Orphanhood	The result of the trial will be disastrous, if against us
21,832	Orphanism	The result is very annoying
21,833	Orphanry .	No better results can be secured
21,834	Orpharion	The result of the inquiry
21,835	Orphic . .	Investigation has had the result of
21,836	Orphrey .	The results of the operations
21,837	Orpiment .	Has had the result of
21,838	Orrery . .	The result is not yet known
21,839	Orrisroot .	What is the result of the negotiations
21,840	Orsedew .	Result of the negotiations as to
21,841	Orthoceras	The result of the action is to
21,842	Orthoclase	**Resulted**
21,843	Orthodox .	Resulted in our having to
21,844	Orthodoxly	Which resulted from

No.	Code Word.	**Resulted** (*continued*)
21,845	Orthodromy	Resulted in the stoppage of
21,846	Orthoepic.	**Resulting**
21,847	Orthogamy	Resulting from our efforts
21,848	Orthogonal	Resulting from the work done
21,849	Orthology.	**Resume**
21,850	Orthometry	(To) resume work on
21,851	Orthopaedy	Shall be able to resume
21,852	Orthophony	Will resume work on the basis of
21,853	Orthopnoea	When can you resume
21,854	Orthopraxy	Cannot resume till
21,855	Orthopter.	When do you resume operations
21,856	Orthose .	Expect to resume operations
21,857	Orthostyle	Resume operations as soon as
21,858	Orthotone.	Cannot resume operations until
21,859	Ortive . .	**Resumed**
21,860	Ortolan .	When you have resumed operations
21,861	Orvietan .	When the men have resumed work
21,862	Oryctology	Resumed work on
21,863	Oryza . .	Men have resumed work
21,864	Oscillancy	**Resumption**
21,865	Oscillator .	The resumption of work
21,866	Oscitantly.	The resumption of our rights
21,867	Osculant .	The resumption of all rights
21,868	Osculating	**Retain(s)**
21,869	Osculatory	Retain the stock till further orders
21,870	Osculatrix.	Retain until further orders
21,871	Osierait .	To retain
21,872	Osierbed .	You can retain
21,873	Osierholt .	Will retain
21,874	Osleoniron	Will not retain
21,875	Osmanlis .	We shall retain
21,876	Osmazome	We shall retain the best man we can get
21,877	Osmelite .	Retain the best expert you can get
21,878	Osmia . .	Retain the best counsel you can get
21,879	Osmiamic.	Retain the possession of
21,880	Osmious .	Retain the services of
21,881	Osmiridium	**Retained**
21,882	Osmometer	Retained a first-class man
21,883	Osmotic .	The man we have retained is a mining expert of
21,884	Osmunda .	Retained at a fee of [great repute
21,885	Osprey. .	The counsel retained has had great experience in
21,886	Osseous .	If not retained promptly [mining cases
21,887	Ossianic .	**Retaining**
21,888	Ossicle. .	By retaining
21,889	Ossified .	We are retaining
21,890	Ossifiage .	**Retarded**
21,891	Ossify . .	Much retarded by
21,892	Ossifying .	Our progress has been greatly retarded by
21,893	Osspringer	Have been retarded owing to the bad weather
21,894	Ossuaries .	Have not retarded the work

No.	Code Word.	
21,895	Ossuary .	**Retired**
21,896	Ostensible	The bill has been retired
21,897	Ostension .	The bill has not been retired
21,898	Ostensive .	Has (have) not retired the bill
21,899	Ostensory .	Have you retired
21,900	Ostentator	If not retired before
21,901	Ostentous .	**Retort(s)**
21,902	Osteoblast	To retort the amalgam
21,903	Osteocolla	**Retorted**
21,904	Osteocope	What is amount of retorted gold
21,905	Osteolepis	Retorted gold
21,906	Osteoma .	**Return(s)**
21,907	Osteomanty	Stop at —— on your return from
21,908	Osteozoa .	You will find letters on your return to
21,909	Ostiolum .	As I (we) return
21,910	Ostler . .	Will return
21,911	Ostleress .	When do you return to
21,912	Ostracism .	Will stop at —— on my (our) return from
21,913	Ostracized	Will return by way of
21,914	Ostracoda	Will return in a few days
21,915	Ostreidae .	You must return home immediately —— is dan-
21,916	Ostrich .	Return immediately [gerously ill
21,917	Ostrya . .	Do not return
21,918	Oswego .	Shall I return
21,919	Oswegotea	You must return home at once —— is dead
21,920	Otacoustic	You must return home at once
21,921	Otalgia .	As soon as you return from
21,922	Otheoscope	As soon as —— return(s)
21,923	Othergates	On the return of
21,924	Otherwards	Cannot return until
21,925	Otherwise .	Shall return by —— steamship (company)
21,926	Otidinae .	Expect —— will return about
21,927	Otiose . .	Do not expect to return before
21,928	Otiosity .	Am (are) waiting for the return of
21,929	Otobafat .	On my return will
21,930	Otoconite .	You had better return by way of
21,931	Otocrane .	The annual return
21,932	Otocyon .	This month's return
21,933	Otography	Next month's return
21,934	Otolithic .	Last month's return
21,935	Otopathy .	The returns show
21,936	Otopteris .	Telegraph the returns for
21,937	Otorrhoea	Returns for this month and next
21,938	Otoscopes	Returns for the next two months
21,939	Ototeal .	Returns for the next three months
21,940	Otozoum .	Future monthly returns
21,941	Ottar . .	Future returns will be much better
21,942	Otterdog .	Future returns will not be so good
21,943	Otterhound	The low return has caused great dissatisfaction
21,944	Ottershell .	What is the cause of the low returns

No.	Code Word.	**Return(s)** (*continued*)
21,945	Otterspear	In order to keep up the returns
21,946	Ottoman .	Do you see your way to better returns
21,947	Ouarine .	See the way to better returns regardless of future
21,948	Oubliette .	Can you make a better return [discoveries
21,949	Oudenodon	Can you make a return as good as
21,950	Ought . .	If the returns do not improve
21,951	Oughtness	Returns should improve consequent upon
21,952	Ouistiti .	We think returns likely to improve
21,953	Oulachon .	Gradual increase in the returns may be expected
21,954	Oulorrhagy	Do not reckon upon a better return
21,955	Ouphen . .	Will not materially affect the return
21,956	Ourebi. .	Will have a most beneficial effect upon the return
21,957	Ouroscopy	What effect are the present developments likely to have upon the returns [better returns
21,958	Ourself .	Considering you are in good ore cannot you make
21,959	Ourselves .	Cannot you work any of the richer ore so as to increase the returns
21,960	Outargue .	Estimated return this month will be
21,961	Outarguing	Estimated return next month will be
21,962	Outbabble	The average returns will be about
21,963	Outbalance	We could show better returns if
21,964	Outbargain	Could otherwise show better returns
21,965	Outbarred.	Instead of steadily increasing returns
21,966	Outbidder	Why should the returns decrease
21,967	Outblaze .	Cannot increase returns without adopting very in-
21,968	Outblazing	With regard to the low returns [judicious means
21,969	Outblown.	Can you fairly exceed the regular returns
21,970	Outblush .	We do not wish the returns increased if you cannot do so without forcing the mine
21,971	Outboard .	Do not pick the eyes of the mine out in order to
21,972	Outbounds	A very poor return [swell the returns
21,973	Outbrag .	A most excellent return
21,974	Outbragged	The return we have just received
21,975	Outbreaks	To equalize the returns
21,976	Outbribe .	Revenue returns
21,977	Outbudding	The returns from the mint
21,978	Outbuilds .	**Returned**
21,979	Outburneth	Has (have) returned from
21,980	Outburst .	Nothing can be done, till —— has (have) returned
21,981	Outcaper .	Has (have) not yet returned
21,982	Outcast .	As soon as —— has (have) returned
21,983	Outcasting	Unless returned at once
21,984	Outcept .	Was returned
21,985	Outcheat .	Was returned by a large majority
21,986	Outclimb .	**Returning**
21,987	Outcome .	Before returning
21,988	Outcompass	After returning
21,989	Outcourts.	When returning
21,990	Outcraft .	**Revenue**
21,991	Outcrier .	Revenue falling off

No.	Code Word.	Revenue (*continued*)
21,992	Outcrop .	Considerable falling off in revenue
21,993	Outcropped	Revenue increasing
21,994	Outcry . .	Revenue has risen from —— to
21,995	Outcrying .	Revenue has risen in —— years
21,996	Outcursing	Revenue accounts
21,997	Outdazzel .	Balance of revenue
21,998	Outdo . .	Revenue derived from
21,999	Outdoeth .	Yield (yielding) a fairly steady revenue
22,000	Outdoors .	To be paid out of revenue
22,001	Outdream .	Has been paid out of revenue
22,002	Outdrink .	All expenses must be paid out of revenue
22,003	Outdwell .	The revenue for the current
22,004	Outermost	The revenue for the past
22,005	Outerplate	The revenue for the next
22,006	Outfawn .	Expenses of —— should be charged to revenue·
22,007	Outfield .	Should not be charged to revenue
22,008	Outfit . .	What part of the expenditure on —— has been
22,009	Outfitters .	Chargeable to revenue [charged to revenue
22,010	Outflanked	Charge to revenue instead of capital
22,011	Outflatter .	To be transferred from revenue to capital
22,012	Outfling .	**Revoke**
22,013	Outflowed .	Revoke power of attorney given to
22,014	Outflowing	We are writing to revoke
22,015	Outfool .	**Revoked**
22,016	Outfuneral	Have revoked
22,017	Outgates .	Have revoked the order
22,018	Outgeneral	Have you revoked
22,019	Outgive .	Have not revoked
22,020	Outgoings	Power of attorney revoked
22,021	Outgrinned	**Revoking**
22,022	Outgrowth	Revoking all powers
22,023	Outguard .	Revoking your power of attorney
22,024	Outhaul .	Have written revoking power
22,025	Outhiss .	**Revolution(s)**
22,026	Outhissing	Revolution has broken out
22,027	Outhouses	Revolution will seriously interfere with
22,028	Outjesting	The revolution has been crushed
22,029	Outjuggle .	The revolution has been successful
22,030	Outkeeper	In consequence of the revolution
22,031	Outknave .	How many revolutions per minute does the ——
22,032	Outlance .	(Making) —— revolutions per minute [make
22,033	Outlandish	Must work at —— revolutions per minute
22,034	Outlaugh .	The revolution in the
22,035	Outlaw .	At a speed of —— revolutions per minute
22,036	Outlawries	The engine works at —— revolutions per minute,
22,037	Outlawry .	**Revolving** [and indicates —— horse-power
22,038	Outleaped	Revolving furnaces
22,039	Outleaping	Revolving barrel(s)
22,040	Outlinear .	Revolving buddles
22,041	Outliving .	**Reward**

£ No.	Code Word.	**Reward** (*continued*)
22,042	Outlooks .	(To) offer a reward
22,043	Outlustre .	As a reward for his services
22,044	Outmantle	Claims a reward
22,045	Outmarch.	Offered a reward for the discovery of
22,046	Outmaster	**Rich**
22,047	Outmeasure	Have struck rich ore in ——, assays give
22,048	Outmounted	Rich strike is reported in —— mine
22,049	Outnoise .	The company is very rich
22,050	Outnumber	The ore is very rich, assaying —— per ton
22,051	Outparish .	Have struck a pocket of very rich ore
22,052	Outpassed.	Shall come to the end of the rich ore in —— days
22,053	Outpatient	The very rich ore only occurs in pockets
22,054	Outpeer .	Rich ore
22,055	Outpeering	Very rich
22,056	Outpenny .	Not very rich
22,057	Outpicket .	Rich chutes of ore
22,058	Outpoising	A chute of rich ore
22,059	Outporch .	**Richer**
22,060	Outposts .	The deeper down the richer the ore
22,061	Outpowered	The ore is richer than ever
22,062	Outprays .	Are now getting into richer ore
22,063	Outpreach	The richer ore we are now getting
22,064	Outprize .	Now milling richer ore
22,065	Outprizing	Richer in some parts than others
22,066	Outquench	**Richest**
22,067	Outrageous	The richest ore we have yet found
22,068	Outrank .	From what part of the mine are you getting the
22,069	Outranking	The richest mine about here [richest ore
22,070	Outrapping	The richest ore is coming from the
22,071	Outrayed .	**Richness**
22,072	Outreached	The richness of the ore
22,073	Outreason.	Of extraordinary richness
22,074	Outreckon	Of great richness, assaying probably [pectation
22,075	Outriding .	The richness of this chute has surpassed our ex-
22,076	Outriggers	**Rid** (of)
22,077	Outright .	Can you get rid of
22,078	Outrival .	Cannot get rid of
22,079	Outroar .	If you can possibly get rid of
22,080	Outromance	We very much wish to get rid of
22,081	Outrooted	If we could get rid of
22,082	Outsail. .	Have at last got rid of
22,083	Outsailing.	**Right(s)**
22,084	Outscented	It would not be right to
22,085	Outscold .	Is everything all right
22,086	Outscorned	You must do whatever you think right
22,087	Outscout .	The right place would be
22,088	Outsentry.	Is in the right place
22,089	Outsettler.	The mill is not in the right place
22,090	Outshine .	The mine has not been worked right
22,091	Outshining	The right thing would be to

No.	Code Word.	**Right(s)** (*continued*)
22,092	Outshoot .	Has (have) no right to
22,093	Outskip .	Is this right
22,094	Outskipped	It is right
22,095	Outskirts .	Should we be right in
22,096	Outslang .	What right had
22,097	Outsoaring	Who has the right
22,098	Outsounded	Our rights must be upheld
22,099	Outspan .	Is within our clear rights
22,100	Outspanned	The only right we can claim
22,101	Outsparkle	Forfeit our rights
22,102	Outspoken	**Rise**
22,103	Outsport .	The vein in the rise
22,104	Outspread	The rise is about —— feet from surface
22,105	Outstay .	Have holed with rise
22,106	Outstepped	Rise up on the vein
22,107	Outstretch	The vein in the rise contains ore
22,108	Outstride .	There is a rise in the vein from the
22,109	Outsubtle .	Samples from rise gave
22,110	Outsuffer .	From the rise
22,111	Outswear .	Fear the water will rise
22,112	Outswelled	Do not think the water will rise
22,113	Outtalked .	The price of the shares will rise
22,114	Outtelling .	Cannot rise unless
22,115	Outthrow .	No. 1 Rise (From Rise No. 1)
22,116	Outtongued	No. 2 Rise (From Rise No. 2)
22,117	Outtop. .	No. 3 Rise (From Rise No. 3)
22,118	Outvalue .	No. 4 Rise (From Rise No. 4)
22,119	Outvenom	No. 5 Rise (From Rise No. 5)
22,120	Outvillain .	Rise No. —— (From Rise No. ——)
22,121	Outvoiced	What prospects are there of a rise in
22,122	Outvote .	Good prospects of a rise
22,123	Outvoting .	No immediate prospects of a rise
22,124	Outwardly	Do not think there will be a rise
22,125	Outwash .	Further rise doubtful for a time
22,126	Outwatched	Expect a rise about
22,127	Outwearied	Buy for me and hold for a rise
22,128	Outweary .	**Risen**
22,129	Outweep .	The water has risen —— ft. in twenty-four hours
22,130	Outweigh .	The price of the shares has risen from —— to
22,131	Outwinded	Has the price of the shares risen
22,132	Outwitting	**Rising**
22,133	Outworking	The water is still rising in shaft
22,134	Outworks .	To stop the water from rising
22,135	Outwrest .	Rising up on the vein
22,136	Outzanied .	After rising to
22,137	Outzany .	**Risk**
22,138	Outzanying	The risk is too great
22,139	Ouviranda	There will be no risk
22,140	Ouzel . .	Is there any risk
22,141	Ova. .	The risk is

No.	Code Word.	**Risk** (*continued*)
22,142	Ovalbumen	Is not worth the risk
22,143	Ovaliform .	There is no risk
22,144	Ovally . .	If it can be done without risk
22,145	Ovariotomy	Will take the risk of
22,146	Ovarium .	The risk is too great for us to undertake
22,147	Ovation .	Think there is serious risk
22,148	Ovenbird .	There is great risk that
22,149	Ovenchyma	At our risk
22,150	Ovenless .	At ——'s risk
22,151	Overabound	Take the risk
22,152	Overact .	**River(s)**
22,153	Overacting	River crosses the property
22,154	Overaction	A river runs close to the property
22,155	Overaffect .	River supplies the water-power
22,156	Overalls .	River is within —— miles
22,157	Overarched	River is within —— of a mile
22,158	Overawed .	There is not much water in the river
22,159	Overawful .	River has flooded the banks
22,160	Overbarren	The river is about —— yards wide
22,161	Overbear .	From the river a plentiful supply of water can be [obtained
22,162	Overbid .	Good site for a mill on the river-side
22,163	Overboard	Water is brought from the —— river
22,164	Overbodied	River nearly dry in summer
22,165	Overbody .	The river affords ample water for power and milling [purposes
22,166	Overboldly	**Road(s)**
22,167	Overbridge	There is a good road to the mine(s)
22,168	Overbright	A good road of —— miles leading from ——, and of very easy grade
22,169	Overbuild .	Is —— miles by good road from
22,170	Overbulked	A very bad road
22,171	Overbuying	Now on the road
22,172	Overcanopy	Owing to the bad state of the roads
22,173	Overcarry .	Teams cannot haul ore, on account of the bad state [of the roads
22,174	Overcatch .	On the road to
22,175	Overcharge	Will this include making roads
22,176	Overcivil .	To include making of roads
22,177	Overclean .	Roads in bad condition
22,178	Overclimb	Roads are impassable after rain
22,179	Overcloud	Roads are too steep for
22,180	Overcoats .	To improve the roads
22,181	Overcolour	Road will have to be made
22,182	Overcometh	Road open
22,183	Overcostly	The road to the mine
22,184	Overcount	The roads are fine natural ones
22,185	Overcritic .	What sort of roads are there
22,186	Overdo .	Roads unfit for anything but mules
22,187	Overdoers	Roads now impassable owing to the rains
22,188	Overdosed	Road impassable, owing to
22,189	Overdress .	Have made a new road
22,190	Overdrink	A road from the mine to the mill

No.	Code Word.	**Road(s)** (*continued*)
22,191	Overdrive .	Road from the mine to
22,192	Overdrown	**Roast**
22,193	Overdry .	It is necessary to roast the ore
22,194	Overdue .	It is not necessary to roast the ore
22,195	Overeager.	Used to roast the ore at one time
22,196	Overeat .	**Roaster(s)**
22,197	Overempty	There is a roaster attached to the mill
22,198	Overenrich	There is no roaster attached to the mill
22,199	Overeye .	Must put up a roaster at the mill
22,200	Overfilled .	**Roasting**
22,201	Overflowed	Roasting furnace
22,202	Overflux .	**Rock(s)**
22,203	Overfondly	Rock-breaker
22,204	Overforce .	Rock-breaker jaws
22,205	Overfrieze	Send —— spare jaws for rock-breaker at once
22,206	Overfront .	Wall rock
22,207	Overgazed	Rock is getting harder
22,208	Overgazing	Rock is getting softer
22,209	Overglided	Send samples of rock from
22,210	Overgloom	The hardness of the rock necessitates
22,211	Overgrace.	Owing to the hardness of the rock
22,212	Overgreat .	It is easy rock to work
22,213	Overgreedy	Owing to the softness of the rock
22,214	Overgross .	Going through hard rock
22,215	Overgrowth	Driving through very hard rock
22,216	Overhanded	Rock-boring
22,217	Overhangs	Rock-boring machinery
22,218	Overhappy	The country rock is
22,219	Overhaste.	The country rock is granite
22,220	Overheated	The country rock is porphyry
22,221	Overheavy	The country rock is slate
22,222	Overhighly	**Rolls**
22,223	Overhold .	Crushing rolls
22,224	Overinform	Send set of Cornish rolls
22,225	Overissue .	Set of Kroms rolls
22,226	Overjoy .	Steel rolls
22,227	Overjump .	**Room(s)**
22,228	Overkind .	Have you room for
22,229	Overlabour	Have plenty of room for
22,230	Overlapped	Want more room
22,231	Overlaps .	Must have more room
22,232	Overlavish	There would be room for
22,233	Overlaying	Reserve room for
22,234	Overliness	What room have you
22,235	Overliver .	**Rope(s)**
22,236	Overloaded	Wire rope —— inches circumference
22,237	Overlong .	Manilla rope —— inches circumference
22,238	Overlooker	Hemp rope —— inches circumference
22,239	Overloop .	Steel rope —— inches circumference
22,240	Overlusty .	Wire rope —— × —— inches, flat

No.	Code Word.	
22,241	Overmanner	**Rough**
22,242	Overmeddle	A rough draft of
22,243	Overmellow	Have sent a rough plan of
22,244	Overmerit.	The men are a rough lot to deal with
22,245	Overmix .	**Route**
22,246	Overmixing	Will be the easier route
22,247	Overmoist	Whichever is the best route
22,248	Overnice .	Go by the shortest route
22,249	Overnicely	Is the best route to take
22,250	Overoffice.	By the quickest route possible
22,251	Overpaint.	The route we shall take will be by
22,252	Overparted	**Rubber**
22,253	Overpeople	Feet of rubber belting —— inches wide
22,254	Overplease	Rubber packing
22,255	Overply .	Rubber hose —— internal diameter
22,256	Overpoised	**Rubbish**
22,257	Overpolish	Are full of rubbish
22,258	Overpower	To clear out all the rubbish
22,259	Overprompt	Have had all the rubbish cleared out
22,260	Overquell .	On account of the shaft being full of rubbish
22,261	Overreach	We are filling in the old workings with rubbish
22,262	Overready	**Rubies**
22,263	Overreckon	The rubies are found in
22,264	Overrefine	The finest rubies
22,265	Overrides .	Are there any rubies in the district
22,266	Overripen.	Rubies are found, but small and poor
22,267	Overruled.	**Ruby**
22,268	Overruling	The ruby mines are situated
22,269	Oversail .	You will have to go to the ruby mines
22,270	Overscent.	Want(s) you to visit the ruby mines of
22,271	Overseers .	To work a ruby mine
22,272	Overshadow	No ruby mine about here
22,273	Overshoot	The only ruby mine near here
22,274	Oversized .	**Ruin(s)**
22,275	Oversleep.	The buildings are in ruins
22,276	Overslow .	The whole place is in ruins
22,277	Oversorrow	**Ruined**
22,278	Overspan .	Was ruined by
22,279	Overspread	Former owners were ruined
22,280	Overspring	Present owners nearly ruined
22,281	Overstock.	The undertaking has been ruined by
22,282	Overstored	Will be ruined unless
22,283	Overstrain	**Ruinous**
22,284	Overstrict .	The whole place is in a most ruinous condition
22,285	Oversubtle	Could only be done at a ruinous cost
22,286	Oversupply	It would be ruinous to attempt to
22,287	Oversure .	**Rumour(s)**
22,288	Overswayed	There is a rumour here that
22,289	Overswell .	Have not heard any rumour of it
22,290	Overswift .	Have heard a rumour that

No.	Code Word.	Rumour(s) (*continued*)
22,291	Overt . .	Is there any truth in the rumour that
22,292	Overtake .	There is no truth in the rumour (that)
22,293	Overtaking	**Run**
22,294	Overtempt	The run for the month
22,295	Overthrow	The run for the next month
22,296	Overthwart	A better run
22,297	Overtilt .	Not so good a run
22,298	Overtitled .	The mill has run —— days
22,299	Overtrade .	After a run of —— days
22,300	Overtrip .	The mill can run —— days
22,301	Overtrust .	What is average run of mill
22,302	Overturner	We shall run the mill
22,303	Overtwine	Run away
22,304	Overvalue .	Run up
22,305	Overview .	The price of shares has run down on account of
22,306	Overvote .	(To) run down the price
22,307	Overvoting	(To) run up the price
22,308	Overwash .	Has run up to
22,309	Overweak .	**Running**
22,310	Overweigh	Running —— stamps
22,311	Overwhelm	Running through
22,312	Overwisely	Mill running well
22,313	Overworked	Mill running badly
22,314	Overwrest .	Running day and night
22,315	Overyeared	Running through good ore
22,316	Overzeal .	Running through low-grade ore
22,317	Ovibos . .	Mill now running on high-grade ore
22,318	Ovicell . .	Mill now running on low-grade ore
22,319	Ovicular .	Mill now running on medium ore
22,320	Ovidian .	Mill now running on ore assaying
22,321	Oviduct .	Mill now running on ore from the
22,322	Ovipara .	Mill now running on selected ore
22,323	Oviparous .	We are now running through
22,324	Oviposit .	We are now running the mill on
22,325	Ovipositor	**Sack(s)**
22,326	Ovisac . .	Ore sacks
22,327	Ovoid . .	A sack sample assayed
22,328	Ovoidal .	Sacks of ore shipped (weighing)
22,329	Ovulation .	Sacks of concentrates
22,330	Ovulum .	Are now shipping —— sacks, weighing
22,331	Owenite .	**Safe**
22,332	Owlery . .	The workings are not safe
22,333	Owleyed .	It would not be safe to
22,334	Owllight .	Quite safe
22,335	Owner . .	Make everything safe
22,336	Ownership	Would it be safe to
22,337	Oxalamide	Workings have now been made quite safe
22,338	Oxalate .	Are they now quite safe
22,339	Oxalic . .	Not safe at present
22,340	Oxalideae .	Do not consider (it) them safe

No.	Code Word.	Safe (*continued*)
22,341	Oxaluria .	It is not safe to work (it)
22,342	Oxalyl . .	Everything has been made safe
22,343	Oxbird .	. Safely
22,344	Oxbiters .	Cannot be safely done
22,345	Oxbow . .	May be safely done
22,346	Oxeyed .	Can be safely worked
22,347	Oxfence .	Cannot be safely worked
22,348	Oxflies . .	Arrived here safely
22,349	Oxfly . .	Safety
22,350	Oxfoot . .	The safety of the men
22,351	Oxfordclay	The safety of the mine
22,352	Oxgang .	Safety lamps
22,353	Oxgoads .	Safety valve of the engine
22,354	Oxidates .	With perfect safety
22,355	Oxidating .	Can be done with perfect safety
22,356	Oxidation .	To ensure safety
22,357	Oxidizable	Said
22,358	Oxidizer .	It was said
22,359	Oxidulated	It has been said by
22,360	Oxlip . .	Something should be said about
22,361	Oxpecker .	What has been said (about)
22,362	Oxpith .	. Sail(s)
22,363	Oxreim .	Expect to sail on the
22,364	Oxstalls .	To sail not later than
22,365	Oxtongue .	To sail on or before
22,366	Oxycoccus	Expected to sail in a few days
22,367	Oxycrate .	Will sail for
22,368	Oxygen .	Cannot sail (before) (until)
22,369	Oxygenacid	Sailed
22,370	Oxygenator	Sailed in the
22,371	Oxygenize	Sailed yesterday
22,372	Oxygenous	Sailed to-day
22,373	Oxygonal .	Salary
22,374	Oxymel .	At a salary of
22,375	Oxymoron	Has (have) engaged —— at a salary of
22,376	Oxymuriate	Salary of —— per annum
22,377	Oxyopia .	At too high a salary
22,378	Oxyphony	What salary does —— get
22,379	Oxyrrhodin	On salary and commission
22,380	Oxysalt .	At a salary of —— and commission of —— per
22,381	Oxytone .	Not sufficient salary [cent.
22,382	Oxytonical	It is not a sufficient salary for a good man
22,383	Oxyuris .	Sale(s)
22,384	Oyez . .	Have completed the sale of
22,385	Oysterbed	The —— mines are for sale
22,386	Oysterling	Have heard the —— mine(s) will shortly be for sale ;
22,387	Oysters .	Have withdrawn the sale [can you do anything
22,388	Oysterwife	Have not completed sale
22,389	Ozokerite .	The property is again for sale
22,390	Ozonation	If the sale is effected

No.	Code Word.	Sale(s) (*continued*)
22,391	Ozone . .	If you hear of a good —— mine for sale for about
22,392	Ozonify .	Effected the sale of [—— telegraph me at once
22,393	Ozonifying	Have heard privately that the —— mines are for sale. Owner(s) would take —— What can you do? It is a great opportunity
22,394	Ozonometry	Will be for sale shortly
22,395	Pabouches	If you can effect a sale of
22,396	Pabular .	We authorize the sale of
22,397	Pabulation	The sale of —— at the price named
22,398	Pacable .	What was paid for last actual sale in
22,399	Pachacamac	Do you advise a sale of
22,400	Pachalic .	In treaty for the sale of
22,401	Pachana .	Sale under the order of the court
22,402	Pachyderm	Sale of the company's property
22,403	Pachyote .	Instructions for sale (of)
22,404	Pacifiable .	Conditions of sale
22,405	Pacifical .	A forced sale
22,406	Pacifying .	Compulsory sale under mortgage
22,407	Packages .	In the event of sale taking place
22,408	Packcloth.	Sale of —— interest in
22,409	Packduck .	The sale produced
22,410	Packers .	Sale of ores
22,411	Packetboat	Sale of concentrates
22,412	Packetday	Sale of granulations, etc.
22,413	Packetship	Sale of machinery
22,414	Packfong .	Loss on sale of
22,415	Packhorse	Sale of property to company will be assured
22,416	Packice .	**Saleable**
22,417	Packingawl	Is it (are they) saleable
22,418	Packingbox	Saleable at present (at)
22,419	Packload .	Is (are) not saleable
22,420	Packmen .	**Salt**
22,421	Packsaddle	Existence of salt at —— ft. has been proved by
22,422	Packsheet .	A salt deposit [boreholes put down
22,423	Packthread	A bed of salt
22,424	Packwares	No salt in the neighbourhood
22,425	Packway .	Can obtain salt easily
22,426	Pacouryava	Mill stopped for want of salt
22,427	Pactolian .	Supply of salt uncertain
22,428	Padalon .	Good supply of salt
22,429	Paddlebeam	There is salt at —— feet down
22,430	Paddlebox	Tons of salt now on hand
22,431	Paddlecock	Quantity of salt consumed
22,432	Paddled .	Difficult to get salt
22,433	Paddlefish	Salt is obtained from the salt-beds (of)
22,434	Paddlehole	Salt is very impure
22,435	Paddlewood	The salt from these beds is very good
22,436	Paddybird	The salt from these beds is very impure
22,437	Padlocked	Salt is very dear
22,438	Padlocking	Rock salt

No.	Code Word.	Salt (*continued*)
22,439	Padnag .	The salt lakes of
22,440	Padowpipe	No salt anywhere nearer than
22,441	Paduasoy .	Salt mines
22,442	Paederia .	**Salted**
22,443	Paganical .	There is no doubt the mine was salted
22,444	Paganish .	The mine had been salted
22,445	Paganize .	The samples were salted
22,446	Paganizing	Sure the mine had not been salted
22,447	Paganly .	Were not the samples salted
22,448	Pageantry .	To prevent their (its) being salted
22,449	Pageants .	Take every precaution that it is not salted
22,450	Pagehood .	**Same**
22,451	Pagellus .	Not the same
22,452	Paginal .	Same as
22,453	Pagination	Same as reported
22,454	Pagodas .	At the same time
22,455	Pagodite .	By the same
22,456	Paguma .	The same people have
22,457	Pagurian .	From the same
22,458	Paideutics	Much the same as
22,459	Paigle . .	About the same
22,460	Pailful . .	Same as before
22,461	Paillasse .	Same quality as
22,462	Painstaker	On the same terms as
22,463	Paintbox .	Not on the same terms
22,464	Painter .	In the same way
22,465	Pairroyal .	If it is not the same
22,466	Pairwise .	If it is all the same to
22,467	Palacecar .	Is it (are they) the same (as)
22,468	Palaces .	In other respects, the same (as)
22,469	Palaeaster	Can you get the same
22,470	Palaeogean	It is the same which we
22,471	Palaeolith .	It is (they are) not the same (as)
22,472	Palaeology	The same, in every respect, as
22,473	Palaeomys	It is the same property
22,474	Palaeophis	It cannot be the same
22,475	Palaeosaur	**Sample(s)**
22,476	Palaeozoic	The samples not to hand
22,477	Palagonite	An average sample assayed
22,478	Palamedea	Average sample from the dump gave
22,479	Palankas .	Impossible to sample the reserves of ore
22,480	Palapterix	A large sample from the whole width of the vein
22,481	Palatable .	An average sample
22,482	Palatial .	An average sample contained no gold
22,483	Palatinate .	Send us, as a sample, about —— lb. of
22,484	Palative .	Sold as sample
22,485	Palaver .	Another sample assayed
22,486	Palavering	Samples from this assayed
22,487	Paleale .	Wait for the assays of samples
22,488	Palebuck .	Send home samples of the ore

No.	Code Word.	Sample(s) *(continued)*
22,489	Paledead .	Samples from outcrop assay [and numbered
22,490	Paleeyed .	Send samples with localities and position marked
22,491	Palefence .	Samples of the tailings
22,492	Paleness .	Samples of concentrates
22,493	Palfrey . .	Samples of the ore (from)
22,494	Palicourea	Samples of the pulp from
22,495	Palimpsest	Samples of diamond drill cores
22,496	Palindrome	Send samples viâ
22,497	Palinodist	Send us some samples as quickly as possible
22,498	Palinody .	Samples, asked for, sent
22,499	Palisading	Samples will be sent
22,500	Palisander	Sealed samples
22,501	Paliurus .	Send some more samples (of)
22,502	Palladium .	The average of —— samples
22,503	Pallbearer . **Sampled**	
22,504	Pallholder	Have been sampled
22,505	Palliated .	Have been sampled, and assayed
22,506	Palliating .	When you have sampled
22,507	Palliative .	Have sampled the ores, and find them
22,508	Palliatory . **Sanction**	
22,509	Pallidly .	You have the sanction of the board
22,510	Pallidness	The board will not give their sanction to
22,511	Pallor . .	You have my (our) sanction to
22,512	Palmacite .	I (we) cannot give my (our) sanction to
22,513	Palmary .	Without the sanction (of)
22,514	Palmata .	With the sanction (of)
22,515	Palmately .	Do you sanction
22,516	Palmatifid . **Sanctioned**	
22,517	Palmbirds .	The court sanctioned the
22,518	Palmbutter	Sanctioned the proposed
22,519	Palmcats .	Has not been sanctioned by
22,520	Palmcolour **Sandstone**	
22,521	Palmelleae	In sandstone
22,522	Palmerworm	Micaceous sandstone
22,523	Palmettes .	Sandstone formation
22,524	Palmhoney **Sapphire(s)**	
22,525	Palmhouse	The sapphires are found in
22,526	Palmigrade	With the sapphires
22,527	Palming .	The finest sapphires
22,528	Palmipeds	Sapphires are said to be found
22,529	Palmisters	Sapphires and rubies found
22,530	Palmistry . **Satisfaction**	
22,531	Palmitic .	To (for) the satisfaction of
22,532	Palmkale .	Can get no satisfaction from
22,533	Palmoil .	Give (given) great satisfaction
22,534	Palmsugar	To our satisfaction
22,535	Palmsunday	(It) has given us the greatest satisfaction
22,536	Palmtree . **Satisfactorily**	
22,537	Palmwine .	Satisfactorily arranged
22,538	Palmworms	The whole thing has been satisfactorily carried out

No.	Code Word.	**Satisfactorily** (*continued*)
22,539	Palmyra .	Has not turned out satisfactorily
22,540	Palpable .	**Satisfactory**
22,541	Palpation .	It would be more satisfactory if
22,542	Palpebral .	Is (are) not satisfactory
22,543	Palpi . .	Is (are) most satisfactory
22,544	Palpicorn .	If you consider it satisfactory
22,545	Palpitate .	May be considered satisfactory
22,546	Palsical .	Satisfactory to both parties
22,547	Palsied .	Not satisfactory to either party
22,548	Palstave .	The most satisfactory arrangement
22,549	Palsy . .	We cannot consider (it) satisfactory
22,550	Palsying .	Not so satisfactory as we could wish
22,551	Palsywort .	**Satisfied**
22,552	Paltock .	Am (are) quite satisfied (with)
22,553	Paltrily .	Am (are) not satisfied (with)
22,554	Paltriness .	If you are satisfied (with)
22,555	Paludament	Is (are) satisfied (with
22,556	Paludina .	Is (are) not satisfied (with)
22,557	Paludinous	Is (are) —— satisfied (with)
22,558	Palustral .	Will not be satisfied (unless)
22,559	**Pampas** .	Requirements cannot be satisfied [results
22,560	Pampascat	Are not satisfied with promises ; want substantial
22,561	Pampering	Hope you (they) will be satisfied that we have done
22,562	Pamphila .	**Satisfy** [all we could
22,563	Pamphlets	In order to satisfy
22,564	Panabase .	We can satisfy you as to
22,565	Panacea .	**Saturday**
22,566	Panamahat	On Saturday
22,567	Pancakes .	Last Saturday
22,568	Panchway .	Next Saturday
22,569	Pancratian	Every Saturday
22,570	Pancreas .	Every other Saturday
22,571	Pancreatic	**Save(s)**
22,572	Pandarism	Could not save
22,573	Pandarized	Did not save
22,574	Pandect .	Will in future save
22,575	Pandemic.	To save the
22,576	Panderly .	Try and save
22,577	Pandowdy	How much do you save (on) (by)
22,578	Panegyrics	Save(s) more than
22,579	Panegyry .	Save(s) as much as
22,580	Panelling .	We can save
22,581	Panelsaw .	Does not save as much as
22,582	Panelwork	Impossible to save (the)
22,583	Pangenesis	Save(s) —— per cent.
22,584	Pangolin .	Do not save more than
22,585	Panicgrass	Do you save anything (by)
22,586	Paniculate	Do not save anything (by)
22,587	Panicum .	**Saved**
22,588	Panivorous	A great deal can be saved by

No.	Code Word.	Saved (*continued*)
22,589	Panmugs .	The amount saved
22,590	Panniered	The value of the —— saved
22,591	Pannierman	Per cent. is saved by
22,592	Panningout	There is nothing saved (by)
22,593	Pannose .	How much is saved by
22,594	Panomphean	**Saving**
22,595	Panophobia	Saving the expense of
22,596	Panoplied.	Will cause a great saving
22,597	Panoply .	By saving the
22,598	Panopticon	Are saving the
22,599	Panoramic	Saving —— per cent. of
22,600	Panorpa .	Have not been saving the
22,601	Panorpidae	Saving —— per month
22,602	Panotype .	Saving the cost of
22,603	Pansclavic.	Not much saving in
22,604	Pansies .	Now saving by the new process
22,605	Panslavism	Effecting a great saving in
22,606	Pansophy .	The saving that has been effected by
22,607	Panspermy	Saving in tailings
22,608	Pantacosm	A very little saving (in) (by)
22,609	Pantagamy	Do not think there would be any saving (in)
22,610	Pantagogue	What saving is there in
22,611	Pantagraph	A saving of about —— per ounce
22,612	Pantalets .	Claim(s) to effect a saving in
22,613	Pantaloons	Will probably effect a saving of
22,614	Pantamorph	**Saw**
22,615	Pantheism	Saw nothing of the
22,616	Pantheon .	Saw everything
22,617	Pantheress	From what I saw
22,618	Pantherine	If —— saw
22,619	Pantile. .	Circular saw
22,620	Pantomime	Frame saw for cutting balks
22,621	Pantonshoe	**Sawmill**
22,622	Pantophagy	There is a sawmill on the property
22,623	Pantopoda	We shall want a sawmill
22,624	Pants . .	Sawmill for cutting
22,625	Panyard .	Sawmill with engine
22,626	Panym. .	For want of a sawmill
22,627	Panzoism .	It would be advisable to erect a sawmill
22,628	Papalists .	**Sawn**
22,629	Papalizing	Sawn lumber
22,630	Papally .	Sawn logs
22,631	Paparchy .	Sawn timbers —— in. × —— in.
22,632	Papaverine	**Say(s)**
22,633	Papaverous	I could not say
22,634	Papawtrees	I should say that
22,635	Papayaceae	Say(s) that there is
22,636	Papboat .	They all say that
22,637	Paperbook	Would you say that
22,638	Paperclip .	It is difficult to say

No.	Code Word.	Say(s) *(continued)*
22,639	Papercoal .	It is impossible to say
22,640	Paperday .	Would say certainly that
22,641	Paperfaced	Say(s) that he (they) will
22,642	Paperknife	Say(s) that he (they) will not
22,643	Papermaker	Says that he
22,644	Papermills	If he says that he can
22,645	Papermoney	If he says that he cannot
22,646	Paperreed	**Scarce**
22,647	Paperruler	Provisions are very scarce
22,648	Papershade	Water is very scarce
22,649	Papescent	Wood is very scarce
22,650	Papess . .	Very scarce on account of
22,651	Papilio . .	Is (are) very scarce
22,652	Papillated	Are scarce and very dear
22,653	Papillous .	Becoming very scarce
22,654	Papion . .	**Scarcely**
22,655	Papistic .	Scarcely any to be found
22,656	Papistical .	Scarcely anything
22,657	Papized .	Scarcely time yet
22,658	Papoose .	Scarcely any —— to be found near
22,659	Pappous .	**Scarcity**
22,660	Pappy . .	Owing to the scarcity of
22,661	Papyrean .	A great scarcity of
22,662	Papyrus .	From the scarcity of
22,663	Parabola .	**Scattered**
22,664	Parabolist .	Very much scattered
22,665	Paraboloid	Scattered here and there
22,666	Parachrose	**Scheme**
22,667	Paracresol .	The scheme is to
22,668	Paradisaic	The scheme has been to
22,669	Paradoxal .	What is the scheme
22,670	Paraffin .	In carrying out this scheme
22,671	Paragoge .	What do you think of the scheme
22,672	Paragram .	Think favourably of the scheme
22,673	Paraiba .	There is a scheme on foot to
22,674	Paralepsy .	If the scheme is carried out
22,675	Parallax .	A very promising scheme
22,676	Parallelly .	Interested in the scheme
22,677	Paralogism	**Schist(s)**
22,678	Paralogize	Talc schists
22,679	Paralyse .	Hornblende schists
22,680	Paralysing	In schists
22,681	Paralytic .	Schist formation
22,682	Paramatta	**Screens**
22,683	Paramentos	Battery screens
22,684	Paramo .	Battery screens —— mesh (wire cloth)
22,685	Paramoudra	Battery screens, No. —— slot
22,686	Paramount	Revolving screens
22,687	Parathine .	**Scrip**
22,688	Paranut .	The scrip is held by

No.	Code Word.	Scrip (*continued*)
22,689	Paranymph	By whom is the scrip held
22,690	Parapegm.	Scrip to be sold at your discretion
22,691	Parapeted.	Scrip to be sold or held at your discretion
22,692	Paraph. .	Scrip must not be sold
22,693	Paraphonia	Scrip must be sold
22,694	Paraphrase	Send our scrip to
22,695	Paraphysis	Scrip only signed by one director
22,696	Paraplegia	Scrip not signed by secretary
22,697	Parapodium	Scrip is informal
22,698	Paraquet .	Scrip has been lost
22,699	Parasang .	Scrip has been found
22,700	Parasceve.	Examine scrip carefully, as many forged certificates
22,701	Parasite .	Scrip not endorsed [are in existence
22,702	Parasitism	Scrip was officially issued
22,703	Parasols .	What date was scrip officially issued
22,704	Parastata .	Bank will not advance more than —— against lodg-
22,705	Paratactic.	Keep the scrip [ments of scrip (as follows)
22,706	Parataxis .	Lodge our scrip in bank against overdraft of
22,707	Parathesis.	Will pay in London if scrip is delivered by you to
22,708	Paratomous	Deliver all our scrip to
22,709	Parboiled .	Draw on me at sight with scrip attached
22,710	Parboiling	Draw on us attaching scrip
22,711	Parbreaked	Bank instructed to cash draft if scrip is attached
22,712	Parbuckle	Send scrip at once
22,713	Parcelbook	**Sea**
22,714	Parceldeaf	It takes —— days to get to —— by sea
22,715	Parcelgilt .	The best way is to go by sea to
22,716	Parcelled .	The mine(s) is (are) situated —— miles from the
22,717	Parcelpoet	And then by sea to [seaport of
22,718	Parcelvan	The sea coast
22,719	Parcenary.	A sea voyage
22,720	Parceners.	All the way by sea
22,721	Parched .	The only way to get to the —— mine(s) is by sea
22,722	Parchments	To the sea [to —— and then
22,723	Pardalotus	**Seam(s)**
22,724	Pardon .	In seams of
22,725	Pardonable	Seams of quartz
22,726	Pardoneth	Showing sulphuret in the seams
22,727	Pardoning	On a thin seam
22,728	Paregmenon	On the seam
22,729	Paregoric .	Thin seams of
22,730	Parelcon .	There is a thin seam of
22,731	Parembole	There is a fine seam of
22,732	Parenchyma	There are some fine seams of
22,733	Parenetic .	**Search**
22,734	Parentage.	Will search
22,735	Parentally	To search for
22,736	Parenthood	A strict search
22,737	Parentless	You must search for
22,738	Parergies .	After a thorough search, found

No.	Code Word.	Search (*continued*)
22,739	Parergy .	The search was fruitless
22,740	Pargeter .	**Searching**
22,741	Pargework	Are now searching for
22,742	Parhelion .	Have been searching for the reef (or vein)
22,743	Pariahs .	Searching for the vein
22,744	Parietal .	**Season(s)**
22,745	Parietaria .	Is now the season (to)
22,746	Parietine .	Will be the season to
22,747	Parisblue .	The only season of the year when
22,748	Parisgreen	During the hot season
22,749	Parisred .	During the hot season there is not enough water
22,750	Paritors .	No change in the seasons
22,751	Parkia . .	Too late in the season
22,752	Parkkeeper	Too early in the season
22,753	Parkleaves	The season is unfavourable for
22,754	Parleyed .	During the winter season
22,755	Parleying .	During the winter season, the mill is shut down
22,756	Parliament	During the winter season, the smelting works at ——
22,757	Parlours .	**Second** [are closed
22,758	Parmelia .	Second-class ore
22,759	Parmesan .	Second telegram
22,760	Parnassia .	Second time
22,761	Parochial .	Second visit
22,762	Parodical .	**Secondary**
22,763	Parodists .	This is a secondary consideration
22,764	Parody . .	**Secret**
22,765	Parodying	Keep this quite secret
22,766	Paronomasy	A secret for the present
22,767	Paronychia	It is kept a profound secret
22,768	Paronyme .	It is no secret here
22,769	Paronymous	We think this should be kept quite secret
22,770	Parotitis .	**Secretary**
22,771	Paroxysm .	Will act as secretary (to)
22,772	Paroxysmal	Declines to act as secretary
22,773	Paroxytone	Who is the secretary (to)
22,774	Parquetage	The secretary of
22,775	Parqueted .	A new secretary
22,776	Parquetry .	Appoint a secretary
22,777	Parrakeets	Appointed secretary to
22,778	Parral . .	Without a secretary at present
22,779	Parrelrope	Have appointed —— as secretary
22,780	Parrhesia .	**Section(s)**
22,781	Parricidal .	Sections of the mine
22,782	Parrotcoal	A section of
22,783	Parroted .	In this section
22,784	Parrotfish .	Plan and section
22,785	Parroting .	With plans and sections
22,786	Parseeism	Plan received, but no section
22,787	Parsimony	Send a section of the mine
22,788	Parsley .	Made in sections not exceeding

No.	Code Word.	Section(s) (*continued*)
22,789	Parsnips .	Made in sections for mule transport
22,790	Parsonbird	If made in sections, extra cost will be
22,791	Parsonic .	No section must exceed —— in weight
22,792	Parsonish .	Weight of each section
22,793	Partake .	In —— sections
22,794	Partaking .	Section of the mine, showing the
22,795	Partheniad	Longitudinal section of
22,796	Parthenope	Horizontal section of
22,797	Partialist .	Cross section of
22,798	Partially .	Vertical section of
22,799	Partibus .	Indicate the square on the section
22,800	Participle .	Square on section marked
22,801	Particular .	See square —— on section
22,802	Parting .	The locality is marked on the section
22,803	Partisans .	Have marked on the section
22,804	Partitive .	**Secure(s)**
22,805	Partnered .	Secure all you can
22,806	Partowner	You must secure control
22,807	Partridge .	Can you secure
22,808	Partsong .	Secure all the time you can
22,809	Parturiate .	Shall secure all we can
22,810	Parturious	If we can secure
22,811	Partyfence	Hope to secure
22,812	Partygold .	Am trying to secure
22,813	Partyism .	Secure it immediately
22,814	Partyjury .	Cannot secure
22,815	Partyman .	To secure the services of
22,816	Partywalls	Cannot secure —— on reasonable terms
22,817	Parulis . .	Am endeavouring to secure a gold property at
22,818	Parvitude .	I can secure a gold property in
22,819	Parvity .	**Secured**
22,820	Pasch . .	Can be secured if deposit of —— paid at once
22,821	Paschal .	Has (have) secured
22,822	Paschalist .	Ought to have secured
22,823	Paschegg .	Has (have) not secured
22,824	Pascuage .	Can be secured
22,825	Pasigraphy	Cannot be secured
22,826	Pasilaly .	Must be well secured
22,827	Pasquil .	Cannot be better secured
22,828	Pasquiller .	Have secured a first-class gold property in
22,829	Pasquinade	Have secured a first-class
22,830	Passages .	Endeavouring to secure a first-class
22,831	Passbook .	Are you secured
22,832	Passboxes .	We are fully secured
22,833	Passcheck	We are partially secured
22,834	Passengers	Has been secured for
22,835	Passerby .	Is (are) not secured
22,836	Passerine .	Think we ought to be secured against
22,837	Passeth .	Secured from further trouble
22,838	Passholder	Have secured a bond upon

No.	Code Word.	Secured (*continued*)
22,839	Passiflora .	Have secured the services of
22,840	Passion .	Security
22,841	Passionary	As a security
22,842	Passionist .	It is good security for
22,843	Passivity .	No security
22,844	Passkey .	No security against
22,845	Passman .	What security have you (that)
22,846	Passover .	The only security
22,847	Passparole	The security is good
22,848	Passport .	What security can —— offer
22,849	Passwords	Offer(s) as security
22,850	Pasteboard	Can —— give security for
22,851	Pastime .	Cannot accept the security offered
22,852	Pastiming .	Must have good security
22,853	Pastinaca .	Want better security
22,854	Pastor . .	Want first-class security
22,855	Pastoral .	Cannot do it without good security
22,856	Pastorless .	If no better security can be got
22,857	Pastorlike .	See(s)
22,858	Pastorly .	You must at once see (to)
22,859	Pastorship	I (we) could not see
22,860	Pastrycook	Am (are) to see
22,861	Pastrymen	See about
22,862	Pasturable	See about this at once
22,863	Pasturage .	See —— immediately
22,864	Patacoon .	See if it is all right
22,865	Patagium .	See what is wrong
22,866	Patavinity .	See what you can arrange
22,867	Patchedly .	If you can see your way
22,868	Patchouli .	You must at once see —— personally, and inform
22,869	Patchwork	See —— at once with regard to {him that
22,870	Patellidae .	Will see to it at once
22,871	Patentable	Have not been able to see
22,872	Patented .	Can you see
22,873	Patenting .	Do you see your way to
22,874	Paternal .	Cannot see my (our) way
22,875	Paternally .	You will see that
22,876	Pathetical .	When you see
22,877	Pathetism .	Have let him (them) see that
22,878	Pathflies .	Seem(s)
22,879	Pathfly . .	It seems to me (us) that
22,880	Pathnage .	Seems to be giving out
22,881	Pathogeny	Seems to be falling off
22,882	Pathognomy	There seems to be
22,883	Pathology .	This seems
22,884	Pathometry	This does not seem
22,885	Pathopoeia	Seem(s) uncertain whether to
22,886	Pathos . .	Seen
22,887	Pathways .	Have you seen
22,888	Patibulary	Has (have) seen

No.	Code Word.	Seen (*continued*)
22,889	Patient .	Has (have) not seen
22,890	Patiently .	Has (have) not yet seen
22,891	Patoncee .	As soon as —— has (have) seen
22,892	Patriarch .	Can be seen
22,893	Patricians .	Cannot be seen
22,894	Patriot . .	Is (are) said to have been seen
22,895	Patriotism.	We do not think any have been seen
22,896	Patrolled .	If any had been seen, we should have heard
22,897	Patrolling .	Gold has been seen
22,898	Patronage .	Has (have) been seen
22,899	Patroness .	From what I have seen
22,900	Patronized	Which has been seen
22,901	Patronymic	Where it was (has been) seen
22,902	Pattemar .	Has been seen in considerable quantities
22,903	Pattypan .	**Seize**
22,904	Pauciloquy	Threaten to seize
22,905	Pauhaugen	Attempted to seize
22,906	Paulianist .	Can seize under the judgment
22,907	Paulician .	Seize the whole of
22,908	Paunchmat	**Seized**
22,909	Paunchy .	Has been seized
22,910	Pauper . .	Has been seized under the judgment
22,911	Pauperism	Seized by order of the court
22,912	Pauperized	Seized by the Government
22,913	Pauropoda	Seized by the custom-house officials
22,914	Paused . .	Seized by the mortgagee
22,915	Pauxi . .	Seized by the creditors
22,916	Pavache .	Seized for the non-payment of
22,917	Pavements	Will be seized if we do not
22,918	Pavesade .	Cannot be seized
22,919	Paviage .	Any risk of being seized
22,920	Pavilions .	Liable to be seized (for)
22,921	Paviors .	Illegally seized
22,922	Pawing .	The whole has been seized
22,923	Pawlbitt .	**Seizure**
22,924	Pawlpost .	The seizure was illegal
22,925	Pawnable .	(To) set aside the seizure
22,926	Pawnbroker	The seizure of the property
22,927	Pawnticket	To avoid seizure
22,928	Paxboards	Have succeeded in setting aside the seizure
22,929	Paxillose .	**Select**
22,930	Paxwax .	Select a good man to
22,931	Payclerk .	Will select
22,932	Paylist . .	Did not select
22,933	Paymaster	You could not select a better
22,934	Payoffice .	Select some of the best
22,935	Payroll . .	**Selected**
22,936	Paysa . .	Selected ore
22,937	Peabeetle .	Selected samples
22,938	Peabugs .	Has (have) selected

No.	Code Word.	Selected (*continued*)
22,939	Peaceably .	Has (have) been selected
22,940	Peaceful .	Was selected
22,941	Peacefully .	Was (were) not selected
22,942	Peacemaker	Had been selected
22,943	Peaceparty	Had not been selected
22,944	Peachafers	Have selected the best
22,945	Peachdown	Have selected some of the best
22,946	Peachick .	**Selection**
22,947	Peachtree .	**Sell**
22,948	Peachwood	On no account sell
22,949	Peacoat .	I (we) have agreed to sell
22,950	Peacocked	Has (have) agreed to sell
22,951	Peacocking	Intend(s) to sell his (their) interest in
22,952	Peacrab .	In order to sell the stock
22,953	Peadove .	Is (are) trying to sell
22,954	Peafowl .	It would be better to sell
22,955	Peagrit. .	Do not sell
22,956	Peagun .	Can you sell
22,957	Peajacket .	If you can sell
22,958	Peakish .	If we now sell
22,959	Peamaggot	If you cannot sell
22,960	Peanisms .	I (we) can sell
22,961	Pearguage	I (we) cannot sell
22,962	Pearifle .	Shall I (we) sell
22,963	Pearl . .	At what price can you sell
22,964	Pearlash .	At what price may I (we) sell
22,965	Pearledge .	You may sell at
22,966	Pearlgrass .	Can you sell any more
22,967	Pearliness .	Can you sell at a fair profit
22,968	Pearlmoth .	We can sell at a profit; shall we do so
22,969	Pearlplant .	Do not sell until they reach
22,970	Pearlsago .	Can most probably sell
22,971	Pearlside .	Will most probably sell
22,972	Pearlspar .	Shall I (we) sell for your account
22,973	Pearlstone	Sell on my account
22,974	Pearlwhite	Sell when they reach
22,975	Pearlwort .	Sell at call best terms
22,976	Pearmain .	Sell the one half
22,977	Pearshaped	Sell the shares on our account
22,978	Peartrees .	Sell as much as you can at
22,979	Peasantry .	Telegraph if you will sell
22,980	Peascod .	Will probably sell better
22,981	Peashell .	Sell at
22,982	Peashooter	Sell at about
22,983	Peasoup .	Sell quickly
22,984	Peatmoss .	Sell quietly
22,985	Peatsoil .	Cannot sell at limit
22,986	Peaty . .	Do not sell any more
22,987	Peaweevil .	Advise you to sell out your holding
22,988	Pebbled .	Sell in your market to arrive at best prices

No.	Code Word.	**Sell** (*continued*)
22,989	Pebrine	I (we) advise you to sell
22,990	Peccadillo	Sell at your discretion
22,991	Peccancies	Sell in your market to arrive
22,992	Peccant .	I (we) advise you not to sell
22,993	Peccantly .	You had better sell ; market will decline
22,994	Peccavi .	Take advantage of present favourable moment to [sell
22,995	Pechurane	We now sell the ore at
22,996	Pecopteris	We now sell the concentrates (to)
22,997	Pecora . .	Sell quickly, prices will be lower
22,998	Pectinated	Sell as soon as you hear of
22,999	Pectized .	Sell at current price
23,000	Pectizing .	Sell by public auction
23,001	Pectorally.	Sell without reserve
23,002	Pectous .	Do you intend to sell
23,003	Peculate .	Sell to arrive
23,004	Peculating	Can sell more at same price
23,005	Peculation	**Sellers**
23,006	Peculiar .	What are sellers asking for
23,007	Peculiarly.	Advise sellers to register sale so as to avoid possible [liability
23,008	Pecuniary .	**Selling**
23,009	Pecunious.	Is (are) selling all his (their) stock
23,010	Pedagogism	Is (are) selling at
23,011	Pedagogued	Stop selling
23,012	Pedagogy .	Selling out
23,013	Pedalbass.	Do you advise selling
23,014	Pedality .	Keep selling
23,015	Pedalnote.	Who is (are) selling
23,016	Pedalorgan	Has (have) stopped selling
23,017	Pedaneous	Selling orders
23,018	Pedantical	Selling orders are in the way
23,019	Pedanticly	**Send**
23,020	Pedantism	Must send to —— (for the ——)
23,021	Pedantized	Have to send the ore to
23,022	Pedantry .	Will send
23,023	Pedatisect.	If he will send
23,024	Peddlery .	If you will send
23,025	Pederast .	If you can send
23,026	Pederastic	Send after deciphering to
23,027	Pedescript	Send the fullest particulars
23,028	Pedestal .	When can you send
23,029	Pedestrian	Can you send
23,030	Pedicelled	Can send
23,031	Pedigree .	Cannot send
23,032	Pediluvy .	Send by telegraph
23,033	Pedimane.	Send to the care of
23,034	Pedimanous	Send bullion
23,035	Pedimental	Do not send
23,036	Pedipalp .	Do not send until (unless)
23,037	Pedireme .	Where am I (are we) to send
23,038	Pedleress .	The best way is to send

No.	Code Word.	Send (*continued*)
23,039	Pedomancy	Will send you to-morrow
23,040	Pedometric	Send by steamer
23,041	Pedomotive	Send by express
23,042	Pedomotors	We shall not send (until)
23,043	Pedotrophy	Cannot send any at present
23,044	Peduncle .	Let us know when you intend to **send**
23,045	Peduncular	Send reply to my (our) last telegram
23,046	Peelhouse	It is impossible to send
23,047	Peeltower .	Can send either to —— or to
23,048	Peepbo .	Send as much as you can
23,049	Peeped .	Send at earliest opportunity
23,050	Peepholes.	Send by quickest route
23,051	Peepshow .	Send a few
23,052	Peerages .	If you will send
23,053	Peerdom .	If you cannot send
23,054	Peeresses .	Have to send the —— **to**
23,055	Peery . .	Send us a competent
23,056	Peevish .	Send as soon as possible
23,057	Peevishly .	When will they send
23,058	Peganum .	We now send to
23,059	Pegasean .	Send —— now, and the balance on
23,060	Peggingawl	Send as much as you can now, the rest can follow
23,061	Pegmatite .	**Sending**
23,062	Pegtankard	Am (are) now sending
23,063	Pegtops .	Are you sending (us)
23,064	Peirameter	Sending all we possibly can
23,065	Pejorative.	We are now sending the ores to
23,066	Pelagic .	We are now sending the concentrates **to**
23,067	Pelagosaur	We are now sending
23,068	Pelecanus.	Sending at first opportunity
23,069	Pelecoids .	Recommend sending
23,070	Pelf. . .	Instead of sending to
23,071	Pelioma .	We propose sending you
23,072	Pellagrins .	What are you sending
23,073	Pellicles .	Recommend sending a competent person
23,074	Pellmell .	**Sent**
23,075	Pellucid .	Have you sent
23,076	Pellucidly.	Has (have) sent (you)
23,077	Peloconite	Has (have) not sent
23,078	Pelorism .	Has (have) —— sent
23,079	Peltate. .	Will be sent
23,080	Peltately .	To be sent to
23,081	Pelted . .	Will be sent as soon as possible
23,082	Peltiform .	Competent person must be sent
23,083	Peltmonger	Competent person is sent
23,084	Peltocaris.	Sent by mail
23,085	Peltrot. .	Sent particulars by letter
23,086	Peltryware	Sent to the care of
23,087	Peltwool .	Was (were) sent
23,088	Pelvic . .	Was (were) sent without

No.	Code Word.	**Sent** (*continued*)
23,089	Pelvimeter	Must be sent to arrive not later than
23,090	Pemmican	Must be sent without fail
23,091	Pemphigus	All goods to be sent viâ
23,092	Penally .	Machinery to be sent viâ
23,093	Penalogist	If you have not yet sent
23,094	Penalties .	If they have not yet been sent
23,095	Penannular	Why have you not sent
23,096	Pencilling.	When will they be sent
23,097	Pencils .	**Separate**
23,098	Pencraft .	Keep the two reports separate
23,099	Pencutter .	Must be kept separate
23,100	Pendentive	It is very difficult to separate
23,101	Pending .	Can you separate
23,102	Pendragon	Cannot separate
23,103	Pendule .	Are keeping them separate
23,104	Pendulous	Do you wish them to be kept separate
23,105	Pendulum.	**Separated**
23,106	Penelope .	Easily separated
23,107	Penetrable	Separated with difficulty
23,108	Penetrancy	Could not be separated
23,109	Penetrated	Separated sects —— classes
23,110	Penfish .	**Separation**
23,111	Penguinery	The separation of the ores
23,112	Penguins .	**Serious**
23,113	Penholder	A serious outbreak has occurred
23,114	Penible .	A serious outbreak is feared
23,115	Peninsula .	Has caused serious damage
23,116	Penitency.	Nothing serious
23,117	Penitent .	Likely to prove serious
23,118	Penknife .	Is it serious
23,119	Penman .	Is (are) in a very serious condition
23,120	Penmanship	Serious disturbances here in consequence of
23,121	Pennached	More serious than at first anticipated
23,122	Pennatula.	Less serious than at first anticipated
23,123	Penniless .	Not of a very serious character
23,124	Pennoncel	The matter is serious, and requires careful handling
23,125	Pennons .	The consequences may be serious
23,126	Pennycress	**Serpentine**
23,127	Pennydog.	In serpentine rock
23,128	Pennygaff.	**Serve**
23,129	Pennyroyal	Will serve
23,130	Pennywise	It will serve to
23,131	Pennyworth	It will serve as a warning
23,132	Penock .	Will serve us faithfully
23,133	Pensility .	Will not serve
23,134	Pension .	Will serve our purpose
23,135	Pensionary	It will not serve our interests
23,136	Pensive .	**Served**
23,137	Pensively .	Served our purpose
23,138	Penslides .	Served us faithfully

No.	Code Word.	**Served** (*continued*)
23,139	Penstocks .	Have been served with a writ
23,140	Pentachord	**Service(s)**
23,141	Pentacles .	Will be of no service
23,142	Pentadesma	Is of no service
23,143	Pentafid .	Will be of great service
23,144	Pentaglot .	To accept service
23,145	Pentagonal	Agent to accept service
23,146	Pentamera	Must appoint an agent to accept service
23,147	Pentamyron	Is of great service
23,148	Pentander.	No longer in the service of the company
23,149	Pentandria	In the service of
23,150	Pentapody	For service rendered
23,151	Pentaptote	What do you consider value of service rendered
23,152	Pentaptych	Consider value of service rendered to be
23,153	Pentarchy.	**Serviceable**
23,154	Pentaspast	Can be made serviceable
23,155	Pentastich	Cannot be made serviceable
23,156	Pentastyle	Can it be made serviceable
23,157	Pentateuch	It is quite serviceable
23,158	Pentecost .	More serviceable
23,159	Pentelican	Is not serviceable
23,160	Penthouses	**Set**
23,161	Pentremite	Has set in
23,162	Pentroof .	Winter has set in
23,163	Pentrough	Has not yet set in
23,164	Penult . .	To be set aside
23,165	Penultima.	Will be set aside
23,166	Penumbral	Have set aside
23,167	Penwiper .	A set of
23,168	Penwoman	A complete set of
23,169	Peonage .	Is set up
23,170	People. .	Will be set up
23,171	Peopling .	Cannot be set up
23,172	Pepper. .	**Settle**
23,173	Pepperbox	To settle
23,174	Peppercake	To settle the matter
23,175	Peppercorn	Can you settle (the)
23,176	Peppermint	Shall I (we) settle on these terms
23,177	Peppermoth	Try and settle the matter amicably
23,178	Peppernel.	Cannot settle
23,179	Peppertree	Will settle
23,180	Pepsine .	Will not settle
23,181	Peptics .	Hope to settle
23,182	Peracted .	Settle the matter, if possible, without going to law
23,183	Peracting .	**Settled**
23,184	Peracute .	Is it settled
23,185	Peragrated	Get it settled
23,186	Perceive .	It is settled
23,187	Perceiving	Not yet settled
23,188	Percentage	Has been settled

No.	Code Word.	**Settled** (*continued*)
23,189	Percept .	Has not been settled
23,190	Perceptive	Getting more settled
23,191	Perch . .	Has (have) settled to
23,192	Perchance	Have you settled
23,193	Perchloric	' Have you settled the contract
23,194	Percidae .	As soon as it is settled
23,195	Percipient	Telegraph what has been settled about
23,196	Percolated	Must have it all settled before
23,197	Perculaced	It was settled before I left (that)
23,198	Percurrent	Must have everything settled
23,199	Percuss .	The dispute will have to be settled by
23,200	Percussing	No prospect of being settled
23,201	Perdifoil .	Cannot be settled as long as
23,202	Perdition.	Will not be settled until
23,203	Perdix. .	The matter can be at once settled (if)
23,204	Perdurance	When once the matter is settled
23,205	Peregrine.	Have settled the terms upon which
23,206	Perempt .	Has been satisfactorily settled
23,207	Peremption	Do not think it can be settled without
23,208	Peremptory	It will have to be settled
23,209	Perendure	**Settlement**
23,210	Perennity .	A settlement of affairs
23,211	Perfect. .	As soon as a settlement has been arranged
23,212	Perfecting.	No settlement
23,213	Perfection.	The settlement of
23,214	Perfectly .	Unless settlement be made by
23,215	Perfervid .	No prospect of a settlement
23,216	Perfidious.	In settlement of claim
23,217	Perfidy .	Have made a settlement with
23,218	Perflable .	Have you made any settlement with
23,219	Perflating .	Do not make any settlement without
23,220	Perfoliate .	**Several**
23,221	Perforator.	It will be several days before
23,222	Perforce .	It will take several weeks
23,223	Perforcing	In several places
23,224	Perfricate .	From several people
23,225	Perfumed .	**Shaft(s).** (See also Sink.)
23,226	Perfuming	Shaft has reached a depth of —— feet
23,227	Pergola .	Shaft is repaired
23,228	Pergunnah	Shaft is being repaired
23,229	Perhaps .	Shaft is timbered, station opened
23,230	Periagua .	Main shaft is down
23,231	Perianth .	The first thing is to sink a vertical shaft
23,232	Periastral .	The shaft has been sunk —— feet
23,233	Peribolos .	Has been prospected by a shaft, to the depth of
23,234	Pericardic.	In the shaft [—— feet
23,235	Perichete .	From the shaft
23,236	Periclase .	Connections must be made to the shaft
23,237	Periclinal .	The vein has been extensively worked by shafts
23,238	Pericope .	At the bottom of the shaft

No.	Code Word.	**Shaft(s)** (*continued*)
23,239	Periderm .	Has (have) seen all the shafts free of water
23,240	Peridiolum	In the bottom of the deepest shaft
23,241	Peridrome	Lack of funds compelled stopping the shaft at
23,242	Perigean .	The shaft must be continued another —— feet
23,243	Perigonium	There are —— shafts on the mine
23,244	Perigord .	There are —— shafts sunk on the vein, varying in
23,245	Perigynous	Old shaft [depth from —— feet to —— feet
23,246	Perihelium	New shaft
23,247	Perilled .	New shaft will be sunk
23,248	Perilous .	Shaft has cut the vein
23,249	Perilously .	In the main shaft
23,250	Perilymph	Sinking shaft
23,251	Perimorph	Inclined shaft
23,252	Period . .	Shaft(s) full of water
23,253	Periodic .	Shaft(s) free of water
23,254	Periodical .	The shaft is on the incline of the vein
23,255	Periosteal .	The shaft is sunk at an angle of ——$^{\circ}$
23,256	Peripetia .	The shaft is perpendicular
23,257	Peripheral	To connect with the shaft
23,258	Periphery .	**Shall**
23,259	Periplast .	Shall I (we)
23,260	Periploca .	When shall you
23,261	Peripteros	I (we) shall not
23,262	Peripyrist .	What shall I (we) do
23,263	Perisarc .	Think we shall
23,264	Periscopic .	Do not think we shall
23,265	Perisheth .	They shall not
23,266	Perisperm .	We shall be able to
23,267	Peristyles .	Shall not be able to
23,268	Peritropal .	Let us know if we shall
23,269	Periwig .	Shall we go on with
23,270	Periwigged	**Share(s)**
23,271	Periwinkle	To share
23,272	Perjenete .	Not to share
23,273	Perjurious	No demand for the shares
23,274	Perjury .	Great demand for the shares
23,275	Perkinism .	Provided the shares are pooled
23,276	Perky . .	How many shares stand in the name of
23,277	Permanable	Shares stand in the name of
23,278	Permanency	No shares stand in the name of
23,279	Permanent	The shares are being largely dealt in
23,280	Permeably	The greater portion of the shares
23,281	Permeated	The greater portion of the shares are held by
23,282	Permiss .	Have disposed of —— shares at —— per share
23,283	Permitting	All shares taken up
23,284	Permixtion	Per share
23,285	Permutably	Who has (have) the largest share in the mine
23,286	Pernicity .	Buy —— shares of
23,287	Perorate .	Shares taken up
23,288	Perorating	Please buy following shares immediately

No.	Code Word.	Share(s) *(continued)*
23,289	Peroration	I am (we are) not open to take shares in the
23,290	Peroxidize	Shares appropriated to you in the —— mine
23,291	Perpend .	Shall I (we) apply for shares for your account (in —— mine)? If so, please telegraph instructions
23,292	Perpetrate	Please apply for —— shares in the —— mine on my (our) account [account
23,293	Perpetual .	I (we) have obtained —— shares in —— for your
23,294	Perpetuity	I (we) have obtained —— shares of
23,295	Perplex .	I (we) have not been able to get any —— shares
23,296	Perplexeth	How many shares have you obtained
23,297	Perplexing	How many shares have you got rid of
23,298	Perplexly .	What is the present buying price per share
23,299	Perquisite .	What is the present selling price per share
23,300	Perruquier .	Present buying price per share is
23,301	Persecute .	Present selling price per share is
23,302	Perseides .	Shares are at —— premium
23,303	Persevered	Shares are at —— discount
23,304	Persicaria .	Shares are at par
23,305	Persiflage .	Sell my (our) shares (in ——)
23,306	Persifleur .	Do not part on any account with —— shares
23,307	Persimmon	Do not part with —— shares under
23,308	Persist . .	Do not buy —— shares above
23,309	Persistent .	Please sell following shares immediately
23,310	Persistive .	Get rid of —— shares at present market prices
23,311	Persolved .	Get rid of —— shares at any price
23,312	Persolving	Have sold —— shares
23,313	Personal .	Cannot sell —— shares
23,314	Personally	You have done well to sell the shares
23,315	Personator	Transfer —— shares to
23,316	Personify .	Transfer all shares to
23,317	Perspicacy	A call of —— per share has been made
23,318	Perspicil .	Please pay the call on my (our) shares
23,319	Perspirate .	To receive —— shares
23,320	Perspiring .	Shares as part payment
23,321	Perstringe .	Total share capital of company will be
23,322	Persuade .	Shares have been allotted to you
23,323	Persuading	Shares allotted to
23,324	Persuasive	Hold shares to my order
23,325	Persuasory	Register shares at once in our name
23,326	Pertaining	Register shares at once in name of
23,327	Perthite .	How many shares is —— on the register for
23,328	Pertinacy .	Shares eagerly sought after here
23,329	Pertingent	Shares not obtainable here
23,330	Pertness .	Shares obtainable here at par
23,331	Perturb .	Shares obtainable here at $\frac{1}{8}$ premium
23,332	Perturbate	Shares obtainable here at $\frac{1}{4}$ premium
23,333	Perturbing	Shares obtainable here at $\frac{3}{8}$ premium
23,334	Peruked .	Shares obtainable here at $\frac{1}{2}$ premium
23,335	Perusal .	Shares obtainable here at $\frac{5}{8}$ premium
23,336	Perusing .	Shares obtainable here at $\frac{3}{4}$ premium

No.	Code Word.	Share(s) *(continued)*
23,337	Pervaded .	Shares obtainable here at $\frac{7}{8}$ premium
23,338	Pervasion .	Shares quoted under par
23,339	Perverse .	Shares not obtainable at all
23,340	Perversely .	No market for shares here
23,341	Perversive	Better buy them, as they are certain to rise
23,342	Perverted .	Shall I buy —— shares
23,343	Perverting	Shall I sell —— shares
23,344	Pervially .	You can buy —— shares
23,345	Pervious .	You can sell —— shares
23,346	Peschito .	Shares falling rapidly
23,347	Peskilly .	Shares likely to fall
23,348	Pessary .	Has (have) sold —— shares and flooded market
23,349	Pessimists .	Has bought —— shares
23,350	Pessimize .	The certificates of (——) shares will be ready on
23,351	Pessomancy	What are —— shares worth
23,352	Pestering .	Shares are worth
23,353	Pesterous .	No —— shares sold for some time, do not know
23,354	Pestiduct .	Do not part with —— shares [value
23,355	Pestilence .	Have you any shares in [both in London and
23,356	Pestled .	Make arrangements that shares can be transferred
23,357	Petal . .	The share list opens
23,358	Petaliform	The share list closes
23,359	Petalism .	Arranged that shares can be transferred both in
23,360	Petaloid .	When does the share list open [London and
23,361	Petardier .	When shall we close the share list
23,362	Petasites .	Close your share list
23,363	Petaurist .	Only allot shares
23,364	Petechial .	Shares only allotted
23,365	Peterboat .	Provisional certificates have been issued for shares
23,366	Peterpence	Share certificates
23,367	Petiolary .	How many shares can you dispose of, and at what
23,368	Petioled .	Sell shares at any price, and wire result [price
23,369	Petiolule .	Shillings per share
23,370	Petitioned	Dollars per share
23,371	Petitory .	What price did the shares realize, and how many
23,372	Petiveria .	I can buy shares in the [did you sell
23,373	Petong . .	Get the brokers to cable to London for shares, as
23,374	Petrean .	Allotted shares [this will benefit us
23,375	Petrescent	We have only got applications from the public for
23,376	Petroleum	Shares have been applied for [—— shares
23,377	Petromyzon	Shares remain unapplied for
23,378	Petronel .	Shares were overapplied for
23,379	Petrosilex .	Subdivision of the shares
23,380	Pettichaps	Shareholders will be entitled to —— new share(s) for
23,381	Petticoat	The new shares will rank [one old share
23,382	Pettifog .	The new shares will rank for dividend *pari passu*
23,383	Pettish. .	Ordinary shares of —— each [with the old shares
23,384	Pettishly .	Preference shares, —— per cent., of —— each
23,385	Pettitoes .	Conducted altogether for share-gambling purposes, with no regard to the future

No.	Code Word.	Share(s) (*continued*)
23,386	Pettyrice .	The present low price of the shares is owing to market speculation, without any reference to their
23,387	Pettywhin .	Founder's shares [intrinsic value
23,388	Petulancy .	Please give permission by wire to buy on my own account shares in [shares in
23,389	Petulant .	You have permission to buy on your own account
23,390	Petulcity .	You have not permission to buy on your own account shares in
23,391	Petulcous .	Large demand has set in for —— shares
23,392	Petunia .	Shares are likely to be dearer
23,393	Petuntze .	Large transactions to-day in —— shares
23,394	Petzite . .	Small parcels to-day in —— shares
23,395	Peucedanin	Procure shares paid up, *ex div.*
23,396	Pewfellow .	Procure shares paid up, *cum div.*
23,397	Pewit . .	Procure shares paid up, *ex* rights
23,398	Pewitgull .	Procure shares paid up, *cum* rights
23,399	Pewopener	Advise buyers buy (shares) *cum* all rights
23,400	Pewterers .	Advise buyers buy (shares) only old issue
23,401	Pewtery .	Advise buyers buy (shares) only new issue
23,402	Peytrel. .	Shares fully paid up, and company registered with
23,403	Peziza . .	The new issue of shares [limited liability
23,404	Pezizoid .	Shares fully paid up ; company not yet registered
23,405	Pezophaps	Shares only —— paid up, but registered
23,406	Phacochere	Shares only —— paid up, but not yet registered
23,407	Phacoid .	Do you advise buying —— shares
23,408	Phacops .	Do you advise a sale of —— shares
23,409	Phaeton .	What are buyers offering for —— shares
23,410	Phagedena	Shares are expected to move —— (good authority)
23,411	Phalaena .	Thoroughly good ; shares are worth
23,412	Phalangium	Buy on my (our) account —— shares
23,413	Phalanxes	Buy on my (our) account —— shares at ——, my
23,414	Phalaris .	Expected fall in —— shares [extreme limit
23,415	Phalarope.	Buy —— shares in your market at best price, and draw through bank, attaching certificates
23,416	Phallic. .	Do not advise sale of —— shares
23,417	Phanerogam	Do not advise buying —— shares
23,418	Phansigar .	What are buyers offering for —— shares
23,419	Phantasm .	Hold the shares or sell at your discretion
23,420	Phantasmal	Buy —— shares at——, and I (we) will take a —— interest in the venture
23,421	Phantastry	Shares must not be sold without our authority
23,422	Phantoms .	Have bought on your account —— shares in —— at
23,423	Pharbitis .	Strongly recommend you to buy —— shares
23,424	Pharisaism	Suggest you buying for your own profit and risk
23,425	Pharisean .	The price of —— shares was —— here
23,426	Pharmacist	The price of —— shares was —— at —— days
23,427	Pharmacon	The price of —— shares was —— to arrive
23,428	Pharmacy.	The price of —— shares was —— to arrive in
23,429	Pharology.	Can you secure for me —— shares at —— limit
23,430	Pharyngeal	Can secure you —— shares at price named

No.	Code Word.	Share(s) (*continued*)
23,431	Pharynx .	Buy for me and hold for a rise —— shares
23,432	Phasma .	Suggest your buying joint risk and profit —— shares
23,433	Phasmidae	Will you join us (risk and profit being divided) in
23,434	Phatagin .	New shares for one old share [—— shares
23,435	Pheasant .	The shares not being delivered as per broker's note, I have refused to accept them
23,436	Pheesy .	We have declined to accept the shares because
23,437	Phenakism	Hold the shares pending further instructions
23,438	Phengite .	Send the shares here, drawing for the amount
23,439	Phenicine .	Is putting on screw and forcing shares into market
23,440	Phenicious	Errors in numbers on share scrip (or certificates)
23,441	Phenogamia	Disquieting rumours here re ——, shares likely to fall
23,442	Phenomenal	Shares have fallen in sympathy with
23,443	Phenylia .	Shares have not been delivered
23,444	Phial . .	Shares were not delivered till
23,445	Phialled .	Decline(s) to accept shares
23,446	Phialling .	Decline(s) to deliver shares
23,447	Phigalian .	Shares have been lodged in bank against overdraft
23,448	Philander .	Cannot get more than —— for (—— shares)
23,449	Philately .	Take over from —— the following share for me (us)
23,450	Philautie .	Sell at once for us —— shares, and apply sum
23,451	Philippic .	**Shareholder(s)** [realized to payment of
23,452	Phillipsia .	One of the largest shareholders in
23,453	Phillyrea .	A shareholder in
23,454	Philogyny .	The shareholders in
23,455	Philologue	One of the largest shareholders
23,456	Philomath	Is (are) not shareholders in
23,457	Philomela .	Shareholders are getting anxious
23,458	Philosophy	Meeting of shareholders
23,459	Phiz . .	Meeting of shareholders will be held
23,460	Phlebitis .	A meeting of the largest shareholders
23,461	Phlebolite .	The English shareholders
23,462	Phlebology	The colonial shareholders
23,463	Phlegm .	The American shareholders
23,464	Phlegmatic	The foreign shareholders
23,465	Phlegmon .	Some of the principal shareholders
23,466	Phloeum .	A committee of shareholders
23,467	Phlogiston	At a conference with some of the principal share-
23,468	Phlogotic .	List of shareholders [holders it was decided
23,469	Phlomis .	A list of shareholders will be sent
23,470	Phlorizin .	Must have a list of shareholders
23,471	Phlox . .	Mail us copy of list of registered shareholders
23,472	Phlyctena .	Is —— on register as a shareholder
23,473	Phoca . .	Shareholders are demanding a committee of inquiry
23,474	Phocacean	Shareholders will not agree to
23,475	Phocidae .	The shareholders are very despondent
23,476	Phoenix .	The shareholders are very angry
23,477	Pholadite .	The shareholders are very cheerful
23,478	Pholadomya	**Ship**
23,479	Phonation	Will ship

No.	Code Word.	**Ship** (*continued*)
23,480	Phonetical	Will not ship
23,481	Phonics .	What quantity of bullion can you ship
23,482	Phonogram	At what rate can you ship
23,483	Phonograph	Can you ship
23,484	Phonolites	How much shall you ship
23,485	Phonometer	If you can ship
23,486	Phonoscope	If you cannot ship
23,487	Phonotype	Ship all you can
23,488	Phormium	Have not been able to ship bullion
23,489	Phoronomy	Will have to ship the ore to
23,490	Phosgene .	Shall not be able to ship more
23,491	Phosphide	**Shipment(s)**
23,492	Phosphor .	Expedite the shipment
23,493	Phosphoric	Next shipment about the
23,494	Phosphuret	Can make no shipment till the
23,495	Photizite .	For the shipment
23,496	Photo . .	Now ready for shipment
23,497	Photogenic	Now in course of shipment
23,498	Photologic	Shipment of ores
23,499	Photophony	Shipment of high-grade ores
23,500	Photopsy .	Shipment of concentrates
23,501	Phragma .	Shipment to smelters
23,502	Phragmites	Statement of all shipments
23,503	Phrasebook	Shipment per first steamer
23,504	Phrased .	Shipment has been delayed by
23,505	Phraseless	In future shipments
23,506	Phraseman	**Shipped**
23,507	Phrasing .	There have been shipped [month
23,508	Phrenesiac	There have been shipped for the first half of the
23,509	Phrensy .	There have been shipped for the latter half of the
23,510	Phryganea	There have been shipped for the month [month
23,511	Phrygian .	In addition to what has been shipped
23,512	Phthisic .	How much ore have you shipped
23,513	Phthisical.	Tonnage and value of all ore shipped
23,514	Phthisicky.	There have been shipped —— tons of ore, of an
23,515	Phycomater	**Shipping** [estimated value of
23,516	Phylacter .	Shipping instructions
23,517	Phylarchy.	Telegraph shipping instructions
23,518	Phyletic .	Shipping dues and charges
23,519	Phyllary .	We are now shipping
23,520	Phyllised .	**Shoot.** (See Chute.)
23,521	Phyllising .	Ore shoot in mine
23,522	Phyllocyst	Ore shoot to mill
23,523	Phyllodium	The old ore shoot
23,524	Phyllogen .	The new ore shoot
23,525	Phylloid .	The ore is sent down a shoot
23,526	Phyllopod.	Timber shoot
23,527	Phyllotaxy	Had to make a new shoot
23,528	Phylloxera	Utilize(d) the shoot
23,529	Phyllula .	**Shore**

No.	Code Word.	Shore (*continued*)
23,530	Phyma . .	Has gone on shore
23,531	Physalite .	On shore off ——; cargo lost
23,532	Physconia .	Is reported on shore
23,533	Physeter .	Run on shore
23,534	Physic . .	Short
23,535	Physical .	Cannot complete in so short a time
23,536	Physically .	The time is too short
23,537	Physicians	Cannot be done in so short a time
23,538	Physicist .	Will be short of
23,539	Physicnuts	Short of money
23,540	Physiology	We are running short of
23,541	Physique .	Has (have) been selling short
23,542	Physnomy	Several have been selling short
23,543	Phyteuma .	If you run short of
23,544	Phytochimy	If you are short of money
23,545	Phytochlor	Shortly
23,546	Phytocrene	Will shortly be
23,547	Phytolacca	Shortly after
23,548	Phytophagy	Shortly before
23,549	Phytozoons	Will shortly have
23,550	Piacular .	If it cannot be done shortly
23,551	Piaculous .	Should
23,552	Pianette .	Should be
23,553	Pianists .	Should not be
23,554	Pianoforte	Should have been
23,555	Pianograph	Should not have been
23,556	Pianos . .	Should have done so (but)
23,557	Piassava .	Should not have done so (but)
23,558	Piazzian .	If you should
23,559	Pibcorn .	If you should not be able
23,560	Pibroch .	If I (we) should
23,561	Picamar .	If I (we) should not
23,562	Picaninny .	Should be done at once
23,563	Picaroons .	Should it
23,564	Picayune .	Should it not
23,565	Piccolo .	Should it not be
23,566	Pickaback	Should it not have been
23,567	Pickaxe .	It should (be)
23,568	Pickerbend	It should not
23,569	Pickerels .	It should not be
23,570	Pickeridge	Unless we should
23,571	Pickery .	Unless you should
23,572	Picklings .	Unless (he) they should
23,573	Picklock .	We should not have
23,574	Pickmeup .	When should
23,575	Pickpocket	Which should
23,576	Pickpurse .	Which should not
23,577	Pickthank .	Why should
23,578	Picktooth .	Why should we
23,579	Pickwick .	Why should we not

No.	Code Word.	**Should** (*continued*)
23,580	Picnicked .	Why should you
23,581	Picnicking	Why should you not
23,582	Picraena .	Why should he (they)
23,583	Picris . .	Why should he (they) not
23,584	Picromel .	You should
23,585	Picrophyll	You should not
23,586	Picrosmine	You should have
23,587	Picrotoxin	You should not have
23,588	Pictorial .	But if not, you should
23,589	Pictoric .	But if not, it should
23,590	Picturable	If so, it should be
23,591	Picture. .	If so, you should
23,592	Picturerod	**Show(s)**
23,593	Picturized .	You can show
23,594	Piecegoods	Do not show
23,595	Piecely. .	Shall I (we) show
23,596	Piecemeal .	Could show
23,597	Piecework	Could not show
23,598	Piedness .	To be able to show
23,599	Pieman .	Cannot show
23,600	Pieplant .	Did not show
23,601	Piepowder	This will show you
23,602	Pierceth .	Who will show you
23,603	Piercing .	Who will show me
23,604	Piercingly .	If ——— will not show
23,605	Pierglass .	Show him (them) my letter of
23,606	Pierian .	Show this to
23,607	Pierides .	This does not show
23,608	Piertable .	Would show
23,609	Pietistic .	There is a good show of
23,610	Pigacia .	The accounts show
23,611	Pigeonhole	If we can show
23,612	Pigeonpea.	If we can show a profit
23,613	Pigeonry .	Shows a profit of
23,614	Pigeonwood	Shows a loss of
23,615	Piggeries .	Shows a balance of
23,616	Pigheaded	**Showed**
23,617	Pigiron. .	Which showed at once
23,618	Pigmentary	This showed
23,619	Pigmentous	**Showing**
23,620	Pigmy . .	Showing plainly that
23,621	Pignerate .	Showing traces of
23,622	Pignuts .	Showing signs of
23,623	Pigpens .	Showing some improvement
23,624	Pigskin .	Not showing such good
23,625	Pigweed .	Upon his (their) showing
23,626	Pigwidgin .	**Shut**
23,627	Pikedevant	Shut down
23,628	Pikestaff .	Shut down for repairs
23,629	Pilastered .	Shut down ——— days for repairs

No.	Code Word.	Shut (*continued*)
23,630	Pilaws . .	Shut down on account of bad weather
23,631	Pilchard .	Shut down —— days on account of bad weather
23,632	Pilcrow .	The mine has been shut down (since)
23,633	Pilecap .	It is reported you have had to shut down
23,634	Pilecarpet.	Shall be compelled to shut down
23,635	Pileclamps	Have shut down the mill
23,636	Piledriver .	Must shut down —— if we cannot
23,637	Pilehoop .	Shut up
23,638	Pileiform .	Everything shut up
23,639	Pilentum .	Will have to shut up
23,640	Pileopsis .	**Sick**
23,641	Pileorhiza.	The men are sick from
23,642	Pileplank .	Constantly sick
23,643	Pileshoe .	Is sick, and unable to attend to business
23,644	Pilewort .	Is sick, and returns to
23,645	Pilfer . .	Is sick, and wishes to return
23,646	Pilfering .	Is sick with
23,647	Pilgarlick .	Sick leave granted
23,648	Pilgrim .	Am sick, and must return
23,649	Pilgrimage	**Sickness**
23,650	Pilidia . .	There is at present much sickness
23,651	Piliferous .	Owing to the prevailing sickness we have a diffi-
23,652	Pillaged .	**Side(s)** [culty in getting labour
23,653	Pillaging .	On the side
23,654	Pillarbox .	On the —— side
23,655	Pillarist .	From the side
23,656	Pillbeetle .	By the side
23,657	Pillboxes .	At the side
23,658	Pillez . .	From one side
23,659	Pillorized .	From the other side
23,660	Pillory . .	On which side
23,661	Pillorying .	On both sides
23,662	Pillowbear	**Sight**
23,663	Pillowlace.	There is no ore in sight
23,664	Pillowslip .	What quantity of ore is there in sight
23,665	Pilltile . .	Estimated amount of ore in sight
23,666	Pilose . .	In sight
23,667	Pilotage .	Not in sight
23,668	Pilotbird .	There is very little ore in sight
23,669	Pilotboats.	Draw at —— days' sight
23,670	Pilotcloth .	Do not draw at sight
23,671	Pilotfish .	Have drawn at —— days' sight
23,672	Pilotjack .	There are now —— tons of ore in sight
23,673	Pilotstar .	Tons of ore blocked out and in sight
23,674	Pilula . .	Bill at —— days' sight
23,675	Pilularia .	Do not lose sight of
23,676	Pilulous .	Have not lost sight of
23,677	Pilumnus .	Will not lose sight of
23,678	Pimelite .	Lost sight of
23,679	Pimelodus	Now in sight between the

No.	Code Word.	
23,680	Pimenta .	**Sign**
23,681	Pimgenet .	To sign
23,682	Pimpernels	Not to sign
23,683	Pimplike .	Will sign
23,684	Pimply. .	Will not sign
23,685	Pimpship .	Shall I (we) sign
23,686	Pinacloth .	Do not sign
23,687	Pinafore .	Do not sign any contract without
23,688	Pinax . .	Do not sign any agreement
23,689	Pinbouke .	Before you sign any agreement
23,690	Pinbuttock	To sign the agreement
23,691	Pincase .	Before —— sign(s) the agreement
23,692	Pinchbeck	Will not sign the contract without
23,693	Pinchfist .	If you sign
23,694	Pinchgut .	Sign nothing
23,695	Pinchingly	**Signature(s)**
23,696	Pinchpenny	Ready for signature
23,697	Pincushion	Signature must be certified by
23,698	Pinda . .	Requires signature
23,699	Pindaric .	Sending deed for your signature
23,700	Pindarical.	Signature guaranteed
23,701	Pindjajap .	Signature notarially attested
23,702	Pindrill .	**Signed**
23,703	Pindrop .	Was signed
23,704	Pindust .	Was not signed
23,705	Pineapple.	Has the agreement been signed
23,706	Pineasters.	The agreement was (has been) **signed and sealed**
23,707	Pinebarren	Could not be signed on account of
23,708	Pinebeetle	Will be signed
23,709	Pinechafer	Will not be signed
23,710	Pineclad .	Has been signed
23,711	Pinefinch .	Has not been signed
23,712	Pineful .	Have signed and returned
23,713	Pineknot .	Nothing has been signed
23,714	Pinemarten	Has (have) not yet signed the
23,715	Pinemast .	When you have signed
23,716	Pineoil .	Unless signed before
23,717	Pinestove .	Not signed by
23,718	Pinetree .	**Significant**
23,719	Pinfeather.	It is very significant
23,720	Pinguefied	**Signified**
23,721	Pinguefy .	Have signified —— intention
23,722	Pinguicula	Have you signified
23,723	Pinguitude	Have they signified
23,724	Pinholes .	Have not signified
23,725	Pinioned .	**Signify**
23,726	Pinionists .	It does not signify
23,727	Pinionwire	**Signing**
23,728	Pinites . .	Before signing
23,729	Pinkeyed .	Before signing the contract

No.	Code Word.	**Signing** (*continued*)
23,730	Pinkneedle	Before signing the agreement
23,731	Pinkroot .	After signing
23,732	Pinkstern .	Immediately upon signing
23,733	Pinmakers	**Silver**
23,734	Pinmoney.	Chloride of silver
23,735	Pinnacling	Native silver
23,736	Pinnated .	Ruby silver
23,737	Pinnatiped	Horn silver
23,738	Pinniform.	Sulphide of silver
23,739	Pinnigrada	The silver is found in
23,740	Pinnipedia	Pockets of silver ore
23,741	Pinnothere	Almost pure silver
23,742	Pinpoint .	Assays give —— ozs. of silver per ton
23,743	Pinrack .	Rich deposits of silver
23,744	Pintails .	Samples show immensely rich in silver
23,745	Pintle . .	Samples show but little silver
23,746	Pintpots .	A vein of silver ore
23,747	Pinworm .	There is no vein, the silver being all in pockets
23,748	Pioneering	Consisting of rich deposits of silver ore in the lime-
23,749	Pioneers .	A silver mine in [stone
23,750	Piophila .	The silver occurs in galena
23,751	Piosoca .	The silver occurs in carbonate of lead
23,752	Piously .	(At) the standard price of silver
23,753	Pipecase .	At 1 dollar 29 c. per oz. for silver
23,754	Pipefish .	(At) the present price of silver
23,755	Pipelayer .	At —— per oz. for silver
23,756	Pipelaying	The market price of silver is now
23,757	Pipemouth	Silver has advanced —— per oz.
23,758	Piperaceae	Silver has fallen —— per oz.
23,759	Piperic. .	Silver is advancing
23,760	Piperidge .	Silver is falling
23,761	Piperolls .	Richer in silver than gold
23,762	Pipestaple.	Not so rich in silver
23,763	Pipestick .	The loss of silver
23,764	Pipestone.	The loss of silver is considerable
23,765	Pipewine .	We estimate the loss of silver at
23,766	Pipingcrow	**Similar**
23,767	Pipings .	Similar in appearance
23,768	Pipistrel .	Not similar to
23,769	Pippinface	Is it similar to
23,770	Pipra . .	In a similar manner
23,771	Pipridae .	In a similar case
23,772	Piquancy .	Very similar to
23,773	Piquant .	**Simplest**
23,774	Pique . .	The simplest way (would be)
23,775	Piquework	**Simplify**
23,776	Piquing .	In order to simplify matters
23,777	Piracy . .	It would tend to simplify
23,778	Piramidig.	This will simplify
23,779	Pirate . .	Will this simplify

No.	Code Word.	**Simplify** (*continued*)
23,780	Piratical .	It would simplify matters, **if**
23,781	Pirating .	Simplify the method
23,782	Piraya . .	Simplify the proceedings
23,783	Pirouetted	**Simplifying**
23,784	Piscation .	By simplifying the accounts
23,785	Piscatory .	By simplifying the process
23,786	Piscidia .	**Simultaneously (with)**
23,787	Piscinal .	**Since**
23,788	Pismire .	Since I (we)
23,789	Pisolite .	Since it has (he has)
23,790	Pisophalt .	Since it is
23,791	Pisselaeum	Since you
23,792	Pistachio .	Since you wrote
23,793	Pistazite .	Since we last heard
23,794	Pistiaceae.	Since we last wrote
23,795	Pistillary .	Since we telegraphed
23,796	Pistolade .	Since we have
23,797	Pistolling .	Since the date of
23,798	Pistols . .	Since when
23,799	Pistonrod .	Since sending
23,800	Pisum . .	Since then
23,801	Pitapat . .	Since the receipt of
23,802	Pitchblack	But since
23,803	Pitchchain	But since you have
23,804	Pitchfield .	(It is) sometime since
23,805	Pitchfork .	It is now ——— since
23,806	Pitchopal .	How long since
23,807	Pitchpine .	Since (he) they
23,808	Pitchpot .	Since commencing
23,809	Pitchstone	Since starting
23,810	Pitchwheel	Not long since
23,811	Pitchy . .	**Sink**
23,812	Pitcoal . .	To sink
23,813	Pitfall . .	Will begin to sink the **shaft**
23,814	Pitfalling .	To sink a shaft
23,815	Pitframe .	Must sink the shaft
23,816	Pithecia .	Do not begin to sink
23,817	Pithecoid .	Since we began to sink
23,818	Pitiable .	Have begun to sink
23,819	Pitiedly .	Shall sink
23,820	Pitiful . .	As soon as ——— can sink the shaft
23,821	Pitifully .	**Sinking**
23,822	Pitiless .	Sinking at the rate of ——— feet per **day**
23,823	Pitilessly .	(To) continue sinking
23,824	Pitsaw . .	Will continue sinking
23,825	Pittacal .	Sinking the shaft
23,826	Pittance .	Sinking the incline shaft
23,827	Pittikins .	Have stopped sinking
23,828	Pituita . .	Am (are) now sinking
23,829	Pituitary .	Shall stop sinking (until)

No.	Code Word.	Sinking (*continued*)
23,830	Pitying . .	Resume sinking
23,831	Pityingly .	Have resumed sinking
23,832	Pityriasis .	A sinking fund
23,833	Pityroid .	To provide a sinking fund
23,834	Piuma . .	**Site(s)**
23,835	Pivot . .	An excellent site for
23,836	Pivotal . .	Mill site
23,837	Pivotguns .	The mill site —— miles from the mine
23,838	Pivotman .	Have selected a site for
23,839	Piwarrie .	The site of
23,840	Pixyled .	There is not a good site for
23,841	Pixyring .	**Situated**
23,842	Pixystool .	Situated as we are
23,843	Placards .	Situated as you have been
23,844	Placate .	How are you situated in respect to
23,845	Placating .	Situated —— miles north of
23,846	Placebo .	Situated —— miles south of
23,847	Placebrick	Situated —— miles east of
23,848	Placement	Situated —— miles west of
23,849	Placenta .	Much better situated than
23,850	Placentary	How is —— situated
23,851	Placeproud	**Situation**
23,852	Placidly .	What is the present situation
23,853	Placidness	The situation of the mines is
23,854	Placitum .	Owing to the bad situation
23,855	Placoderm	Owing to the favourable situation
23,856	Placoidian	Accept the situation
23,857	Plafond .	The financial situation
23,858	Plagal . .	The political situation
23,859	Plagiarism	The situation is
23,860	Plagiarize .	What is the situation
23,861	Plagiary .	**Size**
23,862	Plagiaulax	In size
23,863	Plagueful .	The vast size
23,864	Plagueless	What is the size of
23,865	Plaguemark	The size of
23,866	Plaguesore	What size do you want
23,867	Plaguespot	The size of this vein
23,868	Plaguing .	From the size and appearance of
23,869	Plaiding .	(What is) the largest size
23,870	Plainant .	(What is) the smallest size
23,871	Plainbacks	Owing to the large size of
23,872	Plainchant	**Slate**
23,873	Plainly . .	The formation is clay slate
23,874	Plainness .	Slate formation
23,875	Plainsong .	The hanging-wall is slate
23,876	Plaintful .	The foot-wall is slate
23,877	Plaintiffs .	The country rock is slate
23,878	Plaintive .	Compact slate
23,879	Plainwork .	We are now in slate

R

No.	Code Word.	
23,880	Plaited	**. Slightly**
23,881	Planarian .	Slightly better
23,882	Planarioid.	Slightly worse
23,883	Planched .	Slightly less
23,884	Planeguide	Slightly more
23,885	Planeiron .	Slightly built
23,886	Planera	**. Slip**
23,887	Planerhead	A considerable slip
23,888	Planertree	A slip has occurred
23,889	Planestock	Let the chance slip
23,890	Planetary	**. Slipped**
23,891	Planetule .	Slipped through (in)
23,892	Plangency	Slipped through our **hands**
23,893	Planimetry	Slipped in without being noticed
23,894	Planish	**. Slow**
23,895	Plankroad.	Our progress is very slow owing **to**
23,896	Planksheer	Will not be so slow
23,897	Planorbis .	Slow work
23,898	Plantable .	It is very slow work, as we have had **to**
23,899	Plantago .	Has been very slow on account of
23,900	Plantains .	Progress henceforth will not be so slow
23,901	Plantation	Progress has been very slow owing to the hardness
23,902	Plantcane.	**Slowly** [of the rock
23,903	Planticle .	Slowly at present, but shall make better progress as
23,904	Plantlet .	As slowly as [we get experience in the use of
23,905	Plantlouse	Is (are) now working slowly
23,906	Plashoot .	Too slowly to satisfy us
23,907	Plashwheel	**Sluice(s)**
23,908	Plashy . .	Sluice-boxes
23,909	Plasmatic .	**Small**
23,910	Plasmodium	The vein(s) is (are) **small**
23,911	Plastering	Too small
23,912	Plastery .	Although small
23,913	Plasticity .	Too small to justify
23,914	Plastron .	A small but well-defined **vein**
23,915	Plataazul .	A small loss
23,916	Platalea .	A small profit
23,917	Platanista.	As small as
23,918	Platanus .	As small as possible
23,919	Plataverde	It (there) is a very small
23,920	Plateaux	**. Smaller**
23,921	Plateglass .	Smaller than
23,922	Platelayer.	Very much smaller
23,923	Platemark.	**Smelt**
23,924	Platemetal	To smelt
23,925	Platerack .	Will pay to smelt
23,926	Platessa .	Have to smelt
23,927	Platform .	Not having to smelt
23,928	Platic . .	To smelt the ore
23,929	Platinized .	To smelt —— tons per **day**

No.	Code Word.	
23,930	Platinode .	**Smelted**
23,931	Platinous .	The ore has to be smelted
23,932	Platonical.	The ore will be smelted (at)
23,933	Platonism.	Concentrates will be smelted
23,934	Platymeter	Can only be smelted
23,935	Platyodon	Must be smelted
23,936	Platypus .	Sent to be smelted (at)
23,937	Platyrhina.	**Smelters**
23,938	Platysma .	Price of silver given by smelters
23,939	Platysomes	Price of gold given by smelters
23,940	Plaudit. .	Value by smelters' assays
23,941	Plauditory	Smelters and refiners
23,942	Plausible .	Treatment as smelters
23,943	Playactor .	Per cent. of gold allowed by smelters
23,944	Playbills .	Per cent. of silver allowed by smelters
23,945	Playbook .	Smelters' allow
23,946	Playday .	Smelters charges amount to
23,947	Playdebts .	**Smelting**
23,948	Players .	Smelting ore
23,949	Playfellow	A free-smelting ore
23,950	Playfully .	A free-smelting lead and silver ore
23,951	Playgames	For smelting the ore
23,952	Playgoer .	It is proposed to erect smelting works
23,953	Playgoing.	It will be advisable to erect smelting works
23,954	Playground	Will it be necessary to erect smelting works
23,955	Playhouse	Smelting works
23,956	Playless .	Would not pay for smelting
23,957	Playthings	Rich ore for smelting
23,958	Playtime .	Smelting ore, assaying
23,959	Playwright	The ore is suitable for smelting
23,960	Playwriter	Smelting —— tons per day
23,961	Pleading .	Smelting operations
23,962	Pleasance.	Smelting and refining
23,963	Pleased .	Smelting and refining works
23,964	Pleasedly .	The ore has to be sent to the smelting works at
23,965	Pleasurer .	The ore has been sent to the smelting works at
23,966	Pleasuring	Concentrates to be smelted
23,967	Plebeians .	Ore to be smelted
23,968	Plebeity .	Concentrates to be sent to the smelting works at
23,969	Plebiscite.	Concentrates have been sent to the smelting works at
23,970	Plebs . .	Smelting charges
23,971	Plectrum .	Loss in gold per cent. in smelting
23,972	Pledged .	Loss of silver per cent. in smelting
23,973	Pledgeless.	Smelting furnace
23,974	Pledgery .	It would cost —— to erect smelting works
23,975	Pleiades .	Cost of smelting
23,976	Plenarily .	Profit on smelting
23,977	Plenicorn .	Lead for smelting
23,978	Plenilunar	Flux(es) for smelting
23,979	Plenipos .	**Smoothly**

No.	Code Word.	Smoothly (*continued*)
23,980	Plenished .	The machinery is working very smoothly
23,981	Plenteous .	**Snow**
23,982	Plenties .	Mine cannot be examined on account of snow
23,983	Pleochroic	Have had a heavy fall of snow
23,984	Pleonasm .	On account of the deep snow
23,985	Pleonastic	Great quantities of snow have fallen
23,986	Plerophory	Snow —— feet deep
23,987	Plesiosaur.	Cannot work on account of snow
23,988	Plethora .	Cannot travel to —— on account of snow
23,989	Plethron .	Teams cannot haul ore on account of the snow
23,990	Pleuralgia.	Very little snow has yet fallen
23,991	Pleurisy .	Snow falling heavily ; roads blocked
23,992	Pleuritic .	Railroad blocked by deep snow
23,993	Pleurodont	Thaw set in—snow disappearing
23,994	Pleurotoma	Think snow will soon disappear ; shall then be
23,995	Plevin . .	As soon as the snow disappears [able to
23,996	Plexiform	Snow has stopped all transport
23,997	Pleximeter	If the snow continues
23,998	Pliancy .	Owing to the snow melting
23,999	Pliant . .	**Snowed**
24,000	Pliantness.	Are snowed up every winter for —— months
24,001	Plication	Snowed up
24,002	Pliers . .	Snowed up ; cannot get out
24,003	Plight . .	**Snowing**
24,004	Plighting .	Snowing heavily
24,005	Plinth . .	It has been snowing since
24,006	Pliocene .	**So**
24,007	Pliohippus	So as to
24,008	Pliosaurus.	So that
24,009	Plocaria ´ .	So as not to
24,010	Ploceus .	So as not to hinder
24,011	Plodding .	So far as
24,012	Ploddingly	So far as we have gone
24,013	Plotinist .	So long as
24,014	Plotproof .	So long as it lasts
24,015	Plotters .	So few
24,016	Plough. .	So little
24,017	Ploughable	So much—So very
24,018	Ploughalms	If it is so
24,019	Ploughboys	If it is not so
24,020	Plougheth .	It is not so
24,021	Ploughgate	**Soda**
24,022	Ploughhead	Nitrate of soda
24,023	Ploughing .	Sulphate of soda
24,024	Ploughiron	Carbonate of soda
24,025	Ploughshoe	**Sodium**
24,026	Ploughsock	Sodium amalgam
24,027	Ploughtail	**Soft**
24,028	Pluck . .	The rock is soft and easily worked
24,029	Plucketh .	Rock is so soft here that cost of mining will be low

No.	Code Word.	Soft (*continued*)
24,030	Plucking .	The ore is soft and does not need blasting
24,031	Plugrod .	The rock has become soft
24,032	Plumage .	In soft ground
24,033	Plumassary	**Softer**
24,034	Plumbbob	The rock is softer
24,035	Plumbeous	Running in softer ground
24,036	Plumbery .	Sinking in softer ground
24,037	Plumbic .	**Sold**
24,038	Plumblines	Have you sold
24,039	Plumbosite	If you have sold
24,040	Plumbroth	If you have not sold
24,041	Plumbrule	If not sold before
24,042	Plumcake .	Have sold
24,043	Plumealum	Have already sold
24,044	Plumebird	Had already sold
24,045	Plumeless .	Has (have) been sold
24,046	Plumemaker	Has (have) not been sold
24,047	Plumiped .	Have not sold
24,048	Plumists .	Must be sold
24,049	Plummet .	Must be sold at any price
24,050	Plumosity .	Not to be sold
24,051	Plumparmed	Must not be sold
24,052	Plumpfaced	Not yet sold
24,053	Plumpie .	Cannot be sold
24,054	Plumply .	Could not be sold
24,055	Plumpness	Have sold at your limits
24,056	Plumule .	Have sold above your limits
24,057	Plunder .	Could not be sold at your limits
24,058	Plunderage	Cannot be sold at any price
24,059	Plunderers	Will be sold by
24,060	Plungebath	Sold for
24,061	Plungeon .	Sold privately
24,062	Pluperfect	To be sold at once
24,063	Pluralism .	Telegraph what you have sold
24,064	Plurality .	Telegraph if you have sold
24,065	Pluralized .	Telegraph if he has (they have) sold
24,066	Pluries . .	Telegraph when —— was sold
24,067	Plusher .	Sold for account of
24,068	Plutarchy .	Sold for your account
24,069	Plutocracy	Sold for next account
24,070	Plutocrat .	Sold on joint account
24,071	Plutonian .	Sold as per instructions
24,072	Pluvious .	Sold at good prices
24,073	Plying . .	Sold at better prices
24,074	Pneumatic	Sold at low prices
24,075	Pneumology	Sold at lower prices
24,076	Pneumonia	Sold at sheriff's sale
24,077	Poachiness	Sold by private contract
24,078	Poaching .	Sold for delivery
24,079	Poachy .	Part have been sold

No.	Code Word.	Sold (*continued*)
24,080	Poacite .	Have been sold at a considerable advance
24,081	Pocanbush	Number of shares sold
24,082	Pochard .	Have been sold ; shall sell the rest
24,083	Pockarred	We have sold only
24,084	Pockbroken	Sold the remainder
24,085	Pocketbook	Sold the last at
24,086	Pocketflap	Will probably be sold
24,087	Pocketful .	Sold on speculative account
24,088	Pockethole	A large number have been sold
24,089	Pocketing .	We do not think they should be sold at present
24,090	Pocketlid .	Everything to be sold
24,091	Pockmarks	Sold by auction
24,092	Pockpitted	Will be sold by public auction
24,093	Pockwood	Has been sold by public auction
24,094	Poculent .	Sold before
24,095	Podagra .	Sold before your message arrived
24,096	Podagrical	Sold subject to confirmation by wire
24,097	Podagrous	Have not sold any
24,098	Podargus .	**Solvent**
24,099	Podauger .	Is (are) —— solvent
24,100	Podbit . .	Is (are) quite solvent
24,101	Podetium .	Considered to be quite solvent
24,102	Podgy . .	**Some**
24,103	Podiceps .	Some more
24,104	Podocarp .	Some of it
24,105	Podogynium	Some of them
24,106	Podology .	We shall want some (more)
24,107	Podoscaph	From some
24,108	Podosomata	In some places
24,109	Podosperm	In some ways
24,110	Poduridae	In some places we have found
24,111	Poematic .	By some
24,112	Poephaga .	Some means of
24,113	Poephagous	For some
24,114	Poetasters	Some one
24,115	Poetical .	Send some
24,116	Poetically .	Have sent some
24,117	Poeticules	Have found some
24,118	Poetized .	To find some one to
24,119	Poetizing .	Can you find some one to
24,120	Poetress .	If, in some way
24,121	Poetry . .	Of some importance
24,122	Poetship .	**Something**
24,123	Poetsucker	Something is wrong with
24,124	Poignancy	Something has occurred to
24,125	Poignant .	Unless something is done
24,126	Poikilitic .	Something must be done at once
24,127	Poinciana .	Something must be settled
24,128	Point . .	Cannot something be done (to)
24,129	Pointblank	**Some time (ago)**

No.	Code Word.	**Some time (ago)** (*continued*)
24,130	Pointedly .	For some time past
24,131	Pointlace .	Will take some time
24,132	Pointleted	Some time in the future
24,133	Pointsman	**Somewhat**
24,134	Poisonable	Somewhat better than
24,135	Poisonbulb	Somewhat less than
24,136	Poisoneth .	Somewhat more than
24,137	Poisonfang	**Soon**
24,138	Poisonful .	It is too soon
24,139	Poisonivy .	As soon as possible
24,140	Poisonous	As soon as
24,141	Poisonsome	Very soon
24,142	Poisontree	Not soon enough
24,143	Poisure .	Quite soon enough
24,144	Poitrel .	How soon
24,145	Poitrinal .	How soon can you
24,146	Pokeberry	How soon must I
24,147	Pokebonnet	How soon will you
24,148	Pokelokens	Will be too soon for
24,149	Poker . .	Not at all too soon for
24,150	Pokerish .	How soon could we have
24,151	Pokesleeve	How soon could you send
24,152	Pokeweed	We shall soon
24,153	Pokiest .	We shall soon have
24,154	Poking .	Will soon
24,155	Polacre .	**Sooner**
24,156	Polanisia .	Sooner than
24,157	Polarily .	Cannot be done sooner
24,158	Polaristic .	Can be done sooner (if)
24,159	Polarities .	The sooner the better
24,160	Polarized .	**Sorry**
24,161	Polarizing .	Sorry to learn that
24,162	Poldway .	I am (we are) very sorry
24,163	Poleaxes .	Very sorry to report
24,164	Poleclipt .	Is (are) very sorry that
24,165	Poledavy .	Very sorry you cannot
24,166	Polelathes	Very sorry to have such a bad report (or return)
24,167	Polemarch	Very sorry for your sake
24,168	Polemast .	Very sorry for all concerned
24,169	Polemical .	**Sort(s)**
24,170	Polemics .	(To) sort out the best
24,171	Polemonium	(To) sort the
24,172	Polemy .	Do not sort
24,173	Polestar .	Will sort
24,174	Polewards .	Cannot sort
24,175	Polewig .	**Sorting**
24,176	Policemen	Now sorting out
24,177	Policies .	**Sought**
24,178	Polishable	Sought for
24,179	Polished .	Much sought after

No.	Code Word.	**Sought** (*continued*)
24,180	Polishing .	We have sought in vain for
24,181	Polishment	**Source(s)**
24,182	Polite . .	From its source
24,183	Politeness.	From what source
24,184	Politician .	Traced to its source .
24,185	Politicly .	Have traced it to its source
24,186	Polka . .	The river, from its source
24,187	Pollarchy .	Appears to be the source from which
24,188	Pollarding	**South.** (See Table at end.)
24,189	Pollbook .	**Spare**
24,190	Pollclerk .	Can you spare
24,191	Pollenger .	How much can you spare
24,192	Pollenine .	We can spare
24,193	Pollentube	Cannot possibly spare more (than)
24,194	Pollevil .	If you can spare
24,195	Pollinctor .	As much as you can spare
24,196	Polliwigs .	**Special**
24,197	Pollmoney	Very special
24,198	Pollpick .	Not of special importance
24,199	Pollsilver .	Special information
24,200	Polltax. .	Special power for (to)
24,201	Polluted .	Special reason
24,202	Pollutedly.	A special case .
24,203	Polluting .	Special care
24,204	Pollutions.	For special purpose
24,205	Polo . .	Special reasons are given for
24,206	Polonaise .	Have prepared a special case
24,207	Polonian .	Preparing a special report
24,208	Poltfooted	Have prepared a special report
24,209	Poltronry .	Send a special report
24,210	Poltroon .	Am sending a special report
24,211	Polyadelph	**Specially**
24,212	Polyandry	Specially required for
24,213	Polyanthus	**Specie**
24,214	Polyatomic	Payable in specie
24,215	Polybasic .	Specie sent
24,216	Polycarpon	Specie will be sent
24,217	Polychord	Specie to be forwarded
24,218	Polychrest	Boxes of specie
24,219	Polychrome	Specie will be forwarded by the
24,220	Polyconic .	**Specification**
24,221	Polycracy.	In accordance with the specification
24,222	Polydipsia	Specification and drawings .
24,223	Polyedrous	Contract according to specification
24,224	Polyfoil .	According to specification and drawings
24,225	Polygala .	A complete specification
24,226	Polygaline	Specification is not complete
24,227	Polygamize	Have sent specification
24,228	Polygarchy	Received specification of —— from
24,229	Polygenist	Have you sent specification of .

No.	Code Word.	**Specification** (*continued*)
24,230	Polygeny .	Specification of engine
24,231	Polyglot .	Specification of mill
24,232	Polygonal.	Clause in the specification
24,233	Polygraph	**Specified**
24,234	Polygynia .	In specified terms
24,235	Polygynous	Within the specified time
24,236	Polyhalite.	Specified in the
24,237	Polyhedron	Specified in clause —— of the contract
24,238	Polyhistor.	As specified in the contract (or agreement)
24,239	Polyhymnia	If not otherwise specified
24,240	Polymath .	**Specify(ies)**
24,241	Polymathic	To specify
24,242	Polymerism	You must specify
24,243	Polymnite	Do not specify
24,244	Polymorphy	Does the —— specify
24,245	Polynemus	Will specify
24,246	Polynomial	The contract must specify
24,247	Polyodonta	The contract does not specify (any)
24,248	Polyonymy	The contract specifies
24,249	Polyoptrum	The agreement specifies
24,250	Polyorama	**Specimen(s)**
24,251	Polyparous	Have seen specimens of the
24,252	Polyphagia	Send home specimens of the
24,253	Polyphant.	From specimens
24,254	Polyphonic	No specimens
24,255	Polyphore	The specimens we saw (contained)
24,256	Polypier .	Specimens from the
24,257	Polypifera.'	Some very fine specimens
24,258	Polypodium	**Speculate**
24,259	Polypogon	To speculate
24,260	Polyporite	Do not speculate
24,261	Polypstock	**Speculating**
24,262	Polypterus	Is (are) speculating largely in
24,263	Polypus .	Has (have) been speculating
24,264	Polyscope.	Has (have) not been speculating
24,265	Polyspast .	Speculating, but not to any great extent
24,266	Polysperm	**Speculation**
24,267	Polystyle .	Too risky a speculation
24,268	Polytheism	As a speculation
24,269	Polytheize	A fair prospecting speculation
24,270	Polytomous	Bought for a speculation
24,271	Polytypage	Nothing but a speculation
24,272	Polytyping	Be very careful ; a great deal of speculation is going
24,273	Polyzoa .	Speculation turned out a success [on
24,274	Polyzoary .	Speculation turned out a loss
24,275	Polyzoons.	Will you join me (us) in the speculation
24,276	Pomace .	I (we) will join you in the speculation
24,277	Pomanders	I (we) cannot join you in the speculation
24,278	Pomatum .	I (we) consider this a good speculation
24,279	Pomecitron	**Speculative**

R 2

No.	Code Word.	Speculative (*continued*)
24,280	Pomeroy .	Project is wholly speculative
24,281	Pomeroyal	Project is wholly speculative; so far not a single thing known in the properties worth working
24,282	Pomewater	It is too speculative for us to touch
24,283	Pomey . .	**Speculators**
24,284	Pomfret .	Very few speculators in the market
24,285	Pomiferous	Speculators for the rise
24,286	Pommel .	Speculators for the fall
24,287	Pommelion	Speculators offering
24,288	Pommelled	In the hands of speculators
24,289	Pommelling	**Speed**
24,290	Pomoerium	At what speed are you running the
24,291	Pomotis .	Running at a speed of —— revolutions per minute
24,292	Pomp . .	Running at a speed of —— strokes per minute
24,293	Pompatic .	The highest speed at which (you) (we) can run
24,294	Pompelmous	The speed is too great
24,295	Pempelo .	At a very low rate of speed
24,296	Pompholyx	At a speed of
24,297	Pompillion	**Spend**
24,298	Pompion .	Do not spend so much
24,299	Pomposity	Do not spend more than
24,300	Pompous .	We shall have to spend
24,301	Pompously	You will have to spend
24,302	Pomum .	How much do you propose to spend
24,303	Ponderal .	Now spending at the rate of
24,304	Ponderance	**Spent**
24,305	Ponderers	Spent up to the present
24,306	Pondlily .	Spent more than
24,307	Pondperch	**Split**
24,308	Pondweeds	Much split up
24,309	Pongo . .	Vein splits here
24,310	Poniard .	Will split the difference with you
24,311	Poniarding	**Spread**
24,312	Ponibility .	Spread out
24,313	Pontedeira	Spread over —— months
24,314	Pontific .	Spread over a considerable area
24,315	Pontifical .	**Spreading (out)**
24,316	Pontlevis .	Spreading out over
24,317	Pontonier .	**Spring(s)**
24,318	Pontvolant	This spring
24,319	Poohpooh	Last spring
24,320	Poonawood	Next spring
24,321	Poorbox .	During the spring
24,322	Poorest .	Everything ready to start in the spring
24,323	Poorjohn .	Springs from—Springing from
24,324	Poorlaws .	**Stables**
24,325	Poorly . .	There are good stables for —— horses
24,326	Poorness .	The stables are out of repair
24,327	Poorrate .	Must enlarge the stables
24,328	Popcorn .	**Stage**

No.	Code Word.	**Stage** (*continued*)
24,329	Popdock .	Going on the stage from —— to
24,330	Popedom .	Stage as far as
24,331	Popehood.	Stage thence to
24,332	Popery. .	Please take —— seats in the stage leaving
24,333	Popguns .	**Stamp(s).** (See also Mill.)
24,334	Popinjay .	Must increase the number of stamps
24,335	Poplared .	There is work enough for —— stamps [stamps
24,336	Poplins .	Send the necessary money to put up —— more
24,337	Poplitaeus	Will send the money to put up —— more stamps
24,338	Popliteal .	When will the new stamps be working
24,339	Poplitic .	Shall be working with —— stamps by
24,340	Poppies .	Stamps of —— lbs. each
24,341	Poppyhead	Stamps of 750 lbs. each
24,342	Poppyoil .	Stamps of 850 lbs. each
24,343	Popshop .	Stamps of 950 lbs. each
24,344	Popularize	Stamps of 1000 lbs. each
24,345	Populators	Do you recommend putting up more stamps
24,346	Populosity	I (we) do not recommend putting up more stamps
24,347	Populous .	Stamps are now working on [at present
24,348	Porbeagle.	I (we) recommend putting up —— more stamps at
24,349	Porcated .	How many stamps of the —— mill [once
24,350	Porcelain .	We are employing —— stamps
24,351	Porcellio .	If more stamps are put up
24,352	Porcine .	The large reserves of ore would justify putting up
24,353	Porcupines	By putting up more stamps [more stamps
24,354	Poriferan .	The new stamps are working well
24,355	Poriform .	New stamps will be at work
24,356	Porism. .	Stamps hung up on account of
24,357	Porismatic	Shall have to hang up —— stamps
24,358	Poristical .	Have had to hang up —— stamps
24,359	Porkchop .	Propose to hang up —— stamps
24,360	Porkeater .	**Stand**
24,361	Porklings .	(To) stand
24,362	Porkpie .	(To) stand in the name of
24,363	Porosity .	How do matters stand
24,364	Porous. .	As matters now stand
24,365	Porously .	To take everything as it stands
24,366	Porousness	How do they stand
24,367	Porpentine	How does it stand
24,368	Porphurie .	Stand on our legal rights
24,369	Porphrya .	Stand on the rights acquired by
24,370	Porphyrize	From our standpoint
24,371	Porpoises .	Take our stand upon
24,372	Porraceous	Has (have) no ground to stand upon
24,373	Porridge .	They stand out for
24,374	Portals .	**Standard**
24,375	Portative .	The standard price of
24,376	Portbar .	Is the regular standard for
24,377	Portcannon	Up to the standard required
24,378	Portcluse .	**Standing**

No.	Code Word.	Standing (*continued*)
24,379	Portcrayon	Now standing
24,380	Portcullis .	Standing over
24,381	Portdues .	Standing in the name of
24,382	Portend .	**Standstill**
24,383	Portendeth	Everything is at a standstill
24,384	Portension	Is (are) at a standstill for want of
24,385	Portentous	**Start**
24,386	Porterage .	When will —— start
24,387	Porteress .	Will start
24,388	Porterly .	Do not start
24,389	Portfolio .	Cannot start
24,390	Portglave .	Hope to start
24,391	Portholes .	Expect to start milling on or about
24,392	Porthook .	Prepared to make a start
24,393	Portico .	Will start work in a few days
24,394	Portioned.	If we start now
24,395	Portionist .	If you start now
24,396	Portlast .	If you start on the
24,397	Portlid. .	When do you start
24,398	Portlifter .	Will start early in
24,399	Portliness.	Will start early on that day
24,400	Portmote .	When can you start for
24,401	Portpane .	Can start to examine —— mines on ——, but will await your letter containing full instructions sent [by mail
24,402	Portraits .	**Started**
24,403	Portray .	Has (have) started
24,404	Portrayal .	Has (have) not started
24,405	Portraying	Has (have) —— started
24,406	Portreeve .	The mill has been started
24,407	Portrule .	Will be started
24,408	Portsides .	**Starting**
24,409	Porttackle.	We are now starting
24,410	Portulaca .	**State**
24,411	Portvein .	In a very bad state
24,412	Porwigle .	I am (we are) able to state
24,413	Posaune .	Cannot state
24,414	Posingly .	Can state with certainty
24,415	Positing .	Cannot yet state with certainty
24,416	Positional.	I am (we are) unable to state
24,417	Positions .	State(s) positively that
24,418	Positive .	Is (are) in a very dangerous state
24,419	Positively .	In what state do you find
24,420	Positivism	You do not state
24,421	Possessed .	Please state exactly
24,422	Possessing	In the present state of the
24,423	Possessive	You must state
24,424	Possessory	Owing to the state of
24,425	Possibly .	Please state how you propose to
24,426	Postact .	**Stated**
24,427	Postanal .	As already stated

No.	Code Word.	**Stated** (*continued*)
24,428	Postbill .	As already stated in our letter of
24,429	Postboys .	As already stated in our cable of
24,430	Postcard .	It has been stated that
24,431	Postchaise	What was stated is not true
24,432	Postcoach.	**Statement(s)** [regard to
24,433	Postdating	I (we) can confirm ——'s statements except with
24,434	Postentry .	I (we) cannot confirm ——'s statements with regard
24,435	Posterior .	A statement has been made that [to
24,436	Posters	Please investigate what truth there is in the state-
24,437	Postexist .	Misguided by the statements made by [ment that
24,438	Postfacto .	——'s statements are incorrect and misleading
24,439	Postfixed .	Will draw up a statement
24,440	Postfixing .	Draw up a statement as soon as possible
24,441	Posthaste .	I (we) can confirm ——'s statements in every
24,442	Posthetomy	If you can confirm ——'s statements [respect
24,443	Posthorn .	Unless ——'s statements can be confirmed
24,444	Posthume .	Statement is correct
24,445	Posthumous	Statement is incorrect in some respects
24,446	Posticous .	Statement is entirely false
24,447	Posticum .	Send full statement
24,448	Postilion .	Send a revised statement
24,449	Postilize .	General statement
24,450	Postillate .	Official statement
24,451	Postique .	Cash statement
24,452	Postliminy	Dr. and Cr. statement
24,453	Postmarks	Statement of the company's position
24,454	Postmaster	Statement of claim
24,455	Postmortem	Statement of expenses incurred
24,456	Postnatus .	Statement of estimated
24,457	Postobit .	Statement of all ores shipped
24,458	Postoffice .	Statement you have sent
24,459	Postpaid .	Statement sent by
24,460	Postponeth	Statement made by
24,461	Postponing	With regard to statements
24,462	Postposit .	Statements to be verified
24,463	Postremote	Title and statements to be proved
24,464	Postroad .	Statements must be proved to your satisfaction
24,465	Postcript .	According to (or in accordance with) the statement
24,466	Postulated	The statements are erroneous in many particulars
24,467	Postured .	The statement is in most respects correct
24,468	Posturist .	We require another statement showing
24,469	Potamology	What further statements are required
24,470	Potargo .	Only want an approximate statement
24,471	Potash . .	The statement must be full and complete
24,472	Potassic .	Please send statement(s) of ore shipped
24,473	Potassium	Please send statement of ore milled
24,474	Potations .	Ore statements will be forwarded
24,475	Potato . .	Must have statement for prospectus
24,476	Potbellied.	Must add to your statement list of buildings, plant,
24,477	Potbelly .	A most misleading statement [and machinery

No.	Code Word.	
24,478	Potboiler .	**Station** [—— railway
24,479	Potelot .	The mines are situated at ——, a station on the
24,480	Potentates :	The mines are situated —— miles from ——, a
24,481	Potential .	Pump station [station on the —— Railway
24,482	Potently .	A station will shortly be opened —— miles from
24,483	Potentness	Miles from the nearest station [the mine
24,484	Poterium .	There is a station within
24,485	Pothangers	A station very near the mine
24,486	Potheen .	Now cutting station
24,487	Potherbs .	It will take about —— days to cut station
24,488	Potliquor .	As soon as we have cut station
24,489	Potluck .	Have cut station ; shall now go on sinking
24,490	Potmetal .	Station is now finished
24,491	Potpiece .	**Stationary**
24,492	Potplant .	Stationary engine
24,493	Potstone .	Driven by a stationary engine
24,494	Pottered .	**Stationed (at)**
24,495	Pottering .	Stationed at ——, about —— miles from here
24,496	Potternore	**Statistical**
24,497	Pottlepot .	Statistical abstract (of)
24,498	Pottybaker	**Statistics**
24,499	Potvaliant	From statistics which we have obtained we find that
24,500	Potwaller .	The statistics show that
24,501	Pouched .	Mining statistics
24,502	Pouchmouth	Statistics must be verified
24,503	Poulaine .	Can be proved by the statistics
24,504	Pouldavis .	Published statistics (show)
24,505	Poult . .	Statistics of the
24,506	Poulterer .	Statistics of the United States Mint
24,507	Poultice .	**Stay.** (See also Telegraph.)
24,508	Poulticing .	Would like to stay till
24,509	Poultry .	Can stay till
24,510	Poulverain	Cannot stay till
24,511	Pouncebox	Cannot stay any longer
24,512	Poundcake	Will not stay any longer
24,513	Pounders .	Stay as long as you can
24,514	Poundovert	Stay until you have further instructions
24,515	Poupeton .	Must stay till
24,516	Pouring .	Do not stay any longer
24,517	Pourparler	Stay until
24,518	Pourparty .	How long do you stay at
24,519	Pourpoint .	We shall have to stay
24,520	Poussetted	Stay till we have
24,521	Poverty .	Not stay unless
24,522	Powderbox	Stay proceedings
24,523	Powdercart	If we stay here
24,524	Powdereth	**Staying**
24,525	Powderhorn	Staying all proceedings
24,526	Powdering	Am (are) staying at
24,527	Powdermill	At what hotel will you be staying

No.	Code Word	
24,528	Powderpuff	**Steadily**
24,529	Powderroom	Steadily working (at)
24,530	Powdery .	Steadily working to repair
24,531	Powdike .	Steadily increasing
24,532	Powerful .	Steadily falling (or decreasing)
24,533	Powerfully	Think they will steadily
24,534	Powerpress	We have been steadily
24,535	Powldron .	**Steady**
24,536	Powwows .	Not steady enough
24,537	Poynado .	A good steady man
24,538	Poynette .	Will want good steady man as
24,539	Practician.	Making steady progress (with)
24,540	Practicks .	(With) a steady output
24,541	Practised .	**Steam**
24,542	Practising.	Expect to get steam up
24,543	Praecipe .	Shall not be able to get steam up until
24,544	Praecoces.	Cannot make sufficient steam in the boiler(s)
24,545	Praetexta .	Steam-pipes
24,546	Pragmatism	Steam-power
24,547	Prahu . .	Steam-power to drive
24,548	Prairial .	We recommend steam-power
24,549	Prairiedog	Steam will have to be carried
24,550	Prairiehen	Boiler does not supply sufficient steam for
24,551	Praisably .	Steam for the pans
24,552	Pram . .	Steam to work the pumps
24,553	Pranceth .	Steam to work the hoist
24,554	Prancing .	(Is) steam-power available
24,555	Pranks. .	There is not enough steam-power to drive
24,556	Praseolite.	With additional steam-power
24,557	Prasine .	Heated by steam
24,558	Prater . .	Driven by steam
24,559	Pratincole	**Steamer**
24,560	Pratique .	And thence by steamer to
24,561	Prattled .	There will soon be a daily steamer running to
24,562	Prattling .	A weekly steamer running to
24,563	Prayerbook	A —— steamer running to
24,564	Prayerful .	When will the steamer leave for
24,565	Prayerless.	The steamer(s) leave(s) for —— on
24,566	Prayingly .	The machinery will be sent by steamer from —— on
24,567	Preachers .	By what steamer will machinery be sent
24,568	Preachify .	To be sent by steamer to
24,569	Preachment	By the steamer which leaves
24,570	Preadamic	Sent by steamer leaving
24,571	Preamble .	**Steel**
24,572	Preambling	To be made of steel
24,573	Preappoint	Tool steel
24,574	Prebend .	Steel for drills
24,575	Prebendary	Steel rods
24,576	Precarium	Steel bars
24,577	Precaution	**Steps**

No.	Code Word.	**Steps** (*continued*)
24,578	Precedency	Take immediate steps
24,579	Precellent.	What steps have you taken with regard to
24,580	Preceptial.	Steps must be taken immediately (to)
24,581	Preceptory	Take such steps as you think best
24,582	Precessor .	Has (have) taken steps to
24,583	Precincts .	No steps have been taken to
24,584	Preciously	If steps are not taken without delay
24,585	Precipice .	Steps have been taken
24,586	Precisely .	Take no further steps at present
24,587	Precisian .	Has (have) taken no further steps
24,588	Preclair .	If he takes (they take) any steps
24,589	Precluding	Take steps to prevent
24,590	Preclusive.	Take steps to protect ourselves
24,591	Precociıy .	Have not yet taken any steps
24,592	Precompose	Refuse(s) to take any steps
24,593	Preconceit	Before taking any further steps you had better con-
24,594	Precondemn	**Sterling** [sult
24,595	Preconform	What is the rate of sterling exchange
24,596	Preconquer	The rate of sterling exchange is
24,597	Preconsign	Sterling exchanges—banker's sight
24,598	Precordial	Sterling exchanges—for banker's, 60 days
24,599	Precurse .	Sterling exchanges—for commercial sight
24,600	Precursory	Sterling exchanges—for commercial, 60 days
24,601	Predacean	Sterling exchanges—for commercial, 90 days
24,602	Predaceous	In £ sterling
24,603	Predate .	In £ sterling, at the rate of —— dollars per £
24,604	Predation .	In £ sterling, at the rate of —— rupees per £
24,605	Predecay .	**Stiffening**
24,606	Predeclare	Require(s) stiffening
24,607	Predefined	Are stiffening the
24,608	Predesign .	Prices are stiffening
24,609	Predestiny	**Still**
24,610	Predial .	There are still left
24,611	Predicable	We have still on hand
24,612	Predicant .	Still —— days
24,613	Predicted .	Still —— weeks
24,614	Predictive.	Still working at (on)
24,615	Predispose	Still sinking
24,616	Predoomed	Still driving
24,617	Predorsal .	Still unable to
24,618	Preemploy	Still without
24,619	Preexamine	Still in hand
24,620	Preexist .	What —— have you still on hand
24,621	Preface .	Do you still hold
24,622	Prefacing .	Are you still
24,623	Prefecture	If you can still
24,624	Prefer . .	We still hold
24,625	Preferably.	We still think that
24,626	Preference	We are still
24,627	Preferred .	**Stipulate** (for)

No.	Code Word.	
24,628	Preferring .	**Stipulated**
24,629	Prefident .	As stipulated in the agreement (or contract)
24,630	Prefigure .	As stipulated in clause —— of the contract
24,631	Prefinite .	It is stipulated by —— that
24,632	Prefixed .	It is not stipulated in the agreement
24,633	Prefixing .	It is stipulated that —— is to
24,634	Prefixion .	It must be stipulated that
24,635	Prefool .	**Stipulation(s)**
24,636	Pregaged .	(By) the stipulations of the contract
24,637	Preglacial .	(By) the stipulations of the agreement
24,638	Pregnable .	Insist(s) upon a stipulation that
24,639	Pregravate	Stipulations provide for
24,640	Pregustant	**Stock(s)**
24,641	Prehended	In stock
24,642	Prehensile	Stock(s) on hand
24,643	Prehensory	We have a stock of —— sufficient for
24,644	Prehnite .	What stock have you
24,645	Prejudge .	A fair stock of
24,646	Prejudging	A large stock of
24,647	Prejudical .	Only a small stock of
24,648	Prelacies .	Stocks are rising
24,649	Prelacy .	Stocks are falling
24,650	Prelateity .	The stock of —— is said to be
24,651	Prelatess .	Stocks are diminishing
24,652	Prelatical .	Stocks are accumulating
24,653	Prelatism .	Increase your stock of
24,654	Prelatized .	Purchase of stock
24,655	Prelector .	Have just bought a considerable stock of
24,656	Prelimit .	Our stock amounts to
24,657	Prelook .	If prices favourable, lay in a stock of
24,658	Prelooking	Can do with our present stock until
24,659	Prelude .	Do not allow your stock of —— to get too low
24,660	Preludial .	A necessary stock of
24,661	Preludious	A needlessly large stock of
24,662	Prelumbar	Reduce your stock (of)
24,663	Prelusory .	We are reducing our stock (of)
24,664	Premature	Buy(ing) to increase our stock while prices are low
24,665	Premediate	Stocks in excess of
24,666	Premerited	Stocks are not in excess of
24,667	Premising .	A large stock of —— but little demand for
24,668	Premiums .	There is a great demand for —— but little in stock
24,669	Premna .	Inform us that stocks are likely to
24,670	Premolar .	Have now filled up our stock
24,671	Premonish	A good stock of tools and all appliances
24,672	Premonitor	Stocks are low, and prices rule high
24,673	Premunite	Hold a considerable stock in reserve
24,674	Prenanthes	Advise buyers this stock is not safe to touch except with special guarantee of non-liability beyond
24,675	Prender .	This stock not safe to touch [purchase price
24,676	Prenomen	Thoroughly good stock and likely to rise

No.	Code Word.	**Stock(s)** (*continued*)
24,677	Prenostic .	Indifferent stock of only speculative value
24,678	Preoblige .	Stock Exchange Committee [Committee
24,679	Preobtain .	Case is being brought before the Stock Exchange
24,680	Preoccupy	Case decided against us by Stock Exchange Com-
24,681	Preominate	Stock has large industrial merits [mittee
24,682	Preopinion	Case decided in our favour by Stock Exchange
24,683	Preordered	**Stolen** [Committee
24,684	Prepaid .	Stolen in transit from the mine
24,685	Prepalatal.	Much of the rich ore is stolen
24,686	Preparable	Has (have) stolen
24,687	Preparance	Amalgam has been frequently stolen
24,688	Preparedly	Stolen amalgam
24,689	Preparers .	Stolen bullion
24,690	Prepay .	The stolen —— has been traced
24,691	Prepayment	Recover the stolen
24,692	Prepensing	A good deal of the missing —— was no doubt
24,693	Prepollent	The stolen —— has been recovered [stolen
24,694	Preponder	**Stop**
24,695	Preposed .	Stop payment
24,696	Prepositor	Stop for a time
24,697	Prepossess	To stop the
24,698	Prepotency	Can you stop
24,699	Preremote	Cannot stop
24,700	Prerequire	Will stop
24,701	Preresolve	You must put a stop to (or—on)
24,702	Presageful	In order to put a stop to (or—on)
24,703	Presaging .	Must stop
24,704	Presbyope	Do not stop
24,705	Presbytery	Stop everything till you hear again from
24,706	Presbytia .	Stop the works
24,707	Prescience	Stop working
24,708	Prescind .	Stop all exploratory work
24,709	Prescripts	Stop all operations
24,710	Preselect .	Remove the stop on
24,711	Presented.	Have removed the stop on
24,712	Presential.	Cheque (or bill) has been lost; please stop payment
24,713	Presently .	**Stope(s)**
24,714	Presentoir.	The stopes above Level No.
24,715	Preserving	The next stope will remove
24,716	Preshow .	Begin to stope
24,717	Preshowing	The stopes are looking better
24,718	Presidiary.	The stopes are looking poor
24,719	Presignify.	In the stopes
24,720	Prespinal .	From the stope
24,721	Pressbed .	A sample from the stope above Level —— assayed
24,722	Pressers .	Stopes looking splendid
24,723	Pressfat .	The stopes above Level No. —— have communi-
24,724	Pressgangs	The stope is [cated with the old working above
24,725	Pressitant	When will you begin to stope
24,726	Pressive .	Stopes still hold good in .

No.	Code Word.	Stope(s) (*continued*)
24,727	Pressly .	The stope(s) above
24,728	Pressmoney	The stope in —— chute
24,729	Pressness .	The stope is improving
24,730	Presspack	The stope is not looking so well
24,731	Pressrooms	The stope is yielding ore assaying
24,732	Pressurage	The stope has given out of pay ore
24,733	Presswork .	The stope in —— has communicated with
24,734	Prestable .	The stope has broken into
24,735	Prestige .	Have begun to stope
24,736	Presto . .	From the stope in —— chute
24,737	Presultor .	To connect with the stope
24,738	Presumably	How is the stope looking
24,739	Presuming	Ore from this stope assays [of ore
24,740	Presuppose	From this stope we expect to get a valuable supply
24,741	Presurmise	From this stope we have taken up to now, —— tons
24,742	Pretence .	We hope to get from this stope [of ore
24,743	Pretendant	The old stopes have caved in
24,744	Pretermit .	From the old stopes and workings
24,745	Pretexed .	From some of the old stopes
24,746	Pretibial .	In some of the old stopes
24,747	Pretiosity .	The face of the stope
24,748	Pretorium .	The stopes in the upper part of the mine
24,749	Pretorship	Is the roof in stope good
24,750	Pretorture .	**Stoped**
24,751	Prettiest .	Have stoped
24,752	Prettified .	Stoped out all the ore
24,753	Prettiness .	Ore stoped
24,754	Prettyism .	**Stoping**
24,755	Pretypify .	Commence stoping
24,756	Prevail .	We are now stoping in
24,757	Prevaileth .	Have you begun stoping in
24,758	Prevailing	We have begun stoping in
24,759	Prevalency	When will you begin stoping
24,760	Prevalent .	Stoping above —— Level
24,761	Prevenancy	Stoping in —— Level
24,762	Prevenient	Stoping in the —— chute
24,763	Preventeth	Shall continue stoping
24,764	Prevention	Have discontinued stoping (until) or (on account of)
24,765	Preventive	**Stoppage**
24,766	Previewed	Avoid any stoppage
24,767	Previewing	Why is there a stoppage
24,768	Prevoyant .	In order to prevent any stoppage
24,769	Prewarned	A stoppage has occurred in
24,770	Prewarning	No stoppage has occurred in
24,771	Priapean .	**Stopped**
24,772	Priapism .	Has (have) stopped
24,773	Priceless .	Has not yet stopped
24,774	Prickeared	Have stopped payment (of)
24,775	Prickingup	Has (have) not been stopped
24,776	Pricklouse	Must be stopped

No.	Code Word.	**Stopped** (*continued*)
24,777	Prickly . .	Need not be stopped
24,778	Pricklyash	Why have you stopped
24,779	Prickpost .	As soon as it is stopped
24,780	Prickpunch	Stopped sending
24,781	Prickshaft.	The works have been stopped
24,782	Pricksong.	All operations stopped on account of
24,783	Prickwood	All works must be stopped
24,784	Prideful .	Stopped on account of
24,785	Pridefully .	Stopped for want of
24,786	Priestcap .	It is said that —— have stopped payment
24,787	Priesthood	**Store(s).** (See also Stock.)
24,788	Priestism .	In store
24,789	Priestless .	Stores now in hand
24,790	Priestlike .	Stores in hand at end of month
24,791	Priggery .	You had better reduce stores
24,792	Priggish .	Keep a good store of
24,793	Prill . .	Lay in a good store of
24,794	Prillon .	A large store of
24,795	Primacy .	To supply the necessary stores
24,796	Primage .	Value of stores in hand
24,797	Primality .	Including the stores
24,798	Primatial .	Stores have to be sent —— miles
24,799	Primero .	The company keeps a store which brings a yearly
24,800	Primest .	What store is there of [profit of
24,801	Primestaff.	There is a good store of wood ready stacked at the
24,802	Primetide .	**Stored** [mill
24,803	Primeval .	Have stored
24,804	Primevally	Have been stored
24,805	Primipilar.	Stored on account of
24,806	Primitive .	**Strata (Stratum)**
24,807	Primordial	The strata
24,808	Primprint .	The strata are known to dip towards the
24,809	Primrosed.	The strata are nearly vertical
24,810	Primula .	The strata dip
24,811	Primwort .	Horizontal strata
24,812	Princedom	The strata are well defined
24,813	Princelike.	Well-defined strata
24,814	Princeling.	Strata very irregular
24,815	Princely .	The strata are much distorted
24,816	Princessly.	In a stratum of
24,817	Princewood	**Stratification**
24,818	Principals .	The dip of the vein is the same as the stratification
24,819	Principia .	Stratification nearly perpendicular [of the rock
24,820	Principled.	The peculiar stratification
24,821	Princox .	The pitch of the stratification
24,822	Pringlea .	**Stratified**
24,823	Printfield .	Stratified rocks of the —— period
24,824	Printroom.	**Streak(s)**
24,825	Printshop .	The rich streak is from —— to —— thick
24,826	Printworks	The streak could be followed for some distance

No.	Code Word.	Streak(s) *(continued)*
24,827	Prionidae .	There is a streak of ore in the vein that assays ——
24,828	Prionodon	A streak of ore [per ton
24,829	Prior . .	Rich streaks of ore, carrying
24,830	Priorate .	Thin streaks of
24,831	Prioress .	Traversed by streaks of
24,832	Priorly . .	These streaks are highly charged with
24,833	Priorship .	Stream(s)
24,834	Prisebolt .	A fine stream of water
24,835	Prismoid .	The stream is full only in the winter
24,836	Prismoidial	The stream runs dry in summer
24,837	Prisms . .	The stream is generally dry
24,838	Prisonbase	No streams
24,839	Prisoners .	Several streams
24,840	Prisonment	Constant stream of running water
24,841	Prisonship	There are several small streams
24,842	Prisonvan .	Two or three streams
24,843	Pristinate .	On one side is a good stream of water
24,844	Privacies .	A stream sufficient to drive
24,845	Privateers .	Stream does not supply enough water to
24,846	Privations .	Stream supplies water enough for a mill of ——
24,847	Privileged .	A broad, rapid stream [stamps
24,848	Privily . .	A shallow stream
24,849	Privycoat .	An excellent stream of clear water
24,850	Privypurse	This stream could be dammed up
24,851	Privyseal .	Strength
24,852	Prizeable .	Has not sufficient strength
24,853	Prizecourt	The strength of the vein
24,854	Prized . .	At its full strength
24,855	Prizefight .	Strengthened(ing)
24,856	Prizelist .	Has strengthened considerably
24,857	Prizeman .	We have strengthened the
24,858	Prizemoney	By strengthening the —— it will be safe to work
24,859	Prizerings .	We are strengthening
24,860	Proaulion .	Stress
24,861	Probably .	By stress of weather
24,862	Probal . .	A great stress upon
24,863	Probang .	Strike(s)
24,864	Probative .	A strike is threatened
24,865	Probatory .	A strike has been ordered by the
24,866	Problem .	The reason of the strike
24,867	Proboscis .	The strike is likely to cause us much trouble
24,868	Procacious	How long is strike likely to last
24,869	Procedendo	Do not think strike will last long
24,870	Procedure .	To avert a strike
24,871	Proceedeth	Strike expected among the
24,872	Proceeding	Strike of miners and labourers
24,873	Procellous	Strike will probably last some time
24,874	Proception	Strike has been averted for the present
24,875	Process .	Strike averted, but afraid we shall have trouble by-
24,876	Processive	Strike begun [and-bye

No.	Code Word.	**Strike(s)** (*continued*)
24,877	Prochilus .	Strike continues
24,878	Proclaimed	Strike over ; men have returned to work
24,879	Proclitic .	Strike over ; men gave in
24,880	Proclive .	Strike over ; we have had to
24,881	Proclivity .	Strike over ; terms of payment agreed to
24,882	Proclivous	Workmen on strike
24,883	Procoelia .	Labourers on strike
24,884	Proconsul .	Pitmen on strike
24,885	Procreate .	Miners on strike
24,886	Proctocele	Mill men on strike
24,887	Proctor .	Caused by strike of pitmen
24,888	Proctorage	In the event of a strike
24,889	Proctorial .	Caused by strike of miners
24,890	Procumbent	The strike was caused by
24,891	Procurable	The strike originated with
24,892	Procuracy .	Hope shortly to strike the vein
24,893	Procurator .	If by then we do not strike the vein
24,894	Procured .	To strike the vein
24,895	Procyon .	Telegraph when you strike the vein
24,896	Prodigally .	Expect to strike the vein within —— days
24,897	Prodigated	(To) strike the main vein
24,898	Prodigious	(To) strike side vein
24,899	Prodigy .	Expect to strike the vein at a distance of
24,900	Proditor .	As soon as we strike ore
24,901	Prodromus	Rich strikes have been made at several places
24,902	Producer .	Strike made in
24,903	Producing	Strike reported in
24,904	Productile .	Have made rich strike in
24,905	Production	Strike in —— is not good for much
24,906	Proembryo	Pushing on to strike chute above
24,907	Profanate .	Pushing on to strike chute below
24,908	Professors .	**Stringers**
24,909	Proffer . .	No regular vein, only stringers
24,910	Proffering .	Stringers of rich ore
24,911	Proficient .	Stringers of quartz
24,912	Profilist .	Small stringers of quartz
24,913	Profitable .	**Stromeyrite**
24,914	Profitless .	**Strong**
24,915	Profligacy .	A strong, well-defined vein
24,916	Profluence	Very strong
24,917	Profound .	Not very strong
24,918	Profoundly	The vein is not so strong as we expected
24,919	Profulgent	Not so strong as
24,920	Profusely .	Not strong enough
24,921	Profusion .	Probable strong inquiry at higher prices
24,922	Profusive .	Strong enough to bear the strain
24,923	Progenitor	There is a strong feeling against
24,924	Proglottis .	There is a strong feeling in favour of
24,925	Prognostic	**Stronger**
24,926	Programme	Stronger than it has ever been

No.	Code Word.	**Stronger** (*continued*)
24,927	Progressed	Stronger than we have yet seen it
24,928	Prohibit .	Have made it much stronger; it now works very [well
24,929	Proined .	**Strongest**
24,930	Project .	In the strongest manner possible
24,931	Projectile .	Of the strongest make
24,932	Projection	**Struck**
24,933	Projectors	We have struck
24,934	Prolabium	Have struck a seam of
24,935	Prolapsed .	Telegraph when you have struck ore
24,936	Prolapsing	Telegraph when you have struck the
24,937	Prolapsion	Have not yet struck any ore
24,938	Prolefied .	Struck a rich body of ore
24,939	Proleg . .	The men have struck at the dictation of the Union
24,940	Proleptics .	Miners have struck for less hours
24,941	Proletary .	Miners have struck for higher wages
24,942	Prolicide .	Miners have struck for
24,943	Prolific .	Miners have struck against
24,944	Prolifical .	Miners have struck against proposed reduction in [wages
24,945	Prolixity .	Have struck water in
24,946	Prolixness	Have not struck any water
24,947	Prolocutor	(Have) struck the hanging-wall
24,948	Prologized	(Have) struck the foot-wall
24,949	Prologue .	(Have) struck the lode
24,950	Prolongate	(Have) struck the ore body in
24,951	Prolonging	Have struck pay ore
24,952	Prolongs .	Have struck rich ore
24,953	Promenade	Have struck country rock
24,954	Promerit .	Have not yet struck
24,955	Promerops	Struck by lightning
24,956	Promethean	**Subject (to)**
24,957	Prominency	Subject to the conditions named
24,958	Prominent	Subject to a commission of
24,959	Promiseth .	Subject to a rebate of
24,960	Promisor .	Subject to a return of
24,961	Promissive	Subject to a discount of
24,962	Promissory	Subject to our approval
24,963	Promontory	Subject to approval of
24,964	Promotions	Subject to an early reply
24,965	Promptbook	Subject to the terms of the contract
24,966	Prompting	Subject to confirmation
24,967	Promptly .	Will be subject to
24,968	Promptness	Will not be subject to
24,969	Promptnote	Will always be subject to
24,970	Promptuary	Will —— always be subject to
24,971	Prompture	Will not in future be subject to
24,972	Promulgate	On the subject
24,973	Promulged	The subject referred to in
24,974	Promulging	The subject was fully discussed and it was decided to
24,975	Pronaos .	**Sub-let**
24,976	Prongbuck	Cannot sub-let

No.	Code Word.	**Sub-let** (*continued*)
24,977	Pronghoe .	Have sub-let the contract
24,978	Pronghorns	To sub-let the contract
24,979	Pronominal	Do not allow —— to sub-let the contract
24,980	Pronounced	Have allowed —— to sub-let the contract
24,981	Pronouns .	Will not allow —— to sub-let the contract
24,982	Pronucleus	**Subordinate(d)** [which demand attention
24,983	Pronuncial	(Must be) subordinate to the more important matters
24,984	Proofarm .	Hold (holding) a subordinate position
24,985	Proofhouse	Is a subordinate consideration
24,986	Proofless .	**Subordinating**
24,987	Proofprint	Subordinating everything to this consideration (or
24,988	·Proofsheet	**Subscribe(d)** [object)
24,989	Prooftext .	Will subscribe for —— shares
24,990	Propaganda	Will subscribe to the extent of
24,991	Propagator	Capital already subscribed
24,992	Propago .	How much will you subscribe
24,993	Propagulum	Enough money has been subscribed
24,994	Proparent.	What has been subscribed
24,995	Propel . .	Has (have) subscribed
24,996	Propelled .	Has been subscribed
24,997	Propelling	Capital was subscribed twice over
24,998	Propense .	Capital subscribed —— times over
24,999	Propension	More was subscribed than was wanted
25,000	Propensity	The full amount will be subscribed
25,001	Properates	Was privately subscribed
25,002	Properly .	The total subscribed is
25,003	Properness	Was quickly subscribed
25,004	Propertied	If you have not already subscribed
25,005	Prophasis .	**Subscriber(s)**
25,006	Prophecy .	Put my name down as a subscriber for
25,007	Prophesied	The subscribers to the loan
25,008	Prophet .	The subscribers to the syndicate
25,009	Prophetess	The subscribers to the
25,010	Prophetize	**Subscription(s)**
25,011	Prophoric.	A subscription list has been opened
25,012	Propitiate.	What is the total subscription
25,013	Propitious	Subscription was not well taken up
25,014	Proplasm .	Subscriptions invited for
25,015	Proplastic.	Now inviting subscriptions for
25,016	Propodium	Subscriptions were —— times the amount asked for
25,017	Propolis .	**Submit**
25,018	Proportion	Submit for your consideration
25,019	Proposals .	Submit to arbitration
25,020	Proposedly	Submit to the decision of
25,021	Propounded	Willing to submit it to ——, and abide by his (their)
25,022	Proppage .	Will not submit to [decision
25,023	Propriety .	**Submitted**
25,024	Proproctor	Submitted to the action (of)
25,025	Props . .	Submitted to arbitration
25,026	Propugner	Submitted for consideration

No.	Code Word.	**Submitted** (*continued*)
25,027	Propugning	Submitted to —— for approval
25,028	Propulsory	Has been submitted to
25,029	Propwood	Cannot be submitted to
25,030	Propylaeum	Will be submitted to
25,031	Propylon .	Your proposal was submitted to the Board and
25,032	Proratable	Submitted to —— and approved [approved
25,033	Proroguing	**Subsequent** [your opinion
25,034	Prosaism .	If subsequent investigations cause you to change
25,035	Proscenium	Subsequent events showed that
25,036	Proscolex .	Subsequent investigations have not changed my
25,037	Proscolla .	Subsequent investigations have proved [opinion
25,038	Prosecuted	In a subsequent interview with
25,039	Proselytes.	As read by the light of subsequent events
25,040	Proserpine	**Subsequently**
25,041	Prosimiae .	It was subsequently decided
25,042	Proslavery	(What) occurred subsequently
25,043	Prosodical	**Subsided**
25,044	Prosodist .	As soon as the flood has subsided
25,045	Prosody .	The excitement has subsided
25,046	Prosoma .	Has the excitement subsided (about)
25,047	Prospect .	Flood has subsided
25,048	Prospector	Flood has subsided; not much damage done
25,049	Prospered.	Flood has subsided; have re-started
25,050	Prospering	**Substance**
25,051	Prosperous	Was, in substance, correct
25,052	Prostatic .	Give the substance of
25,053	Prosthesis.	The substance of what was said
25,054	Prostrated	**Substantially**
25,055	Prostyle .	Substantially built
25,056	Protamoeba	Has been substantially rebuilt
25,057	Protandry.	Substantially as reported upon
25,058	Protasis .	**Substitute(d)**
25,059	Proteacea .	May we substitute
25,060	Protean .	You may substitute
25,061	Proteanly .	For the old —— we shall substitute
25,062	Protected .	Substituted for
25,063	Protective.	**Substituting**
25,064	Protectrix.	We are now substituting
25,065	Proteidae .	**Substitution**
25,066	Proteinous	By the substitution of
25,067	Proteles .	**Succeed(s)**
25,068	Protervity	Hope you will succeed in
25,069	Protestant	In order to succeed
25,070	Protests .	Will succeed
25,071	Prothallus.	Will not succeed
25,072	Prothorax.	Can you succeed in
25,073	Prothyrum	Fear I (we) shall not succeed in
25,074	Protocol .	Hope to succeed in
25,075	Protogyny	If we succeed
25,076	Protopapas	We shall succeed

No.	Code Word.	Succeed (*continued*)
25,077	Protophyte	Cannot succeed in
25,078	Protoplast	Hope —— will succeed in
25,079	Protopteri	Fear —— will not succeed in
25,080	Protornis .	Succeed in obtaining
25,081	Protosalt .	Cannot succeed except by (or unless)
25,082	Protospore	If you succeed in doing this
25,083	Prototypes	If we do not succeed
25,084	Protoxide.	If we do not succeed it will be on account of
25,085	Protozoic .	If you do not succeed
25,086	Protract .	If —— does not succeed
25,087	Protractor	Did not succeed in
25,088	Protrude .	Likely to succeed
25,089	Protruding	**Succeeded**
25,090	Protrusile.	Have you succeeded
25,091	Protutors .	Has (have) succeeded
25,092	Proudish .	Has (have) not succeeded
25,093	Proudly .	To be succeeded by
25,094	Proudness	Have succeeded beyond our expectations
25,095	Proudpied	Succeeded in obtaining
25,096	Provably .	Have not yet succeeded, but still hope to do so
25,097	Provant .	As soon as we have succeeded
25,098	Proveditor	Succeeded in finding the vein
25,099	Provedores	**Success**
25,100	Provender	A great success
25,101	Proverbial	Success now assured
25,102	Proverbs .	Success is doubtful
25,103	Provided .	(We) can make it a great success
25,104	Providence	Must make it a success
25,105	Providing.	Has been a great success
25,106	Provinces.	What success have you had in
25,107	Provision .	Have had a great success
25,108	Provisory .	If it should not be a success
25,109	Provokable	Is going to be a great success
25,110	Provoke .	Success may be considered certain
25,111	Provoking	Success is very doubtful
25,112	Provost .	Success is assured, if
25,113	Prowess .	Confident of success
25,114	Prowled .	In order to make it a success
25,115	Prowleries	There is a good prospect of success
25,116	Prowling .	May expect moderate success
25,117	Proxenet .	We think that future success is certain
25,118	Proximal .	**Successful**
25,119	Proximity.	Has (have) been very successful.
25,120	Proximo .	Have always been successful
25,121	Proximous	Was (were) not at first successful
25,122	Proxy . .	Was (were) ultimately successful
25,123	Proxying .	Has (have) not been successful
25,124	Proxyship.	Has (have) —— been successful (in)
25,125	Prudential	Not likely to be successful
25,126	Prudently.	Not so successful (as)

No.	Code Word.	Successful (*continued*)
25,127	Prudery .	Was very successful in the past
25,128	Prudishly .	The old owners were very successful
25,129	Pruinate .	If not successful at present
25,130	Prunelet .	We have at last been successful in
25,131	Pruner . .	If you are not successful
25,132	Prunings .	**Successfully**
25,133	Pruriency .	Was successfully carried out (or completed)
25,134	Prurient .	**Such**
25,135	Prurigo .	From such
25,136	Pruritus .	By such
25,137	Prussiate .	From such a source
25,138	Prutenic .	Such is (are)
25,139	Pryest . .	Is (are) such the
25,140	Prytaneum	Such as
25,141	Prytany .	In such a case
25,142	Psalm . .	Just such another as
25,143	Psalmist .	If such is the case
25,144	Psalmodic	Such as may
25,145	Psalterium	**Suction**
25,146	Psaltery .	Suction valves and seats
25,147	Psamma .	Suction valve-box
25,148	Psammitic	Suction lift
25,149	Psammodus	**Sudden(ly)**
25,150	Psarolite .	There has been a sudden change
25,151	Psellismus	The sudden disappearance of
25,152	Psephism .	A sudden change in the weather
25,153	Pseudobulb	Cannot account for the sudden
25,154	Pseudodox	Very suddenly (or)—Very sudden
25,155	Pseudology	Died suddenly last
25,156	Pseudonym	**Sue**
25,157	Pseudopus	To sue (for)
25,158	Pseudova .	Intend to sue
25,159	Pshaw . .	Do not sue
25,160	Pshawing .	Must sue
25,161	Psidium .	Sue at once
25,162	Psilothron	To sue them for damages
25,163	Psittacid .	Will sue
25,164	Psittacula .	Will not sue
25,165	Psoas . .	Shall sue them in the —— court
25,166	Psophia .	Shall sue for substantial damages
25,167	Psoralea .	Shall sue for a re-hearing of the case
25,168	Psoriasis .	If he sues (they sue) us
25,169	Psoric . .	If we sue him (them)
25,170	Psychal .	**Sued**
25,171	Psychiatry	Expect to be sued for
25,172	Psychism .	As to which we are sued
25,173	Psychology	**Suing**
25,174	Psychopomp	Now suing —— for
25,175	Psychosis .	Suing for heavy damages
25,176	Psychotria.	**Sufficient(ly)**

No.	Code Word.	**Sufficient(ly)** (*continued*)
25,177	Psylla . .	There is sufficient
25,178	Psyllidae .	Not in sufficient quantities
25,179	Ptarmic .	Sufficient money in hand (to)
25,180	Ptarmigan	Not sufficient money in hand (to)
25,181	Pteraspis .	A sufficient quantity
25,182	Pteroceras	Reserving only sufficient
25,183	Pterodon .	There is sufficient wood to last for years
25,184	Pteroma .	Is it (are they) sufficient
25,185	Pteromys .	It is (they are) sufficient
25,186	Pteropidae	It is (they are) not sufficient
25,187	Pteropus .	The quantity is sufficient to
25,188	Pterosaur .	The quantity is not sufficient to
25,189	Pterygium	Have you sufficient
25,190	Pterygoid .	Have sufficient
25,191	Pterygotus	Have sufficient —— to last for
25,192	Ptinidae .	Have not sufficient
25,193	Ptisan . .	Not sufficient
25,194	Ptochogony	There not being sufficient
25,195	Ptolemaic.	If not sufficient
25,196	Ptyalin. .	Sufficient for the purpose (of)
25,197	Ptychodus	Sufficient to justify
25,198	Puberulent	Not sufficient to justify
25,199	Pubescency	Producing sufficient for
25,200	Pubescent	Milling sufficient ore to keep up
25,201	Public . .	The output is sufficient to keep the mill going
25,202	Publicans .	There will not be sufficient
25,203	Publicist .	We have not sufficient —— for more than
25,204	Publicly .	Sufficient to carry us over
25,205	Publicness	Sufficient to enable us to
25,206	Publish .	Sufficiently large to
25,207	Publishers	Sufficiently satisfactory
25,208	Puccinia .	**Suggest**
25,209	Puceron .	What would you suggest
25,210	Puchapat .	Can you suggest any means (or cause)
25,211	Puckball .	Cannot suggest
25,212	Puckered .	Cannot suggest any better way
25,213	Puckering.	We suggest that you should
25,214	Puckfists .	If you can suggest any better way
25,215	Puddening	**Suggested**
25,216	Puddingbag	It has been suggested to us (that)
25,217	Puddingpie	We suggested to him (them)
25,218	Puddleball	**Suggestion(s)**
25,219	Puddlepoet	Acted on your suggestion
25,220	Puddly .	At our suggestion
25,221	Pudendous	At —— suggestion
25,222	Pudicity .	Do not approve of —— suggestion
25,223	Puerile. .	Approve(s) of —— suggestion
25,224	Puerilely .	Suggestion is impracticable
25,225	Puerperal.	Cannot act on your suggestion
25,226	Puerperous	Glad to receive any suggestion

No.	Code Word.	**Suggestion(s)** (*continued*)
25,227	Puff . . .	Think the suggestion a good one
25,228	Puffadder .	The suggestion will be considered
25,229	Puffbird .	Your suggestion shall have our best consideration
25,230	Puffery. .	Suggestions for you to consider
25,231	Puffiness .	Suggestions for you and —— to consider
25,232	Puffiingly .	Will consider your suggestions
25,233	Puffpaste .	Our joint suggestions
25,234	Pugdog .	Your joint suggestions
25,235	Pugfaced .	**Suit(s)**
25,236	Pugh . .	To suit
25,237	Pugilism .	Will suit
25,238	Pugilistic .	Will not suit
25,239	Pugmill .	Anything to suit
25,240	Pugnacious	Nothing that will suit
25,241	Pugnacity .	Will it suit
25,242	Pugnosed .	Will suit very well
25,243	Pugpiles .	Would suit much better (if)
25,244	Pugpiling .	Would not suit so well
25,245	Puisny . .	In the suit of —— versus
25,246	Puissance .	The suit of the company versus
25,247	Puissantly.	The suit of —— versus the company
25,248	Pulex . .	(It is) a suit for damages
25,249	Pulicaria .	(It is) a suit for trespass
25,250	Pulicene .	At the suit of
25,251	Pulings .	This suit will be heard
25,252	Pullback .	You need have no fear about the suit
25,253	Pulleys .	The suit will not come on until
25,254	Pullicat .	The hearing of the suit
25,255	Pullmancar	If the suit is heard
25,256	Pullulated.	Who are the plaintiffs in the suit
25,257	Pulmograda	Who are the defendants in the suit
25,258	Pulmonary	Are the plaintiffs ; —— are the defendants in the suit
25,259	Pulmonata	The plaintiff in the suit
25,260	Pulmonicks	The defendant in the suit
25,261	Pulmonifer	The result of the suit
25,262	Pulpers .	If it will suit
25,263	Pulpited .	If it will not suit
25,264	Pulpitical .	If it suits our purpose, we will
25,265	Pulpitish .	It suits our purpose
25,266	Pulpitry .	Suits —— purpose
25,267	Pulpous .	When it suits
25,268	Pulsated .	We shall not move until it suits us to do so
25,269	Pulsatilla .	**Suitable**
25,270	Pulsations.	Would it be suitable
25,271	Pulsative .	Suitable in every way
25,272	Pulsatory .	Not suitable for the work
25,273	Pulseglass.	Not at all suitable
25,274	Pulseless .	Telegraph if anything suitable offers
25,275	Pulsific. .	Nothing suitable offered
25,276	Pulsion .	Think it would be very suitable

No.	Code Word.	Suitable (*continued*)
25,277	Pultaceous	If nothing suitable offers
25,278	Pulverable	Nothing more suitable to offer
25,279	Pulverate .	More suitable for our purpose
25,280	Pulverized	In other respects, it is very suitable
25,281	Pulverous .	Might be made suitable
25,282	Pulvillo .	Is there any suitable
25,283	Pulvinuli .	Find anything suitable
25,284	Pulwar . .	Cannot find a suitable
25,285	Pumicose .	Would be very suitable for
25,286	Pumpbarrel	**Sulphide (of)**
25,287	Pumpbits .	Sulphide of silver
25,288	Pumpbox .	Sulphide of copper
25,289	Pumpbrake	Sulphide of zinc
25,290	Pumpchain	**Sulphur**
25,291	Pumpdale.	**Sulphurets**
25,292	Pumpetball	Cannot treat the sulphurets
25,293	Pumphandle	These sulphurets can only be treated by
25,294	Pumphood	In the sulphurets
25,295	Pumprooms	**Sum(s)**
25,296	Pumpspear	A sum of
25,297	Pumpstock	Too large a sum
25,298	Pumpwell .	Not too large a sum
25,299	Punawind .	Do you think it too large a sum to pay for the mine
25,300	Punch . .	I (we) consider it a fair sum
25,301	Punchbowl	A lump sum of
25,302	Punchcheck	Offer him (them) a lump sum down
25,303	Puncheon .	A lump sum offered
25,304	Punchladle	Large sums have been spent
25,305	Punctuated	**Summer**
25,306	Punctators	This summer
25,307	Punctiform	Last summer
25,308	Punctilio .	Next summer
25,309	Punctist .	During the summer
25,310	Punctual .	Very hot in summer
25,311	Punctually	**Summit**
25,312	Punctulate	On the summit
25,313	Puncturing	From the summit
25,314	Pundum .	Over the summit
25,315	Pungently.	At the summit
25,316	Pungled .	The summit of
25,317	Punica . .	Near the summit
25,318	Punish . .	**Sunday**
25,319	Punishable	On Sunday
25,320	Punishing .	Last Sunday
25,321	Punishment	Next Sunday
25,322	Punitory .	Every Sunday
25,323	Punnets .	Every other Sunday
25,324	Punsters .	**Sunk**
25,325	Puparial .	Have sunk
25,326	Pupilage .	Sunk —— feet on the vein

No.	Code Word.	**Sunk** (*continued*)
25,327	Pupilarity .	Sunk on the vein
25,328	Pupils . .	Have sunk —— feet
25,329	Pupivora .	Have not sunk far enough
25,330	Pupivorous	After they had sunk —— feet
25,331	Puppetish .	How far have you sunk
25,332	Puppetly .	Have sunk —— feet for water
25,333	Puppetman	Shaft sunk
25,334	Puppetplay	**Superficial**
25,335	Puppetshow	Could only make superficial examination
25,336	Puppyism .	From a superficial examination
25,337	Purblind .	Have made only a superficial examination
25,338	Purblindly	The superficial measurement
25,339	Purchase .	**Superfluous**
25,340	Purchasing	Would be quite superfluous
25,341	Purflewed .	**Superintend(s)**
25,342	Purgament	Who will superintend
25,343	Purgation .	Will superintend
25,344	Purgatives.	Will not superintend
25,345	Purgatory .	Cannot superintend
25,346	Purgingnut	You must superintend yourself
25,347	Purify . .	Has (have) sent proper person to superintend
25,348	Purists . .	Shall we send proper person to superintend
25,349	Puritan .	Send a proper person to superintend
25,350	Puritanism	Must have efficient man to superintend
25,351	Puritanize.	**Superintended**
25,352	Purlieus .	Has been superintended by
25,353	Purloin .	Who superintended
25,354	Purloining	**Superintendence**
25,355	Purparties.	The chief superintendence of
25,356	Purplefish.	**Superintendent**
25,357	Purplewood	The mine superintendent
25,358	Purpling .	The underground superintendent
25,359	Purport .	The superintendent of the
25,360	Purporting	Surface superintendent
25,361	Purposeful	**Superintending**
25,362	Purposely .	Superintending everything
25,363	Purposive .	**Superior**
25,364	Purpurate .	Superior to any we have yet seen
25,365	Purreic. .	Superior to all others
25,366	Pursecrab .	If not superior
25,367	Purseful .	A much superior kind
25,368	Pursemouth	Do not think it superior to
25,369	Pursenet .	**Supersede(s)**
25,370	Pursepride	Intended to supersede
25,371	Purseproud	Will it supersede the old
25,372	Purslane .	Think it will supersede
25,373	Pursuable .	**Superseded**
25,374	Pursuance.	Have (has) superseded
25,375	Pursueth .	Has been superseded (by)
25,376	Pursuivant	**Supervision**

No.	Code Word.	**Supervision** (*continued*)
25,377	Purtenance	Under good supervision
25,378	Purulency.	There has been no proper supervision
25,379	Purulent .	To be placed under proper supervision
25,380	Purvey. .	Greatly in need of supervision
25,381	Purveyance	Give this your personal supervision
25,382	Purveyors.	Without —— personal supervision
25,383	Purview .	Under the personal supervision of
25,384	Puseyism .	Will give his personal supervision
25,385	Puseyistic.	Most careful supervision
25,386	Pushed .	**Supplement**
25,387	Pushpins .	In order to supplement the supply of
25,388	Pussmoth .	Supplement what is wanted to complete the
25,389	Pustular .	**Supplementary (to)**
25,390	Putamen .	A supplementary report
25,391	Putchock .	**Supplemented** (by)
25,392	Puteal . .	**Supplied**
25,393	Putloghole	Supplied with
25,394	Putorius .	Badly supplied (with)
25,395	Putrefied .	Well supplied (with)
25,396	Putrid . .	Has supplied us with
25,397	Putridity .	Was (were) supplied
25,398	Putterout .	Must be supplied with
25,399	Puttyers .	Are you well supplied (with)
25,400	Puttyeye .	We are now well supplied (with)
25,401	Puttyfaced	Shall be well supplied with
25,402	Puttyknife	Supplied with tools and other appliances
25,403	Puttyroot .	Have supplied all that is wanted
25,404	Puzzle . .	New —— have been supplied
25,405	Puzzledom	Supplied to replace the old
25,406	Puzzlement	**Supplies**
25,407	Puzzling .	Supplies are very short
25,408	Puzzolite .	Supplies easily hauled from the seaport of
25,409	Pyaemia .	Supplies hauled with great difficulty from
25,410	Pycnidium	Milling and mining supplies
25,411	Pycnite .	Large supplies
25,412	Pycnodont	Small supplies
25,413	Pycnogonum	Supplies coming in slowly
25,414	Pycnostyle	Supplies coming in quickly
25,415	Pyebald .	Large supplies expected
25,416	Pygarg. .	Supplies have to be hauled
25,417	Pygathrix .	Supplies coming to hand daily
25,418	Pygmean .	**Supply**
25,419	Pygmiest .	Can —— supply
25,420	Pygopus .	Can you supply
25,421	Pyjama .	If you cannot supply
25,422	Pyloridea .	Can supply
25,423	Pyogenesis	Cannot supply
25,424	Pyogenia .	Must supply
25,425	Pyoning .	Supply exceeds the demand
25,426	Pyracanth.	Supply falls short of demand

No.	Code Word.	**Supply** (*continued*)
25,427	Pyracid .	Supply us with
25,428	Pyramidal.	Will furnish a regular supply (for) (of)
25,429	Pyramidion	Will supply
25,430	Pyramises.	Will not supply
25,431	Pyreneite .	Would supply
25,432	Pyrethrum	We must have a good supply of
25,433	Pyretic. .	Has (have) now a regular supply
25,434	Pyretology	The supply is irregular (of)
25,435	Pyrexial .	Has (have) a good supply in hand (of)
25,436	Pyritized .	Sufficient supply of
25,437	Pyritizing .	Short supply of
25,438	Pyritous .	An abundant supply of
25,439	Pyroacetic	Have bought a large supply of
25,440	Pyrochlore	The supply appears inexhaustible
25,441	Pyrochroa.	A constant supply of
25,442	Pyrogallic.	A supply of new
25,443	Pyrogenous	Can you supply the extra —— needed to
25,444	Pyrognomic	Will supply us with enough ore to last
25,445	Pyrolaceae	**Supplying**
25,446	Pyrolatry .	Am (are) supplying the mill with ore from the
25,447	Pyroleter .	Supplying all materials
25,448	Pyrolithic .	Supplying what is necessary for
25,449	Pyrologist .	**Support(s)**
25,450	Pyrolusite.	The support of the government
25,451	Pyromantic	Promise of support (from)
25,452	Pyrope. .	The authorities will support
25,453	Pyrophorus	Will require support from
25,454	Pyrorthite.	In order to support
25,455	Pyrosoma .	To support
25,456	Pyrotechny	Will support
25,457	Pyrouric .	Will not support .
25,458	Pyroxene .	Must support
25,459	Pyroxyline	Get —— to support you
25,460	Pyrrhicist .	The supports
25,461	Pyrrholite.	No support can be expected from
25,462	Pyrrhonean	The supports have given way
25,463	Pyrrhonic .	(To) have the support of
25,464	Pyrrhula .	(To) replace the old supports
25,465	Pyrulinae .	The old supports are quite decayed
25,466	Pythagoric	Can no longer support
25,467	Pythiad .	Must put in new supports to make all safe
25,468	Python .	New supports
25,469	Pythoness.	Support the strain thrown upon it
25,470	Pyxineae .	You will have the support of
25,471	Pyxis . .	Agreed to support
25,472	Quabird .	(To) support the proposition
25,473	Quackery .	**Supported**
25,474	Quackhood	Is (are) entirely supported by
25,475	Quackish .	Must be supported
25,476	Quackled .	Unless supported by

S

No.	Code Word.	
25,477	Quackling.	**Supposed**
25,478	Quadra .	Supposed to have been
25,479	Quadragene	Supposed not to have been
25,480	Quadrangle	Is it supposed that
25,481	Quadrantal	It was supposed to be (that)
25,482	Quadrated	It is supposed to be (that)
25,483	Quadrating	**Supposition**
25,484	Quadratrix	On the supposition that
25,485	Quadrature	**Sure**
25,486	Quadricorn	Are you sure
25,487	Quadrics .	To be quite sure
25,488	Quadrifid .	I am (we are) quite sure
25,489	Quadrilles.	Is (are) sure
25,490	Quadriloge	I am (we are) not sure
25,491	Quadrivium	Is (are) not sure
25,492	Quadroon .	Make quite sure of your ground
25,493	Quadroxide	We have made quite sure of
25,494	Quadrumana	Quite sure we are right
25,495	Quadruped	Quite sure it was the right thing to **do**
25,496	Quaesta .	Do not feel quite sure
25,497	Quaff . .	If you are not sure, you should
25,498	Quaffing .	**Surface.** (See also Indications.)
25,499	Quagmire .	On the surface
25,500	Quags . .	From the surface
25,501	Quahaug .	Along the surface
25,502	Quailcall .	Below the surface
25,503	Quailed .	Much nearer the surface
25,504	Quailpipe .	Near the surface
25,505	Quaint . .	Surface examination impossible on account of snow
25,506	Quaintly .	Surface indications point to
25,507	Quaintness	The surface works consist of
25,508	Quakegrass	Surface workings
25,509	Quaker .	(By) surface workings and trial shafts
25,510	Quakerbird	Many surface holes from —— to —— feet deep
25,511	Quakergun	One or two surface holes
25,512	Quakerism	Surface trials made
25,513	Quaketail .	Development consists of surface-openings only
25,514	Quakingly.	From these surface openings
25,515	Qualified .	The whole surface of the claims
25,516	Qualify .	The surface of the entire district
25,517	Qualifying	Here and there along the surface
25,518	Qualm . .	Picked up on the surface
25,519	Qualming .	The surface features of the country
25,520	Qualmish .	From the surface down to
25,521	Qualmishly	**Surplus**
25,522	Quamoclit.	The surplus to be applied (in)
25,523	Quandaried	What surplus will there be
25,524	Quandary .	The surplus that will be left
25,525	Quantitive	Will leave a surplus of
25,526	Quantum .	Have a surplus of

No.	Code Word.	Surplus (*continued*)
25,527	Quarantine	Any surplus there may be
25,528	Quarrel .	With the surplus
25,529	Quarrelled	**Surprised**
25,530	Quarriable	Have been surprised to hear
25,531	Quarriers .	Have been surprised at receiving
25,532	Quarry. .	Was (were) surprised
25,533	Quarrying.	Was (were) not surprised
25,534	Quarryman	Have been surprised to find
25,535	Quarterday	You will be surprised to learn that
25,536	Quartering	**Surrounded**
25,537	Quartettes.	The property is surrounded by
25,538	Quarto. .	**Surrounding(s)**
25,539	Quartpots.	From the surroundings
25,540	Quartridge	The surrounding country
25,541	Quartzite .	The surrounding mines
25,542	Quartzmill	**Survey(s)**
25,543	Quartzoid.	Have a survey made (of)
25,544	Quartzrock	Make an accurate survey
25,545	Quartzy .	Have finished the survey
25,546	Quaschi .	Have not yet finished the survey
25,547	Quasifee .	Has any survey been made
25,548	Quasimodo	Will make a survey and report upon
25,549	Quassation	A survey of all the company's property
25,550	Quassia .	An accurate survey of
25,551	Quaterfoil	(From) a survey of the property [cession
25,552	Quaternate	(From) a survey of the district covered by the con-
25,553	Quatorzain	An engineer has been appointed to make the neces-
25,554	Quatrain .	The surveys required by [sary surveys
25,555	Quatuor .	As soon as the surveys are completed
25,556	Quavered .	The surveys will take —— to complete
25,557	Quayage .	Estimated cost of the surveys
25,558	Quayberth	If no surveys are made within the specified time,
25,559	Quaying .	Survey contracts [the concession will be forfeited
25,560	Queached.	Go to —— to commence surveys
25,561	Queasily .	Quarterly survey(s)
25,562	Quebecoak	Available for surveys
25,563	Queenapple	Surveys will cost
25,564	Queenbee.	Arrangements have been made to survey
25,565	Queencake	Surveys in the various districts, for which legal
25,566	Queencraft	**Surveyed** [notices have been complied with
25,567	Queendom	Has (have) surveyed
25,568	Queengold	Has (have) not surveyed
25,569	Queenhood	Has (have) the mine(s) been properly surveyed
25,570	Queening .	Has (have) been thoroughly surveyed
25,571	Queenlike.	Has (have) never been properly surveyed
25,572	Queenly .	Expect to have —— surveyed
25,573	Queenpost	Have had —— surveyed
25,574	Queenship	Not yet surveyed
25,575	Queensware	Must be surveyed before
25,576	Queentruss	The lands (or claims) to be surveyed

No.	Code Word.	
25,577	Queerish .	**Surveying**
25,578	Quell . .	When can you commence surveying
25,579	Quemeful .	Surveying the old workings
25,580	Quenchable	Will commence surveying
25,581	Quenchcoal	Am (are) now surveying
25,582	Quenching	The expense of surveying
25,583	Quenchless	Surveying instruments
25,584	Quercetic .	**Surveyors**
25,585	Quercitron	Surveyors will be sent
25,586	Quercus .	Surveyors have been sent
25,587	Queried .	When will surveyors commence
25,588	Querimony	Surveyors' report
25,589	Querists .	Have arranged with surveyors
25,590	Querulous.	**Suspect**
25,591	Questers .	Do you suspect (any) (that)
25,592	Questioned	I (we) suspect that
25,593	Questman.	I (we) have reason to suspect
25,594	Questuary.	We suspect that the cause was
25,595	Quibble .	Have no reason to suspect
25,596	Quibbling .	**Suspected**
25,597	Quickbeam	Has (have) been suspected (of) (that)
25,598	Quicken .	Was (were) never suspected (of) (that)
25,599	Quickeneth	Has (have) —— ever been suspected of
25,600	Quickening	I (we) suspected (that)
25,601	Quickhatch	Having suspected (that)
25,602	Quickhedge	I (we) never suspected (that)
25,603	Quicklime.	**Suspecting**
25,604	Quickmarch	Suspecting that all was not right
25,605	Quickness.	**Suspend**
25,606	Quicksands	To suspend
25,607	Quickstep.	Likely to suspend
25,608	Quidam .	Not likely to suspend
25,609	Quiddany.	Suspend further action
25,610	Quiddative	Suspend work until
25,611	Quidnunc .	Will suspend
25,612	Quiesced .	We shall suspend all work
25,613	Quiescency	We shall have to suspend
25,614	Quiescing .	Suspend all operations
25,615	Quietened.	Have had to suspend work in —— on account of
25,616	Quietism .	Have had to suspend work in —— in order to put
25,617	Quietistic .	**Suspended**
25,618	Quietous .	Has (have) suspended
25,619	Quietously	Has (have) not suspended
25,620	Quietsome	Must be suspended
25,621	Quietude .	Cannot be suspended
25,622	Quillnibs .	All operations have been suspended in consequence
25,623	Quillwork.	Been partly suspended [of the severe frost
25,624	Quiltings .	Been suspended for —— days
25,625	Quinceseed	Has (have) suspended payment
25,626	Quincetree	**Suspending**

No.	Code Word.	Suspending (*continued*)
25,627	Quincewine	We are suspending operations
25,628	Quincunx .	Cause for suspending
25,629	Quinine .	**Suspense**
25,630	Quininism.	Have been kept in suspense
25,631	Quinoa .	In a state of suspense
25,632	Quinology	**Suspension**
25,633	Quinquevir	The suspension has been announced of
25,634	Quinquina	**Suspicion**
25,635	Quinsywort	Is (are) under grave suspicion of having
25,636	Quintals .	No suspicion attaches to
25,637	Quinteron.	Is there any suspicion of
25,638	Quintilian.	Great suspicion
25,639	Quintuple.	No suspicion (of)
25,640	Quirinalia.	We had no suspicion that
25,641	Quiristers .	Should be regarded with suspicion
25,642	Quirky . .	**Suspicious**
25,643	Quisqualis.	I am (we are) very suspicious (of)
25,644	Quitclaim .	Are you suspicious
25,645	Quitrents .	Is (are) —— suspicious
25,646	Quittance .	No cause to be suspicious
25,647	Quiver . .	Every reason to be suspicious
25,648	Quivering .	**Sustained**
25,649	Quixotic .	Credit must be sustained
25,650	Quiz . .	Great loss will be sustained
25,651	Quizzed .	Sustained no damage (or loss)
25,652	Quizzical .	Sustained very little damage (or loss)
25,653	Quizzism .	What loss (damage) has been sustained
25,654	Quoddy .	**Sustaining**
25,655	Quodlibet.	We are sustaining
25,656	Quondam .	Sustaining heavy loss
25,657	Quorum .	**Swindle(d)**
25,658	Quota . .	The mine(s) is (are) a regular swindle
25,659	Quotable .	The whole thing is a regular swindle
25,660	Quotations	Do you think it a swindle
25,661	Quoteless .	If you think it a swindle
25,662	Quotidian .	Was (has been) completely swindled
25,663	Quotient .	**Syenite**
25,664	Quoting .	The hanging-wall is syenite
25,665	Raasch. .	The foot-wall is syenite
25,666	Rabbetted.	The country rock is syenite
25,667	Rabbinical	**Syndicate**
25,668	Rabbinism	A syndicate has been formed
25,669	Rabbitfish.	From a syndicate
25,670	Rabbiting .	In the hands of a syndicate
25,671	Rabbits .	Ought to be bought by a syndicate, who would
25,672	Rabblement	If it were in the hands of a syndicate [develop it
25,673	Rabblerout	Syndicate with a capital of —— in —— shares
25,674	Rabdoidal.	Is suitable for a syndicate
25,675	Rabdology	A small syndicate
25,676	Rabdomancy	A syndicate to provide, say

No.	Code Word.	**Syndicate** (*continued*)
25,677	Rabidly .	Obtained by the syndicate
25,678	Rabidness.	Endeavouring to syndicate
25,679	Raccahout	A powerful syndicate
25,680	Raccoons .	Represent a syndicate of
25,681	Racecourse	Who are in the (——) syndicate
25,682	Raceginger	Fresh syndicate has been formed
25,683	Raceground	The syndicate gets
25,684	Racehorses	The syndicate gets in cash and shares respectively
25,685	Racemation	The syndicate gets no cash, only shares
25,686	Racemic .	The syndicate gets no shares, only cash
25,687	Racemous	Syndicate with a small capital
25,688	Racemulose	A small syndicate might be formed
25,689	Racer . .	A syndicate to find means to have it reported on
25,690	Racetrack.	Syndicate being formed to [and developed
25,691	Raceway .	Syndicate has taken
25,692	Rachialgia	It would be best to form a syndicate
25,693	Rachidian	Syndicate will be formed
25,694	Rachilla .	Syndicate has expended
25,695	Rachis . .	Syndicate has sent out
25,696	Rachitome	Syndicate to take up a land concession in
25,697	Racial . .	Syndicate has bought the concession
25,698	Rackbars .	Syndicate has obtained the concession
25,699	Rackblock	Syndicate to explore
25,700	Rackingcan	Share in the syndicate
25,701	Rackpin .	The syndicate has called up
25,702	Rackrail .	Syndicate must send out
25,703	Rackrenter	Syndicate must be prepared to
25,704	Racksaw .	**System**
25,705	Rackstick.	Must have a better system
25,706	Rackwork.	Must alter the system of
25,707	Racodium.	The system is bad
25,708	Racoonda	The system is very good
25,709	Racovian .	A better system of
25,710	Racquets .	Cannot adopt a better system
25,711	Radevore .	There has been an entire want of system
25,712	Radiance .	The entire system of work wants thorough revision
25,713	Radiantly .	Have adopted a new system
25,714	Radiating .	Under the old system
25,715	Radiative .	The change from the old system to the new
25,716	Radiator .	Cost of the respective systems
25,717	Radical .	By the new system we expect to save
25,718	Radicalism	The saving effected by —— system
25,719	Radically .	Is the system adapted for our
25,720	Radicose .	The system is not adapted for
25,721	Radicular .	**Table(s)**
25,722	Radiolaria	Refer to the table of
25,723	Radiolite .	By the tables which we send
25,724	Radious .	Shaking tables
25,725	Radix . .	**Tabular**
25,726	Raduliform	Tabular statement(s)

TAB

519

No.	Code Word.	**Tabular** (*continued*)
25,727	Raffish . .	Tabular statement of results
25,728	Rafflenet .	Tabular statement of monthly expenses
25,729	Rafflers .	Tabular statement of work done
25,730	Rafflesia .	Tabular statement of ore extraction
25,731	Raftbridge	Tabular statement will show
25,732	Raftdog .	Information arranged in a tabular form
25,733	Raftering .	**Tailings**
25,734	Raftports .	Average sample of tailings (give)
25,735	Ragabash .	The tailings would be worth working
25,736	Ragamuffin	Lost in the tailings
25,737	Ragbolts .	Assay of tailings
25,738	Ragcarpet.	Tailings will pay to work ·
25,739	Ragdust .	Tailings will not pay to work
25,740	Ragerie .	Estimate tailings on hand to be —— tons
25,741	Ragfair .	Estimate tailings on hand to be worth
25,742	Ragingly .	Tons of tailings worth
25,743	Ragmanroll	The tailings from the —— mill
25,744	Ragounce .	Tailings from vanners
25,745	Ragpicker.	How much can you save in the tailings
25,746	Ragshop .	The best way of treating the tailings
25,747	Ragstone .	Cost of treating the tailings
25,748	Ragwheel .	Treating tailings by the —— process
25,749	Ragwort .	Can treat —— tons of tailings every twenty-four
25,750	Railcar .	By this process of treating the tailings [hours
25,751	Railfence .	Slum mill for treating tailings
25,752	Railjoint .	Assay value of tailings now being worked
25,753	Raillery .	Have tried —— process for treating the tailings
25,754	Railroad .	Tried —— process of treating the tailings, but **it**
25,755	Railway .	**Take** [was not successful
25,756	Raiment .	Take all you can get
25,757	Rainbirds .	Take all the
25,758	Rainbowed	(To) take up
25,759	Rainclouds	(To) take out
25,760	Raindrop .	(To) take over
25,761	Rainfall .	Not to take
25,762	Raingauges	Will take
25,763	Raininess .	Will you take
25,764	Rainprint .	Will he (they) take
25,765	Raintight .	Will it take
25,766	Rainwater.	Will not take
25,767	Rainy . .	Will not take less **(than)**
25,768	Raise . .	Can take
25,769	Raisingbee	Can take any quantity
25,770	Raisins .	Can you take
25,771	Rajahship .	Cannot take
25,772	Rakestale .	Should take
25,773	Rakevein .	Ought not to take
25,774	Rakish . .	Ought not to take more **(than)**
25,775	Rakishly .	It will take too long
25,776	Rakishness	It will take some time

No.	Code Word.	**Take** (*continued*)
25,777	Rallidae .	It will not take much
25,778	Rallus . .	How much will it take to
25,779	Ramagious	In order to take out —— tons per day
25,780	Ramberges	How long will it take to
25,781	Rambler .	Will take —— subject to
25,782	Rambling .	Will be compelled to take
25,783	Rambootan	Necessary to take
25,784	Ramekin .	When it suits —— to take
25,785	Ramhead .	If we take it
25,786	Ramified .	If we take any part of
25,787	Ramiparous	If we do not take it
25,788	Rampage .	We must take
25,789	Rampallian	Have had to take out
25,790	Rampancy	**Taken**
25,791	Ramphastos	Very little has been taken (from)
25,792	Rampion .	A great deal has been taken (from)
25,793	Rampired .	Has much been taken (from)
25,794	Ramrods .	How much has it taken
25,795	Ramsagul .	How much has been taken
25,796	Ramtilla .	It has taken
25,797	Ramulous .	It has taken more than
25,798	Ramuscule	It has not taken
25,799	Rancescent	Have you taken
25,800	Ranched .	I (we) have taken
25,801	Rancidity .	Has (have) taken
25,802	Rancorous	Has (have) not taken
25,803	Rancour .	Has (have) not taken more tha
25,804	Randanite	Taken from
25,805	Randletree	Taken from the dump
25,806	Random .	Sample(s) taken from —— assay(s)
25,807	Randomly	Ore taken from —— assays
25,808	Rangement	Have taken out the old
25,809	Rangership	Have taken some of
25,810	Raniceps .	Have not taken any of
25,811	Rankest .	No more has been taken
25,812	Rankling .	**Taking**
25,813	Rankriding	In taking these samples
25,814	Ransacked	Taking out —— tons per day
25,815	Ransacking	Has (have) been taking
25,816	Ransomable	Not taking
25,817	Ransombill	Taking out sufficient to
25,818	Ransomfree	Is (are) taking out
25,819	Ransoming	We are now taking out more than
25,820	Ransomless	Not taking out as much as
25,821	Ranters .	We are now taking out [new
25,822	Ranterism	Taking out the old timbering and replacing it by
25,823	Rantipole .	**Talc**
25,824	Ranula . .	**Talk**
25,825	Ranunculus	There is a talk of
25,826	Rapaces .	There is no talk of

No.	Code Word.	**Talk** (*continued*)
25,827	Rapadura .	Is there any talk of
25,828	Rapecake .	**Tank(s)**
25,829	Rapeoil .	Cast iron tank(s) to hold —— gallons
25,830	Raperoot .	Wrought iron tank(s) to hold —— gallons
25,831	Rapeseed .	Square wooden tank(s)
25,832	Rapewine .	Round wooden tank(s)
25,833	Raphaelism	Tank(s) —— long, —— wide, —— deep
25,834	Raphaelite	Round tanks —— diameter, —— deep
25,835	Raphanus .	Tank excavated in the rock
25,836	Raphe . .	Tank must have a capacity of —— gallons
25,837	Raphides .	Tank burst
25,838	Rapids . .	Tank(s) holding —— gallons
25,839	Rapierfish .	**Tariff**
25,840	Rapinous .	The new tariff will come into operation
25,841	Rapparee .	Duty under the new tariff
25,842	Raptorial .	The new tariff will be prejudicial
25,843	Raptorious	New tariff imposes heavy duties upon
25,844	Rapturist .	New tariff will make all material much dearer
25,845	Rapturized	New tariff will make everything cheaper
25,846	Rarebit .	**Tax(es)**
25,847	Rareeshow	A heavy tax
25,848	Rarefiable .	To pay the taxes on
25,849	Rarefy . .	What taxes will there be
25,850	Rarefying .	To meet this tax
25,851	Rascaldom	Intend(s) to levy a tax on
25,852	Rascality .	To cover the taxes
25,853	Rascallike.	The tax on
25,854	Rascallion	Must get the tax reduced
25,855	Rascals .	Intend(s) to do away with the tax on
25,856	Rashly . .	The taxes must be paid
25,857	Rashness .	Taxes and insurance
25,858	Raskolnik .	The tax is levied on
25,859	Raspatory.	Let us know the amount of taxes
25,860	Raspberry.	The taxes amount to
25,861	Rasure . .	If taxes are not paid by —— the property will be [seized
25,862	Ratability .	Income tax on profits
25,863	Ratable .	Income tax levied on
25,864	Ratcatcher	Valuation for taxes
25,865	Ratch . .	What is the valuation upon which taxes will be [assessed
25,866	Ratebooks	**Taxing**
25,867	Ratepayer	Without taxing too heavily
25,868	Ratetithe .	Improperly taxing
25,869	Ratholite .	**Team(s)**
25,870	Rathripe .	Ore team broke down owing to
25,871	Ratifia . .	With one team —— tons per day can be hauled
25,872	Ratifiers .	Will necessitate having another team
25,873	Rationally.	Cannot get teams at present
25,874	Rationing .	Including a team of —— horses and mules
25,875	Ratitate .	Team working steadily
25,876	Ratmara .	With one team

No.	Code Word.	**Team(s)** *(continued)*
25,877	Ratpit . .	With two teams
25,878	Ratsbane .	Another team
25,879	Ratsnake .	Team can haul
25,880	Rattailed .	What can team haul in one day
25,881	Ratteen .	Team(s) cannot haul ore on account of
25,882	Rattlebox .	Team(s) at work again
25,883	Rattlecap .	**Technical**
25,884	Rattlehead	In the technical accounts
25,885	Rattlepate	The technical accounts to be sent
25,886	Rattlewort	The technical accounts
25,887	Rattraps .	Technical objections to
25,888	Rauchwacke	From a purely technical point of view
25,889	Ravaged .	Raising technical objections to
25,890	Ravaging .	There are technical difficulties to be overcome
25,891	Ravehook .	Technical difficulties have been overcome
25,892	Ravelment	**Telegram(s).** (See also Date.)
25,893	Raveningly	Your telegram of to-day's date [to
25,894	Ravenously	Cannot understand your telegram ; repeat from ——
25,895	Ravens .	Must prepare report for the public on your telegrams
25,896	Ravenstone	In a telegram from
25,897	Raviney .	On receipt of telegram
25,898	Ravisher .	Has (have) received telegram (from)
25,899	Ravishing .	Has (have) not received any telegram (from)
25,900	Rawboned	Unless you get a telegram from
25,901	Rawest. .	Wait for telegram
25,902	Rawport .	Wait for telegram giving result of assays
25,903	Rayah . .	Waiting reply to my telegram
25,904	Razorable .	Not receiving any telegram
25,905	Razorback	Have you had any telegram from
25,906	Razorbill .	This is our —— telegram of to-day
25,907	Razorfish .	This refers to our —— telegram yesterday
25,908	Razors . .	Referring to your telegram (of)
25,909	Razorshell	Referring to our telegram (of)
25,910	Razorstone	We confirm our telegram of
25,911	Razorstrop	The —— word of telegram should read
25,912	Razzia . .	Repeat —— word of your telegram
25,913	Reabsorb .	See —— telegram to
25,914	Reaccess .	Telegrams and letters to be sent to
25,915	Reaccused	You will find telegram at [advices
25,916	Reaccusing	Do not act on telegram of —— but await further
25,917	Reachable	Telegram did not reach me (us) until
25,918	Reacted .	Have no telegram from you (since)
25,919	Reactions .	Reply to our telegram immediately
25,920	Reactive .	Report fully and send telegram to
25,921	Reactively	You have not yet replied to our telegram
25,922	Readably .	Your telegram to hand
25,923	Readept .	Telegram came to hand too late
25,924	Readepting	Telegram is not satisfactory
25,925	Readeption	This telegram is by A B C Code
25,926	Readiest .	This telegram is by Moreing's Code, second edition

No.	Code Word.	Telegram(s) (continued)
25,927	Readingboy	This telegram is by Moreing's New General and
25,928	Readjourn	Send a telegram by A B C Code [Mining Code
25,929	Readjusted	Send a telegram by Moreing's Code, second edition
25,930	Readmitted	Send a telegram by Moreing's New General and Mining Code
25,931	Readopted	**Telegraph.** (See also Cable—Wire.)
25,932	Readorn .	Telegraph by A B C Code; —— has no copy of Moreing's New General and Mining Code
25,933	Readorning	Telegraph by Moreing's Code, second edition
25,934	Readvance	Telegraph by Moreing's Code, second edition; —— has no copy of Moreing's New General and Mining Code
25,935	Readymade	Telegraph by Moreing's New General and Mining
25,936	Readymoney	Telegraph how long you stay at [Code
25,937	Reaffirm .	Telegraph when you can leave for
25,938	Reafforest .	Telegraph width of vein
25,939	Reagreeing	Telegraph to me at
25,940	Realgar .	Telegraph fuller report
25,941	Realism .	Telegraph as soon as possible
25,942	Realistic .	Telegraph any new strike at once
25,943	Realities .	Telegraph what has been done with regard to
25,944	Reality .	Telegraph at my (our) expense
25,945	Realizable	Telegraph all particulars
25,946	Realized .	Telegraph the latest information about
25,947	Realizing .	Telegraph any information of interest relating to the
25,948	Reallege .	Telegraph what you have done [developments
25,949	Realleging	Telegraph what you are doing in
25,950	Realliance	Telegraph your opinion as to the present appearance
25,951	Realm . .	Telegraph the result of [of the
25,952	Realmless .	Telegraph if there is any improvement
25,953	Reanimate	Telegraph how the new workings are looking
25,954	Reannex .	Telegraph the substance of
25,955	Reannexing	Telegraph how much you can send
25,956	Reanoint .	Telegraph whether you can
25,957	Reanswer .	Telegraph a definite answer
25,958	Reapers .	Telegraph clearly what is wanted
25,959	Reapparel	Telegraph if it is not in order
25,960	Reappear .	Please telegraph to —— to pay me
25,961	Reapplied	Telegraph reply to reach us without fail to-morrow
25,962	Reapply .	Please telegraph reply not later than [morning
25,963	Reapplying	We require this information by telegraph not later
25,964	Reapproach	Will telegraph again [than
25,965	Reardorse .	Will telegraph
25,966	Rearfront .	Cannot telegraph
25,967	Rearguard	Will telegraph instructions
25,968	Reargue .	Please telegraph me at once
25,969	Rearguing	Shall not be able to telegraph
25,970	Rearingbit	Could not telegraph sooner on account of
25,971	Rearmost .	Do not answer by telegraph; answer by letter
25,972	Rearrange	Reply by telegraph immediately

No.	Code Word.	
25,973	Rearrank	**Telegraphed**
25,974	Rearvault .	Have telegraphed —— to pay you
25,975	Rearwards	Telegraphed you yesterday
25,976	Reascend .	Telegraphed you this morning
25,977	Reason .	Have telegraphed to
25,978	Reasonings	Have telegraphed full instructions to
25,979	Reasonless	**Telegraphic**
25,980	Reassemble	Telegraphic communication
25,981	Reassign .	There is no telegraphic communication with
25,982	Reassumed	Is the nearest telegraphic station
25,983	Reattach .	All telegraphic communications stopped on account
25,984	Reattained	Telegraphic communication resumed [of
25,985	Reattempt	Telegraphic communication between —— and ——
25,986	Reaumuria	**Telegraphing** [is interrupted
25,987	Reavowed	When telegraphing in future (use)
25,988	Reavowing	**Tell**
25,989	Reawake .	Please tell —— to
25,990	Reawaking	Can you tell
25,991	Rebaptism	Cannot tell
25,992	Rebaptize .	Cannot tell how it occurred
25,993	Rebatement	Cannot tell who it was
25,994	Rebeck .	Cannot tell exactly
25,995	Rebeccaism	If you can tell us
25,996	Rebeccaite	Do not tell
25,997	Rebeldom	**Telluride**
25,998	Rebelled .	The gold is found as a telluride
25,999	Rebellious	**Temporary**
26,000	Rebellowed	As a temporary measure
26,001	Rebloom .	A temporary way of
26,002	Reblooming	Only temporary
26,003	Reblossom	Can you make any temporary arrangement
26,004	Reboant .	Do you think the improvement is only temporary
26,005	Reboation	Think improvement is only temporary
26,006	Rebounding	Is this temporary or permanent
26,007	Rebrace .	Temporary arrangements
26,008	Rebracing.	A temporary advance
26,009	Rebreathe.	A temporary advantage
26,010	Rebucous.	**Tend**
26,011	Rebuff. .	This will tend to
26,012	Rebuffeted	Will this tend to
26,013	Rebuffing .	This will not tend to
26,014	Rebuilder.	**Tendency**
26,015	Rebuilding	With a tendency to
26,016	Rebuilt .	No tendency to
26,017	Rebukable	Is there any tendency
26,018	Rebuked .	There is a strong tendency to
26,019	Rebukeful.	If you notice any tendency to
26,020	Rebuking .	**Tender(s)**
26,021	Rebukingly	Will send in a tender
26,022	Rebuoy .	Have sent in a tender

No.	Code Word.	**Tender(s)** (*continued*)
26,023	Rebuoying	Send in a tender
26,024	Reburied .	Has (have) not sent any tender
26,025	Rebus . .	The tender must include
26,026	Rebuttal .	Tender is the lowest
26,027	Recadency	The highest tender was
26,028	Recall . .	The lowest tender was
26,029	Recallable	Tenders for sinking the shaft
26,030	Recalling .	Tenders too high
26,031	Recallment	Has (have) called for tenders
26,032	Recaptor .	Intend(s) to call for tenders
26,033	Recaptured	Inviting tenders from
26,034	Recarnify .	From whom shall we invite tenders
26,035	Recarried .	Tenders from the best two or three makers
26,036	Recarry .	Have received tenders from
26,037	Recasting .	Cannot accept your tender
26,038	Receipting	Tender accepted for
26,039	Receiptor .	**Tendered**
26,040	Receivable	Tendered for the work
26,041	Receivers .	Tendered for the supply of
26,042	Recent. .	**Term(s)**
26,043	Recentred	For a term of —— years
26,044	Receptacle	Term has nearly expired
26,045	Recessed .	What terms can you get
26,046	Recession.	Terms are impossible
26,047	Recharged	Impossible to come to terms with
26,048	Rechasten	Are you likely to come to terms with
26,049	Rechauffes	The best terms that I can get
26,050	Recheat .	Cannot accept the terms offered
26,051	Recherche	Accept the terms offered
26,052	Rechoose .	If these terms are accepted
26,053	Rechoosing	On the same terms as before
26,054	Recidivate	Hope —— will accept these terms
26,055	Recidivous	Will you accept these terms
26,056	Recipe. .	Would advise accepting these terms
26,057	Recipiency	On satisfactory terms
26,058	Recipient .	Make the best terms possible
26,059	Reciprocal	The best terms possible have been made
26,060	Reciprok .	If to be had on easy terms
26,061	Recital .	Terms will be made to suit
26,062	Recitation	Terms are too high
26,063	Reciters .	Terms are very low
26,064	Reckless .	Let us know the lowest terms
26,065	Recklessly	Let us know the best terms you can obtain
26,066	Reckon .	Try to obtain better terms
26,067	Reckoneth	Try to come to terms on the matter
26,068	Reckoning	The best terms I (we) can offer
26,069	Reclaimant	According to the terms of the
26,070	Reclaimed	Have offered better terms
26,071	Reclinate .	Have been offered better terms
26,072	Reclosed .	Would like to come to terms

No.	Code Word.	**Term(s)** (*continued*)
26,073	Reclosing .	If you cannot obtain better terms
26,074	Reclothe .	Obtain satisfactory terms
26,075	Reclusely .	More favourable terms
26,076	Recognitor	Terms are not stated
26,077	Recognized	Better terms should be got, considering
26,078	Recoiled .	Terms have been modified
26,079	Recoinage	Terms must be modified
26,080	Recollect .	Terms must be distinctly stated
26,081	Recolonize	Have it on the most favourable terms
26,082	Recombine	On the following terms
26,083	Recomfort	**Terminate**
26,084	Recommend	(To) terminate the affair
26,085	Recommit	(To) terminate the engagement
26,086	Recompact	**Terminated**
26,087	Recompense	The difficulty was terminated by
26,088	Recompile	Was terminated by
26,089	Reconciled	**Terminating**
26,090	Recondite	Thus terminating the
26,091	Reconduct	**Termination**
26,092	Reconfirm	The termination of the strike
26,093	Reconjoin .	The termination of the difficulty
26,094	Reconquest	The termination of the agreement (or contract)
26,095	Reconsider	**Territory**
26,096	Recontinue	In the territory of
26,097	Reconvey .	A vast territory
26,098	Recordance	The extent of the territory
26,099	Recorder .	The territory covered by the concession
26,100	Recouching	**Tertiary**
26,101	Recoup .	Tertiary formation
26,102	Recoupment	**Test.** (See Mill Test.)
26,103	Recovering	A fair test of
26,104	Recreancy	Can you test (will you test)
26,105	Recreating	Can you not test
26,106	Recross .	Subjected to a thorough test
26,107	Recrossing	Will give it a fair test
26,108	Recruiting	Have made a fair test of
26,109	Rectangle .	**Tested**
26,110	Rectified .	Have tested
26,111	Rectify. .	Has (have) been thoroughly tested
26,112	Rectifying	Not yet been tested
26,113	Rectitude .	Have not yet tested
26,114	Rectorship	The statements can be tested
26,115	Rectory .	**Testify (Testifies)**
26,116	Rectrix. .	Shall have to testify
26,117	Recumbence	Testifies to
26,118	Recuperate	**Testimony**
26,119	Recureful .	Testimony will be required
26,120	Recurrency	Testimony in support of
26,121	Recurvated	Shall be prepared to bear testimony
26,122	Recurve .	Most important testimony

No.	Code Word.	
26,123	Recurvity .	**Than**
26,124	Recurvous	Than we thought
26,125	Recusancy	More than
26,126	Recusant .	More than was expected
26,127	Recusative	More than ever
26,128	Redacting.	Not more than
26,129	Redbelly .	Less than
26,130	Redbird .	Not less than
26,131	Redbreast.	Better than
26,132	Redcaps .	Lower than
26,133	Redcoat .	Higher than
26,134	Redcross .	Than that
26,135	Redden .	Cannot be less than
26,136	Reddening	Cannot be more than
26,137	Reddendo.	Than any
26,138	Reddish .	Than any yet
26,139	Reddition.	Than it could have been
26,140	Redditive .	**Thank(s)**
26,141	Redecorate	Has (have) begged me to thank you
26,142	Rededicate	Please to thank —— on our behalf
26,143	Redeem .	Thank him for his services
26,144	Redeemeth	Thanks for your services
26,145	Redeeming	Thank all for what they have done
26,146	Redeemless	Accept my thanks for
26,147	Redelivery	A vote of thanks was passed
26,148	Redemand	Passed a vote of thanks to you and staff
26,149	Redemise .	Our thanks are due to
26,150	Redemising	Accept our best thanks
26,151	Redemption	**That**
26,152	Redemptive	That is (are)
26,153	Redemptory	That is (are) not
26,154	Redented .	Is that
26,155	Redeposit.	Is that so
26,156	Redfire .	All that can be
26,157	Redgame .	If there is any that
26,158	Redhanded	In that case
26,159	Redhot .	Is that all that
26,160	Redigest .	That will be
26,161	Rediminish	That will do
26,162	Redisburse	That will not be
26,163	Rediscover	That will not do
26,164	Redisposed	If that is so
26,165	Redisseize	If that is not so
26,166	Redissolve	Unless that is done
26,167	Redivide .	Unless that
26,168	Redividing	If that can be
26,169	Redlattice.	If that cannot be
26,170	Redletter .	That is not the case
26,171	Redlooked	When that is done
26,172	Redly . .	**Their**

No.	Code Word.	Their (*continued*)
26,173	Rednose .	Is not their own
26,174	Redolency	By their own
26,175	Redolent .	Their own
26,176	Redoubled	**Their stock**
26,177	Redoubling	When their
26,178	Redoubt .	**Is it their own**
26,179	Redowa .	It is their
26,180	Redperch .	It is their fault
26,181	Redpole .	Their view of the matter
26,182	Redraft .	In their case
26,183	Redrafting	Their contention is
26,184	Redressal .	In their opinion
26,185	Redressing	Their opinion is
26,186	Redressive	Their —— will be
26,187	Redriven .	How is their
26,188	Redshank .	What is their
26,189	Redsilver .	Who is their
26,190	Redskin .	He is their
26,191	Redsorrel .	Their friends
26,192	Redstart .	Their agent(s)
26,193	Redstreak	Their services
26,194	Redtapery	They want their
26,195	Redtapism	Will want for their
26,196	Redthroat .	**Them**
26,197	Redtop .	To them
26,198	Redub . .	By them
26,199	Redubber .	For them
26,200	Reducibly .	For them and their friends
26,201	Redundance	Against them
26,202	Redwood .	From them
26,203	Reebok .	Let them know
26,204	Reechoed .	Inform them that
26,205	Reechoing	**Themselves**
26,206	Reechy .	For themselves
26,207	Reedbirds .	By themselves
26,208	Reedbuck .	**Then**
26,209	Reedgrass .	Will know by then
26,210	Reedified .	If, by then
26,211	Reedify .	Then you can
26,212	Reedifying	Then you will
26,213	Reedling .	Will then
26,214	Reedmace .	Will not then
26,215	Reedorgan	Then if you find
26,216	Reedplane	You can then proceed
26,217	Reedstop .	**Theodolite**
26,218	Reefbands	Please send theodolite as soon as possible
26,219	Reefknot .	Have sent theodolite
26,220	Reefline .	**Theory**
26,221	Reefpoint .	If this theory is correct
26,222	Reeftackle	Upon this theory

No.	Code Word.	
26,223	Reefties	. **There**
26,224	Reelcotton	There is (are)
26,225	Reelection	There is (are) not (no)
26,226	Reelevate .	If there are
26,227	Reeligible .	If there are not
26,228	Reelstand .	In case there is (are)
26,229	Reembark	Is (are) there (any)
26,230	Reembattle	There will be
26,231	Reembodied	There will not be
26,232	Reembody	Will there be
26,233	Reembraced	Is there likely to be
26,234	Reemerging	There could not be (have been)
26,235	Reenaction	There may be (may have been)
26,236	Reendow .	There might be (might have been)
26,237	Reengraved	If there might not be (not have been)
26,238	Reenjoyed	There ought to be (to have been)
26,239	Reenjoying	There was (were)
26,240	Reenkindle	There were not (any)
26,241	Reenlisted	There would not be (not have been)
26,242	Reenslave	If there should be (should have been)
26,243	Reenstamp	Unless there are
26,244	Reentering	When there are
26,245	Reentrance	**Thereby**
26,246	Reentry .	Can tell thereby
26,247	Reermouse	Cannot tell thereby
26,248	Reexamined	Can judge thereby
26,249	Reexchange	**Thermometer**
26,250	Reexhibit .	Very cold weather set in; thermometer down to ——
26,251	Reexpel .	Thermometer has been down to ——°
26,252	Reexpelled	Thermometer has been up to ——°
26,253	Reexport .	Thermometer ——° below zero
26,254	Reextent .	The heat intense; thermometer at ——°
26,255	Refashion .	**They**
26,256	Refastened	They are
26,257	Refectorer	They are not
26,258	Referees .	They can
26,259	Referrible .	They cannot
26,260	Refigured .	If they can
26,261	Refinedly .	If they are
26,262	Refinement	If they are not
26,263	Refineries .	They will
26,264	Refinery .	They will not
26,265	Refit . .	They will try
26,266	Refitted .	They could
26,267	Reflecting .	They might
26,268	Reflexed .	Unless they
26,269	Reflexible .	Until they
26,270	Reflexity .	When they
26,271	Reflexive .	**Thick**
26,272	Reflourish	Is very thick

No.	Code Word.	
		Thick (*continued*)
26,273	Reflower .	A thick seam of
26,274	Refluency.	**Thickness**
26,275	Refluent .	What is the thickness of
26,276	Refomented	The thickness of the seam
26,277	Reformers	**Thin**
26,278	Reformist .	The vein is thin and ill defined
26,279	Reformly .	Where the vein is thin
26,280	Refortify .	Very thin
26,281	Refract .	A very thin streak of
26,282	Refragable	A thin line (of)
26,283	Refrain .	Too thin to work
26,284	Refraineth	**Think(s)**
26,285	Refraining	As you think best
26,286	Reframed .	What do you think of
26,287	Refresher .	I think you had better
26,288	Refreshful	Telegraph if you think
26,289	Refreshing	Think(s) it advisable to
26,290	Refringent	I (we) think
26,291	Refuge . .	I (we) do not think
26,292	Refugeeism	Think(s) that
26,293	Refulgent .	Think it will
26,294	Refunded .	Think it will not
26,295	Refunding	Do(es) not think that
26,296	Refundment	If you think that
26,297	Refurnish .	Think it prudent
26,298	Refutably .	Hardly think that
26,299	Refutation	There is reason to think
26,300	Refutatory	Do not think it advisable (to)
26,301	Regaining.	Think you ought to
26,302	Regality .	When do you think of
26,303	Regardant	Do you think it well to
26,304	Regarded .	**Third(s)**
26,305	Regardful .	One-third (of)
26,306	Regardless	Two-thirds (of)
26,307	Regather .	**This**
26,308	Regattas .	By this you will see
26,309	Regency .	This does not prove
26,310	Regenesis .	Does this mean
26,311	Regentbird	This is all I can find out
26,312	Regentship	This is a (the)
26,313	Regicidal .	This is an attempt to
26,314	Regifugium	This may be (so)
26,315	Regild . .	If this is so
26,316	Regilding .	If this is not so
26,317	Regimental	This is not the
26,318	Regional .	In this case
26,319	Registered	This or the other
26,320	Registrary	This must not
26,321	Regium .	This has not
26,322	Regma . .	This must be so

No.	Code Word.	**This** (*continued*)
26,323	Regnicides	This was
26,324	Regranted	This was not
26,325	Regrators .	This can
26,326	Regret . .	This cannot
26,327	Regretful .	This could
26,328	Regretting	What has this to do with
26,329	Regrowth .	After this
26,330	Reguerdon	Before this
26,331	Regularize	During this
26,332	Regularly .	**Thoroughly**
26,333	Regulated	A thoroughly good
26,334	Regulation	Went most thoroughly into every detail
26,335	Regulus .	You must thoroughly investigate every matter con- [nected with
26,336	Rehearsers	Have thoroughly
26,337	Rehelm .	Have not thoroughly
26,338	Rehelming	You must thoroughly
26,339	Rehibition	**Thought**
26,340	Rehibitory	Has (have) thought it necessary to
26,341	Reillumine	Has (have) not thought it necessary
26,342	Reimburse	Without any thought of the future
26,343	Reimmerged	Have thought it advisable
26,344	Reimplant	Have not thought it advisable
26,345	Reimported	It is thought here that
26,346	Reimpose .	It is not thought much of here
26,347	Reimposing	It is thought highly of here
26,348	Reimpress	Have thought the matter over
26,349	Reimprison	**Thoughtless(ness)**
26,350	Reincense	Is too thoughtless
26,351	Reincite .	It was very thoughtless of
26,352	Reincrease	From sheer thoughtlessness
26,353	Reincurred	**Thousand(s)**
26,354	Reindeer .	There are —— thousand tons of ore
26,355	Reinduced	Many thousand tons
26,356	Reinducing	At a cost of several thousands (of)
26,357	Reinfect .	Have spent thousands (of)
26,358	Reinforced	**Threaten(s)**
26,359	Reinfused .	Threaten(s) to bring an action
26,360	Reinhabit .	Threaten(s) us with
26,361	Reinquire .	Now threaten(s) to
26,362	Reinserted	Threaten(s) to oppose us
26,363	Reinspect .	**Threatened**
26,364	Reinspired	Is (are) threatened with
26,365	Reinstall .	Has (have) threatened to
26,366	Reinstates	Have been threatened with
26,367	Reinstruct	**Through**
26,368	Reinsured.	Through the
26,369	Reinterred	To get through
26,370	Reinthrone	Will soon be through
26,371	Reinundate	Cannot get through
26,372	Reinvest .	We are not yet through

No.	Code Word.	
		Through (*continued*)
26,373	Reinvolve.	Through the ledge
26,374	Reissuable	Carry it through
26,375	Reissue	We can carry it through
26,376	Reissuing .	Cannot carry it through
26,377	Reiterant .	Through all the workings
26,378	Rejectable	Have been through all
26,379	Rejective .	Through the whole length of
26,380	Rejoice .	Through our
26,381	Rejoicing.	Through them
26,382	Rejoinders	Through their
26,383	Rejointed.	Through what was
26,384	Rejolt . .	Through —— exertions
26,385	Rejudged .	Through —— influence
26,386	Rejuvenate	**Throughout**
26,387	Relapsing .	Throughout the country
26,388	Relational	Throughout the mine
26,389	Relatively.	Throughout the district
26,390	Relators .	Throughout the summer
26,391	Relatrix .	**Through-rate(s)**
26,392	Relax . .	Through-rate to
26,393	Relaxable.	The lowest through-rate we can get
26,394	Relaxation	The nearest place to which we can get a through-rate
26,395	Relaxative	**Thrown**
26,396	Relaxeth .	Has (have) been thrown away
26,397	Relbun .	It is simply good money thrown away
26,398	Releasable	Has (have) been thrown
26,399	Release .	Thrown on the market
26,400	Releasing .	**Throwing**
26,401	Relegated.	Are now throwing away
26,402	Relegating	It is throwing money away (to)
26,403	Relenting .	**Thursday**
26,404	Relentment	Next Thursday
26,405	Relessor .	Last Thursday
26,406	Relevancy	On Thursday
26,407	Relevant .	Every Thursday
26,408	Reliably .	Every other Thursday
26,409	Relicly. .	**Tight**
26,410	Relics . .	Can be made tight
26,411	Reliefless .	It has to be made tight
26,412	Relieved .	Cannot make it tight
26,413	Religious .	**Timber(s)**
26,414	Relinquish	Plentiful supply of timber
26,415	Reliqua .	Timber is scarce [district
26,416	Reliquary.	There is no timber for mining purposes in the
26,417	Relishable	Timber has to be brought a long distance to the mine
26,418	Reloaded .	Timber is plentiful for all mining purposes, and for
26,419	Reloading	Is there plenty of timber near the mine(s) [fuel
26,420	Relocating	The shaft(s) require(s) fresh timbers
26,421	Relodge .	Timbers rotten in the shaft
26,422	Relodging	Estimate cost of new timbers in the shaft at

No.	Code Word.	**Timber(s)** (*continued*)
26,423	Reluctancy	What do you estimate new timbers in shaft will cost
26,424	Reluctant .	All timber has to be imported
26,425	Relumined	Plenty of valuable timber
26,426	Remanation	The right of cutting timber
26,427	Remanent	There is plenty of timber on the property
26,428	Remarkably	There is no timber on the property
26,429	Remarriage	The timbers are still sound
26,430	Remarrying	The timbers are quite rotten
26,431	Remeasure	The old timbers are not safe; we must put new ones in
26,432	Remediably	Now putting in new timbers
26,433	Remedial .	Time required to put in new timbers
26,434	Remediless	Have shored the old workings with heavy timbers
26,435	Remedy .	Size of timbers required
26,436	Remedying	Timbers: length —— feet —— inches × ——
26,437	Remelted .	Timbers —— inches square [inches
26,438	Remelting	Timbers —— × —— inches
26,439	Remember	The extreme length of timbers
26,440	Rememorate	New timbers throughout
26,441	Remercied	Timbers for framing
26,442	Remigable	Timber cribs
26,443	Remiges .	**Timbered**
26,444	Remigrate	Have timbered the
26,445	Remijia .	**Timbering**
26,446	Reminders	Timbering commenced
26,447	Remissful .	Timbering must be done at once
26,448	Remissible	Estimate cost of timbering at
26,449	Remission	What will be the cost of timbering
26,450	Remissly .	Does your estimate include timbering
26,451	Remissness	My estimate includes timbering
26,452	Remitment	Timbering throughout
26,453	Remittal .	Have finished timbering
26,454	Remix . .	Timbering will be a heavy item in the cost, owing
26,455	Remnants	**Time(s)** [to scarcity of wood
26,456	Remodel .	You must get the time extended
26,457	Remodelled	Will —— extend the time to
26,458	Remodified	Time extended to
26,459	Remodify .	Will extend the time to
26,460	Remontoir	Will not extend the time beyond
26,461	Remordency	There is not sufficient time allowed
26,462	Remorsed .	Impossible to do anything in so short a time
26,463	Remotely .	Cannot get to —— and examine the mine in the
26,464	Remoteness	So that no time may be lost [time allowed
26,465	Remould .	Shall not have time to
26,466	Remoulding	It is now a good time (to)
26,467	Remounted	It is now a bad time (to)
26,468	Removable	To be in time
26,469	Remove .	Will you have time to
26,470	Removing	Last time
26,471	Remugient	There will be time enough
26,472	Remunerate	Time is too short

No.	Code Word.	**Time(s)** (*continued*)
26,473	Remurmur	What time will you
26,474	Rename .	For some time to come
26,475	Renaming	For some time past
26,476	Renascent	This time
26,477	Renascible	Next time
26,478	Renavigate	In the time
26,479	Renculus .	In a week's time
26,480	Rendering	In a month's time
26,481	Rendezvous	In —— months' time
26,482	Rendrock .	Another time
26,483	Reneague .	At all times of the year
26,484	Renegation	Can only be worked at times
26,485	Renerving	There are times when the mine is completely shut off
26,486	Renewable	The mine can be worked at all times [from
26,487	Renneted .	Can the mine(s) be worked at all times of the year
26,488	Rennetwhey	Many times
26,489	Renovate .	At such times
26,490	Renovating	Within the specified time
26,491	Renovation	Time must be limited
26,492	Renowned	Is (are) in good time
26,493	Renownedly	Can you extend the time
26,494	Renownful	Cannot get time extended
26,495	Renownless	A short time ago
26,496	Rentarrear	Until the time has expired
26,497	Rentcharge	Unless —— can get time extended
26,498	Rentday .	At what time
26,499	Rentroll .	How much time can you give
26,500	Rentseck .	Time occupied in
26,501	Renverse .	A great waste of time
26,502	Renvoy .	A most anxious time
26,503	Reobtained	Time is all important
26,504	Reoccupied	The best time
26,505	Reopen .	It is now the best possible time to
26,506	Reopening	The worst time
26,507	Reopposed	If time permits
26,508	Reopposing	Hardly expect to be in time
26,509	Reordained	Expect to be in time for
26,510	Reordering	Will be in time
26,511	Reorganize	Will not be in time
26,512	Reorient .	It will take some time to
26,513	Repacified	Cannot fix a time
26,514	Repacify .	What time did our message reach you
26,515	Repacking	What time do our messages arrive
26,516	Repairer .	Have plenty of time to
26,517	Repairment	If at any time
26,518	Repandous	**Tin**
26,519	Reparably	Tin has been found
26,520	Reparative	Tin ore
26,521	Repartee .	Stream tin
26,522	Repass. .	**Title(s)**

No.	Code Word.	Title(s) (*continued*)
26,523	Repassage	Have the titles examined by some competent person
26,524	Repatriate	Searching examination of titles must be made
26,525	Repayable	The title is good
26,526	Repeal .	The title is not good
26,527	Repealers .	Titles to property are perfect
26,528	Repeatedly	Titles to property are not perfect
26,529	Repellency	Titles certificated
26,530	Repellent .	The titles are all right
26,531	Repentance	Is the title to the property good
26,532	Repenteth	Will give a proper title
26,533	Repeopling	Cannot get a good title
26,534	Repercuss	Unless you can get a better title
26,535	Reperusal .	Have had the titles examined ; they are all right
26,536	Reperusing	Have had the titles examined ; they are not good
26,537	Repetend .	Owing to the titles being wrong (or defective)
26,538	Repetitive	Examine into the title of ——, and furnish us with
26,539	Repiningly	The title is disputed [the best legal authority
26,540	Replacing .	There is a flaw in the title
26,541	Replaited .	Trouble about titles will cause delay
26,542	Replead .	Owing to defective title must decline the property
26,543	Repledge .	Title deed not on record
26,544	Repledging	Title not patented
26,545	Replegiare	The title must be valid in every respect
26,546	Replenish .	Title must be patented
26,547	Replevin .	Title must be recorded
26,548	Replevisor	Title not yet patented (but)
26,549	Replevy .	Is title in order
26,550	Replevying	Title has not yet been recorded
26,551	Replicated	It is a question of title
26,552	Replunged	Lawyers state title is in order
26,553	Replyer .	An indefeasible title
26,554	Repoison .	Under what title is the property held
26,555	Repolished	The title is held direct from
26,556	Reportable	Title held under U.S. patent
26,557	Reportage	The validity of the titles is questioned
26,558	Reporting	Cannot do anything until we are satisfied as to title
26,559	Reposance	Title in perfect order ; transfer deeds registered (or recorded)
26,560	Reposedly	Title bad, do not touch it
26,561	Reposeful .	**To**
26,562	Repository	In order to
26,563	Reposure .	So as to
26,564	Repour .	Not to be
26,565	Represent .	So as not to be
26,566	Repress .	Are not to
26,567	Reprieve .	Is not to
26,568	Reprieving	**To-day**
26,569	Reprimand	Arrived here to-day
26,570	Reprinted .	Leaving here to-day for
26,571	Reprinting	Will send to-day

No.	Code Word.	To-day. (*continued*)
26,572	Reprized .	Have sent to-day
26,573	Reproach .	Have written to-day
26,574	Reprobacy	Have heard to-day that
26,575	Reprobance	To-day at —— a.m.
26,576	Reprobated	To-day at —— p.m.
26,577	Reproduce	Meeting held to-day
26,578	Reproof .	We have to-day
26,579	Reprovably	**Together**
26,580	Reproval .	Both together
26,581	Reptile .	Together with
26,582	Reptilian .	Not together
26,583	Reptonize .	**To-morrow.** (See also Telegraph.)
26,584	Republican	Will leave to-morrow for
26,585	Republish .	Send me (us) to-morrow without fail
26,586	Repudiated	Will send to-morrow
26,587	Repugnancy	Cannot send to-morrow
26,588	Repugnant	To-morrow morning
26,589	Repugned	To-morrow evening
26,590	Repulpit .	Meeting will be held to-morrow
26,591	Repulsing .	We want reply early to-morrow
26,592	Repulsive .	The day after to-morrow
26,593	Repurchase	**Ton(s).** (See Tables at end.)
26,594	Reputeless	Value per ton
26,595	Requested	Per ton
26,596	Requesting	Hauling —— tons per day
26,597	Requiem .	How many tons can you haul in twenty-four hours
26,598	Requietory	Weighing —— tons
26,599	Requirable	About —— tons of quartz
26,600	Required .	What is the value per ton (of)
26,601	Requiring .	Per ton of 2000 lbs.
26,602	Requisite .	Tons of 2000 lbs.
26,603	Requitals .	Per ton of 2240 lbs.
26,604	Reredos .	Tons of 2240 lbs.
26,605	Rerefine .	Yield per ton
26,606	Rerefining	Grains per ton
26,607	Resailed .	Ounces per ton
26,608	Resaunt .	How much per ton
26,609	Rescinded	How many tons of
26,610	Rescinding	How many tons have you
26,611	Rescissory	How many tons of ore are you getting from
26,612	Rescuable	How many tons of ore are you milling
26,613	Rescue .	Producing —— tons of ore
26,614	Rescueless	Have given us —— tons of ore
26,615	Rescussor .	Shall require about —— tons
26,616	Research .	From this chute we have got —— tons
26,617	Reseda .	Tons crushed
26,618	Reseizure .	Tons milled
26,619	Resemblant	Tons smelted
26,620	Reseminate	Tons, valued at
26,621	Resentful .	Dry tons

No.	Code Word.	Ton(s) (continued)
26,622	Reserate .	Wet tons
26,623	Reservance	**Tonnage**
26,624	Reservoir .	Tonnage and value of ore
26,625	Resettable	Tonnage and value of ore reserves
26,626	Resettle .	What is the tonnage and value of
26,627	Resettling	What is the tonnage of
26,628	Reshape .	Gross tonnage
26,629	Reshaping	Net tonnage
26,630	Reshipment	**Too**
26,631	Reshipped	Too much (many)
26,632	Residenter	Not too much
26,633	Residuum .	Too little
26,634	Resignant .	Not too little
26,635	Resilient .	Too great
26,636	Resinbush	Too few
26,637	Resinoid .	**Top**
26,638	Resinously	At the top
26,639	Resiny . .	Near the top
26,640	Resisteth .	At the top of the shaft
26,641	Resistible .	At the top of the hill
26,642	Resistless .	**Total**
26,643	Resolder .	I estimate the total expense of working the ore at
26,644	Resoluble .	I estimate the total expense of erecting machinery at
26,645	Resolution	I estimate the total value of mine, machinery, and [plant at
26,646	Resolutive	Total expense
26,647	Resolve .	Total amount
26,648	Resolvedly	What do you estimate is total value of mine and [machinery
26,649	Resonancy	Total expense of mill this month is
26,650	Resonator	Total expense of mine this month is
26,651	Resorb .	What has been the total expense of
26,652	Resorbent	Total amount of ore hauled in —— was
26,653	Resorcine .	Total amount of ore for the year —— was
26,654	Resource .	What has been total amount of ore hauled
26,655	Respectful	What is the total
26,656	Respecting	What will be the total
26,657	Respection	At a total cost of
26,658	Respirable	Telegraph total amount
26,659	Respirator	Total amount of freight
26,660	Respired .	Total amount of stores
26,661	Respiting .	Total amount of invoice
26,662	Respond .	Total amount of expenditure
26,663	Respondeth	Total sales amount to
26,664	Responding	Total purchases amount to
26,665	Responsal	The total will be about
26,666	Responsion	The total —— so far, is
26,667	Responsive	**Touch**
26,668	Responsory	Do not touch it
26,669	Restagnate	Will not touch it
26,670	Restaur .	It is not worth while to touch it
26,671	Restaurant	Advise you not to touch it on any account

No.	Code Word.	Touch (*continued*)
26,672	Restem .	Keep in touch with
26,673	Restemming	Near enough to touch
26,674	Restfully .	**Towards**
26,675	Restharrow	Towards the
26,676	Resthouse	Now heading towards the
26,677	Restiaceae	**Town**
26,678	Restiff . .	No town nearer than ——, —— miles away
26,679	Restiform .	The mine is situated —— miles from the town **of**
26,680	Resting .	Near the town of
26,681	Restitute .	At the town of
26,682	Restively .	**Trace(s)**
26,683	Restorable	A trace of
26,684	Restorator	No trace of
26,685	Restored .	To trace
26,686	Restrained	Can trace
26,687	Resubject .	Cannot trace
26,688	Resublime	Can you trace
26,689	Result . .	We found traces of
26,690	Resultance	On assaying, traces were found of
26,691	Resulting .	Cannot find any traces of
26,692	Resummons	If you can trace
26,693	Resumption	If you cannot trace
26,694	Resumptive	Destroyed all traces of
26,695	Resupplied	**Traced**
26,696	Resupply .	The vein can be traced for
26,697	Resurgence	The vein cannot be traced beyond
26,698	Resurprise	Can the vein be traced for any distance
26,699	Resurrect .	Can be traced
26,700	Resurvey .	Cannot be traced
26,701	Retailers .	Has been traced for miles
26,702	Retainal .	Was traced for a short distance
26,703	Retaliate .	Has (have) been traced (to)
26,704	Retard . .	Has (have) not been traced
26,705	Retarding .	The lode has been traced for
26,706	Retecious .	**Trachyte**
26,707	Retepora .	**Tracing(s)**
26,708	Rethor . .	Send tracing(s) of plan(s) as quickly as possible
26,709	Retiarius .	Tracings of the plans
26,710	Retiary .	Tracings of plan and section
26,711	Reticency .	Will send tracings
26,712	Reticle . .	**Tract**
26,713	Reticulosa	Comprises a large tract of country
26,714	Reticulum	Covering a large tract of country
26,715	Retinalite .	The tract of country included in the
26,716	Retinervis .	Has granted a large tract
26,717	Retinitis .	A small tract
26,718	Retiped .	**Trade**
26,719	Retiracy .	Trade is brisk
26,720	Retirade .	Trade is dull and likely to continue so
26,721	Retiredly .	Trade is dull, but likely to improve shortly

No.	Code Word.	**Trade** (*continued*)
26,722	Retorsion .	Do (does) a large trade in
26,723	Retorted .	**Train**(s)
26,724	Retortive .	Trains delayed by
26,725	Retoss . .	No trains running, owing to
26,726	Retossing .	By train to —— and thence by
26,727	Retouch .	Everything in train (for)
26,728	Retractate	Put everything in train for
26,729	Retraxit .	Send by passenger train
26,730	Retreated .	Send by goods train
26,731	Retreatful.	Sent by passenger train
26,732	Retribute .	Sent by goods train
26,733	Retrieval .	**Tram**
26,734	Retrimmed	A tram line
26,735	Retrimming	**Tramway**
26,736	Retroact .	Wire tramway
26,737	Retrocede	Tramway —— miles long
26,738	Retrochoir	Tramway of —— inch gauge
26,739	Retroflex .	Conveyed by tramway
26,740	Retrofract	Must have a tramway for
26,741	Retrograde	A tramway from the mine to
26,742	Retrogress	Tramway from the shaft to the mill
26,743	Retrorsely	Tramway from tunnel to the mill
26,744	Retrospect	There is a tramway between
26,745	Retrovert .	Overhead wire tramway
26,746	Returnable	Tramway on the —— system
26,747	Returnball	Inclined tramway from
26,748	Returnday	Length of tramway
26,749	Returner .	Estimated cost of tramway
26,750	Returnless	Waggons for tramway
26,751	Reunion .	Tramway rails —— lbs. per yard
26,752	Reunitedly	**Transact**
26,753	Reuniting .	To transact
26,754	Reurge .	Will transact
26,755	Reurging .	Cannot transact
26,756	Revalue .	To transact the business
26,757	Revaluing	Has (have) no authority to transact
26,758	Revamp .	Give full authority to —— to transact
26,759	Revamping	**Transacted**
26,760	Revealable	As soon as the business is transacted
26,761	Revealed .	Has (have) transacted
26,762	Revealment	Has (have) not transacted
26,763	Revegetate	Has (have) —— transacted
26,764	Reveland .	Has (have) been successfully transacted
26,765	Revelation	Cannot be transacted
26,766	Revelatory	**Transacting**
26,767	Revelrout .	The business we are now transacting
26,768	Revenge .	Is transacting considerable business
26,769	Revengeful	**Transaction**(s)
26,770	Revenging	Large transactions in
20,771	Reverence	Small transactions in

No.	Code Word.	**Transactions** (*continued*)
26,772	Reverently	No transactions in
26,773	Reveries .	What transactions have there been in
26,774	Reversal .	Transactions have been mostly of a speculative
26,775	Reversedly	In all our transactions with [character
26,776	Reversible	There have been some heavy transactions
26,777	Revertant .	**Transfer**
26,778	Revestiary	Please to execute transfer
26,779	Revesture .	Do not transfer to
26,780	Revibrate .	The transfer of the mine is effected
26,781	Reviction .	Will transfer the deeds at once
26,782	Revictual .	Will not transfer the deeds till
26,783	Review .	When will transfer of deeds take place
26,784	Reviewage	The transfer of the property
26,785	Revigorate	Transfer of the shares
26,786	Revileth .	Transfer of the —— shares
26,787	Revised .	Transfer —— shares
26,788	Revisions .	Transfer —— shares standing in the name of
26,789	Revisory .	Have telegraphed —— to transfer
26,790	Revivalism	·Transfer at once
26,791	Revivers .	Do not give transfer
26,792	Revivified	We will transfer
26,793	Revivify .	Will not give transfer
26,794	Revocably	Transfer is obtained
26,795	Revocated	Transfer is not obtained
26,796	Revocatory	Do not want transfer
26,797	Revoke .	To give money against transfer
26,798	Revokingly	Until transfer is completed
26,799	Revolted .	Transfer will be registered
26,800	Revolting .	Transfer has been registered
26,801	Revolvency	Transfer is irregular
26,802	Revolvers .	Transfer not properly executed
26,803	Rewardably	Transfer cannot be registered
26,804	Rewrite .	Transfer books will be closed from —— to
26,805	Rewriting .	Transfer books closed
26,806	Reynard .	As soon as transfer is completed
26,807	Rhabdolith	Undertake(s) to transfer
26,808	Rhamadan	A telegraph transfer
26,809	Rhamnaceae	**Transferred**
26,810	Rhamnus .	Has (have) been transferred to
26,811	Rhapsode .	Has (have) not been transferred to
26,812	Rhapsodist	Has (have) the —— been transferred
26,813	Rheafibre .	Have the —— transferred to
26,814	Rheic . .	As soon as the —— is transferred
26,815	Rheinberry	As soon as deeds are transferred
26,816	Rhenish .	As soon as shares are transferred
26,817	Rheochord	Until —— have been transferred
26,818	Rheometer	Has (have) transferred
26,819	Rheometric	Has (have) not yet transferred
26,820	Rheoscope	Cannot be transferred
26,821	Rheostat .	**Transferring** .

No.	Code Word.	
26,822	Rheotome	**Transhipment**
26,823	Rhesus .	The transhipment of the
26,824	Rhetor. .	**Transhipped**
26,825	Rhetoric .	Had to be transhipped at
26,826	Rhetorical	Must be transhipped
26,827	Rhetorized	Can be transhipped
26,828	Rheuma .	**Transit**
26,829	Rheumatism	Is it (are they) for transit
26,830	Rhexia .	Has been lost in the transit
26,831	Rhigolene.	In course of transit
26,832	Rhinal. .	Overland transit
26,833	Rhinanthus	**Transmitted**
26,834	Rhindmart	Power transmitted by
26,835	Rhinoceros	**Transmitting**
26,836	Rhinoscopy	**Transpired**
26,837	Rhipiptera	It has transpired that
26,838	Rhipsalis .	(It) has not yet transpired
26,839	Rhizanth .	**Transport**
26,840	Rhizodont	Arrange for transport
26,841	Rhizodus .	Have arranged transport
26,842	Rhizogen .	Transport will cost —— per ton
26,843	Rhizoid .	What means of transport have you (are there)
26,844	Rhizomania	No means of transport except by
26,845	Rhizomys .	Mule transport
26,846	Rhizophaga	Transport by ox waggon
26,847	Rhizopod .	Transport from —— to
26,848	Rhizostoma	**Transporting**
26,849	Rhizotaxis	**Treasury**
26,850	Rhodaloze	Cash in the treasury
26,851	Rhodanic .	No money in the treasury
26,852	Rhodanthe	What is there in the treasury
26,853	Rhodeswood	**Treat**
26,854	Rhodiola .	To treat the ore
26,855	Rhodium .	To treat with
26,856	Rhodonite	Cannot treat the ore
26,857	Rhodymenia	Cannot treat with
26,858	Rhomb .	In order to treat the ore
26,859	Rhomboid	In order to treat with
26,860	Rhomboidal	To enable us to treat with
26,861	Rhombspar	To treat the tailings
26,862	Rhombus .	To treat the concentrates
26,863	Rhoncal .	To treat —— by the —— process
26,864	Rhopalic .	**Treated**
26,865	Rhopalodon	By which means the ore can be treated
26,866	Rhubarb .	How is the ore treated
26,867	Rhumbline	The ore is treated
26,868	Rhyme .	Can only be treated (by)
26,869	Rhymeless	**Treating**
26,870	Rhymeroyal	(By) treating the ore
26,871	Rhymery .	Have been treating the ore

No.	Code Word.	**Treating** (*continued*)
26,872	Rhymist' .	The method of treating
26,873	Rhymster .	By this method of treating
26,874	Rhynchops	To see the new way of treating the
26,875	Rhysimeter	By the —— method of treating
26,876	Rhythm .	We are treating the —— (by)
26,877	Rhythmical	**Treaty**
26,878	Rhythming	In treaty for
26,879	Rhythmus	Under the treaty
26,880	Rhytidoma	A new treaty
26,881	Rhytina .	**Trespass**
26,882	Ribadoquin	The alleged trespass
26,883	Ribald . .	It is a trespass case
26,884	Ribaldish .	**Trespassing**
26,885	Ribaldrous	To prevent any trespassing on our claim
26,886	Ribaldry .	Is (are) trespassing on
26,887	Ribandweed	Not trespassing
26,888	Ribbands .	Allege that we are trespassing on
26,889	Ribbonfish	**Trial**
26,890	Ribbonism	The trial is fixed for
26,891	Ribbonmaps	The trial is postponed till
26,892	Ribbonsaw	Till the trial comes off
26,893	Ribbonworm	Trial of machinery
26,894	Ribroast .	Give it a trial
26,895	Ribwort .	What is the result of the trial
26,896	Ricciaceae	The trial is coming on in a day or two
26,897	Ricebird .	The trial will take place (on) (or about)
26,898	Ricecorn .	Trial will not take place until
26,899	Ricedust .	Trial concluded ; the result is
26,900	Riceflour .	The trial is not yet over
26,901	Riceglue .	As a result of the trial
26,902	Ricemilk .	Will a trial be given to
26,903	Ricepaper	A —— trial will be given
26,904	Riceshell .	A —— trial will not be given
26,905	Ricesoup .	Have given it a fair trial
26,906	Riceweevil	Will give it a fair trial
26,907	Rich . .	The result of the trial is very satisfactory
26,908	Richardia .	The result of the trial is very unsatisfactory
26,909	Richly . .	We expect the trial will
26,910	Ricinolic .	Will pay for expense of a trial
26,911	Ricinus .	**Tribute**
26,912	Rickers .	The men working on tribute
26,913	Ricketish .	No men working on tribute
26,914	Rickstand .	Can let on tribute
26,915	Ricochet .	At what tribute
26,916	Riddance .	Tribute of —— per cent.
26,917	Riddled .	Some men who have been working on tribute
26,918	Riderless .	Have you any men working on tribute
26,919	Riderroll .	Advisable to have men working on tribute
26,920	Ridgebands	Not advisable to have men working on tribute
26,921	Ridgelet .	How many men have you on tribute

No.	Code Word.	**Tribute** (*continued*)
26,922	Ridgeling .	Working on a tribute of —— per cent.
26,923	Ridgepiece	Offer to take the work on tribute
26,924	Ridgeplate	Have let it on tribute
26,925	Ridgepoles	**Tried**
26,926	Ridgeroof .	Has (have) tried
26,927	Ridgetile .	Has (have) not tried
26,928	Ridiculed .	Have you tried
26,929	Ridiculing	I (we) have tried
26,930	Ridiculize	Tried again, but with no better success
26,931	Ridiculous	Have tried the new process, and found it
26,932	Riding . .	**Tropics**
26,933	Ridingday	In the tropics
26,934	Ridinghood	**Trouble**
26,935	Ridingrods	To avoid trouble in the future
26,936	Ridingwhip	There is great trouble in
26,937	Riffraff .	There will be no trouble in
26,938	Rifleball .	Will there be any trouble in
26,939	Riflebirds .	The trouble has been to
26,940	Riflecorps .	What is the trouble
26,941	Rifleman .	We expect to have considerable trouble with
26,942	Riflepit .	We had trouble at first with the
26,943	Rifted . .	The machinery gave some trouble at first, but all [goes well now
26,944	Rigadoon .	Has not given any trouble
26,945	Rigafir . .	Trouble likely to be caused by
26,946	Rightabout	Has been the cause of much trouble
26,947	Rightdrawn	All the trouble might have been avoided
26,948	Righteous	Has given great trouble
26,949	Rightfully .	**Troublesome**
26,950	Righthand	A very troublesome business
26,951	Rightwhale	Is likely to prove troublesome
26,952	Rightwise .	**True**
26,953	Rigid . .	Is it true (that)
26,954	Rigidly .	It is true
26,955	Rigidness .	If it is true
26,956	Rigmarole	It is not true
26,957	Rigorism .	If what you say is true
26,958	Rigorous .	What was stated is quite true
26,959	Rigorously	What was stated is not true
26,960	Rimbase .	**Trust**
26,961	Rimlock .	If you can trust
26,962	Rinderpest	Can you trust
26,963	Ringarmour	Can trust
26,964	Ringbolts .	Cannot trust
26,965	Ringchuck	You must not trust
26,966	Ringcourse	Do not trust the report of
26,967	Ringdial .	We trust that you will
26,968	Ringdoves	We trust to your judgment
26,969	Ringfence	In trust for
26,970	Ringfinger	The trust deed
26,971	Ringformed	The trust deed has been registered

No.	Code Word.	**Trust** (*continued*)
26,972	Ringgauge	Trust deed not registered
26,973	Ringings .	Send copy of trust deed
26,974	Ringleader	**Trusted**
26,975	Ringleted .	Can —— be trusted (with)
26,976	Ringman .	Can be implicitly trusted
26,977	Ringmaster	Cannot be trusted
26,978	Ringmoney	May be trusted to any amount
26,979	Ringouzel .	May be trusted to the extent of
26,980	Ringrope .	Has (have) trusted —— to
26,981	Ringsaws .	Can be trusted to look after our interests
26,982	Ringshaped	**Trustee(s)**
26,983	Ringstand	As our trustee(s)
26,984	Ringtailed	As trustee(s) for
26,985	Ringwall .	In the hands of trustees
26,986	Ringworm	To name as trustee(s)
26,987	Riolite. .	Will act as trustee(s)
26,988	Riotous .	Will not act as trustee(s)
26,989	Riotously .	The property has been vested in trustees
26,990	Riparian .	Would —— (and ——) act as trustee(s)
26,991	Ripeness .	As trustee for the estate of
26,992	Ripening .	As trustee for the bondholders
26,993	Riphean .	**Truth**
26,994	Rippingsaw	Is there any truth in the report that
26,995	Ripplemark	There is no truth in the report
26,996	Ripplingly	The real truth is
26,997	Ripply. .	What is the real truth
26,998	Riprap .	**Try**
26,999	Risibility .	You must try to
27,000	Risible. .	Will try to
27,001	Risky . .	Will not try to
27,002	Risotto .	Useless to try to
27,003	Rissoles .	Try to do without
27,004	Risus . .	Try elsewhere
27,005	Ritely . .	Will try elsewhere
27,006	Ritual . .	Try what you can do
27,007	Ritualism .	We shall try to
27,008	Ritually .	**Trying**
27,009	Rivalling .	Has (have) been trying to
27,010	Rivalry .	Has (have) not been trying to
27,011	Rivals . .	Has (have) —— been trying to
27,012	Rivalship .	After trying
27,013	Riverbed .	We are trying to
27,014	Rivercrab .	We are trying by all possible means
27,015	Riverhog .	Trying every method to
27,016	Riverhorse	**Tuesday**
27,017	Riverplain	On Tuesday
27,018	Riverside .	Last Tuesday
27,019	Riversnail .	Next Tuesday
27,020	Riverwall .	Every Tuesday
27,021	Riverwater	Every other Tuesday

No.	Code Word.	
27,022	Rivetboy .	**Tunnel(s)**
27,023	Riveter .	The rock is soft to tunnel through
27,024	Riveting .	The rock is difficult to tunnel through
27,025	Rivetjoint.	The rock is hard to tunnel through
27,026	Rivose. .	A tunnel driven on the vein for —— feet
27,027	Rivulet .	Has been taken from this tunnel
27,028	Rixation .	A blast put in the end of the tunnel
27,029	Rixatrix .	Tunnel No. ——
27,030	Rixdollar .	The tunnel is in
27,031	Rizom . .	From the end of the tunnel to the surface
27,032	Rizzered .	Tunnels have been driven
27,033	Roadbed .	In the tunnel
27,034	Roadbook	From the tunnel
27,035	Roadharrow	The tunnel has caved for the entire length
27,036	Roadmetal	Tunnel being driven at rate of —— feet per day
27,037	Roadroller	The only development a tunnel —— feet in length
27,038	Roadsteads	By a tunnel
27,039	Roadsulky	To drive a tunnel
27,040	Roadway .	Have driven a tunnel
27,041	Roadworthy	Tunnel stopped for lack of funds
27,042	Roamer .	What progress do you make with the tunnel
27,043	Roantree .	Tunnel is in —— feet; no ore
27,044	Roareth .	Tunnel is in —— feet; shows ore all along
27,045	Roaringly .	Tunnel is flooded by water
27,046	Roasting .	Feet from the end of tunnel
27,047	Robbercrab	Have begun tunnel
27,048	Robemaker	Tunnel mouth
27,049	Roberdsman	From the mouth of the tunnel
27,050	Robertine .	When shall you begin to tunnel
27,051	Robingroom	We wish you to drive a tunnel
27,052	Robinia .	Tunnel to intercept
27,053	Robinwake	Tunnel to cut vein
27,054	Roborant .	Tunnel to cut vein —— feet below surface
27,055	Roboration	Tunnel to cut vein —— feet below mouth of shaft
27,056	Roboreous	Tunnel now being driven
27,057	Robust .	Tunnel now being driven through country rock
27,058	Robustious	Tunnel now being driven through granite
27,059	Robustly .	Tunnel now being driven through slate
27,060	Robustness	Tunnel now being driven through quartz
27,061	Rocambole	Tunnel now being driven through
27,062	Roccella .	Tunnel has been driven
27,063	Rochelime	Tunnel has been driven —— feet
27,064	Rockalum	Tunnel to meet —— level
27,065	Rockaway	Tunnel to cut vein below water-level
27,066	Rockbasin	Tunnel has cut vein below water-level
27,067	Rockbound	Tunnel has cut
27,068	Rockbutter	Tunnel has cut vein
27,069	Rockcork .	Tunnel has cut vein —— feet from
27,070	Rockdove .	Tunnel will be driven
27,071	Rockdrill .	Tunnel will be driven with all possible speed

T

No.	Code Word.	Tunnel(s) *(continued)*
27,072	Rocketcase	How many feet has tunnel been driven
27,073	Rocklimpet	How long will it take to complete tunnel
27,074	Rockling .	Cost of driving tunnel
27,075	Rockmaple	Have discontinued tunnel (on account of)
27,076	Rockmilk .	We wish you to stop tunnel
27,077	Rockmoss	Tunnel stopped for the present
27,078	Rockoil .	Tunnel stopped until
27,079	Rockpigeon	Tunnel stopped until further orders
27,080	Rockplant	Propose to tunnel
27,081	Rockrabbit	Proposed tunnel
27,082	Rockribbed	**Tunnelled**
27,083	Rockruby.	Have tunnelled through
27,084	Rocksalt .	Must be tunnelled through
27,085	Rockshaft .	**Tunnelling**
27,086	Rockshell.	Tunnelling to reach the
27,087	Rockslater	**Turbine**
27,088	Rocksnake	High-pressure turbine
27,089	Rocksoap .	Low-pressure turbine
27,090	Rockstaffs	Turbine of —— h.p. for a fall of
27,091	Rocktar .	Driven by a turbine
27,092	Rocktemple	A turbine can advantageously be employed
27,093	Rocky . .	Recommend a turbine being put down
27,094	Rodiron .	**Turning**
27,095	Rodomel .	Turning out satisfactorily
27,096	Rodomont	Turning out badly
27,097	Roebuck .	**Ultimately**
27,098	Roedeer .	Will be able ultimately
27,099	Roffia . .	Can you ultimately
27,100	Rogation .	Hope ultimately to
27,101	Rogerian .	Think the mine will ultimately turn out successful
27,102	Rogueries ·	Will ultimately prove to be worthless
27,103	Roguery .	May be ultimately worked up into a paying concern
27,104	Rogueship	Ultimately will pay
27,105	Roguishly .	**Ultimo**
27,106	Roignous .	Refer to our letter of the —— ultimo
27,107	Roister	Refer to our telegram of the —— ultimo
27,108	Roisterly .	Refer to your letter of the —— ultimo
27,109	Rokelay .	Refer to your telegram of the —— ultimo
27,110	Rokette .	On or about the —— ultimo
27,111	Rollable .	**Unable (to)**
27,112	Rollabout .	Unable at present to
27,113	Rollcall .	We have been unable to
27,114	Rolled . .	Unable to do better
27,115	Rollerbolt.	Unable to do what is required
27,116	Rolleyway	Unable to reply at present
27,117	Rollick .	Are you unable to
27,118	Rollicking	If you are unable to
27,119	Rollingpin	If he is (they are) unable to
27,120	Rolypoly .	Unable to do more
27,121	Romancer	Cable if you are unable to

No.	Code Word.	
27,122	Romancing	**Unaccountable**
27,123	Romanesque	Quite unaccountable
27,124	Romanized	It is quite unaccountable how
27,125	Romantic .	**Unadvisable**
27,126	Romanticly	Consider it most unadvisable
27,127	Romantype	A most unadvisable proceeding
27,128	Romanwhite	**Unanimous**
27,129	Romanza .	Unanimous in supporting
27,130	Romaunt .	Unanimous in opposing
27,131	Rombowline	(By) a unanimous vote
27,132	Romepenny	(By) the unanimous expression of
27,133	Romeshot.	Unanimous in opinion
27,134	Rompingly	**Unanimously**
27,135	Rompish .	Unanimously agreed to
27,136	Rondeau .	Resolution passed unanimously
27,137	Rondeletia	Resolution unanimously rejected
27,138	Rondure .	**Unanswered**
27,139	Ronyon .	My communication of —— is unanswered
27,140	Roodarch.	Why is my (our) telegram of —— unanswered
27,141	Roodbeam	Why is my (our) letter of —— unanswered
27,142	Roodfree .	**Unappropriated**
27,143	Roodloft .	The unappropriated balance
27,144	Roodscreen	The unappropriated funds
27,145	Roodtower	If entirely unappropriated
27,146	Roofing .	The unappropriated portion of the
27,147	Rooflet .	**Unassessable**
27,148	Roofy . .	Shares unassessable
27,149	Rooklers .	**Unauthorized**
27,150	Rookpie .	Is (are) unauthorized
27,151	Roomily .	Was (were) unauthorized
27,152	Roominess	In a most unauthorized manner
27,153	Roomridden	**Unavoidable**
27,154	Roomsome	The delay is unavoidable
27,155	Roomth .	Unavoidable on account of
27,156	Roomthier	Is it unavoidable
27,157	Roosaoil .	**Unaware**
27,158	Roostcock	Unaware of what was done
27,159	Rootbuilt .	Unaware of any such thing
27,160	Rootcrop .	**Uncertain**
27,161	Rooteater .	Very uncertain whether
27,162	Rootleaf .	Uncertain what to do
27,163	Rootmildew	Uncertain what is best to be done about
27,164	Rooyebok.	In the present uncertain state of affairs
27,165	Ropebark .	Uncertain how long it may last
27,166	Ropedancer	The supply of ore is uncertain, owing to
27,167	Ropeladder	It is very uncertain (when)
27,168	Ropemaker	Very uncertain whether
27,169	Ropemaking	Still uncertain
27,170	Ropemat .	Am (are) uncertain
27,171	Ropeporter	Is (are) uncertain

No.	Code Word.	**Uncertain** (*continued*)
27,172	Ropepump	Owing to the uncertain way in which
27,173	Ropeshaped	**Uncertainty**
27,174	Ropetrick .	The uncertainty which we feel as to
27,175	Ropewalker	Owing to the great uncertainty
27,176	Ropeyarn .	The uncertainty of the market
27,177	Ropily . .	If there is any uncertainty
27,178	Roquelaure	**Unchanged**
27,179	Rorifluent.	Aspect of affairs is unchanged
27,180	Rorulent .	Character of rock still unchanged
27,181	Rosaceous	Remain(s) unchanged
27,182	Rosalina .	If still unchanged
27,183	Rosaniline	**Uncommon**
27,184	Rosary . .	Not uncommon in this district
27,185	Roseacacia	Very uncommon
27,186	Rosebeetle	A very uncommon occurrence
27,187	Rosebuds .	**Uncompromising**
27,188	Rosebushes	Uncompromising hostility to
27,189	Rosechafer	Uncompromising resistance to
27,190	Rosecold .	Shows a most uncompromising attitude
27,191	Rosecolour	**Unconditional(ly)**
27,192	Rosecross .	An unconditional surrender
27,193	Rosedrop .	Unconditionally agreed to
27,194	Roseelder .	**Undecided**
27,195	Roseengine	Undecided whether
27,196	Rosefaced	**Under**
27,197	Rosefever .	Under the agreement (of)
27,198	Rosefish .	Under the system in use
27,199	Roseflies .	Not under any
27,200	Rosefly .	From under
27,201	Rosegall .	Driving under
27,202	Rosehued .	Appear to lie under
27,203	Roseknot .	If it (they) come(s) under
27,204	Roselathe .	Under the Act of Parliament
27,205	Roselips .	Under the Alien Act
27,206	Rosemadder	Under the Act of Congress
27,207	Rosemallow	Under the laws of
27,208	Rosemary .	Under the local laws or regulations
27,209	Rosenoble	Under the terms of
27,210	Roseola .	All that are under
27,211	Rosepink .	Under great pressure
27,212	Rosequartz	Under the heavy strain upon it
27,213	Roserash .	**Undergone**
27,214	Rosered .	Has (have) undergone
27,215	Roselum .	Has undergone great changes
27,216	Rosewater	**Underground**
27,217	Rosewindow	Surveying underground
27,218	Rosewood	(In) the underground workings
27,219	Rosintin .	An underground foreman
27,220	Rosmarine	The underground superintendent's report
27,221	Rosoglio .	Underground survey(s)

No.	Code Word.	**Underground** (*continued*)
27,222	Rosselly .	Plan of the underground workings
27,223	Rostellum	Section of the underground workings
27,224	Rostriform	Plan and section of the underground workings
27,225	Rosytinted	Both at surface and underground
27,226	Rotacism .	**Underhand**
27,227	Rotaclub .	Owing to the underhand way in which
27,228	Rotaeform	Has (have) behaved all through in an underhand
27,229	Rotascope	The underhand manner of [way
27,230	Rotated .	Some underhand work
27,231	Rotating .	Underhand stoping
27,232	Rotational	**Underlie(s)**
27,233	Rothernail	The vein underlies
27,234	Rothoffite .	**Underlying**
27,235	Rottenness	The underlying rocks
27,236	Rottolo .	The underlying strata (stratum)
27,237	Rotular .	Underlying the entire area
27,238	Rotund .	Underlying the whole of
27,239	Rotundate	**Understand**
27,240	Rotundious	Did not understand (why)
27,241	Rotundity .	I (we) understand (that)
27,242	Rotundness	Understand from
27,243	Roturier .	Do you understand
27,244	Rouged .	As far as I (we) can understand
27,245	Roughcast	We understand that you have
27,246	Roughclad	We are given to understand that
27,247	Roughdraft	Cannot understand your instructions
27,248	Roughdried	If you do not understand
27,249	Roughdry .	If he does (they do) not understand
27,250	Roughens	Do you clearly understand ? if not, wire
27,251	Roughhewer	Cannot understand what is wanted; please wire
27,252	Roughness	Understand from your letter of [more definitely
27,253	Roughrider	Understand from your telegram of
27,254	Roughshod	Understand from your remarks
27,255	Roughtree	Understand, that unless
27,256	Roughwork	Understand, that until
27,257	Rouging .	Are we to understand that
27,258	Rounceval	With respect to the part you do not understand
27,259	Roundabout	Do not quite understand your telegram
27,260	Roundalls .	Do not understand that part of your telegram
27,261	Roundelay	We are sorry to understand [referring to
27,262	Roundfish	We are pleased to understand
27,263	Roundhand	In order to understand
27,264	Roundhouse	**Understanding**
27,265	Roundridge	According to our understanding of (or with)
27,266	Roundrobin	The understanding between us
27,267	Roundshot	Must have a clear understanding before we can .
27,268	Roundtower	By an understanding with
27,269	Roundtrade	Is there any understanding between
27,270	Roundworm	There has been an understanding between
27,271	Rousant .	There has been no understanding (between)

No.	Code Word.	**Understanding** (*continued*)
27,272	Routcakes	Cannot come to any understanding with
27,273	Routinary .	**Understood**
27,274	Routinist .	I (we) clearly understood from —— that
27,275	Rovingshot	Understood that (on)
27,276	Rowboats .	Understood to be
27,277	Rowdies .	It was an understood thing that
27,278	Rowdy. .	Understood by both parties
27,279	Rowdydowdy	Let it be clearly understood
27,280	Rowdyism	Was not understood to be
27,281	Rowel . .	Never understood this
27,282	Rowelhead	Was understood to refer to
27,283	Rowelling .	**Undertake(s)**
27,284	Rowleyragg	Will —— undertake (to)
27,285	Rowlocks .	Will undertake (to)
27,286	Royalarch	Will not undertake (to)
27,287	Royalets .	Do not undertake
27,288	Royally .	Can you undertake (to)
27,289	Royalmast	Cannot possibly undertake to
27,290	Royalties .	Must undertake to
27,291	Royalyard	Must undertake at his own cost to
27,292	Royena .	Must not undertake
27,293	Roysterer .	Must undertake not to
27,294	Rubadub .	Will not undertake it, unless
27,295	Rubbishing	If you can undertake to
27,296	Rubblewall	If you cannot undertake to
27,297	Rubblework	Unless —— undertake(s) to
27,298	Rubbly .	If —— do(es) not undertake to
27,299	Rubellite .	Undertake to do it before
27,300	Rubeola .	Can —— undertake
27,301	Rubeoloid	**Undertaking**
27,302	Rubezahl .	Before undertaking
27,303	Rubiacine .	No hesitation in undertaking
27,304	Rubicative	An arduous undertaking
27,305	Rubicelle .	Do not care about undertaking
27,306	Rubicund .	**Undertook**
27,307	Rubific .	He (they) undertook to
27,308	Rubiginose	I (we) undertook to
27,309	Rubigo .	**Undervaluation**
27,310	Rubrician .	Owing to the undervaluation of the products
27,311	Rubrics .	Owing to the undervaluation of the reserves
27,312	Rubsencake	**Undervalue**
27,313	Ruby . .	Think you undervalue the
27,314	Rubyblende	I (we) do not undervalue the
27,315	Rubysilver	**Undervalued**
27,316	Rucheing .	Was (were) undervalued
27,317	Rudderband	Have undervalued the
27,318	Ruddercase	In the statement of —— I (we) undervalued
27,319	Ruddercoat	Owing to —— having been undervalued
27,320	Rudderfish	**Underwrite**
27,321	Rudderhole	Will vendors underwrite any portion of the capital

No.	Code Word.	**Underwrite** (*continued*)
27,322	Ruddernail	Vendors will not underwrite any part of the capital
27,323	Ruddied .	Vendors will underwrite
27,324	Ruddleman	Will underwrite
27,325	Ruddocks	Agreed to underwrite
27,326	Rudenture	**Underwriters**
27,327	Ruderary .	The underwriters have withdrawn
27,328	Rudesby .	**Underwritten**
27,329	Rudiment .	All the capital is underwritten
27,330	Rudimental	Has been underwritten
27,331	Rudmasday	The remainder will be underwritten
27,332	Rudolphine	Can you get —— underwritten
27,333	Ruebargain	Can get —— underwritten
27,334	Rued . .	Underwritten at a commission of
27,335	Rueful . .	Not all underwritten
27,336	Ruefully .	**Undesirable**
27,337	Ruefulness	It is most undesirable to
27,338	Ruellbone	**Uneasiness**
27,339	Ruellia ! .	(There is) no cause for uneasiness as yet ; we hope to be able to give reassuring information
27,340	Rufescent .	Much uneasiness as to the political condition of the [country
27,341	Ruffian .	These things cause much uneasiness
27,342	Ruffianage	**Uneasy**
27,343	Ruffianism	Shareholders feel very uneasy about the position of [affairs
27,344	Ruffianly .	We are very uneasy in respect to
27,345	Ruffinous .	There is no reason for you to feel uneasy
27,346	Ruffled .	**Unexpected**
27,347	Ruffleless .	Quite unexpected
27,348	Rufflement	Happened in a most unexpected manner
27,349	Rufterhood	**Unexplored**
27,350	Rugged .	Entirely unexplored hitherto
27,351	Ruggedly .	Unexplored ground to the north
27,352	Ruggedness	Unexplored ground to the south
27,353	Rugging .	Unexplored ground to the east
27,354	Ruggown .	Unexplored ground to the west
27,355	Rugheaded	The unexplored claims
27,356	Rugosa .	**Unfair**
27,357	Ruination .	It is a most unfair arrangement
27,358	Ruiniform	It is most unfair
27,359	Ruinously	Is it unfair
27,360	Rulers . .	An unfair advantage
27,361	Rumex .	It is unfair to saddle us with
27,362	Ruminantia	Was most unfair to us
27,363	Ruminantly	Would be very unfair
27,364	Ruminated	**Unfairness**
27,365	Ruminators	The unfairness is evident
27,366	Rummaged	**Unfavourable**
27,367	Rummaging	I (we) have formed a very unfavourable opinion of
27,368	Rummy .	The reports are very unfavourable
27,369	Rumour .	If your report is unfavourable
27,370	Rumpfed .	Should your report be unfavourable

No.	Code Word.	**Unfavourable** (*continued*)
27,371	Rumpless .	Report is very unfavourable
27,372	Rumpling.	Prospects are unfavourable
27,373	Rumpsteak	Counsel's opinions are unfavourable
27,374	Rumpus .	A most unfavourable time for
27,375	Rumshrub	The weather has been very unfavourable
27,376	Rumswizzle	Unfavourable opinions have been expressed about
27,377	Runagate .	Very unfavourable
27,378	Runaway .	Conditions are not of an unfavourable character
27,379	Rundle .	Has (have) taken a most unfavourable turn
27,380	Runecraft .	Under most unfavourable circumstances
27,381	Runologist	**Unfavourably**
27,382	Rupellary .	Sorry to have to report unfavourably of the position
27,383	Rupicapra	Spoke very unfavourably of [and prospects
27,384	Rupicola .	**Unfinished**
27,385	Ruppia .	Owing to the unfinished state of
27,386	Ruptuary .	Left unfinished
27,387	Ruralist .	Had to be left unfinished
27,388	Rurality .	**Unfit**
27,389	Ruralized .	Unfit for the position
27,390	Ruralizing	Quite unfit for
27,391	Ruralness .	Unfit for use
27,392	Ruricolist .	**Unforeseen**
27,393	Rurigenous	Unless something unforeseen occurs
27,394	Ruscus .	Was quite unforeseen at the time
27,395	Rushcandle	Unforeseen events which have since occurred
27,396	Rushlights	**Unfounded**
27,397	Rushlike .	The report is quite unfounded
27,398	Rushy . .	Quite unfounded
27,399	Russet . .	**Unhealthy**
27,400	Russeting .	The district is most unhealthy
27,401	Russophile	The mine(s) is (are) situated in a very unhealthy
27,402	Rustical .	Climate unhealthy for Europeans [part
27,403	Rustically.	Unhealthy at certain seasons
27,404	Rusticated	**Unhurt**
27,405	Rusticity .	All unhurt except
27,406	Rustling .	The men escaped unhurt
27,407	Rustydab .	Hope all have escaped unhurt
27,408	Rutabaga .	**Unimportant**
27,409	Rutelidae .	Otherwise, the changes are unimportant
27,410	Rutha . .	I think it is quite unimportant
27,411	Ruthenium	An unimportant matter
27,412	Ruthless .	**Uninformed**
27,413	Ruthlessly	Why are we kept uninformed
27,414	Rutilant .	Have been kept quite uninformed
27,415	Rutterkin .	Was (were) quite uninformed
27,416	Ryacolite .	Complain that they are kept uninformed about
27,417	Ryotwar .	**Unintelligible**
27,418	Rypecks .	Your telegram unintelligible ; repeat
27,419	Sabadilla .	The following words in your telegram unintelligible ;
27,420	Sabathian .	**Unless** [please repeat them

No.	Code Word.	**Unless** (*continued*)
27,421	Sabbath .	Unless you can
27,422	Sabbaton .	Will not unless
27,423	Sabbire .	Unless he (they) will
27,424	Sabella .	Unless I (we) can
27,425	Sabellaria .	Do not go unless
27,426	Sabianism .	Do not agree unless
27,427	Sablemouse	Will go unless
27,428	Sabling .	Will agree unless
27,429	Sabretache	Will not go unless
27,430	Saccade .	Will not agree unless
27,431	Saccharate	Not unless
27,432	Saccharify	Unless I (we) have
27,433	Saccharoid	Unless he has (they have)
27,434	Sacciform .	Unless it saves
27,435	Saccomys .	Unless you can do better
27,436	Saccosoma	Unless you cannot do better
27,437	Sacculated	Unless there is a change for the better
27,438	Sacellum .	Unless you will
27,439	Sacerdotal	Unless he (they) can
27,440	Sachemdom	Unless there is
27,441	Sachemship	Unless there has been
27,442	Sacheverel	Unless we
27,443	Sackbarrow	Unless you
27,444	Sackbuts .	Unless they
27,445	Sackdoudle	Unless this can be done
27,446	Sackful .	Unless you can do this
27,447	Sacking .	Unless it is absolutely necessary
27,448	Sackless .	Unless prevented by
27,449	Sackposset	**Unlikely**
27,450	Sacrament	It is very unlikely (that)
27,451	Sacrarium .	Is it unlikely (that)
27,452	Sacrated .	It is not unlikely (that)
27,453	Sacredly .	**Unlimited**
27,454	Sacredness	Unlimited time
27,455	Sacrific .	With unlimited means
27,456	Sacrificial .	An unlimited supply
27,457	Sacrilege .	An unlimited demand (for)
27,458	Sacrist .	**Unloading**
27,459	Sacristans .	Including unloading
27,460	Sacrosanct	Including unloading and delivery at mine
27,461	Sacrum .	Unloading at station and delivery at mine
27,462	Saddened .	Unloading at port of delivery
27,463	Saddest .	Upon unloading
27,464	Saddlebow	The ship is now unloading
27,465	Saddlegall .	**Unlucky**
27,466	Saddleroof	It is very unlucky
27,467	Saddlerugs .	Very unluckily for us
27,468	Saddlery .	**Unnecessary**
27,469	Saddlesick	Quite unnecessary
27,470	Saddletree	**Unobtainable**

No.	Code Word.	**Unobtainable** (*continued*)
27,471	Sadducaic	Unobtainable at any price
27,472	Sadducism	Unobtainable at present
27,473	Sadducized	**Unpaid**
27,474	Sadeyed .	Has (have) been left unpaid
27,475	Sadfaced .	Remain(s) unpaid
27,476	Sadhearted	Unpaid accounts at end of month
27,477	Sadirons .	Unpaid accounts on the
27,478	Safeguards	Amount due, still unpaid
27,479	Safely . .	**Unprofitable**
27,480	Safepledge	So far, has been unprofitable
27,481	Safeties .	Too unprofitable to work
27,482	Safetyarch	**Unpromising**
27,483	Safetybelt .	The state of affairs is unpromising
27,484	Safetybuoy	The state of the mine is unpromising
27,485	Safetycage	**Unprovided**
27,486	Safetyfuse .	Quite unprovided for
27,487	Safetylamp	**Unreasonable**
27,488	Safetypin .	Demands are quite unreasonable
27,489	Safetyplug	It is unreasonable to expect
27,490	Safflower .	Do you think it unreasonable (to ask)
27,491	Saffron . .	I do not think it unreasonable
27,492	Saffroning .	**Unreliable**
27,493	Sagacious .	The sample was unreliable
27,494	Sagacity .	The assays were (are) unreliable
27,495	Sagamore .	We think they are unreliable
27,496	Sagapen .	Is (are) very unreliable
27,497	Sagapenum	Unreliable and disappointing
27,498	Sagathy .	The indications are unreliable
27,499	Sagenitic .	Are unreliable as affording any indications of
27,500	Sagina . .	You may consider it (them) unreliable
27,501	Sagittal .	**Unsafe** [mine
27,502	Sago . .	It is unsafe to work in the present condition of the
27,503	Saguerus .	The workings were unsafe, and therefore abandoned
27,504	Sahib . .	Becoming unsafe
27,505	Sahlite . .	Too unsafe (to be)
27,506	Sailborne .	It would be unsafe to leave them as they are
27,507	Sailcloth .	Is (are) unsafe
27,508	Sailhook .	Is it unsafe to
27,509	Sailloft . .	The old workings, which were very unsafe, have been
27,510	Sailmaker .	Was (were) too unsafe to work in [filled in
27,511	Sailneedle .	Think it very unsafe to
27,512	Sailorlike .	Would be unsafe to
27,513	Saintdom .	Unsafe to rely upon
27,514	Saintesses .	Workings are very unsafe
27,515	Sainthood .	**Unsatisfactory**
27,516	Saintism .	Most unsatisfactory
27,517	Saintlike .	A most unsatisfactory state of affairs
27,518	Saintly . .	The account you give is most unsatisfactory
27,519	Saintship .	The unsatisfactory result is owing to
27,520	Sakur . .	The output is very unsatisfactory

No.	Code Word.	**Unsatisfactory** (*continued*)
27,521	Salaaming	The results are very unsatisfactory
27,522	Salaams .	How do you account for such unsatisfactory
27,523	Salad . .	**Unsettled**
27,524	Saladoil .	The unsettled condition of affairs
27,525	Salalberry .	Everything is in a most unsettled state
27,526	Salamander	Owing to the unsettled state of the
27,527	Salamba .	**Unsuccessful**
27,528	Salamstone	Has (have) been very unsuccessful
27,529	Salaried .	If —— should be unsuccessful
27,530	Salary . .	**Unsuitable**
27,531	Salarying .	Quite unsuitable for the purpose
27,532	Salep . .	Find it unsuitable for
27,533	Saleratus .	Is considered unsuitable for
27,534	Saleroom .	**Until**
27,535	Salework .	Nothing can be done until
27,536	Saliaunce .	Until it is
27,537	Salicetum .	Until it has been
27,538	Salicornia .	Until it has been proved
27,539	Salicylic .	Until we can
27,540	Salifiable .	Until you can
27,541	Salimeter .	Until he (they) can
27,542	Salineness .	Until —— are
27,543	Saliniform .	Until —— have
27,544	Salinity .	Until this is done
27,545	Salinous .	Until we see our way to
27,546	Salique .	Until you see your way to
27,547	Salisburia .	But not until
27,548	Salivary .	Will not go until
27,549	Salivated .	Until the last
27,550	Salivating .	Until (I) we hear from you
27,551	Salleeman .	I waited until
27,552	Salleting .	Must ask you to wait until
27,553	Sallow . .	**Unusual(ly)**
27,554	Sallowish .	It is very unusual
27,555	Sallylun .	It is not unusual
27,556	Sallyport .	Has been unusually
27,557	Salmagundy	**Unwell**
27,558	Salmiac .	Am (is) very unwell
27,559	Salmonet .	Too unwell to attend to any business
27,560	Salmonidae	Too unwell to attend to his duties
27,561	Salmonoid	Has been very unwell
27,562	Salomonian	**Unwilling(ly)**
27,563	Saloons .	Shall be very unwilling to
27,564	Salopian .	If —— is unwilling to
27,565	Salpidae .	Is (are) very unwilling to
27,566	Salpinx .	Did it most unwillingly
27,567	Salsilla . .	Shall have most unwillingly (to)
27,568	Salsoacid .	**Unwise**
27,569	Saltant . .	I think it would be very unwise to
27,570	Saltatory .	Do you think it would be unwise to

No.	Code Word.	**Unwise** (*continued*)
27,571	Saltbox .	Very unwise to
27,572	Saltbutter .	**Up**
27,573	Saltcake .	Not quite up to
27,574	Saltcellar .	Up to the present time
27,575	Saltduty .	Upcast shaft
27,576	Saltern .	**. Uphill**
27,577	Saltfish .	It is very uphill work
27,578	Saltfoot .	**Upper**
27,579	Saltgreen .	The upper part of the mine
27,580	Saltholder	The upper workings
27,581	Salticus .	The ore from the upper workings
27,582	Saltigrade .	**Upon**
27,583	Saltishly .	Upon the whole
27,584	Saltjunk .	Upon what basis have you
27,585	Saltlick .	Upon what grounds do you
27,586	Saltly . .	Upon the basis of
27,587	Saltmarsh .	**Upraised**
27,588	Saltmine .	Have upraised to
27,589	Saltness .	Have upraised to level above
27,590	Saltpan .	Have upraised to No. —— level
27,591	Saltpetre .	Have upraised to winze in
27,592	Saltrakers .	**Upraise**
27,593	Saltrheum .	The upraise in —— level
27,594	Saltsea . .	The upraise from
27,595	Saltspring .	The upraise from —— is now completed
27,596	Saltwater .	**Upraising**
27,597	Salubrious	Upraising to
27,598	Salubrity .	Upraising to —— so as to get better ventilation
27,599	Salutarily .	**Upset**
27,600	Salutation .	This will entirely upset my plans
27,601	Saluted .	Has completely upset my calculations
27,602	Salvadora .	**Urge**
27,603	Salvage .	Strongly urge you to
27,604	Salve . .	Urge the importance of
27,605	Salvific . .	We cannot too strongly urge
27,606	Samaritan .	**Urgent**
27,607	Samaroid .	Is it urgent
27,608	Sambucus .	It is most urgent that you should
27,609	Sambuke .	Am (are) in most urgent need of money
27,610	Samolus .	Very urgent demands have been made
27,611	Samoyedic.	Urgent necessity for
27,612	Sampans .	The matter is urgent ; let it have immediate attention
27,613	Samphire .	In case the matter is urgent
27,614	Sampleroom	The urgent demands upon
27,615	Samshoo .	There is urgent need of your presence
27,616	Samyda .	If anything urgent should necessitate
27,617	Samydaceae	There is no urgent necessity for
27,618	Sanability .	**Us**
27,619	Sanatorium	For us and those interested with us
27,620	Sanctebell .	For us and you

No.	Code Word. **Us** (*continued*)	
27,621	Sanctified .	Us and our friends
27,622	Sanctify .	Acting with us
27,623	Sanctimony **Use**	
27,624	Sanctioned	Necessitating the use of
27,625	Sanctitude	Now in use
27,626	Sanctuary .	Use your own discretion in
27,627	Sanctum .	Use your influence to
27,628	Sandalled .	May I use my own discretion in
27,629	Sandals .	If you can use
27,630	Sandalwood	What is the use of
27,631	Sandbanks	Will use
27,632	Sandbath .	No use for
27,633	Sandbed .	No use now
27,634	Sandblast .	Of no use whatever
27,635	Sandboy .	Will have to use
27,636	Sandcanal.	Will be of some use
27,637	Sandcorn .	Use other means
27,638	Sandcrack	Use the wire (cable) freely
27,639	Sanddrift .	There is no need to use the wire (cable) so freely
27,640	Sandflag .	Can you make any use of
27,641	Sandfluke .	Cannot make any use of
27,642	Sandgall .	(To) use in case of need
27,643	Sandgrouse	Use more
27,644	Sandheat .	To use it to advantage
27,645	Sandhills . **Usage**	
27,646	Sandhopper	According to usage and custom
27,647	Sandiver . **Used**	
27,648	Sandjet .	Has heen used as
27,649	Sandlance	Used for a long time
27,650	Sandlizard	Is used for
27,651	Sandmartin **Using**	
27,652	Sandmole .	Has (have) been using
27,653	Sandmyrtle	Is (are) using
27,654	Sandnecker	We are now using
27,655	Sandoricum	Using all the influence —— possesses (as)
27,656	Sandpipe . **Useless**	
27,657	Sandprey .	It is useless to
27,658	Sandpride.	Is it useless to
27,659	Sandsmelt.	Will be quite useless (to)
27,660	Sandstones	The machine is quite useless
27,661	Sandstorm **Usual**	
27,662	Sandtubes.	It is usual
27,663	Sandwasp .	Is it usual
27,664	Sandwich .	It is not usual
27,665	Sandwort .	(Subject to) the usual conditions
27,666	Sandyx .	(Subject to) the usual terms
27,667	Sangaree .	(Subject to) the usual commission
27,668	Sangfroid.	What are the usual terms (conditions)
27,669	Sanglier .	More than usual
27,670	Sangreal .	Less than usual

No.	Code Word.	
27,671	Sanguifier .	**Usually**
27,672	Sanguify .	(As) is usually the case
27,673	Sanguinary	If usually done
27,674	Sanguined .	**Utmost**
27,675	Sanguisuga	Am (are) doing my (our) utmost
27,676	Sanhedrim	Is (are) doing his (their) utmost
27,677	Sanidine .	Do your utmost to
27,678	Sanitary .	Utmost limit
27,679	Sanitist .	The utmost care has been taken
27,680	Sanjak . .	The utmost care must be taken
27,681	Sansappel .	The utmost caution must be observed
27,682	Sanseviera	Will do our utmost (to)
27,683	Sanskrit .	Will do their (his) utmost to
27,684	Santalum .	The utmost that can be done
27,685	Santhee .	Have done our utmost
27,686	Saouari .	To the utmost point
27,687	Sapball . .	The utmost point we have reached
27,688	Sapcolour .	**Vacancy**
27,689	Sapfaggot .	A vacancy has occurred
27,690	Sapgreen .	If a vacancy occurs
27,691	Saphead .	**Vacant**
27,692	Saphenous	The vacant spaces have been filled in
27,693	Sapidless .	The vacant space will be used for
27,694	Sapience .	**Vague(ly)**
27,695	Sapiential .	Vague rumours are afloat that
27,696	Sapiently .	Your instructions are too vague
27,697	Sapium .	Very vaguely worded
27,698	Saponacity	**Valid**
27,699	Saponaria .	Quite valid
27,700	Saponified.	Is it valid
27,701	Saponify .	Title is quite valid
27,702	Saponine .	Valid powers conferred upon
27,703	Saporosity	Is the deed valid
27,704	Sapota . .	Valid in any court of law
27,705	Sapotaceae	Has been pronounced to be quite valid
27,706	Sapphic .	If —— give(s) a valid title
27,707	Sapphirine	Is (are) not valid
27,708	Sappodilla	The concession is quite valid
27,709	Saproller .	**Validity**
27,710	Saprophyte	The validity has been established
27,711	Saprot . .	Validity unimpeachable
27,712	Sapsago .	The validity of our title
27,713	Sapskull .	The validity of our claims
27,714	Sapsucker .	The validity of the deed
27,715	Sapwood .	Concession of doubtful validity
27,716	Sapygidae .	The validity is doubtful
27,717	Sarabaite .	**Valuable**
27,718	Saracenic .	Have developed a most valuable mine
27,719	Sarasin . .	Will become very valuable
27,720	Sarcasmous	Is likely to become more valuable

No.	Code Word.	**Valuable** (*continued*)
27,721	Sarcastic .	Do you think the mine sufficiently valuable
27,722	Sarcilis . .	———'s services would be very valuable in
27,723	Sarcinula .	Likely to turn out very valuable
27,724	Sarcled .	Considered to be very valuable
27,725	Sarcobasis	Very valuable concession
27,726	Sarcocarp .	Most valuable
27,727	Sarcocolla .	Least valuable
27,728	Sarcoderm	The portions alienated least valuable
27,729	Sarcodic .	The portions alienated most valuable
27,730	Sarcolemma	Very valuable stocks of
27,731	Sarcolobe .	Most valuable deposits of
27,732	Sarcology .	The most valuable part of the
27,733	Sarcoma .	Is not as valuable as stated
27,734	Sarcophaga	The least valuable part of
27,735	Sarcophile	The least valuable ores assay from ——— to ——— per
27,736	Sarcoptes .	Chiefly valuable as containing [ton
27,737	Sarcosperm	**Valuation**
27,738	Sardachate	Purchaser(s) to take over at a valuation
27,739	Sardonyx .	Have set too high a valuation on
27,740	Sargassum	To be taken over at a valuation
27,741	Sarmentous	The valuation of the mine
27,742	Sarmentum	What do you think the true valuation of the mine
27,743	Sarplier .	Taken at a low valuation
27,744	Sarracenia	Our valuation is based upon
27,745	Sarsa . .	Could be realized at our valuation
27,746	Sarsenet .	If realized at our valuation
27,747	Sartorius .	Your valuation too high
27,748	Sashbar .	At a nominal valuation
27,749	Sashchisel	**Value.** (See also Estimate.)
27,750	Sashdoors	Of no value
27,751	Sasheries .	The value of
27,752	Sashery .	Estimated value (of)
27,753	Sashframe	The full value of
27,754	Sashgate .	State as near as you can the value
27,755	Sashsaws .	What is the value of
27,756	Sashsluice .	Of great value
27,757	Sassanage .	Of greater value than estimated
27,758	Sassenath .	Value of bullion
27,759	Sassoline .	Value of bullion in hand and in transit
27,760	Sassorol .	Value of concentrates
27,761	Sassorolla .	Value of concentrates in hand and in transit
27,762	Sasstea . .	Of no practical value
27,763	Satanical .	Of very little value
27,764	Satchel .	Concession of very little value
27,765	Satellites .	The greater part of the value is in
27,766	Satiable .	May be of very great value
27,767	Satinbird .	The nominal value is ———; doubt if it would realize
27,768	Satinpaper	To the value of [this
27,769	Satinspar .	Of the greatest value at the present moment
27,770	Satinstone	The actual realized or realizable value

No.	Code Word.	**Value** (*continued*)
27,771	Satinturk .	The value has been proved by repeated tests and
27,772	Satinwood	Value of the stocks in hand [assays
27,773	Satiny . .	Value of stores in hand
27,774	Satirized .	Reduces its value very much
27,775	Satirizing .	Reduced its value
27,776	Satisfier .	Not considered of any value
27,777	Satisfy . .	**Valueless**
27,778	Satisfying .	Absolutely valueless
27,779	Satrap . .	**Variation**
27,780	Satrapical .	Without much variation
27,781	Satrapies .	There is great variation
27,782	Saturable .	**Vary**
27,783	Saturated .	Vary from day to day
27,784	Saturating.	Do not (does not) vary very much
27,785	Saturnalia .	The appearance varies daily
27,786	Saturnine .	Has not varied
27,787	Satyr . .	**Varying**
27,788	Satyriasis .	Varying very much in quality
27,789	Satyrical .	Varying very much in appearance
27,790	Satyrion .	Varying very much in quantity and quality
27,791	Saucealone	Working with varying success
27,792	Sauceboat.	Varying from —— to
27,793	Saucepans	Varying in width from —— feet to —— feet
27,794	Saucer . .	Has been worked by the process with varying results
27,795	Sauciest .	**Vein(s)**
27,796	Saucily . .	Have found the vein
27,797	Saucybark .	The vein in rise No. ——
27,798	Saultree .	The vein in winze No. ——
27,799	Saunterer .	On the vein
27,800	Sauntering	Vein is well defined
27,801	Saurian .	Vein is fully —— feet wide
27,802	Saurillus .	A true fissure vein
27,803	Saurodon .	A contact vein
27,804	Sauropsida	A gash vein
27,805	Saurypike .	The vein averages in value
27,806	Sausage .	The vein is barren everywhere
27,807	Sausefleme	On the course of the vein
27,808	Saussurite .	The vein has been driven upon —— feet [show
27,809	Savagely .	Assays made of the vein matter taken promiscuously
27,810	Savageness	There is (are) —— vein(s) embraced in the property
27,811	Savagism .	Vein is very much broken and disturbed
27,812	Saveall .	Vein is compact and well defined
27,813	Saveloy .	Sink upon the vein
27,814	Savingly .	The vein is between walls of
27,815	Savingness	Vein has all the appearance of being permanent
27,816	Savorous .	The vein runs north and south
27,817	Savory . .	The vein runs east and west
27,818	Savourily .	Vein is almost horizontal
27,819	Savouring .	Vein is vertical
27,820	Savourless	The vein dips at angle of ——° to the

No.	Code Word.	**Vein(s)** (*continued*)
27,821	Sawarranut	The vein is very irregular
27,822	Sawdust .	Vein is very narrow, not over —— inches wide
27,823	Sawfile. .	Vein is not over —— feet wide
27,824	Sawframe.	Telegraph width of vein
27,825	Sawgate .	The vein can be seen on the surface, coursing the
27,826	Sawmill .	The veins are [whole length of the claim
27,827	Sawsash .	The vein is irregular and much mixed with country
27,828	Sawtoothed	The vein is [rock
27,829	Sawwhetter	Average width of vein is —— feet
27,830	Sawwrest .	Telegraph the character of the vein
27,831	Saxatile .	The vein is composed of
27,832	Saxhorn .	A quartz vein (containing)
27,833	Saxicavous	A vein of lead ore
27,834	Saxicola .	A vein of copper ore
27,835	Saxifraga .	A vein of iron ore
27,836	Saxon . .	Vein has faulted
27,837	Saxonblue.	Vein runs north-east and south-west
27,838	Saxondom	Vein runs north and south, dipping east
27,839	Saxongreen	Vein runs north and south, dipping west
27,840	Saxonisms	Vein runs east and west, dipping north
27,841	Saxophone	Vein runs east and west, dipping south
27,842	Saxotromba	A belt of veins running
27,843	Saxtuba .	A belt of veins dipping
27,844	Sayings .	The veins are of various widths
27,845	Scabbard .	The veins are of various widths, the largest averaging
27,846	Scabies .	The veins are of various widths, the smallest
27,847	Scabiosa .	The mother vein [averaging
27,848	Scabredity	The side vein
27,849	Scabrous .	Subsidiary veins
27,850	Scabwort .	Veins running parallel to the main vein
27,851	Scaffold .	The veins intersect
27,852	Scaffraff .	The vein is improving in depth
27,853	Scaglia. .	The vein is getting poorer in depth
27,854	Scalable .	The two veins appear to merge into each other
27,855	Scalaria .	The two veins diverge
27,856	Scaldfish .	Veins crossing at an angle of ——°
27,857	Scaldhead.	The vein at present yields a very low-grade ore
27,858	Scaldic .	Is located on the same vein
27,859	Scalebeam	Distinct mineral veins
27,860	Scaleboard	In addition to smaller veins
27,861	Scalefern .	The veins contain many chutes and chimneys of ore
27,862	Scaleless .	The vein was worked in former days
27,863	Scalemoss	Is one of the best veins in the district
27,864	Scalenus	**Vendor(s)**
27,865	Scalestone	The vendor(s) would take
27,866	Scalingbar	You must see ——, the vendor of the —— mine
27,867	Scaliola .	The vendor(s) to retain an interest
27,868	Scallion .	If the vendor(s) will
27,869	Scallop .	Ascertain from the vendor(s)
27,870	Scalloping	Shall soon see ——, the vendor(s) of the —— mine

No.	Code Word.	**Vendor(s)** *(continued)*
27,871	Scalpel .	Have seen the vendor(s)
27,872	Scalpellum	From the vendor(s)
27,873	Scalplock .	To the vendor(s)
27,874	Scalprum .	The vendor(s) will [tially correct
27,875	Scammonia	The representations of the vendor(s) are substan-
27,876	Scamp. .	You must see vendor(s) and arrange with them per-
27,877	Scampered	The vendor(s) offer(s) to [sonally
27,878	Scampering	The vendor(s) will not agree to a lower price than
27,879	Scampish .	The vendor(s) will not agree (to) [that
27,880	Scandalize	If the vendor(s) still insist upon the terms, suggest
27,881	Scandalous	Will the vendor(s) allow us to go to allotment for
27,882	Scandent .	The vendor(s) to [the working capital alone
27,883	Scanning .	The vendor(s) of the property is (are) going to
27,884	Scansion .	The vendor will pay towards promotion the sum of
27,885	Scansores .	Will vendor pay expenses of examination and report
27,886	Scansorial.	Vendor will not bear the expense of examination and
27,887	Scantiest .	Will vendor pay the expense of [report
27,888	Scantilone	Vendor will pay the expense of
27,889	Scantily .	Vendor will bear the expense of examination and
27,890	Scantiness	Vendor will not pay the expense of [report
27,891	Scantle .	Vendor wishes examination to be made immediately
27,892	Scantling .	Vendor will reduce the amount of purchase money
27,893	Scapegoat.	Vendor declines to reduce the amount of purchase
27,894	Scapegrace	Vendor will accept [money
27,895	Scapellus .	**Ventilated**
27,896	Scapewheel	Mine is ventilated by
27,897	Scapha. .	Mine is ventilated by upcast and downcast shafts
27,898	Scaphander	Mine is ventilated by a —— fan
27,899	Scaphidium	**Ventilating**
27,900	Scaphism .	Ventilating fan and blower
27,901	Scaphoid .	Ventilating fan has broken down
27,902	Scapiform.	**Ventilation**
27,903	Scapolite .	For purposes of ventilation
27,904	Scapulary.	To improve the ventilation of the mine
27,905	Scapus .	Owing to insufficient ventilation
27,906	Scarab. .	The proper ventilation of the mine
27,907	Scarabaeus	**Venture**
27,908	Scaramouch	It is too risky a venture
27,909	Scarbroite	Will venture
27,910	Scarcely .	Do not venture
27,911	Scarcement	Would not venture
27,912	Scarceness	It is worth while to venture
27,913	Scarebabe	**Verdict**
27,914	Scarecrow.	The verdict is in our favour
27,915	Scarfjoint .	The verdict is against us
27,916	Scarfskin .	The verdict is in favour of
27,917	Scarfweld .	The verdict is against
27,918	Scarifiers .	Telegraph as soon as you know the verdict
27,919	Scarifying.	The verdict is for us ; —— pays all costs [costs
27,920	Scariose .	The verdict is against us ; we have to pay all the

No.	Code Word.	Verdict (*continued*)
27,921	Scarlatina .	Verdict will be given on
27,922	Scarleted .	**Verge**
27,923	Scarmage .	On the verge of
27,924	Scarnbee .	**Verification**
27,925	Scarred .	Statements require verification
27,926	Scatebrous	In full verification of —— report
27,927	Scatheful .	**Verified**
27,928	Scathly .	Statements verified
27,929	Scattereth	Must be verified
27,930	Scaturient.	Must be verified and legally attested
27,931	Scauper .	If statements are not verified
27,932	Scavage .	**Verify**
27,933	Scavenger.	If you can verify ——'s statements
27,934	Scavenging	I can verify
27,935	Scazon . .	I cannot verify
27,936	Scelerat .	Can verify all statements made by
27,937	Scelides .	Can verify, in the main, his reports
27,938	Sceneman.	**Version**
27,939	Sceneries .	According to his (their) version of the matter
27,940	Scenery .	According to your version of the matter
27,941	Scenework	Think this is the true version
27,942	Scenograph	**Vertical**
27,943	Scented .	Vertical shaft —— feet deep
27,944	Scentful .	To sink a vertical shaft
27,945	Sceptic .	The vein is nearly vertical
27,946	Sceptical .	**Very**
27,947	Scepticism	Very few men at work
27,948	Scepticize.	Very few to be got
27,949	Sceptring .	Could only get very few
27,950	Schalstein.	Without very great trouble
27,951	Schapziger	Very nearly
27,952	Schediasm	Very great
27,953	Schedule .	Very high
27,954	Scheduling	Very low
27,955	Scheelite .	Of very little account
27,956	Scheelium	**Vessel**
27,957	Scheellead	Send by sailing vessel
27,958	Schema .	Have sent by sailing vessel
27,959	Schematism	Vessel will sail shortly
27,960	Schematize	Vessel expected to sail about the
27,961	Schemeful	Send them by the first vessel leaving for
27,962	Scheming .	**Vested**
27,963	Schemingly	Is vested in the hands of trustees
27,964	Schererite .	To be vested in
27,965	Scherzando	Ought to be vested in
27,966	Scherzo .	If not vested in
27,967	Schinus .	**Vicinity**
27,968	Schireman	In the immediate vicinity
27 969	Schirrhus .	Visited the mines in the immediate vicinity
27,970	Schism .	Owing to the near vicinity of

No.	Code Word.	**Vicinity** (*continued*)
27,971	Schismatic	Can be got in the vicinity of the mines
27,972	Schismless	The country in the immediate vicinity
27,973	Schistous .	**View(s)**
27,974	Schizandra	Nothing in view
27,975	Schizodus.	Not much in view
27,976	Schizopod	In view of
27,977	Schlich .	What is your view of
27,978	Schmelze .	What have you in view
27,979	Schnapps .	Keep in view
27,980	Schoenus .	If this meets —— views
27,981	Scholar .	Have nothing in view at present
27,982	Scholarity.	Do not let any one know your views about
27,983	Scholastic .	Have (has) other views
27,984	Scholiasts .	Fall in with —— views
27,985	Scholiaze .	In accordance with —— view
27,986	Scholical .	Our views agree with
27,987	Scholiums	A favourable view of
27,988	Scholy . .	An unfavourable view of
27,989	School . .	Are your views still unchanged
27,990	Schoolbook	I (we) cannot carry out your views
27,991	Schoolboys	Views still unchanged
27,992	Schoolbred	Meet their views to the extent of [property
27,993	Schooldame	Directors anxious to have your views upon the ——
27,994	Schoolgirl.	**Vigorously**
27,995	Schooling .	If vigorously undertaken, the work will soon be done
27,996	Schoolmaid	If vigorously set about, can soon
27,997	Schoolroom	Vigorously prosecuted
27,998	Schoolship	Have vigorously proceeded with the
27,999	Schorist .	**Vigour**
28,000	Schorlite .	With the utmost vigour
28,001	Schottish .	**Vindicated**
28,002	Schrode .	Has fully vindicated our claim
28,003	Sciaenurus	Has fully vindicated our character
28,004	Sciamachy	**Vindication**
28,005	Sciatheric.	In vindication of our character
28,006	Sciatic . .	In vindication of our claim to priority
28,007	Sciatical .	In vindication of all our rights
28,008	Scienced .	**Violation**
28,009	Sciential .	In violation of the agreement
28,010	Scientific .	In violation of the terms laid down
28,011	Scientism .	In violation of the law
28,012	Scilicet .	**Visible**
28,013	Scilla . .	Visible to the eye
28,014	Scillitine .	Not visible
28,015	Scincoid .	Is visible throughout
28,016	Scincus .	Rich in visible free gold
28,017	Scinque .	Samples taken from where the gold was not visible
28,018	Scintilla .	**Visit** [gave
28,019	Sciography	I cannot visit —— at present
28,020	Sciolism .	Visit —— as soon as you can

No.	Code Word.	Visit (*continued*)
28,021	Sciolous .	After the next visit to
28,022	Sciomancy	If you have time, visit the —— mine
28,023	Scion . .	Should like to visit
28,024	Sciopticon	Was (were) unable to visit
28,025	Scirewyte .	Can you visit large mines (at)
28,026	Scirrhoid .	I cannot visit the place
28,027	Scirrosity .	I will visit the place when opportunity offers
28,028	Scised . .	I will visit the place at once, and let you have
28,029	Scissible .	I will visit [required report by
28,030	Scissors .	To visit and report upon the place will take me
28,031	Scissure .	I do not consider it wise to visit or report upon
28,032	Sciurine .	On the occasion of each visit
28,033	Sclavonic .	If possible, will visit the
28,034	Scleragogy	Had not time to visit the
28,035	Sclerema .	**Visited**
28,036	Scleriasis .	After having visited the —— mine(s) (of)
28,037	Sclerites .	After you have visited
28,038	Sclerobase	I have visited
28,039	Scleroderm	Have never visited the place
28,040	Sclerogen .	Have you visited the —— mine
28,041	Scleroid .	**Void(ed)**
28,042	Sclerotic .	Will become void
28,043	Sclerous .	Will make void
28,044	Scobina .	Will this make void the
28,045	Scoff . .	Will not make void
28,046	Scoffery .	This makes the agreement perfectly void
28,047	Scoffingly .	The sale is void on account of
28,048	Scoffings .	The purchase is void owing to
28,049	Scold . .	Has become void by lapse of time .
28,050	Scolders .	To become void
28,051	Scolecida .	If the —— is void
28,052	Scolite . .	Has voided the agreement
28,053	Scolloped .	Has voided the concession
28,054	Scolopax .	**Volcanic**
28,055	Scolymus .	Of volcanic origin
28,056	Scomber .	In volcanic rock
28,057	Scomberoid	**Vouch**
28,058	Scooping .	To vouch for
28,059	Scoopnet .	Cannot vouch for
28,060	Scoopwheel	Will vouch for
28,061	Scopelidae	Will —— vouch for
28,062	Scopiped .	Will vouch for the truth of the statement (made by)
28,063	Scorbutic .	**Voucher(s)**
28,064	Scorched .	The voucher(s)
28,065	Scorching .	Bring all the vouchers with you
28,066	Scordium .	Send all the vouchers
28,067	Scorers. .	Will bring all the vouchers with me
28,068	Scoriac .	Have sent on all the vouchers
28,069	Scorious .	Voucher(s) will follow by next mail
28,070	Scornful .	Want the vouchers for

No.	Code Word.	
28,071	Scornfully .	**Voyage**
28,072	Scorodite .	Voyage safely over
28,073	Scorpaena	A voyage of —— days
28,074	Scorpions .	A long stormy voyage
28,075	Scorpiurus	Had a very long voyage
28,076	Scorza . .	Had a very quick voyage
28,077	Scorzonera	Had a rough voyage
28,078	Scotchhop	Had a very pleasant voyage
28,079	Scoterduck	**Wages**
28,080	Scotfree .	Wages amount to
28,081	Scotist . .	Owing to the high wages
28,082	Scotodinia	Miners are getting very high wages
28,083	Scotograph	To reduce the wages
28,084	Scotomy .	Wages have been reduced to
28,085	Scotoscope	Low wages
28,086	Scotticism	Could not wages be reduced
28,087	Scotticize .	What do the wages amount to
28,088	Scoundrel .	The wages of competent miners are —— per day
28,089	Scourged .	Wages per day
28,090	Scourging .	What is the rate of wages
28,091	Scout . .	Are demanding more wages
28,092	Scovanlode	About —— when wages are low
28,093	Scowerer .	Wages are very high at present
28,094	Scowlingly	Expect to have to advance wages very soon
28,095	Scraffito .	Must have —— to pay the wages of
28,096	Scraffle. .	It is not a wages' question
28,097	Scrag . .	**Wait**
28,098	Scraggily	Wait to see
28,099	Scramblers	Wait to hear
28,100	Scrannel .	Not to wait
28,101	Scrapbook	Shall I (we) wait (for)
28,102	Scraped .	Must wait here for
28,103	Scrapiana .	Shall have to wait for
28,104	Scrapingly	Will wait here for
28,105	Scrapings .	Will not wait any longer
28,106	Scrapiron .	Cannot wait any longer
28,107	Scrapmetal	Will wait till I hear from you
28,108	Scrappy .	Will wait
28,109	Scratch .	Wait for a further decline
28,110	Scratchpan	Wait for better prices
28,111	Scratchwig	Wait for result of assays
28,112	Scrawled .	Do not wait
28,113	Screable .	Wait for a few days
28,114	Screechowl	Wait till you receive my (our) letter(s) of
28,115	Screechy .	Wait for
28,116	Screenings	Wait till
28,117	Screwbolts	Do not wait any longer
28,118	Screwbox .	How long must —— wait
28,119	Screwcap .	Agree(s) to wait
28,120	Screwclamp	Have no objection to wait

No.	Code Word.	**Wait** (*continued*)
28,121	Screwdock	Wait for further orders
28,122	Screwjacks	Wait on the chance of
28,123	Screwkey .	Can you wait (until)
28,124	Screwnail .	Cannot wait an indefinite time
28,125	Screwpile .	You had better wait for
28,126	Screwplate	How long shall we have to wait for
28,127	Screwpost.	How long can they wait
28,128	Screwpress	Wait till you receive our letter of —— before send-
28,129	Screwshell	If you will wait [ing report
28,130	Screwstone	Wait a little longer (before)
28,131	Screwtree .	If I (we) have to wait (for)
28,132	Screwvalve	What is there to wait for
28,133	Scribbet .	No reason to wait longer
28,134	Scribblers .	**Waited**
28,135	Scriber. .	Waited as long as possible
28,136	Scribism .	Waited in hope of
28,137	Scrimmages	Waited in expectation of being able to announce
28,138	Scrimply .	**Waiting**
28,139	Scrimpness	Am (are) kept waiting
28,140	Scriptory .	Waiting for
28,141	Scriptural.	Waiting for your reply to my (our) telegram of
28,142	Scrivello .	Waiting for you to
28,143	Scrivener .	Waiting for reply to my (our) letter of
28,144	Scrofula .	Whilst waiting for
28,145	Scrofulous	Why are you waiting
28,146	Scroll . .	Am (are) all ready; only waiting for
28,147	Scrollhead	There is no good in waiting longer
28,148	Scrollsaws .	Waiting to hear from
28,149	Scrollwork	Am (are) not waiting
28,150	Scroop. .	Still waiting (for)
28,151	Scrotiform	Waiting for orders
28,152	Scrotocele	Waiting for remittance
28,153	Scroyles .	Waiting for funds
28,154	Scrubbed .	Waiting until the road is clear
28,155	Scrubbing.	Waiting until the frost has broken up
28,156	Scrubbyish	**Waive**
28,157	Scruboak .	Will you waive the right
28,158	Scrubrace.	Will not waive the right
28,159	Scrubstone,	Agree to waive the right
28,160	Scruff . .	If they will not waive their rights
28,161	Scruple .	**Wall(s)**
28,162	Scrupling .	Between walls of
28,163	Scrupulize	Without walls
28,164	Scrutable .	On the hanging-wall
28,165	Scrutation.	With plainly defined hanging and foot walls (of)
28,166	Scrutators.	Having well-defined walls
28,167	Scrutineer.	Against the foot-wall
28,168	Scrutinize .	Samples taken from the hanging-wall
28,169	Scrutinous	Samples taken from the foot-wall
28,170	Scrutiny .	**Want**

No.	Code Word.	**Want** (*continued*)
28,171	Scrutoire .	I (we) want
28,172	Sculleries .	He (they) want(s)
28,173	Scullery .	Do you want
28,174	Sculptor .	Want(s) you to
28,175	Sculptress.	Do(es) not want
28,176	Sculptured	In want of
28,177	Scummings	From want of
28,178	Scummy .	Want(s) too much
28,179	Scurfiness .	Want(s) immediately
28,180	Scurrile .	Will want
28,181	Scurrility .	Will not want
28,182	Scurrilous.	What —— do you want
28,183	Scurvy . .	Let me (us) know exactly what you want
28,184	Scutage .	If you want me (us) to
28,185	Scutellate .	Do not want any more
28,186	Scuttles .	Do you want me to go to
28,187	Scybala .	Do you want me to
28,188	Scyllarian .	In case you want
28,189	Scyllarus .	We shall want
28,190	Scymitar .	We shall want more (than)
28,191	Scymnidae	We still want —— to make up
28,192	Scyphiform	What did —— want
28,193	Scyphulus .	Greatly in want of
28,194	Scythed :	Want all we (you) can get
28,195	Scytheman	Want(s) very much to
28,196	Scythewhet	Want(s) —— to
28,197	Scything .	Will want —— to
28,198	Scythrops .	**Wanted**
28,199	Seaacorn .	Wanted him (them) to
28,200	Seaadder .	Wanted you to
28,201	Seaanemone	Wanted me (us) to
28,202	Seabarrow	Not wanted
28,203	Seabasket .	Wanted immediately
28,204	Seabass .	Will be wanted
28,205	Seabeaten.	Will not be wanted
28,206	Seabirds .	Will soon be wanted
28,207	Seabiscuit .	You are wanted immediately in
28,208	Seabound .	What is principally wanted is
28,209	Seabreach.	Let us know what is wanted
28,210	Seabreeze .	If we know what is wanted
28,211	Seabugloss	Unless we know what is wanted
28,212	Seabuilt .	**War**
28,213	Seacabbage	War threatening between —— and
28,214	Seacalf . .	Since the war
28,215	Seacaptain	Owing to the war
28,216	Seacarps .	Nothing can be done owing to the rumour of war
28,217	Seacatgut .	War between —— and —— averted
28,218	Seachange	War is over
28,219	Seacoast .	Continually at war with
28,220	Seacompass	In the event of war breaking out

No.	Code Word.	**War** (*continued*)
28,221	Seacraft	The war scare caused a panic
28,222	Seacrow	The war scare has subsided
28,223	Seadevil	A war must seriously affect us
28,224	Seadog	War has broken out
28,225	Seadottrel	There are rumours of war
28,226	Seadragon	The seat of war
28,227	Seaducks	Expect war will soon be declared
28,228	Seaeagle	Expect war will soon be over
28,229	Seaegg	A renewal of the war is imminent
28,230	Seafarer	Do you think there will be any war
28,231	Seafaring	War has been declared between
28,232	Seafennel	To support the war a tax has been imposed
28,233	Seafern	Intend to levy a war-tax
28,234	Seafight	**Warn**
28,235	Seaforthia	I think it right to warn you
28,236	Seafoxes	Must warn —— not to
28,237	Seagirdle	Warn —— against
28,238	Seagirkin	Will warn
28,239	Seagirt	**Warned**
28,240	Seagoddess	Has (have) been warned not to
28,241	Seagoing	Has (have) been warned against
28,242	Seagrape	Was (were) not warned
28,243	Seagreen	Be warned in time
28,244	Seagudgeon	**Warrant**
28,245	Seagulls	There is nothing to warrant
28,246	Seaheath	Sufficient to warrant the
28,247	Seaholly	Do you think there is sufficient —— to warrant
28,248	Seaholm	There is a warrant out against
28,249	Seaisland	To warrant me (us)
28,250	Sealark	To warrant such a course
28,251	Sealegs	**Was**
28,252	Sealeopard	I was
28,253	Sealetters	I was not
28,254	Sealingwax	He was
28,255	Sealskin	He was not
28,256	Sealwax	Was he
28,257	Seamaid	Was he not
28,258	Seamanship	Was not
28,259	Seamarge	It was
28,260	Seamblast	It was not
28,261	Seamlace	If it was
28,262	Seamonster	If it was not
28,263	Seamrent	Was it not
28,264	Seamroller	There was
28,265	Seamstress	There was not
28,266	Seaneedle	Was it so
28,267	Seanymph	Why was it not
28,268	Seaonion	Was there any
28,269	Seaooze	There was not one (any)
28,270	Seaorb	When was

No.	Code Word.	**Was** (*continued*)
28,271	Seaotter .	Where was
28,272	Seapad .	Which was
28,273	Seaparrot .	**Washing**
28,274	Seaperch .	In washing
28,275	Seapieces .	By washing the stuff, we found
28,276	Seaplant .	Washing for gold
28,277	Seapoacher	**Waste(s)**
28,278	Seapudding	Great waste of
28,279	Seapurse .	Waste rock
28,280	Seaquake .	There is no waste of
28,281	Searadish .	To stop the waste of
28,282	Searaven .	The waste in
28,283	Searchable	There has been a great waste of
28,284	Searchless	Can you stop the waste of
28,285	Searisque .	Waste no time
28,286	Searobbers	Causing a great waste of time
28,287	Searobin .	Waste of money
28,288	Searocket .	Causing a great waste of money
28,289	Searoom .	From the waste of
28,290	Searoving .	Entails a considerable waste of
28,291	Seasalt. .	It would be a positive waste of time (to)
28,292	Seascape .	There is considerable waste of
28,293	Seaserpent	Considerable waste of quicksilver
28,294	Seasharks .	Why is there such a waste of
28,295	Seashore .	**Watch**
28,296	Seasick .	Watch carefully
28,297	Seaside .	Watch the market
28,298	Seaslater .	Watch the mine
28,299	Seasleeve .	You must watch
28,300	Seaslug .	Will watch
28,301	Seasnail .	Cannot watch
28,302	Seasnakers	To watch over
28,303	Seasnipe .	Watch carefully what is being done
28,304	Seasonably	Watch very closely what he is (they are) doing
28,305	Seasonage	Watch the case in our interest
28,306	Seaspider .	**Watched**
28,307	Seasquirt .	Have carefully watched
28,308	Seastock .	If not watched
28,309	Seasurgeon	**Watchful(ness)**
28,310	Seaswallow	Extreme watchfulness is necessary
28,311	Seaswine .	**Watching**
28,312	Seathief .	Watching for the opportunity
28,313	Seathong .	Watching the proceedings
28,314	Seatitling .	**Water**
28,315	Seatossed .	There is always a great scarcity of water
28,316	Seaturtles .	From the scarcity of water
28,317	Seaunicorn	Water very scarce
28,318	Seaurchin .	Water plentiful
28,319	Seaviews .	Water is diminishing
28,320	Seawalled .	Water getting low

No.	Code Word.	**Water** (*continued*)
28,321	Seawater .	Water increasing
28,322	Seaweed .	Water will not interfere with the working of the mine
28,323	Seawife .	The water from the mine
28,324	Seawolds .	Could see nothing, the mine being full of water
28,325	Seaworthy.	It will take —— days and cost —— to get water
28,326	Seawrack .	Full of water [out of mine
28,327	Sebacic .	Free of water
28,328	Sebastes .	The mine being always free of water
28,329	Sebestan .	Water-power is not enough for all demands
28,330	Sebiparous	Water-power is ample for all demands
28,331	Sebka . .	Water tank(s)
28,332	Sebundy .	Do not go to —— unless you hear the water has been taken out of the mine
28,333	Seceder .	Has seen the mine free of water
28,334	Seceding .	Succeeded in lowering the water to
28,335	Secern . .	Sufficient water to (or for)
28,336	Secernment	Not sufficient water to (or for)
28,337	Secluded .	Will get the water out of the mine
28,338	Secludedly	Mine flooded with water, and will require ——
28,339	Secluding .	Water to work the [pumps to cope with it
28,340	Seclusion .	Water will be out of mine in —— days
28,341	Second .	This creek is always full of water [pany at
28,342	Secondary	Water in any quantity obtainable from a water com-
28,343	Secondbest	Water costs —— per miner's inch
28,344	Secondhand	Amount of water at command is —— inches with a
28,345	Secondrate	From want of water [head of —— feet
28,346	Secrecies .	What is the head of water and quantity
28,347	Secrecy .	In order to keep the water under [hours
28,348	Secret . .	There are —— gallons of water in the twenty-four
28,349	Secretage .	Water is available for [miles
28,350	Secretions.	The water is brought in pipes a distance of ——
28,351	Secretists .	The water-pipes will have to be renewed at a cost of
28,352	Secretness.	Water-power is ample to run —— stamps
28,353	Secretory .	Sufficient water available to run all the machinery
28,354	Sectarist .	Available water-power ample for all demands, but will require —— miles ditch, costing
28,355	Sectional .	Is there plenty of water on the place
28,356	Sectionize .	What fall of water have you
28,357	Sectism .	Fall of water will be —— feet
28,358	Sectmaster	The water has risen in
28,359	Secularism	The water still rising [season
28,360	Secularity .	Calculating at the lowest flow of water in the dry
28,361	Secundated	Calculating at the present good flow of water
28,362	Secundine.	The property has no water upon it
28,363	Securable .	The property is far from water
28,364	Securely .	The property is close to water
28,365	Secureness	The property has water on it
28,366	Securifer .	For how many stamps have you water
28,367	Securiform	What depth of water is there
28,368	Securitan .	Depth of high-water spring-tides

No.	Code Word.	**Water** (*continued*)
28,369	Securities .	Depth of high-water neap-tides
28,370	Sedanchair	The water is not sufficient for motive power,.but is
28,371	Sedately .	Water-ditch [enough for the plates of a battery
28,372	Sedateness	There is always plenty of water in the creek
28,373	Sedentary .	Flume for conveying water
28,374	Sederunt .	Gives us the command of water
28,375	Sedgebird .	Plenty of water to be obtained
28,376	Sedgy . .	Water is now in fork
28,377	Sedilia . .	Water can be kept down without difficulty
28,378	Seditious .	Water is giving much trouble
28,379	Seducingly	Water is not giving much trouble
28,380	Seductive .	A considerable stream of water
28,381	Seductress	A considerable stream of water is constantly flowing
28,382	Sedulously	Under a full head of water
28,383	Seedbasket	Under a —— feet head of water
28,384	Seedbud .	Water is highly charged with acid
28,385	Seedcakes.	Water rapidly corrodes the pumps
28,386	Seedcoat .	Is there much water in the mine
28,387	Seedfield .	Is there much water in the shaft
28,388	Seedgarden	Is the mine free from water
28,389	Seedgrain .	When will the mine be free from water
28,390	Seedleaf .	Mine flooded with water
28,391	Seedlobe .	Workings flooded with water
28,392	Seedoil . .	Water will be out of the mine on
28,393	Seedpearl .	Water in any quantity obtainable
28,394	Seedplot .	Amount of water at command is —— inches
28,395	Seedsheet .	A head of water of —— feet can be obtained
28,396	Seedsman .	Water is available from the
28,397	Seedtime .	Water has to be brought a distance of
28,398	Seedvessel	Water would have to be brought from a distance of
28,399	Seedy . .	The water is brought in pipes (from)
28,400	Seeker . .	The water is brought in a leat from
28,401	Seeksorrow	Water can be brought
28,402	Seemingly.	The water-pipes will have to be renewed
28,403	Seemlily .	Water-pipes required —— inches diameter
28,404	Seemliness	Available water to run
28,405	Seemlyhood	Available water power ample for all demands, but
28,406	Seerhand .	Water-pipe [will require —— pipes, costing
28,407	Seership .	Water to be lifted, estimated at —— gallons per hour
28,408	Seersucker	Gallons of water per hour
28,409	Seesaw . .	Gallons of water per minute to be lifted —— feet
28,410	Seesawing.	Water-power estimated at —— h.p.
28,411	Seethe . .	Stopped for want of water
28,412	Sefatian .	In order to keep the water under, we shall require
28,413	Seghol . .	**Watered**
28,414	Segholate .	The stock has been heavily watered
28,415	Segment .	The watered stock is worth about
28,416	Segmenting	What is the watered stock worth
28,417	Segmentsaw	**Water level**
28,418	Segregated.	Below the water level

No.	Code Word.	**Water level** (*continued*)
28,419	Seguidilla .	At the water level
28,420	Seignorage	From the water level
28,421	Seignory .	Above the water level
28,422	Seineboat .	Water level has been reached
28,423	Seismology	To reach the water level
28,424	Seisura. .	Have you reached the water level
28,425	Seizable .	Have not yet reached the water level
28,426	Seize . .	Shall sink until we reach the water level
28,427	Seizing. .	Expect to reach water level (in)
28,428	Sejunction	Continue sinking until you reach water level
28,429	Sejungible	From the water level downwards
28,430	Selachian .	**Water-right(s)**
28,431	Selbite . .	Water-right surveyed
28,432	Seldom .	The property has no water-right
28,433	Seldomness	The property has an excellent water-right
28,434	Seldshown	Get a water-right
28,435	Select . .	Water-rights have been secured
28,436	Selectedly .	Water-rights must be secured
28,437	Selecting .	Water-rights can be bought for
28,438	Selectman .	Water-rights have been bought for
28,439	Selectness .	Secure water-rights
28,440	Selectors .	Survey water-rights
28,441	Selenitic .	Cannot secure water-rights
28,442	Selenium .	Water-rights fully protected
28,443	Seleniuret .	By getting these water-rights we avoid risk of
28,444	Selenology	All the available water-rights [litigation
28,445	Selfabased	Litigation respecting water-rights
28,446	Selfacting .	Dispute about water-rights
28,447	Selfaction .	The respective water-rights
28,448	Selfborn .	Water-rights extend
28,449	Selfbounty	**Water-wheel(s)**
28,450	Selfbreath .	Pelton water-wheel
28,451	Selfdanger	Overshot water-wheel
28,452	Selfdeceit .	Undershot water-wheel
28,453	Selfdenial .	Breast water-wheel
28,454	Selfesteem	Water-wheel —— feet diameter, × —— feet wide
28,455	Selffaced .	Wooden water-wheel
28,456	Selffed . .	Iron water-wheel
28,457	Selffeeder .	The water-wheel drives
28,458	Selfhelp .	**Way**
28,459	Selfishly .	Either way
28,460	Selfkilled .	Either one way or the other
28,461	Selfleft . .	Which way
28,462	Selflike. .	Whichever way may be chosen
28,463	Selfloving .	In every way possible
28,464	Selfmade .	In that way
28,465	Selfmettle .	It is the only way (to)
28,466	Selfmotion	Is it the only way possible
28,467	Selfmoved .	The best way to
28,468	Selfmurder	Do not consider it the best way

No.	Code Word.	**Way** (*continued*)
28,469	Selfpity .	On the way to
28,470	Selfpraise .	Out of the way
28,471	Selfprofit .	In a fair way to
28,472	Selfrolled .	In a fair way to recovery
28,473	Selfruined .	We see our way clearer
28,474	Selfsame .	Do you see your way clearer
28,475	Selfscorn .	Now on the way to
28,476	Selfseeker .	**We**
28,477	Selfslain .	We are
28,478	Selfstyled .	We are not
28,479	Selftaught .	We can
28,480	Selftrust .	We cannot
28,481	Selfview .	We have
28,482	Selfwill . .	We have not
28,483	Selfwrong .	If we are
28,484	Seltzogene	If we have
28,485	Semaphore	If we can
28,486	Semblable .	We may be
28,487	Semblance	We may have
28,488	Semblative	**Weakened**
28,489	Semecarpus	Much weakened by
28,490	Semeiotics	**Wealthy**
28,491	Semencine	In the hands of wealthy people
28,492	Semiacid .	The owners are wealthy
28,493	Semiangle .	Very wealthy and important firm
28,494	Semiannual	Reputed to be wealthy
28,495	Semibreve .	Was (were) once very wealthy
28,496	Semibull .	Has (have) become very wealthy
28,497	Semichorus	Is (are) not so wealthy as represented
28,498	Semicircle .	**Weather**
28,499	Semicolons	Frosty weather
28,500	Semicolumn	Cannot leave owing to the bad weather
28,501	Semicubium	Cannot go to —— on account of bad weather
28,502	Semidome	Are subject to sudden and great changes of weather
28,503	Semidouble	Weather permitting
28,504	Semifable .	In bad weather
28,505	Semiflexed	In fine weather
28,506	Semifluid .	Weather too rough to do anything
28,507	Semiformed	There is a favourable change in the weather
28,508	Semiliquid	The mine(s) is (are) inaccessible in bad weather
28,509	Semilunar .	Every appearance of present weather lasting
28,510	Semimetal .	During the hot weather
28,511	Semiminim	The weather is exceedingly hot
28,512	Seminality ..	Severe weather; snow and gales
28,513	Seminarian	Severe weather; heavy blizzard
28,514	Seminary .	Severe weather; the snow has completely blocked
28,515	Seminating	Owing to the severity of the weather [the roads
28,516	Semiopal .	As soon as weather permits
28,517	Semiopaque	Experienced bad weather during the voyage
28,518	Semiped .	Experienced fine weather during the voyage

No.	Code Word.	Weather (*continued*)
28,519	Semipedal .	Weather has become much milder; think we shall soon be free from snow
28,520	Semispinal	Weather milder; frost breaking up
28,521	Semisteel .	The bad weather has made it impossible (for)
28,522	Semitism .	Weather very much against my (our)
28,523	Semitonic .	Weather unusually severe
28,524	Semivocal .	**Wednesday**
28,525	Semolella .	On Wednesday
28,526	Semolina .	Last Wednesday
28,527	Sempervive	Next Wednesday
28,528	Semuncia .	Every Wednesday
28,529	Senatus .	Every other Wednesday
28,530	Senecaoil .	**Week(s)**
28,531	Senecaroot	During the week
28,532	Senecio .	Last week
28,533	Senega . .	Next week
28,534	Senescence	The week after next
28,535	Senile . .	For the week ending
28,536	Senility .	This week
28,537	Senior . .	One week
28,538	Seniority .	Two weeks
28,539	Seniorized.	Three weeks
28,540	Sennachy .	Four weeks
28,541	Sensation .	Five weeks
28,542	Senseless .	Six weeks
28,543	Sensibly .	Seven weeks
28,544	Sensific .	Eight weeks
28,545	Sensitive .	For —— weeks
28,546	Sensitory .	In —— weeks
28,547	Sensorium	For the next few weeks
28,548	Sensual .	For the past few weeks
28,549	Sensualism	For the next —— weeks
28,550	Sensuality.	This week's run
28,551	Sensuous .	In a week or ten days
28,552	Sensuously	Early next week
28,553	Sentenced.	Towards the end of next week
28,554	Sentencing	Every week regularly
28,555	Sentiency .	The work can easily be done in a week
28,556	Sentiment.	The work can easily be done in —— weeks
28,557	Sentinels .	Can you get through in a week
28,558	Sentries .	About the beginning of the week
28,559	Sentry . .	About the middle of the week
28,560	Sentrybox.	About the end of the week
28,561	Sepalous .	**Weekly**
28,562	Separately.	Weekly statements
28,563	Separation	Weekly shipments
28,564	Separatist .	(To) show what has been done weekly
28,565	Sepawn .	**Weigh(s)**
28,566	Sepia . .	No piece weighs more than
28,567	Sepiolite .	The whole of the machinery weighs

No.	Code Word.	**Weigh(s)** (*continued*)
28,568	Sepometer.	Do (does) not weigh more than
28,569	Septaria .	Must not weigh more than
28,570	Septemvir .	**Weighing**
28,571	Septenary .	Weighing less than
28,572	Septennate	Weighing more than
28,573	Septennium	**Weight**
28,574	Septentrio .	The weight of each piece to be kept under —— lbs.
28,575	Septette .	What is the weight
28,576	Septfoil .	The weight is —— lbs.
28,577	Septically .	The estimated weight
28,578	Septicidal .	What weight is required
28,579	Septicity .	The average weight
28,580	Septiform .	Weight of ore
28,581	Septimal .	Weight of dry ore
28,582	Septuagint	Weight of concentrates
28,583	Septulate .	The greatest weight
28,584	Septuor .	Broken down under the weight of
28,585	Sepulchral	Give the size and weight
28,586	Sepulture .	Loss of weight
28,587	Sequarious	Loss of weight may be taken at
28,588	Sequels .	Loss of weight —— per cent
28,589	Sequence .	The invoice weight
28,590	Sequential	The net weight
28,591	Sequester .	The gross weight
28,592	Sequestrum	What are the net and gross weights respectively
28,593	Seraglio .	Net weight —— gross weight
28,594	Seralbumen	The weight of the heaviest piece (is)
28,595	Seraph . .	**Well(s)**
28,596	Seraphic .	It would be as well to
28,597	Seraphical	Not so well
28,598	Seraskier .	Is (are) well known
28,599	Serbonian .	Is (are) not well known
28,600	Serenade .	Very well
28,601	Serenading	Is (are) —— well known
28,602	Serenely .	Can you get water by a well
28,603	Serenitude	A well has been put down —— feet
28,604	Serf . . .	Wells have been put down
28,605	Serfdom .	Artesian wells supply ample water
28,606	Serfhood .	Abundance of well water
28,607	Sergeancy .	From wells
28,608	Sergeant .	From a well
28,609	Seriatim .	By a well —— feet deep
28,610	Sericeous .	It will be necessary to sink —— well(s)
28,611	Sericulus .	Is (are) —— well
28,612	Seriema .	Is (are) quite well
28,613	Seriocomic	I am quite well
28,614	Seriously .	I am not well enough to
28,615	Serjania .	Not being very well
28,616	Sermon .	Is (are) not well enough to
28,617	Sermonical	Is (are) looking well

No.	Code Word.	Well(s) (*continued*)
28,618	Sermoning	Doing very well
28,619	Sermonish	In other respects, all is going well
28,620	Sermonized	Generally, all is going well
28,621	Serotine .	**Were**
28,622	Serotinous	If we were
28,623	Serpent .	If we were not
28,624	Serpentary	If you were
28,625	Serpentize.	If you were not
28,626	Serpigo .	If he (they) were
28,627	Serpolet .	If he (they) were not
28,628	Serranus .	We were
28,629	Serratula .	We were not
28,630	Serricorn .	You were
28,631	Sertularia .	You were not
28,632	Servable .	They were
28,633	Servantess	They were not
28,634	Servantry .	There were
28,635	Service. ..	There were not
28,636	Serviceage	There were not any
28,637	Servilely .	Unless there were
28,638	Servingman	Unless we were
28,639	Servitium .	Unless you were
28,640	Servitors .	Unless they were
28,641	Sesame .	Were there
28,642	Sesamoid .	If there were
28,643	Sesamoidal	If there were any chance of
28,644	Sesbania .	**West.** (See Table at end.)
28,645	Seseli . .	Going west
28,646	Sesleria .	Driving west
28,647	Sesqui . .	At the west end
28,648	Sesquisalt .	West cross-cut
28,649	Session. .	West drift
28,650	Sessional .	West vein
28,651	Sestertius .	To the west (of)
28,652	Setback .	West of north
28,653	Setbolt .	On the west side of (the)
28,654	Setdown .	Miles west of
28,655	Setfair . .	**Wet**
28,656	Setigera .	The mine(s) is (are) very wet
28,657	Setigerous.	The mill is constructed for wet crushing
28,658	Setireme	Very wet
28,659	Setline. . ·	Becoming wetter
28,660	Setoff . .	Not so wet as it has been
28,661	Setscrew .	Wet crushing
28,662	Settersoff .	Battery for wet crushing
28,663	Setterup .	Tons of wet ore
28,664	Setterwort	It is an unusually wet season
28,665	Settlebed .	Much delayed by the wet season
28,666	Settlement	During the wet season
28,667	Settlers .	**Wharf**

U

No.	Code Word.	**Wharf** (*continued*)
28,668	Setulose .	From the company's wharf
28,669	Setwall . .	A private railway to the wharf
28,670	Sevenfold .	The company have a tram line to their wharf
28,671	Sevennight	As the company owns a wharf at
28,672	Sevensome	It is essential to have a wharf at
28,673	Seventhly .	The stuff can be run down at small cost to the wharf
28,674	Several .	To our own wharf [by means of a tramway
28,675	Severalize .	Delivered free on the wharf at
28,676	Severally .	The company's wharf
28,677	Severeness	Now lying on the wharf (at)
28,678	Sevoeja .	Have been lying on the wharf since
28,679	Sewage .	**Wharfage**
28,680	Sewers . .	Wharfage charges
28,681	Sexagenary	Wharfage and lighterage
28,682	Sexagesima	**Wharves**
28,683	Sexangled .	The company owns wharves at
28,684	Sexangular	**What**
28,685	Sexdecimal	What is (are)
28,686	Sexennial .	What are you
28,687	Sexhindman	What are they
28,688	Sexifid . .	What shall I
28,689	Sexivalent .	What should I
28,690	Sexless . .	What will
28,691	Sexlocular .	What is the
28,692	Sexly . .	What would
28,693	Sextain .	What do you think of
28,694	Sextile . .	What he (they)
28,695	Sextillion .	What he (they) may do
28,696	Sextonry .	What you may do
28,697	Sextonship	What had we better do
28,698	Sextuple .	What was
28,699	Sexualized	What we
28,700	Sexually .	What were
28,701	Shabbiest .	What is to be done
28,702	Shabbily .	And if so, what
28,703	Shabrack .	**Whatever**
28,704	Shackatory	Whatever you think best
28,705	Shackbolt .	Whatever course is best
28,706	Shackle .	Whatever you think right
28,707	Shacklebar	Whatever may be done
28,708	Shackling .	Whatever —— may do
28,709	Shacklock .	Whatever it was possible to do
28,710	Shadbush .	**Wheel(s)**
28,711	Shaded .	Wheel —— feet diameter
28,712	Shadefish .	Fly-wheel broken
28,713	Shadfrog .	Trolley wheels and axles
28,714	Shadings .	**When(ever)**
28,715	Shadoof .	When are
28,716	Shadowing	When can
28,717	Shadowish	When is

No.	Code Word.	When(ever) *(continued)*
28,718	Shadowless	When we are
28,719	Shaft . .	When you are
28,720	Shaftalley .	When are you
28,721	Shafthorse	When you can
28,722	Shaftman .	When can you
28,723	Shagbark .	When you do
28,724	Shagged .	When do you
28,725	Shagginess	When I can
28,726	Shagging .	When can I (we)
28,727	Shaghaired	When will you
28,728	Shagreen .	When do you think
28,729	Shakedown	When you think
28,730	Shakefork .	When we
28,731	Shakerag .	When we were
28,732	Shakerism.	When you
28,733	Shalloon .	When you were
28,734	Shallowest	When they
28,735	Shallowly .	When they were
28,736	Sham . .	When there were
28,737	Shamanism	When were you
28,738	Shambles .	When were they
28,739	Shambling	And if so, when
28,740	Shamefaced	When —— is (are) likely to
28,741	Shameful .	Cable when
28,742	Shamefully	When and where
28,743	Shameless.	Whenever you
28,744	Shameproof	Whenever it can be done
28,745	Shamfight .	**Where(ever)**
28,746	Shamoying	Where are
28,747	Shampoo .	Where is
28,748	Shampooing	Where was
28,749	Shamrock .	From where
28,750	Shandrydan	Where can I
28,751	Shandygaff	Where you can
28,752	Shankbeer	Where can you
28,753	Shantyman	Where there is
28,754	Shapeable .	Where we
28,755	Shapely .	Where you
28,756	Shapesmith	Where did you
28,757	Shapournet	Where he (they)
28,758	Shardborne	Where —— is (are) likely to
28,759	Sharebeam	And where
28,760	Shareline .	Wherever it occurs (they occur)
28,761	Sharelists .	Is (are) found, wherever
28,762	Sharkray .	**Whether**
28,763	Sharpcut .	Whether we
28,764	Sharpen .	Whether you
28,765	Sharpeneth	Whether he (they)
28,766	Sharpening	Whether or not
28,767	Sharpness .	Whether it is probable

No.	Code Word.	**Whether** (*continued*)
28,768	Sharptail .	Whether it is possible
28,769	Shathmont	Cable us whether or not you
28,770	Shattering .	**Which(ever)**
28,771	Shavegrass	Which are
28,772	Shawfowl .	Which have
28,773	Shayaroot .	Which have been
28,774	Shearbill .	Which can be
28,775	Shearhogs.	Which must
28,776	Shearman .	From which
28,777	Shearwater	To which
28,778	Sheathclaw	Which of the
28,779	Sheathing .	From which of the
28,780	Sheathless.	Which we
28,781	Sheavehole	Which you
28,782	Shebander	Which he (they)
28,783	Shebeen .	Whichever is best
28,784	Shebeening	**While**
28,785	Shed . .	While I am (we are)
28,786	Shedroof .	While you are
28,787	Sheenly .	While it is being
28,788	Sheepberry	While —— is there
28,789	Sheepbiter	While I am there
28,790	Sheepcote.	While —— is (are) here
28,791	Sheepdip .	While he is (they are)
28,792	Sheepdogs	For a long while past
28,793	Sheepfaced	For a long while to come
28,794	Sheepfold .	While that is being done
28,795	Sheephooks	Cannot be done while
28,796	Sheepish .	**Whim(s)**
28,797	Sheepishly	Horse whim
28,798	Sheeplouse	By a whim
28,799	Sheeprun .	By a horse whim
28,800	Sheepsbane	Put up a whim
28,801	Sheepseye.	The horse whim is in course of erection
28,802	Sheepshank	The whim won't work
28,803	Sheepshead	Whim erected and working well
28,804	Sheepskins	**Who**
28,805	Sheepsplit	Who are
28,806	Sheeptick .	Who are not
28,807	Sheepwalk	Who is
28,808	Sheerhulk.	Who is not
28,809	Sheermould	Who is it
28,810	Sheerplan .	Who was
28,811	Sheetcable	Who was not
28,812	Sheetglass.	Who were
28,813	Sheetiron .	Who were not
28,814	Sheetlead .	Who will
28,815	Sheetpile .	Who will not
28,816	Shekarry .	Who must
28,817	Shekel . .	Who should

No.	Code Word.	Who *(continued)*
28,818	Sheldrakes	Who can
28,819	Shelfy . .	Who cannot
28,820	Shellac .	Who have
28,821	Shellapple	Who have not (no)
28,822	Shellbark .	Who have been
28,823	Shellbit .	Who was it that
28,824	Shellboard	Who are our
28,825	Shellcameo	Who gave
28,826	Shellfish .	The man (men) who
28,827	Shellgum .	**Whole**
28,828	Shelllime .	On the whole
28,829	Shellmarl .	From the whole
28,830	Shellmeat .	To get the whole
28,831	Shellproof	The whole of the
28,832	Shellroad .	**Whom**
28,833	Shellsand .	From whom
28,834	Sheltered .	By whom
28,835	Sheltering .	With whom
28,836	Shemitic .	Without whom
28,837	Shendfully	To whom have you
28,838	Shepherd .	To whom did you
28,839	Shepherdly	To whom we are indebted
28,840	Sherardia .	To whom are we (am I) indebted for
28,841	Sherbet .	To whom do they (does it) beiong
28,842	Sheriff . .	**Whose**
28,843	Sherkers .	From whose
28,844	Sheslips .	By whose
28,845	Shewbread	With whose
28,846	Sheworld .	Without whose
28,847	Shibboleth	By whose instructions
28,848	Shield . .	By whose orders
28,849	Shieldeth .	By whose exertions
28,850	Shieldfern .	**Why**
28,851	Shielding .	Why is (are)
28,852	Shieldless .	Why is not
28,853	Shiftable .	Why are not
28,854	Shiftiness .	Why is this
28,855	Shiftingly .	Why did
28,856	Shillelah .	Why did you
28,857	Shimmer .	Why did you not
28.858	Shimmering	And if so, why
28,859	Shimplough	Why then was it
28,860	Shinbone .	Why then did you
28,861	Shineth .	Why will
28,862	Shingle .	Why will not
28,863	Shintoism .	Why will he (they) not
28,864	Shipboard	Why do you
28,865	Shipboy .	Why do you not
28,866	Shipcanal .	Why does he (do they)
28,867	Shipfever .	**Width**

No.	Code Word.	**Width** (*continued*)
28,868	Shipful . .	In width
28,869	Shipholder	The width of the vein
28,870	Shipless .	Width not yet determined
28,871	Shipletter .	Varying in width
28,872	Shipman .	Varying in width from —— feet to —— feet
28,873	Shipmoney	From the width
28,874	Shiprigged	At its greatest width
28,875	Shipshape .	The whole width (of the)
28,876	Shipworm .	What is the width
28,877	Shipwreck	**Will**
28,878	Shipwright	Will you
28,879	Shipyard .	Will you not
28,880	Shireclerk	Will he (they)
28,881	Shiremote	I (we) will
28,882	Shirereeve	I (we) will not
28,883	Shiretown .	He (they) will
28,884	Shirewick .	He (they) will not
28,885	Shirk . .	If he (they) will
28,886	Shirtfront .	If he (they) will not
28,887	Shiver . .	Will it
28,888	Shivereth .	Will it not
28,889	Shiverspar	It will not
28,890	Shoadpit .	When will you be able
28,891	Shoadstone	When will he (they)
28,892	Shoal . .	If you will
28,893	Shoaliness	If you will not
28,894	Shoalwise .	Will not
28,895	Shockdog .	Will not be
28,896	Shockhead	You will not
28,897	Shocking .	It will be
28,898	Shockingly	It will not be
28,899	Shoddy .	In that case, will
28,900	Shoddymill	Will be here
28,901	Shoeblack .	Will be with you
28,902	Shoeboy .	Will be with you about
28,903	Shoebrush	Will not, in any case, be
28,904	Shoebuckle	We will at once
28,905	Shoefactor	Will, if you wish
28,906	Shoehammer	Hope you will at once
28,907	Shoehorn .	**Willing**
28,908	Shoeknife .	Would —— be willing to
28,909	Shoemaker	Am (are) willing to
28,910	Shoemaking	Is (are) willing to
28,911	Shoepeg .	Is (are) not willing to
28,912	Shoeshave	Perfectly willing to
28,913	Shoestones	Perfectly willing to leave it **to**
28,914	Shoestrap .	If you are willing to
28,915	Shoestring	If you are not willing to
28,916	Shoetie .	If he is (they are) willing
28,917	Shooters .	If he is (they are) not willing

No.	Code Word.	
28,918	Shopbill	. **Wind**
28,919	Shopbook	To wind up
28,920	Shopgirls	Resolution passed to wind up
28,921	Shopkeeper	**Winding-up**
28,922	Shoplifter	Voluntary winding-up
28,923	Shopmaid	Winding-up under inspection
28,924	Shoppish	. **Winding engine.** (See Hauling.)
28,925	Shopshift	Winding engine is wanted for
28,926	Shopwalker	A small winding engine of about —— h.p.
28,927	Shopwoman	Putting down a winding engine
28,928	Shoreweed	Winding engine will be placed
28,929	Shortage	Winding engine will be started
28,930	Shortarmed	Winding engine started; going well
28,931	Shortbread	Winding engine for —— shaft
28,932	Shortcake	Winding engine for the tramway
28,933	Shortdated	**Windlass**
28,934	Shortdrawn	A steam windlass capable of hoisting (or hauling)
28,935	Shortener	Can do at present with a hand windlass
28,936	Shorthand	Now hoisting with a small steam windlass
28,937	Shortlived	Using at present hand windlass
28,938	Shortly	. **Winter**
28,939	Shortness	Mine cannot be examined in winter
28,940	Shortrib	Must finish before winter sets in or it will have to
28,941	Shortsight	In the winter [stand over until the spring
28,942	Shotanchor	Through the winter
28,943	Shotbelt	The winter begins in
28,944	Shotclog	The winter is not over before
28,945	Shotgauge	During the winter months
28,946	Shotgun	Cannot work in the winter
28,947	Shothole	The winter has set in with great severity
28,948	Shotlocker	(If) we cannot get done before the winter sets in
28,949	Shotmetal	Cannot get mill up before the winter sets in
28,950	Shotplug	Must finish before winter sets in
28,951	Shotpouch	Winter has begun in earnest; ground is frozen hard
28,952	Shotproof	Looking for the end of the winter
28,953	Shotrack	Winter nearly over; frost is breaking up
28,954	Shotsilk	. **Winze(s)**
28,955	Shottower	Winze No. 1
28,956	Shotwindow	Winze No. 2
28,957	Shoulder	Winze No. 3
28,958	Shoutings	Winze No. 4
28,959	Shovegroat	Winze No. 5
28,960	Shovelard	Winze No. 6
28,961	Shovelful	Winze No. 7
28,962	Shovelhat	Winze No. 8
28,963	Showboxes	Winze No. 9
28,964	Showcards	Winze No. 10
28,965	Showcase	Winze No. ——
28,966	Showerbath	In winze No. ——
28,967	Showerless	Samples from winze No. —— assayed

No.	Code Word.	**Winze(s)** (*continued*)
28,968	Showery .	Estimate the cost of sinking winze at
28,969	Showglass .	The vein in winze No. ——
28,970	Showiest .	From winze
28,971	Showplace	By a winze
28,972	Showrooms	Connected by a winze
28,973	Shrank. .	Have commenced winze No. —— in —— level
28,974	Shredhead	Winze from —— level down —— feet
28,975	Shredless .	Winze (in)
28,976	Shreetalum	Sinking a winze (in)
28,977	Shrew . .	Sinking a winze to connect with
28,978	Shrewdest .	Winze has been sunk
28,979	Shrewdly .	Winze has been sunk to a depth of
28,980	Shrewdness	Shall sink a winze (in)
28,981	Shrewishly	Shall sink a winze from —— (level)
28,982	Shrewmole	Shall sink a winze to connect with
28,983	Shriek . .	Winze form —— level should be sunk
28,984	Shrieking .	Winze form —— level has been sunk
28,985	Shriekowl .	Winze has made connection with the upraise in
28,986	Shrieval .	How deep has winze been sunk in
28,987	Shrievalty .	How deep has winze been sunk in ——, and what is the appearance and assay of ore therefrom
28,988	Shrill . .	Winze from —— level
28,989	Shrillness .	Winze from —— has been sunk to connect with
28,990	Shrimp .	Winze down —— feet, ore therefrom assays
28,991	Shrimpers .	Have sunk a winze (in)
28,992	Shrimpnet	Have commenced sinking a winze
28,993	Shrinkage .	To connect with —— winze
28,994	Shrivelled .	Continue sinking winze No. ——
28,995	Shroff . .	Have discontinued sinking winze (from)
28,996	Shroffage .	Winze has not been sunk during the month
28,997	Shroud .	Do not propose to sink winze any further
28,998	Shrouding	Propose to sink a winze (in or from)
28,999	Shroudless	**Wire**
29,000	Shroudrope	Feet of steel wire rope, —— inches circumference
29,001	Shrovetide	Wire tramway
29,002	Shrubbery .	Feet of steel wire flat rope, —— inches × ——
29,003	Shrugs . .	Wire rope [inch
29,004	Shudder .	Wire rope for tramway
29,005	Shuddereth	Wire rope for hoist
29,006	Shuddering	Copper wire
29,007	Shufflecap	**Wish(es)**
29,008	Shufflers .	Wish you to
29,009	Shuntguns	Wish you not to
29,010	Shunting .	Wish(es) that
29,011	Shuttlebox	Do(es) not wish
29,012	Shwanpan	When do you wish
29,013	Shyly . .	I (we) do not wish
29,014	Shyness .	If you wish
29,015	Siagush .	If you do not wish
29,016	Sialagogue	Do you wish (for)(to)

No.	Code Word.	**Wish(es)** *(continued)*
29,017	Sialidae .	Do they (does he) wish (for) (to)
29,018	Sibbaldia .	Do not wish (for) (to)
29,019	Siberite .	Unless you wish me (us) to
29,020	Sibilancy .	Wish that you would
29,021	Sibilant .	Sorry we cannot comply with your wish
29,022	Sibilating .	Impossible to do what —— wish(es)
29,023	Sibilatory .	We wish to avoid
29,024	Sibilous .	Wish you could (would)
29,025	Sibthorpia	Wish he (they) could (would)
29,026	Sibylline .	Wish him (them) not to
29,027	Sibyllists .	Your wishes shall have our best attention
29,028	Siccated .	Wishes shall be attended to at once
29,029	Siccating .	Have acted in accordance with your wishes
29,030	Siccation .	Has (have) not acted in accordance with our wishes
29,031	Siccific .	**Withdraw**
29,032	Siccity . .	I would advise you to withdraw your offer
29,033	Siciliana .	Better to withdraw from the matter
29,034	Sickbay .	Wish(es) to withdraw his (their) offer
29,035	Sickbed .	Do not withdraw
29,036	Sickberth .	Will withdraw
29,037	Sickerly .	Will not withdraw
29,038	Sickfallen .	Withdraw —— letter
29,039	Sickleman .	Withdraw if possible
29,040	Sicklewort	Withdraw the claim
29,041	Sicklist . .	Withdraw from the case
29,042	Sicknesses	Withdraw all objections
29,043	Sidearms .	Cannot now withdraw
29,044	Sideaxe .	**Withdrawing**
29,045	Sidebar .	**Withdrawn**
29,046	Sideboxes .	Has (have) withdrawn from
29,047	Sidechain .	Has (have) not withdrawn from
29,048	Sidecut .	Have you withdrawn
29,049	Sidedish .	Have withdrawn the offer
29,050	Sideglance	Withdrawn our claim
29,051	Sidehead .	Withdrawn their claim
29,052	Sidehook .	Withdrawn the petition
29,053	Sidelever .	Has been withdrawn (from)
29,054	Sidelights .	Have withdrawn our account from
29,055	Sidelong .	All objections have been withdrawn
29,056	Sidepipe .	Has not been withdrawn
29,057	Sideplane .	Has (have) withdrawn his (their) offer
29,058	Sidepost .	**Within**
29,059	Sideration .	Within the time
29,060	Sidereal .	Within the space
29,061	Siderolite .	Within the limit
29,062	Siderostat .	Within the limited space
29,063	Siderotype	Within the limited time
29,064	Sidesaddle	Within reasonable limits
29,065	Sideslip .	Within the specified amount
29,066	Sidestitch .	Within the specified time

No.	Code Word.	**Within** (*continued*)
29,067	Sidetaking	Cannot keep within
29,068	Sidetimber	Not within
29,069	Sideview .	**Without**
29,070	Sidewalk .	Without the
29,071	Sideways .	Without our consent
29,072	Sidewinds.	Without our knowledge
29,073	Sidled . .	Without prejudice
29,074	Siegetrain .	Without any warning
29,075	Sigaultian .	Without any (more)
29,076	Sigheth .	Not without reason
29,077	Sighing .	Not without
29,078	Sighingly .	Has been done without
29,079	Sightdraft.	Cannot do without
29,080	Sighthole .	Have had to do without
29,081	Sightseer .	Run(s) without a hitch
29,082	Sightshot .	**Witness(es)**
29,083	Sightsman	To witness
29,084	Sigillaria .	Can witness
29,085	Sigmoid .	Will witness
29,086	Sigmoidal.	The witness
29,087	Signable .	As witness(es)
29,088	Signal . .	Can you call witnesses to prove
29,089	Signalbox .	Can call plenty of witnesses to prove
29,090	Signalfire .	Witnesses can prove
29,091	Signality .	Witnesses cannot prove
29,092	Signalized.	**Won**
29,093	Signallamp	Has (have) won the case
29,094	Signalling .	Telegraph who has won
29,095	Signalman	Has (have) won the
29,096	Signalpost	Have won the case with costs allowed
29,097	Signation .	The other side have won the case
29,098	Signature .	The other side have won ; we shall appeal
29,099	Signboards	We have won ; our opponents intend to appeal
29,100	Signers .	**Wood(s)**
29,101	Signetring.	Good wood
29,102	Signified .	Poor wood
29,103	Signify .	Plenty of wood
29,104	Signiorize .	There is a fair quantity of wood in the neighbourhood of the mines
29,105	Signmanual	A scarcity of wood
29,106	Signposts .	Wood costs per cord, delivered at works
29,107	Silence .	Wood only costs for the hauling
29,108	Silencing .	Estimate there is enough wood to last —— years
29,109	Silentious .	Teams have to haul wood —— miles
29,110	Silently .	What is the yearly cost of wood
29,111	Silentness .	What is the cost of wood at the mine
29,112	Silesian .	Engine(s) take(s) —— cords of wood per day
29,113	Silicated .	Mill takes —— cords of wood per day
29,114	Silicify . .	Large woods in the neighbourhood
29,115	Silicon . .	**Work(s)**

No.	Code Word.	**Work(s)** (*continued*)
29,116	Silicula .	There is nothing to prevent the work of mining and milling proceeding throughout the year
29,117	Siliginose .	Prevents the regular work of mining and milling from being carried on throughout the year
29,118	Silingdish .	The necessary work for developing the mine
29,119	Siliquae .	A great deal of dead work has to be done
29,120	Siliquella .	Do you consider —— sufficient for opening mine
29,121	Siliquous .	Work is being pushed ahead [for full work
29,122	Silkcotton.	Work suspended (owing to)
29,123	Silkened .	Work will be resumed
29,124	Silkfowl .	Have stopped work
29,125	Silkhen .	Work will be commenced
29,126	Silkmercer	Will not work
29,127	Silkmill .	Stop all work upon
29,128	Silkweaver	Men refuse to work
29,129	Silkweed .	Men refuse to work at a reduction
29,130	Silkworms	Men agree to work
29,131	Sillabub .	Agree to work
29,132	Silliest . .	Work day and night
29,133	Silly . .	Work only eight hours a day
29,134	Silting . .	Too much work
29,135	Siluridan .	No work is being done on the place
29,136	Silvanite .	No work is being done in the mine
29,137	Silverbell .	A great amount of surface work has been done
29,138	Silverbush	The work done on the place is of the most meagre description and of no practical value
29,139	Silverfox .	Work(s) well
29,140	Silvergray.	Work(s) badly
29,141	Silvering .	I believe these works will open up a mine worth
29,142	Silverized .	The first works to be done are
29,143	Silverleaf .	The works will be completed about
29,144	Silverless .	Can only work —— months in the year
29,145	Silvertree .	Must have —— to do the necessary work
29,146	Silverweed	A great deal of work will have to be done before
29,147	Simblot .	A great deal of the work has been done
29,148	Simiadae .	Can work —— to better advantage
29,149	Similar .	Work has been resumed
29,150	Similarity .	Work has been begun upon
29,151	Similitude	Will start work (upon)
29,152	Simious .	Have begun work
29,153	Simoniac .	Have not yet begun (to) work
29,154	Simoniacal	The necessary work(s)
29,155	Simonists .	The whole of the work to be done
29,156	Simony .	The whole of the work that has been done
29,157	Simoom .	The works will include
29,158	Simper. .	The only work at present carried on
29,159	Simpering	Cannot now work to a profit
29,160	Simpleness	Work had progressed to a distance of
29,161	Simpleton	Work done upon this lode
29,162	Simplician	Upon this work (these works) alone have been spent

No.	Code Word.	
29,163	Simplicity .	**Worked**
29,164	Simplified .	The mine is being worked
29,165	Simplify .	Has (have) been largely worked
29,166	Simplistic .	The mine is not being worked
29,167	Simploce .	The mine is worked out
29,168	Simulachre	The mine will soon be worked out
29,169	Simulacrum	Have worked out
29,170	Simulated	Which was (were) worked
29,171	Simulating	The whole of the upper levels (above No. ——)
29,172	Simulation	Not worked [worked out
29,173	Simulatory	Extensively worked
29,174	Simulium .	Very little worked
29,175	Simulties .	Only worked on the surface
29,176	Simulty .	Neither of the mines are being worked
29,177	Simurg . .	None of the mines are being worked
29,178	Sinaitic .	The mine(s) has (have) been badly worked
29,179	Sinapism .	The mine(s) has (have) been well worked
29,180	Sinborn ,	Owing to the mine(s) having been so badly worked
29,181	Sinbred .	Worked —— tons
29,182	Sincere .	Worked —— hours
29,183	Sincerely .	Can the (——) mine be worked profitably
29,184	Sincipital .	Can the property be worked to a profit
29,185	Sinciput .	The whole of the —— is worked out
29,186	Sineater .	Worked by steam-power
29,187	Sinecural .	Worked by water-power
29,188	Sinecurism	How is (are) the —— worked
29,189	Sinewiness	Has not been worked since
29,190	Sinewous .	When it was last worked
29,191	Sinewy . .	Was (were) formerly worked
29,192	Sinful . .	When worked by the previous owner
29,193	Sinfully .	The mine was worked in former days
29,194	Sinfulness .	It is said that the mine when formerly worked
29,195	Singeress .	The mine, as now worked [yielded
29,196	Singingman	The greatest depth to which the mine has been
29,197	Singlecut .	Can be worked to better advantage [worked
29,198	Singletree.	Cannot be worked to better advantage
29,199	Singly . .	(If) worked under a better system
29,200	Singular .	Worked in a very imperfect way
29,201	Singularly.	Has been successfully worked
29,202	Singultous	The mine was worked by the Spaniards
29,203	Sinister .	Was never properly worked
29,204	Sinisterly .	From the way in which it has been worked
29,205	Sinistral .	Has been worked
29,206	Sinistrous.	Has not been worked
29,207	Sinkhole .	Not recently worked
29,208	Sinktrap .	Most energetically worked
29,209	Sinlessly .	Have worked hard in order to
29,210	Sinologist.	Worked continuously for
29,211	Sinology .	Now being worked
29,212	Sinople .	Not being worked at present

No.	Code Word.	**Worked** (*continued*)
29,213	Sintoc . .	Worked to its (their) utmost capacity
29,214	Sinuose .	Cannot be worked more economically
29,215	Sinuosity .	Must be worked more economically
29,216	Siphonage	**Working(s)**
29,217	Siphoncup	In the old workings
29,218	Siphonida .	The old workings have caved, and are inaccessible
29,219	Siphonifer	Out of the old workings
29,220	Siphuncle .	Very extensive old workings
29,221	Sipunculus	The workings were consequently abandoned
29,222	Sirenical .	To start the workings up again
29,223	Sirenize .	Cost of working is about —— per ton
29,224	Sirenizing .	The ore from these workings
29,225	Siriasis . .	Cheap working
29,226	Siricidae .	Great facilities exist for working the ore cheaply
29,227	Sirius . .	For working the ore
29,228	Sirloins .	Working day and night
29,229	Sirocco .	Working —— hours
29,230	Siruped .	Working day of eight hours
29,231	Sisalgrass .	Working day of —— hours
29,232	Sisalhemp	Working hard to obtain
29,233	Siserara .	There are considerable old workings on the property
29,234	Siskiwit .	How are the workings looking
29,235	Sismometer	Cable how the new workings are looking
29,236	Sison . .	The new workings and developments
29,237	Sisterhood	The workings above the
29,238	Sisterly .	The upper workings
29,239	Sisymbrium	The workings in the upper part of the mine
29,240	Sisyphean	The lower workings
29,241	Sitfast . .	The workings in the lower part of the mine
29,242	Sitomania .	The lowest workings
29,243	Sitophobia	The workings below the [encouraging
29,244	Situate . .	The development in this working has been most
29,245	Situations .	Conditions and prospects of the workings are most [encouraging
29,246	Sitzbath .	Are you still working on
29,247	Sium . .	Have you anything to report upon the workings in
29,248	Sixfold . .	The workings in the —— level
29,249	Sixshooter	From these workings we expect to obtain
29,250	Sizar . .	We are still working on the ore from the
29,251	Sizarship .	To connect (connecting) these workings with
29,252	Sizeroll .	The caves in the old workings
29,253	Sizestick .	Owing to the —— workings having caved in
29,254	Skaddon .	The old workings have caved on the surface
29,255	Skainsmate	(From) the ore body we are now working
29,256	Skater . .	(From) the face of the workings
29,257	Skating .	The presence of —— in these workings
29,258	Skayle . .	No working facilities whatever
29,259	Skedaddle	Working facilities very inadequate
29,260	Skegger .	Working fairly, but do not appear to understand what they have to deal with
29,261	Skegshore	(From) the surface workings

No.	Code Word.	**Working(s)** (*continued*)
29,262	Skeins . .	Now working on the —— vein
29,263	Skelder .	The workings on the —— vein
29,264	Skeletal .	Working with little hope of success
29,265	Skeletons .	Working with good prospects of success
29,266	Skelloch .	This working (these workings) have so far not
29,267	Skellum .	As soon as the workings are opened up [yielded
29,268	Skerry . .	Until the workings have been opened up
29,269	Sketching .	These workings have opened up
29,270	Sketchy .	The workings have opened out well
29,271	Skewarch .	The old workings, from which were extracted
29,272	Skewbald .	Immense bodies of ore in sight in these workings
29,273	Skewbridge	Ore from these workings has been treated
29,274	Skewcorbel	The workings are very extensive
29,275	Skewed .	The recent workings and developments
29,276	Skewering	Have taken the men out of these workings
29,277	Skewfillet .	Discontinue operations in these workings
29,278	Skewplane	Workings in —— have been abandoned
29,279	Skewput .	The extent and depth of the workings
29,280	Skidpan .	More extensive and deeper workings
29,281	Skied . .	These workings have been carried on for
29,282	Skilful . .	The workings promise to develop large ore bodies
29,283	Skilfully .	Give us some information about the workings in
29,284	Skimington	Cable the latest information about the workings and
29,285	Skimmilk .	Workings very discouraging [developments (in)
29,286	Skinbound	Anxious to know what you are doing in the work-
		ings in, or (in these workings) [to keep on
29,287	Skindeep .	The workings are more encouraging, and we intend
29,288	Skinflint .	There is no change in the workings (since)
29,289	Skinwool .	There is no material change in the workings (since)
29,290	Skipjacks .	Is there any change in the workings
29,291	Skipkennel	Is there any change in the workings or develop-
29,292	Skipping .	Workings and developments [ments (since)
29,293	Skippingly	Workings very discouraging, and we intend to stop
29,294	Skirmisher	The workings show an inclination to develop into
		a substantial pay chute [a good pay chute
29,295	Skirret . .	The workings show no inclination to develop into
29,296	Skittish .	No probability of finding in these workings any ore
29,297	Skittishly .	Width and size of ore body we (you) are now work-
29,298	Skolecite .	Must stop working if you cannot [ing on
29,299	Skuagull .	Must stop working unless we are provided with
29,300	Skulked .	Have had to stop working [more funds
29,301	Skulkingly	Stop working until
29,302	Skullcap .	The expense of working the
29,303	Skulless .	Delay in working caused by
29,304	Skulpin .	Restarted working upon (in)
29,305	Skunkbirds	The cost of working is too great
29,306	Skunkish .	The poor results obtaining from (the) working
29,307	Skunkweed	In working the mill
29,308	Skyblue .	Working as cheaply as
29,309	Skyborn .	**Worn**

No.	Code Word.	**Worn** (*continued*)
29,310	Skycolour	Quite worn out
29,311	Skydrain .	Quite worn out, and must replace them
29,312	Skydyed .	Replace the worn
29,313	Skyhigh .	**Worse**
29,314	Skyish . .	Do you think the mine looks worse
29,315	Skylark .	The mine is looking worse
29,316	Skylarking	The mine is not looking worse
29,317	Skyless .	Getting worse
29,318	Skylights .	Matters growing worse every day
29,319	Skyparlour	Appear to be getting worse
29,320	Skyplanted	Is worse, come home at once
29,321	Skyrocket.	Is worse, we do not expect he will live
29,322	Skyroofed.	Is (are) no worse
29,323	Skysail . .	Worse and worse
29,324	Skyscraper	Worse than when we last wrote
29,325	Skyward .	The last accounts we had were much worse
29,326	Slabline .	The ore is much worse and will not pay expenses
29,327	Slabsided .	Much worse, will hardly pay
29,328	Slack . .	Worse than we anticipated
29,329	Slackening	If matters grow worse
29,330	Slackjaw .	Cannot be much worse
29,331	Slackly .	Expect things will be worse yet
29,332	Slackness .	**Worst**
29,333	Slakeless .	If it comes to the worst
29,334	Slamkin .	The worst that could happen
29,335	Slammed .	The worst that we have had to contend with
29,336	Slammerkin	The worst of all
29,337	Slamming .	The worst results that have been obtained
29,338	Slander .	The worst place possible (for)
29,339	Slandering	The worst that could be found
29,340	Slanderous	The worst predictions have been verified
29,341	Slantingly.	The worst of the matter is, that
29,342	Slantwise .	Has (have) realized our worst fears
29,343	Slapbang .	One of the worst points about it
29,344	Slapdash .	Put the worst possible construction upon
29,345	Slapup . .	**Worth**
29,346	Slashing .	What do you think the mine is worth
29,347	Slateaxe .	I estimate the mine is worth
29,348	Slateclay .	Think it is worth
29,349	Slattern .	Not worth more than
29,350	Slatternly .	Worth altogether
29,351	Slaty . .	I estimate the mine and mill are worth
29,352	Slaveborn .	Is it worth while to
29,353	Slaveforks	It is hardly worth while to
29,354	Slavegrown	Hardly worth the expense
29,355	Slavelike .	Worth much more than
29,356	Slavery .	Worth much less than
29,357	Slaveship .	Will be worth more when
29,358	Slavetrade	The whole is not worth more than
29,359	Slavocracy	Worth considerably more than was given for it

No.	Code Word.	**Worth** (*continued*)
29,360	Sleavesilk .	Think it is well worth
29,361	Sleaziness .	Think it is well worth while to try
29,362	Sleazy . .	Is not worth altogether
29,363	Sledging .	Says it is fully worth
29,364	Sleekstone	May be worth —— when
29,365	Sleepers .	What is —— really worth
29,366	Sleepful .	Would cost more than it is worth
29,367	Sleepiest .	Much more than it is actually worth
29,368	Sleepily .	**Worthless**
29,369	Sleepwaker	Quite worthless
29,370	Sleeress .	Is it worthless
29,371	Sleevefish .	Would be worthless
29,372	Sleevehand	Worthless at present owing **to**
29,373	Sleeveknot	**Would**
29,374	Sleeveless .	Would it be
29,375	Sleevelink	Would it not be
29,376	Sleigh . .	Would be
29,377	Sleighbell .	Would be better
29,378	Sleighing .	Would be better not
29,379	Slenderly .	Would you
29,380	Slept . .	Would there be any
29,381	Slibowitz .	Would he (they) do this
29,382	Slicebar .	It would be
29,383	Slicer . .	It would not be
29,384	Slidegroat.	I (we) would
29,385	Sliderail .	I (we) would not
29,386	Sliderest .	He (they) would
29,387	Sliderod .	He (they) would not
29,388	Sliderpump	How would
29,389	Slidevalve.	How would you wish it
29,390	Slighten .	If you would
29,391	Sliminess .	If he (they) would
29,392	Slingcart .	If it would be
29,393	Slingdog .	If it would not be too much
29,394	Slipclutch .	If it would not inconvenience (you)
29,395	Slipdock .	Unless he (they) would
29,396	Slipkiln .	**Wreck(ed)**
29,397	Slipknot .	Saved from the wreck
29,398	Sliplink .	Is a total wreck
29,399	Slipon . .	Totally wrecked (by)
29,400	Slippered .	Has been wrecked by a gang of swindlers
29,401	Slipperily.	Has been wrecked by gross mismanagement
29,402	Sliprope .	Has been wrecked by
29,403	Slipshod .	**Writ**
29,404	Slipsloppy	A writ has been issued
29,405	Slipstring .	Has (have) been served with a writ
29,406	Slipthrift .	Has (have) issued a writ against
29,407	Slitdeal .	The writ issued
29,408	Slobbered .	**Write(s)** [what is being done in
29,409	Slobbering	Write me full particulars regarding appearance, and

No.	Code Word.	**Write(s)** (*continued*)
29,410	Slogan . .	Write me at
29,411	Sloop . .	Write fully by first post
29,412	Slopbasin .	Write immediately full particulars
29,413	Slopbowl .	Write immediately following particulars
29,414	Sloped . .	Write promptly if any change occurs
29,415	Slopewise .	Write to me, care of —— at
29,416	Slopingly .	Write all you know about
29,417	Sloppail .	Write fully to —— at once saying
29,418	Sloppiness	Write for particulars
29,419	Slopping .	Write what has been done
29,420	Sloppy . .	Write full particulars next mail what has been done
29,421	Sloproom .	Write full instructions [about
29,422	Slopseller .	Write regularly
29,423	Slopshops .	Write as soon as you
29,424	Slopwork .	Will write
29,425	Sloshy . .	Will write to you to explain
29,426	Slothful .	Will write you fully
29,427	Slothfully .	Will write him (them) fully
29,428	Slothound	Writes us saying
29,429	Slouch . .	Will write you as soon as
29,430	Slouchhat .	Promise(d) to write
29,431	Slouching .	No use to write, you must telegraph
29,432	Slovenly .	**Writing**
29,433	Slowback .	Writing you by to-day's post
29,434	Slowgaited	Writing you by next mail
29,435	Slowlemur	Writing to you at
29,436	Slowness .	Writing to ask why (about)
29,437	Slowpaced	Have it all in writing
29,438	Slowwinged	Will be sure to have it all in writing
29,439	Sluerope .	Not in writing
29,440	Slugabed .	Am writing you fully
29,441	Sluggard .	Am writing him (them)
29,442	Sluggish .	Better to put it in writing
29,443	Sluggishly .	If you are writing
29,444	Slughorn .	When writing you last, we omitted to send —— it
29,445	Slugsnail .	In writing [will be sent with our next
29,446	Sluice . .	When writing
29,447	Sluicegate .	After writing
29,448	Sluiceway .	**Written**
29,449	Sluicing .	Have written you to care of —— at
29,450	Slumberer	Has (have) written that
29,451	Slungshot .	Have you written to
29,452	Slurred .	Have both written and wired
29,453	Slyboots .	Have both written and wired, but can get no reply
29,454	Smalkaldic	Has been written to on the subject
29,455	Small . .	All that has been written
29,456	Smallage .	If all has been written is true
29,457	Smallarms	Have written fully explaining
29,458	Smallbeer .	**Wrong**
29,459	Smallcoal .	Is anything wrong

No.	Code Word.	**Wrong** (*continued*)
29,460	Smallcraft .	Quite wrong
29,461	Smallfry .	Is (are) wrong
29,462	Smallhand	The accounts are wrong
29,463	Smallish .	There is nothing wrong (with)
29,464	Smallpica .	If anything is wrong (with) (about)
29,465	Smallpox .	There appears to be something wrong
29,466	Smallstuff.	Working in a wrong direction
29,467	Smalltalk .	Has gone wrong
29,468	Smallwares	Something has gone wrong with
29,469	Smaltine .	Have done wrong
29,470	Smaragd .	Have found nothing wrong as yet
29,471	Smaragdine	Whatever has been done wrong
29,472	Smartened	You have a wrong idea
29,473	Smartly .	It is a wrong assumption
29,474	Smartmoney	Put a wrong construction upon
29,475	Smartness.	Has (have) not been far wrong
29,476	Smartweed	Are they not wrong
29,477	Smashers .	The plan is quite wrong
29,478	Smatterers	I am (we are) not wrong
29,479	Smearcase	There must be something wrong with
29,480	Smeardab.	The entire system is wrong
29,481	Smectite .	Wrong from beginning to end
29,482	Smeddum .	**Wrongly**
29,483	Smegmatic	Wrongly kept
29,484	Smellfeast.	Wrongly detained
29,485	Smelling .	Wrongly accused
29,486	Smellless .	Has been greatly wronged by
29,487	Smelltrap .	**Yard(s)**
29,488	Smeltery .	Cubic yard(s)
29,489	Smerlin .	Square yard(s)
29,490	Smilaceae.	Cost per yard
29,491	Smilax . .	What is the cost of driving per yard
29,492	Smilingly .	Cost of driving is —— per yard
29,493	Smirching	What is the cost of sinking per yard
29,494	Smirked .	Cost of sinking is —— per yard
29,495	Smithcraft	Per lineal yard
29,496	Smitheries	Per cubic yard
29,497	Smithery .	Per square yard.
29,498	Smock . .	Cost of —— is —— per cubic yard
29,499	Smockfaced	What is the cost per cubic yard
29,500	Smockfrock	**Year(s)**
29,501	Smockmill	In one year
29,502	Smokable .	For one year
29,503	Smokearch	In —— years
29,504	Smokeball	For —— years
29,505	Smokeblack	In the year
29,506	Smokeboard	For the year
29,507	Smokebox	For the past year
29,508	Smokecloud	For the current year
29,509	Smokedried	Next year

No.	Code Word.	Year(s) (*continued*)
29,510	Smokedry .	Last year
29,511	Smokejack	Early in the year
29,512	Smokemoney	During the year
29,513	Smokepenny	During last year
29,514	Smokeplant	During the last —— years
29,515	Smoketight	During next year
29,516	Smoketree	About the beginning of next year
29,517	Smokiest .	About the middle of the year
29,518	Smokily .	About the end of the year
29,519	Smokiness	About the end of last year
29,520	Smoking .	About the end of next year
29,521	Smokingcap	All the year round
29,522	Smooth . .	The average of the last year
29,523	Smoothbore	The average of the last —— years
29,524	Smoothing	Towards the end of the year
29,525	Smoothly .	A year or two hence
29,526	Smoothness	It will take at least a year
29,527	Smothered	It will take several years
29,528	Smotherfly	Cannot last more than —— years
29,529	Smothering	In the course of the year
29,530	Smouched	How many years (has)
29,531	Smouldered	Over a number of years
29,532	Smudge .	For several years past
29,533	Smudgecoal	**Yearly**
29,534	Smugboat .	A yearly income of
29,535	Smugfaced	A yearly profit of
29,536	Smuggle .	Yearly accounts
29,537	Smuggling	Yearly statements
29,538	Smugly .	Forward the yearly accounts as soon as possible
29,539	Smutball .	Have not yet received yearly accounts
29,540	Smuttiest .	Have sent the yearly accounts
29,541	Smuttily .	**Yes**
29,542	Smuttiness	Yes, certainly
29,543	Smyrnium	Yes, you may do it
29,544	Snacks . .	Yes, do what you propose
29,545	Snaffle . .	Yes, act as you think best
29,546	Snafflebit .	Referring to your —— our reply is—Yes
29,547	Snaffling .	**Yesterday**
29,548	Snagtooth .	Arrived here yesterday
29,549	Snailfish .	Left yesterday for
29,550	Snaillike .	Wrote yesterday
29,551	Snailpaced	Went yesterday
29,552	Snailplant .	Sent yesterday
29,553	Snailshell .	We received yesterday
29,554	Snailslow .	We wrote you yesterday
29,555	Snakebird .	When writing yesterday
29,556	Snakeboat	The day before yesterday
29,557	Snakeflies .	Early yesterday morning
29,558	Snakefly .	Too late yesterday to reply
29,559	Snakegourd	**Yet**

No.	Code Word.	**Yet** (*continued*)
29,560	Snakehead	Not yet
29,561	Snakemoss	Cannot yet
29,562	Snakenut .	Will not yet
29,563	Snakepiece	It is yet too soon (to)
29,564	Snakestone	But yet there may be
29,565	Snakeweed	**Yield.** (See also Ton.)
29,566	Snakewort	The gross yield being
29,567	Snapbug .	Estimate the yield at
29,568	Snapdragon	I believe this will increase immensely the yield of
29,569	Snaphance	Will yield
29,570	Snaplock .	Will not yield
29,571	Snappish .	Can yield
29,572	Snappishly	Cannot yield
29,573	Snaredrum	What will be the probable yield of
29,574	Snarlknot.	What is the average yield
29,575	Snatched .	The yield has averaged
29,576	Snattocks .	The yield from —— has been
29,577	Sneakcup .	The yield has averaged —— gold and —— silver
29,578	Sneakiness	The average yield in gold and silver
29,579	Sneaking .	The yield from all the mines
29,580	Sneerful .	Will yield a good return
29,581	Sneeringly	Will yield enough to
29,582	Sneers . .	**Yielded**
29,583	Sneeshin .	The mines have yielded
29,584	Sneezed .	The mines have not yielded much lately
29,585	Sneezewood	Yielded —— gold
29,586	Sneezings .	Yielded —— silver
29,587	Sniff . .	Yielded —— ozs. of gold and —— ozs. silver
29,588	Sniggering	Yielded —— copper
29,589	Snipebill .	Yielded —— per cent. of
29,590	Snipefish .	**Yielding**
29,591	Snippety .	Not yielding any profit
29,592	Snipsnap .	The mines are yielding a net profit of
29,593	Snivel . .	What are the mines yielding
29,594	Snivelly .	The mines are yielding —— tons per month
29,595	Snob . .	**You**
29,596	Snobbery .	Do you advise
29,597	Snobbish .	Do you admit
29,598	Snobbishly	What do you
29,599	Snoblings .	If you
29,600	Snobocracy	When you
29,601	Snortings .	When do you
29,602	Snowapple	**Your**
29,603	Snowballs .	By your
29,604	Snowberry	From your
29,605	Snowboot .	To your
29,606	Snowbreak	To your care
29,607	Snowbroth	From your letters
29,608	Snowdrift .	By your instructions
29,609	Snowdrops	**Yourself** (**selves**)

No.	Code Word.	**Yourself (selves)** (*continued*)
29,610	Snoweyes .	By yourself
29,611	Snowfed .	Not by yourself
29,612	Snowfield .	You have only yourself to blame (if)
29,613	Snowflakes	It rests with yourself
29,614	Snowfleck .	If you, yourself, cannot make it pay, who can
29,615	Snowflood	You have, yourself
29,616	Snowgoose	**Zeal(ous)**
29,617	Snowhut .	Has (have) shown great zeal in
29,618	Snowish .	Has (have) not shown much zeal in
29,619	Snowlight .	Has shown more zeal than discretion
29,620	Snowlimbed	Is very zealous to do what he can
29,621	Snowmould	Zealously working to effect (or) (carry out)
29,622	Snowplant	**Zero**
29,623	Snowplough	——° below zero [——° below zero
29,624	Snowshoes	Through the winter the thermometer at night averages
29,625	Snowskate	The thermometer falling ——° below zero
29,626	Snowslip .	**Zinc**
29,627	Snowstorm	Zinc ore
29,628	Snowwater	Veins of zinc ore
29,629	Snowwhite	Sulphide of zinc
29,630	Snowwreath	Carbonate of zinc
29,631	Snowy . .	Zinc blende occurs with
29,632	Snubbed	Silicate of zinc

No.	Code Word.
29,633	Snubnosed
29,634	Snubpost
29,635	Snuffbox .
29,636	Snuffdish .
29,637	Snuffers .
29,638	Snuffmill .
29,639	Snufftaker
29,640	Snuggeries
29,641	Snuggery .
29,642	Snugness .
29,643	Snush . .
29,644	Soak . .
29,645	Soakings .
29,646	Soapboiler
29,647	Soapbubble
29,648	Soapcerate
29,649	Soapengine
29,650	Soaphouse
29,651	Soappan .
29,652	Soapplant .
29,653	Soapstone.
29,654	Soapsuds .
29,655	Soaptest .
29,656	Soapworks
29,657	Soapy . .

No.	Code Word.
29,658	Soarant
29,659	Soarfalcon
29,660	Sobeit . .
29,661	Soberized .
29,662	Soberly .
29,663	Soberness .
29,664	Soboles .
29,665	Sobrieties .
29,666	Sobriety .
29,667	Socalled .
29,668	Socdolager
29,669	Sociably .
29,670	Socialists .
29,671	Socialized .
29,672	Socialness .
29,673	Societary .
29,674	Socinian .
29,675	Sociologic .
29,676	Socket . .
29,677	Socketbolt
29,678	Sockplate .
29,679	Socmanry
29,680	Socome .
29,681	Socotrine .
29,682	Socratical .

No.	Code Word.
29,683	Socratism .
29,684	Soda . .
29,685	Sodaalum .
29,686	Sodaash .
29,687	Sodalime .
29,688	Sodality .
29,689	Sodapowder
29,690	Sodasalt .
29,691	Sodawater
29,692	Sodburning
29,693	Sodden .
29,694	Soddening
29,695	Sodomapple
29,696	Sofabed .
29,697	Sofism . .
29,698	Soften . .
29,699	Softeneth .
29,700	Softening .
29,701	Softeyed .
29,702	Softgrass .
29,703	Softheaded
29,704	Softhorn .
29,705	Softish . .
29,706	Softspoken
29,707	Soil . . .

No.	Code Word.
29,708	Soilless .
29,709	Soilpipe .
29,710	Soirees. .
29,711	Sojourn .
29,712	Sojourneth
29,713	Sojourning
29,714	Sokereeve.
29,715	Solace . .
29,716	Solacement
29,717	Solacing .
29,718	Solangoose
29,719	Solania .
29,720	Solatiums .
29,721	Soldanel .
29,722	Soldanella
29,723	Soldanry .
29,724	Soldieress .
29,725	Soldierly .
29,726	Solecisms .
29,727	Solecistic .
29,728	Solecizing.
29,729	Solely . .
29,730	Solemn .
29,731	Solemnity .
29,732	Solemnizer

No.	Code Word.
29,733	Solemnness
29,734	Solenacea .
29,735	Solenodon
29,736	Solenoid .
29,737	Soleship .
29,738	Solfanaria .
29,739	Solicit . .
29,740	Soliciteth .
29,741	Soliciting .
29,742	Solicitous .
29,743	Solicitude .
29,744	Solidago .
29,745	Solidarity .
29,746	Solidate .
29,747	Solidified .
29,748	Solidism .
29,749	Solidness .
29,750	Solidus .
29,751	Solifidian .
29,752	Soliform .
29,753	Soliloquy .
29,754	Solipedal .
29,755	Solitaires .
29,756	Solitarian .
29,757	Solitary .

No.	Code Word.
29,758	Solitudes .
29,759	Solivagant
29,760	Solivagous
29,761	Sollerets .
29,762	Sollunar .
29,763	Solograph .
29,764	Soloists .
29,765	Solos . .
29,766	Solpuga .
29,767	Solpugidae
29,768	Solstice .
29,769	Solstitial .
29,770	Solubility .
29,771	Solutions .
29,772	Solutive .
29,773	Solvency .
29,774	Solvent .
29,775	Somateria .
29,776	Somatical .
29,777	Somatics .
29,778	Somatocyst
29,779	Somatology
29,780	Somatome
29,781	Sombre .
29,782	Sombrely .

No.	Code Word.
29,783	Sombreness
29,784	Sombrerite
29,785	Sombrous .
29,786	Sombrously
29,787	Somebody
29,788	Somedeal ,
29,789	Somegate .
29,790	Somehow .
29,791	Somersault
29,792	Something
29,793	Sometimes
29,794	Somewhat
29,795	Somewhile
29,796	Somnambule
29,797	Somnial .
29,798	Somniative
29,799	Somniatory
29,800	Somnific .
29,801	Somniloquy
29,802	Somnipathy
29,803	Somnolency
29,804	Somnolent
29,805	Somnolism
29,806	Sonance .
29,807	Sondeli .

No.	Code Word.
29,808	Songbird .
29,809	Songcraft .
29,810	Songful .
29,811	Songless .
29,812	Songster .
29,813	Songstress
29,814	Songthrush
29,815	Soniferous
29,816	Soninlaw .
29,817	Sonneteer .
29,818	Sonnetists
29,819	Sonnetize .
29,820	Sonometer
29,821	Sonorous.
29,822	Sonorously
29,823	Sonship .
29,824	Soochong .
29,825	Soofeeism .
29,826	Sooterkin .
29,827	Sootflake .
29,828	Soothed .
29,829	Soothfast .
29,830	Soothingly
29,831	Soothsay .
29,832	Sooty . .

No.	Code Word.
29,833	Soph . .
29,834	Sophic . .
29,835	Sophical .
29,836	Sophisms .
29,837	Sophistry .
29,838	Sophomoric
29,839	Sophora .
29,840	Sopiting .
29,841	Sopition .
29,842	Soporose .
29,843	Sopra . ,
29,844	Sopranist ,
29,845	Sorbapple
29,846	Sorbate ,
29,847	Sorbonical
29,848	Sorbonist .
29,849	Sorcerous .
29,850	Sorcery .
29,851	Sordidly -
29,852	Soredia :
29,853	Sorghum .
29,854	Sororal , .
29,855	Sororicide.
29,856	Sororizing .
29,857	Sorriest .

No.	Code Word.
29,858	Sorrily . .
29,859	Sorrowed .
29,860	Sorrowful .
29,861	Sorrowless
29,862	Sorters .
29,863	Sortilegy .
29,864	Sortments.
29,865	Sotadean .
29,866	Sothiac .
29,867	Sottishly .
29,868	Soulamea .
29,869	Soulbells ,
29,870	Soulcurer }
29,871	Soulfoot .
29,872	Soulshot .
29,873	Soulsick .
29,874	Soundable
29,875	Soundboard
29,876	Soundbow
29,877	Soundest .
29,878	Soundings
29,879	Soundly .
29,880	Soundness
29,881	Soundpost
29,882	Soupcon .

No.	Code Word.
29,883	Soupmaigre
29,884	Soupticket
29,885	Source . .
29,886	Sourcrout .
29,887	Sourdine .
29,888	Sourdock .
29,889	Sourgourd .
29,890	Sourmilk .
29,891	Soursop .
29,892	Sourtree .
29,893	Sourwood .
29,894	Souterrain
29,895	Southdown
29,896	Southerner
29,897	Southernly
29,898	Southmost
29,899	Southward
29,900	Souvenance
29,901	Souvenir .
29,902	Soverainly
29,903	Sovereign .
29,904	Sowans .
29,905	Sowthistle .
29,906	Soyled . .
29,907	Spa . . .

No.	Code Word.
29,908	Spaced
29,909	Spaceline
29,910	Spacerule
29,911	Spacially
29,912	Spadassins
29,913	Spadebone
29,914	Spadeful
29,915	Spadeiron
29,916	Spadiceous
29,917	Spadicose
29,918	Spadix
29,919	Spadroon
29,920	Spagyric
29,921	Spagyrical
29,922	Spalpeen
29,923	Spancel
29,924	Spancelled
29,925	Spandogs
29,926	Spandrels
29,927	Spanemy
29,928	Spangled
29,929	Spanishelm
29,930	Spanishfly
29,931	Spanishnut
29,932	Spanishred

No.	Code Word.
29,933	Spanlong
29,934	Spanpiece
29,935	Spanroof
29,936	Spansaw
29,937	Spanworm
29,938	Sparadrap
29,939	Spardeck
29,940	Sparely
29,941	Sparerib
29 942	Sparganium
29,943	Sparhawk
29,944	Sparidae
29,945	Sparingly
29,946	Sparkful
29,947	Sparkish
29,948	Sparklets
29,949	Sparkling
29,950	Sparlyre
29,951	Sparrow
29,952	Sparry
29,953	Sparseth
29,954	Sparsedly
29,955	Sparseness
29,956	Sparsim
29,957	Sparterie

No.	Code Word.
29,958	Spartina
29,959	Spasmodic
29,960	Spasmology
29,961	Spastic
29,962	Spasticity
29,963	Spatangus
29,964	Spatchcock
29,965	Spathal
29,966	Spathella
29,967	Spathiform
29,968	Spathodea
29,969	Spathous
29,970	Spathulate
29,971	Spatiated
29,972	Spatularia
29,973	Spavin
29,974	Spawnings
29,975	Spayade
29,976	Speakable
29,977	Speakhouse
29,978	Speared
29,979	Spearfoot
29,980	Speargrass
29,981	Spearhand
29,982	Spearmint

No.	Code Word.
29,983	Spearwort
29,984	Specht
29,985	Specialist
29,986	Specific
29,987	Specifical
29,988	Specifying
29,989	Specimens
29,990	Speciology
29,991	Specked
29,992	Spectacle
29,993	Spectant
29,994	Spectation
29,995	Spectators
29,996	Spectatrix
29,997	Spectral
29,998	Spectrally
29,999	Spectrebat
30,000	Spectrums

Code Word.	s.	d.	Code Word.	s.	d.	Code Word.	s.	d.
Specular	0	0 1/16	Sphereborn	0	3 3/16	Spinellane	1	2 1/4
Speculated	0	0 1/8	Spherical	0	3 1/4	Spinescent	1	3
Speculum	0	0 3/16	Sphericity	0	3 5/16	Spinnaker	1	3 1/2
Speechday	0	0 1/4	Spherics	0	3 3/8	Spinnery	1	4
Speechful	0	0 5/16	Sphering	0	3 7/16	Spinous	1	4 1/2
Speechify	0	0 3/8	Spheroidic	0	4	Spinozism	1	5
Speeching	0	0 7/16	Spherulate	0	4 1/8	Spinstress	1	5 1/2
Speechless	0	0 1/2	Sphery	0	4 1/4	Spinstry	1	6
Speedfully	0	0 9/16	Sphex	0	4 3/8	Spinthere	1	6 1/2
Speediest	0	0 5/8	Sphincter	0	4 1/2	Spinulose	1	7
Speedily	0	0 11/16	Sphinxes	0	4 5/8	Spirally	1	7 1/2
Speediness	0	0 3/4	Sphingidae	0	4 3/4	Spiranthy	1	8
Speedwell	0	0 13/16	Sphragide	0	4 7/8	Spirifer	1	8 1/2
Speiss	0	0 7/8	Sphrigosis	0	5	Spiritduck	1	9
Spelaean	0	0 15/16	Sphygmic	0	5 1/8	Spiritedly	1	9 1/2
Speldron	0	1	Spials	0	5 1/4	Spiritful	1	10
Spellable	0	1 1/16	Spica	0	5 3/8	Spiritism	1	10 1/2
Spellbound	0	1 1/8	Spicated	0	5 1/2	Spiritlamp	1	11
Spellful	0	1 3/16	Spiccato	0	5 5/8	Spiritless	1	11 1/2
Spellwork	0	1 1/4	Spicebush	0	5 3/4	Spiritoso	2	0
Spelter	0	1 5/16	Spicenut	0	5 7/8	Spiritroom	2	0 1/2
Spelunc	0	1 3/8	Spicewood	0	6	Spiritual	2	1
Spendall	0	1 7/16	Spiciform	0	6 1/4	Spirituous	2	1 1/2
Spendeth	0	1 1/2	Spicknel	0	6 1/2	Spirometer	2	2
Spenserian	0	1 9/16	Spicose	0	6 3/4	Spirorbis	2	2 1/2
Spermaceti	0	1 5/8	Spicosity	0	7	Spirula	2	3
Spermagone	0	1 11/16	Spidercrab	0	7 1/4	Spirulidae	2	3 1/2
Spermarium	0	1 3/4	Spiderfly	0	7 1/2	Spissated	2	4
Spermary	0	1 13/16	Spiderlike	0	7 3/4	Spissitude	2	4 1/2
Spermatize	0	1 7/8	Spiderwort	0	8	Spitbox	2	5
Spermatoid	0	1 15/16	Spigelia	0	8 1/4	Spitfire	2	5 1/2
Spermcell	0	2	Spignet	0	8 1/2	Spitously	2	6
Spermic	0	2 1/16	Spigurnel	0	8 3/4	Spitpoison	2	6 1/2
Spermoderm	0	2 1/8	Spikelet	0	9	Spittly	2	7
Spermoil	0	2 3/16	Spikenard	0	9 1/4	Spittoons	2	7 1/2
Spermwhale	0	2 1/4	Spikeoil	0	9 1/2	Spitvenom	2	8
Spetches	0	2 5/16	Spikeplant	0	9 3/4	Spitzdog	2	8 1/2
Spetum	0	2 3/8	Spiketeam	0	10	Splachnei	2	9
Sphacel	0	2 7/16	Spiketub	0	10 1/4	Splanchnic	2	9 1/2
Sphacelism	0	2 1/2	Spilanthes	0	10 1/2	Splashed	2	10
Sphaereda	0	2 9/16	Spilehole	0	10 3/4	Splashwing	2	10 1/2
Sphaerodus	0	2 5/8	Spilikin	0	11	Splatter	2	11
Sphagnei	0	2 11/16	Spilth	0	11 1/4	Splayfoot	2	11 1/2
Sphagnous	0	2 3/4	Spilus	0	11 1/2	Splaymouth	3	0
Sphagnum	0	2 13/16	Spinaceous	0	11 3/4	Spleen	3	0 1/2
Sphenodon	0	2 7/8	Spinach	1	0	Spleenful	3	1
Sphenogram	0	2 15/16	Spinacidae	1	0 1/2	Spleening	3	1 1/2
Sphenoid	0	3	Spinal	1	1	Spleenish	3	2
Sphenoidal	0	3 1/16	Spindrift	1	1 1/2	Spleenless	3	2 1/2
Spheral	0	3 1/8	Spineless	1	2	Spleenwort	3	3

Code Word.	s.	d.	Code Word.	s.	d.	Code Word.	£	s.	d.
Splenalgia	3	3½	Sponsion	5	9	Sprigbolt	0	9	11
Splendent	3	4	Sponsorial	5	10	Spriggy	0	10	0
Splendidly	3	4½	Sponsors	5	11	Sprighted	0	10	3
Splendour	3	5	Spoolstand	6	0	Sprightful	0	10	6
Splendrous	3	5½	Spoonbills	6	1	Springal	0	10	9
Splenetic	3	6	Spoonbit	6	2	Springback	0	11	0
Splenocele	3	6½	Spoondrift	6	3	Springbox	0	11	3
Splenology	3	7	Spoonfuls	6	4	Springcart	0	11	6
Splentcoal	3	7½	Spoongouge	6	5	Springers	0	11	9
Splenule	3	8	Spoonily	6	6	Springfeed	0	12	0
Splice	3	8½	Spoonmeat	6	7	Springgun	0	12	3
Splicing	3	9	Spoonworm	6	8	Springhaas	0	12	6
Splintbone	3	9½	Sporadic	6	9	Springhalt	0	12	9
Splintery	3	10	Sporadical	6	10	Springlet	0	13	0
Splitcloth	3	10½	Sporangium	6	11	Springlock	0	13	3
Splitnew	3	11	Sporecase	7	0	Springpins	0	13	6
Splitpease	3	11½	Sporid	7	1	Springrye	0	13	9
Splotchy	4	0	Sporidiola	7	2	Springstay	0	14	0
Spluttered	4	0½	Sporidium	7	3	Springtail	0	14	3
Spodomancy	4	1	Sporocarp	7	4	Springtide	0	14	6
Spodumene	4	1½	Sporocyst	7	5	Sprinkled	0	14	9
Spoffish	4	2	Sporogen	7	6	Sprinkling	0	15	0
Spoilbank	4	2½	Sporophore	7	7	Sprintrace	0	15	3
Spoilers	4	3	Sporosac	7	8	Sprites	0	15	6
Spoilfive	4	3½	Sporozoid	7	9	Spritsail	0	15	9
Spoilful	4	4	Sportfully	7	10	Sprouting	0	16	0
Spoilsport	4	4½	Sportingly	7	11	Sprucefir	0	16	3
Spokeshave	4	5	Sportive	8	0	Sprucely	0	16	6
Spokesman	4	5½	Sportively	8	1	Spruceness	0	16	9
Spoliary	4	6	Sportling	8	2	Spumescent	0	17	0
Spoliate	4	6½	Sportsman	8	3	Spuminess	0	17	3
Spoliating	4	7	Sportulary	8	4	Spungold	0	17	6
Spoliation	4	7½	Sportule	8	5	Spunsilk	0	17	9
Spoliatory	4	8	Spotlens	8	6	Spunsilver	0	18	0
Spondaic	4	8½	Spotlessly	8	7	Spunyarn	0	18	3
Spondaical	4	9	Spousage	8	8	Spurgall	0	18	6
Spondyl	4	9½	Spouseless	8	9	Spurgear	0	18	9
Spondylus	4	10	Spoutfish	8	10	Spurgeflax	0	19	0
Sponge	4	10½	Spouthole	8	11	Spurgewort	0	19	3
Spongecake	4	11	Spoutshell	9	0	Spurious	0	19	6
Spongecrab	4	11½	Spragged	9	1	Spuriously	0	19	9
Spongelet	5	0	Spragging	9	2	Spurning	1	0	0
Spongeous	5	1	Sprains	9	3	Spurrier	1	0	3
Spongetree	5	2	Sprawl	9	4	Spurroyal	1	0	6
Spongida	5	3	Sprawling	9	5	Spurtled	1	0	9
Spongiform	5	4	Spraydrain	9	6	Spurway	1	1	0
Sponginess	5	5	Sprayey	9	7	Spurwheel	1	1	3
Spongiole	5	6	Spread	9	8	Sputation	1	1	6
Spongoid	5	7	Spreadeth	9	9	Sputative	1	1	9
Sponsible	5	8	Spreading	9	10	Sputtering	1	2	0

Code Word.	£	s.	d.	Code Word.	£	s.	d.	Code Word.	£	s.	d.
Spyal	1	2	3	Squierie	1	14	9	Staidly	2	7	3
Spycraft	1	2	6	Squiggle	1	15	0	Stainand	2	7	6
Spyglass	1	2	9	Squillagee	1	15	3	Stainers	2	7	9
Spyism	1	3	0	Squillitic	1	15	6	Staircase	2	8	0
Spymoney	1	3	3	Squinancy	1	15	9	Stairfoot	2	8	3
Spyred	1	3	6	Squinny	1	16	0	Stairhead	2	8	6
Squabashed	1	3	9	Squinnying	1	16	3	Stairrods	2	8	9
Squabbish	1	4	0	Squinteyed	1	16	6	Stairway	2	9	0
Squabbled	1	4	3	Squiralty	1	16	9	Stairwire	2	9	3
Squabbling	1	4	6	Squireage	1	17	0	Staith	2	9	6
Squabby	1	4	9	Squirearch	1	17	3	Staithman	2	9	9
Squabchick	1	5	0	Squireen	1	17	6	Staithwort	2	10	0
Squabpie	1	5	3	Squirehood	1	17	9	Stakenet	2	10	3
Squacco	1	5	6	Squirelets	1	18	0	Stalactic	2	10	6
Squadroned	1	5	9	Squireling	1	18	3	Stalagmite	2	10	9
Squalid	1	6	0	Squireship	1	18	6	Stalemate	2	11	0
Squalidity	1	6	3	Squirrels	1	18	9	Stalkeyed	2	11	3
Squallings	1	6	6	Stabbing	1	19	0	Stalking	2	11	6
Squalor	1	6	9	Stabbingly	1	19	3	Stallation	2	11	9
Squama	1	7	0	Stableboy	1	19	6	Stallboard	2	12	0
Squamated	1	7	3	Stablished	1	19	9	Stallinger	2	12	3
Squamella	1	7	6	Staccato	2	0	0	Stalwart	2	12	6
Squamiform	1	7	9	Stackcover	2	0	3	Stambha	2	12	9
Squamipen	1	8	0	Stadium	2	0	6	Stamened	2	13	0
Squamoid	1	8	3	Staffangle	2	0	9	Stamfortis	2	13	3
Squamosal	1	8	6	Staffbead	2	1	0	Stamina	2	13	6
Squamulose	1	8	9	Staffhole	2	1	3	Stamineous	2	13	9
Squander	1	9	0	Staffiers	2	1	6	Staminidia	2	14	0
Square	1	9	3	Staffman	2	1	9	Stampact	2	14	3
Squarefile	1	9	6	Staffsling	2	2	0	Stampduty	2	14	6
Squarely	1	9	9	Stafftree	2	2	3	Stampede	2	14	9
Squareness	1	10	0	Stagbeetle	2	2	6	Stamphead	2	15	0
Squarerig	1	10	3	Stagdance	2	2	9	Stamping	2	15	3
Squareroof	1	10	6	Stagebox	2	3	0	Stampnote	2	15	6
Squaresail	1	10	9	Stagecoach	2	3	3	Stanchions	2	15	9
Squaretoed	1	11	0	Stagedoors	2	3	6	Stanchless	2	16	0
Squarish	1	11	3	Stageplay	2	3	9	Standage	2	16	3
Squarrose	1	11	6	Stagevil	2	4	0	Standards	2	16	6
Squashbug	1	11	9	Stagewagon	2	4	3	Standcrop	2	16	9
Squasher	1	12	0	Stagey	2	4	6	Standerby	2	17	0
Squeak	1	12	3	Stageyness	2	4	9	Standersup	2	17	3
Squeakers	1	12	6	Staggered	2	5	0	Standeth	2	17	6
Squeasy	1	12	9	Staghound	2	5	3	Standpoint	2	17	9
Squeezable	1	13	0	Stagnancy	2	5	6	Standrest	2	18	0
Squeezed	1	13	3	Stagnated	2	5	9	Standstill	2	18	3
Squelch	1	13	6	Stagnating	2	6	0	Standup	2	18	6
Squelching	1	13	9	Stagnation	2	6	3	Stanielry	2	18	9
Squenched	1	14	0	Stagworm	2	6	6	Stannotype	2	19	0
Squeteague	1	14	3	Stagyrite	2	6	9	Stannous	2	19	3
Squib	1	14	6	Stahlian	2	7	0	Stanza	2	19	6

Code Word.	£	s.	d.	Code Word.	£	s.	d.	Code Word.	£	s.	d.
Stanzaic	2	19	9	Statua	3	12	3	Steelpens	4	4	9
Stapedius	3	0	0	Statuaries	3	12	6	Steelplate	4	5	0
Stapelia	3	0	3	Statuesque	3	12	9	Steelyards	4	5	3
Staphyle	3	0	6	Statuettes	3	13	0	Steepdown	4	5	6
Staphyloma	3	0	9	Statured	3	13	3	Steepness	4	5	9
Stapled	3	1	0	Statutable	3	13	6	Steerage	4	6	0
Starapples	3	1	3	Statutecap	3	13	9	Steersman	4	6	3
Starblind	3	1	6	Statutory	3	14	0	Steganopod	4	6	6
Starched	3	1	9	Staurolite	3	14	3	Stegnotic	4	6	9
Starconner	3	2	0	Stauropus	3	14	6	Steinbock	4	7	0
Starfinch	3	2	3	Stavesacre	3	14	9	Stelechite	4	7	3
Starfort	3	2	6	Stavewood	3	15	0	Stellaria	4	7	6
Starfruit	3	2	9	Staybar	3	15	3	Stellated	4	7	9
Stargazer	3	3	0	Staybolts	3	15	6	Stellerine	4	8	0
Stargazing	3	3	3	Staybusk	3	15	9	Stelliform	4	8	3
Stargrass	3	3	6	Staylaces	3	16	0	Stellulate	4	8	6
Starjelly	3	3	9	Stayplough	3	16	3	Stemleaf	4	8	9
Starlight	3	4	0	Stayrods	3	16	6	Stemlet	4	9	0
Starlike	3	4	3	Staytackle	3	16	9	Stemmata	4	9	3
Starmonger	3	4	6	Staywedge	3	17	0	Stenwinder	4	9	6
Starnose	3	4	9	Steadier	3	17	3	Stenches	4	9	9
Starosty	3	5	0	Steadyrest	3	17	6	Stenchtrap	4	10	0
Starpagoda	3	5	3	Stealth	3	17	9	Stencil	4	10	3
Starproof	3	5	6	Stealthily	3	18	0	Stenciller	4	10	6
Starshake	3	5	9	Steamboats	3	18	3	Stentorian	4	10	9
Starshoot	3	6	0	Steambrake	3	18	6	Stepchild	4	11	0
Starslough	3	6	3	Steamchest	3	18	9	Stepdame	4	11	3
Starter	3	6	6	Steamcock	3	19	0	Stepfather	4	11	6
Starveling	3	6	9	Steamcoil	3	19	3	Stepgrate	4	11	9
Statarian	3	7	0	Steamdome	3	19	6	Stephanite	4	12	0
Statary	3	7	3	Steamed	3	19	9	Stepladder	4	12	3
Stateball	3	7	6	Steamgas	4	0	0	Stepmother	4	12	6
Statebarge	3	7	9	Steamguage	4	0	3	Stepparent	4	12	9
Statebed	3	8	0	Steampipe	4	0	6	Stepper	4	13	0
Statecraft	3	8	3	Steamport	4	0	9	Stepsister	4	13	3
Statedly	3	8	6	Steampower	4	1	0	Stepsons	4	13	6
Statehouse	3	8	9	Steampress	4	1	3	Stercorate	4	13	9
Stateliest	3	9	0	Steamship	4	1	6	Stercorist	4	14	0
Statements	3	9	3	Steamtight	4	1	9	Stercory	4	14	3
Statepaper	3	9	6	Steamtilt	4	2	0	Sterculia	4	14	6
Stateroom	3	9	9	Steamtugs	4	2	3	Stereobate	4	14	9
Statesman	3	10	0	Steamwheel	4	2	6	Stereogram	4	15	0
Statesword	3	10	3	Steamwinch	4	2	9	Stereotomy	4	15	3
Statetrial	3	10	6	Stearinery	4	3	0	Stereotype	4	15	6
Statically	3	10	9	Stearyl	4	3	3	Sterilcoal	4	15	9
Stationary	3	11	0	Steatocele	4	3	6	Sterilized	4	16	0
Stationers	3	11	3	Steatoma	4	3	9	Sternalgia	4	16	3
Statism	3	11	6	Steatopyga	4	4	0	Sternboard	4	16	6
Statistics	3	11	9	Steelbow	4	4	3	Sternchase	4	16	9
Statoblast	3	12	0	Steelclad	4	4	6	Sternfast	4	17	0

Code Word.	£	s.	d.	Code Word.	£	s.	d.	Code Word.	£	s.	d.
Sternframe	4	17	3	Stingbull	6	19	0	Stonebrash	9	9	0
Sternidae	4	17	6	Stingfish	7	0	0	Stonebreak	9	10	0
Sternknee	4	17	9	Stingiest	7	1	0	Stonebucks	9	11	0
Sternmost	4	18	0	Stingily	7	2	0	Stonecast	9	12	0
Sternness	4	18	3	Stinginess	7	3	0	Stonechats	9	13	0
Sternway	4	18	6	Stingo	7	4	0	Stonecold	9	14	0
Stertorious	4	18	9	Stingray	7	5	0	Stonecoral	9	15	0
Stevedore	4	19	0	Stinkard	7	6	0	Stonecrop	9	16	0
Steward	4	19	3	Stinkhorn	7	7	0	Stonedeaf	9	17	0
Stewardess	4	19	6	Stinkingly	7	8	0	Stoneater	9	18	0
Stewardly	4	19	9	Stinkstone	7	9	0	Stonefern	9	19	0
Stewing	5	0	0	Stinkwood	7	10	0	Stoneflies	10	0	0
Stewpans	5	1	0	Stintance	7	11	0	Stonefly	10	1	0
Sthenic	5	2	0	Stipend	7	12	0	Stonefruit	10	2	0
Stibial	5	3	0	Stippled	7	13	0	Stonegall	10	2	6
Stibialism	5	4	0	Stipula	7	14	0	Stonegrig	10	3	0
Stibnite	5	5	0	Stipulates	7	15	0	Stonehawk	10	4	0
Stichic	5	6	0	Stirabout	7	16	0	Stonelily	10	5	0
Stichidium	5	7	0	Stirring	7	17	0	Stonemason	10	6	0
Sticketh	5	8	0	Stitchwort	7	18	0	Stoneochre	10	7	0
Sticklac	5	9	0	Stoats	7	19	0	Stoneoil	10	7	6
Sticklebag	5	10	0	Stochastic	8	0	0	Stonepine	10	8	0
Sticklers	5	11	0	Stockades	8	1	0	Stonepitch	10	9	0
Sticky	5	12	0	Stockdoves	8	2	0	Stoneseed	10	10	0
Sticta	5	13	0	Stockfish	8	3	0	Stonesnipe	10	11	0
Stiffbit	5	14	0	Stockgold	8	4	0	Stonewalls	10	12	0
Stiffborne	5	15	0	Stockinged	8	5	0	Stonewort	10	12	6
Stiffen	5	16	0	Stocklock	8	6	0	Stoolball	10	13	0
Stiffening	5	17	0	Stockman	8	7	0	Stoolend	10	14	0
Stiffish	5	18	0	Stockpot	8	8	0	Stoopeth	10	15	0
Stiffly	5	19	0	Stockpurse	8	9	0	Stopcocks	10	16	0
Stiffneck	6	0	0	Stockstill	8	10	0	Stopgaps	10	17	0
Stiflebone	6	1	0	Stockyards	8	11	0	Stopless	10	17	6
Stifling	6	2	0	Stocial	8	12	0	Stopmotion	10	18	0
Stigmaria	6	3	0	Stoicism	8	13	0	Stoppages	10	19	0
Stigmas	6	4	0	Stoicity	8	14	0	Stoppering	11	0	0
Stigmatic	6	5	0	Stokehole	8	15	0	Stopplank	11	1	0
Stigmatose	6	6	0	Stokers	8	16	0	Stopvalve	11	2	0
Stiletto	6	7	0	Stolen	8	17	0	Stopwatch	11	2	6
Stillbirth	6	8	0	Stolidity	8	18	0	Storehouse	11	3	0
Stillhouse	6	9	0	Stomacace	8	19	0	Storepay	11	4	0
Stillicide	6	10	0	Stomach	9	0	0	Storeship	11	5	0
Stillingia	6	11	0	Stomachful	9	1	0	Storify	11	6	0
Stilllife	6	12	0	Stomachous	9	2	0	Storifying	11	7	0
Stillness	6	13	0	Stomapod	9	3	0	Storksbill	11	7	6
Stillroom	6	14	0	Stomatitis	9	4	0	Stormbeat	11	8	0
Stillstand	6	15	0	Stoneaxe	9	5	0	Stormbirds	11	9	0
Stilton	6	16	0	Stoneblue	9	6	0	Stormcock	11	10	0
Stimulants	6	17	0	Stoneborer	9	7	0	Stormcone	11	11	0
Stimuli	6	18	0	Stonebows	9	8	0	Stormdoor	11	12	0

Code Word.	£	s.	d.	Code Word.	£	s.	d.	Code Word.	£	s.	d.
Stormdrum	11	12	6	Strawhat	13	14	0	Stringless	15	16	0
Stormfinch	11	13	0	Strawhouse	13	15	0	Stripleaf	15	17	0
Stormglass	11	14	0	Strawpaper	13	16	0	Striplings	15	17	6
Storminess	11	15	0	Strawplait	13	17	0	Strippet	15	18	0
Stormproof	11	16	0	Strawrope	13	17	6	Stritchel	15	19	0
Stormsail	11	17	0	Strawworm	13	18	0	Strobiline	16	0	0
Stormstead	11	17	6	Straying	13	19	0	Strokal	16	1	0
Stormwind	11	18	0	Streaked	14	0	0	Strokeoar	16	2	0
Stormy	11	19	0	Stream	14	1	0	Strolled	16	2	6
Storybook	12	0	0	Streamers	14	2	0	Strolling	16	3	0
Storying	12	1	0	Streamful	14	2	6	Stromb	16	4	0
Storyposts	12	2	0	Streamice	14	3	0	Strombidae	16	5	0
Storyrod	12	2	6	Streamtin	14	4	0	Strombus	16	6	0
Stoutbuilt	12	3	0	Streamwort	14	5	0	Stromnite	16	7	0
Stoutest	12	4	0	Streetarab	14	6	0	Strong	16	7	6
Stouthrief	12	5	0	Streetdoor	14	7	0	Stronghold	16	8	0
Stoutly	12	6	0	Streetward	14	7	6	Strongish	16	9	0
Stoutmade	12	7	0	Strelitzia	14	8	0	Strongknit	16	10	0
Stoutness	12	7	6	Stremma	14	9	0	Strongly	16	11	0
Stowage	12	8	0	Strengthen	14	10	0	Strongroom	16	12	0
Stowaway	12	9	0	Strenuity	14	11	0	Strongset	16	12	6
Stowwood	12	10	0	Strenuous	14	12	0	Strongylus	16	13	0
Strabismus	12	11	0	Streperous	14	12	6	Strontian	16	14	0
Strabotomy	12	12	0	Strephon	14	13	0	Strontites	16	15	0
Strachy	12	12	6	Strepitoso	14	14	0	Strophe	16	16	0
Stragglers	12	13	0	Stress	14	15	0	Strophiole	16	17	0
Straighten	12	14	0	Stretcheth	14	16	0	Strophulus	16	17	6
Straightly	12	15	0	Strewed	14	17	0	Strossers	16	18	0
Strainable	12	16	0	Strewment	14	17	6	Struck	16	19	0
Strained	12	17	0	Striating	14	18	0	Structural	17	0	0
Stramash	12	17	6	Striation	14	19	0	Strumatic	17	1	0
Stramazoun	12	18	0	Stricken	15	0	0	Strumiform	17	2	0
Stramonium	12	19	0	Strict	15	1	0	Strummed	17	2	6
Stramony	13	0	0	Strictest	15	2	0	Strumous	17	3	0
Strangeful	13	1	0	Strictly	15	2	6	Strumpet	17	4	0
Strangury	13	2	0	Strictness	15	3	0	Strumstrum	17	5	0
Strapheads	13	2	6	Stricture	15	4	0	Strumulose	17	6	0
Strappado	13	3	0	Stridor	15	5	0	Strutbeam	17	7	0
Strapper	13	4	0	Stridulate	15	6	0	Struthio	17	7	6
Strapping	13	5	0	Stridulous	15	7	0	Struthiola	17	8	0
Stratagems	13	6	0	Strifeful	15	7	6	Struthious	17	9	0
Strategist	13	7	0	Striga	15	8	0	Strutters	17	10	0
Strategy	13	7	6	Strigidae	15	9	0	Strychnine	17	11	0
Stratified	13	8	0	Strigilose	15	10	0	Strychnos	17	12	0
Stratiotes	13	9	0	Strigops	15	11	0	Stubbiness	17	12	6
Stratonic	13	10	0	Striketh	15	12	0	Stubblefed	17	13	0
Stratum	13	11	0	Strikingly	15	12	6	Stubbly	17	14	0
Strawberry	13	12	0	Stringency	15	13	0	Stubborn	17	15	0
Strawboard	13	12	6	Stringendo	15	14	0	Stubbornly	17	16	0
Strawbuilt	13	13	0	Stringent	15	15	0	Stubnails	17	17	0

Code Word.	£	s.	d.	Code Word.	£	s.	d.	Code Word.	£	s.	d.
Stucco	17	17	6	Stylohyoid	19	19	o	Subclavian	44	10	o
Stuccoing	17	18	o	Styloid	20	o	o	Subconcave	45	o	o
Stuccowork	17	19	o	Stylometer	20	10	o	Subconical	45	10	o
Stuckup	18	o	o	Stylopod	21	o	o	Subcordate	46	o	o
Studbolt	18	1	o	Stylops	21	10	o	Subcostal	46	10	o
Studbooks	18	2	o	Stylospore	22	o	o	Subcranial	47	o	o
Studentry	18	2	6	Styptic	22	10	o	Subdeacon	47	10	o
Students	18	3	o	Styptical	23	o	o	Subdean	48	o	o
Studhorse	18	4	o	Stypticity	23	10	o	Subdeanery	48	10	o
Studiedly	18	5	o	Styracine	24	o	o	Subdecanal	49	o	o
Studious	18	6	o	Styrax	24	10	o	Subdented	49	10	o
Studiously	18	7	o	Styrole	25	o	o	Subdeposit	50	o	o
Studwork	18	7	6	Suability	25	10	o	Subdialect	51	o	o
Study	18	8	o	Suage	26	o	o	Subdilated	52	o	o
Studying	18	9	o	Suaging	26	10	o	Subdivides	53	o	o
Stufa	18	10	o	Suasible	27	o	o	Subdolous	54	o	o
Stuffgown	18	11	o	Suasion	27	10	o	Subdual	55	o	o
Stuffy	18	12	o	Suasively	28	o	o	Subduce	56	o	o
Stuke	18	12	6	Suasory	28	10	o	Subducing	57	o	o
Stultifier	18	13	o	Suavified	29	o	o	Subduement	58	o	o
Stultify	18	14	o	Suavifying	29	10	o	Subduers	59	o	o
Stumbled	18	15	o	Suaviloquy	30	o	o	Subdulcid	60	o	o
Stumpy	18	16	o	Subacetate	30	10	o	Subduple	61	o	o
Stupefied	18	17	o	Subacrid	31	o	o	Subeditor	62	o	o
Stupefy	18	17	6	Subact	31	10	o	Subequal	63	o	o
Stupefying	18	18	o	Subacting	32	o	o	Suberic	64	o	o
Stupendous	18	19	o	Subaction	32	10	o	Subfamily	65	o	o
Stupeous	19	o	o	Subaerial	33	o	o	Subfibrous	66	o	o
Stupid	19	1	o	Subagency	33	10	o	Subfossil	67	o	o
Stupidity	19	2	o	Subagent	34	o	o	Subfuscous	68	o	o
Stupidness	19	2	6	Subaided	34	10	o	Subgeneric	69	o	o
Stupor	19	3	o	Subalate	35	o	o	Subgenus	70	o	o
Stuprate	19	4	o	Subalmoner	35	10	o	Subglacial	71	o	o
Stuprating	19	5	o	Subalpine	36	o	o	Subglobose	72	o	o
Stupration	19	6	o	Subaltern	36	10	o	Subgroup	73	o	o
Stupulose	19	7	o	Subangular	37	o	o	Subinduced	74	o	o
Sturdiest	19	7	6	Subapical	37	10	o	Subinfer	75	o	o
Sturdily	19	8	o	Subaquatic	38	o	o	Subitane	76	o	o
Sturdiness	19	9	o	Subaqueous	38	10	o	Subjacent	77	o	o
Sturionian	19	10	o	Subarctic	39	o	o	Subjected	78	o	o
Sturnus	19	11	o	Subastral	39	10	o	Subjecting	79	o	o
Stygian	19	12	o	Subaud	40	o	o	Subjection	80	o	o
Stylidium	19	12	6	Subbeadle	40	10	o	Subjic.ble	81	o	o
Styliform	19	13	o	Subblush	41	o	o	Subjoin	82	o	o
Styling	19	14	o	Subbourdon	41	10	o	Subjoinder	83	o	o
Styliscus	19	15	o	Subbreed	42	o	o	Subjoining	84	o	o
Stylish	19	16	o	Subcaudal	42	10	o	Subjugate	85	o	o
Stylishly	19	17	o	Subcentral	43	o	o	Subkingdom	86	o	o
Stylistic	19	17	6	Subchanter	43	10	o	Sublapsary	87	o	o
Stylobite	19	18	o	Subclass	44	o	o	Sublation	88	o	o

Code Word.	£	Code Word.	£	Code Word.	£
Sublative	89	Subresin	360	Subtonic	775
Sublessee	90	Subrigid	370	Subtorrid	780
Subletting	91	Subrogated	375	Subtracted	790
Sublimable	92	Subrotund	380	Subtrahend	800
Sublimated	93	Subsaline	390	Subtrifid	810
Sublime	94	Subsalt	400	Subtriple	820
Sublimely	95	Subscribe	410	Subtrude	825
Sublingual	96	Subscripts	420	Subtruding	830
Sublunary	97	Subsecute	425	Subtutor	840
Submammary	98	Subsellium	430	Subtypical	850
Submarine	99	Subsequent	440	Subuliform	860
Submarshal	100	Subserous	450	Subulipalp	870
Submental	105	Subsesqui	460	Subumbonal	875
Submerge	110	Subsessile	470	Subunguial	880
Submerging	115	Subsidence	475	Suburban	890
Submersion	120	Subsidiary	480	Subvariety	900
Submiss	125	Subsidize	490	Subvene	910
Submissive	130	Subsign	500	Subvening	920
Submissly	135	Subsigning	510	Subvention	925
Submitted	140	Subsisted	520	Subversed	930
Submonish	145	Subsizar	525	Subversive	940
Submucous	150	Subsoiling	530	Subvertant	950
Subnexed	155	Subsoils	540	Subverters	960
Subnexing	160	Subsolary	550	Subways	970
Subnormal	165	Subspecies	560	Subworker	975
Subnuvolar	170	Substant	570	Succedanea	980
Subobtuse	175	Substernal	575	Succeed	990
Suboctave	180	Substitute	580	Succeedant	1000
Suboctuple	185	Substratum	590	Succeeders	1050
Subocular	190	Substructs	600	Succentor	1100
Subofficer	195	Substylar	610	Success	1150
Suborbital	200	Subsultive	620	Successary	1200
Suborder	210	Subsultory	625	Successful	1250
Suborn	220	Subsultus	630	Succession	1300
Suborning	225	Subsumed	640	Successive	1350
Suboval	230	Subsuming	650	Successors	1400
Suboxide	240	Subtangent	660	Succiduous	1450
Subplinth	250	Subtenants	670	Succinated	1500
Subpoena	260	Subtend	675	Succinct	1550
Subpolar	270	Subtepid	680	Succinctly	1600
Subprefect	275	Subterfuge	690	Succory	1650
Subprior	280	Subterrany	700	Succotash	1700
Subpubic	290	Subterrene	710	Succoured	1750
Subrameal	300	Subtile	720	Succuba	1800
Subramose	310	Subtilely	725	Succulency	1850
Subreader	320	Subtiliate	730	Succulent	1900
Subrector	325	Subtilism	740	Succumbed	1950
Subregions	330	Subtilized	750	Succumbing	2000
Subreption	340	Subtleness	760	Succursal	2050
Subreptive	350	Subtly	770	Suchwise	2100

Code Word.	£	Code Word.	£	Code Word.	£
Suckener	2150	Suggestive	6,500	Sunbeams	14,600
Sucrose	2200	Suggil	6,600	Sunbonnet	14,800
Suction	2250	Suggilate	6,700	Sunburner	15,000
Suctoria	2300	Suicidal	6,750	Sunburning	15,250
Suctorious	2350	Suicidally	6,800	Sunburst	15,500
Sudak	2400	Suicidism	6,900	Sunclad	15,750
Sudamina	2450	Suilline	7,000	Sundart	16,000
Sudatory	2500	Suist	7,100	Sundered	16,250
Sudorific	2600	Suitably	7,200	Sundering	16,500
Sufferably	2700	Suitbroker	7,250	Sunderment	16,750
Sufferance	2800	Suited	7,300	Sundials	17,000
Sufferer	2900	Suithold	7,400	Sundown	17,250
Sufferings	3000	Suitings	7,500	Sundried	17,500
Sufficeth	3100	Suitors	7,600	Sundrily	17,750
Sufficient	3200	Suitress	7,700	Sundryman	18,000
Suffix	3300	Sulcated	7,800	Sunflowers	18,250
Suffixion	3400	Sulkily	7,900	Sunhemp	18,500
Sufflate	3500	Sullen	8,000	Sunkfence	18,750
Sufflating	3600	Sullenly	8,200	Sunlight	19,000
Sufflation	3700	Sullenness	8,400	Sunlit	19,250
Suffocated	3800	Sullevate	8,600	Sunniah	19,500
Suffossion	3900	Sullying	8,800	Sunniness	19,750
Suffragant	4000	Sulphamate	9,000	Sunny	20,000
Suffrages	4100	Sulphatic	9,200	Sunnywarm	20,250
Suffrutex	4200	Sulphides	9,400	Sunopal	20,500
Suffumige	4300	Sulphoacid	9,600	Sunpicture	20,750
Suffuse	4400	Sulphosalt	9,800	Sunplant	21,000
Suffusing	4500	Sulphur	10,000	Sunproof	21,250
Sugarbaker	4600	Sulphurous	10,200	Sunrise	21,500
Sugarbean	4700	Sulphydric	10,400	Sunrising	21,750
Sugarberry	4800	Sulpitian	10,600	Sunsetting	22,000
Sugarbush	4900	Sultanate	10,800	Sunshades	22,250
Sugarcamp	5000	Sultanship	11,000	Sunshiny	22,500
Sugarcandy	5100	Sumach	11,200	Sunspurge	22,750
Sugarcanes	5200	Sumbul	11,400	Sunstar	23,000
Sugared	5250	Summarized	11,600	Sunstroke	23,250
Sugarhouse	5300	Summary	11,800	Sunup	23,500
Sugaring	5400	Summation	12,000	Sunward	23,750
Sugarless	5500	Summerduck	12,200	Sunworship	24,000
Sugarloaf	560c	Summered	12,400	Sunyear	24,250
Sugarmaple	5700	Summerstir	12,600	Superable	24,500
Sugarmills	5750	Summertree	12,800	Superaltar	24,750
Sugarmite	5800	Summists	13,000	Superb	25,000
Sugarmould	5900	Summitless	13,200	Superbly	25,250
Sugarplum	6000	Summon	13,400	Superbness	25,500
Sugartongs	6100	Summoning	13,600	Supercargo	25,750
Sugartrees	6200	Summonses	13,800	Superchery	26,000
Sugescent	6250	Sumpters	14,000	Supercilia	26,250
Suggest	6300	Sumptuary	14,200	Superexalt	26,500
Suggesting	6400	Sumptuous	14,400	Superfine	26,750

Code Word.	£	Code Word.	£	Code Word.	£
Superflux	27,000	Surbet	49,000	Surroyal	120,000
Superheat	27,250	Surceasing	49,500	Sursanure	122,500
Superhuman	27,500	Surcharge	50,000	Surseance	125,000
Superior	27,750	Surcoats	51,000	Sursolid	127,500
Superiorly	28,000	Surcrease	52,000	Surtout	130,000
Supernal	28,250	Surcrew	53,000	Survenue	132,500
Superplant	28,500	Surculous	54,000	Surveyed	135,000
Superposed	28,750	Surculi	55,000	Surveying	137,500
Superregal	29,000	Surditas	56,000	Survivancy	140,000
Supersede	29,250	Surefooted	57,000	Survivors	142,500
Supersolar	29,500	Surement	58,000	Susceptors	145,000
Supertonic	29,750	Suresbyes	59,000	Suscipient	147,500
Supertotus	30,000	Surety	60,000	Suscitate	150,000
Supervened	30,500	Suretyship	61,000	Suspectful	152,500
Supervisal	31,000	Surface	62,000	Suspecting	155,000
Supination	31,500	Surfaceman	63,000	Suspenders	157,500
Supine	32,000	Surfacing	64,000	Suspense	160,000
Supinely	32,500	Surfboat	65,000	Suspension	162,500
Supineness	33,000	Surfduck	66,000	Suspensory	165,000
Suppage	33,500	Surfeited	67,000	Suspicable	167,500
Supperless	34,000	Surfeiting	68,000	Suspicious	170,000
Suppertime	34,500	Surgeful	69,000	Suspiral	172,500
Supplejack	35,000	Surgeless	70,000	Suspiring	175,000
Supplely	35,500	Surgeoncy	71,000	Sustain	177,500
Supplement	36,000	Surgery	72,000	Sustaineth	180,000
Suppletive	36,500	Surgical	73,000	Sustaining	182,500
Suppletory	37,000	Surinamine	74,000	Sustenance	185,000
Supplial	37,500	Surmark	75,000	Susurrant	187,500
Suppliance	38,000	Surmisal	76,000	Susurrous	190,000
Supplicant	38,500	Surmisings	77,000	Sutteeism	192,500
Suppliers	39,000	Surmounted	78,000	Suzerainty	195,000
Supplying	39,500	Surmullet	79,000	Swaggerer	197,500
Support	40,000	Surnames	80,000	Swainish	200,000
Supporteth	40,500	Surnominal	82,000	Swainmote	202,500
Supportful	41,000	Surpass	84,000	Swallowed	205,000
Supporting	41,500	Surpasseth	86,000	Swampoak	207,500
Supposable	42,000	Surpliced	88,000	Swamppink	210,000
Supposure	42,500	Surplusage	90,000	Swampwood	212,500
Suppress	43,000	Surprises	92,000	Swampy	215,000
Suppressor	43,500	Surprising	94,000	Swanflower	217,500
Suppurated	44,000	Surquedous	96,000	Swanlike	220,000
Supralunar	44,500	Surquedrie	98,000	Swanmark	222,500
Suprarenal	45,000	Surquedy	100,000	Swanneries	225,000
Supremacy	45,500	Surrebut	102,500	Swannery	227,500
Supreme	46,000	Surrejoin	105,000	Swansdown	230,000
Suradanni	46,500	Surrenal	107,500	Swanshot	232,500
Surbased	47,000	Surrenders	110,000	Swanskin	235,000
Surbating	47,500	Surrogatum	112,500	Swarth	237,500
Surbedded	48,000	Surround	115,000	Swarthiest	240,000
Surbedding	48,500	Surroy	117,500	Swarthily	242,500

Code Word.	£	Code Word.	£	Code Word.	£
Swartstar	245,000	Swimmable	440,000	Sybarite	15,000,000
Swartzia	247,500	Swimmeret	445,000	Sybaritism	16,000,000
Swashbank	250,000	Swimmingly	450,000	Sycamore	17,000,000
Swashplate	252,500	Swindlers	455,000	Sycite	18,000,000
Swashway	255,000	Swinebread	460,000	Sycoma	19,000,000
Swathbond	257,500	Swinecrue	465,000	Syconus	20,000,000
Swathed	260,000	Swinedrunk	470,000	Sycophancy	
Swathing	262,500	Swinegrass	475,000	Sycophant	
Swaybacked	265,000	Swineherd	480,000	Syenitic	
Swayful	267,500	Swineoat	485,000	Syllabaria	
Sweaters	270,000	Swinepipe	490,000	Syllabical	
Sweatiness	272,500	Swinepox	495,000	Syllabify	
Sweepbar	275,000	Swinestone	500,000	Syllable	
Sweepnet	277,500	Swinesty	505,000	Syllabling	
Sweeps	280,000	Swingbeam	510,000	Sylleptic	
Sweepstake	282,500	Swingboats	515,000	Syllogisms	
Sweetapple	285,000	Swingism	520,000	Syllogize	
Sweetbay	287,500	Swingknife	525,000	Sylph	
Sweetbriar	290,000	Swingtree	530,000	Sylphish	
Sweetcorn	292,500	Swingwheel	535,000	Sylvan	
Sweetened	295,000	Swinishly	540,000	Sylvatic	
Sweetening	297,500	Swiple	545,000	Sylviadae	
Sweetest	300,000	Switch	550,000	Symbolic	
Sweetfern	305,000	Switchman	555,000	Symbolical	
Sweetflag	310,000	Swiveleyed	560,000	Symbolisms	
Sweetgale	315,000	Swivelgun	565,000	Symbolized	
Sweetgum	320,000	Swivelhook	570,000	Symbology	
Sweetheart	325,000	Swizzled	575,000	Symmetrist	
Sweetjohn	330,000	Swizzling	580,000	Symmetry	
Sweetly	335,000	Swobbers	585,000	Sympathise	
Sweetmeats	340,000	Swollen	590,000	Sympathy	
Sweetoil	345,000	Swooning	595,000	Symphonies	
Sweetroot	350,000	Swooningly	600,000	Symphonize	
Sweetrush	355,000	Swooped	700,000	Symphony	
Sweetsop	360,000	Swordarm	800,000	Symphyseal	
Sweetwater	365,000	Swordbelt	900,000	Symphysis	
Swelldom	370,000	Swordblade	1,000,000	Symphytum	
Swellish	375,000	Swordcanes	2,000,000	Symplesite	
Swellmob	380,000	Swordcut	3,000,000	Symplocos	
Swellorgan	385,000	Sworddance	4,000,000	Symposaic	
Swelting	390,000	Swordfight	5,000,000	Symposium	
Swertia	395,000	Swordgrass	6,000,000	Symptoms	
Swerved	400,000	Swordhand	7,000,000	Synagogal	
Swiftfoot	405,000	Swordknot	8,000,000	Synagogues	
Swiftly	410,000	Swordless	9,000,000	Synalepha	
Swiftness	415,000	Swordlily	10,000,000	Synanthous	
Swig	420,000	Swordmats	11,000,000	Synanthy	
Swigged	425,000	Swordplay	12,000,000	Synapta	
Swilley	430,000	Swordstick	13,000,000	Synartesis	
Swillings	435,000	Swung	14,000,000	Syncarpous	

Code Word.	Cents.	Code Word.	Dol. Cts.	Code Word.	Dol. Cts
Synchronal	1	Syringed	51	Tacca	1.05
Synchrony	2	Syringing	52	Tachylite	1.10
Syncladei	3	Syrinx	53	Tachypetes	1.15
Synclinal	4	Syrma	54	Tacitly	1.20
Syncopated	5	Syrupy	55	Taciturn	1.25
Syncopize	6	Systaltic	56	Taciturnly	1.30
Syncratism	7	Systasis	57	Tackduty	1.35
Syncretic	8	System	58	Tackled	1.40
Syncrisis	9	Systematic	59	Tackling	1.45
Syndactyl	10	Systemizer	60	Tacksman	1.50
Syndic	11	Systemless	61	Tackspins	1.55
Syndicate	12	Systyle	62	Tacktackle	1.60
Syndrome	13	Syzygies	63	Tactic	1.65
Synecdoche	14	Syzygium	64	Tactical	1.70
Synechia	15	Syzygy	65	Tactically	1.75
Synedrous	16	Tabachir	66	Tacticians	1.80
Synergetic	17	Tabanidae	67	Tactility	1.85
Synergism	18	Tabanus	68	Tadorna	1.90
Synergy	19	Tabard	69	Tadpoledom	1.95
Syngenesia	20	Tabbinet	70	Tadpoles	2.00
Syngnathus	21	Tabbycat	71	Tafelspath	2.05
Syngraph	22	Tabellion	72	Taffata	2.10
Synochus	23	Tabering	73	Taffrail	2.15
Synocreate	24	Tabernacle	74	Tafilet	2.20
Synod	25	Tabetic	75	Tagbelt	2.25
Synodal	26	Tablature	76	Tagetes	2.30
Synodists	27	Tableaux	77	Tagrag	2.35
Synoecious	28	Tablebeer	78	Taguan	2.40
Synomosy	29	Tablebells	79	Taguicati	2.45
Synonym	30	Tablebook	80	Tailage	2.50
Synonymal	31	Tablecloth	81	Tailblock	2.55
Synonymize	32	Tablecover	82	Tailboards	2.60
Synonymous	33	Tableknife	83	Tailend	2.65
Synopsis	34	Tablelinen	84	Tailorbird	2.70
Synoptical	35	Tableman	85	Tailrace	2.75
Synovia	36	Tablements	86	Tailvalve	2.80
Synovitis	37	Tablemoney	87	Tailvices	2.85
Syntactic	38	Tableshore	88	Tailwater	2.90
Syntax	39	Tablespar	89	Taintless	2.95
Synthermal	40	Tablespoon	90	Taintworm	3.00
Synthesis	41	Tabletalk	91	Tajassu	3.05
Synthetize	42	Tablets	92	Takeoff	3.10
Syntomy	43	Tablinum	93	Takingly	3.15
Synzygia	44	Tabooed	94	Takingness	3.20
Syphilitic	45	Tabourets	95	Talbotype	3.25
Synphilize	46	Tabula	96	Talcite	3.30
Syphiloid	47	Tabularize	97	Talcky	3.35
Syphon	48	Tabulated	98	Talcous	3.40
Syphonic	49	Tabulating	99	Talcschist	3.45
Syriacism	50	Tacamahac	1.00	Talcslate	3.50

Code Word.	Dol. Cts.	Code Word	Dol. Cts.	Code Word	Dol. Cts.
Talebearer	3.55	Tangibly	7.10	Taraxacine	15.25
Talemaster	3.60	Tanhouse	7.20	Tarboggin	15.50
Talented	3.65	Tanist	7.30	Tarboosh	15.75
Taleteller	3.70	Tankards	7.40	Tardation	16.00
Talisman	3.75	Tankengine	7.50	Tardied	16.25
Talismanic	3.80	Tankia	7.60	Tardigrada	16.50
Talkative	3.85	Tankiron	7.70	Tardo	16.75
Talked	3.90	Tankworm	7.80	Tardying	17.00
Tallness	3.95	Tanmill	7.90	Targeted	17.25
Tallower	4.00	Tannadar	8.00	Targetiers	17.50
Tallowface	4.05	Tannery	8.10	Targum	17.75
Tallowtree	4.10	Tannings	8.20	Targumist	18.00
Tallwood	4.15	Tannometer	8.30	Tariffing	18.25
Tally	4.20	Tanpickle	8.40	Tariffs	18.50
Tallyman	4.25	Tanpits	8.50	Tarlatan	18.75
Tallytrade	4.30	Tanspud	8.60	Tarnish	19.00
Talmigold	4.35	Tansy	8.70	Tarnishers	19.25
Talmud	4.40	Tantalism	8.80	Taroc	19.50
Talmudic	4.45	Tantalized	8.90	Tarpan	19.75
Talmudical	4.50	Tantamount	9.00	Tarpaulin	20.00
Talook	4.55	Tantivy	9.10	Tarquinish	20.25
Talookah	4.60	Tantrums	9.20	Tarragon	20.50
Talookdars	4.65	Tanvats	9.30	Tarsiatura	20.75
Talus	4.70	Tanystome	9.40	Tarsius	21.00
Tamandua	4.75	Tanzimat	9.50	Tarsotomy	21.25
Tamarinds	4.80	Tapcinder	9.60	Tartaric	21.50
Tamarisk	4.85	Tapeism	9.70	Tartarous	21.75
Tambac	4.90	Tapeline	9.80	Tartrate	22.00
Tambour	4.95	Taperingly	9.90	Tartuffe	22.25
Tambourine	5.00	Taperness	10.00	Tartuffish	22.50
Tambreet	5.10	Tapers	10.25	Tasimeter	22.75
Tameness	5.20	Tapestried	10.50	Tasimetric	23.00
Tammin	5.30	Tapestry	10.75	Taskmaster	23.25
Tammuz	5.40	Tapeworms	11.00	Taskwork	23.50
Tamperers	5.50	Taphole	11.25	Tasmanite	23.75
Tampingbar	5.60	Tapinage	11.50	Tasselgent	24.00
Tamtams	5.70	Tapioca	11.75	Tasselling	24.25
Tanacetum	5.80	Tapiridae	12.00	Tastable	24.50
Tanagra	5.90	Tapiroid	12.25	Tasteful	24.75
Tanagrinae	6.00	Tapirus	12.50	Tastefully	25.00
Tanballs	6.10	Taplash	12.75	Tasteless	25.50
Tanbeds	6.20	Tapnets	13.00	Tattle	26.00
Tandem	6.30	Tappice	13.25	Tattling	26.50
Tangencies	6.40	Taproom	13.50	Tattooer	27.00
Tangency	6.50	Taprooted	13.75	Tattooing	27.50
Tangential	6.60	Taquanut	14.00	Taught	28.00
Tangents	6.70	Tarafern	14.25	Taunting	28.50
Tangerine	6.80	Tarantass	14.50	Tauridor	29.00
Tangfish	6.90	Tarantella	14.75	Tauriform	29.50
Tanghinia	7.00	Taraquira	15.00	Taurocoll	30.00

Code Word	Dol. Cts.	Code Word.	Dollars.	Code Word.	Dollars.
Tauromachy	30.50	Teasing	61	Tellina	111
Taustaff	31.00	Teaspoon	62	Tellinidae	112
Tautaug	31.50	Teataster	63	Telltale	113
Tautolite	32.00	Teatray	64	Telltroth	114
Tautologic	32.50	Teaurns	65	Telluric	115
Taverner	33.00	Teazehole	66	Tellurions	116
Tavernings	33.50	Teazetenon	67	Temeration	117
Tavernman	34.00	Tebbad	68	Temerity	118
Tawdries	34.50	Technic	69	Temperancy	119
Tawdriness	35.00	Technicals	70	Temperate	120
Tawdry	35.50	Technicist	71	Tempestive	121
Tawdrylace	36.00	Technology	72	Tempests	122
Taxability	36.50	Tectrices	73	Templars	123
Taxably	37.00	Tecumfibre	74	Templed	124
Taxaceae	37.50	Tedder	75	Templeless	125
Taxatively	38.00	Tedious	76	Templinoil	126
Taxcart	38.50	Tediously	77	Tempo	127
Taxel	39.00	Tedium	78	Temporal	128
Taxfree	39.50	Teemeth	79	Temporally	129
Taxiarch	40.00	Teemful	80	Temporist	130
Taxicorn	40.50	Teeming	81	Temporized	131
Taxidermic	41.00	Teemless	82	Temptable	132
Taxing	41.50	Teesdalia	83	Temptation	133
Taxless	42.00	Teetotal	84	Tempting	134
Taxodites	42.50	Teetotally	85	Temptingly	135
Taxodium	43.00	Teetotums	86	Temsebread	136
Taxology	43.50	Tegmentum	87	Temulency	137
Taxonomic	44.00	Teguexin	88	Tenable	138
Taxpayer	44.50	Tegumina	89	Tenacity	139
Teaboard	45.00	Teidae	90	Tenaculum	140
Teacaddies	45.50	Teiltrees	91	Tenaillon	141
Teacaddy	46.00	Teinds	92	Tenancy	142
Teacake	46.50	Teinoscope	93	Tenant	143
Teachest	47.00	Telamones	94	Tenanteth	144
Teacups	47.50	Telarly	95	Tenantless	145
Teadealer	48.00	Telegraphy	96	Tenantsaw	146
Teadrinker	48.50	Telemeter	97	Tench	147
Teagarden	49.00	Teleophyte	98	Tendance	148
Teak	49.50	Teleosaur	99	Tendencies	149
Teakettle	50.00	Teleost	100	Tendency	150
Teaktree	51.00	Teleostean	101	Tenderest	151
Tealead	52.00	Teleozoon	102	Tenderling	152
Teamsters	53.00	Telephone	103	Tenderloin	153
Teamwork	54.00	Telephorus	104	Tenderly	154
Teaoil	55.00	Telerpeton	105	Tenderness	155
Teapots	56.00	Telescopes	106	Tendotome	156
Teardrop	57.00	Telesia	107	Tendrac	157
Teasaucer	58.00	Telestick	108	Tendriled	158
Teased	59.00	Telic	109	Tendsome	159
Teaservice	60.00	Tellership	110	Tenebrific	160

Code Word.	Dollars.	Code Word.	Dollars.	Code Word.	Dollars.
Tenebrious	161	Terminator	211	Testify	305
Tenebrose	162	Terminist	212	Testimony	310
Tenement	163	Terminthus	213	Testiness	315
Tenemental	164	Termitary	214	Testobject	320
Tenendas	165	Termites	215	Testpaper	325
Tenesmic	166	Termitidae	216	Testplate	330
Tennantite	167	Termly	217	Testpump	335
Tennisball	168	Termpiece	218	Testril	340
Tenonaugur	169	Ternately	219	Testtube	345
Tenorino	170	Terneplate	220	Testudinal	350
Tenotomy	171	Ternion	221	Testudo	355
Tenpounder	172	Terpodion	222	Testy	360
Tensionrod	173	Terraced	223	Tetanoid	365
Tensity	174	Terracing	224	Tetanus	370
Tentacular	175	Terracotta	225	Tethys	375
Tentative	176	Terraneous	226	Tetracerus	380
Tenterhook	177	Terrapene	227	Tetrachord	385
Tenthredo	178	Terraquean	228	Tetracolon	390
Tentmaker	179	Terreen	229	Tetrad	395
Tentories	180	Terremote	230	Tetradite	400
Tentorium	181	Terreplein	231	Tetraedal	405
Tentory	182	Terrestre	232	Tetragonal	410
Tentstitch	183	Terribly	233	Tetragram	415
Tentwort	184	Terricolae	234	Tetragynia	420
Tenuate	185	Terrific	235	Tetrameter	425
Teocallis	186	Terrifical	236	Tetramorph	430
Tepejilote	187	Terrifying	237	Tetrander	435
Tephrosia	188	Territory	238	Tetrandria	440
Tepidarium	189	Terrorism	239	Tetraonid	445
Tepidity	190	Terrorized	240	Tetrapla	450
Tepidness	191	Terseness	241	Tetrapody	455
Teratical	192	Tertenant	242	Tetraptote	460
Teratogeny	193	Tertiated	243	Tetrarch	465
Teratology	194	Tertiating	244	Tetraspore	470
Terbium	195	Teruncius	245	Tetrastic	475
Tercellene	196	Terutero	246	Tetrastoon	480
Tercemajor	197	Terzarima	247	Tetrastyle	485
Tercine	198	Terzetto	248	Tetratomic	490
Terebate	199	Tessellar	249	Tetricity	495
Terebella	200	Tessera	250	Tetricous	500
Terebic	201	Tesseraic	255	Tetryl	505
Terebinth	202	Testacea	260	Tetrylene	510
Teredina	203	Testaceous	265	Tetterous	515
Teretous	204	Testaments	270	Tettigonia	520
Tergal	205	Testamur	275	Teucrium	525
Termagancy	206	Testation	280	Teuthidans	530
Termagant	207	Testators	285	Teutlose	535
Termfee	208	Testatrix	290	Teutonized	540
Terminable	209	Testglass	295	Textbook	545
Terminal	210	Testified	300	Texthand	550

Code Word.	Dollars.	Code Word.	Dollars.	Code Word.	Dollars.
Textorial	555	Thenardite	805	Theurgic	1110
Textpen	560	Thencefrom	810	Theurgical	1120
Textrine	565	Theobroma	815	Theurgists	1130
Textualist	570	Theocracy	820	Thewy	1140
Textually	575	Theocrat	825	Thibaudia	1150
Textuists	580	Theocratic	830	Thicketty	1160
Textury	585	Theodicean	835	Thickeyed	1170
Teyne	590	Theodicy	840	Thickhead	1180
Thalamus	595	Theodolite	845	Thicklips	1190
Thalarctos	600	Theogonism	850	Thicknee	1200
Thalessema	605	Theologers	855	Thickness	1210
Thalian	610	Theologian	860	Thickset	1220
Thalictrum	615	Theologue	865	Thickskin	1230
Thallious	620	Theomancy	870	Thickskull	1240
Thallogen	625	Theopathy	875	Thickstuff	1250
Thamnium	630	Theophanic	880	Thieftaker	1260
Thanage	635	Theorbo	885	Thieved	1270
Thanatici	640	Theorems	890	Thievery	1280
Thanedom	645	Theoretic	895	Thievish	1290
Thanehood	650	Theorists	900	Thievishly	1300
Thaneland	655	Theorize	905	Thigh	1310
Thaneship	660	Theorizing	910	Thighbone	1320
Thanked	665	Theory	915	Thimbleful	1330
Thankful	670	Theosophy	920	Thimblerig	1340
Thankfully	675	Theotheca	925	Thimbles	1350
Thankless	680	Theowman	930	Thingumbob	1360
Thanksgive	685	Thereanent	935	Thinkable	1370
Thanus	690	Thereaway	940	Thinking	1380
Thapsia	695	Therefore	945	Thinkingly	1390
Thatchers	700	Thereinto	950	Thinly	1400
Thatchtree	705	Thereof	955	Thinned	1410
Thawed	710	Thereunder	960	Thinspun	1420
Thawing	715	Therewhile	965	Thirdpenny	1430
Theandric	720	Therfbread	970	Thirdrate	1440
Thearchies	725	Theriac	975	Thirdsman	1450
Thearchy	730	Thermal	980	Thirlage	1460
Theatine	735	Thermally	985	Thirst	1470
Theatres	740	Thermidor	990	Thirsteth	1480
Theatrical	745	Thermogen	995	Thirstily	1490
Thebaia	750	Thermology	1000	Thirstless	1500
Thecal	755	Thermopile	1010	Thisness	1510
Thecaphore	760	Thermostat	1020	Thistle	1520
Thecidae	765	Thermotics	1030	Thither	1530
Thecodont	770	Thermotype	1040	Thlaspi	1540
Theft	775	Thesaurus	1050	Thlipsis	1550
Theftbote	780	Thesicle	1060	Tholepin	1560
Theiform	785	Thesis	1070	Tholobate	1570
Theistical	790	Thesium	1080	Tholus	1580
Themselves	795	Thespesia	1090	Thomaism	1590
Thenadays	800	Theta	1100	Thomasite	1600

Code Word.	Dollars	Code Word.	Dollars.	Code Word.	Dollars.
Thomean	1610	Thrillant	2110	Thumbstall	2625
Thomsonian	1620	Thrilled	2120	Thunder	2650
Thong	1630	Thrincia	2130	Thunderfit	2675
Thorax	1640	Thrips	2140	Thundering	2700
Thorinum	1650	Thrissops	2150	Thunderous	2725
Thornapple	1660	Thrivers	2160	Thunderrod	2750
Thornback	1670	Thrivingly	2170	Thurible	2775
Thornbush	1680	Throatband	2180	Thurifers	2800
Thornbut	1690	Throated	2190	Thurified	2825
Thornhedge	1700	Throatpipe	2200	Thurify	2850
Thorntail	1710	Throatwort	2210	Thurifying	2875
Thorny	1720	Throbbed	2220	Thuringite	2900
Thoroughly	1730	Throbbings	2230	Thurrok	2925
Thorow	1740	Thrombosis	2240	Thuytes	2950
Thorowwax	1750	Thrombus	2250	Thwack	2975
Thoughted	1760	Throneless	2260	Thwarter	3000
Thoughtful	1770	Throngful	2270	Thwarting	3050
Thrackscat	1780	Throngly	2280	Thwartly	3100
Thraldom	1790	Thronize	2290	Thwartness	3150
Thralling	1800	Thronizing	2300	Thwartship	3200
Thralllike	1810	Throstling	2310	Thyine	3250
Thrasaetus	1820	Throttled	2320	Thylacinus	3300
Threadbare	1830	Through	2330	Thylacoleo	3350
Threadcell	1840	Throughout	2340	Thymele	3400
Threaders	1850	Throve	2350	Thymus	3450
Threadlace	1860	Throwcock	2360	Thymy	3500
Threadworm	1870	Throwing	2370	Thyroid	3550
Thready	1880	Throwoff	2380	Thyroideal	3600
Threat	1890	Throwster	2390	Thyrse	3650
Threatened	1900	Thrum	2400	Thyrsiform	3700
Threatful	1910	Thrumming	2410	Thysanoura	3750
Threatless	1920	Thrumwort	2420	Thyself	3800
Threeaged	1930	Thrushes	2430	Tiberts	3850
Threecoat	1940	Thrust	2440	Tibia	3900
Threefold	1950	Thrusthoe	2450	Tibicinate	3950
Threeman	1960	Thrymsa	2460	Ticement	4000
Threepiled	1970	Thuban	2470	Tichorhine	4050
Threeply	1980	Thud	2480	Tickbean	4100
Threescore	1990	Thuggism	2490	Ticketday	4150
Threnetic	2000	Thuja	2500	Ticketed	4200
Threnode	2010	Thulite	2510	Ticketing	4250
Threnodial	2020	Thumbband	2520	Ticklish	4300
Thresh	2030	Thumbblue	2530	Tickseed	4350
Threshing	2040	Thumbcleat	2540	Ticktack	4400
Threshold	2050	Thumbed	2550	Ticorea	4450
Thridace	2060	Thumbkins	2560	Ticpolonga	4500
Thridacium	2070	Thumblatch	2570	Tidbits	4550
Thrifallow	2080	Thumbmarks	2580	Tidecoach	4600
Thriftiest	2090	Thumbnut	2590	Tideday	4650
Thrifty	2100	Thumbscrew	2600	Tidedial	4700

Code Word.	Dollars.	Code Word.	Dollars.	Code Word.	Dollars.
Tideful	4750	Timarcha	7250	Tinplate	9750
Tidegate	4800	Timbal	7300	Tinsaw	9800
Tidegauge	4850	Timberhead	7350	Tinselled	9850
Tidelock	4900	Timberlode	7400	Tinsmiths	9900
Tiderip	4950	Timbermare	7450	Tinstone	9950
Tidetables	5000	Timberson	7500	Tintamar	10,000
Tidewaiter	5050	Timbertree	7550	Tinternell	10,100
Tidewave	5100	Timberwork	7600	Tinttool	10,200
Tidewheel	5150	Timbrelled	7650	Tintype	10,300
Tidily	5200	Timbrels	7700	Tinworms	10,400
Tidiness	5250	Timebook	7750	Tipcart	10,500
Tidingwell	5300	Timecandle	7800	Tipcheese	10,600
Tidying	5350	Timefuse	7850	Tippling	10,700
Tiebeam	5400	Timegun	7900	Tipsified	10,800
Tiercel	5450	Timeist	7950	Tipsifying	10,900
Tierods	5500	Timekeeper	8000	Tipsily	11,000
Tiewig	5550	Timelessly	8050	Tipstaff	11,100
Tiffany	5600	Timeliness	8100	Tipsycake	11,200
Tigellate	5650	Timenoguy	8150	Tiptilted	11,300
Tigellus	5700	Timepieces	8200	Tiptoeing	11,400
Tigercat	5750	Timeserver	8250	Tiptop	11,500
Tigercowry	5800	Timid	8300	Tipulary	11,600
Tigerism	5850	Timocratic	8350	Tipulidae	11,700
Tigerlily	5900	Timoneer	8400	Tiresome	11,800
Tigermoth	5950	Timonize	8450	Tiresomely	11,900
Tigershell	6000	Timonizing	8500	Tironian	12,000
Tigerwolf	6050	Timorous	8550	Tirralirra	12,100
Tighten	6100	Timorously	8600	Tissue	12,200
Tightness	6150	Timorsome	8650	Titanate	12,300
Tightrope	6200	Timpani	8700	Titaness	12,400
Tigrisoma	6250	Timwhiskey	8750	Titanitic	12,500
Tikus	6300	Tinamidae	8800	Titanium	12,600
Tiledrain	6350	Tinamotis	8850	Titanshorl	12,700
Tileearth	6400	Tinamou	8900	Tithable	12,800
Tilefield	6450	Tinctorial	8950	Tithe	12,900
Tileore	6500	Tinctured	9000	Tithefree	13,000
Tilepin	6550	Tinderbox	9050	Titheless	13,100
Tiletea	6600	Tinderlike	9100	Tithepig	13,200
Tiliaceae	6650	Tinewald	9150	Tithing	13,300
Tillandsia	6700	Tinfloor	9200	Tithingman	13,400
Tillerhead	6750	Tinfoil	9250	Tithonic	13,500
Tillerrope	6800	Tinglass	9300	Tithymal	13,600
Tillers	6850	Tinker	9350	Titillated	13,700
Tilleyseed	6900	Tinkering	9400	Titivating	13,800
Tillman	6950	Tinkerly	9450	Titlark	13,900
Tiltboat	7000	Tinkerman	9500	Titledeed	14,000
Tilthammer	7050	Tinliquor	9550	Titleleaf	14,100
Tiltmill	7100	Tinmordant	9600	Titlepage	14,200
Tiltyard	7150	Tinnient	9650	Titlerole	14,300
Timalia	7200	Tinnitus	9700	Titmice	14,400

Code Word.	Dollars.	Code Word.	Dollars.	Code Word.	Dollars.
Titmouse	14,500	Toilworm	19,500	Tonguepad	24,500
Titrating	14,600	Tokened	19,600	Tongueshot	24,600
Tittering	14,700	Tokening	19,700	Tonguesore	24,700
Tittlebat	14,800	Tolerably	19,800	Tonguester	24,800
Titubation	14,900	Tolerance	19,900	Tonguetied	24,900
Titular	15,000	Tolerated	20,000	Tongueworm	25,000
Titupping	15,100	Tolerating	20,100	Tonguing	25,100
Tituppy	15,200	Toleration	20,200	Tonnage	25,200
Tityretu	15,300	Tolibants	20,300	Tonous	25,300
Toadeater	15,400	Toll	20,400	Tonsilar	25,400
Toadeating	15,500	Tollbars	20,500	Tonsilitic	25,500
Toadfish	15,600	Tollbooth	20,600	Tonsorial	25,600
Toadflax	15,700	Tollbridge	20,700	Tonsured	25,700
Toadies	15,800	Tollcorn	20,800	Toolpost	25,800
Toadlets	15,900	Tolldish	20,900	Toolstock	25,900
Toadseye	16,000	Tollgates	21,000	Toothache	26,000
Toadspit	16,100	Tollhop	21,100	Toothback	26,100
Toadstool	16,200	Tollhouse	21,200	Toothbrush	26,200
Toadyism	16,300	Tolsester	21,300	Toothedge	26,300
Toasted	16,400	Tolsey	21,400	Toothful	26,400
Toastrack	16,500	Toluol	21,500	Toothkeys	26,500
Toastwater	16,600	Tolutree	21,600	Toothless	26,600
Tobacco	16,700	Tomahawk	21,700	Toothpicks	26,700
Tobaccobox	16,800	Tomalline	21,800	Toothshell	26,800
Tobaccoman	16,900	Tomato	21,900	Toothsome	26,900
Tobacconer	17,000	Tomboys	22,000	Toothwort	27,000
Tobagocane	17,100	Tombstone	22,100	Toothy	27,100
Tobine	17,200	Tomcats	22,200	Toparch	27,200
Toboggan	17,300	Tomcod	22,300	Toparmour	27,300
Tockay	17,400	Tomedes	22,400	Topaz	27,400
Tocsin	17,500	Tomentose	22,500	Topazolite	27,500
Tocussa	17,600	Tomfool	22,600	Topbeam	27,600
Toddalia	17,700	Tomfoolery	22,700	Topblock	27,700
Toddle	17,800	Tomfoolish	22,800	Topboots	27,800
Toddling	17,900	Tomifarous	22,900	Topchain	27,900
Toddybird	18,000	Tomjohn	23,000	Topcloth	28,000
Todidae	18,100	Tommied	23,100	Topdress	28,100
Tofana	18,200	Tommying	23,200	Topfilled	28,200
Tofieldia	18,300	Tommyshop	23,300	Topful	28,300
Tofore	18,400	Tommystore	23,400	Topgallant	28,400
Toftman	18,500	Tomnoddy	23,500	Tophaceous	28,500
Toga	18,600	Tompoker	23,600	Tophamper	28,600
Togated	18,700	Tomrig	23,700	Topheavy	28,700
Together	18,800	Tomtits	23,800	Tophonour	28,800
Togglebolt	18,900	Tondino	23,900	Topiarian	28,900
Toiletset	19,000	Toned	24,000	Topiary	29,000
Toilful	19,100	Tongabean	24,100	Topic	29,100
Toilinette	19,200	Tongkang	24,200	Topical	29,200
Toilsome	19,300	Tongs	24,300	Topically	29,300
Toilsomely	19,400	Tongueless	24,400	Topknot	29,400

Code Word.	Dollars.	Code Word.	Dollars.	Code Word.	Dollars.
Toplantern	29,500	Torrefy	52,500	Toughly	77,500
Toplight	29,600	Torrelite	53,000	Toumbeki	78,000
Toplining	29,700	Torrentbow	53,500	Toupettit	78,500
Topmasts	29,800	Torrential	54,000	Tour	79,000
Topmaul	29,900	Torrentine	54,500	Touraco	79,500
Topography	30,000	Torridity	55,000	Tourism	80,000
Topolatry	30,500	Torrock	55,500	Touristic	80,500
Toponomy	31,000	Torrontes	56,000	Tourmaline	81,000
Toppingest	31,500	Torsion	56,500	Tournament	81,500
Toppingly	32,000	Torsional	57,000	Tourneries	82,000
Toppled	32,500	Torteaux	57,500	Tourniquet	82,500
Topproud	33,000	Tortfeasor	58,000	Tournure	83,000
Toprails	33,500	Tortile	58,500	Towardness	83,500
Toprim	34,000	Tortoise	59,000	Towards	84,000
Topsawyer	34,500	Tortozon	59,500	Towelgourd	84,500
Topshell	35,000	Tortrix	60,000	Towelhorse	85,000
Topsiturn	35,500	Tortulous	60,500	Towerlet	85,500
Topsoiling	36,000	Tortuosity	61,000	Towingpath	86,000
Topsyturvy	36,500	Torturable	61,500	Towingpost	86,500
Toptackle	37,000	Torturer	62,000	Towingrope	87,000
Toptimber	37,500	Torturing	62,500	Towline	87,500
Toque	38,000	Torulacei	63,000	Townbox	88,000
Torchdance	38,500	Torulose	63,500	Townclerk	88,500
Torched	39,000	Torvity	64,000	Towncrier	89,000
Torchlight	39,500	Torvulae	64,500	Townhouse	89,500
Torchrace	40,000	Toscarock	65,000	Townland	90,000
Torchstaff	40,500	Tossed	65,500	Townless	90,500
Torcular	41,000	Tossily	66,000	Townmajor	91,000
Tordylium	41,500	Tosspot	66,500	Townships	91,500
Toreutic	42,000	Tossup	67,000	Towntalk	92,000
Torgant	42,500	Totalize	67,500	Towntop	92,500
Torilis	43,000	Totalizing	68,000	Townwards	93,000
Tormented	43,500	Totally	68,500	Towpath	93,500
Tormentful	44,000	Totemism	69,000	Towrope	94,000
Tormenting	44,500	Totterers	69,500	Towser	94,500
Tormentors	45,000	Touch	70,000	Toxicant	95,000
Tormina	45,500	Touchable	70,500	Toxicology	95,500
Torminous	46,000	Touchbox	71,000	Toxoceras	96,000
Tornadoes	46,500	Touchhole	71,500	Toxotes	96,500
Torpedo	47,000	Touchiness	72,000	Toyman	97,000
Torpescent	47,500	Touching	72,500	Tozy	97,500
Torpidly	48,000	Touchingly	73,000	Trabeated	98,000
Torpidness	48,500	Touchpan	73,500	Trabeation	98,500
Torpified	49,000	Touchpaper	74,000	Trabecula	99,000
Torpifying	49,500	Touchpiece	74,500	Traceably	99,500
Torpitude	50,000	Touchstone	75,000	Traceries	100,000
Torpor	50,500	Touchwood	75,500	Tracheal	200,000
Torporific	51,000	Toughened	76,000	Tracheitis	300,000
Torqued	51,500	Toughening	76,500	Trachoma	400,000
Torreador	52,000	Toughish	77,000	Trachytic	500,000

Code Word.	Dollars.	Code Word.	Dollars.	Code Word.	Dollars.
Trackroad	600,000	Trammelled	5,600,000	Trapesing	13,000,000
Trackscout	700,000	Tramontane	5,700,000	Trapezate	13,500,000
Trackways	800,000	Trampled	5,800,000	Trapeziums	14,000,000
Tractarian	900,000	Trampling	5,900,000	Trapezoid	14,500,000
Tractators	1,000,000	Tramppick	6,000,000	Trapholes	15,000,000
Tractatrix	1,100,000	Tramroads	6,100,000	Trappean	15,500,000
Traded	1,200,000	Tramway	6,200,000	Trappists	16,000,000
Tradeful	1,300,000	Trancedly	6,300,000	Trappy	16,500,000
Tradehall	1,400,000	Trangram	6,400,000	Traprock	17,000,000
Trademarks	1,500,000	Trankum	6,500,000	Trapair	17,500,000
Tradeprice	1,600,000	Tranlaced	6,600,000	Trapstick	18,000,000
Tradesale	1,700,000	Trannel	6,700,000	Traptree	18,500,000
Tradesfolk	1,800,000	Tranquil	6,800,000	Traptufa	19,000,000
Tradesmen	1,900,000	Tranquilly	6,900,000	Trashiness	19,500,000
Tradewind	2,000,000	Transact	7,000,000	Traulism	20,000,000
Trading	2,100,000	Transactor	7,100,000	Traumatism	21,000,000
Traditores	2,200,000	Transcend	7,200,000	Traunter	22,000,000
Traducible	2,300,000	Transcribe	7,300,000	Travail	23,000,000
Traducing	2,400,000	Transearth	7,400,000	Travailous	24,000,000
Traduct	2,500,000	Transenna	7,500,000	Travelled	25,000,000
Traduction	2,600,000	Transepts	7,600,000	Travelling	26,000,000
Traductive	2,700,000	Transfer	7,700,000	Traverser	27,000,000
Traffic	2,800,000	Transflux	7,800,000	Traversing	28,000,000
Trafficker	2,900,000	Transforms	7,900,000	Travestied	29,000,000
Tragacanth	3,000,000	Transfrete	8,000,000	Travesty	30,000,000
Tragalism	3,100,000	Transgress	8,100,000	Travis	31,000,000
Tragedian	3,200,000	Tranship	8,200,000	Trawlbeams	32,000,000
Tragedy	3,300,000	Transiency	8,300,000	Trawlboat	33,000,000
Tragetour	3,400,000	Transient	8,400,000	Trawlhead	34,000,000
Tragic	3,500,000	Transiting	8,500,000	Trawlwarp	35,000,000
Tragical	3,600,000	Transition	8,600,000	Traytip	36,000,000
Tragically	3,700,000	Transitory	8,700,000	Treachery	37,000,000
Tragicomic	3,800,000	Transluce	8,800,000	Treachour	38,000,000
Tragopan	3,900,000	Translunar	8,900,000	Treacly	39,000,000
Tragopogon	4,000,000	Transmoved	9,000,000	Treadeth	40,000,000
Tragulidae	4,100,000	Transmuter	9,100,000	Treadmill	41,000,000
Trailnets	4,200,000	Transpired	9,200,000	Treadwheel	42,000,000
Trainband	4,300,000	Transplace	9,300,000	Treasonous	43,000,000
Trainers	4,400,000	Transposal	9,400,000	Treasures	44,000,000
Trainoil	4,500,000	Transprint	9,500,000	Treasuring	45,000,000
Traipse	4,600,000	Transtra	9,600,000	Treasury	46,000,000
Traitor	4,700,000	Transude	9,700,000	Treatably	47,000,000
Traitorism	4,800,000	Transumpt	9,800,000	Treatiser	48,000,000
Traitorly	4,900,000	Transverse	9,900,000	Treatments	49,000,000
Traitorous	5,000,000	Transview	10,000,000	Trebled	50,000,000
Traject	5,100,000	Transvolve	10,500,000	Trebleness	60,000,000
Trajecting	5,200,000	Trapballs	11,000,000	Trebling	70,000,000
Trajectory	5,300,000	Trapbat	11,500,000	Trebucket	80,000,000
Tralucency	5,400,000	Trapdoor	12,000,000	Treebeard	90,000,000
Tralucent	5,500,000	Trapelus	12,500,000	Treecrabs	100,000,000

Code Word.	No.	Code Word.	No.	Code Word.	No.
Treefern	1	Triad	51	Tridacna	101
Treefrog	2	Triadic	52	Tridactyle	102
Treegoose	3	Triakenium	53	Tridented	103
Treehair	4	Trialday	54	Triduan	104
Treejobber	5	Trialfire	55	Trientalis	105
Treeless	6	Triality	56	Trierarch	106
Treelouse	7	Trialogue	57	Trieterics	107
Treemallow	8	Trialtrips	58	Trifacial	108
Treeonion	9	Triander	59	Trifid	109
Treepigeon	10	Triandrous	60	Triflers	110
Treeship	11	Triangles	61	Triflingly	111
Treeshrike	12	Triangular	62	Trifloral	112
Treesorrel	13	Trianthema	63	Triflorous	113
Treetoad	14	Triarchy	64	Trifolium	114
Treewool	15	Triarian	65	Trifoly	115
Trehala	16	Triassic	66	Triformity	116
Trehalose	17	Triatomic	67	Trifurcate	117
Treillages	18	Tribalism	68	Trigamist	118
Trekked	19	Tribometer	69	Trigastric	119
Trektow	20	Triboulet	70	Trigemini	120
Trellised	21	Tribrach	71	Trigintal	121
Trellising	22	Tribunals	72	Triglidae	122
Tremarctos	23	Tribune	73	Triglochin	123
Trematoda	24	Tributary	74	Triglyph	124
Tremblers	25	Tricennial	75	Triglyphic	125
Tremella	26	Trichechus	76	Trigonella	126
Tremellini	27	Trichiasis	77	Trigonia	127
Tremelloid	28	Trichidium	78	Trigrammic	128
Tremendous	29	Trichilia	79	Trigraph	129
Tremolando	30	Trichinous	80	Trigyn	130
Tremolite	31	Trichocyst	81	Trigynous	131
Tremor	32	Trichodon	82	Trihedron	132
Tremulent	33	Trichogyne	83	Trihilate	133
Trenchcart	34	Trichonema	84	Trijugous	134
Trenched	35	Trichords	85	Trilaminar	135
Trenchmore	36	Trichotomy	86	Trilinear	136
Trend	37	Trichroism	87	Trilingual	137
Trentsand	38	Trickery	88	Triliteral	138
Trepan	39	Trickiness	89	Trilith	139
Trepanized	40	Trickishly	90	Trilithic	140
Trepanners	41	Trickment	91	Trilithons	141
Trepeget	42	Tricksome	92	Trillibub	142
Trephining	43	Tricksters	93	Trills	143
Tresayle	44	Triclinate	94	Trilobed	144
Trespassed	45	Tricoccae	95	Trilobitic	145
Tressful	46	Tricoccous	96	Trilogy	146
Tressour	47	Tricolour	97	Trimembral	147
Trevat	48	Triconodon	98	Trimera	148
Triable	49	Tricostate	99	Trimester	149
Triaconter	50	Tricycles	100	Trimmed	150

NUMBERS.

63 5

Code Word.	No.	Code Word.	No.	Code Word.	No.
Trimorphic	151	Trisulcate	201	Trotcosy	251
Trimyarian	152	Tritheism	202	Trothring	252
Trinervate	153	Triticum	203	Trotters	253
Trinerved	154	Tritone	204	Troubador	254
Trinity	155	Tritonidae	205	Troublable	255
Triniunity	156	Tritorium	206	Trouble	256
Trinketed	157	Tritozooid	207	Troubledly	257
Trinketry	158	Triturable	208	Troublous	258
Trinoctial	159	Triturated	209	Trousered	259
Trinoda	160	Tritylene	210	Trousering	260
Trinominal	161	Triumph	211	Trousseau	261
Triobolar	162	Triumphal	212	Troutful	262
Trioctile	163	Triumphing	213	Troutlet	263
Trioecious	164	Triumviry	214	Troutling	264
Triolet	165	Trivalent	215	Trouveur	265
Trionyx	166	Trivalve	216	Trowest	266
Triosteum	167	Trivial	217	Trowsed	267
Triparted	168	Trivialism	218	Truage	268
Tripaschal	169	Trivially	219	Truancy	269
Tripedal	170	Triweekly	220	Truanted	270
Tripeman	171	Trocar	221	Truanting	271
Tripennate	172	Trochanter	222	Truantship	272
Triphammer	173	Trochilics	223	Trubtail	273
Triphane	174	Trochiscus	224	Truce	274
Triphasia	175	Trochisk	225	Truckage	275
Triphthong	176	Trochleary	226	Trucking	276
Triphyline	177	Trochoid	227	Trucklebed	277
Triplex	178	Trochoidal	228	Truckler	278
Triplicity	179	Trodden	229	Truculency	279
Tripmadam	180	Troglodyte	230	Trudgeman	280
Tripodian	181	Trogonidae	231	Trudging	281
Tripods	182	Trolley	232	Trueblue	282
Tripping	183	Trollol	233	Trueborn	283
Tripsis	184	Trollolled	234	Truebred	284
Tripterous	185	Trollopish	235	Truelove	285
Triptich	186	Trollopy	236	Truepenny	286
Triptote	187	Trollplate	237	Truetable	287
Triptychon	188	Trombone	238	Truisms	288
Tripudiary	189	Trompour	239	Truismatic	289
Tripyramid	190	Troopbird	240	Truly	290
Triquetra	191	Troopers	241	Trumpery	291
Trireme	192	Troopmeal	242	Trumpeters	292
Trisagion	193	Troopship	243	Trumpetfly	293
Trisected	194	Tropaeolum	244	Trumpeting	294
Trisecting	195	Trophies	245	Truncal	295
Trisection	196	Trophonian	246	Truncating	296
Trismus	197	Trophosome	247	Truncheons	297
Trispaston	198	Trophy	248	Trundling	298
Tristfully	199	Tropicbird	249	Trunk	299
Trisulc	200	Tropologic	250	Trunkfish	300

Code Word.	No.	Code Word.	No.	Code Word.	No.
Trunkhose	301	Tugiron	351	Turbinate	401
Trunkline	302	Tuille	352	Turbinidae	402
Trunkwork	303	Tuition	353	Turbot	403
Trunnioned	304	Tuitionary	354	Turbulency	404
Trusion	305	Tulametal	355	Turbulent	405
Trusshoop	306	Tulchan	356	Turcism	406
Trustdeeds	307	Tulipist	357	Turdus	407
Trustingly	308	Tuliptree	358	Turf	408
Trustless	309	Tulipwood	359	Turfclad	409
Truthful	310	Tumblebug	360	Turfdrain	410
Truthfully	311	Tumbledown	361	Turfhedge	411
Truthlover	312	Tumblehome	362	Turfhouse	412
Tryable	313	Tumblerful	363	Turfknife	413
Trycock	314	Tumblers	364	Turfmoss	414
Trygon	315	Tumefy	365	Turfplough	415
Tryma	316	Tumescence	366	Turfspade	416
Trysted	317	Tumidly	367	Turgesce	417
Trysting	318	Tummals	368	Turgescing	418
Tubberman	319	Tumoured	369	Turgid	419
Tubbing	320	Tumpline	370	Turgidity	420
Tubbish	321	Tumulated	371	Turgidous	421
Tubby	322	Tumulosity	372	Turinnut	422
Tubeplate	323	Tumult	373	Turio	423
Tubeplugs	324	Tumulters	374	Turiones	424
Tubepouch	325	Tumultuary	375	Turkeybird	425
Tuberacei	326	Tumultuate	376	Turkeycock	426
Tubercle	327	Tumultuous	377	Turkeyhone	427
Tubercular	328	Tunably	378	Turkeyred	428
Tuberosity	329	Tunbellied	379	Turkised	429
Tubesheet	330	Tunefully	380	Turkishly	430
Tubewell	331	Tuneless	381	Turkscap	431
Tubfast	332	Tungstate	382	Turkshead	432
Tubful	333	Tungstenic	383	Turlupins	433
Tubicolae	334	Tungusic	384	Turmeric	434
Tubicolous	335	Tunhoof	385	Turmoil	435
Tubiporite	336	Tunicate	386	Turmoiling	436
Tubivalve	337	Tuning	387	Turnbench	437
Tubman	338	Tuningfork	388	Turnbroach	438
Tubthumper	339	Tuningkey	389	Turncoats	439
Tubularian	340	Tunnelhead	390	Turncock	440
Tubule	341	Tunnelling	391	Turndown	441
Tubulicole	342	Tunnelpit	392	Turnip	442
Tuburcinia	343	Tupaia	393	Turnipfly	443
Tubwheel	344	Tupaiadae	394	Turnkeys	444
Tuckahoe	345	Turacine	395	Turnout	445
Tucknet	346	Turbaned	396	Turnover	446
Tuftaffeta	347	Turbantop	397	Turnpike	447
Tufthunter	348	Turbary	398	Turnplate	448
Tufty	349	Turbidly	399	Turnscrew	449
Tugboats	350	Turbillion	400	Turnsick	450

Code Word.	No.	Code Word.	No.	Code Word.	No.
Turnsole	451	Twicetold	505	Typewriter	810
Turnspit	452	Twiddled	510	Typhaceae	820
Turnstiles	453	Twiggy	515	Typhfever	830
Turnstones	454	Twigrush	520	Typhline	840
Turntable	455	Twilight	525	Typhlops	850
Turntippet	456	Twinborn	530	Typhoid	860
Turpentine	457	Twinflower	535	Typhomania	870
Turpeth	458	Twinges	540	Typhoons	880
Turquoise	459	Twinkler	545	Typhpoison	890
Turreted	460	Twinkling	550	Typifiers	900
Turretship	461	Twinleaf	555	Typify	910
Turribant	462	Twinscrew	560	Typifying	920
Turrilite	463	Twinsister	565	Typocosmy	930
Turritella	464	Twiscar	570	Typolite	940
Turritis	465	Twisting	575	Typology	950
Turtleback	466	Twitched	580	Tyrannical	960
Turtledove	467	Twitching	585	Tyrannidae	970
Turtler	468	Twittered	590	Tyranning	980
Turtlesoup	469	Twittingly	595	Tyrannous	990
Turtling	470	Twocelled	600	Tyrants	1000
Turves	471	Twocleft	605	Tyrocinium	1050
Tusked	472	Twodecker	610	Tyrociny	1100
Tussehsilk	473	Twoedged	615	Tyronism	1150
Tussicular	474	Twofaced	620	Uberty	1200
Tussle	475	Twofold	625	Ubication	1250
Tussocky	476	Twohand	630	Ubiquarian	1300
Tutenague	477	Twoheaded	635	Ubiquist	1350
Tutmouthed	478	Twolipped	640	Ubiquitous	1400
Tutnose	479	Twoness	645	Ubiquity	1450
Tutor	480	Twopetaled	650	Udal	1500
Tutorage	481	Tworanked	655	Udaller	1550
Tutorism	482	Twosome	660	Udalman	1600
Tutorly	483	Twotongued	665	Uddered	1650
Tutworkman	484	Twovalved	670	Udderless	1700
Tuwhits	485	Twyfoil	675	Udometer	1750
Tuyere	486	Twyforked	680	Ugliest	1800
Twaddling	487	Tyall	685	Uglifies	1850
Twaddly	488	Tyburntree	690	Uglify	1900
Twaincloud	489	Tychonic	695	Uglifying	1950
Twang	490	Tydy	700	Ugliness	2000
Twanging	491	Tyer	710	Ugric	2050
Twangled	492	Tylophora	720	Ugsome	2100
Twankay	493	Tymp	730	Ugsomeness	2150
Twayblade	494	Tympanitic	740	Ukase	2200
Tweed	495	Tympanized	750	Ulcer	2250
Tweedledum	496	Tympanum	760	Ulcerable	2300
Tweezers	497	Tyndaridae	770	Ulcerated	2350
Twelfthday	498	Typemetal	780	Ulcerating	2400
Twentyfold	499	Types	790	Ulceration	2450
Twibill	500	Typesetter	800	Ulcerous	2500

Code Word.	No.	Code Word.	No.	Code Word.	No.
Ulcerously	2550	Unacquired	7100	Unavenged	29,000
Ulcuscle	2600	Unactable	7200	Unavoided	30,000
Ulema	2650	Unacted	7300	Unawakened	31,000
Uletree	2700	Unactive	7400	Unbacked	32,000
Ullage	2750	Unactuated	7500	Unbaffled	33,000
Ullmannite	2800	Unadjusted	7600	Unbagging	34,000
Ulmaceous	2850	Unadmitted	7700	Unbailable	35,000
Ulmic	2900	Unadorned	7800	Unbalanced	36,000
Ulmous	2950	Unadvised	7900	Unbank	37,000
Ulodendron	3000	Unaffable	8000	Unbaptized	38,000
Ulorrhagia	3100	Unaffected	8100	Unbashful	39,000
Ulotrichan	3200	Unafraid	8200	Unbearably	40,000
Ulsters	3300	Unalarmed	8300	Unbearded	41,000
Ulterior	3400	Unalarming	8400	Unbearing	42,000
Ultimate	3500	Unalist	8500	Unbeaten	43,000
Ultimately	3600	Unallayed	8600	Unbeavered	44,000
Ultimation	3700	Unalliable	8700	Unbecome	45,000
Ultimatums	3800	Unallied	8800	Unbecoming	46,000
Ultimo	3900	Unaltered	8900	Unbedinned	47,000
Ultraism	4000	Unamazed	9000	Unbefool	48,000
Ultroneous	4100	Unamusive	9100	Unbegotten	49,000
Ululating	4200	Unanchor	9200	Unbeguiled	50,000
Ulvaceae	4300	Unangular	9300	Unbegun	51,000
Umbellated	4400	Unanimity	9400	Unbeheld	52,000
Umbellifer	4500	Unanimous	9500	Unbeknown	53,000
Umbellule	4600	Unannoyed	9600	Unbelief	54,000
Umbery	4700	Unanointed	9700	Unbeliever	55,000
Umbilic	4800	Unanxious	9800	Unbeloved	56,000
Umbilical	4900	Unappalled	9900	Unbelted	57,000
Umbonulate	5000	Unapparent	10,000	Unbenign	58,000
Umbracle	5100	Unapplied	10,500	Unbenumb	59,000
Umbraculum	5200	Unapproved	11,000	Unbereaven	60,000
Umbrageous	5300	Unaptly	11,500	Unbereft	61,000
Umbrated	5400	Unargued	12,000	Unbeseem	62,000
Umbratical	5500	Unarm	13,000	Unbesought	63,000
Umbrating	5600	Unarming	14,000	Unbespeak	64,000
Umbrellas	5700	Unartful	15,000	Unbestowed	65,000
Umbrose	5800	Unartfully	16,000	Unbetide	66,000
Umbrosity	5900	Unartistic	17,000	Unbetrayed	67,000
Umlaut	6000	Unascried	18,000	Unbewailed	68,000
Umpirage	6100	Unaspiring	19,000	Unbewares	69,000
Umpires	6200	Unassuming	20,000	Unbewitch	70,000
Umpireship	6300	Unassured	21,000	Unbiassed	71,000
Unabashed	6400	Unatonable	22,000	Unbiassing	72,000
Unableness	6500	Unatoned	23,000	Unbigoted	73,000
Unabridged	6600	Unattached	24,000	Unbinding	74,000
Unabsurd	6700	Unattested	25,000	Unbishop	75,000
Unabundant	6800	Unattire	26,000	Unblamable	76,000
Unaccented	6900	Unattiring	27,000	Unblasted	77,000
Unaccorded	7000	Unavailing	28,000	Unbleeding	78,000

Code Word.	No.	Code Word.	Inches.	Code Word.	Ft.	Ins.
Unblended	79,000	Uncertain	$\frac{1}{16}$	Uncongeal	2	0
Unblighted	80,000	Uncessant	$\frac{1}{8}$	Unconjugal	2	3
Unblissful	81,000	Unchain	$\frac{3}{16}$	Unconning	2	6
Unblooded	82,000	Unchaplain	$\frac{6}{16}$	Unconstant	2	9
Unblushing	83,000	Uncharged	$\frac{3}{8}$	Uncontrite	3	0
Unboastful	84,000	Unchariot	$\frac{7}{16}$	Uncordial	3	3
Unbodkined	85,000	Uncharity	$\frac{9}{16}$	Uncouple	3	6
Unbolt	86,000	Uncharnel	$\frac{5}{8}$	Uncoupling	3	9
Unbolting	87,000	Unchaste	$\frac{11}{16}$	Uncourted	4	0
Unbonneted	88,000	Unchastely	$\frac{13}{16}$	Uncouth	4	3
Unbookish	89,000	Unchecked	$\frac{7}{8}$	Uncouthly	4	6
Unborrowed	90,000	Uncheerful	$\frac{15}{16}$	Uncovering	4	9
Unbosomer	91,000	Uncheery	$1\frac{1}{4}$	Uncowled	5	0
Unbosoming	92,000	Unchild	$1\frac{1}{2}$	Uncrafty	6	0
Unbounded	93,000	Unchildish	$1\frac{3}{4}$	Uncredit	7	0
Unbowable	94,000	Uncholeric	$2\frac{1}{4}$	Uncrippled	8	0
Unbreast	95,000	Unchristen	$2\frac{1}{2}$	Uncritical	9	0
Unbreathed	·96,000	Unchurch	$2\frac{3}{4}$	Uncrossed	10	0
Unbreech	97,000	Unciatim	$3\frac{1}{4}$	Uncrown	11	0
Unbribable	98,000	Uncipher	$3\frac{1}{2}$	Uncrowning	12	0
Unbridled	99,000	Uncivil	$3\frac{3}{4}$	Unctuosity	13	0
Unbridling	100,000	Uncivilly	$4\frac{1}{4}$	Unctuous	14	0
Unbroken	200,000	Unclaimed	$4\frac{1}{2}$	Uncular	15	0
Unbruised	300,000	Unclasped	$4\frac{3}{4}$	Unculpable	16	0
Unbuckling	400,000	Unclassic	$5\frac{1}{4}$	Unculture	17	0
Unbudded	500,000	Unclean	$5\frac{1}{2}$	Uncumbered	18	0
Unbuoyed	600,000	Uncleanly	$5\frac{3}{4}$	Uncurl	19	0
Unburdened	700,000	Unclench	$6\frac{1}{4}$	Uncurrent	20	0
Unburnt	800,000	Unclerical	$6\frac{1}{2}$	Uncustomed	21	0
Unburrow	900,000	Uncloak	$6\frac{3}{4}$	Undaunted	22	0
Unburthen	1,000,000	Uncloaking	$7\frac{1}{4}$	Undawning	23	0
Unburying	2,000,000	Unclogged	$7\frac{1}{2}$	Undeadly	24	0
Unbuttoned	3,000,000	Unclose	$7\frac{3}{4}$	Undeaf	25	0
Unbuxom	4,000,000	Unclothed	$8\frac{1}{4}$	Undecagon	30	0
Uncadenced	5,000,000	Uncloudy	$8\frac{1}{2}$	Undecaying	35	0
Uncage	6,000,000	Unclutched	$8\frac{3}{4}$	Undeceived	40	0
Uncaging	7,000,000	Uncoached	$9\frac{1}{4}$	Undecency	45	0
Uncalled	8,000,000	Uncoffined	$9\frac{1}{2}$	Undecide	50	0
Uncalm	9,000,000	Uncoif	$9\frac{3}{4}$	Undeck	60	0
Uncamped	10,000,000	Uncoloured	$10\frac{1}{4}$	Undecking	70	0
Uncandid	15,000,000	Uncombine	$10\frac{1}{2}$	Undecreed	80	0
Uncanny	20,000,000	Uncomely	$10\frac{3}{4}$	Undefaced	90	0
Uncanonize	25,000,000	Uncommixed	$11\frac{1}{4}$	Undefended	100	0
Uncanopied	30,000,000	Uncommon	$11\frac{1}{2}$	Undefining	150	0
Uncaptious	35,000,000	Uncommonly	$11\frac{3}{4}$	Undeify	200	0
Uncareful	40,000,000	Unconcern	19	Undeluged	250	0
Uncaria	45,000,000	Uncondited	20	Undeniably	300	0
Uncastled	50,000,000	Unconform	21	Undeplored	350	0
Uncensured	60,000,000	Unconfound	22	Undepraved	400	0
Uncentre	70,000,000	Unconfused	23	Underagent	500	0

* See also under Inch in Alphabetical Code.

Code Word.	Weight.	Code Word.	Weight.	Code Word.	Weight.
Underaided	lbs. 1	Undershoot	lbs. 51	Undulative	lbs. 101
Underboard	,, 2	Undershrub	,, 52	Undulatory	,, 102
Underbrace	,, 3	Undersign	,, 53	Undull	,, 103
Underbrush	,, 4	Undersized	,, 54	Undumpish	,, 104
Underbuy	,, 5	Undersky	,, 55	Undusted	,, 105
Underchaps	,, 6	Undersoil	,, 56	Unduteous	,, 106
Underclay	,, 7	Underspend	,, 57	Undutiful	,, 107
Undercliff	,, 8	Underspore	,, 58	Unearth	,, 108
Undercoat	,, 9	Understock	,, 59	Unearthing	,, 109
Undercroft	,, 10	Undersuit	,, 60	Unearthly	,, 110
Underdelve	,, 11	Undertaxed	,, 61	Uneasily	,, 111
Underdo	,, 12	Underturn	,, 62	Uneasiness	cwt. 1
Underdosed	,, 13	Undervalue	,, 63	Uneclipsed	,, 2
Underdrain	,, 14	Underverse	,, 64	Unedge	,, 3
Underfeed	,, 15	Underwater	,, 65	Unedging	,, 4
Underfoot	,, 16	Underwear	,, 66	Unedified	,, 5
Undergird	,, 17	Underwings	,, 67	Unedifying	,, 6
Undergoing	,, 18	Underworld	,, 68	Uneducate	,, 7
Undergone	,, 19	Underwrite	,, 69	Unelected	,, 8
Undergowns	,, 20	Undesired	,, 70	Unelegant	,, 9
Undergroan	,, 21	Undesiring	,, 71	Unembodied	,, 10
Undergrub	,, 22	Undesirous	,, 72	Unemphatic	,, 11
Underheave	,, 23	Undevilled	,, 73	Unemployed	,, 12
Underhung	,, 24	Undevout	,, 74	Unemptied	,, 13
Underjaw	,, 25	Undiademed	,, 75	Unendeared	,, 14
Underkind	,, 26	Undid	,, 76	Unendly	,, 15
Underlaid	,, 27	Undigested	,, 77	Unendowed	,, 16
Underlayer	,, 28	Undight	,, 78	Unenglish	,, 17
Underleaf	,, 29	Undiocesed	,, 79	Unenjoyed	,, 18
Underlies	,, 30	Undirect	,, 80	Unenjoying	,, 19
Underlip	,, 31	Undirectly	,, 81	Unenlarged	tons 1
Underlock	,, 32	Undisclose	,, 82	Unenslaved	,, 2
Underlying	,, 33	Undismayed	,, 83	Unentangle	,, 3
Undermatch	,, 34	Undisputed	,, 84	Unentering	,, 4
Undermeal	,, 35	Undivine	,, 85	Unentombed	,, 5
Undermines	,, 36	Undivorced	,, 86	Unenviable	,, 6
Undermirth	,, 37	Undivulged	,, 87	Unenvied	,, 7
Undermost	,, 38	Undoer	,, 88	Unequal	,, 8
Underntide	,, 39	Undomestic	,, 89	Unequally	,, 9
Underparts	,, 40	Undoubtful	,, 90	Unerringly	,, 10
Underpight	,, 41	Undoubting	,, 91	Unespied	,, 11
Underpin	,, 42	Undoubtous	,, 92	Unethes	,, 12
Underplay	,, 43	Undreaded	,, 93	Uneven	,, 13
Underprize	,, 44	Undress	,, 94	Unevenly	,, 14
Underprop	,, 45	Undressing	,, 95	Unevenness	,, 15
Underpull	,, 46	Undriven	,, 96	Uneventful	,, 16
Underrated	,, 47	Undrooping	,, 97	Unevident	,, 17
Underroof	,, 48	Undueness	,, 98	Unexacted	,, 18
Underscore	,, 49	Undulant	,, 99	Unexamined	,, 19
Undersell	,, 50	Undulation	,, 100	Unexampled	,, 20

Code Word.	Tons.	Code Word.	Tons.	Code Word.	Tons.
Unexcised	21	Unfraught	71	Unhailed	1900
Unexempt	22	Unfreezed	72	Unhallows	2000
Unexpected	23	Unfreezing	73	Unhampered	2100
Unexplored	24	Unfret	74	Unhandily	2200
Unextorted	25	Unfriendly	75	Unhandled	2300
Unfadable	26	Unfrock	76	Unhandsome	2400
Unfaded	27	Unfrocking	77	Unhappiest	2500
Unfairly	28	Unfruitful	78	Unhappily	2600
Unfairness	29	Unfurnish	79	Unhaps	2700
Unfaith	30	Ungainful	80	Unhardy	2800
Unfaithful	31	Ungallant	81	Unharmful	2900
Unfalcated	32	Ungartered	82	Unharness	3000
Unfamiliar	33	Ungauged	83	Unhatted	3100
Unfastened	34	Ungenerous	84	Unhattings	3200
Unfatherly	35	Ungenial	85	Unhazarded	3300
Unfathomed	36	Ungenteel	86	Unhealthy	3400
Unfatigued	37	Ungetting	87	Unheard	3500
Unfaulty	38	Ungifted	88	Unheededly	3600
Unfearful	39	Ungirds	89	Unhelm	3700
Unfeasible	40	Ungive	90	Unhelped	3800
Unfeather	41	Ungiving	91	Unhelpful	3900
Unfeatured	42	Ungkaputi	92	Unhidden	4000
Unfeeling	43	Unglazed	93	Unhinged	4100
Unfeigned	44	Unglazing	94	Unhitched	4200
Unfeigning	45	Unglorify	95	Unhitching	4300
Unfellow	46	Ungloved	96	Unhoarding	4400
Unfeminine	47	Unglue	97	Unholiness	4500
Unfenced	48	Unglutted	98	Unholy	4600
Unfertile	49	Ungodded	99	Unhonest	4700
Unfetter	50	Ungodly	100	Unhonestly	4800
Unfigured	51	Ungorgeous	150	Unhonoured	4900
Unfilial	52	Ungotten	200	Unhoped	5000
Unfilially	53	Ungoverned	250	Unhopeful	5500
Unfinished	54	Ungown	300	Unhorse	6000
Unfirm	55	Ungraced	350	Unhospital	6500
Unfix	56	Ungraceful	400	Unhostile	7000
Unfixing	57	Ungracious	500	Unhumanize	7500
Unflagging	58	Ungravely	600	Unhumbled	8000
Unfledged	59	Ungreable	700	Uniate	9000
Unfleshing	60	Unground	800	Uniaxal	10,000
Unfleshly	61	Ungrudging	900	Unicameral	11,000
Unflower	62	Ungual	1000	Unicity	12,000
Unfluent	63	Unguarded	1100	Uniclinal	13,000
Unforbade	64	Unguentary	1200	Unicorn	14,000
Unforeseen	65	Unguentous	1300	Unicornous	15,000
Unforetold	66	Unguical	1400	Unicostate	16,000
Unforgiven	67	Unguicular	1500	Unideal	17,000
Unforgot	68	Unguidably	1600	Unifacial	18,000
Unforsaken	69	Unguiform	1700	Unific	19,000
Unframed	70	Unhabile	1800	Unifilar	20,000

Code Word.	Assays (or Assaying).	Code Word.	Assay (or Assaying).
Uniflorous	1 dwt. per ton	Universe	4 ozs. 15 dwts. per ton
Unifoil	2 ,, ,,	University	5 ,, 0 ,, ,,
Unifoliate	3 ,, ,,	Univocal	5 ,, 5 ,, ,,
Uniformity	4 ,, ,,	Univocally	5 ,, 10 ,, ,,
Uniforms	5 ,, ,,	Unjealous	5 ,, 15 ,, ,,
Unigenous	6 ,, ,,	Unjointing	6 ,, 0 ,, ,,
Unijugate	7 ,, ,,	Unjoyful	6 ,, 5 ,, ,,
Unilabiate	8 ,, ,,	Unjoyous	6 ,, 10 ,, ,,
Unilateral	9 ,, ,,	Unjustly	6 ,, 15 ,, ,,
Unillumed	10 ,, ,,	Unjustness	7 ,, 0 ,, ,,
Unillusory	11 ,, ,,	Unkempt	7 ,, 5 ,, ,,
Unilocular	12 ,, ,,	Unkindly	7 ,, 10 ,, ,,
Unimagined	13 ,, ,,	Unkindness	7 ,, 15 ,, ,,
Unimmortal	14 ,, ,,	Unkindred	8 ,, 0 ,, ,,
Unimparted	15 ,, ,,	Unkinglike	8 ,, 5 ,, ,,
Unimposing	16 ,, ,,	Unkingship	8 ,, 10 ,, ,,
Unimproved	17 ,, ,,	Unknelled	8 ,, 15 ,, ,,
Uninclosed	18 ,, ,,	Unknightly	9 ,, 0 ,, ,,
Unindented	19 ,, ,,	Unknotty	9 ,, 5 ,, ,,
Uninflamed	20 ,, ,,	Unknow	9 ,, 10 ,, ,,
Uninspired	21 ,, ,,	Unknowable	9 ,, 15 ,, ,,
Unintitled	22 ,, ,,	Unknowing	10 ,, 0 ,, ,,
Uninvented	23 ,, ,,	Unlaboured	10 ,, 10 ,, ,,
Uninvite	24 ,, ,,	Unladylike	11 ,, 0 ,, ,,
Uninviting	25 ,, ,,	Unlapping	11 ,, 10 ,, ,,
Uninvolved	26 ,, ,,	Unlashing	12 ,, 0 ,, ,,
Union	27 ,, ,,	Unlavished	12 ,, 10 ,, ,,
Unionidae	28 ,, ,,	Unlawful	13 ,, 0 ,, ,,
Unionism	29 ,, ,,	Unlawfully	13 ,, 10 ,, ,,
Unionistic	30 ,, ,,	Unleashed	14 ,, 0 ,, ,,
Unionjack	31 ,, ,,	Unlectured	14 ,, 10 ,, ,,
Unionjoint	32 ,, ,,	Unlettered	15 ,, 0 ,, ,,
Uniparous	33 ,, ,,	Unlicensed	15 ,, 10 ,, ,,
Uniplicate	34 ,, ,,	Unlikely	16 ,, 0 ,, ,,
Unique	35 ,, ,,	Unlikeness	16 ,, 10 ,, ,,
Uniquely	36 ,, ,,	Unlimber	17 ,, 0 ,, ,,
Uniqueness	37 ,, ,,	Unlimited	17 ,, 10 ,, ,,
Uniseptate	38 ,, ,,	Unlineal	18 ,, 0 ,, ,,
Uniserial	39 ,, ,,	Unliquored	18 ,, 10 ,, ,,
Unisonance	2 ozs. 0 dwts. per ton	Unload	19 ,, 0 ,, ,,
Unisonous	2 ,, 5 ,, ,,	Unloading	19 ,, 10 ,, ,,
Unitable	2 ,, 10 ,, ,,	Unlocated	20 ,, 0 ,, ,,
Unitarian	2 ,, 15 ,, ,,	Unlocketh	20 ,, 10 ,, ,,
Unitary	3 ,, 0 ,, ,,	Unlodge	21 ,, 0 ,, ,,
Unitedly	3 ,, 5 ,, ,,	Unlogical	21 ,, 10 ,, ,,
Unition	3 ,, 10 ,, ,,	Unloosened	22 ,, 0 ,, ,,
Unitively	3 ,, 15 ,, ,,	Unloving	22 ,, 10 ,, ,,
Unitizing	4 ,, 0 ,, ,,	Unlucent	23 ,, 0 ,, ,,
Univalent	4 ,, 5 ,, ,,	Unluckiest	23 ,, 10 ,, ,,
Universal	4 ,, 10 ,, ,,	Unlucky	24 ,, 0 ,, ,,

Code Word.	Assays (or Assaying).	Code Word.	Assays (or Assaying).
Unluminous	24 ozs. 10 dwts. per ton	Unnamed	150 to 160 ozs. per ton
Unlustrous	25 ,, 0 ,, ,,	Unnative	160 ,, ,,
Unlute	25 ,, 10 ,, ,,	Unnatural	160 to 170 ,, ,,
Unmade	26 ,, 0 ,, ,,	Unnaturing	170 ,, ,,
Unmaidenly	26 ,, 10 ,, ,,	Unnervate	170 to 180 ,, ,,
Unmakable	27 ,, 0 ,, ,,	Unnestled	180 ,, ,,
Unmanacle	27 ,, 10 ,, ,,	Unnestling	180 to 190 ,, ,,
Unmanaged	28 ,, 0 ,, ,,	Unniggard	190 ,, ,,
Unmanlike	28 ,, 10 ,, ,,	Unnobly	190 to 200 ,, ,,
Unmarked	29 ,, 0 ,, ,,	Unnoticed	200 ,, ,,
Unmarried	29 ,, 10 ,, ,,	Unobscured	200 to 250 ,, ,,
Unmarrying	30 ,, 0 ,, ,,	Unobserved	250 ,, ,,
Unmartyr	30 to 35 ozs. ,,	Unobvious	250 to 300 ,, ,,
Unmasking	35 ,,	Unoccupied	300 ,, ,,
Unmatched	35 to 40 ,, ,,	Unoften	300 to 350 ,, ,,
Unmeaning	40 ,, ,,	Unopposed	350 ,, ,,
Unmeant	40 to 45 ,, ,,	Unorder	350 to 400 ,, ,,
Unmeasured	45 ,, ,,	Unordering	400 ,, ,,
Unmeetly	45 to 50 ,, ,,	Unorderly	400 to 450 ,, ,,
Unmeetness	50 ,, ,,	Unoriginal	450 ,, ,,
Unmellowed	50 to 55 ,, ,,	Unorthodox	450 to 500 ,, ,,
Unmerciful	55 ,, ,,	Unowned	500 ,, ,,
Unmerited	55 to 60 ,, ,,	Unpack	500 to 600 ,, ,,
Unmew	60 ,, ,,	Unpacking	600 ,, ,,
Unminded	60 to 65 ,, ,,	Unpaint	600 to 700 ,, ,,
Unmindful	65 ,, ,,	Unparadise	700 ,, ,,
Unmiry	65 to 70 ,, ,,	Unparroted	700 to 800 ,, ,,
Unmitre	70 ,, ,,	Unpastor	800 ,, ,,
Unmitring	70 to 75 ,, ,,	Unpay	800 to 900 ,, ,,
Unmixed	75 ,, ,,	Unpayable	900 ,, ,,
Unmodified	75 to 80 ,, ,,	Unpeace	900 to 1000 ,, ,,
Unmodishly	80 ,, ,,	Unpeaceful	1000 ,, ,,
Unmoist	80 to 85 ,, ,,	Unpeerable	1000 to 1100 ,, ,,
Unmolested	85 ,, ,,	Unpen	1100 ,, ,,
Unmoneyed	85 to 90 ,, ,,	Unpenitent	1100 to 1200 ,, ,,
Unmonkish	90 ,, ,,	Unpenning	1200 ,, ,,
Unmoor	90 to 95 ,, ,,	Unpeople	1200 to 1300 ,, ,,
Unmooring	95 ,, ,,	Unpeopling	1300 ,, ,,
Unmortised	95 to 100 ,, ,,	Unperegal	1300 to 1400 ,, ,,
Unmosaic	100 ,, ,,	Unperjured	1400 ,, ,,
Unmotherly	100 to 110 ,, ,,	Unperplex	1400 to 1500 ,, ,,
Unmoulded	110 ,, ,,	Unpervert	1500 ,, ,,
Unmoulding	110 to 120 ,, ,,	Unpierced	1500 to 1600 ,, ,,
Unmovable	120 ,, ,,	Unpillared	1600 ,, ,,
Unmuffling	120 to 130 ,, ,,	Unpiloted	1600 to 1700 ,, ,,
Unmurmured	130 ,, ,,	Unpinion	1700 ,, ,,
Unmuscular	130 to 140 ,, ,,	Unpitied	1700 to 1800 ,, ,,
Unmusical	140 ,, ,,	Unpitiful	1800 ,, ,,
Unmuzzled	140 to 150 ,, ,,	Unpitous	1800 to 1900 ,, ,,
Unmystery	150 ,, ,,	Unpitying	2000 ,, ,,

Code Word.			Code Word.		
Unplagued	% loss		Unrazored	11 % loss	
Unplausive	% profit		Unreality	11½	,,
Unpleasant	% premium		Unrealized	12	,,
Unpliant	% discount		Unreason	12½	,,
Unplucked	% higher than		Unrecuring	13	,,
Unpoetic	% lower than		Unredeemed	13½	,,
Unpoetical	2½ % higher		Unrefined	14	,,
Unpoisoned	5	,,	Unreformed	14½	,,
Unpolicied	7½	,,	Unregarded	15	,,
Unpolish	10	,,	Unrejoiced	15½	,,
Unpolitely	12½	,,	Unrelated	16	,,
Unpolitic	15	,,	Unrelative	16½	,,
Unpolluted	17½	,,	Unremedied	17	,,
Unpopular	20	,,	Unremitted	17½	,,
Unposted	2½ % lower		Unremoved	18	,,
Unpowerful	5	,,	Unrenewed	18½	,,
Unpraising	7½	,,	Unrepaid	19	,,
Unpreached	10	,,	Unrepealed	19½	,,
Unpredict	12½	,,	Unrequited	20	,,
Unprepared	15	,,	Unresisted	20½	,,
Unpressed	17½	,,	Unresolve	21	,,
Unpretty	20	,,	Unrespited	21½	,,
Unpriest	1/16 % loss		Unrestful	22	,,
Unpriestly	1/8	,,	Unresty	22½	,,
Unprince	1/4	,,	Unreverend	23	,,
Unprincely	3/8	,,	Unrevoked	23½	,,
Unprincing	1/2	,,	Unriddle	24	,,
Unprizable	5/8	,,	Unrigging	24½	,,
Unprized	3/4	,,	Unrighting	25	,,
Unprolific	7/8	,,	Unrioted	25½	,,
Unpromise	1	,,	Unripened	26	,,
Unprompted	1½	,,	Unrobed	26½	,,
Unproposed	2	,,	Unrobing	27	,,
Unpropped	2½	,,	Unromantic	27½	,,
Unprovided	3	,,	Unroofing	28	,,
Unpublic	3½	,,	Unroyal	28½	,,
Unpunctual	4	,,	Unruffled	29	,,
Unpunished	4½	,,	Unruinable	29½	,,
Unpurged	5	,,	Unruinated	30	,,
Unpurified	5½	,,	Unruliment	30½	,,
Unqualify	6	,,	Unruliness	31	,,
Unqueen	6½	,,	Unrumple	31½	,,
Unquelled	7	,,	Unrumpling	32	,,
Unquick	7½	,,	Unsaddens	32½	,,
Unquieting	8	,,	Unsaddle	33	,,
Unquietly	8½	,,	Unsaddling	33½	,,
Unquietude	9	,,	Unsadness	34	,,
Unraised	9½	,,	Unsafely	35	,,
Unraptured	10	,,	Unsalaried	36	,,
Unravelled	10½	,,	Unsanitary	37	,,

Code Word.			Code Word.		
Unsavoury	38	% loss	Unsluice	15½	% profit
Unsaying	39	,,	Unsluicing	16	,,
Unscanned	40	,,	Unsmirched	16½	,,
Unschooled	41	,,	Unsmitten	17	,,
Unscience	42	,,	Unsmoked	17½	,,
Unscorched	43	,,	Unsmooth	18	,,
Unscoured	44	,,	Unsociably	18½	,,
Unscreened	45	,,	Unsocial	19	,,
Unscrew	46	,,	Unsolders	19½	,,
Unscrewing	47	,,	Unsolemn	20	,,
Unseasoned	48	,,	Unsolid	20½	,,
Unseconded	49	,,	Unsoundly	21	,,
Unsecret	50	,,	Unspeak	21½	,,
Unsecular	1/16	% profit	Unspeaking	22	,,
Unseduced	1/8	,,	Unspeedy	22½	,,
Unseemly	1/4	,,	Unspike	23	,,
Unseized	3/8	,,	Unspirited	23½	,,
Unseldom	1/2	,,	Unspoil	24	,,
Unselfish	5/8	,,	Unsportful	24½	,,
Unservice	3/4	,,	Unspotted	25	,,
Unsettling	7/8	,,	Unsquared	25½	,,
Unsevered	1	,,	Unsqueezed	26	,,
Unsexed	1½	,,	Unsquiring	26½	,,
Unshackle	2	,,	Unstabled	27	,,
Unshadowed	2½	,,	Unstarch	27½	,,
Unshakable	3	,,	Unstartled	28	,,
Unshape	3½	,,	Unstating	28½	,,
Unsheathed	4	,,	Unsteadily	29	,,
Unshed	4½	,,	Unstirred	29½	,,
Unshielded	5	,,	Unstooping	30	,,
Unshook	5½	,,	Unstop	30½	,,
Unshowered	6	,,	Unstopped	31	,,
Unshriven	6½	,,	Unstormed	31½	,,
Unshrouded	7	,,	Unstrewed	32	,,
Unshrubbed	7½	,,	Unstringed	32½	,,
Unshut	8	,,	Unstruck	33	,,
Unshutting	8½	,,	Unstudied	33½	,,
Unsightly	9	,,	Unsubdued	34	,,
Unsimple	9½	,,	Unsubject	35	,,
Unsinew	10	,,	Unsuccess	36	,,
Unsinewing	10½	,,	Unsucked	37	,,
Unsinned	11	,,	Unsuitably	38	,,
Unsisterly	11½	,,	Unsuited	39	,,
Unsizable	12	,,	Unsuiting	40	,,
Unskilful	12½	,,	Unsullied	41	,,
Unskill	13	,,	Unsunny	42	,,
Unslain	13½	,,	Unsupple	43	,,
Unslaked	14	,,	Unsupplied	44	,,
Unsleek	14½	,,	Unsuspect	45	,,
Unsling	15	,,	Unswayed	46	,,

Code Word.		Code Word.	
Unswept	47 % profit	Unturn	11 per cent
Unswerving	48 ,,	Untutored	$11\frac{1}{2}$,, ,,
Unswilled	49 ,,	Untwisting	12 ,, ,,
Unsworn	50 ,,	Ununited	$12\frac{1}{2}$,, ,,
Untaken	$\frac{1}{16}$ per cent	Unusage	13 ,, ,,
Untalented	$\frac{1}{8}$,, ,,	Unusual	$13\frac{1}{2}$,, ,,
Untappice	$\frac{3}{16}$,, ,,	Unusually	14 ,, ,,
Untasted	$\frac{1}{4}$,, ,,	Unuttered	$14\frac{1}{2}$,, ,,
Unteach	$\frac{5}{16}$,, ,,	Unvalued	15 ,, ,,
Unteaching	$\frac{3}{8}$,, ,,	Unvarying	$15\frac{1}{2}$,, ,,
Untemper	$\frac{1}{2}$,, ,,	Unveiledly	16 ,, ,,
Untenant	$\frac{5}{8}$,, ,,	Unveileth	$16\frac{1}{2}$,, ,,
Untendered	$\frac{3}{4}$,, ,,	Unvenomous	17 ,, ,,
Untearific	$\frac{7}{8}$,, ,,	Unveracity	$17\frac{1}{2}$,, ,,
Unthankful	1 ,, ,,	Unverdant	18 ,, ,,
Unthawed	$1\frac{1}{8}$,, ,,	Unvicar	$18\frac{1}{2}$,, ,,
Unthinker	$1\frac{1}{4}$,, ,,	Unvicaring	19 ,, ,,
Unthinking	$1\frac{3}{8}$,, ,,	Unviolable	$19\frac{1}{2}$,, ,,
Unthorny	$1\frac{1}{2}$,, ,,	Unvirtuous	20 ,, ,,
Unthread	$1\frac{5}{8}$,, ,,	Unvitiated	$20\frac{1}{2}$,, ,,
Unthrift	$1\frac{3}{4}$,, ,,	Unvizard	21 ,, ,,
Unthroned	$1\frac{7}{8}$,, ,,	Unvote	$21\frac{1}{2}$,, ,,
Untidiest	2 ,, ,,	Unvulgar	22 ,, ,,
Untidy	$2\frac{1}{8}$,, ,,	Unwarily	$22\frac{1}{2}$,, ,,
Untillable	$2\frac{1}{4}$,, ,,	Unwariness	23 ,, ,,
Untimely	$2\frac{3}{8}$,, ,,	Unwatchful	$23\frac{1}{2}$,, ,,
Untimeous	$2\frac{1}{2}$,, ,,	Unwavering	24 ,, ,,
Untoiling	$2\frac{5}{8}$,, ,,	Unweary	$24\frac{1}{2}$,, ,,
Untold	$2\frac{3}{4}$,, ,,	Unwearying	25 ,, ,,
Untombed	$2\frac{7}{8}$,, ,,	Unwebbed	$25\frac{1}{2}$,, ,,
Untongue	3 ,, ,,	Unweighed	26 ,, ,,
Untonguing	$3\frac{1}{4}$,, ,,	Unwelcome	$26\frac{1}{2}$,, ,,
Untooth	$3\frac{1}{2}$,, ,,	Unwhipped	27 ,, ,,
Untoothing	$3\frac{3}{4}$,, ,,	Unwhole	$27\frac{1}{2}$,, ,,
Untouched	4 ,, ,,	Unwieldy	28 ,, ,,
Untoward	$4\frac{1}{4}$,, ,,	Unwilling	$28\frac{1}{2}$,, ,,
Untowardly	$4\frac{1}{2}$,, ,,	Unwisdom	29 ,, ,,
Untrading	$4\frac{3}{4}$,, ,,	Unwise	$29\frac{1}{2}$,, ,,
Untragic	5 ,, ,,	Unwisely	30 ,, ,,
Untrampled	$5\frac{1}{2}$,, ,,	Unwished	$30\frac{1}{2}$,, ,,
Untrifling	6 ,, ,,	Unwithered	31 ,, ,,
Untrimmed	$6\frac{1}{2}$,, ,,	Unwittily	$31\frac{1}{2}$,, ,,
Untrodden	7 ,, ,,	Unwomaning	32 ,, ,,
Untrouble	$7\frac{1}{2}$,, ,,	Unwomanly	$32\frac{1}{2}$,, ,,
Untruisms	8 ,, ,,	Unwonder	33 ,, ,,
Untruly	$8\frac{1}{2}$,, ,,	Unwontedly	$33\frac{1}{2}$,, ,,
Untrussers	9 ,, ,,	Unworldly	34 ,, ,,
Untruthful	$9\frac{1}{2}$,, ,,	Unworth	$34\frac{1}{2}$,, ,,
Untruths	10 ,, ,,	Unworthily	35 ,, ,,
Unturbaned	$10\frac{1}{2}$,, ,,	Unwrapping	$35\frac{1}{2}$,, ,,

Code Word.			Code Word.		
Unwrinkle	36 per cent		Uplooking	170 per cent	
Unwrung	36½ ,,	,,	Upper	180 ,,	,,
Unyielding	37 ,,	,,	Upperhand	190 ,,	,,
Unyoked	37½ ,,	,,	Uppermost	200 ,,	,,
Unyoking	38 ,,	,,	Upperworld	225 ,,	,,
Upas	38½ ,,	,,	Uppishness	250 ,,	,,
Upastree	39 ,,	,,	Upplough	275 ,,	,,
Upbind	39½ ,,	,,	Uppluck	300 ,,	,,
Upblow	40 ,,	,,	Uppricked	325 ,,	,,
Upborne	40½ ,,	,,	Upprop	350 ,,	,,
Upbraideth	41 ,,	,,	Uppropping	375 ,,	,,
Upbraiding	41½ ,,	,,	Upputting	400 ,,	,,
Upbreak	42 ,,	,,	Uprearing	450 ,,	,,
Upbreaking	42½ ,,	,,	Upridge	500 ,,	,,
Upbringing	43 ,,	,,	Upright	600 ,,	,,
Upbrought	43½ ,,	,,	Uprightly	700 ,,	,,
Upbuoyance	44 ,,	,,	Uprisings	800 ,,	,,
Upburst	44½ ,,	,,	Uproar	900 ,,	,,
Upcast	45 ,,	,,	Uproarious	1000 ,,	,,
Upcaught	45½ ,,	,,	Uprooted		
Upcheer	46 ,,	,,	Uproused		
Upcheering	46½ ,,	,,	Uprush		
Upclimb	47 ,,	,,	Uprushing		
Upclimbing	47½ ,,	,,	Upseek		
Upcoiled	48 ,,	,,	Upsend		
Upcoiling	48½ ,,	,,	Upsetteth		
Updrawn	49 ,,	,,	Upsetting		
Upfilled	49½ ,,	,,	Upshoot		
Upflowing	50 ,,	,,	Upshooting		
Upgather	55 ,,	,,	Upskip		
Upgaze	60 ,,	,,	Upsoared		
Upgazing	65 ,,	,,	Upsoaring		
Upgrowth	70 ,,	,,	Upspear		
Uphang	75 ,,	,,	Upsprung		
Uphanging	80 ,,	,,	Upstairs		
Upheaped	85 ,,	,,	Upstand		
Upheavals	90 ,,	,,	Upstanding		
Uphill	95 ,,	,,	Upstir		
Uphoarded	100 ,,	,,	Upstroke		
Upholdeth	105 ,,	,,	Upswarm		
Upholding	110 ,,	,,	Upswell		
Upholstery	115 ,,	,,	Upswelling		
Uplander	120 ,,	,,	Uptake		
Uplandish	125 ,,	,,	Uptaking		
Uplay	130 ,,	,,	Uptear		
Uplaying	135 ,,	,,	Upthrow		
Upled	140 ,,	,,	Upthrowing		
Uplifting	145 ,,	,,	Upthunder		
Uplocked	150 ,,	,,	Uptown		
Uplook	160 ,,	,,	Uptrace		

JANUARY.	FEBRUARY.	MARCH.	
Code Word.	Code Word.	Code Word.	
Uptracing	Usable	Vacated	——
Uptrained	Usager	Vacating	1st of
Upturning	Usance	Vaccinator	2nd of
Upupa	Usefully	Vaccinists	3rd of
Upupidae	Uselessly	Vaccinium	4th of
Upwafted	Usherance	Vachery	5th of
Upwards	Usherdom	Vacillancy	6th of
Upwhirl	Ushereth	Vacillated	7th of
Upwhirling	Ushership	Vacoa	8th of
Upwinding	Usnea	Vacuity	9th of
Uraemia	Usquebaugh	Vacuolated	10th of
Uralian	Ustilago	Vacuumpan	11th of
Uranic	Ustulation	Vacuumpump	12th of
Uranium	Usualness	Vacuums	13th of
Uranmica	Usucaption	Vacuumtube	14th of
Uranochre	Usufruct	Vademecum	15th of
Uranolite	Usurarious	Vadimony	16th of
Uranology	Usuring	Vagabond	17th of
Uranoscopy	Usuriously	Vagabondry	18th of
Uranous	Usurp	Vagal	19th of
Uranutan	Usurpant	Vagantes	20th of
Urban	Usurpation	Vagarish	21st of
Urbanists	Usurpatory	Vaginant	22nd of
Urbanity	Usurpature	Vaginule	23rd of
Urbanized	Usurpers	Vagrantly	24th of
Urbanizing	Usurpingly	Vaguely	25th of
Urbiculous	Utensils	Vagueness	26th of
Urceola	Utilities	Vaguest	27th of
Urceolaria	Utility	Vainglory	28th of
Urchin	Utilized	Valentines	29th of
Uredo	Utilizing	Valeriana	30th of
Urged	Utopian	Valeric	31st of
Urgently	Utopianism	Valerole	For the month of
Urgewonder	Utraquist	Valet	During the month of
Urginea	Utricle	Valhalla	Early in
Urinary	Utriculate	Validated	About the middle of
Urinators	Utriculoid	Validating	Towards the end of
Urinometer	Utterable	Valinch	The first week in
Urnful	Utterances	Valises	The second week in
Urodela	Utterly	Valkyrian	The third week in
Urogenital	Uveous	Vallatory	The fourth week in
Uromastix	Uvrou	Valleys	The last week in
Uroplania	Uvularly	Vallicula	From 1st to 15th
Urox	Uwarowite	Valonia	From 15th to 30th (31st)
Ursidae	Uxorial	Valorous	From 1st to 10th
Ursuline	Uxoricide	Valorously	From 10th to 20th
Ursus	Uxorious	Valour	From 20th to 30th (31st)
Urtica	Vaagmar	Valuable	For first half of
Urticaria	Vacancies	Valuators	For second half of
Urtication	Vacancy	Valueless	Month(s) ending 30th (31st)

APRIL.	MAY.	JUNE.	
Code Word.	Code Word.	Code Word.	——
Valvasor	Variety	Veiled	
Valvecage	Variformed	Veilless	1st of
Valvegear	Variola	Veinlets	2nd of
Valveshell	Variolaria	Veinstone	3rd of
Valvestems	Variolitic	Veinstuff	4th of
Valvular	Varioloid	Veiny	5th of
Vambrace	Variorum	Velar	6th of
Vampirebat	Varix	Velarium	7th of
Vampirism	Vartabed	Velatura	8th of
Vamplet	Varvicite	Velella	9th of
Vanadous	Vascular	Velellidae	10th of
Vancourier	Vasculares	Veliferous	11th of
Vanda	Vasculose	Velivolant	12th of
Vandalism	Vasculum	Vellicate	13th of
Vandellia	Vaseline	Vellon	14th of
Vandyke	Vasiform	Velloped	15th of
Vandyking	Vasomotor	Vellozia	16th of
Vanellus	Vassalages	Vellumy	17th of
Vanessa	Vassalry	Veloce	18th of
Vanfoss	Vastate	Velociman	19th of
Vangloe	Vastators	Velocipede	20th of
Vanguards	Vastitude	Velumen	21st of
Vanillin	Vaticanism	Velveteen	22nd of
Vanished	Vaticide	Velvetleaf	23rd of
Vanishment	Vaticinal	Velvetmoss	24th of
Vanquished	Vatted	Velvetpile	25th of
Vapidly	Vatting	Velvety	26th of
Vap	Vault	Venalities	27th of
Vaporable	Vaultage	Venality	28th of
Vaporation	Vaulting	Venatorial	29th of
Vaporific	Vaunteth	Vendace	30th of
Vaporized	Vauntful	Vendetta	31st of
Vaporizing	Vauntingly	Vendibly	For the month of
Vaporose	Vauntmure	Vendition	During the month of
Vapourbath	Vauqueline	Vendors	Early in
Vapoured	Vavasory	Veneered	About the middle of
Vapouring	Vedanga	Veneering	Towards the end of
Vapourish	Vedantists	Veneermoth	The first week in
Vaquero	Vedettes	Venefical	The second week in
Varangian	Vegetables	Venenosa	The third week in
Variably	Vegeta	Venerably	The fourth week in
Variance	Vegetality	Veneraceae	The last week in
Variations	Vegetarian	Venerate	From 1st to 15th
Varicella	Vegetating	Veneration	From 15th to 30th (31st)
Variciform	Vegetative	Venereous	From 1st to 10th
Varicocele	Vegetous	Venetians	From 10th to 20th
Varicose	Vehemently	Veneys	From 20th to 30th (31st)
Varicosity	Vehicled	Vengeance	For first half of
Varied	Vehiculary	Vengeful	For second half of
Variegate	Vehmic	Vengefully	Month(s) ending 30th (31st)

JULY.	AUGUST.	SEPTEMBER.	
Code Word.	Code Word.	Code Word.	
Vengement	Verjuice	Vestryman	————
Venison	Vermelet	Vestryroom	1st of
Venomed	Vermetus	Vesuvians	2nd of
Venomous	Vermicelli	Vetchling	3rd of
Venomously	Vermiceous	Vetchy	4th of
Venosity	Vermicide	Veterinary	5th of
Ventanna	Vermiform	Vetiver	6th of
Ventbit	Vermifugal	Veto	7th of
Ventfield	Vermilion	Vetoist	8th of
Ventiduct	Vermin	Vexatious	9th of
Ventilator	Vermuth	Vexillary	10th of
Ventousing	Vernacular	Vexillum	11th of
Ventpegs	Vernage	Vexing	12th of
Ventpiece	Vernicle	Vexingly	13th of
Ventplug	Vernier	Viable	14th of
Ventricose	Vernonia	Viaduct	15th of
Venturers	Verruca	Viand	16th of
Venulose	Verrucidae	Viandry	17th of
Verandah	Versatile	Viatecture	18th of
Veratric	Versemaker	Viatic	19th of
Verb	Verseman	Viaticum	20th of
Verbal	Versified	Viators	21st of
Verbalisms	Versify	Vibex	22nd of
Verbalized	Versifying	Vibices	23rd of
Verbascum	Versionist	Vibraculum	24th of
Verbatim	Vertebra	Vibrant	25th of
Verbena	Vertexes	Vibrating	26th of
Verberated	Verticity	Vibrations	27th of
Verbiage	Vertumnus	Vibratory	28th of
Verbose	Vervain	Vibrio	29th of
Verbosely	Vesania	Vibrissae	30th of
Verdancy	Vesicated	Vibriscope	31st of
Verdant	Vesicating	Viburnum	For the month of
Verdeawine	Vesicatory	Vicarage	During the month of
Verderor	Vesiculate	Vicarial	Early in
Verdict	Vesiculous	Vicarious	About the middle of
Verdigris	Vesperal	Vicarship	Towards the end of
Verditure	Vesperbell	Vicebitten	The first week in
Verdoy	Vespertine	Viceconsul	The second week in
Verdured	Vespiary	Vicegerent	The third week in
Verdurous	Vespidae	Viceking	The fourth week in
Verecund	Vespillo	Vicenary	The last week in
Vergaloo	Vessels	Vicennial	From 1st to 15th
Vergeboard	Vessignon	Viceregal	From 15th to 30th (31st)
Vergency	Vestiarian	Vicinity	From 1st to 10th
Vergers	Vestibule	Viciously	From 10th to 20th
Veridical	Vestiges	Vicontiel	From 20th to 30th (31st)
Veriest	Vestments	Victim	For first half of
Verifiable	Vestries	Victimate	For second half of
Verifiers	Vestry	Victimized	Month(s) ending 30th (31st)

OCTOBER.	NOVEMBER.	DECEMBER.	
Code Word.	Code Word.	Code Word.	
Victorine	Violaceae	Visitation	——
Victory	Violaceous	Visitress	1st of
Victualled	Violascent	Vismia	2nd of
Vicugna	Violating	Visnomy	3rd of
Vidame	Violation	Vistas	4th of
Videlicet	Violators	Visuality	5th of
View	Violently	Visualized	6th of
Viewhalloo	Violetwood	Vital	7th of
Viewing	Violinist	Vitalism	8th of
Viewless	Violone	Vitalized	9th of
Viewly	Viper	Vitalizing	10th of
Vigesimal	Viperina	Vitally	11th of
Vigilantly	Viperish	Vitellary	12th of
Vigils	Viperous	Vitellicle	13th of
Vignette	Viraginity	Vitiated	14th of
Vigour	Virago	Vitiating	15th of
Vile	Virent	Viticula	16th of
Viliaco	Vireonidae	Vitrify	17th of
Vilified	Virgated	Vitriol	18th of
Vilipend	Virgilia	Vitriolize	19th of
Villagery	Virgin	Vitriolous	20th of
Villainize	Virginal	Vitrotype	21st of
Villainous	Virginborn	Vitruvian	22nd of
Villainy	Virginhood	Vituline	23rd of
Villakin	Virgoleuse	Vituperate	24th of
Villanage	Virgularia	Vivace	25th of
Villarsia	Virgulate	Vivacity	26th of
Villatic	Viripotent	Vivandière	27th of
Vinaigrous	Viroled	Vivarium	28th of
Vinatico	Virtueless	Vivency	29th of
Vincentian	Virtuosity	Viverra	30th of
Vincible	Virtuous	Viverridae	31st of
Vincture	Virulency	Vivianite	For the month of
Vinculum	Virulent	Vivid	During the month of
Vindemial	Virus	Vivifical	Early in
Vindicable	Visaged	Vivifying	About the middle of
Vindicator	Visaging	Viviparity	Towards the end of
Vindictive	Visavis	Viviparous	The first week in
Vineal	Viscach	Vixen	The second week in
Vinegar	Visceraa	Vixenish	The third week in
Vinegareel	Viscerated	Vixenly	The fourth week in
Vinegrub	Viscosity	Vizament	The last week in
Vinemildew	Viscum	Vizierial	From 1st to 15th
Vineyards	Visibly	Vizoring	From 15th to 30th (31st)
Viniferae	Visigothic	Vlackevark	From 1st to 10th
Vinometer	Vision	Vocable	From 10th to 20th
Vintagers	Visionary	Vocabulary	From 20th to 30th (31st)
Vintnery	Visionists	Vocabulist	For first half of
Viola	Visitable	Vocalic	For second half of
Violable	Visitant	Vocality	Month(s) ending 30th (31st)

OUR LETTER	YOUR LETTER	HAVE RECEIVED YOUR LETTER	
Code Word.	Code Word.	Code Word.	
Vocalized	Voluminous	Voyaging	Of the ——
Vocalness	Volumists	Vugg	,, 1st
Vocations	Volunteers	Vulcanizer	,, 2nd
Vocative	Voluptuary	Vulgar	,, 3rd
Vochyaceae	Voluptuous	Vulgarisms	,, 4th
Vociferant	Voluspa	Vulgarity	,, 5th
Vociferous	Volutidae	Vulgarness	,, 6th
Voglite	Volvox	Vulnerable	,, 7th
Vogue	Volvulus	Vulnerary	,, 8th
Voiced	Vomerine	Vulnerose	,, 9th
Voiceful	Vomica	Vulnific	,, 10th
Voiceless	Vomicnut	Vulnifical	,, 11th
Voicing	Vomited	Vulpecular	,, 12th
Voiders	Vomiting	Vulpes	,, 13th
Voidness	Vomitory	Vulpicide	,, 14th
Voiture	Voracious	Vulpinism	,, 15th
Volatility	Voracity	Vulturish	,, 16th
Volatilize	Vortex	Vulturous	,, 17th
Volauvent	Vortexring	Vulviform	,, 18th
Volcanian	Vortically	Wabbly	,, 19th
Volcanoes	Vorticella	Wabronleaf	,, 20th
Volitient	Votarists	Wadded	,, 21st
Volitional	Votary	Waddling	,, 22nd
Volkameria	Votively	Waddlingly	,, 23rd
Volleying	Vouchers	Wadhook	,, 24th
Voltaic	Vouchment	Wadingbird	,, 25th
Voltairism	Vouchsafe	Wadsetter	,, 26th
Voltameter	Voussoir	Waferers	,, 27th
Voltaplast	Vowbreak	Waferirons	,, 28th
Voltatype	Vowelish	Waferwoman	,, 29th
Voltigeurs	Vowfellows	Wafture	,, 30th
Voltzia	Voxhumana	Wagering	,, 31st
Volubilate	Voyage	Wagework	Referring to
Volubly	Voyageable	Waggoner	In answer to
Volumetric	Voyageurs	Waghalter	

REFER TO OUR LETTER	REFER TO YOUR LETTER	HAVE YOU RECEIVED OUR LETTER	
Code Word.	Code Word.	Code Word.	
Wagnerite	Wallerite	Wapatoo	Of the ——
Wagon	Walleteer	Wapenshaw	„ 1st
Wagonage	Walleyed	Wapentake	„ 2nd
Wagonettes	Wallfruit	Wapp	„ 3rd
Wagontrain	Wallmoss	Warblers	„ 4th
Wagtails	Wallnewts	Warblingly	„ 5th
Waif	Wallopped	Warcraft	„ 6th
Wainbote	Wallpapers	Warcries	„ 7th
Wainhouse	Wallpepper	Warcry	„ 8th
Wainman	Wallpie	Wardance	„ 9th
Wainropes	Wallplate	Wardcorn	„ 10th
Wainscot	Wallrocket	Wardecorps	„ 11th
Waistband	Wallsided	Wardenpie	„ 12th
Waistbelt	Walltent	Wardenry	„ 13th
Waistcoat	Walltrees	Wardmote	„ 14th
Waitingly	Wallwort	Wardpenny	„ 15th
Waitress	Walnutoil	Wardrobes	„ 16th
Waived	Walnuts	Wardroom	„ 17th
Waiwode	Waltron	Wardship	„ 18th
Wakeful	Waltzer	Wardsman	„ 19th
Wakefully	Waltzing	Wardstaff	„ 20th
Wakened	Wambais	Warehouses	„ 21st
Wakerife	Wampum	Warfare	„ 22nd
Wakerobin	Wandereth	Warfaring	„ 23rd
Walchowite	Wandering	Warfield	„ 24th
Waldgrave	Wanderoo	Warflame	„ 25th
Waleknot	Wanghee	Wargarron	„ 26th
Walepiece	Wangtooth	Wargear	„ 27th
Walk	Wanhope	Warhorse	„ 28th
Walkable	Wantonings	Wariangle	„ 29th
Walking	Wantonize	Wariment	„ 30th
Walkmill	Wantonly	Warish	„ 31st
Wallaba	Wantonness	Warlike	Referring to
Wallbox	Wantwit	Warlockry	In answer to
Wallcress	Wapacut	Warmarked	

Our Cable	Your Cable	Have received your Cable	
Code Word.	Code Word.	Code Word.	
Warmheaded	Waspflies	Waterborne	Of the ——
Warmingpan	Waspfly	Waterbreak	„ 1st
Warmness	Waspish	Waterbug	„ 2nd
Warmonger	Waspishly	Watercarts	„ 3rd
Warmsided	Wassail	Watercask	„ 4th
Warmth	Wassailcup	Waterclock	„ 5th
Warpaint	Wassailers	Watercraft	„ 6th
Warpath	Wastage	Watercress	„ 7th
Warped	Wasteth	Waterdrain	„ 8th
Warping	Wastebooks	Waterdrops	„ 9th
Warplume	Wastegate	Waterelder	„ 10th
Warproof	Wastegood	Waterfall	„ 11th
Warragal	Wastelcake	Waterflood	„ 12th
Warrantize	Wastepaper	Waterflies	„ 13th
Warrantors	Wasteweir	Waterfly	„ 14th
Warrener	Watchbells	Waterfowls	„ 15th
Warrior	Watchbox	Waterfox	„ 16th
Warrioress	Watchcase	Waterguage	„ 17th
Warsong	Watchdogs	Waterglass	„ 18th
Warthog	Watched	Watergruel	„ 19th
Warthought	Watchful	Waterhemp	„ 20th
Wartwort	Watchfully	Waterinch	„ 21st
Warwasted	Watchglass	Waterlaid	„ 22nd
Warwearied	Watchguard	Waterleaf	„ 23rd
Warwhoop	Watchhouse	Waterlevel	„ 24th
Warwolf	Watchkeys	Waterlily	„ 25th
Washable	Watchlight	Waterman	„ 26th
Washballs	Watchpaper	Watermarks	„ 27th
Washhouses	Watchtower	Watermelon	„ 28th
Washiba	Watchwords	Watermill	„ 29th
Washing	Waterage	Waternewts	„ 30th
Washoff	Wateraloes	Waternixie	„ 31st
Washpots	Waterapple	Waternymph	Referring to
Washstand	Wateravens	Waterousel	In answer to
Washtub	Waterbirds	Waterpipe	

REFER TO OUR CABLE	REFER TO YOUR CABLE	HAVE YOU RECEIVED OUR CABLE	
Code Word.	Code Word.	Code Word.	
Waterplant	Wavellite	Waywardens	Of the ——
Waterpoise	Waveloaf	Waywiser	„ 1st
Waterpower	Wavemotion	Weakeneth	„ 2nd
Waterproof	Waveringly	Weakening	„ 3rd
Waterqualm	Waveshell	Weakfish	„ 4rd
Waterrails	Waveson	Weakheaded	„ 5th
Waterrams	Waxbasket	Weakness	„ 6th
Waterscape	Waxbill	Wealdclay	„ 7th
Watersheds	Waxcandle	Wealsman	„ 8th
Watershrew	Waxcloth	Wealth	„ 9th
Waterside	Waxdolls	Wealthiest	„ 10th
Watersnail	Waxflower	Weanlings	„ 11th
Watersnake	Waxinsect	Weapon	„ 12th
Waterspout	Waxlight	Weaponless	„ 13th
Waterstead	Waxmoth	Weaponry	„ 14th
Watertabby	Waxmyrtle	Weariable	„ 15th
Watertable	Waxpalm	Weariest	„ 16th
Watertanks	Waxpaper	Weariful	„ 17th
Watertap	Waxtree	Wearifully	„ 18th
Waterthief	Waxwork	Weariless	„ 19th
Waterthyme	Waxworkers	Wearisome	„ 20th
Watertick	Waybaggage	Weasand	„ 21st
Watertight	Waybennet	Weaselcoot	„ 22nd
Watertwist	Waybread	Weather	„ 23rd
Watertwyer	Waydoor	Weatherbow	„ 24th
Watervole	Wayfarer	Weatherspy	„ 25th
Waterweed	Waygoose	Weaverbird	„ 26th
Waterwing	Waylay	Weaverfish	„ 27th
Waterworm	Waylaying	Weaving	„ 28th
Watery	Wayleaves	Webeye	„ 29th
Wattlebird	Waymark	Webfoot	„ 30th
Wattled	Waypost	Webpress	„ 31st
Waveborne	Wayside	Websterite	Referring to
Wavelength	Waystation	Weddingday	In answer to
Waveless	Wayward	Wedge	

NORTH.	SOUTH.	
Code Word.	Code Word.	
Weedery	Wellread	
Weedgrown	Wellrooms	The —— boundary (of)
Weekday	Wellseeing	„ —— boundaries (of)
Weep	Wellsinker	„ —— drift (of)
Weepable	Wellspent	„ —— end (of)
Weepful	Wellspoken	„ —— end line (of)
Weepingash	Wellspring	„ —— extremity (of)
Weevil	Wellsweep	„ —— level (or the level —— of)
Weevilled	Welltook	„ —— limit (of)
Wehrgelt	Welltrod	„ —— lode or vein (of)
Wehrwolf	Wellwater	„ —— side (of)
Weigh	Wellwiller	„ —— side line (of)
Weighable	Wellwished	„ —— line of demarcation (of)
Weighage	Wellwon	„ extreme —— limit (of)
Weighboard	Welshers	bounded on the —— by
Weighhouse	Weltered	lies to the —— (or lying to ——)
Weighlock	Weltering	at the extreme ——
Weighshaft	Wencher	extends —— (or to the ——)
Weighted	Wenchlike	to the —— (of)
Weightiest	Wendish	at the —— (of)
Weightily	Wertherian	at the —— end (of)
Weightnail	Wesleyan	(in) a —— direction
Weird	Westerner	the direction of the lode is ——
Weirdness	Westmost	the vein runs ——
Weissite	Westringia	from —— to
Welaway	Westwardly	the hanging wall is to the ——
Welcome	Wetdock	the footwall is to the ——
Welcomely	Wetfinger	the vein dips ——
Welcoming	Wetnurse	dipping to the ——
Welkin	Wetshod	due ——
Wellarmed	Whack	a bend to the ——
Wellboat	Whalebird	the vein has taken a bend to ——
Wellborer	Whalebone	probable continuation to ——
Wellbucket	Whalecalf	driving ——
Wellchosen	Whalefin	have begun driving ——
Welldoing	Whalelouse	have stopped driving ——
Welldrain	Whaleshot	have driven ——
Wellfamed	Whallabee	has been driven ——
Wellgraced	Wharf	if you drive ——
Wellhole	Wharfage	why not drive ——
Wellknit	Whartboat	—— end of this drift
Wellknown	Wharfinger	—— face of this drift is now in ore
Wellliking	Wharves	—— of the shaft
Wellloved	Whatlike	driving a cross-cut to the ——
Wellmeaner	Whatnots	have cross-cut to the ——
Wellmet	Whatsoever	have begun to cross-cut ——
Wellminded	Whealworm	why not cross-cut to the ——
Wellness	Wheatears	you should cross-cut to the ——
Wellnigh	Wheaten	cross-cut to the ——
Wellpaid	Wheatflies	from the ——

EAST.	WEST.	
Code Word.	Code Word.	
Wheatfly	Whimsey	
Wheatgrass	Whimshaft	The —— boundary (of)
Wheatmidge	Whimsical	„ —— boundaries (of)
Wheatmoth	Whimsies	„ —— drift (of)
Wheedle	Whimwham	„ —— end (of)
Wheel	Whinaxe	„ —— end-line (of)
Wheelbands	Whinchat	„ —— extremity (of)
Wheelboat	Whining	„ —— level (or the level —— of)
Wheelbug	Whinnied	„ —— limit (of)
Wheelchair	Whinnying	„ —— lode or vein (of)
Wheelhorse	Whinyard	„ —— side (of)
Wheelless	Whipcat	„ —— side line (of)
Wheellock	Whipcord	„ —— line of demarcation (of)
Wheelman	Whipgraft	„ extreme —— limit (of)
Wheelore	Whiphand	bounded on the —— by
Wheelrace	Whipjack	lies to the —— (or lying to ——)
Wheelropes	Whiplash	at the extreme ——
Wheelswarf	Whipmaker	extends —— (or to the ——)
Wheeltires	Whipperin	to the —— (of)
Wheelworks	Whipray	at the —— (of)
Wheezing	Whipsaws	at the —— end (of)
Whelky	Whipshaped	(in) a —— direction
Whelm	Whipsnake	the direction of the lode is ——
Whelped	Whipstaff	the vein runs ——
Whence	Whipstalk	from —— to
Whenever	Whipster	the hanging-wall is to the ——
Whereabout	Whipstitch	the foot-wall is to the ——
Whereness	Whipstocks	the vein dips ——
Whereout	Whirlblast	dipping to the ——
Whereso	Whirler	due ——
Whereunto	Whirlicote	a bend to the ——
Wherry	Whirligigs	the vein has taken a bend to ——
Wherryman	Whirlpits	probable continuation to ——
Whethering	Whirlpool	driving ——
Wheyey	Whirlpuff	have begun driving ——
Wheyface	Whirlwater	have stopped driving ——
Wheyish	Whirlwinds	have driven ——
Whichever	Whirlybat	has been driven ——
Whiff	Whirrick	if you drive ——
Whiffing	Whiskered	why not drive ——
Whiffled	Whiskey	—— end of this drift
Whig	Whiskified	—— face of this drift is now in ore
Whigamore	Whisking	—— of the shaft
Whiggarchy	Whiskyjack	driving a cross-cut to the ——
Whiggishly	Whisperers	have cross-cut to the ——
Whilom	Whistled	have begun to cross-cut ——
Whimbrel	Whiteant	why not cross-cut to the ——
Whimmy	Whitebay	you should cross-cut to the ——
Whimper	Whitebeard	cross-cut to the ——
Whimpering	Whiteblaze	from the ——

N.E.	N.W.	
Code Word.	Code Word.	
Whitebrant	Whortles	
Whitecaps	Whoso	The —— boundary (of)
Whitecedar	Whyles	„ —— boundaries (of)
Whitefaced	Wicked	„ —— drift (of)
Whitefilm	Wickedest	„ —— end (of)
Whiteflaw	Wickedly	„ —— end line (of)
Whitefriar	Wickedness	„ —— extremity (of)
Whitegum	Wickentree	„ —— level of (or the level —— of)
Whiteiron	Wickerwork	„ —— limit (of)
Whitelead	Wicketgate	„ —— lode or vein (of)
Whitelie	Wicking	„ —— side (of)
Whitelight	Wicliffite	„ —— side line (of)
Whitelily	Wicopy	„ —— line of demarcation (of)
Whitelimed	Wideawake	„ extreme —— limit (of)
Whitemeats	Widely	bounded on the —— by
Whitemetal	Widened	lies to the —— (or lying to ——)
Whitemoney	Widening	at the extreme ——
Whiteoak	Widespread	extends —— (or to the ——)
Whitepine	Widewhere	to the —— (of)
Whitepoppy	Widgeons	at the —— (of)
Whitepot	Widow	at the —— end (of)
Whiterent	Widowbench	(in) a —— direction
Whiterope	Widowbird	the direction of the lode is ——
Whitesalt	Widowers	the vein runs ——
Whiteshark	Widowhood	from —— to
Whitesmith	Widowly	the hanging-wall is to the ——
Whitespur	Widowmaker	the foot-wall is to the ——
Whitesters	Widowwail	the vein dips ——
Whitestone	Width	dipping to the ——
Whitetail	Widual	due ——
Whitethorn	Wieldable	a bend to the ——
Whitewash	Wielding	the vein has taken a bend to ——
Whitewhale	Wieldsome	probable continuation to ——
Whitewitch	Wieldy	driving ——
Whitewood	Wifehood	have begun driving ——
Whitingmop	Wiferidden	have stopped driving ——
Whitish	Wigblock	have driven ——
Whitsour	Wigless	has been driven ——
Whitsunale	Wigreve	if you drive ——
Whittawer	Wigwams	why not drive ——
Whitybrown	Wigweavers	—— end of this drift
Whizzingly	Wildbasil	—— face of this drift is now in ore
Whoa	Wildbeast	—— of the shaft
Whobub	Wildboar	driving a cross-cut to the ——
Wholeness	Wildbrains	have cross-cut to the ——
Wholesome	Wildcat	have begun to cross-cut ——
Wholly	Wildcherry	why not cross-cut to the ——
Whoop	Wildduck	you should cross-cut to the ——
Whopping	Wilderedly	cross-cut to the ——
Whopper	Wilderment	from the ——

S.E.	S.W.	
Code Word.	Code Word.	
Wilderness	Windowsash	————————
Wildfowl	Windowseat	The ——— boundary (of)
Wildgoose	Windowsill	,, ——— boundaries (of)
Wildhoney	Windowtax	,, ——— drift (of)
Wildish	Windpipe	,, ——— end (of)
Wildland	Windplant	,, ——— end line (of)
Wildlichen	Windpump	,, ——— extremity (of)
Wildmare	Windrose	,, ——— level (or the level ——— of)
Wildrice	Windsails	,, ——— limit (of)
Wildswan	Windshaken	,, ——— lode or vein (of)
Wildtansy	Windshock	,, ——— side (of)
Wildthyme	Windside	,, ——— side line (of)
Wildwood	Windsucker	,, ——— line of demarcation (of)
Wileful	Windtrunk	,, extreme ——— limit (of)
Wilfully	Windup	bounded on the ——— by
Wilfulness	Windway	lies to the ——— (or lying to ———)
Willowgall	Winebibber	at the extreme ———
Willowherb	Winecasks	extends ——— (or to the ———)
Willowish	Winecellar	to the ——— (of)
Willowlark	Winecooler	at the ——— (of)
Willowmoth	Wineglass	at the ——— end (of)
Willowoaks	Winegrower	(in) a ——— direction
Willowwren	Wineheated	the direction of the lode is ———
Wimbled	Winemaking	the vein runs ———
Wimbling	Winepalm	from ——— to
Windage	Winepress	the hanging-wall is to the ———
Windbags	Winesap	the foot-wall is to the ———
Windbeam	Wineskin	the vein dips ———
Windbill	Winevault	dipping to the ———
Windbound	Winewhey	due ———
Windbreak	Wingcovert	a bend to the ———
Windbroach	Wingfooted	the vein has taken a bend to ———
Windbroken	Winglet	probable continuation to ———
Windchest	Wingshell	driving ———
Winddropsy	Wingstroke	have begun driving ———
Windegg	Wingswift	have stopped driving ———
Winders	Winnowing	have driven ———
Windfallen	Winsome	has been driven ———
Windflower	Winterclad	if you drive ———
Windgall	Wintercrop	why not drive ———
Windgauge	Winterkill	——— end of this drift
Windguns	Winterlove	——— face of this drift is now in ore
Windhatch	Wintermew	——— of the shaft
Windhover	Wintermoth	driving a cross-cut to the ———
Windiness	Winterpear	have cross-cut to the ———
Windlasses	Wintertide	have begun to cross-cut ———
Windlift	Winterweed	why not cross-cut to the ———
Windowbars	Wintrous	you should cross-cut to the ———
Windowed	Wintry	cross-cut to the ———
Windowless	Wirecloth	from the ———

Code Word.

Wiredrawer	.	Dynamo
Wireedge .	.	Series dynamo
Wirefence.	.	Shunt　　„
Wiregauze	.	Compound dynamo
Wiregrub .	.	Alternating　„
Wireguards	.	Dynamo to give —— volts
Wirepuller	.	„　　　„　—— ampères
Wireropes.	.	„　　　„　—— alternations per second
Wiretwist .	.	„　gives —— volts
Wirewheel	.	„　　„　—— ampères
Wireworker	.	„　　„　—— alternations per second
Wireworm	.	„　armature faulty
Wiry . .	.	„　broken down
Wisdom .	.	„　for electric refining
Wiseacres.	.	„　for —— vats
Wiselike .	.	„　for —— lbs. per hour
Wiselings .	.	What is speed of dynamo
Wisest .	.	Dynamo speed is —— revs. per minute
Wishbone .	.	Spare armature wanted
Wishedly .	.	Electrical efficiency of dynamo
Wishers .	.	„　　„　　„　is —— %
Wishfully .	.	Commercial efficiency of dynamo
Wishingcap	.	„　　„　　„　is —— %
Wishywashy	.	Electric Light
Wistaria .	.	Series arc system
Witchcraft	.	Series incandescence system
Witcheries	.	No. of lamps
Witchery .	.	No. of lamps of —— volts
Witchmeal	.	Lamps of —— volts
Witchtree .	.	Lamps of 8 c.p.
Witcracker	.	„　16　„
Withal . .	.	„　32　„
Withamite	.	„　50　„
Withdraw .	.	„　100　„
Withdrawal	.	„　200　„
Witherband	.	„　300　„
Withereth.	.	„　500　„
Witherling	.	„　——　„
Withernam	.	Parallel incandescence system
Witherod .	.	„　arc system
Withheld .	.	Alternating current system
Withholden	.	Alternating current
Withinside	.	Volts in primary circuit
Without .	.	Volts in secondary circuit
Withstood	.	Accumulators
Withwind .	.	Storage cells
Witjar . .	.	Accumulators (or storage cells) for —— ampère hours
Witmonger	.	E.M.F. of storage —— volts
Witnesseth	.	Electric installation to be carried out
Witnessing	.	„　　„　is carried out

Code Word.

Witsnapper .	Electrical transmission of power
Witstarved .	We require to transmit —— h.p.
Witticisms .	Distance —— yards
Wittiest . .	Water power to be used for driving generators
Wittified . .	Steam „ „ „ „ „
Wittily. . .	What kind of power to be used for driving generators
Wittiness . .	Generators to be driven direct
Witwanton .	„ „ „ by gearing
Wives . . .	„ to give —— volts
Wizardly . .	„ „ —— ampères
Wizen . . .	„ for —— E.H.P.
Woadwaxen .	„ to run —— revs. per min.
Wodegeld. .	„ to be —— wound
Woebegone .	„ to have a spare armature
Woful . . .	„ require „ „
Wofully . .	Electro-motor (or motors) to give —— B.H.P.
Wofulness. .	„ „ „ —— yards from generator
Wolf . . .	„ „ for —— volts
Wolfdog . .	„ „ for —— ampères
Wolffian . .	„ „ „ —— inch fan
Wolffish . .	„ „ „ —— „ exhauster
Wolfishly . .	„ „ „ pumping
Wolflings . .	„ „ „ —— stamps
Wolfnet . .	„ „ to be coupled direct
Wolfram . .	Transmit power from motor by belt
Wolfsbane .	„ „ „ „ by worm
Wolfsclaw. .	„ „ „ „ by spur-wheel
Wolfskin . .	Series electro-motor
Wolverine .	Shunt „ „
Woman . .	What B.H.P. do you require
Womanborn .	„ voltage do you require
Womanfully .	„ system of conductors
Womangrown	„ fall in E.M.F. —— % allowed
Womanguard	„ speed do you require to run
Womanhater .	„ size pulley required
Womanhood .	Overhead conductors
Womanish .	Underground conductors
Womankind .	—— % fall in E.M.F. allowed
Womanless .	Pulley —— inches diameter
Womanlike .	„ —— inches face
Womanly . .	„ to be flat
Womanpost .	„ „ crowned
Womantired .	„ „ plated
Wonderful .	„ „ shrouded
Wonderland .	Gearing to be —— to
Wondermaze.	We require a Telpher line
Wonderment.	What distance of Telpher line required
Wonderwork.	Telpher line —— yards
Wondrous. .	„ „ —— miles
Wondrously .	Cars to carry —— cwt. each

Code Word.

Wontedness .	Agra Bank, Limited.
Woodacid. .	Alliance Bank, Limited.
Woodapple .	Anglo-Argentine Bank, Limited.
Woodashes .	Anglo-Californian Bank, Limited.
Woodbine .	Anglo-Foreign Banking Co., Limited.
Woodbirds .	Banco-Nacional de Brazil.
Woodboring .	Banco-Nacional de Mexico.
Woodbound .	Bank of Adelaide.
Woodcarpet .	Bank of Africa, Limited.
Woodchoir .	Bank of Australasia.
Woodchucks .	Bank of British Columbia.
Woodcraft .	Bank of British North America.
Woodculver .	Bank of California.
Woodcut . .	Bank of Egypt, Limited.
Woodcutter .	Bank of England.
Wooddove. .	Bank of France.
Wooddrink .	Bank of Montreal.
Woodduck .	Bank of New South Wales.
Wooded . .	Bank of New Zealand.
Woodenly. .	Bank of Scotland.
Woodfall . .	Bank of South Australia, Limited.
Woodgas . .	Bank of Tarapaca and London, Limited.
Woodgods. .	Bank of Victoria, Limited (Australia).
Woodgrouse .	Barclay, Bevan, Tritton, Ransom, Bouverie, & Co.
Woodhole. .	Baring, Brothers, & Co., Limited.
Woodibis . .	British Linen Co.
Woodkern. .	Brooks & Co.
Woodlarks .	Brown (Alexander) & Co.
Woodlayer .	Brown, Brothers, & Co. (New York).
Woodmite. .	Brown, Janson, & Co.
Woodmonger.	Brown, Shipley, & Co.
Woodnotes .	Capital and Counties' Bank, Limited.
Woodnymph .	Chartered Bank of India, Australia, and China.
Woodoil . .	Chartered Mercantile Bank of India, London, and China.
Woodopal. .	Child & Co.
Woodpaper .	City Bank, Limited.
Woodpie . .	Clydesdale Bank, Limited.
Woodpigeon .	Cocks, Biddulph, & Co.
Woodreeve .	Colonial Bank of New Zealand.
Woodrush. .	Commercial Bank of South Australia.
Woodsage. .	Commercial Banking Company of Sydney.
Woodscrew .	Commercial Bank of Australia, Limited.
Woodsia . .	Comptoir National d'Escompte de Paris (Paris).
Woodslaves .	Comptoir National d'Escompte de Paris (London).
Woodsman .	Consolidated Bank, Limited.
Woodsoot . .	Coutts & Co.
Woodsorrel .	Cox & Co.
Woodspirit .	Credit Lyonnais (Paris).
Woodspite. .	Credit Lyonnais (London).
Woodstamp .	Cunliffe (Roger), Sons, & Co.
Woodtar . .	Delhi and London Bank, Limited.

Code Word.

Woodwasps	Deutsche Bank (Berlin).
Woodwork	Dimsdale, Fowler, Barnard, & Dimsdales.
Woodwren	D'Obree, Samuel, & Sons.
Wooers	Drexel, Morgan, & Co. (New York).
Woolburler	Drummond, Messrs.
Woolcomber	Eives and Allen.
Woolder	English Bank of the River Plate, Limited.
Wooldriver	English Bank of Rio de Janeiro, Limited.
Wooldyed	Erlanger (Emile) & Co. (Paris).
Woolgrower	Erlanger (Emile) & Co. (London).
Woolliness	Fuller, Banbury, Nix, & Co.
Woollybut	Gibbs (Anthony) & Sons.
Woollyhead	Glyn, Mills, Currie, & Co.
Woolmill	Grant & Co.
Woolpacker	Herries, Farquhar, & Co.
Woolsack	Hoare, Messrs.
Woolsey	Hong-Kong and Shanghai Banking Corporation.
Woolshears	Imperial Bank, Limited.
Woolsorter	Imperial Bank of Persia.
Woolstaple	Imperial Ottoman Bank.
Woolstocks	King (Henry S.) & Co.
Woolwinder	Knowles & Foster.
Woorali	Kountze Brothers (New York).
Wootz	Land Mortgage Bank of India, Limited.
Wordbook	Land Mortgage Bank of Victoria, Limited.
Wordish	Lloyds, Barnett's, & Bosanquet's Bank, Limited.
Wordsquare	London & Brazilian Bank, Limited.
Workaday	London & County Banking Co., Limited.
Workbag	London & Provincial Bank, Limited.
Workboxes	London & River Plate Bank, Limited.
Workers	London & San Francisco Bank, Limited.
Workfellow	London & South Western Bank, Limited.
Workfolks	London & Westminster Bank, Limited.
Workful	London Bank of Mexico & South America, Limited.
Workhouse	London Chartered Bank of Australia.
Workingday	London Joint Stock Bank, Limited.
Workmanly	Manchester & Liverpool District Banking Co., Limited.
Workmaster	Martin & Co.
Workmen	Matheson & Co.
Workpeople	Melville, Evans, & Co.
Workshops	Merchant Banking Company of London.
Worktables	Morgan, J. S., & Co.
Workwoman	C. de Murrieta & Co., Limited.
Worldlings	National Bank, Limited.
Worldwide	National Bank of Australasia.
Wormbark	National Bank of India, Limited.
Wormcast	National Bank of New Zealand, Limited.
Wormeaten	National Bank of Scotland, Limited.
Wormfence	National Provincial Bank of England, Limited.
Wormfever	New London & Brazilian Bank.

BANKERS.

Code Word.

Code Word	Banker
Wormgear	New Oriental Bank Corporation, Limited.
Wormgrass	Peabody (Henry W.) & Co.
Wormholes	Praeds & Co.
Wormlike	Prescott & Co.
Wormoil	Queensland National Bank, Limited.
Wormpowder	Redfern, Alexander, & Co.
Wormseed	Robarts, Lubbock, & Co.
Wormshaped	Rothschild, N. M., & Sons.
Wormshell	Royal Bank of Scotland.
Wormsocket	Robinson South African Banking Co.
Wormwheels	Schuster, Son, & Co.
Wormwood	Scott (Sir Samuel), Bart., & Co.
Wornout	Seligmann & Co. (New York).
Worried	Seyd & Co.
Worry	Silver, S. W., & Co.
Worrying	Smith, Payne, & Smiths.
Worryingly	Société Générale.
Worsening	Standard Bank of South Africa, Limited.
Worship	Twining, R., & Co.
Worshipful	Union Bank of Australia, Limited.
Worshipped	Union Bank of London, Limited.
Worst	Union Bank of Scotland, Limited.
Worthier	Union Bank of Spain & England, Limited.
Worthily	Union Discount Co., Limited.
Worthless	Wells, Fargo, & Co.
Worthy	Western Australian Bank.
Woundwort	Williams, Deacon, and Manchester & Salford Bank, Limited.

Code Word.

Woundy . .	African Gold Share Investment Company, Limited.
Woven . .	African Investment Corporation, Limited.
Wracked . .	Agency and Exploration Company of Australasia, Limited.
Wrackgrass .	Alliance Trust and Investment Company, Limited.
Wrainbolt .	Assets Realization Company, Limited.
Wraith . .	British and African Investment Company.
Wranglers .	British North Borneo Company.
Wrap . . .	British South Africa Company.
Wrappages .	Commercial Trust and Agency.
Wraprascal .	Debenture Corporation, Limited.
Wrasse . .	Empire of India Corporation, Limited.
Wrathfully .	Exploration Company, Limited.
Wrathily . .	Financial Trust Corporation.
Wreathes . .	Financial Trust, Limited.
Wreckage . .	Foreign and Colonial Debenture Corporation, Limited.
Wreckers . .	Hudson's Bay Company.
Wreckfree .	Imperial British East Africa Company.
Wreckful . .	Industrial and General Trust, Limited.
Wren . . .	International Trustee, Assets, & Debenture Corporation, Ltd
Wrencheth .	London and Australasian Debenture Corporation, Limited
Wrenching .	London and Colonial Finance Corporation, Limited.
Wrestler . .	London and Hamburg Gold Recovery Company, Limited.
Wrestling .	London Share and Debenture Company, Limited.
Wretch . .	London and South African Exploration Company, Limited.
Wretchcock .	Merchants' Trust, Limited.
Wretchedly .	Metropolitan Contract Corporation.
Wretchless .	Mexican Association, Limited.
Wriggle . .	Mexican Company of London, Limited.
Wriggling . .	Mexican Explorations, Limited.
Wrightia . .	Mexican and General Concessions Company, Limited.
Wringstaff .	Mexican General Land, Mortgage, & Investment Co., Ltd.
Wrinkles . .	Mexican Investment Corporation, Limited.
Wristdrop .	Mines Company, Limited.
Writable . .	Mines Contract Company, Limited.
Writerling .	Mines Trust, Limited.
Writers . .	Mining and Financial Trust Syndicate, Limited.
Writership .	Mining Shares Investment Company, Limited.
Writhled . .	Mosambique Company.
Writingink .	National Financial Corporation.
Wroken . .	Phœnix Trust Company.
Wrongdoer .	South African Gold Trust and Agency Company, Limited
Wrongdoing .	South African Prospecting and Mortgage Corporation, Ltd
Wrongful . .	South African Loan, Mortgage, & Mercantile Agency, Ltd
Wrongfully .	South African Trust and Finance Company.
Wronghead .	South American and Mexican Company, Limited.
Wronging .	Transvaal Mortgage, Loan, and Finance Company.
Wrongless .	Trust and Agency Company of Australasia, Limited.
Wrongly . .	Trustees Executors & Securities Insurance Corporation, Ltd.
Wrongous .	United Kingdom & Foreign Investment & Finance Co., Ltd.
Wroth . , .	United States & South American Investment Trust Co., Ltd.
Wrung . . .	United States Debenture Corporation, Limited.

Code Word.
Wryly

Wrymouthed

Wryneck

Wurrus

Wychelm

Wychhazel

Wychwaller

Wyvern

Xanthate

Xantheine

Xanthium

Xantho

Xanthocon

Xanthopous

Xanthosis

Xenelasia

Xenium

Xenodochy

Xenops

Xenotime

Xerasia

Xeroderma

Xerodes

Xeromyrum

Xerophagy

Code Word.
Xiphias

Xiphidium

Xiphodon

Xiphoidian

Xiphosura

Xylanthrax

Xylene

Xylobius

Xylocopa

Xylograph

Xyloidine

Xylol

Xylophagus

Xylophilan

Xylophylla

Xylopia

Xyloretine

Xylotile

Xyridaceae

Xyst

Xystarch

Xysters

Yaccawood

Yacht

Yachtclub

Code Word.

Yachting

Yachtsman

Yaffingale

Yagger

Yanked

Yankeeism

Yanolite

Yaourt

Yapocks

Yapon

Yardarm

Yardland

Yardstick

Yarely

Yarn

Yarnuts

Yawls

Yawneth

Yawning

Yeanling

Yearbook

Yeared

Yearlily

Yearningly

Yeast

Code Word.

Yeastiness

Yeastplant

Yell

Yelling

Yellow

Yellowbird

Yellowgum

Yellowish

Yellowjack

Yellowlegs

Yellowness

Yellowpine

Yellowtop

Yellowwood

Yenite

Yeoman

Yeomanly

Yerbamate

Yergas

Yestereve

Yestermorn

Yestern

Yewbow

Yewen

Yewtree

Code Word.
Yieldance

Yieldingly

Yieldless

Yirdhouse

Yoicks

Yokeage

Yokefellow

Yokeline

Yokemates

Yokerope

Yoketh

Yoky

Yolk

Yonder

Young

Youngeyed

Youngish

Younglings

Youngly

Youngness

Youngster

Youth

Youthful

Youthfully

Youthhood

Code Word.
Youthly

Youthsome

Yucca

Yulan

Yuleblock

Yulelog

Yuletide

Zabism

Zacchean

Zamang

Zamia

Zanied

Zanonia

Zantewood

Zany

Zanying

Zaphara

Zapotilla

Zayat

Zeal

Zealful

Zealless

Zealotical

Zealotism

Zealotry

Code Word.		
Zebeck		
Zebra		
Zebraplant		
Zebrawolf		
Zebrawood		
Zechin		
Zechstein		
Zedoary		
Zeine		
Zemindary		
Zenana		
Zenith		
Zenithal		
Zeolitic		
Zephyrs		
Zero		
Zested		
Zetetics		
Zeticula . .	3 H.P. vertical portable hoisting engine—Hindley (see advertisement, facing inside back cover).
Zeuglodon .	4 H.P. ditto.
Zeugma . .	6 H.P. ditto.
Zeugmatic .	8 H.P. ditto.
Zeuxite . .	Circular saw bench to take 48″ saw, with self-acting rope-feed, variable while working—Hindley.
Zibeth . .	Combined slow-speed engine and dynamo for electric lighting. State number of arc or incandescent lamps plant required for.
Zibethum .	Combined high-speed engine coupled direct to dynamo. State number of arc or incandescent lamps plant required for.
	If boiler required for electric light plant, add code word "Bearish."

Code Word.
Ziega

Zigzag

Zigzaggery	.	Send — of Charleton's Cost-sheet Forms, Reference Order No.
Zincamyl	.	Send — of Charleton's Report-Book for Mining Engineers.
Zincblende	.	Repeat last order for Incandescent Lamp (Ediswan).
Zincbloom	.	Send Price Lists of Electrical Plant (Ediswan).
Zincethyl	.	Send address of Nearest Agent (Ediswan).
Zincky	. .	Send complete set of Catalogues (Ediswan).
Zincoid	. .	Send particulars of Mine and Mill fittings (Ediswan).
Zincolysis	.	White's System of Steel Portable Railway complete.
Zincolyte	.	White's Patent Portable Aërial Wire Ropeway.
Zincopolar	.	White's Patent Aërial Wire Ropeway, to carry 5-ton loads with electric block system of haulage.
Zincwhite	.	White's Patent Aërial Wire Ropeway, to carry 5-ton loads for endless rope haulage.
Zingel	. .	White's Patent Automatic Level Crossing Gates, opened and closed by the passing train.
Zingho	. .	White's Patent Steam Pulsating Pump.
Zingian	. .	Send Hindley's Price List of Boilers and Engines.
Zingiber	. .	How much Gold, Bullion, or Specie have you banked ?
Zinkenite	. .	Send Report, &c., to The Financial News.
Zircon	. .	Ship per first steamer.
Zirconium	.	Tailings pump (add one word giving size of pump, as "four," "eight," or whatever it may be).
Zithern	. .	35 feet cast-iron piping for pump ordered, with foot-valve bend, and jointing material.
Zizania	. .	60 feet cast-iron piping for pump ordered, with foot-valve, bend, and jointing material.

z

Code Word.

Zizyphus.	.	Tailings pumping-engine (add one word giving size of pump, as "four," "eight," or whatever it may be).
Zoadulae.	.	35 feet cast-iron piping for pumping-engine ordered, with foot-valve, bend, and jointing material.
Zoantharia	.	60 feet cast-iron piping for pumping-engine ordered, with foot-valve, bend, and jointing material.
Zoanthidae	.	Send Catalogue Surveying Instruments (Stanley's).
Zoanthropy	.	Send Mining Dial A 97 (Stanley's).
Zocco	. .	Send A 42 Improved Level (Stanley's).
Zodiac	. .	Send A 2 with A 4 Theodolite for Mining (Stanley's).
Zodiacal	. .	Send A 175 Patent 6-ft. Mining Staff (Stanley's).
Zeotrope	. .	Send A 130 Mining Aneroid Barometer (Stanley's).
Zohar	. .	Bullivant's Flexible Steel Wire Rope.
Zoilean	. .	Bullivant's Special Flexible Steel Wire Rope.
Zoilism	. .	Bullivant's Crucible Steel Wire Mining Rope.
Zomboruks	.	Bullivant's Plough Steel Wire Mining Rope.
Zonulet	. .	Bullivant's Special Steel Wire Tramway Cable.
Zoocarp	. .	Bullivant's Aerial Wire Tramway Plant.
Zoochemy	.	Composite Frame Engine (Robey & Co., Limited, Lincoln).
Zoogen	. .	Undertype Robey Winding Engine, W.I.F. (Robey).
Zoogenic	. .	Vertical Hoisting Engine (Robey).
Zoography	.	Steel Frame Ore Crusher (Robey).
Zooid	. . .	10 Stamp Mill (Robey;
Zoolatry	. .	Undertype Robey Compound Engine (Robey).
Zoolite	. .	Send Spons' latest Catalogue of Engineering books.

Code Word.

Zoologer . . Send copy of Moreing and Neal's Mining Code, price one guinea.

Zoological . Harling's special surveying and mining aneroid, giving readings to single feet of altitude. 3½ inches diameter.

Zoomorphic . No. 1087, ditto, ditto, same as above but 5 inches diameter, giving more open divisions, £6.

Zoophaga . Harling's 6-inch transit theodolite, divided on silver to 20 seconds. In case, with tripod stand, price £26 10s.

Zoophagous . Harling's portable anemometer, in box, with two dials, reading to 1000 feet, with disconnector, price £2 5s.

Zoophilist . Harling's ditto, ditto, with six dials, reading to 10,000,000 feet, £2 12s. 6d.

Zoophily . . No. 1, pulsometer with foot and back pressure valves, short length of suction pipe and strainer, steam valve, and set of spare parts.

Zoophoric . No. 2, ditto, ditto.

Zoophytes . No. 3, ditto, ditto.

Zoophytoid . No. 4, ditto, ditto.

Zoosperm . No. 5, ditto, ditto.

Zootheca . . No. 6, ditto, ditto.

Zootic . . No. 7, ditto, ditto.

Zootomist . No. 7½, ditto, ditto.

Zootomy . . No. 8, ditto, ditto.

Zoozoo . . No. 9, ditto, ditto.

Zopilote . . No. 10, ditto, ditto.

Zopissa . . No. 11½, ditto, ditto.

Zosterite . . If pump is wanted with the Pulsometer Engineering Company's special wrought-iron flanged pipe for 80 feet total lift, add this word.

Zosterops . If pulsometer is required with boiler, add this word.

Code Word.

Zouaves . . "Deane" double-plunger sinking pump with one 7-in. and
one 5-in. plunger, 10-in. cylinder, 16-in. stroke, capable of
raising 4500 gallons per hour, 150 feet high, with 50 lbs.
of steam.

Zoutch . . "Deane," ditto, ditto, for 10,000 gallons per hour.

Zygomatic . Please send out the Pulsometer Engineering Company's
lists.

Zygosis . . Ask the Pulsometer Engineering Company, 63, Queen
Victoria Street, E.C., to give you quotation for pumps
suitable for the work.

www.ingramcontent.com/pod-product-compliance
Lightning Source LLC
LaVergne TN
LVHW012209040326
832903LV00003B/203